Nonlinear Dynamics

Advanced Texts in Physics

This program of advanced texts covers a broad spectrum of topics which are of current and emerging interest in physics. Each book provides a comprehensive and yet current and emerging interest in physics. Each book provides a comprehensive and yet accessible introduction to a field at the forefront of modern research. As such, these texts are intended for senior undergraduate and graduate students at the MS and PhD level; however, research scientists seeking an introduction to particular areas of physics will also benefit from the titles in this collection.

Springer

Berlin
Heidelberg
New York
Hong Kong
London
Milan
Paris
Tokyo

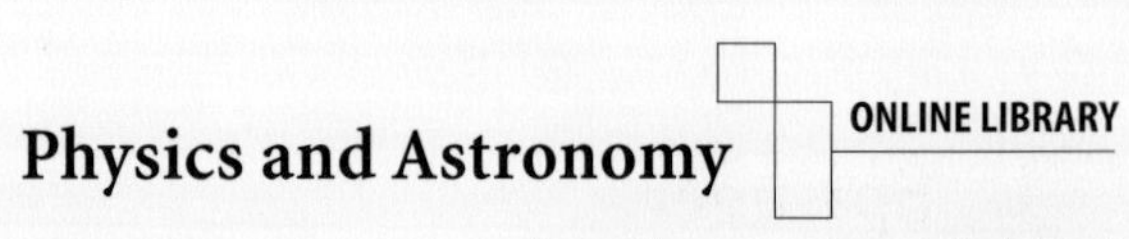

M. Lakshmanan S. Rajasekar

Nonlinear Dynamics

Integrability,
Chaos, and Patterns

With 193 Figures, 17 Tables,
Numerous Examples and Problems

Springer

Professor M. Lakshmanan
Department of Physics
and Center for Nonlinear Dynamics
Bharathidasan University
Tiruchirapalli – 620 024
India

Dr. S. Rajasekar
Department of Physics
Manonmaniam Sundaranar University
Tirunelveli – 627 012
India

Library of Congress Cataloging-in-Publication Data.
Lakshmanan, M. (Muthuswamy)
Nonlinear dynamics : integrability, chaos, and patterns / M. Lakshmanan; S. Rajasekar.
p. cm – (Advanced texts in physics, ISSN 1439-2674)
Includes bibliographical references and index.
ISBN 3540439080 (acid-free paper)
1. Dynamics. 2. Nonlinear theories. I. Rajasekar, S. (Shanmuganathan), 1963- II. Title. III Series.
QA845.L25 2002 531'.11-dc21 2002030441

ISSN 1439-2674

ISBN 3-540-43908-0 Springer-Verlag Berlin Heidelberg New York

Springer-Verlag Berlin Heidelberg New York
a member of BertelsmannSpringer Science+Business Media GmbH

http://www.springer.de

Typesetting: Data prepared by the authors using a Springer TEX macro package

Final page layout: EDV-Beratung Frank Herweg, Leutershausen
Cover design: *design & production* GmbH, Heidelberg

Printed on acid-free paper SPIN 10884430 57/3141/tr 5 4 3 2 1 0

To our parents

Preface

Nonlinearity is ubiquitous and all pervading in the physical world. For a long time nonlinear systems were essentially studied under linear approximations, barring a few exceptions. However, the famous Fermi-Pasta-Ulam numerical experiments of the year 1955 on energy sharing between modes in anharmonic lattices triggered the golden era of modern nonlinear dynamics. Several path-breaking discoveries followed in the subsequent decades. Two of these developments during the 1960s stand apart as having radically changed our outlook on nonlinear systems and the underlying dynamics.

In the year 1963, E.N. Lorenz numerically integrated a simplified system of the three coupled first-order nonlinear equations of the fluid convection model describing the atmospheric weather conditions. The bounded nonperiodic trajectories of the equations started from two nearby initial states diverged exponentially until they become completely uncorrelated resulting in unpredictability of future state in a fully deterministic dynamical system. Such a solution became known as chaotic and, with this discovery, the field of chaotic dynamics was born.

Not much later, in 1965, Zabusky and Kruskal numerically analysed the initial value problem of the Korteweg–de Vries (KdV) equation, which represents a nonlinear dispersive system. They observed a phenomenon completely opposite to that of chaos. In their experiments solitary waves interacted among themselves and re-emerged unchanged in form and speed. Because of the particle-like nature of the collision of solitary waves, Zabusky and Kruskal coined the name soliton to describe such a solitary wave. Kruskal and coworkers went on to develop a completely analytic procedure called the inverse scattering transform (IST) to solve the initial value problem of the KdV equation. This marked the advent of the modern era of integrable nonlinear systems.

Independent of these developments, various important studies on nonlinear diffusive and dissipative systems and the underlying patterns were pursued during this period. Subsequently, starting in the 1970s, the study of nonlinear dynamical systems experienced an explosive growth. Analytical and numerical tools were developed and fascinating results obtained. In recent times increasing attention has been focussed on exploring real technological applications of nonlinear dynamics: Controlling of chaos, synchronization of chaos and secure communication, cryptography, optical-soliton-based communication, magnetoelectronics, spatio-temporal patterns to name but a few.

Applications of nonlinear dynamics have been found throughout the realms of physics, engineering, chemistry and biology. Numerous mathematical ideas and techniques have been used to study nonlinear systems, and these, in turn, have enriched the field of mathematics itself. The field of nonlinear dynamics has hence emerged as a highly interdisciplinary endeavour.

Considering its multidisciplinary nature and important practical applications, the topic of nonlinear dynamics has now been introduced as part of advanced level undergraduate, graduate and masters-level courses in many countries and the number of researchers on this field is continuing to grow. The present authors have spent a major part of their academic careers in studying and understanding the many-faceted and facinating features of nonlinear dynamics. They have also been giving courses on various aspects of nonlinear dynamics to masters-level and research students in physics and mathematics for several years. The authors have now endeavoured to develop their teaching materials into an advanced level text book containing 16 chapters and 10 appendices, hopefully catering to the needs of advanced undergraduate, graduate and masters-level course students in physics, mathematics and engineering, who aspire to obtain a sound basic knowledge of nonlinear dynamics. The material covered includes in a rather unified way the three major themes of nonlinear dynamics: Chaos, integrability (including solitons) and spatio-temporal patterns. Ideally, the material can be fully covered in a two semester course: One possibility is to have one semester on chaos and one semester on integrability and patterns. Another is to have a basic course on chaos and integrability (Chaps. 1–5, 7, 11–13) and an advanced course on the remaining materials (covered in Chaps. 6, 8–10, 14–16). A single semester course can also be devised by including relevant chapters and the lecturer will be able to identify the suitable parts very easily since the text is clearly structured.

Regarding the content, the first nine chapters are concerned with bifurcations and chaos. Starting from the basic notions of nonlinearity and nonlinear dynamical systems we have introduced the concept of attractors, bifurcations, chaos and characterization of regular and chaotic motions. These aspects are analysed in simple dissipative and conservative systems and in electronic circuits. Additional compact discussions of specialized and advanced topics on chaos are presented in Chap. 9. Chapters 10–14 deal with integrability and integrable systems. In particular, Chap. 10 is devoted to integrable finite dimensional nonlinear systems. Starting from the notion of integrability, we describe the analytical methods used to identify integrability and employ them to study typical nonlinear systems. Chapters 11–14 are concerned with the study of solitonic systems. After introducing the notion of solitary waves and solitons, the inverse scattering transform method, Hirota's bilinearization and the Bäcklund transformation techniques are presented and the notion of completely integrable infinite-dimensional nonlinear dynamical systems are elucidated. These methods are applied to many nonlinear wave equations of contemporary interest. Then in Chap. 15, spatio-temporal patterns in nonlinear reaction-diffusion systems are presented. Some of the potential technological applications of chaos and solitons are discussed in Chap. 16. The book also contains ten appendices which deal with several important concepts and methods which could not be presented in the main text without disrupting the continuity and style of presentation of the topics concerned.

The book also contains numerous Exercises and Problems which the the authors hope will enhance the understanding of the subject to the level of present-day active research. In our classification the 'Exercises' are essentially meant to augment the discussion and derivation in the text, while the 'Problems' are generally of a more advanced nature. Some of the problems even relate to current research investigations and we hope that they will motivate the student to become involved in contemporary research in nonlinear dynamics. They will also help students and teachers to identify suitable projects for further investigations. Necessary references to the literature are included chapter-by-chapter (towards the end of the book), among them publications that will be helpful for tackling the Exercises and Problems.

During the preparation of this book we have received considerable support from many colleagues, students and friends. In particular we are grateful to Prof. A. Kundu, Dr. K.M. Tamizhmani, Dr. R. Sahadevan, Dr. S.N. Pandey, Dr. S. Parthasarathy, Dr. M. Senthil Velan, Dr. K. Murali, Dr. P. Muruganandam, and Dr. R. Sankaranarayanan among others for their critical reading and suggestions on various parts of the book. It is a pleasure to thank Dr. A. Venkatesan, Dr. P. Philominathan, Dr. V. Chinnathambi, Mr. K. Thamilmaran, Mr. T. Kanna, Mr. P. Palaniyandi, Ms. P.S. Bindu, Mr. C. Senthil Kumar, Mr. D.V. Senthil Kumar and Mr. V.K. Chandrasekar for their assistance in the preparation of some of the figures, checking the derivations and problems, etc. However, the authors are solely responsible for any shortcomings, errors or misconceptions that remain. A few of the illustrations are reproduced from other sources and appropriate references are given at the relevant places. We sincerely thank the respective authors and publishers for granting us permission to use these figures. It is also a pleasure for us to record our thanks to the Department of Science of Technology, Government of India for providing support under various research projects, which enabled us to undertake this task. We also appreciate very much the various suggestions of Springer–Verlag, especially Dr. Angela Lahee of the Physics Editorial Department, on the manuscript. Finally, we thank our family members for their unflinching support and encouragement during the course of this project, without which it would not have been possible to complete it.

Tiruchirapalli,
July 2002

M. Lakshmanan
S. Rajasekar

Contents

1. What is Nonlinearity?

As time goes by all things in nature change: Animals are born, they grow, live and die, so do plants. Planets, stars, etc. move around all the time changing their positions continuously. Oceans, rivers, clouds etc. again change their state continuously. Crystals grow and chemicals interact. Even inanimate objects like furniture, buildings, sculptures, etc. change their physical state, perhaps more slowly and over a longer period of time. Change is inevitable in nature (though often on a finite time scale a system may be considered stationary). This change of state of physical systems as a function of time is their *evolution*, whose study constitutes the subject of *dynamics*. How do such changes arise? Obviously changes take place due to the interplay of forces, simple and complex, which act on the systems. Think of any natural system, it is always being acted upon by one or more forces.

Examples:

System	Type of dominant force
Pendulum	restoring force
Planetary system	gravitational force
Moving charges	electromagnetic force
Stationary charges	electrostatic force
Atomic nucleus	nuclear force

The nature of evolution of different physical systems depends upon the nature of the forces acting on them and on their initial state. In this chapter we will distinguish between *linear* and *nonlinear* forces, thereby leading to the study of *linear* and *nonlinear dynamical systems* respectively, especially with reference to systems with finite number of degrees of freedom. Their relevance to systems with infinite number of degrees of freedom will be taken up later in the book (in Chaps. 10–16).

1.1 Dynamical Systems: Linear and Nonlinear Forces

Newton's laws form the foundation for the description of evolution of physical systems. Based on these laws suitable mathematical formulations can be developed in the form of differential (ordinary/partial) equations, difference equations and even integral equations or combinations of these. For example, concentrating on conservative systems (recall the definition or see Sect. 4.4) with finite number of degrees of freedom, it is well known that at least the three standard but equivalent descriptions given in Table 1.1 are available to describe the dynamics [1,2], provided the initial conditions are suitably prescribed and the nature of the forces are known. The forces can be derived from potentials. In Table 1.1 overdot refers to differentiation with respect to time t. $\boldsymbol{F}_i$'s are the forces, V is the potential, and T is the kinetic energy, L and H are respectively the Lagrangian and Hamiltonian of the system. It is assumed here that no constraints are present. Suitable extensions can be made for nonconservative systems and constrained systems also. It is thus clear from the form of the dynamical equations or *equations of motion*, the behaviour of physical systems is essentially determined by the form of the force (or by the form of the potentials) acting upon them. Some of the familiar physical systems are

(1) linear harmonic oscillator
(2) Kepler's planetary system
(3) spinning top
(4) simple pendulum

and so on.

Table 1.1. Three equivalent formulations of dynamics when the forces are conservative

Type of formulation	State variables	Basic quantity	Equation of motion
Newtonian	position $\boldsymbol{r}_i$, velocity $\dot{\boldsymbol{r}}_i$	$\boldsymbol{F}_i$	$m_i\ddot{\boldsymbol{r}}_i = \boldsymbol{F}_i, \quad i=1,2,...,N$
Lagrangian	generalized coordinates and velocities $(q_i, \dot{q}_i)$	$L = T - V$	$\frac{\mathrm{d}}{\mathrm{d}t}\left(\frac{\partial L}{\partial \dot{q}_i}\right) - \frac{\partial L}{\partial q_i} = 0$
Hamiltonian	generalized coordinates and momenta (q_i, p_i)	$H = T + V$	$\dot{q}_i = \frac{\partial H}{\partial p_i}, \quad \dot{p}_i = -\frac{\partial H}{\partial q_i}$

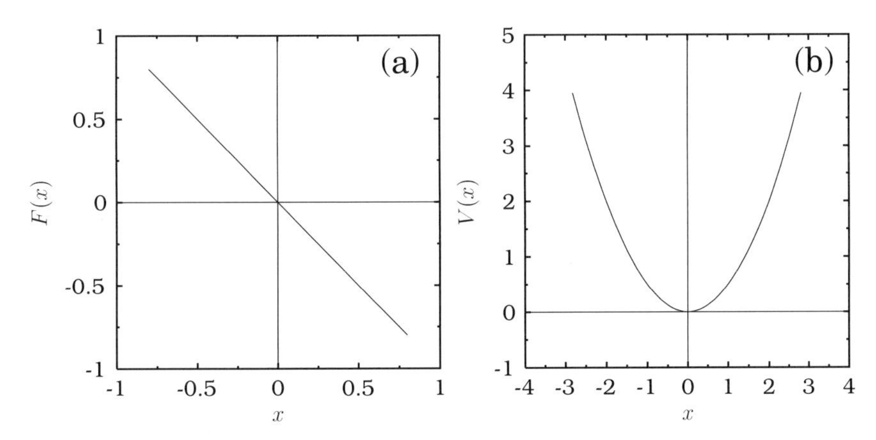

Fig. 1.1. Graphs of (**a**) the linear force $F = -kx$ with $k = 1$ and (**b**) the corresponding potential function $V(x)$

Examples:

(1) *Linear Harmonic Oscillator*

For a linear harmonic oscillator the restoring force F is given by

$$F = -kx, \quad k = \text{constant} > 0 \tag{1.1a}$$

and is directly proportional to the displacement x, that is F is a linear function of x. This linear relationship is depicted graphically in Fig. 1.1a. Correspondingly, the potential $V(x)$ is a quadratic function of x (Fig. 1.1b),

$$V = -\int_0^x F\mathrm{d}x' = \frac{1}{2}kx^2 . \tag{1.1b}$$

(2) *Kepler Problem*

Consider the Kepler problem, namely the motion of a planet around the sun. The gravitational force $\boldsymbol{F}$ is given by

$$\boldsymbol{F} = \frac{-k}{r^2}\hat{\boldsymbol{r}} , \tag{1.2}$$

where r is the radial coordinate and $\hat{\boldsymbol{r}}$ is a radial unit vector. Here the magnitude of the force is not directly proportional to the radial displacement (but it is inversely proportional to r), see Fig. 1.2a. So $|\boldsymbol{F}|$ is not a linear function of r, but it is a nonlinear function of r, or in other words, $\boldsymbol{F}$ is a *nonlinear force*. The corresponding potential is proportional to $1/r$ and is shown in Fig. 1.2b.

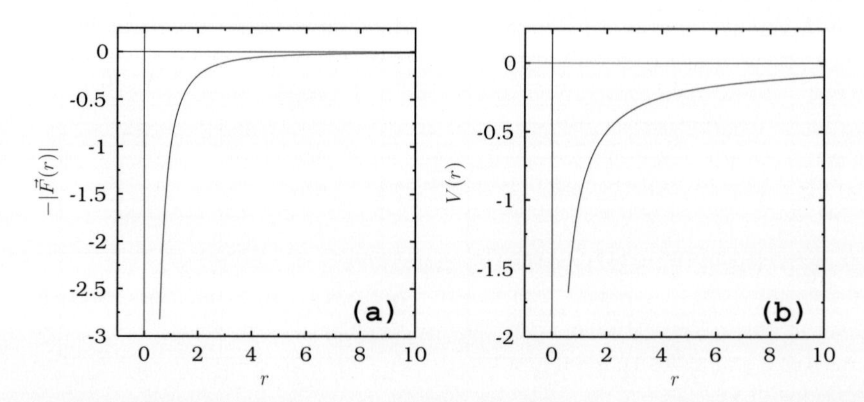

Fig. 1.2. Graphs of (**a**) the negative of the magnitude of the nonlinear force given by (1.2) and (**b**) the corresponding potential function $V(r)$ for the Kepler problem. Here k is chosen as 1

(3) *Anharmonic Oscillator*

Consider now a cubic anharmonic oscillator with a force

$$F = -kx - \alpha x^3 \, . \tag{1.3a}$$

It is definitely nonlinear. Its potential is a quartic function of x and its form depends upon the sign of the constant parameters k and α. Figures 1.3a and 1.3b depict the form of the force $F(x)$ and the corresponding potential function

$$V(x) = \frac{1}{2}kx^2 + \frac{\alpha}{4}x^4 \, , \tag{1.3b}$$

respectively, for $k = -1$ and $\alpha = 1$.

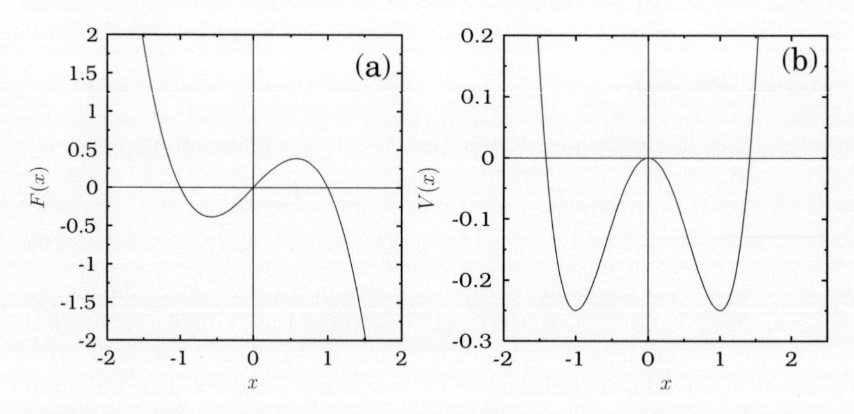

Fig. 1.3. Graphs of (**a**) the nonlinear force given by (1.3) and (**b**) the corresponding double-well potential function $V(x)$. The values of k and α are chosen as -1 and $+1$ respectively

Why should one worry about whether the force is linear or nonlinear? Does it make much of a difference in the physical behaviour of the system? Recent studies have clearly shown that it does make a profound difference depending on whether the force is linear or nonlinear. These studies have lead to revolutionary new concepts (when the force is nonlinear), and this field of study is generally called *nonlinear dynamics*.

1.2 Mathematical Implications of Nonlinearity

As noted above, the dynamics (that is, the evolution) of an N-particle system of masses m_i and position vectors $\boldsymbol{r}_i, i = 1, 2, ..., N$, is described by Newton's equations of motion (or by equivalent Lagrange's or Hamilton's equations of motion),

$$m_i \frac{\mathrm{d}^2 \boldsymbol{r}_i}{\mathrm{d}t^2} = \boldsymbol{F}_i \left(\boldsymbol{r}_1, \boldsymbol{r}_2, ..., \boldsymbol{r}_N, \frac{\mathrm{d}\boldsymbol{r}_1}{\mathrm{d}t}, \frac{\mathrm{d}\boldsymbol{r}_2}{\mathrm{d}t}, ..., \frac{\mathrm{d}\boldsymbol{r}_N}{\mathrm{d}t}, t \right), \quad i = 1, 2, ..., N \tag{1.4}$$

with prescribed $6N$ initial conditions, say $\boldsymbol{r}_i(0)$ and $\dot{\boldsymbol{r}}_i(0) \equiv (\mathrm{d}\boldsymbol{r}_i/\mathrm{d}t)\,|_{t=0}$. Here, for a given i, $\boldsymbol{F}_i$ is the total force acting on the ith particle of the system. We assume that $\boldsymbol{F}_i$'s are sufficiently well behaved for there to exist a unique and nonsingular solution $\boldsymbol{r}_i(t)$, $i = 1, 2, ..., N$, of (1.4). Equations (1.4) constitute a system of $3N$ coupled second-order ordinary differential equations of *deterministic* type, as long as the forces $\boldsymbol{F}_i$ are deterministic (that is no noise or randomness or stochasticity is present). So if one solves the initial value problem corresponding to (1.4), one expects that the future ($\boldsymbol{r}_i(t)$ and $\dot{\boldsymbol{r}}_i(t)$) can be predicted with required accuracy, provided quantum effects, relativistic effects and statistical aspects can be neglected.

However, there arises an important difference in the mathematical nature of the differential equations (1.4) depending upon the functional form of $\boldsymbol{F}_i$'s, that is whether $\boldsymbol{F}_i$'s are linear or not in the dependent variables $\boldsymbol{r}_j$ and $\mathrm{d}\boldsymbol{r}_j/\mathrm{d}t$, $j = 1, 2, ..., N$. (Note that t is an independent variable). To appreciate this difference, we may proceed as follows.

1.2.1 Linear and Nonlinear Systems

(A) If all the $\boldsymbol{F}_i$'s are linear in all the $\boldsymbol{r}_j$'s and $\mathrm{d}\boldsymbol{r}_j/\mathrm{d}t$'s (that is, $\boldsymbol{F}_i$'s are linearly proportional to $\boldsymbol{r}_j$'s and $\mathrm{d}\boldsymbol{r}_j/\mathrm{d}t$'s), then the system of differential equations (1.4) constitutes a system of *linear differential equations*.

Examples:

(1) ***Linear Harmonic Oscillator***

Force:

$$F = -kx\,. \tag{1.5}$$

Equation of motion:

$$m\frac{\mathrm{d}^2x}{\mathrm{d}t^2} = -kx \,. \tag{1.6}$$

(2) *Three-Dimensional Isotropic Harmonic Oscillator*

Force:

$$\boldsymbol{F} = -k\boldsymbol{r} \,. \tag{1.7}$$

Equation of motion:

$$m\frac{\mathrm{d}^2\boldsymbol{r}}{\mathrm{d}t^2} = -k\boldsymbol{r} \,. \tag{1.8}$$

(3) *Damped Harmonic Oscillator*

Force:

$$F = -b\frac{\mathrm{d}x}{\mathrm{d}t} - kx \,. \tag{1.9}$$

Equation of motion:

$$m\frac{\mathrm{d}^2x}{\mathrm{d}t^2} = -b\frac{\mathrm{d}x}{\mathrm{d}t} - kx \,. \tag{1.10}$$

In all the above examples, the forces are linear and the corresponding equations of motion are also linear differential equations.

(B) If any one (or more) of the $\boldsymbol{F}_i$'s is not linear even in one of the $\boldsymbol{r}_j$'s or $\mathrm{d}\boldsymbol{r}_j/\mathrm{d}t$'s or contains products of them, then we have essentially a system of *nonlinear ordinary differential equations* as the equations of motion.

Examples:

(1) *Anharmonic Oscillator*

Force:

$$F = -kx - \lambda x^3 \,. \tag{1.11}$$

Equation of motion:

$$m\frac{\mathrm{d}^2x}{\mathrm{d}t^2} = -kx - \lambda x^3 \,. \tag{1.12}$$

(2) *Kepler Problem*

Force:

$$\boldsymbol{F} = -\frac{k}{r^2}\hat{\boldsymbol{r}}, \quad \hat{\boldsymbol{r}} = \frac{\boldsymbol{r}}{r} . \tag{1.13}$$

Equation of motion (in polar coordinates):

$$mr^2\frac{\mathrm{d}\theta}{\mathrm{d}t} = l = \text{constant} , \tag{1.14}$$

$$m\frac{\mathrm{d}^2 r}{\mathrm{d}t^2} = -\frac{\mathrm{d}}{\mathrm{d}r}\left(-\frac{k}{r} + \frac{1}{2}\frac{l^2}{mr^2}\right) . \tag{1.15}$$

Equations (1.12) and (1.14–15) are nonlinear differential equations and the corresponding forces are nonlinear as well. Generally, we call physical systems subjected to linear forces as *linear dynamical systems* while systems driven by nonlinear forces as *nonlinear dynamical systems*.

Thus, we find that the set of equations of motion becomes a set of either linear or nonlinear differential equations depending upon whether the force is linear or nonlinear. The point is that the properties of nonlinear differential equations can drastically differ from that of linear differential equations, so that there can be characteristic differences in the mathematical and so the physical properties exhibited by these two kinds of systems. Correspondingly the mathematical analysis of them can also become widely different: for linear systems there are well defined analytical methods such as Laplace, Fourier and other transforms, Green's function, spectral and other methods, while for nonlinear differential equations no systematic analytic tools are in general available. As a result, very little is known about the behaviour of nonlinear dynamical systems and only very recently some unique general phenomena underlying these systems have been realized.

1.2.2 Linear Superposition Principle

Linear systems and so linear differential equations possess certain characteristic properties which make them easy to analyse. In particular, linear systems admit the so-called *linear superposition principle*. That is, if u_1 and u_2 are any two linearly independent solutions of a given homogeneous differential equation

$$Lu = 0 , \tag{1.16}$$

then the *linear* combination (or linear superposition)

$$u = au_1 + bu_2 , \tag{1.17}$$

where L is any given linear differential operator and a and b are arbitrary complex constants, is also a solution.

Proof:

Since $Lu = 0$, $Lu_1 = 0$, $Lu_2 = 0$

$$\Longrightarrow L(au_1 + bu_2) = a(Lu_1) + b(Lu_2) = 0 . \tag{1.18}$$

Exercise:

1. For the linear harmonic oscillator, equation (1.6), verify that the linear combination of the two independent solutions $\cos \omega t$ and $\sin \omega t$ namely $A \cos \omega t + B \sin \omega t$ is also a solution, where $\omega = \sqrt{k/m}$ and A, B are arbitrary real constants.

Thus for the nth order *linear homogeneous* differential equation of the form

$$\frac{\mathrm{d}^n u}{\mathrm{d}x^n} + a_1(x)\frac{\mathrm{d}^{n-1} u}{\mathrm{d}x^{n-1}} + \ldots + a_{n-1}(x)\frac{\mathrm{d}u}{\mathrm{d}x} + a_n(x)u = 0 , \tag{1.19}$$

where $a_i(x)$'s, $i = 1, 2, \ldots, n$, are functions of x, the general solution is given by the *linear superposition* of n linearly independent solutions, $u_i(x)$, $i = 1, 2, \ldots, n$, as

$$u(x) = C_1 u_1(x) + C_2 u_2(x) + \ldots + C_n u_n(x) , \tag{1.20}$$

where $C_1, C_2, \ldots, C_n$ are arbitrary constants. On comparing (1.19) with (1.16), the linear differential operator is

$$L = \frac{\mathrm{d}^n}{\mathrm{d}x^n} + a_1(x)\frac{\mathrm{d}^{n-1}}{\mathrm{d}x^{n-1}} + \ldots + a_{n-1}(x)\frac{\mathrm{d}}{\mathrm{d}x} + a_n(x) . \tag{1.21}$$

Now what happens in the case of a *linear inhomogeneous* differential equation

$$Lu = f(x) , \tag{1.22}$$

where L is given by (1.21) and $f(x)$ is a given function of x? One can verify that in this case the general solution can be written as

$$u(x) = \alpha_1(x)u_1(x) + \alpha_2(x)u_2(x) + \ldots + \alpha_n(x)u_n(x) , \tag{1.23}$$

where the unknown functions $\alpha_i(x)$, $i = 1, 2, \ldots, n$, may be obtained by quadratures (direct integrations) involving the linearly independent functions u_i, $i = 1, 2, \ldots, n$, of the linear homogeneous equation (1.19).

Proof:

By substituting the solution (1.23) into (1.22), one can choose

$$u_1\frac{\mathrm{d}\alpha_1}{\mathrm{d}x} + u_2\frac{\mathrm{d}\alpha_2}{\mathrm{d}x} + \ldots + u_n\frac{\mathrm{d}\alpha_n}{\mathrm{d}x} = 0 ,$$

$$u_1' \frac{d\alpha_1}{dx} + u_2' \frac{d\alpha_2}{dx} + \ldots + u_n' \frac{d\alpha_n}{dx} = 0 ,$$

$$\ldots$$

$$u_1^{(n-1)} \frac{d\alpha_1}{dx} + u_2^{(n-1)} \frac{d\alpha_2}{dx} + \ldots + u_n^{(n-1)} \frac{d\alpha_n}{dx} = f(x) , \qquad (1.24)$$

where $u_i' = du_i/dx$ and $u_i^{(n-1)} = d^{(n-1)}u_i/dx^{n-1}$. These equations may be treated as a system of n linear equations for the derivatives of the n unknown functions α_1, α_2, ..., α_n. Since u_1, u_2, ..., u_n are linearly independent, one can solve for $d\alpha_1/dx$, $d\alpha_2/dx$, ..., $d\alpha_n/dx$ uniquely as

$$\frac{d\alpha_1}{dx} = \gamma_1(x), \quad \frac{d\alpha_2}{dx} = \gamma_2(x), \quad \ldots, \quad \frac{d\alpha_n}{dx} = \gamma_n(x), \qquad (1.25)$$

where $\gamma_1, \gamma_2, \ldots, \gamma_n$ are known functions of u_i $(i = 1, 2, \ldots, n)$ and their derivatives. Integrating each member of (1.25), the unknown functions α_1, α_2, ..., α_n can be determined.

One can also easily prove that the above procedure is also equivalent to saying that the general solution of the linear inhomogeneous equation can be written as the sum of the complimentary functions and a particular integral, where the complimentary function is a linear superposition of the linearly independent solutions of the homogeneous part $Lu = 0$ and the particular integral is a particular solution of the inhomogeneous equation $Lu = f$.

Naturally for any nonlinear system (nonlinear differential equation) the linear superposition principle fails completely. Linear combinations of two independent solutions are no longer solutions. This leads to the consequence that unlike the case of a linear oscillator the frequency (or period) of a nonlinear oscillator depends on the amplitude of oscillation (see next chapter). Or in other words the dynamics of a nonlinear system depends heavily on the initial conditions while it is not so for linear systems.

Exercises:

2. For the following systems write the form of the forces and the corresponding equations of motion. Identify which of them are linear and which of them are nonlinear:
   a) Pendulum in free space
   b) Rigid body with one point stationary
   c) Linearly damped and periodically driven harmonic oscillator
   d) The Atwood's machine
   e) Particle in an exponential potential $V(\boldsymbol{r}) = V_0\, e^{-r/a}$
   f) Particle in a screened Coulomb potential $V(\boldsymbol{r}) = V_0 \frac{e^{-r/a}}{r}$
3. Verify that for the anharmonic oscillator equation (1.12), $x = ax_1 + bx_2$, where x_1 and x_2 are two linearly independent solutions, is not a solution.

4. Verify that for a linear second-order differential equation

$$\frac{\mathrm{d}^2u}{\mathrm{d}x^2} + a_1(x)\frac{\mathrm{d}u}{\mathrm{d}x} + a_2(x)u = f(x) ,$$

the general solution can be written as

$$u(x) = au_1(x) + bu_2(x) - u_1 \int \frac{f(x)u_2(x)}{W(x)}\mathrm{d}x + u_2 \int \frac{f(x)u_1(x)}{W(x)}\mathrm{d}x ,$$

where u_1 and u_2 are the two linearly independent solutions of the homogeneous part, a and b are arbitrary constants and the Wronskian $W(x) = (u_1\mathrm{d}u_2/\mathrm{d}x - u_2\mathrm{d}u_1/\mathrm{d}x)$. Identify the complimentary function and particular integral. Show that W is a constant for $a_1(x) = 0$.

1.3 Working Definition of Nonlinearity

From the above discussions, we can introduce a working definition to distinguish between linear and nonlinear differential equations. This can then be naturally extended to any other nonlinear system (like difference equations, partial differential equations, etc.).

If each of the terms of a given differential equation, *after rationalization*, has a total degree either 1 or 0 in the dependent variables and their derivatives, then, it is a linear differential equation. Even if one of the terms has a degree different from 0 or 1 in the dependent variables (and their derivatives), then it is nonlinear. Note that the presence of the independent variable does not affect the linearity/nonlinearity nature.

Examples:

A. ***Linear Differential Equations***

(1) $\mathrm{d}x/\mathrm{d}t + \omega_0^2 x = f\cos\omega t$

(2) $\mathrm{d}^2x/\mathrm{d}t^2 + \alpha\mathrm{d}x/\mathrm{d}t + \omega_0^2 x = f\cos\omega t$

(3) $\mathrm{d}^2x/\mathrm{d}t^2 + \mathrm{d}x/\mathrm{d}t + t^2 = 0$

(4) $\mathrm{d}^2x/\mathrm{d}t^2 + \mathrm{d}x/\mathrm{d}t + \mathrm{e}^{-t} = 0$

(5) $\mathrm{d}^2x/\mathrm{d}t^2 + t^2\mathrm{d}x/\mathrm{d}t + (a + b\cos\omega t)x = f\cos\Omega t$

(6) $\mathrm{d}x/\mathrm{d}t = ax + by$,
$\mathrm{d}y/\mathrm{d}t = cx + f\cos\omega t$

(7) $\mathrm{d}^2\boldsymbol{r}/\mathrm{d}t^2 + \alpha\mathrm{d}\boldsymbol{r}/\mathrm{d}t + k\boldsymbol{r} = \boldsymbol{f}\sin\omega t$

(8) $\sqrt{\mathrm{d}^2x/\mathrm{d}t^2 + \alpha\mathrm{d}x/\mathrm{d}t + \omega_0^2 x} = t$

B. ***Nonlinear Differential Equations***

(1) $\mathrm{d}x/\mathrm{d}t - x + x^3 = 0$

(2) $\mathrm{d}^2x/\mathrm{d}t^2 + \alpha \mathrm{d}x/\mathrm{d}t + \omega_0^2 x + \beta x^3 = f \cos\omega t$

(3) $\mathrm{d}^2x/\mathrm{d}t^2 + (\mathrm{d}x/\mathrm{d}t)^2 + t^2 = 0$

(4) $\mathrm{d}^2x/\mathrm{d}t^2 + \mathrm{d}x/\mathrm{d}t + \sin x = 0$

(5) $\mathrm{d}^2x/\mathrm{d}t^2 + \mathrm{d}x/\mathrm{d}t + \mathrm{e}^{-x} = 0$

(6) $\sqrt{\mathrm{d}^2x/\mathrm{d}t^2 + \alpha \mathrm{d}x/\mathrm{d}t + \omega_0 x} = x$

(7) $\mathrm{d}^2\boldsymbol{r}/\mathrm{d}t^2 + \mathrm{d}\boldsymbol{r}/\mathrm{d}t + \boldsymbol{r}/r^3 = 0$

(8) $\mathrm{d}x/\mathrm{d}t = a + \dfrac{1}{b+x} + y\,,$
$\mathrm{d}y/\mathrm{d}t = x - xy$

(9) $\mathrm{d}x/\mathrm{d}t = A + x^2y - Bx - x\,,$
$\mathrm{d}y/\mathrm{d}t = Bx - x^2y$

(10) $\mathrm{d}x/\mathrm{d}t = \sigma(y - x)\,,$
$\mathrm{d}y/\mathrm{d}t = rx - y - xz\,,$
$\mathrm{d}z/\mathrm{d}t = xy - bz$

1.4 Effects of Nonlinearity

Before we begin our detailed analysis of nonlinear systems, it is pertinent to ask ourselves whether the effects of nonlinearity on natural phenomena are so prominent as to warrant a separate study. To convince the readers that it is indeed so, let us briefly discuss some common effects of nonlinearity [3], which one encounters in our every day life, so as to appreciate the ubiquitous nature of nonlinear phenomena.

First of all, linear systems are generally gradual and gentle; their smooth and regular behaviour is met in various physical phenomena such as slowly flowing streams, small vibrations of a pendulum, electrical circuits that operate under normal conditions, engines working at low power, slowly reacting chemicals and so on. In contrast, nonlinear systems can exhibit regular as well as complicated and irregular behaviours depending upon various factors. Some examples are given below.

(i) Consider a pendulum placed in air medium (Fig. 1.4a). The restoring force is proportional to $\sin\theta$ which is nonlinear in θ. Its equation of motion is

$$\frac{\mathrm{d}^2\theta}{\mathrm{d}t^2} + \alpha\frac{\mathrm{d}\theta}{\mathrm{d}t} + \frac{g}{L}\sin\theta = 0\,, \tag{1.26}$$

where α is the damping coefficient. For small displacements $\sin\theta \approx \theta$ and the pendulum is a linear system. In this approximation its equation of motion is a linear differential equation given by

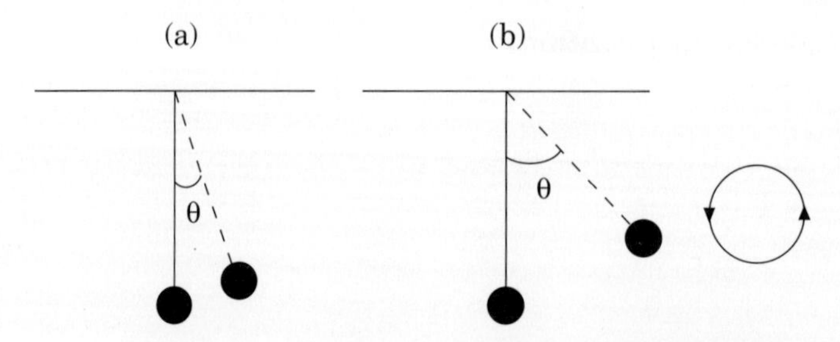

Fig. 1.4. (**a**) A pendulum executing small oscillations. (**b**) Pendulum executing circular motion for a sufficiently large initial displacement

$$\frac{\mathrm{d}^2\theta}{\mathrm{d}t^2} + \alpha\frac{\mathrm{d}\theta}{\mathrm{d}t} + \frac{g}{L}\,\theta = 0\,. \tag{1.27}$$

When the bob is disturbed from its equilibrium position, as time increases the amplitude of oscillation decreases and finally it comes to the rest state. When it is subjected to a weak periodic external force, in the limit $t \to \infty$ the bob exhibits periodic oscillations with the frequency of the applied force. However, if the initial displacement is sufficiently large then one has to analyse the full nonlinear equation (1.26) with an additional term representing the external force. It not only shows oscillatory behaviour but also shows rotational (moving in a circle) motion (Fig. 1.4b), and other complicated periodic and irregular oscillations as well, depending upon the strength of the applied force (see later chapters for more details on such motions).

(ii) When a metal bar with one end fixed is loaded with weight M, the bar bends (Fig. 1.5a). The bending depends on the magnitude M and the

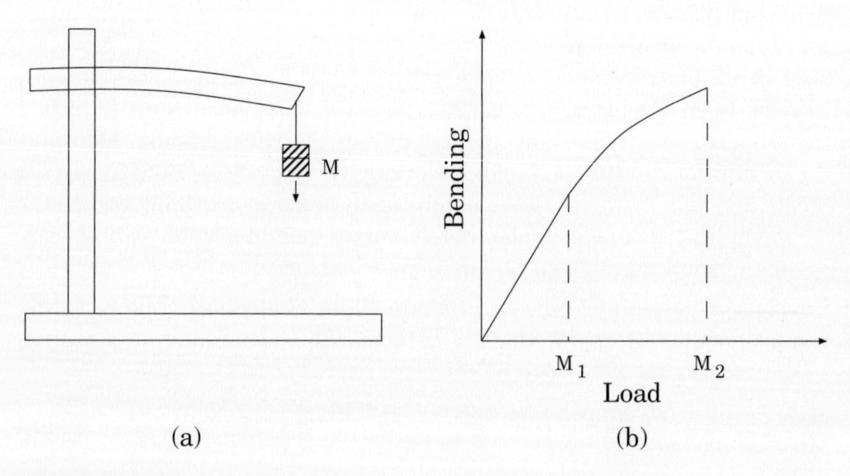

Fig. 1.5. (**a**) A metal bar with a load M. (**b**) Bending versus load. For load $M < M_1$, the bending is almost linear, while for sufficiently large M, say $M_1 < M < M_2$, the bending is nonlinear. At $M = M_2$, the bar breaks due to overload

stiffness, k, of the material. For a small load the metal bar is deformed slightly and will return to its original shape when the load is removed. For small weights the bending versus load is a linear curve. However, for large load the curve is no longer linear and becomes nonlinear and at a critical load the metal bar fractures or breaks (Fig. 1.5b).

(iii) Imagine the possible motions of a particle in the quadratic potential well (Fig. 1.1b) and double-well potential (Fig. 1.3b). For the quadratic potential well as seen earlier the force is linear. The particle exhibits oscillatory motion about the minimum of the potential. What will happen to the particle in the double-well potential where the force is nonlinear? Different oscillatory motions are possible. For example, the particle can exhibit

(a) oscillatory motion in the left well alone (Fig. 1.6a)

(b) oscillatory motion in the right well alone (Fig. 1.6a)

(c) jump between the two wells (Fig. 1.6b).

These types of different dynamics occur only in systems with nonlinear force. Fuller details are discussed in Chap. 5.

(iv) The propagation of sound waves in air or water appears to be a linear phenomenon, since they are usually modelled by the linear wave equation. But it is not difficult to create noisy and nonperiodic acoustic effects in a trumpet, clarinet or a wind instrument. These arise essentially due to the combined effects of the nonlinearities in the medium, the acoustic generator, the reflection and the impedance or reception of the acoustic waves. One may observe nonperiodic modulation in an organ-pipe generator of sound when a nonlinear mechanical impedance is placed at the open end of the meter-long pipe.

(v) Nonlinear effects can be clearly seen in certain chemical reactions. When a small amount of salt is added to water we get a salt solution; when sugar is added to coffee nothing unexpected happens. Cane sugar ($C_{12}H_{22}O_{11}$) and water (H_2O) in the presence of H^+ (catalyst) react to form glucose as follows

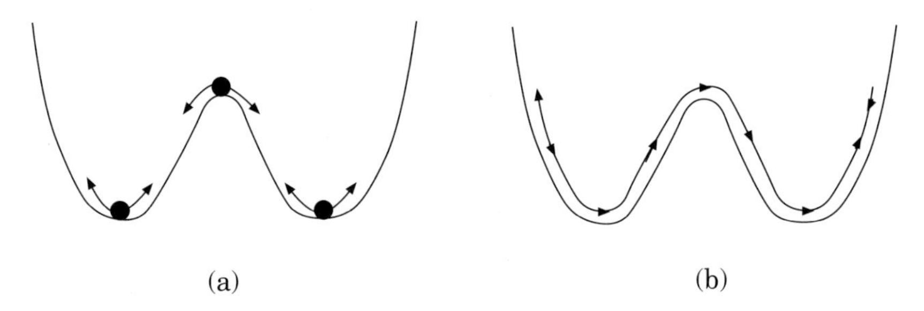

Fig. 1.6. Some of the possible motions of a particle in a double-well potential. (**a**) Particle motion is confined to the left or right potential well alone. (**b**) Hoping motion between the left and right wells

$$C_{12}H_{22}O_{11} + H_2O \xrightarrow{H^+} 2C_6H_{12}O_6 \ .$$

The rate equation describing the time variation of concentration $x(t)$ of glucose is given by

$$\frac{dx}{dt} = k(\alpha - x) \ , \tag{1.28}$$

where k is a positive constant for reaction and α is the initial concentration of cane sugar. Equation (1.28) is a linear differential equation. However, there are chemical reactions, governed by nonlinear differential equations, which show complicated variation of concentrations of chemical compounds. Nitric oxide (NO) and oxygen (O_2) react to form NO_2 as

$$2NO + O_2 \longrightarrow 2NO_2 \ .$$

The rate equation describing the time variation of concentration of NO_2 (denoted by $x(t)$) is given by the nonlinear differential equation

$$\frac{dx}{dt} = k(\alpha - x)^2(\beta - x) \ , \tag{1.29}$$

where k is a positive constant for the reaction, α and β are the initial concentrations of NO and O_2 respectively and $x(0) = 0$. The concentration of SO_3 in the reaction

$$2SO_2 + O_2 \longrightarrow 2SO_3$$

satisfies again a nonlinear differential equation

$$\frac{dx}{dt} = \frac{k(\alpha - x)}{\sqrt{x}} \ , \tag{1.30}$$

where k is again a positive constant for the reaction, α is the initial concentration of SO_2 and $x(0) = 0$. Another example of nonlinear phenomenon is the so-called Belousov–Zhabotinsky reaction [4,5]. In this reaction an organic molecule (malonic acid) is oxidised by bromate ions, the process being catalyzed by a Ce^{4+}/Ce^{3+} ion. The basic reactants are $Ce_2(SO_4)_3$, $NaBrO_3$, $CH_2(COOH)_2$ and H_2SO_4 to which a colour indicator is added. For a certain range of temperature the concentration of Ce^{4+} and that of other chemicals oscillate periodically which can be identified by periodic changes of the colour of the solution when an indicator is added. Above a certain critical temperature, a complicated variation of concentration of the chemicals occurs.

(vi) Pressing the accelerator of a car results in a smooth increase of speed. At first this increase lies in a linear region and everything is regular. However, as the car approaches a certain critical power output nonlinear effects become important. A small additional pressure on the accelerator may cause the car to vibrate violently or the engine to overheat and seize up. Likewise, a small turn of the volume knob of a stereo will produce a linear response from a speaker, but if the volume knob is turned too far then the nonlinearity in the electrical circuitry produces a marked distortion.

(vii) Rings of Saturn, stock market fluctuations, complicated population growth and uncertainties in weather forecast are essentially due to the nonlinearities present in the systems.

Numerous other examples can be cited where one can clearly realize the effects of nonlinearity.

Thus one of the main tasks in nonlinear dynamics is to consider typical equations of motion or evolution equations which describe physical as well as other natural systems and investigate the characteristic features underlying them. Through these studies one can try to bring out general behaviours underlying these systems. For this purpose, it is useful to study very simple but realistic models and extract the required informations. This is what we will do in the following chapters.

2. Linear and Nonlinear Oscillators

The linear superposition principle which is valid for linear differential equations is no longer valid for nonlinear ones. A physical consequence is that when the given dynamical system admits oscillatory motion, the associated frequency of oscillation is in general amplitude-dependent in the case of nonlinear systems, while it is not so in the case of linear systems. Particularly, this can have dramatic consequences in the case of forced and damped nonlinear oscillators, leading to nonlinear *resonances* and *jump (hysteresis) phenomenon* for low strengths of nonlinearity parameters. Such behaviours can be analysed using various approximation methods. However, as the *control parameter* varies further, the nonlinear systems can enter into more and more complex motions through different *routes*, where detailed numerical analysis and possible analog simulations using electronic circuits can be of much help to analyse them. In this chapter, we will introduce some basic features associated with nonlinear oscillations and postpone the discussions on more complex motions to later chapters. However, before discussing the nature of such nonlinear oscillations, we shall first discuss briefly the salient features associated with a damped and driven linear oscillator in order to compare its properties with nonlinear oscillators.

2.1 Linear Oscillators and Predictability

As we have discussed in Chap. 1, physical systems whose motions are described by linear differential equations are called linear systems. If they are associated with oscillatory behaviour, then they are designated as *linear oscillators*. The characteristic features of such linear systems are their *insensitiveness to infinitesimal changes in initial conditions* and at the most constant separation of nearby trajectories in phase space. As a consequence the future behaviour becomes completely predictable. To illustrate these ideas, let us consider the simple example of a linear harmonic oscillator of mass m, damped by a viscous drag force, and acted upon by an external periodic force. Such a model represents a very large number of physical systems ranging from forced oscillations in an LCR circuit to electron oscillations in an electromagnetic field. The oscillations are then described by an inhomogeneous, linear, second-order differential equation of the form

$$m\frac{\mathrm{d}^2x}{\mathrm{d}t^2}+\alpha'\frac{\mathrm{d}x}{\mathrm{d}t}+\omega_0'^2x=F\sin\omega t\ ,$$

or equivalently

$$\frac{\mathrm{d}^2x}{\mathrm{d}t^2}+\alpha\frac{\mathrm{d}x}{\mathrm{d}t}+\omega_0^2x=f\sin\omega t\ ,\quad \alpha=\frac{\alpha'}{m},\quad f=\frac{F}{m}\ ,\quad \omega_0^2=\frac{\omega_0'^2}{m}\ ,\tag{2.1}$$

where $x(t)$ is the displacement of the system, subjected to suitable initial conditions. We choose the initial conditions for convenience to be $x(0)=A$, $\dot{x}(0)=0$. In (2.1), $\omega_0/2\pi$ corresponds to the natural frequency, α is the strength of the damping, while f and ω stand for the forcing amplitude and angular frequency, respectively, of the external force. We will first consider the special cases of (2.1) before looking at the full equation.

2.1.1 Free Oscillations

When the damping and external forcing are absent, $\alpha=0$, $f=0$, the system (2.1) is essentially a free linear harmonic oscillator. Its solution corresponding to the initial conditions $x(0)=A$, $\dot{x}(0)=0$ is

$$x(t)=A\cos\omega_0 t\ .\tag{2.2}$$

This solution represents simple harmonic vibrations with period $T=2\pi/\omega_0$, which is obviously *independent* of the amplitude A. A typical plot of x versus t called a *trajectory plot* resulting from the initial condition $(x,\dot{x})=(1,0)$ at time $t=0$ is given in Fig. 2.1a. The solution can be displayed in the $x-\dot{x}$ plane as well, which we call the *phase plane* or *phase space*. A point in the phase plane, with coordinates $(x(t),\ \dot{x}(t))$ at certain time t, is called a *phase point*. In general, for increasing t, a phase point shall move through the phase plane. A plot of the solution of the system in the phase plane is called

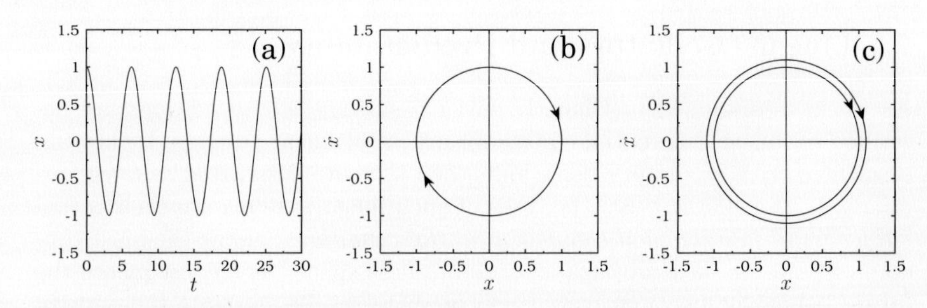

Fig. 2.1. (**a**) Solution curve and (**b**) phase portrait of the free linear harmonic oscillator equation (2.1) ($\alpha=0$, $f=0$). The subplot (**c**) depicts the nearby periodic orbits with initial conditions $(1,0)$ and $(1.1,0)$

a *phase trajectory/flow/orbit*. More often, the plot of x versus $\dot{x}$ is called a *phase portrait*[1].

The phase portrait of the system (2.1) with $\alpha = 0$, $f = 0$, and for $(x(0), \dot{x}(0)) = (1, 0)$, is shown in Fig. 2.1b. The motion is represented by a closed ellipse as shown in the figure, and the state value or the phase point $(x, \dot{x})$ moves continuously on this closed orbit as time progresses. What will happen if we change the starting position (initial condition) by a small amount? This is illustrated in Fig. 2.1c where two periodic orbits resulting from slightly different starting values of $(x, \dot{x})$ run in step with just a small difference in the amplitude and phase. They continue to run nicely in step for all times because the period of oscillations of the two motions are the same. Each of the phase trajectory is in fact characterized by the constant energy $E = \frac{1}{2}(\dot{x}^2 + \omega_0^2 x^2) = \frac{1}{2}\omega_0^2 A^2$.

2.1.2 Damped Oscillations

Consider again the system (2.1), now with the damping present and the forcing absent, $\alpha \neq 0$, $f = 0$. The explicit solution of this linear differential equation can be readily written. It has the form

$$x(t) = A_1 \exp(m_1 t) + A_2 \exp(m_2 t) \ , \tag{2.3a}$$

where

$$m_{1,2} = \frac{1}{2}\left[-\alpha \pm \sqrt{\alpha^2 - 4\omega_0^2}\right] \tag{2.3b}$$

and A_1, A_2 are integration constants. We now have three possibilities:

(1) Under damping: $0 < \alpha < 2\omega_0$

(2) Critical damping: $\alpha = 2\omega_0$

(3) Over damping : $\alpha > 2\omega_0$

For $0 < \alpha < 2\omega_0$ the solution (2.3) becomes

$$x(t) = (A\omega_0/C)\exp(-\alpha t/2)\cos(Ct - \delta) \ , \tag{2.4a}$$

where

$$C = \sqrt{\omega_0^2 - (\alpha^2/4)} \ , \quad \delta = \tan^{-1}(\alpha/2C) \ . \tag{2.4b}$$

It represents exponentially decaying sinusoidal solution. Figure 2.2a shows the phase trajectory for $\alpha = 0.25$ and $\omega_0^2 = 1$. The trajectory spirals about the origin and reaches it in the asymptotic limit $t \to \infty$, Fig. 2.2a. For $\alpha > 2\omega_0$ the solution is nonoscillatory and approaches the origin exponentially. This

[1] More generally, we will later on call the x versus p plot, where p is the canonically conjugate momentum to x, as the phase portrait. In the present case $p = m\dot{x}$, with m taken as the unit mass.

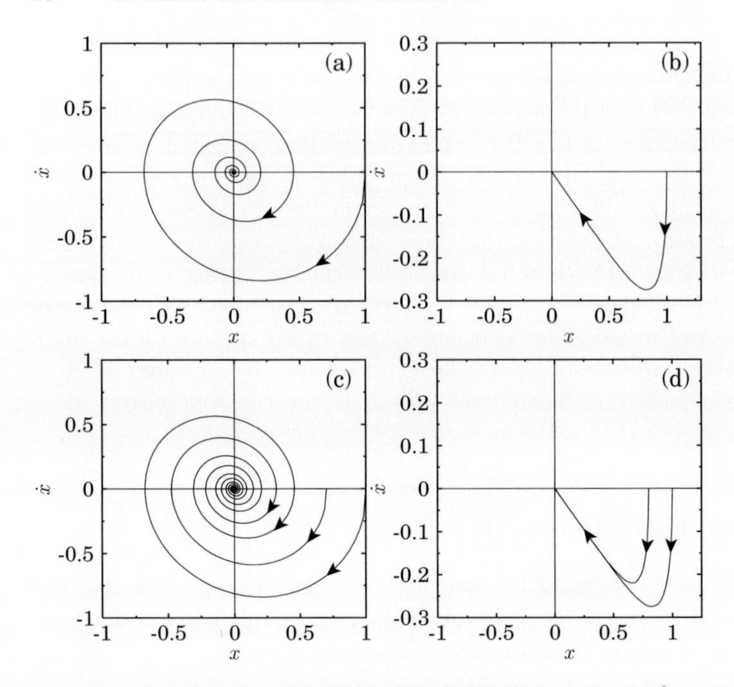

Fig. 2.2. Phase trajectories of the system (2.1) with $f = 0$, $\omega_0^2 = 1$ for (**a**) $\alpha = 0.25$ and (**b**) $\alpha = 3$. Evolution for two different initial conditions for the same set of parameters for (**c**) $\alpha = 0.25$ and (**d**) $\alpha = 3$

is shown in Fig. 2.2b for $\alpha = 3$ and $\omega_0^2 = 1$. Further in both the cases the trajectories starting from different initial conditions asymptotically reach the origin as shown in Figs. 2.2c and 2.2d. Thus the motion is insensitive to the initial conditions.

Exercise:

1. Discuss the motion of the damped linear harmonic oscillator for $\alpha = 2\omega_0$ and $\alpha > 2\omega_0$.

2.1.3 Damped and Forced Oscillations

The full system (2.1), $\alpha \neq 0$, $f \neq 0$, can also be easily integrated. In the underdamped case, it reads

$$x(t) = (A_t\omega_0/C)\ \mathrm{e}^{-\alpha t/2}\cos(Ct - \delta) + A_p\cos(\omega t - \gamma)\,, \tag{2.5a}$$

where

$$A_p = \frac{f}{(\omega_0^2 - \omega^2)^2 + \alpha^2\omega^2}\,, \quad \gamma = \tan^{-1}\left(\frac{\omega^2 - \omega_0^2}{\alpha\omega}\right)\,, \tag{2.5b}$$

and C is given by (2.4b). Here the constants A_t and δ are chosen so as to satisfy the initial conditions $x(0) = A$ and $\dot{x}(0) = 0$. There are several interesting observations that we can make about the solution. For example, if $\alpha > 0$, the first term in the solution (2.5) which is independent of f and ω falls off exponentially fast to zero while the second term oscillates periodically with time. Thus, the first term is the *transient* and the second term is the dominant component of the solution. For large t, the frequency of oscillation is $\omega/2\pi$ and its amplitude is A_p. At the *resonance* value $\omega = \sqrt{\omega_0^2 - (\alpha^2/2)}$ ($\approx \omega_0$, if the damping coefficient α is sufficiently small), the amplitude of oscillation A_p takes a maximum value. The corresponding solution curve and the phase trajectory are depicted in Figs. 2.3a and 2.3b for $\alpha = 0.1$, $\omega_0^2 = 1$, $\omega = 1$ and $f = 0.1$. Here also one can check that the motion is insensitive to initial conditions.

Exercise:

2. Obtain the solution (2.5) for the equation (2.1). Show that A_p takes its maximum value at $\omega^2 = \omega_0^2 - (\alpha^2/2)$. What is the value of $(A_p)_{\max}$?

2.2 Damped and Driven Nonlinear Oscillators

As a prototype to understand the effect of nonlinearity on resonant linear oscillations described in the previous section, we ask the question as to how the dynamics of the system (2.1) gets modified when typical nonlinear spring forces are included. As a standard example, we include a cubic nonlinear force to the left hand side of (2.1) so that the equation of motion becomes

$$\ddot{x} + \alpha\dot{x} + \omega_0^2 x + \beta x^3 = f \sin \omega t \,, \quad (\dot{} = \mathrm{d}/\mathrm{d}t\,) \tag{2.6}$$

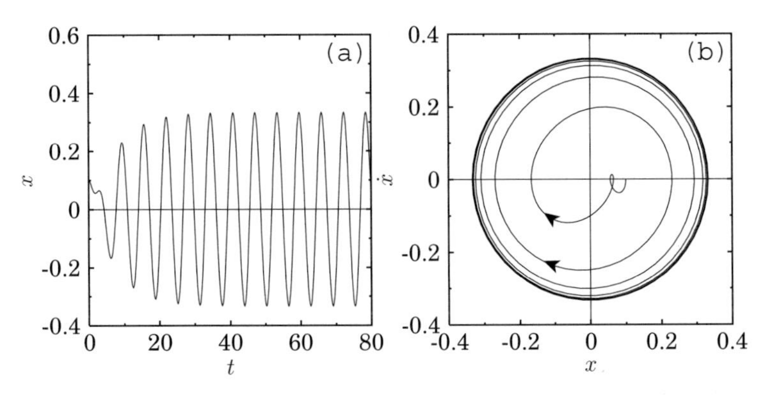

Fig. 2.3. (**a**) Solution curve and (**b**) phase trajectory of the system (2.1) for $\alpha = 0.1$, $\omega_0^2 = 1$, $\omega = 1$ and $f = 0.1$

where β is the strength of nonlinearity. Equation (2.6) is a ubiquitous nonlinear system called the Duffing oscillator about which we will have much more to say in later chapters. Here we will confine ourselves to some of its elementary dynamical properties alone.

2.2.1 Free Oscillations

For the undamped and unforced case ($\alpha = 0,\ f = 0$), again one has bounded and periodic solutions but they are now described by Jacobian elliptic functions (see appendix A). For example, when $\omega_0^2 > 0,\ \beta > 0$, with the initial conditions $x(0) = A,\ \dot{x}(0) = 0$ we have the solution [1,2] (Fig. 2.4a)

$$x(t) = A\mathrm{cn}(\overline{\omega}t; k), \quad \overline{\omega} = \sqrt{\omega_0^2 + \beta A^2}\,, \tag{2.7}$$

where cn is the Jacobian elliptic function of modulus $k = \sqrt{\beta A^2/2\,(\omega_0^2 + \beta A^2)}$. Note here the important feature that unlike the case of linear oscillator the angular frequency $\bar{\omega}$ is now *dependent* on the amplitude A, which is the characteristic feature of nonlinear oscillations. This means that the dynamics is sensitive to the amplitude or initial condition in the present nonlinear system, a theme which will recur frequently in our discussions further in this book. Again the phase trajectories are concentric curves with the origin being an equilibrium point (as shown in Fig. 2.4b, for two different values of amplitude A corresponding to two different initial conditions). They are characterized by the energy $E = \frac{1}{2}\left(\dot{x}^2 + \omega_0^2 x^2 + \frac{\beta}{2}x^4\right)$, which is a constant.

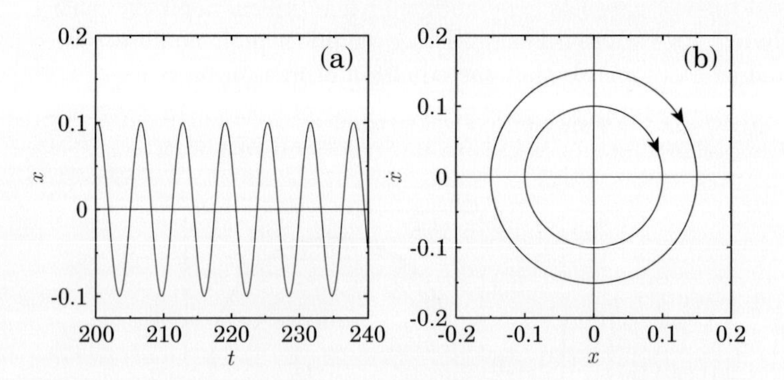

Fig. 2.4. (**a**) Trajectory plot and (**b**) phase portrait of the undamped and unforced (free) nonlinear oscillator equation (2.6) ($\alpha = 0,\ f = 0$)

Exercise:

3. An even more illustrative example which shows very clearly the amplitude dependance of frequency or period of nonlinear oscillations is the following velocity dependent potential system,

$$\left(1+\lambda x^2\right)\ddot{x} - \lambda x\dot{x}^2 + \omega_0^2 x = 0\ , \quad \lambda : \text{constant}\ . \tag{2.8}$$

This equation can be directly integrated (try!) to give the exact solution for the initial conditions $x(0) = A$, $\dot{x}(0) = 0$,

$$x(t) = A\cos\Omega t, \quad \Omega = \frac{\omega_0}{\sqrt{1+\lambda A^2}} \tag{2.9}$$

so that the angular frequency Ω or period $T = 2\pi/\Omega$ is amplitude dependent [3].

2.2.2 Damped Oscillations

When $\alpha > 0$ and $f = 0$ in (2.6), in the physically interesting under-damped case $\alpha < 2\omega_0$, again we have damped oscillatory solution corresponding to an inwardly spiralling trajectory towards the equilibrium point at the origin (Fig. 2.5). Though in the general case no explicit solution can be given, in the special case $\alpha = \pm(3/\sqrt{2})\omega_0$, and $\beta > 0$, one can give the exponentially decaying oscillatory solution [4] as (see again appendix A)

$$\begin{aligned} x(t) &= \left(\omega_0/\sqrt{\beta}\right) A\exp\left(-\omega_0 t/\sqrt{2}\right)\mathrm{cn}(Av;k)\ , \\ v &= -\sqrt{2}\exp\left(-\omega_0 t/\sqrt{2}\right) - v_0\ , \end{aligned} \tag{2.10}$$

where the modulus $k = 1/\sqrt{2}$ and A and v_0 are constants.

2.2.3 Forced Oscillations – Primary Resonance and Jump Phenomenon (Hysteresis)

One of the most interesting aspects of nonlinear oscillators of the form (2.6) is that, even for very small β, its behaviour can be qualitatively different

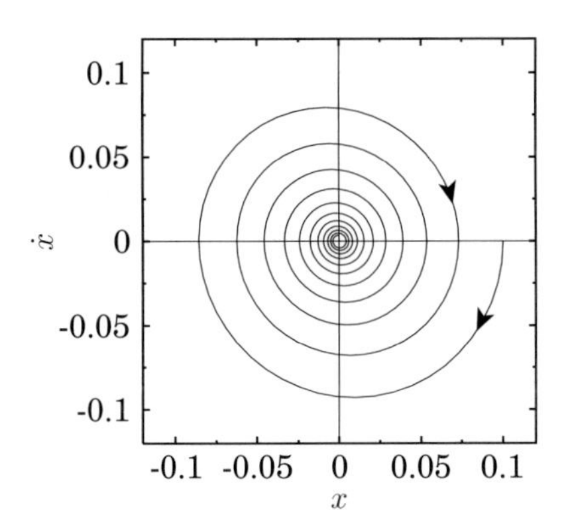

Fig. 2.5. Damped oscillatory solution of the nonlinear equation (2.6)) for $f = 0$, $\alpha = 1$, $\beta = 1$ and $\omega_0^2 = 1$

from that of the linear oscillator discussed in Sect. 2.1. This can be seen very easily as discussed below.

Following our discussion on resonant linear oscillations in Sect. 2.1, we can assume the solution of (2.6) for $0 < \beta \ll 1$ and $\omega \approx \omega_0$ (primary resonance) in the lowest order approximation in β (valid for large t) to be of the form

$$x(t) = A\sin(\omega t + \delta) , \tag{2.11}$$

where δ is a phase constant to be fixed. Substituting (2.11) into (2.6) and equating the coefficients of $\sin\omega t$ and $\cos\omega t$ to zero separately (neglecting higher harmonic terms), we obtain

$$\left[-A\omega^2 + \omega_0^2 A + \frac{3}{4}\beta A^3\right] \cos\delta - \alpha A\omega \sin\delta = f , \tag{2.12a}$$

$$\left[-A\omega^2 + \omega_0^2 A + \frac{3}{4}\beta A^3\right] \sin\delta + \alpha A\omega \cos\delta = 0 . \tag{2.12b}$$

Multiplying (2.12a) by $\cos\delta$ and (2.12b) by $\sin\delta$ and adding the two we get

$$\left[-A\omega^2 + \omega_0^2 A + \frac{3}{4}\beta A^3\right] = f\cos\delta . \tag{2.13a}$$

Similarly multiplying (2.12a) by $\sin\delta$ and (2.12b) by $\cos\delta$ and subtracting the two we get

$$-\alpha A\omega = f\sin\delta . \tag{2.13b}$$

Squaring and adding the two equations (2.13a) and (2.13b) we obtain

$$\left[(\omega_0^2 - \omega^2)A + \frac{3}{4}\beta A^3\right]^2 + (\alpha A\omega)^2 = f^2 . \tag{2.14}$$

Equation (2.14) is an algebraic equation for the amplitude A of the response and it defines a functional relation between A and the parameters in (2.6). It is often called the *frequency-response equation* for obvious reasons. The plot of A as a function of ω for given values of f, ω_0, α and β is called a *frequency-response curve*. It is not symmetrical as in the linear case, (2.5b), but it leans to the right of $\omega = \omega_0$ for $\beta > 0$. A typical form of it is shown in Fig. 2.6 for $\beta > 0$. We note that (2.14) reduces to the previous relation (2.5b) in the limit $\beta = 0$, if we choose $A_p = A$. Once again, near $\omega = \omega_0$, we have the resonance but now in the nonlinear case. For this reason (2.14) is also called a *resonance curve*.

Exercise:

4. Plot the frequency-response curve for $\beta < 0$ and show that it leans to the left of $\omega = \omega_0$.

It can be observed from Fig. 2.6 that for a range of values of the frequency ω the amplitude A has multiple values. The multivaluedness of the response

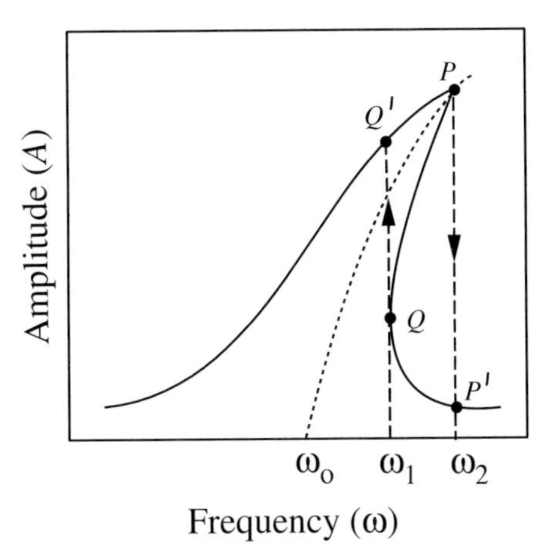

Fig. 2.6. A typical frequency-response curve of the nonlinear equation (2.6) for $\beta > 0$

curve due to the nonlinearity has a significance because it leads to *jump* or *hysteresis* phenomenon [5–7], as explained below.

When the angular frequency ω of the periodic driving force is gradually increased (Fig. 2.6) while keeping its magnitude f constant, the amplitude of oscillation A will be found to increase steadily and it reaches a maximum value at P. Further increase of frequency results in a markedly different behaviour as the amplitude of oscillation makes a discontinuous jump from P to P'. Similarly, an amplitude jump from Q to Q' will occur if the frequency is gradually lowered from a value beyond P'. The maximum amplitude corresponding to the point P is attainable only when approached from a lower frequency. The portion of the response curve between the points P and Q is unstable and hence cannot be produced in numerical or experimental simulation.

Suppose one solves the equation (2.6) numerically for a specific value of ω lying in the region ω_1 and ω_2. What would be the amplitude of the resulting motion? In this case the amplitude of the response depends on the initial conditions. For a set of initial conditions one may get a steady state solution with amplitude lying in the lower portion $P'Q$. Solution with amplitude corresponding to the upper portion PQ' will be realized for another set of initial conditions. Thus, in contrast with linear systems, the *nature* of the steady state *solution of a nonlinear system can depend on the initial conditions*.

2.2.4 Secondary Resonances (Subharmonic and Superharmonic)

Apart from the jump phenomenon and primary resonance discussed above, another characteristic behaviour of nonlinear systems of the form (2.6) is the existence of *secondary resonances* [5–7]. For example, when $\omega \approx 3\omega_0, \beta \ll 1$, the solution to (2.6) can be approximated by

$$x(t) = A\sin\left(\frac{\omega t}{3} + \delta\right) + \frac{1}{\omega_0^2 - \omega^2} f\sin\omega t\,, \tag{2.15}$$

which is the exact solution of (2.6) for $\beta = 0$, $\alpha = 0$, $\omega = 3\omega_0$ and A and δ are integration constants. To see this one can proceed as follows. The solution of (2.6) without the nonlinear term consists of the complementary function (x_c) or free oscillation which is the solution with $f = 0$, $\alpha = 0$ and the particular solution (x_p) given by

$$x_p = \frac{1}{\mathrm{D}^2 + \omega_0^2} f\sin\omega t\,, \tag{2.16}$$

where $\mathrm{D}^2 = \mathrm{d}^2/\mathrm{d}t^2$. From $\mathrm{D}^2\sin\omega t = -\omega^2\sin\omega t$ we find that $\mathrm{D}^2 = -\omega^2$. Thus

$$x_p = \frac{1}{-\omega^2 + \omega_0^2} f\sin\omega t\,. \tag{2.17}$$

The free oscillatory solution ($f = 0$, $\alpha = 0$) is

$$x_c = A\sin(\omega_0 t + \delta)\,. \tag{2.18}$$

Adding (2.17) and (2.18) and substituting $\omega_0 = \omega/3$ we obtain the solution (2.15). For $\alpha \neq 0$ and $\beta \ll 1$ we can assume the solution of (2.6) as (2.15) and obtain the algebraic equation for the amplitude A of the response as was done previously for the primary resonance. We note that the frequency of the particular solution is the same as that of the excitation while the frequency of free oscillation is one-third of the frequency of the excitation. For this reason the associated resonance is called *one-third subharmonic resonance* . This subharmonic resonance is a consequence of the nonlinearity. For more details see for example, A.H. Nayfeh and D.T. Mook [6]. Figures 2.7 show the synthesis of a steady state subharmonic response. Figure 2.7a is the steady state free oscillation term. Figure 2.7b is the particular solution of the linearized governing equation. Figure 2.7c is the first approximation of the actual response.

For $\omega \approx \omega_0/3$, $\beta \ll 1$, one can obtain the approximate solution

$$x(t) = A\sin(3\omega t + \delta) + \frac{1}{\omega_0^2 - \omega^2} f\sin\omega t\,, \tag{2.19}$$

in which the first term corresponds to a *superharmonic resonance of order three* or *overtones*. In Fig. 2.8 the three curves show how the response is

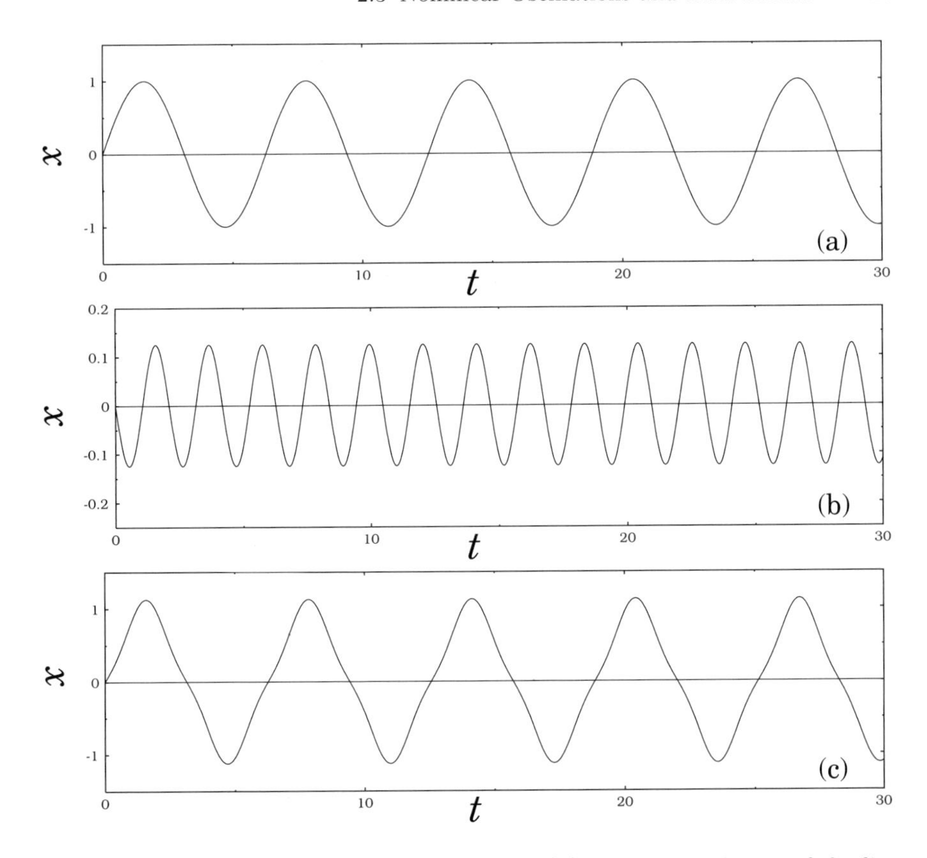

Fig. 2.7. Plots of (**a**) free oscillatory solution, (**b**) particular solution of the linearized governing equation and (**c**) approximate actual response of the system in the subharmonic resonance region for (2.6). Here $f = 1$, $\omega_0 = 1$, $\omega = 3$, $A = 1$ and $\delta = 0$

formed from the particular solution and the free oscillation term. The subharmonic and superharmonic resonances constitute the *secondary resonances* phenomena.

2.3 Nonlinear Oscillations and Bifurcations

The complete dynamics of the nonlinear oscillators of the form (2.6) is too complicated to be described in terms of resonant oscillations of the type mentioned above alone. The equation of motion is in general not solvable exactly. Qualitative and quantitative ideas on the types of oscillations and their stability for small strength of the nonlinearity parameter can be obtained by making use of one of the several perturbation methods available in the literature (For details, see appendix B).

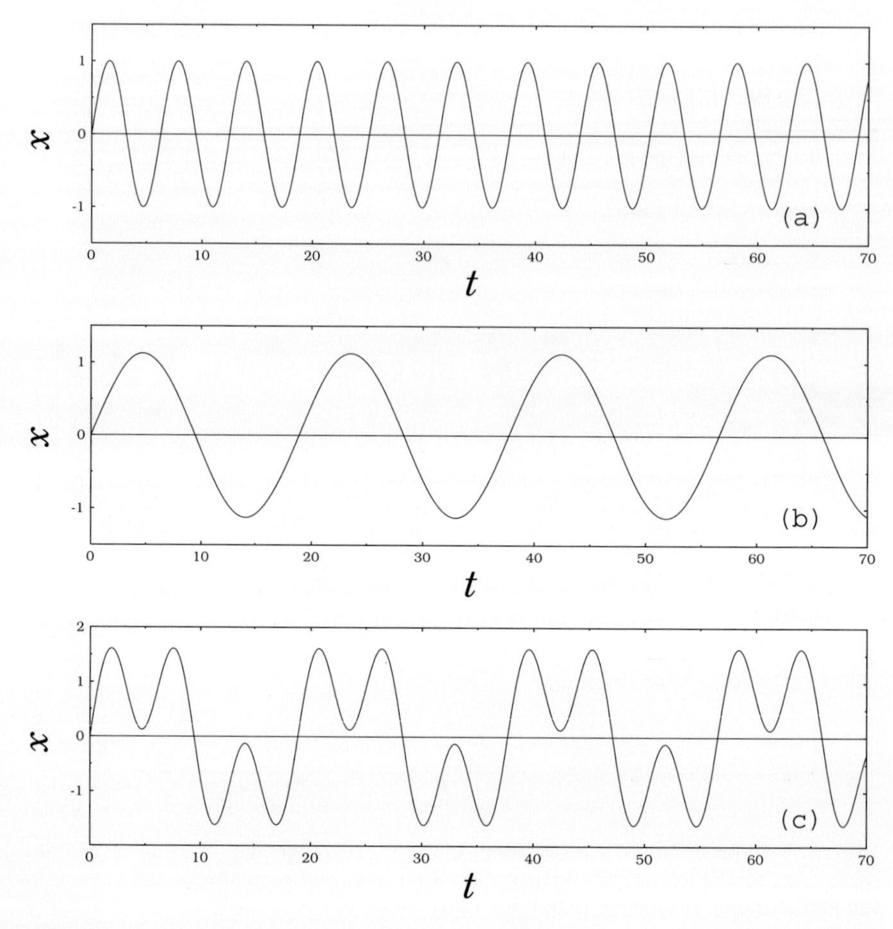

Fig. 2.8. Same as Figs. 2.7 but now in the superharmonic resonance region. Here $f = 1$, $\omega_0 = 1$, $\omega = 1/3$, $A = 1$ and $\delta = 0$

However, a more complete picture of the dynamics can be obtained essentially by a straightforward and detailed numerical analysis using any one of the available standard numerical algorithms, like the fourth-order Runge–Kutta integration method (see appendix C). The result is that one obtains a rich variety of *bifurcation* phenomena, namely the successive qualitative (and quantitative) changes in the nature of oscillations at critical values of the nonlinearity parameter (or another control parameter for a fixed nonlinearity parameter) as it is varied smoothly. One finds that the dynamical behaviour of nonlinear systems can very sensitively depend on initial conditions. Detailed analysis of the Duffing oscillator system will be undertaken in Chap. 5.

Problems

Linear Oscillators

1. Obtain the solution of the undamped, forced linear harmonic oscillator
$$\ddot{x} + \omega_0^2 x = f \sin \omega t$$
and identify the nature of resonance.
2. Analyse the dynamics of the damped linear harmonic oscillator driven by the two combined periodic external forcings of different frequencies,
$$\ddot{x} + \alpha\dot{x} + \omega_0^2 x = f_1 \sin \omega_1 t + f_2 \sin \omega_2 t\ , \quad \alpha > 0\ .$$
3. Oscillator systems instead of being driven by external forces can also be excited by parametric modulation. Consider the sinusoidal parametric excitation of the linear harmonic oscillator described by the Mathieu equation
$$\ddot{x} + \omega_0^2(1 + \epsilon \cos \omega t)x = 0\ , \quad \epsilon \ll 1\ .$$
Discuss the dynamics underlying the system (see also Ref. [6]).

Nonlinear Oscillators

4. Obtain the frequency-response relations and draw the primary resonance curves for the following nonlinear oscillators. Also discuss the nature of the secondary resonances. Compare the dynamics between the various systems.
   a) Quadratic Duffing oscillator:
   $$\ddot{x} + \alpha\dot{x} + \omega_0^2 x + \beta x^2 = f \sin \omega t$$
   b) Periodically driven elastic oscillator:
   $$\ddot{x} + \alpha\dot{x} + \omega_0^2 x + \beta x^2 + \gamma x^3 = f \sin \omega t$$
   c) Driven van der Pol oscillator:
   $$\ddot{x} - \alpha(1 - x^2)\dot{x} + \omega_0^2 x = f \sin \omega t$$
   d) Duffing–van der Pol oscillator:
   $$\ddot{x} - \alpha(1 - x^2)\dot{x} + \omega_0^2 x + \beta x^3 = f \sin \omega t$$
   e) Driven velocity-dependent oscillator:
   $$\left(1 + \lambda x^2\right)\ddot{x} - \lambda x\dot{x}^2 + \alpha\dot{x} + \omega_0^2 x = f \sin \omega t$$
5. Consider the following parametrically modulated nonlinear oscillators and investigate the underlying *parametric resonances* and frequency-response relations.
   a) Parametrically driven Duffing oscillator:
   $$\ddot{x} + \alpha\dot{x} + \omega_0^2(1 + \eta \sin \omega t)x + \beta x^3 = 0\ ,$$
   where η is a parameter.

b) Parametrically driven van der Pol oscillator:

$$\ddot{x} - \alpha\left(1 - x^2\right)\dot{x} + \omega_0^2(1 + \eta\sin\omega t)x = 0$$

c) Parametrically driven Duffing–van der Pol oscillator:

$$\ddot{x} - \alpha\left(1 - x^2\right)\dot{x} + \omega_0^2(1 + \eta\sin\omega t)x + \beta x^3 = 0$$

6. a) Consider the undamped Duffing oscillator when simultaneously excited by means of two different external periodic forces of differing frequencies and satisfying the equation of motion

$$\ddot{x} + \omega_0^2 x + \beta x^3 = f_1\cos\omega_1 t + f_2\cos\omega_2 t\ .$$

By using any one of the approximation procedures, show that the forced oscillations contain terms with frequencies $(\omega_1 \pm 2\omega_2)$ and $(\omega_2 \pm 2\omega_1)$ at second-order and obtain the corresponding frequency-amplitude relation. What is the effect of damping?
b) Extend the above study to the other nonlinear oscillators mentioned under problem 4.
c) Analyse the effect of an external periodic driving to the parametrically excited nonlinear oscillators in problem 5 in the case of (i) identical frequencies and (ii) different frequencies.

3. Qualitative Features

We shall now turn our attention to the study of the general behaviour of nonlinear systems. Already we have mentioned that there are no general methods to solve these systems, even though some interesting problems like the dynamics of the pendulum, the Kepler system and the rigid body are solvable exactly (see for example the book of H. Goldstein: *Classical Mechanics* [1]). Recent progress in identifying such solvable or integrable dynamical systems will be discussed in more detail in Chap. 10. However, even when the problem is not exactly solvable, one can start the analysis from the simplest possible states or structures admitted by the system, which one may be able to identify by inspection or by simple analysis. For example, one can easily identify the *equilibrium/ static/ stationary states (fixed points)* and then ask the question how stable are these states for small disturbances or perturbations. Or in other words, what is the nature of the dynamics in the vicinity of these equilibrium states? Physically, an equilibrium state is a steady or homogeneous state and it may mean, for example, no motion (rest state) of a pendulum, constant population density of a species and so on. The question is whether the system will continue to remain so for all times under practical situations.

Example 1:

A pendulum has two distinct vertical equilibrium states (one stable and the other unstable).

Example 2:

A particle moving in the double-well potential (Fig. 1.3b) has three equilibrium states: The origin (the local maximum of the potential) is unstable and the other two (the two minima of the potential) are stable.

The next interesting and familiar states exhibited by these systems are the *periodic motions* such as the oscillatory motion of a pendulum, bounded orbits of a Kepler particle and so on. In simple dynamical systems they can also be identified and analysed rather easily, but in more complicated systems even this may require considerable effort. Often dynamical systems may admit much more *complex motions*. Can one develop a systematic but at least qualitative description of such motions?

Now, the equations of motion describing a finite number of physical variables in most situations form a system of second-order differential equations (recall equation (1.4)). In addition, in dynamics one frequently comes across evolution equations which are higher-order in nature and also systems of first-order differential equations, depending upon the nature of the system under consideration.

Examples:

For a linear harmonic oscillator the equation of motion is a second-order differential equation. For a two body problem it is a system of six coupled second-order ordinary differential equations and so on. The rate equations describing chemical reactions are sets of first-order differential equations.

However, it is interesting to note that any differential equation (of any order) can always be written as a *system of first-order* differential equations, that too without any explicit time (independent variable) dependence, as demonstrated in the following sections. This allows one to study the *qualitative features* of dynamical systems through combined *geometrical* and *analytical* methods. Such qualitative studies also lead to the notions of *attractors* (and *repellers*) and *basins of attractions* associated with them. Stable equilibrium points are the simplest of such attractors. Limit cycles (which we have already come across in the last chapter) are bit more sophisticated attractors. There are even more exotic attractors (quasiperiodic, strange and so on) and other orbits which we will study in this and later chapters.

3.1 Autonomous and Nonautonomous Systems

To start with we will introduce the notion of *autonomous* and *nonautonomous dynamical systems*. Dynamical systems whose equations of motion have no explicit dependence on time are called *autonomous systems*. If they do have explicit dependence on time, they are called *nonautonomous systems*.

Examples:

In the following, $\boldsymbol{r} = (x, y, z)$ are the coordinate variables, t is the time variable, all the other quantities are control parameters of the corresponding systems.

A. *Autonomous Systems:*

1. Damped harmonic oscillator

$$\mathrm{d}^2x/\mathrm{d}t^2 + \alpha \mathrm{d}x/\mathrm{d}t + \omega_0^2 x = 0$$

2. An exponential oscillator

$$d^2x/dt^2 + e^x = 0$$

3. Kepler problem

$$\mu d^2\boldsymbol{r}/dt^2 = -k\boldsymbol{r}/r^3$$

4. Brusselator equations
(a model chemical reaction of two species)

$$\dot{x} = a - x - bx + x^2y ,$$
$$\dot{y} = bx - x^2y .$$

Here and in the following overdot denotes differentiation with respect to t.

5. Lorenz equations
(a simplified model system of atmospheric convection)

$$\dot{x} = \sigma(y - x) ,$$
$$\dot{y} = rx - y - xz ,$$
$$\dot{z} = xy - bz .$$

6. Rössler equations
(a model system of chemical reactions in a stirred tank)

$$\dot{x} = -(y + z) ,$$
$$\dot{y} = x + ay ,$$
$$\dot{z} = b + z(x - c) .$$

B. *Nonautonomous Systems:*

1. Driven linear oscillator

$$d^2x/dt^2 + \alpha dx/dt + \omega_0^2 x = f \cos \omega t$$

2. Driven nonlinear oscillator (Duffing oscillator)

$$d^2x/dt^2 + \alpha dx/dt + \omega_0^2 x + \beta x^3 = f \sin \omega t$$

3. Parametrically driven nonlinear oscillator

$$d^2x/dt^2 + \alpha dx/dt + \omega_0^2(1 + f \sin \omega t)x + \beta x^3 = 0$$

4. Driven van der Pol oscillator (a relaxation oscillator)

$$d^2x/dt^2 - b\left(1 - x^2\right) dx/dt + \omega_0^2 x = f \cos \omega t , \quad b > 0$$

5. Driven pendulum

$$d^2x/dt^2 + \alpha dx/dt + \sin x = f \sin \omega t$$

6. Driven Morse (exponential) oscillator

$$\mathrm{d}^2x/\mathrm{d}t^2 + \alpha\mathrm{d}x/\mathrm{d}t + \beta\mathrm{e}^{-x}\left(1 - \mathrm{e}^{-x}\right) = f\cos\omega t$$

The above kind of classification of dynamical systems into autonomous and nonautonomous types is often convenient as they can correspond to different physical situations in which, respectively, external forcing (including modulation) is present or absent.

3.2 Dynamical Systems as Coupled First-Order Differential Equations: Equilibrium Points

Considering any of the above two types of finite dimensional dynamical systems, namely autonomous and nonautonomous and whether linear or nonlinear, they can always be written as a *system* of first-order autonomous differential equations. Let us first consider a few examples.

1. *Linear harmonic oscillator*
   Equation of motion:

$$\mathrm{d}^2x/\mathrm{d}t^2 + \omega_0^2 x = 0\ .$$

   Equivalent first-order differential equations:

$$\dot{x} = y\ ,$$
$$\dot{y} = -\omega_0^2 x\ ,$$

   where overdot denotes differentiation with respect to t.

2. *Duffing oscillator*
   Equation of motion:

$$\ddot{x} + \alpha\dot{x} + \omega_0^2 x + \beta x^3 = f\cos\omega t\ .$$

   Equivalent first-order differential equations (nonautonomous):

$$\dot{x} = y\ ,$$
$$\dot{y} = -\alpha y - \omega_0^2 x - \beta x^3 + f\cos\omega t\ ,$$

   or equivalently (autonomous):

$$\dot{x} = y\ ,$$
$$\dot{y} = -\alpha y - \omega_0^2 x - \beta x^3 + f\cos z\ ,$$
$$\dot{z} = \omega\ .$$

3. *A system of coupled anharmonic oscillators*
   Equations of motion:

$$\ddot{x} + 2Ax + 4\alpha x^3 + 2\delta xy^2 = 0\ ,$$
$$\ddot{y} + 2By + 4\beta y^3 + 2\delta x^2 y = 0\ .$$

Equivalent first-order differential equations:

$$\begin{aligned}
\dot{x} &= u\ , \\
\dot{u} &= -2Ax - 4\alpha x^3 - 2\delta x y^2\ , \\
\dot{y} &= v\ , \\
\dot{v} &= -2By - 4\beta y^3 - 2\delta x^2 y\ .
\end{aligned}$$

4. *Rigid body equations*

$$\begin{aligned}
I_1\dot{\omega}_1 - \omega_2\omega_3\,(I_2 - I_3) &= N_1\ , \\
I_2\dot{\omega}_2 - \omega_3\omega_1\,(I_3 - I_1) &= N_2\ , \\
I_3\dot{\omega}_3 - \omega_1\omega_2\,(I_1 - I_2) &= N_3\ .
\end{aligned}$$

5. *Lotka–Volterra equations*
(describing the population of a system of two competing species)

$$\begin{aligned}
\dot{x} &= ax - xy\ , \\
\dot{y} &= xy - by\ .
\end{aligned}$$

Similarly all other examples given in Sect. 3.1 can also be written as equivalent first-order equations.

In the above examples, we see that there are systems such as the Lorenz, rigid body and Lotka–Volterra equations which are inherently sets of first-order coupled differential equations. On the other hand, equations of motion of dynamical systems with n degrees of freedom are n coupled second-order differential equations which can be equivalently written as $2n$ first-order coupled differential equations. For example, it is a second-order differential equation for a linear harmonic oscillator and for the different single anharmonic oscillators, while it is a set of two coupled second-order differential equations for the two coupled anharmonic oscillators system. However, all these equations of motion can also be rewritten as suitable sets of first-order coupled differential equations as shown above. Similarly, even for nonautonomous cases also, one can formally introduce a new *dependent* variable and rewrite the system as a set of first-order *autonomous* system (see the example 2 above).

Thus, a general dynamical system can always be described by a system of n first-order equations for n dynamical variables. For the special case of Hamiltonian or conservative systems, these dynamical variables often will be the (position)coordinates and (canonically conjugate)momenta. But other choices are also possible (like energy, angular momentum, etc.). We can designate the state variables collectively as the *state* $X = (x_1, x_2, ..., x_n)^{\mathrm{T}} \in R^n$, which is an n-dimensional vector. Then the equation of motion can be written in the general form

$$\dot{X} = F(X)\ , \tag{3.1a}$$

or equivalently

$$\dot{x}_1 = F_1(x_1, x_2, \ldots, x_n) ,$$
$$\dot{x}_2 = F_2(x_1, x_2, \ldots, x_n) ,$$
$$\ldots$$
$$\ldots$$
$$\dot{x}_n = F_n(x_1, x_2, \ldots, x_n) , \tag{3.1b}$$

where $F = (F_1, F_2, \ldots, F_n)^{\mathrm{T}}$. Examples are as given above.

One obvious physically important and relevant state of the system (as discussed in the introduction of this chapter) is the *equilibrium* or *static* or *stationary state* (that is the state which does not change in time),

$$\dot{X} = 0 = F(X) , \tag{3.2}$$

provided it exists. Any admissible solution of $F(X) = 0$, which we call as X^*, then gives an *equilibrium* or *fixed point* of the system (for reasons explained in the next section). This is because if X^* is a solution of (3.2) at a given time, it continues to be so for all times, and so it is a fixed point or equilibrium point. It is also sometimes called a *singular point* (as $\mathrm{d}x_1/\mathrm{d}x_2$, etc. are not defined at this state. See Sect. 3.4.1 also.). This means that the equilibrium points, if they exist, correspond to the condition

$$F_1(x_1^*, x_2^*, \ldots, x_n^*) = F_2 = \ldots = F_n = 0 . \tag{3.3}$$

Examples:

1. For the linear harmonic oscillator, $X^* = (x^*, y^*) = (0, 0)$ is the equilibrium point.
2. For the Lorenz equations, there are three equilibrium points: $(x^*, y^*, z^*) = (0,0,0)$, $\left(\sqrt{b(r-1)}\,,\ \sqrt{b(r-1)}\,,\ r-1\right)$, $\left(-\sqrt{b(r-1)}\,,\ -\sqrt{b(r-1)}\,,\ r-1\right)$.

Exercise:

1. Find the fixed points of the other examples given in Sects. 3.1 and 3.2.

3.3 Phase Space/Phase Plane and Phase Trajectories: Stability, Attractors and Repellers

In Chap. 2 we have seen that the dynamics of the various subcases of the Duffing oscillator can be studied qualitatively by plotting the *trajectories* in the two-dimensional $(x - \dot{x})$ *phase plane*. In fact such studies can be carried out in a geometrical framework for the more general system (3.1) also by introducing the n-dimensional *phase space* corresponding to the coordinates $(x_1, x_2, \ldots, x_n)$.

Thus, it is instructive and advantageous to represent geometrically the trajectory of (or the path followed by) the *phase point* $X(t)$, representing the set of state variables and starting from the initial state $X(t_0)$, in the n-dimensional phase space defined by coordinate axes $(x_1, x_2, \ldots, x_n)$. In particular, for a single particle system like the harmonic oscillator we have the phase space coordinates (x_1, x_2) or $(x, y = \dot{x})$, defining a two-dimensional phase plane[1]. The trajectories are then curves in this two-dimensional phase space. The equilibrium points are then just fixed points in the phase plane. As noted above, we have already seen several examples of such two-dimensional systems in Chap. 2. One can obtain or investigate the type of trajectories in the neighbourhood of the given equilibrium/fixed points, which will also indicate the type of stability of the equilibrium point (examples are discussed in the next section).

One can say that the fixed point or equilibrium point X^* is *stable* if the neighbouring trajectories approach X^* asymptotically (as $t \to \infty$), so that it is an *attractor*. Small deviations about the stable equilibrium point do not alter the state of the system as it returns to its original state after the transients die down (like a Thanjavur doll). On the other hand, X^* is *unstable*, if the neighbouring trajectories move away from it as $t \to \infty$, so that it is a *repeller*. If neither of these happens, then X^* is neutrally stable. One can make a finer distinction of stability as *orbitally stable, asymptotically stable, Lyapunov stable* and so on states. However, we do not consider these finer distinctions in this book and we consider only the above type of asymptotic stability[2] in our further discussions. For more sophisticated mathematical treatment see for example P. Glendenning: *Stability, Instability and Chaos* [2].

Examples:

Consider the dynamics of a simple pendulum and that of a wall-clock pendulum in air medium. In both the systems if the bobs are at rest initially, they continue to remain so for all times. So the rest state corresponds to an equilibrium point in the associated phase plane. When the simple pendulum bob is given a small initial displacement, it starts to exhibit damped oscillations and ultimately comes to a rest state as time progresses. On the

[1] For a single variable system like the population density of species, or concentration density of a chemical substance, satisfying a single first-order differential equation, the phase space will be just one-dimensional.

[2] More specifically, asymptotic stability can be defined as follows: Let X^* be a stable solution for $t \geq t_0$. Further, if there exists a $\eta(t_0) > 0$ such that $||X(t_0) - X^*|| < \eta$ implies $\lim_{t \to \infty} ||X(t) - X^*|| = 0$, then the solution X^* is asymptotically stable.

other hand, if the initial displacement is sufficiently large then the wall-clock pendulum does not come to the rest state asymptotically (until the energy supplied by the tensional spring or the battery is ceased), instead it exhibits a periodic oscillation, namely the limit cycle. Here the limit cycle motion is due to the balance between self-excitation and damping. Due to the air resistance the amplitude of the oscillation decreases with time. On the other hand, the anchor clip controlling the rotation of the escape wheel delivers periodic impulses to the pendulum bob which in turn increases the amplitude of the oscillation. A dynamic balance between these two effects leads to the limit cycle motion of the pendulum bob. For the simple pendulum the equilibrium point is stable, while for the wall-clock pendulum it is unstable.

With these introductory remarks let us consider a general two-dimensional system and see how the equilibrium points can be analysed for their stability property rigorously.

3.4 Classification of Equilibrium Points: Two-Dimensional Case

How does one develop definitive mathematical criterion for the type of (asymptotic) stability defined in the previous section? One way is to infinitesimally disturb or perturb the system near the given equilibrium point *linearly* and analyse under what conditions the perturbation will die down or grow exponentially fast. This will then give the required classification of equilibrium points based on the linear stability analysis. We will carry out such a formulation for the general two-dimensional system in this section. and extend it to the general n-dimensional system later in this chapter.

3.4.1 General Criteria for Stability

Consider a two-dimensional dynamical system

$$\dot{x} = P(x,y) \,, \tag{3.4a}$$

$$\dot{y} = Q(x,y) \,, \quad (\dot{} = \mathrm{d}/\mathrm{d}t) \tag{3.4b}$$

where P and Q are some well defined functions of x and y (as in the examples given above in Sect. 3.2). Let

$$X = X^* = X_0 \equiv (x_0, y_0) \tag{3.5}$$

be the equilibrium point[3] of (3.4), so that $P(x_0,y_0) = Q(x_0,y_0) = 0$. In order to determine the stability of this equilibrium point we slightly disturb it (or we say that we infinitesimally *perturb* it),

[3] Since $\mathrm{d}x/\mathrm{d}y = P(x,y)/Q(x,y)$, it becomes indeterminate at the equilibrium point (x_0,y_0). For this reason the equilibrium point is also sometimes called a *singular point*, while all the other points are *regular points*.

$$x = x_0 + \xi(t) , \tag{3.6a}$$

$$y = y_0 + \eta(t) , \quad \xi, \eta \ll 1 \tag{3.6b}$$

and ask the question what the resultant motion is. For this purpose we make a Taylor expansion about the equilibrium point,

$$\begin{aligned} P(x, y) &= P(x_0 + \xi, y_0 + \eta) \\ &= P(x_0, y_0) + \partial P/\partial x\,|_{x_0,y_0} \cdot \xi + \partial P/\partial y\,|_{x_0,y_0} \cdot \eta \\ &\quad + \text{ higher-order terms in } (\xi, \eta) , \end{aligned} \tag{3.7a}$$

$$\begin{aligned} Q(x, y) &= Q(x_0 + \xi, y_0 + \eta) \\ &= Q(x_0, y_0) + \partial Q/\partial x\,|_{x_0,y_0} \cdot \xi + \partial Q/\partial y\,|_{x_0,y_0} \cdot \eta \\ &\quad + \text{ higher-order terms in } (\xi, \eta) . \end{aligned} \tag{3.7b}$$

Here $\partial P/\partial x\,|_{x_0,y_0}$, etc. mean that the derivatives are evaluated at (x_0, y_0).

Now using the fact that at the equilibrium point $P(x_0, y_0) = Q(x_0, y_0) = 0$, and that the higher-order terms on the right hand sides can be neglected in most cases (see below), equation (3.4) can be rewritten (or approximated) as a system of two first-order *linear* coupled differential equations,

$$\dot{\xi} = a\xi + b\eta , \tag{3.8a}$$

$$\dot{\eta} = c\xi + d\eta , \tag{3.8b}$$

where

$$\begin{aligned} a &= \partial P/\partial x\,|_{x_0,y_0}, \quad b = \partial P/\partial y\,|_{x_0,y_0} , \\ c &= \partial Q/\partial x\,|_{x_0,y_0}, \quad d = \partial Q/\partial y\,|_{x_0,y_0} . \end{aligned} \tag{3.8c}$$

Differentiating (3.8a) with respect to t and eliminating the η dependence, we obtain a second-order differential equation for the variable ξ:

$$\ddot{\xi} - (a + d)\dot{\xi} + (ad - bc)\xi = 0 . \tag{3.9}$$

Its solution is given by (using standard method)

$$\xi(t) = A \exp(\lambda_1 t) + B \exp(\lambda_2 t) , \tag{3.10a}$$

where

$$\lambda_{1,2} = \frac{1}{2}\left[(a + d) \pm \sqrt{(a + d)^2 - 4(ad - bc)}\,\right], \quad ad - bc \neq 0 , \tag{3.11}$$

and A, B are in general complex integration constants. Note that in general the eigenvalues can be complex. Substituting (3.10a) into (3.8a) we have

$$\eta(t) = C \exp(\lambda_1 t) + D \exp(\lambda_2 t) , \tag{3.10b}$$

where

$$C = A(\lambda_1 - a)/b, \quad D = B(\lambda_2 - a)/b . \tag{3.10c}$$

Thus the solution of (3.8) is given by (3.10). From (3.11) it is easy to check that λ_1 and λ_2 are the eigenvalues of the matrix $M = \begin{pmatrix} a & b \\ c & d \end{pmatrix}$, $\det M \neq 0$.

When $\det M = 0$, that is $ad - bc = 0$, we find from (3.11) that at least one of the eigenvalues must be zero and in this case the dynamics of the system near the equilibrium point is governed by the omitted nonlinear terms on the right hand side of (3.7) (see below).

An equilibrium point $X_0 = (x_0, y_0)$ of the system (3.4) is dynamically stable (asymptotically), if all the permissible perturbations (ξ, η) which solve (3.8) decay in time exponentially fast and that no admissible perturbation can grow in time. This means that the stability of the equilibrium point (x_0, y_0) can be studied from (3.10) using (3.11) and that it depends solely on the numerical values (magnitude and sign) of the eigenvalues λ_1 and λ_2. Thus for an equilibrium point to be stable, unstable or neutrally stable, the following conditions are to be satisfied:

Stability nature of equilibrium point	Conditions on the eigenvalues
Stable	Real parts of both the eigenvalues are negative
Unstable	Real part of even one of the two eigenvalues is positive
Neutral	Real parts of both the eigenvalues are zero

3.4.2 Classification of Equilibrium (Singular) Points

Based on the above criteria for stability or instability of the equilibrium points the following broad classification of them can be made, depending upon the nature of the eigenvalues.

Case 1: $\lambda_1 \leq \lambda_2 < 0$ – stable node/star

When both λ_1 and λ_2 are real and less than zero, it is clear from (3.10) that both $\xi(t)$ and $\eta(t) \to 0$ as $t \to \infty$. Thus any trajectory starting from the neighbourhood of the equilibrium point (x_0, y_0) in the $(x - y)$ phase plane approaches it exponentially in the long time limit. In order to study how the trajectories reach the equilibrium point in the $(x-y)$ phase space we consider the slope $\mathrm{d}y/\mathrm{d}x$ of the trajectories. Shifting the origin to the equilibrium point (x_0, y_0) for convenience (which means replacing $x \to x - x_0$ and $y \to y - y_0$ in (3.6) and then rewriting the equation for the new variables (x, y)), from (3.10) and (3.6) we obtain

$$\frac{\mathrm{d}y}{\mathrm{d}x} = \frac{C\lambda_1 \mathrm{e}^{(\lambda_1 - \lambda_2)t} + D\lambda_2}{A\lambda_1 \mathrm{e}^{(\lambda_1 - \lambda_2)t} + B\lambda_2} . \tag{3.12}$$

Here we can distinguish two cases: (i) $\lambda_1 = \lambda_2 = \lambda$ and (ii) $\lambda_1 \neq \lambda_2$.

(i) When $\lambda_1 = \lambda_2 = \lambda$ we have

$$\frac{\mathrm{d}y}{\mathrm{d}x} = \frac{(C+D)\lambda}{(A+B)\lambda} = \text{constant} = m \,. \tag{3.13}$$

Integration of the above equation yields

$$y = mx + s \,, \tag{3.14}$$

where s is a constant. Using the fact that the origin is the fixed point we may choose $s = 0$ and so equation (3.14) is a straight line passing through the origin (the fixed point). Thus, in the $(x-y)$ phase space trajectories approach the equilibrium point along straight line paths as shown in Fig. 3.1a. Here (and in all the other Figs. 3.1) the origin corresponds to the point (x_0, y_0). An equilibrium point of this type is called a *stable star* (because of the starlike structure).

(ii) For $\lambda_1 \neq \lambda_2$, the slope of the trajectories vary with time and it decreases exponentially to the value D/B since $\lambda_1 < \lambda_2$. One can easily show that the trajectories in the $(x-y)$ plane approach the equilibrium point along parabolic paths. (For details see appendix D). Figure 3.1b depicts the phase trajectories in the neighbourhood of the equilibrium point after a suitable rotation of the coordinate axes. The equilibrium point of this type is called a *stable node*.

Case 2: $\boldsymbol{\lambda_1 \geq \lambda_2 > 0}$ – *unstable node/star*

Next, when λ_1 and λ_2 are real and positive, $|\xi(t)|, |\eta(t)| \to \infty$ as $t \to \infty$. Therefore, the trajectories starting from a neighbourhood of the equilibrium point diverge from it and so the equilibrium point is unstable. For $\lambda_1 = \lambda_2 = \lambda$ the slope $\mathrm{d}y/\mathrm{d}x$ is given again by (3.13) and the trajectories now diverge along straight lines. The equilibrium point is said to be an *unstable star*. When $\lambda_1 \neq \lambda_2$ it is an *unstable node*. The corresponding phase trajectories are illustrated in Figs. 3.1c and 3.1d.

Case 3: $\boldsymbol{\lambda_1, \lambda_2}$ *complex conjugates – stable/unstable focus*

Let λ_1 and λ_2 be complex conjugates given by

$$\lambda_1 = \alpha + \mathrm{i}\beta \,, \quad \lambda_2 = \alpha - \mathrm{i}\beta \,, \tag{3.15}$$

where α, β are real constants and $\beta > 0$ (for convenience). Now the solution of (3.8) becomes

$$\xi(t) = |A| \mathrm{e}^{\alpha t} \cos(\beta t + \phi) \,, \tag{3.16a}$$

$$\eta(t) = |D| \mathrm{e}^{\alpha t} \sin(\beta t + \phi') \,, \tag{3.16b}$$

where

$$A = A_r + \mathrm{i}A_i \,, \quad D = D_r + \mathrm{i}D_i \,,$$

$$\phi = \tan^{-1}(A_i/A_r) \,, \quad \phi' = \tan^{-1}(D_r/D_i) \,. \tag{3.16c}$$

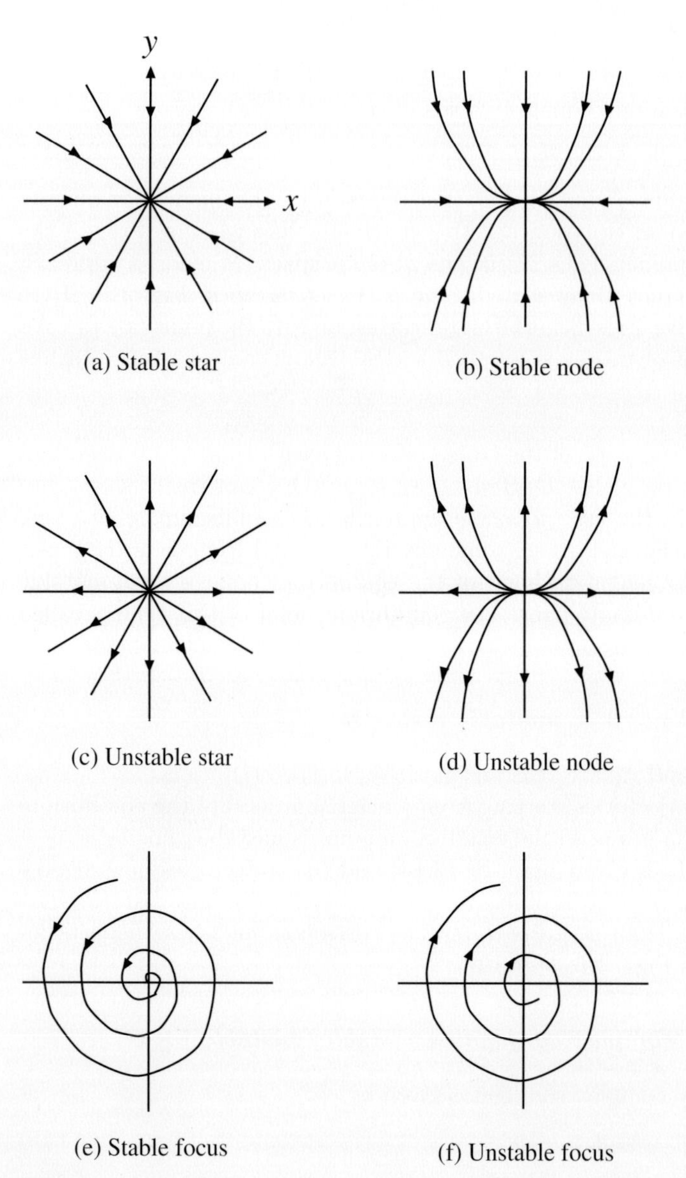

Fig. 3.1. (**a–h**) Classification of equilibrium points: Phase trajectories near the equilibrium points. In all the figures the equilibrium point is at the origin. (**i–l**) Phase trajectories of a linear system for (**i**) $M = \begin{pmatrix} 0 & 0 \\ 0 & 0 \end{pmatrix}$; $\lambda_1 = \lambda_2 = 0$, (**j**) $M = \begin{pmatrix} 0 & 1 \\ 0 & 0 \end{pmatrix}$; $\lambda_1 = \lambda_2 = 0$, (**k**) $M = \begin{pmatrix} 1 & 0 \\ 0 & 0 \end{pmatrix}$; $\lambda_1 = 1$, $\lambda_2 = 0$, and (**l**) $M = \begin{pmatrix} -1 & 0 \\ 0 & 0 \end{pmatrix}$; $\lambda_1 = -1$, $\lambda_2 = 0$

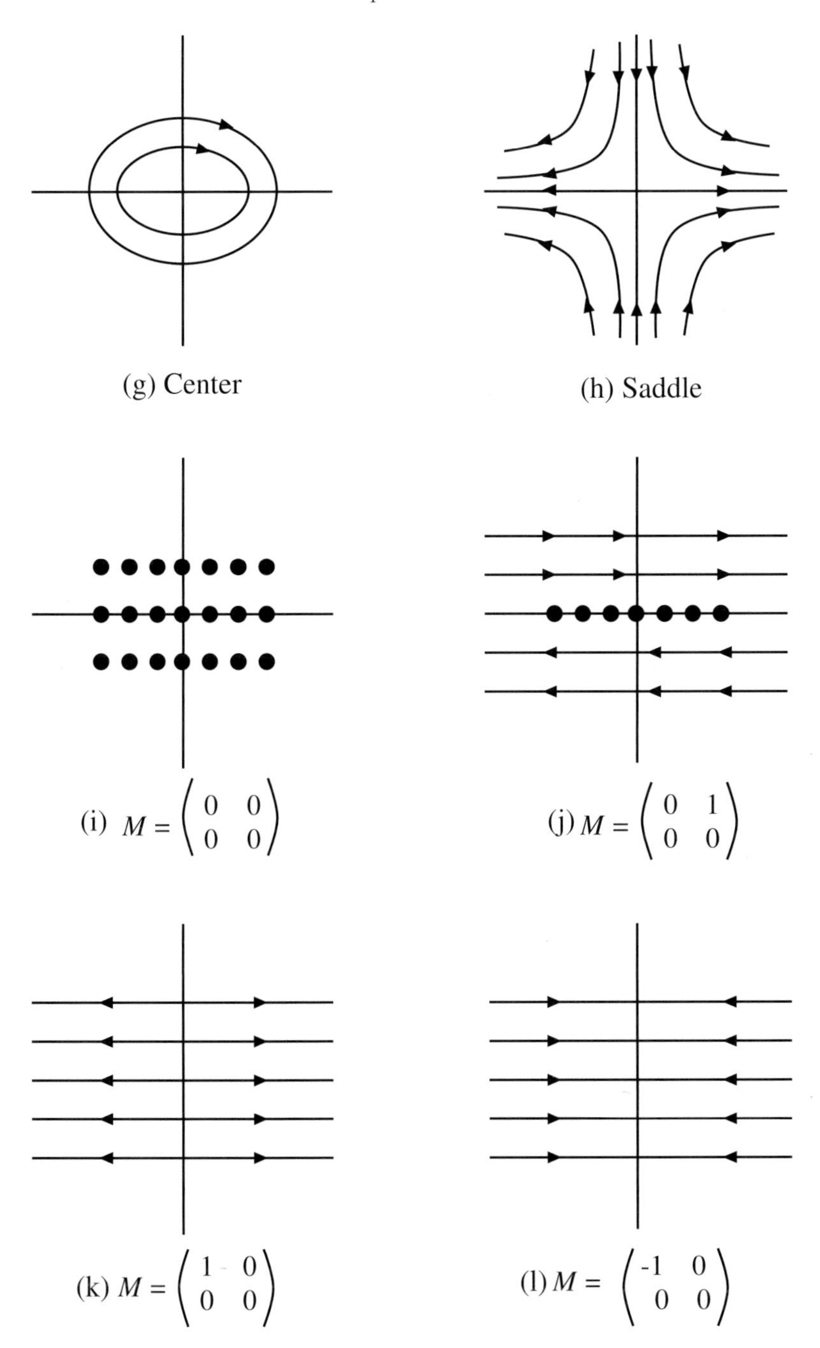

Fig. 3.1. (continued)

Here $A(= B^*)$ is an integration constant while $D(= C^*)$ is given by (3.10c). For $\alpha < 0$, both $\mid \xi \mid$ and $\mid \eta \mid \rightarrow 0$ as $t \rightarrow \infty$. The equilibrium point is thus stable. Figure 3.1e depicts the trajectories near the equilibrium point. The trajectories wind around the equilibrium point a number of times before reaching it asymptotically. In this case, the equilibrium point is called a *stable spiral point* or *stable focus*. For $\alpha > 0$, the trajectories move away from the equilibrium point along spiral paths (Fig. 3.1f). The equilibrium point is an *unstable focus* in this case.

Case 4: λ_1 *and* λ_2 *pure imaginary and complex conjugates – center/elliptic*

For $\lambda_{1,2} = \pm\, i\beta(\alpha = 0)$, from (3.16) we can rewrite,

$$\xi(t) = |A| \cos(\beta t + \phi)\ , \tag{3.17a}$$

$$\eta(t) = |D| \sin(\beta t + \phi')\ . \tag{3.17b}$$

The perturbation neither decays to zero nor diverges to infinity but it varies periodically with time. In this case the trajectories form closed orbits about the equilibrium point as shown in Fig. 3.1g. The trajectories do not approach the equilibrium point as $t \rightarrow \pm\infty$. The equilibrium point is called *center type* or *elliptic equilibrium point* and is neutrally stable. (To study its true stability properties, one has to consider the effect of nonlinear terms which were neglected in (3.8)).

Case 5: $\lambda_1 < 0 < \lambda_2$ *– saddle/hyperbolic*

For $\lambda_1 < 0$ and $\lambda_2 > 0$ (or vice versa), the first terms in (3.10a) and (3.10b) approach zero as $t \rightarrow \infty$ while the second terms diverge to ∞ as $t \rightarrow \infty$. When both the terms are considered we have $|\xi|, |\eta| \rightarrow \infty$ as $t \rightarrow \infty$ and the solution curves are hyperbolic (for details see appendix D). However, for certain initial conditions, namely for which $B = D = 0$, we have $|\xi|$, $|\eta| \rightarrow 0$ as $t \rightarrow \infty$ and hence the trajectories approach the equilibrium point. Figure 3.1h depicts the trajectories in the neighbourhood of the equilibrium point after a suitable rotation of the coordinate axes as discussed in appendix D. In this figure we note that trajectories reach the equilibrium point along two directions only and in all other directions the trajectories diverge from it. Therefore we can say that generally the trajectories diverge from the equilibrium point. This type of equilibrium point is called a *saddle* which is unstable. It is also called the *hyperbolic equilibrium point*.

Case 6: Degenerate Cases (λ_1 *or* $\lambda_2 = 0$; *both* $\lambda_1 = \lambda_2 = 0$)

In the above classification of equilibria we have excluded the choices (a) $\lambda_1 > 0, \lambda_2 = 0$, (b) $\lambda_1 = 0, \lambda_2 < 0$, and (c) $\lambda_1 = \lambda_2 = 0$, where in (a) and (b) λ_1 and λ_2 are ordered as $\lambda_1 > \lambda_2$. We note that when $\det M = 0$, $M = \begin{pmatrix} a & b \\ c & d \end{pmatrix}$, at least one of the eigenvalues of M is zero. In this case the equilibrium point is degenerate. That is, the equilibrium point is not an isolated point and there are infinitely many equilibrium points in its neighbourhood. In this situation the linear system exhibits the following behaviour.

(i) $M = \begin{pmatrix} 0 & 0 \\ 0 & 0 \end{pmatrix}$: $\lambda_1 = \lambda_2 = 0$.

For this case the linear system (3.8) becomes $\dot{\xi} = 0$, $\dot{\eta} = 0$ with the solution $\xi(t) = \xi(0)$, $\eta(t) = \eta(0)$. The solution is independent of time and is a function of initial conditions only. As t increases $\xi(t)$ and $\eta(t)$ remain as $\xi(0)$ and $\eta(0)$ respectively and do not change with time. That is, every orbit is an equilibrium point as shown in the Fig. 3.1i.

(ii) $M = \begin{pmatrix} 0 & 1 \\ 0 & 0 \end{pmatrix}$: $\lambda_1 = \lambda_2 = 0$.

For M of the form given above, the corresponding linear equation may be written as $\dot{\xi} = \eta$, $\dot{\eta} = 0$ whose solution is $\xi(t) = \eta(0)t + \xi(0)$, $\eta(t) = \eta(0)$. For a given initial nonzero values of $\xi(0)$ and $\eta(0)$, $\eta(t)$ remains as $\eta(0)$ whereas $\xi(t) \to \infty$ linearly for $\eta(0) > 0$ and $\xi(t) \to -\infty$ for $\eta(0) < 0$ as $t \to \infty$. The trajectories are parallel to the x-axis. When $\eta(0) = 0$ then $\xi(t) = \xi(0)$. In this case all points on the x-axis are equilibrium points. The corresponding phase portrait is depicted in the Fig. 3.1j.

(iii) $M = \begin{pmatrix} 1 & 0 \\ 0 & 0 \end{pmatrix}$: $\lambda_1 = 1$, $\lambda_2 = 0$.

In this case the linear system may be written as $\dot{\xi} = \xi$, $\dot{\eta} = 0$. Its solution is $\xi(t) = \xi(0)\mathrm{e}^t$, $\eta(t) = \eta(0)$. For $\xi(0) < 0$, $\xi(t) \to -\infty$ and for $\xi(0) > 0$, $\xi(t) \to \infty$ as $t \to \infty$ while $\eta(t)$ remains as $\eta(0)$. Further, when $\xi(0) = 0$ we have $\xi(t) = 0$ and $\eta(t) = \eta(0)$. That is, all points on the y-axis are equilibrium points and are unstable as shown in the Fig. 3.1k.

(iv) $M = \begin{pmatrix} -1 & 0 \\ 0 & 0 \end{pmatrix}$: $\lambda_1 = -1$, $\lambda_2 = 0$.

For this case the linear system is $\dot{\xi} = -\xi$, $\dot{\eta} = 0$. Its solution is $\xi(t) = \xi(0)\mathrm{e}^{-t}$, $\eta(t) = \eta(0)$. From the solution we find that all points on the y-axis are stable equilibrium points (Fig. 3.1l).

The above four cases represent the dynamics of the linearized system only. The actual dynamics of the nonlinear system near the equilibrium point is influenced by nonlinear term(s). Now to determine the stability of the equilibrium point one has to consider higher-order terms in the Taylor series expansion in (3.7). Such an analysis is more involved and beyond our scope.

(For details about nonlinear stability the reader may refer to J. Hale, H. Kocak: *Dynamics and Bifurcations* [3]).

From the above analysis we can appreciate that the linear stability of equilibrium points can be tested without actually solving the differential equation but by knowing the eigenvalues of the matrix M alone. Let us now consider some simple examples to illustrate the above ideas.

Examples:

1. Damped Cubic Anharmonic Oscillator

The equation of motion of the damped (double-well) cubic anharmonic oscillator (see Sect. 2.2) is

$$\ddot{x} + d\dot{x} - ax + bx^3 = 0 \, , \tag{3.18}$$

where d, a and b are positive constants (see Chap. 5 for more details). Equation (3.18) can be rewritten as

$$\dot{x} = y \, , \tag{3.19a}$$

$$\dot{y} = -dy + ax - bx^3 \, . \tag{3.19b}$$

Now we will find the equilibrium points and investigate their stability.

(a) ***Equilibrium Points:***

The equilibrium points are obtained by substituting $\dot{x} = 0, \ \dot{y} = 0$, so that

$$y = 0 \, , \quad ax - bx^3 = 0 \, . \tag{3.20}$$

The roots of the second equation are $x = 0, \pm\sqrt{a/b}$. Thus, there are three equilibrium points, namely

$$(x^*, y^*) = (0, 0) \, , \ \left(\sqrt{a/b}\,, 0\right) \, , \ \left(-\sqrt{a/b}\,, 0\right) \, . \tag{3.21}$$

(b) ***Stability of Equilibrium Points:***

The stability determining eigenvalues are obtained (see below (3.10c)) from

$$\det(M - \lambda I) = \begin{vmatrix} -\lambda & 1 \\ a - 3bx^{*2} & -d - \lambda \end{vmatrix} = 0 \, . \tag{3.22}$$

Expanding the determinant, we obtain

$$\lambda^2 + d\lambda + \left(-a + 3bx^{*2}\right) = 0 \, . \tag{3.23}$$

Then the eigenvalues are

$$\lambda_\pm = \frac{1}{2}\left[-d \pm \left(d^2 + 4a - 12bx^{*2}\right)^{1/2}\right] \, . \tag{3.24}$$

(i) For the equilibrium point $(x^*, y^*) = (0,0)$, $\lambda_\pm$ are given by

$$\lambda_\pm = \frac{1}{2}\left[-d \pm \sqrt{d^2+4a}\right] . \tag{3.25}$$

Since $\sqrt{d^2+4a}$ is always greater than d we have $(\lambda_- < 0 < \lambda_+)$ and hence the origin is a saddle equilibrium point, which is of course unstable.
(ii) For both the remaining equilibrium points $(x^*, y^*) = \left(\pm\sqrt{a/b}\,, 0\right)$, the eigenvalues are

$$\lambda_\pm = \frac{1}{2}\left[-d \pm \sqrt{d^2-8a}\right] . \tag{3.26}$$

For $d^2 > 8a$, λ_+ and λ_- are negative and both the equilibrium points are stable nodes. For $d^2 < 8a$,

$$\lambda_\pm = \frac{1}{2}\left[-d \pm \mathrm{i}\,\sqrt{\mid d^2-8a \mid}\right] . \tag{3.27}$$

The eigenvalues are complex conjugates with negative real parts and the equilibrium points are stable foci. When $d^2 = 8a$, both the eigenvalues are negative real and identical. The equilibrium points are thus stable stars.

To summarize, the oscillator (3.18) has three real equilibrium points. The equilibrium point $(0,0)$ is always a saddle. The remaining two equilibrium points $\left(\pm\sqrt{a/b}\,, 0\right)$ are stable nodes for $d^2 > 8a$, stable stars for $d^2 = 8a$ and stable foci for $d^2 < 8a$. Numerically computed phase trajectories for $a = b = 1$, $d = 3$, $(d^2 > 8a)$ and $a = b = 1$, $d = 0.4$, $(d^2 < 8a)$ are shown in Figs. 3.2a and 3.2b respectively. In these figures, trajectories which start in the neighbourhood of the equilibrium points $\left(\pm\sqrt{a/b}\,, 0\right)$ approach them asymptotically. That is, these equilibrium points attract nearby trajectories and are also called *attractors* . Generally, an orbit (including an equilibrium point) which attracts nearby trajectories is called an attractor. Specifically, in this problem the attractors discussed here are *point attractors*. The initial part of the trajectory approaching the attractor is called a *transient*. In Figs. 3.2, the transient evolution of a trajectory can be clearly seen and is simply the portion of the trajectory excluding the equilibrium point on which it ends. A trajectory which starts directly on the attractor does not have a transient, it is recurrent.

In the Figs. 3.2 we observe that only along two directions the trajectories reach the origin which is a saddle equilibrium point of the system (3.18). That is, trajectories starting with initial conditions on the curves F_1A and F_1B alone end up on the origin. All other trajectories are repelled away from the origin. Thus the origin in this case is a *repeller*. Further in the Fig. 3.2a we clearly see that the trajectories to the left of the curve AF_1B are attracted to the equilibrium point F_2 while those starting from the initial states to the right of AF_1B are attracted to the other equilibrium point F_3. This is due to the consequence of noncrossing of trajectories in the phase space (why?). The

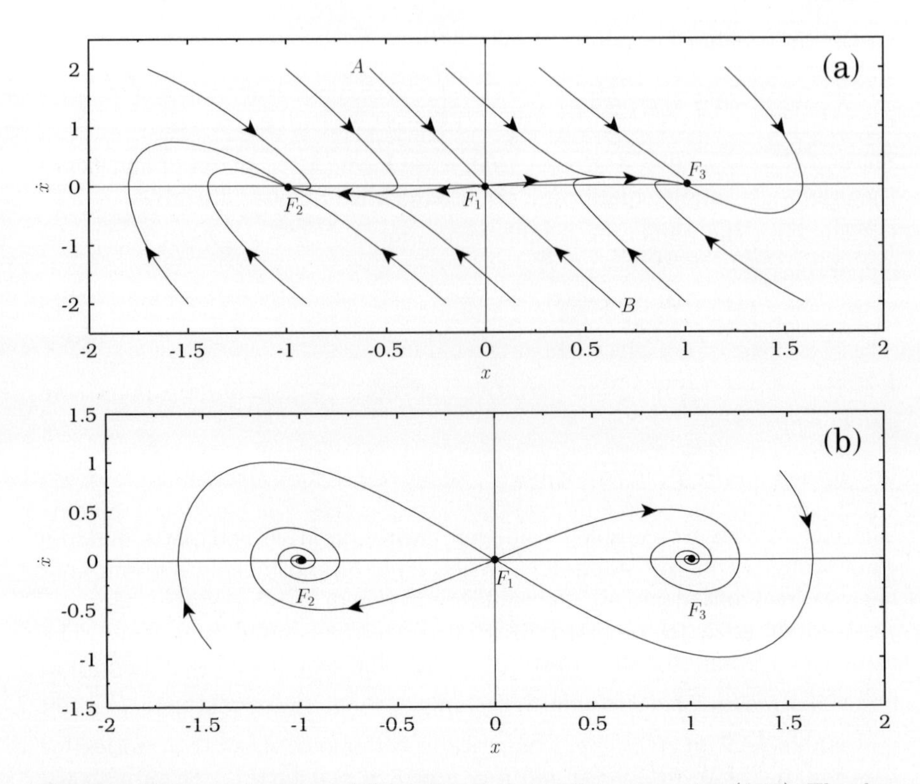

Fig. 3.2. Phase trajectories of the anharmonic oscillator equation (3.18). The chosen values of the parameters are $a = b = 1$. (**a**) $d = 3$, the equilibrium points F_2 and F_3 are stable nodes while F_1 is a saddle. (**b**) $d = 0.4$, the previously stable nodes become stable foci, while F_1 is still a saddle

set of initial states of trajectories ending on a particular attractor is called the *basin of attraction* [4,5] of that attractor. The regions to the left and right of the curve AF_1B are the basins of attraction of the equilibrium points F_2 and F_3, respectively. The curve AF_1B is actually the (*basin*) *boundary* separating the basins of attraction of the two stable equilibrium points.

Exercise:

2. Identify the basins of attraction of the equilibrium points F_2 and F_3 in the phase space diagram (3.2b).

---

2. Undamped Pendulum Equation

The equation of motion of the undamped pendulum (after suitable scaling) is

$$\ddot{x} + \sin x = 0 \,, \tag{3.28}$$

where x here stands for the angle of displacement of the pendulum bob from its equilibrium position or rest state. It can be rewritten as

$$\dot{x} = y \,, \tag{3.29a}$$

$$\dot{y} = -\sin x \,. \tag{3.29b}$$

Its equilibrium points are

$$(x^*, y^*) = (n\pi, 0) \,, \quad n = 0, \pm 1, \pm 2, \ldots \tag{3.30}$$

The system has an infinite number of equilibrium points in the interval $-\infty < x < \infty$. The corresponding stability determining eigenvalues are

$$\lambda_{\pm} = \pm\sqrt{-\cos x^*} \,. \tag{3.31}$$

For the equilibrium points with even values of n the eigenvalues are $\lambda = \pm \mathrm{i}$, which are pure imaginary and complex conjugates. The equilibrium points are thus centers. The eigenvalues of the equilibrium points with odd values of n are $\lambda = \pm 1$ and are saddles. Numerically computed phase trajectories near the equilibrium points $(x^*, y^*) = (-\pi, 0), (0, 0)$ and $(\pi, 0)$ are shown in Fig. 3.3, which correspond to saddle, center and saddle, respectively. The system exhibits three types of motion. In the neighbourhood of the center the trajectories are closed ellipses (denoted by A and B in Fig. 3.3) and they correspond to *bounded periodic motions*. The orbits marked as D and E represent *unbounded aperiodic motions*. These two types of motion are separated by the solution curve C called *separatrix*. The separatrix solution

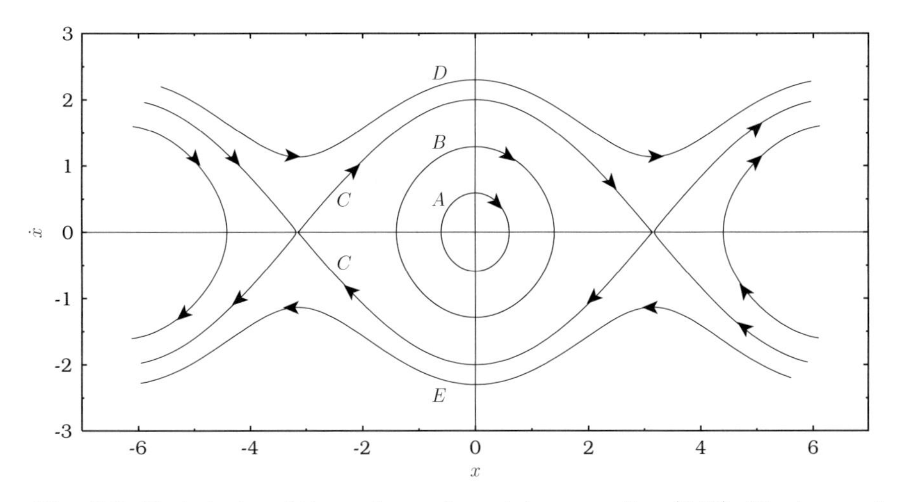

Fig. 3.3. Trajectories of the undamped pendulum equation (3.28). The two equilibrium points $(\pm\pi, 0)$ are saddles, while the origin is a center. For a more detailed description see the text

connects the two saddles. For further details regarding periodic solutions see appendix A.

Exercise:

3. Solve the equation (3.28) explicitly in terms of Jacobian elliptic function (see appendix A) and verify the above assertions.

3.5 Limit Cycle Motion – Periodic Attractor

In addition to the point attractors (or stable equilibrium points), two-dimensional dynamical systems can have isolated but closed stable orbits called (*stable*) *limit cycles* when certain conditions are satisfied. Let $X(t) = (x(t), y(t))$ be a solution of (3.4). If the solution $X(t)$ for an initial condition $X(0)$ is a closed orbit, then for a least unique value of T, $0 < T < \infty$,

$$X(t) = X(t+T) \ . \tag{3.32}$$

Then the solution is said to be periodic with period T. Further, if no other periodic solutions sufficiently close to it exist so that the periodic orbit is an isolated one, then it is called a *limit cycle*. Every trajectory beginning sufficiently near a limit cycle approaches it either for $t \to \infty$ or $t \to -\infty$. If all nearby trajectories approach a limit cycle as $t \to \infty$, it is said to be *stable*. If they approach it as $t \to -\infty$ (that is, deviate from it as $t \to \infty$), it is called *unstable*. A stable limit cycle is a *period-T attractor* or a *periodic attractor*.

Example 1:

A simple example of a limit cycle is that of the wall-clock pendulum motion. Here the cyclic motion is due to the balance between self-excitation and damping (air resistance).

Example 2:

Another example is provided by the van der Pol oscillator (which represents the van der Pol model of an electrical circuit with a triode valve, the resistance property of which changes with current),

$$\dot{x} = y = f_1(x, y) \ , \tag{3.33a}$$

$$\dot{y} = b\left(1 - x^2\right) y - x = f_2(x, y) \ , \tag{3.33b}$$

where b is the damping coefficient. The equilibrium point is the origin. The stability determining eigenvalues are given by

$$\lambda_\pm = \frac{1}{2}\left[b \pm \sqrt{b^2 - 4} \right] . \tag{3.34}$$

Now we can investigate the nature of the singular point $E : (0, 0)$ as a function of the control parameter b in the range $(-\infty, \infty)$ by analysing the form of the eigenvalues $\lambda_\pm$ given by (3.34). The following identifications can be easily made in different regions of b by making use of our classification of equilibrium points discussed earlier in Sect. 3.4.

Sl. No.	Range of b	Nature of eigenvalues	Type of attractor/ repeller
1.	$-\infty < b < -2$	$\lambda_\pm < 0, \quad \lambda_+ \neq \lambda_-$	stable node
2.	$b = -2$	$\lambda_+ = \lambda_- < 0$	stable star
3.	$-2 < b < 0$	$\lambda_\pm = \alpha \pm \mathrm{i}\beta, \quad \alpha < 0$	stable focus
4.	$b = 0$	$\lambda_\pm = \pm\,\mathrm{i}\beta$	center
5.	$0 < b < 2$	$\lambda_\pm = \alpha \pm \mathrm{i}\beta, \quad \alpha > 0$	unstable focus
6.	$b = 2$	$\lambda_+ = \lambda_- > 0$	unstable star
7.	$2 < b < \infty$	$\lambda_\pm > 0, \quad \lambda_+ \neq \lambda_-$	unstable node

In the above table, of particular interest is the region $0 < b < \infty$, where the equilibrium point is unstable. So what happens to the diverging trajectories starting from the neighbourhood of the origin. We note that the trajectories have different characteristics in the two regions $| x |< 1$ and $| x |> 1$. The quantity $\nabla \cdot \boldsymbol{F} = \partial f_1/\partial x + \partial f_2/\partial y$ for the equation (3.33) is a measure as to specify whether the system is dissipative or nondissipative. $\nabla \cdot \boldsymbol{F} < 0$ and > 0 represent dissipative and (phase space) area expanding characters, respectively, of the system (more details will be given in Sect. 3.8). For the equation (3.33), $\nabla \cdot \boldsymbol{F} = b(1 - x^2)$ which is negative for $| x |> 1$ and positive for $| x |< 1$. That is, we have a positive damping (area expanding/energy generation) when $| x |< 1$, negative damping (area contraction/energy dissipation) when $| x |> 1$. Consequently, the trajectories in the phase space near the equilibrium point move outward for $| x |< 1$ but for $| x |> 1$ there is positive damping which begins to suppress the divergence of the trajectory. The resulting dynamics is that of a closed orbit due to the balance between positive and negative damping. Figure 3.4 shows the limit cycle of the van der Pol oscillator for $b = 0.4$.

From the above classification, we note that in different regions of the parameter b, we have different types of equilibrium states and, in particular, in the case of the unstable equilibrium point for $0 < b < \infty$, the system enters into a limit cycle motion. Further we observe that at three critical parameter values of b, sudden qualitative changes, namely *bifurcations* in the dynamical behaviour occur, when the parameter b is slowly changed. They are

(1) at $b = -2$, a bifurcation from a stable node to stable focus occurs.

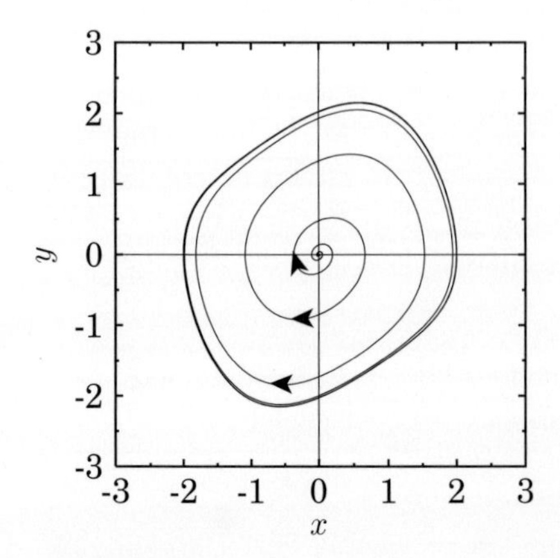

Fig. 3.4. Phase trajectories of the van der Pol equation (3.33) for $b = 0.4$ asymptotically approaching the limit cycle attractor

(2) At $b = 0$, a bifurcation from a stable focus to a limit cycle motion occurs.

(3) At $b = 2$, a bifurcation from an unstable focus to an unstable node occurs.

In particular at $b = 0$, we find that the real parts of the eigenvalues $\lambda_\pm$ change from negative to positive values, crossing the imaginary axis at the critical value, as the parameter b increases through zero, leading to a bifurcation from stable focus to stable limit cycle. Such a bifurcation is called a *Hopf bifurcation*. (More details about such bifurcations will be given in the following chapters).

3.5.1 Poincaré–Bendixson Theorem

An important theorem concerning the existence of limit cycles and other possible motions in two-dimensional systems is the so-called Poincaré–Bendixson theorem. Without going into the technical details, the statement of the theorem [6] can be given as follows.

If a solution curve C:$[x(t), y(t)]$ of the two-dimensional system (3.4) remains in a domain of the $(x - y)$ phase space in which P and Q have continuous first partial derivatives for all $t \geq \tau$, for some τ, without approaching equilibrium points, then there exists a limit cycle in the domain, and either C is a limit cycle or it approaches a limit cycle as $t \to \infty$.

Thus, the possible attractors of two-dimensional systems are point attractors and limit cycle attractors. No other attractors are in general possible. For systems with dimension greater than two other types of motions such as quasiperiodic, chaotic, etc. are possible. These will be studied in Sect. 3.7 and in later chapters.

There is also a theorem given by Bendixson which gives the condition for the nonexistence of a limit cycle motion in a given two-dimensional system. The theorem [6] can be stated as follows:

For the system (3.4), if $s = \partial P/\partial x + \partial Q/\partial y$ *does not change sign or vanish identically in a region* D *of the* $(x-y)$ *phase space enclosed by a single closed curve, then no limit cycles lie entirely in* D.

For fuller details we may refer to Ref. [6].

Example:

For the damped and unforced cubic anharmonic oscillator system

$$\dot{x} = y \,, \tag{3.35a}$$

$$\dot{y} = -dy - x + x^3 \tag{3.35b}$$

the quantity s is $-d$ and thus it does not change sign in any region of the phase space for a given value of d. That is, the Bendixson's theorem predicts that a closed orbit does not occur in the above system. Numerical work also confirms the absence of a limit cycle. On the other hand, for the van der Pol equation (3.33) the quantity s is $(1-x^2)b$. It is positive for $|\, x \,| < 1$ and negative for $|\, x \,| > 1$. So s changes sign in the phase space. The system may or may not have a limit cycle. However, as shown above a limit cycle motion does occur due to Hopf bifurcation.

Exercise:

4. Applying Bendixson's nonexistence theorem, discuss the possibility of occurrence of a limit cycle in the following systems. Also investigate analytically/numerically the nature of solutions in these systems (see also Chaps. 4 and 5 for more details).
   a) Bonhoeffer–van der Pol oscillator
   $$\dot{x} = x - x^3/3 - y + A_0 \,,$$
   $$\dot{y} = c(x + a - by) \,. \quad (A_0,\ a,\ b,\ c: \text{ constants})$$
   b) Brusselator equations
   $$\dot{x} = a - x - bx + x^2 y \,,$$
   $$\dot{y} = bx - x^2 y \,.$$
   c) Duffing–van der Pol oscillator
   $$\ddot{x} - b(1 - x^2)\dot{x} + ax + bx^3 = 0 \,.$$
   d) $$\dot{x} = y - x\left(x^2 + y^2 - \alpha\right) \,,$$
   $$\dot{y} = -x - y\left(x^2 + y^2 - \alpha\right) \,, \quad \alpha: \text{ constant}$$

3.6 Higher Dimensional Systems

The analysis of equilibrium points discussed in Sect. 3.4 can be extended in principle to higher dimensional systems without much difficulty. First, we briefly consider the linear stability analysis of the equilibrium points for the equation of motion (3.1) of the n-dimensional system, and then consider an example.

Similar to the two-dimensional system, we substitute in (3.1)

$$X = X^* + \delta X \,, \quad \| \delta X \| \ll 1 \,, \tag{3.36}$$

where $X^* = (x_1^*, x_2^*, ..., x_n^*)$ is the equilibrium point and $\delta X = (\xi_1, \xi_2, ..., \xi_n)^{\mathrm{T}}$ is a small perturbation and $\|\delta X\| = \sqrt{\xi_1^2 + \xi_2^2 + ... + \xi_n^2}$. Then proceeding as in Sect. 3.4 we obtain the linearised equation,

$$\delta \dot{X} = \left. \frac{\partial F}{\partial X} \right|_{X = X^*} \delta X \equiv A \delta X \,, \tag{3.37a}$$

where A is the Jacobian matrix with constant elements

$$A_{ij} = \left. \frac{\partial F_i}{\partial x_j} \right|_{X = X^*} . \tag{3.37b}$$

The formal matrix solution to (3.37) can be given by

$$\delta X(t) = \mathrm{e}^{At} \delta X(0) \,, \tag{3.38}$$

where $\delta X(0)$ is the initial disturbance from the equilibrium point X^*. Now if the matrix A is diagonalized, then it satisfies the eigenvalue problem

$$A \hat{e}_i = \lambda_i \, \hat{e}_i \,, \quad i = 1, 2, \ldots, n \tag{3.39}$$

where λ_i's are the eigenvalues corresponding to the eigenvectors $\hat{e}_i$'s. When the n eigenvectors are linearly independent then one can write

$$\delta X(0) = \sum_{i=1}^{n} C_i \hat{e}_i \tag{3.40}$$

and from (3.38), using (3.39), that

$$\delta X(t) = \sum_{i=1}^{n} C_i \exp{(\lambda_i t)} \, \hat{e}_i \,, \tag{3.41}$$

where we have used the operator identity

$$\exp(At) = I + tA + \frac{t^2 A^2}{2!} + \ldots + \frac{t^n A^n}{n!} + \ldots \tag{3.42}$$

If $\delta X(t) \to 0$ as $t \to \infty$, then X^* is stable. This is true only if the real parts of all the λ_i's, namely $\mathrm{Re}\lambda_i$, are less than zero for all i. If $\mathrm{Re}\lambda_i > 0$ for at least one value of i then $\| \delta X(t) \|$ diverges as t increases and the equilibrium point is unstable. Thus the stability of an equilibrium point can be studied without explicitly calculating all the eigenvalues but by determining the signs of the real parts of the eigenvalues.

According to Routh–Hurwitz's criterion [7] all the roots of the equation

$$a_0\lambda^n + a_1\lambda^{n-1} + \ldots + a_{n-1}\lambda + a_n = 0\,, \quad a_0 \neq 0 \tag{3.43}$$

have negative real parts *if and only if the quantities* $T_0, T_1, T_2, ..., T_n$ *are all positive*, where

$$T_0 = a_0, \quad T_1 = a_1, \quad T_2 = \begin{vmatrix} a_1 & a_0 \\ a_3 & a_2 \end{vmatrix}, \quad T_3 = \begin{vmatrix} a_1 & a_0 & 0 \\ a_3 & a_2 & a_1 \\ a_5 & a_4 & a_3 \end{vmatrix}, \ \ldots \tag{3.44}$$

This is true if and only if all the a_i's and either all even-numbered T_k's or all odd-numbered T_k's are positive. In order to use this criterion, one can write down the characteristic equation associated with the stability matrix A, $\det|A - \lambda I| = 0$, in the form of (3.43) and check whether all the roots have negative real parts from (3.44). Then one may conclude whether the given equilibrium point is stable or not.

One can try to identify different possible equilibrium points based on stability considerations for $n > 2$ also and classify them accordingly as in the two-dimensional case. However, such analysis become too involved and laborious. While for $n = 3$ it is possible to make such a classification (see for example [7]), it becomes too unwieldy for higher dimensional cases. More meaningfully, one can deal with individual problems as such using the above analysis. In the following, we will treat a typical example, namely the Lorenz equations, for $n = 3$.

3.6.1 Example: Lorenz Equations

The Lorenz equations [8] are given by

$$\dot{x} = \sigma(y - x)\,, \tag{3.45a}$$

$$\dot{y} = rx - y - xz\,, \tag{3.45b}$$

$$\dot{z} = xy - bz\,, \tag{3.45c}$$

where the parameters σ, r and b are positive constants. For some more details see Sect. 5.2. We continue in the usual way by finding the equilibrium points and analysing their linear stability. Substituting $\dot{x} = \dot{y} = \dot{z} = 0$, we obtain

$$y^* = x^*\,, \quad rx^* - y^* - z^*x^* = 0\,, \quad x^*y^* - bz^* = 0\,. \tag{3.46}$$

Solving these, we obtain the three equilibrium points C_1, C_2, and C_3:

$$C_1 : (x^*, y^*, z^*) = (0, 0, 0)$$

$$C_2,\ C_3 : (x^*, y^*, z^*) = \left(\pm\sqrt{b(r-1)}\,,\ \pm\sqrt{b(r-1)}\,,\ r-1\right)\,.$$

For $r < 1$, the origin is the only real equilibrium point. For $r > 1$, there are three real equilibrium points. The stability determining eigenvalues for all of them are obtained from

$$\begin{vmatrix} -\sigma-\lambda & \sigma & 0 \\ r-z^* & -1-\lambda & -x^* \\ y^* & x^* & -b-\lambda \end{vmatrix} = 0 \,. \tag{3.47}$$

Expanding the above equation we obtain

$$\lambda^3 + a_1\lambda^2 + a_2\lambda + a_3 = 0 \,, \tag{3.48a}$$

where

$$\begin{aligned} a_1 &= \sigma + 1 + b \,, \quad a_2 = \sigma\left(1 + b - r + z^*\right) + b + x^{*2} \,, \\ a_3 &= \sigma\left(b - br + x^{*2} + x^* y^* + z^* b\right) \,. \end{aligned} \tag{3.48b}$$

***Case 1:* $(x^*, y^*, z^*) = (0,0,0)$**

For this equilibrium point the values of a_1, a_2 and a_3 in (3.48b) become

$$a_1 = \sigma + 1 + b > 0 \,, \quad a_2 = \sigma(1 + b - r) + b \,, \quad a_3 = \sigma b(1 - r) \,. \tag{3.49}$$

It is easy to check that $\lambda_1 = -b$ is one solution. Then we solve the eigenvalue equation for the remaining two eigenvalues. We obtain

$$\lambda_{2,3} = \frac{1}{2}\left[-(\sigma+1) \pm \sqrt{(\sigma+1)^2 \;+\; 4\sigma(r-1)} \right] . \tag{3.50}$$

If $r > 1$, then $\lambda_2 > 0$. The equilibrium point is unstable. For $r < 1$, all the λ's are negative and hence the equilibrium point is stable. One can arrive at the same conclusion by using Routh–Hurwitz's criterion without calculating the eigenvalues. The Routh–Hurwitz's conditions for the roots of (3.48a) to have negative real parts (with $a_0 = 1$) are

$$a_1 > 0 \,, \quad a_2 > 0 \,, \quad a_3 > 0 \quad \text{and} \quad a_1 a_2 - a_3 > 0 \,. \tag{3.51}$$

From (3.49) we find

$$a_1 a_2 - a_3 = \sigma^2(1 + b - r) + \sigma\left(2b + 1 - r + b^2\right) + b(b+1) \,. \tag{3.52}$$

For $r < 1$, all the four conditions given by (3.51) are satisfied and the equilibrium point is stable. Figure 3.5 shows the trajectories in the neighbourhood of the equilibrium point for $r = 0.5$. When $r > 1$, we have $a_3 < 0$ and so the Routh–Hurwitz's criterion is violated and therefore the equilibrium point is unstable.

***Case 2:* $(x^*, y^*, z^*) = \left(\pm\sqrt{b(r-1)}\;,\; \pm\sqrt{b(r-1)}\;,\; r-1\right).$**

Now the coefficients a_1, a_2 and a_3 in (3.48) are given by

$$a_1 = \sigma + b + 1 \,, \quad a_2 = b(\sigma + r) \,, \quad a_3 = 2\sigma b(r-1) \,. \tag{3.53}$$

The above two equilibrium points exist only for $r \geq 1$. For $r > 1$, the first three conditions in the above equation (3.51) are satisfied without any restriction on the parameters σ, b and r. However, the fourth condition in (3.51) is satisfied only if

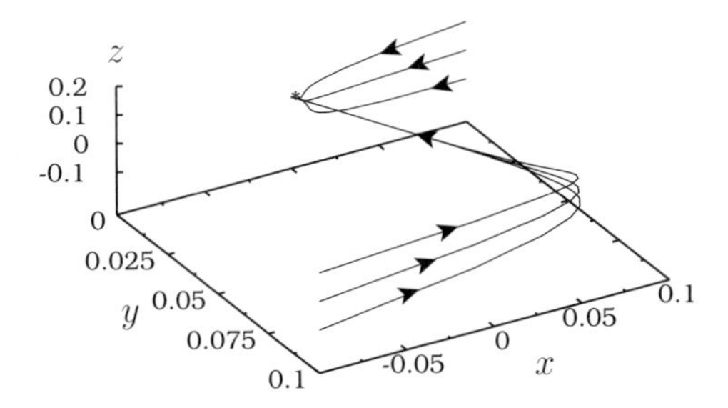

Fig. 3.5. Trajectories in the neighbourhood of the equilibrium point $(x^*, y^*, z^*) = (0, 0, 0)$ of the Lorenz equations (3.45) for $r = 0.5$, $b = 8/3$ and $\sigma = 10$. The equilibrium point is a stable node

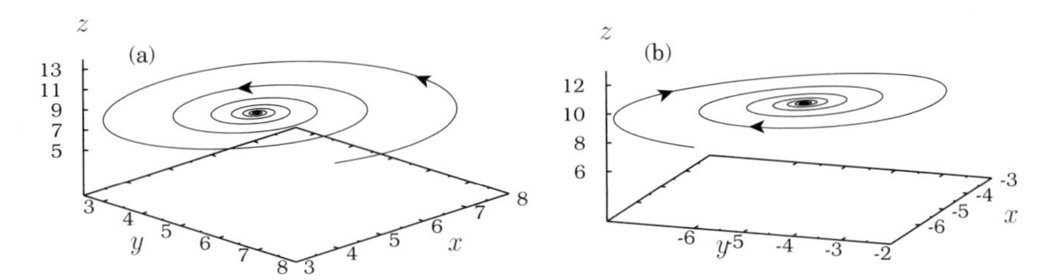

Fig. 3.6. Typical trajectories in the neighbourhood of the stable foci (**a**) $(2\sqrt{6}\,,\ 2\sqrt{6}\,,\ 9)$ and (**b**) $(-2\sqrt{6}\,,\ -2\sqrt{6}\,,\ 9)$ for $\sigma = 10$, $b = 8/3$ and $r = 10$

$$r < r_c = \frac{\sigma(\sigma + b + 3)}{(\sigma - b - 1)} \,. \tag{3.54}$$

Therefore both the equilibrium points are stable only for $r < r_c$. When $\sigma = 10$, $b = 8/3$ (the values used by Lorenz in his numerical study), r_c is $470/19 \approx 24.74$. The equilibrium points are thus stable for $1 < r < r_c \approx 24.74$. Figure 3.6 shows the numerically computed phase trajectories in the neighbourhood of the equilibrium points for $\sigma = 10$, $b = 8/3$ and $r = 10$. For $r > r_c$, two of the eigenvalues are complex conjugate to each other with positive real parts, while the third is negative and hence the equilibrium points are unstable.

What is the nature of dynamics for $r > r_c$? This is where the new developments in nonlinear dynamics begin [8]. These will be discussed later in Chap. 5.

Exercises:

5. Show that the Routh–Hurwitz's necessary conditions for an equilibrium point of the Rössler equations

$$\dot{x} = -(y + z) \,,$$
$$\dot{y} = x + ay \,,$$
$$\dot{z} = b + z(x - c) \,.$$

to be stable [9] are

$$a_1 = c - a - x^* > 0 \, ,$$
$$a_3 = c - x^* - az^* > 0 \, ,$$
$$a_1a_2 - a_3 = (c - x^*)\, z^* - a\,(c - x^*)^2 + a^2\,(c - x^*) - a > 0 \, .$$

6. For the FitzHugh–Nagumo equations

$$\ddot{V} = -V + V^3/3 + R - u\dot{V} \, ,$$
$$\dot{R} = -(c/u)(V + a - bR)$$

show that the threshold condition for Hopf bifurcation to occur is $a_{\pm} = -V_{0\pm}(3 - 3b + bV_{0\pm}^2)/3$, where $V_{0\pm} = \pm[1 - bc + (b^2c^2 - c)/u^2]^{1/2}$ with V_0 being the V component of the equilibrium point. Draw the bifurcation curve in the $a - u$ parameter space. For $a = 0.6$, $b = 0.5$ and $c = 0.1$ calculate the value of $u(u_H)$, at which Hopf bifurcation occurs. Numerically verify the occurrence of limit cycle for values of u near u_H [10].

7. The equilibrium points of the system

$$\dot{x} = y \, ,$$
$$\dot{y} = (\sin(\cos x))z^2 - \sin x - \mu y \, ,$$
$$\dot{z} = K(\cos x - \rho)$$

with $\mu > 0$, $K > 0$ and $0 < \rho < 1$ are given by

$$(x_1^*, y_1^*, z_1^*) = \left(\cos^{-1}\rho, 0,\ 1/\sqrt{\rho}\right) \, ,$$
$$(x_2^*, y_2^*, z_2^*) = \left(\cos^{-1}\rho, 0,\ -1/\sqrt{\rho}\right) \, .$$

Applying the Routh–Hurwitz criterion show that
   a) the equilibrium point (x_1^*, y_1^*, z_1^*) is stable for $\mu > \mu_c = 2K\rho^{3/2}$, undergoes Hopf bifurcation at $\mu = \mu_c$ and
   b) the equilibrium point (x_2^*, y_2^*, z_2^*) is unstable for the parameter range of interest [11].

3.7 More Complicated Attractors

In the earlier section we have described two attractors namely the point and limit cycle attractors. These two are the only attractors which can arise in a two-dimensional dynamical system. However, the situation can be different in higher dimensional systems: In addition to the above two types of attractors, nonlinear systems can have other forms of attractors such as quasiperiodic and chaotic ones. As a prelude to our further discussions in the next chapters, we give a brief account of these two attractors in this section. We will study their characteristic properties in the next two chapters. Quasiperiodic motion of an n-dimensional system takes place on a n-dimensional *torus* T^n. Therefore, first we describe the notion of a two torus T^2.

3.7.1 Torus

Consider a system of n-uncoupled oscillators whose Hamiltonian is

$$H = \frac{1}{2} \sum_{i=1}^{n} \left(p_i^2 + \omega_i^2 q_i^2 \right) , \tag{3.55}$$

where q_i, p_i and ω_i's are the position, momentum and natural frequency of ith oscillator, respectively. Introducing the change of variables

$$q_i = \sqrt{2P_i/\omega_i} \, \sin Q_i \, , \tag{3.56a}$$

$$p_i = \sqrt{2P_i\omega_i} \, \cos Q_i \, , \quad i = 1, 2, \ldots, n \tag{3.56b}$$

the Hamiltonian given by (3.55) can be transformed into

$$H = \sum_{i=1}^{n} \omega_i P_i \, , \tag{3.57}$$

so that Q_i are cyclic variables and they are defined modulo 2π. From (3.57), the equation of motion of the system in the new variables (P_i, Q_i) (which are of course the action and angle variables, respectively) can be written as

$$\frac{\mathrm{d}P_i}{\mathrm{d}t} = -\frac{\partial H}{\partial Q_i} = 0 \, , \tag{3.58a}$$

$$\frac{\mathrm{d}Q_i}{\mathrm{d}t} = \frac{\partial H}{\partial P_i} = \omega_i \, , \quad i = 1, 2, \ldots, n \, . \tag{3.58b}$$

Integrating the above equations, we obtain

$$P_i(t) = \text{constant} = C_i \, , \tag{3.59a}$$

$$Q_i(t) = \omega_i t + D_i \, , \tag{3.59b}$$

where C_i and D_i are integration constants. We note that if we consider the motion of the ith oscillator and represent it by plotting the corresponding position and momentum with proper scalings, namely, using $\sqrt{\omega_i} q_i$ and $p_i/\sqrt{\omega_i}$ as coordinates, then all the trajectories are circles as shown in Fig. 3.7. For a given trajectory the radius of the circle is $\sqrt{2P_i}$ and the angle measured in the counter-clockwise direction from the vertical axis is Q_i.

Consider first the case $n = 1$. A trajectory is represented by a segment of length 2π on the Q_1 axis (Fig. 3.8a). The representative point moves uniformly towards the right with a velocity $\omega_1 (> 0)$ and when it reaches the point $Q_1 = 2\pi$ it jumps back to $Q_1 = 0$. This discontinuity can be avoided by representing the trajectory as a circle with radius $\sqrt{2P_1}$ and Q_1 as angular coordinate of the representative point (Fig. 3.8b).

Next, we consider the case $n = 2$. In the (Q_1, Q_2) plane, the motion takes place inside a square with sides of length 2π (Fig. 3.9a). The point moves with a constant velocity, whose components are (ω_1, ω_2). When it hits a side of the square, it reappears on the opposite side. Here again, these discontinuities can be avoided by sewing first the left side of the square to the right side,

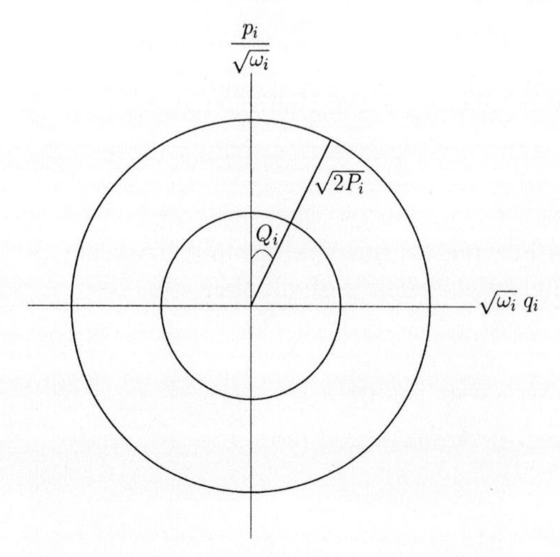

Fig. 3.7. Circular motions of the ith oscillator of the system (3.58), after rescaling

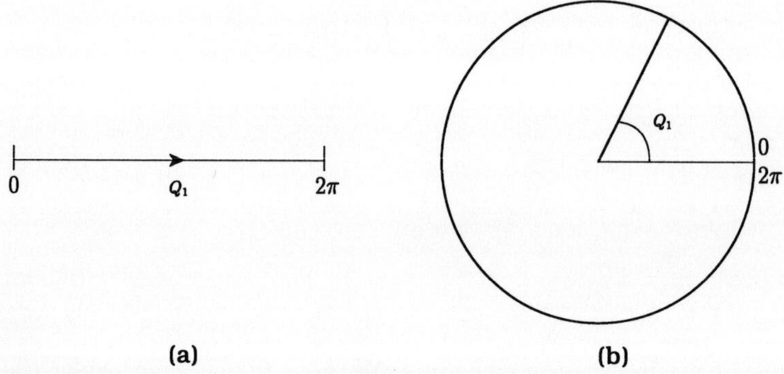

Fig. 3.8. (**a**) A trajectory of the system (3.58) on the one-dimensional axis Q_1. (**b**) Alternative representation of the trajectory on a circular path with Q_1 as angular coordinate

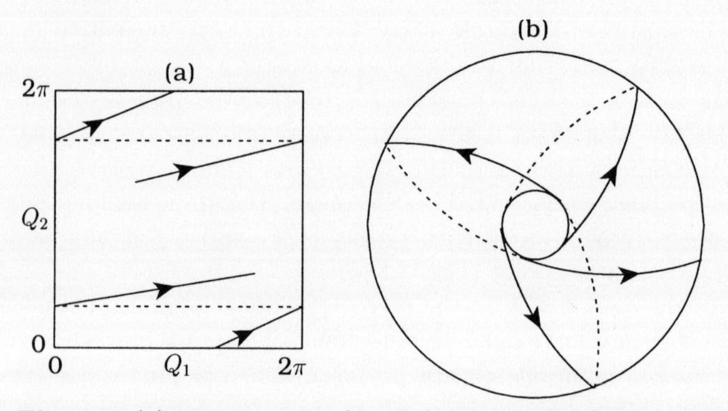

Fig. 3.9. (**a**) Motion in the (Q_1, Q_2) plane and (**b**) on a two-dimensional torus

and then the lower side to the upper side. The surface thus obtained is an ordinary torus (Fig. 3.9b) with $\sqrt{2P_1}$ and $\sqrt{2P_2}$ as the radii of inner and outer circles, respectively. Think of the familiar examples such as a ring tennis, an inflated cycle tube or a scooter tube or a doughnut. Q_1 and Q_2 are now the two angular coordinates which specify the position of a point on the torus.

More generally, for an n degrees of freedom system, the motion can be pictured mentally as taking place either inside an n-dimensional hypercube, in the cartesian space $(Q_1, Q_2, ..., Q_n)$, with discontinuities at the sides or on an n-torus, with the Q_i's as angular coordinates and P_i's as radial coordinates, and without any discontinuity in the motion.

For $n = 1$, the trajectory simply coincides with its supporting 1−torus (circle). For $n = 2$, if the ratio ω_1/ω_2 is a rational number p/q, where p and q are integers, then the trajectory closes back upon itself and is periodic. If ω_1/ω_2 is an irrational number, that is, it is not in the ratio of integers (for example, if $\omega_1/\omega_2 = \sqrt{2}$, π, etc.), the trajectory never closes by itself and it will fill densely the square or the surface of the torus. The trajectory in this case is called a *quasiperiodic* trajectory. Generalization to the arbitrary case (n-torus) can be made as in the following subsection.

The above analysis holds good not only for coupled linear oscillators but for coupled nonlinear oscillators as well provided suitable canonical transformations to action and angle variables exist. This is indeed the case for the so-called integrable systems about which we will discuss in Chaps. 7, 10 and 13. A well known example of nonlinear system for which action and angle variables can be found explicitly is the Kepler particle moving in the attractive gravitational force field (see Goldstein, Ref. [1]).

Exercises:

8. Starting from an initial condition (Q_{10}, Q_{20}), prove that the orbit on a torus closes back upon itself when ω_1/ω_2 is rational and does not close when it is irrational.
9. For the one-dimensional pendulum

$$H_0 = \frac{p^2}{2m} - g\cos x = E_0 \ ,$$

show that the canonical transformation from the variables (x, p) to the angle and action variables (θ, J) can be given by

$$x = 2\sin^{-1}\left[k \ \mathrm{sn}\left(\frac{2K(k)\theta}{\pi}, k\right)\right],$$

$$p = 2k\sqrt{mg}\ \mathrm{cn}\left(\frac{2K(k)\theta}{\pi}, k\right) ,$$

where sn$u$ and cn$u$ are Jacobian elliptic functions (see also appendix A) and $K(k)$ is the complete elliptic integral of the first kind with modulus $k = \sqrt{(2E_0 + g)/(2g)}$ and $J = 8\sqrt{mg}\ \left[E(k) - k'^2 K(k)\right]$. Here $E(k)$ is the complete elliptic integral of the second kind. Find the form of $H_0(J)$ (see Ref. [12]).

3.7.2 Quasiperiodic Attractor

More specifically, a quasiperiodic orbit is one which can be written as

$$X(t) = \phi(\omega_1 t, \omega_2 t, \ldots, \omega_n t) , \tag{3.60}$$

where ϕ is periodic, of period 2π, in each of the $n(n > 1)$ arguments. Further, the frequencies $(\omega_1, \omega_2, \ldots, \omega_n)$ must have the following properties:

(1) They are linearly independent, that is, there does not exist a nonzero set of integers $(l_1, l_2, \ldots, l_n)$ such that

$$l_1\omega_1 + l_2\omega_2 + \ldots + l_n\omega_n = 0 . \tag{3.61}$$

(2) For each i, $\omega_i \neq | l_1\omega_1 + l_2\omega_2 + \ldots + l_n\omega_n |$ for some integers $(l_1, l_2, \ldots, l_n)$, that is, there is no rational relation between the ω's. (For example, for $n = 2$, this implies that the ratio ω_1/ω_2 is irrational).

In other words, a *quasiperiodic orbit* is the sum of periodic orbits each of whose frequency is one of the various sums and differences of the finite set of frequencies $(\omega_1, \omega_2, \ldots, \omega_n)$. Such a quasiperiodic motion takes place on a n-dimensional torus T^n. Suppose that this torus is an attracting set, that is nearby trajectories get attracted to it, then we say that T^n is a *quasiperiodic attractor*.

As a simple example for the realization of quasiperiodic motion let us assume that the approximate solution of a dynamical system

$$\ddot{x} = F(x, \dot{x}) + f \cos \omega_1 t , \tag{3.62}$$

where F is nonlinear, is given by

$$x(t) = b_1 \cos \omega_1 t + b_2 \cos \omega_2 t . \tag{3.63}$$

The motion is periodic if the frequency ratio ω_1/ω_2 is a rational number. Consequently, the solution curve in the $(x, y = \dot{x})$ phase space is a simple closed orbit. If ω_1/ω_2 is irrational then the motion is not periodic. Figure 3.10a shows the phase portrait for $b_1 = 1$, $b_2 = 1$, $\omega_1 = 1$, $\omega_2 = \sqrt{2}$. The solution is clearly not a closed orbit and is *quasiperiodic* or almost periodic in the sense that trajectory returns arbitrarily close to its starting point infinitely often.

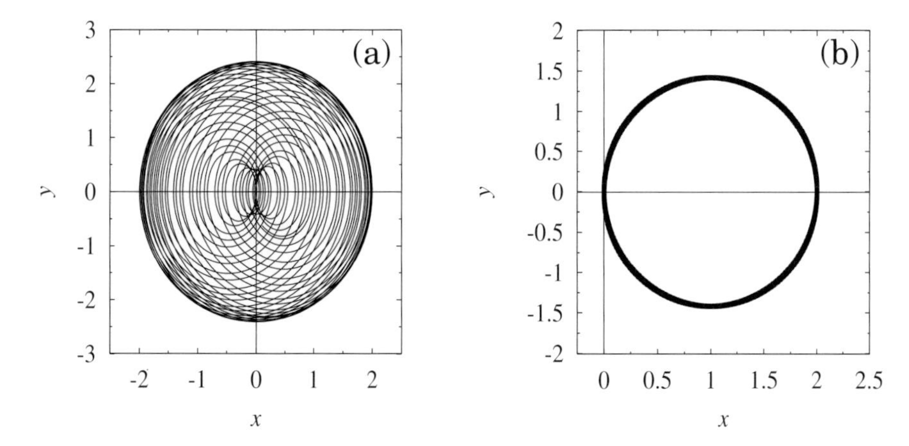

Fig. 3.10. (**a**) Phase portrait and (**b**) Poincaré map of the quasiperiodic solution (3.63)

3.7.3 Poincaré Map

Generally, the complicated structure of an attractor will be more transparent if we look at the dynamics at suitably defined discrete times or intervals of time. In this case the motion will appear as a sequence of points, a *Poincaré map*, in the phase space. If a system has a periodic evolution, the Poincaré map will consist of a repeating set of points. For a quasiperiodic attractor the Poincaré map will appear as a closed orbit. (More details about the Poincaré map will be discussed in the following chapters). To clearly visualize quasiperiodic motion we collect the data at $t_n = 2\pi n/\omega_1$, $n = 0, 1, 2, \ldots$ and denote them as x_n. Then from the above solution we obtain

$$x_n = b_1 + b_2 \cos(2\pi n\omega_2/\omega_1) \, , \tag{3.64a}$$

$$\dot{x}_n = y_n = -b_2\omega_2 \sin(2\pi n\omega_2/\omega_1) \, . \tag{3.64b}$$

The set of points (x_n, y_n) form the Poincaré map of the solution (3.64). Figure 3.10b shows the Poincaré map of the quasiperiodic attractor shown in Fig. 3.10a. As n increases, the points x_n and y_n move around an ellipse in the Poincaré map. This is clear because one can check that from (3.64), x_n's and y_n's satisfy the relation $(x_n - b_1)^2 + (y_n/\omega_2)^2 = b_2^2$, which obviously defines an ellipse.

An example of a dynamical system exhibiting quasiperiodic motion is

$$\dot{x} = y \, , \tag{3.65a}$$

$$\dot{y} = -0.5 \sin y - x + f \cos \omega t \, . \tag{3.65b}$$

Figures 3.11 show the phase portrait and Poincaré map of the attractor for $f = 0.532$ and $\omega = 0.8$, which was obtained by numerical integration of (3.65) (for details see Chap. 5). The trajectory in the phase portrait moves

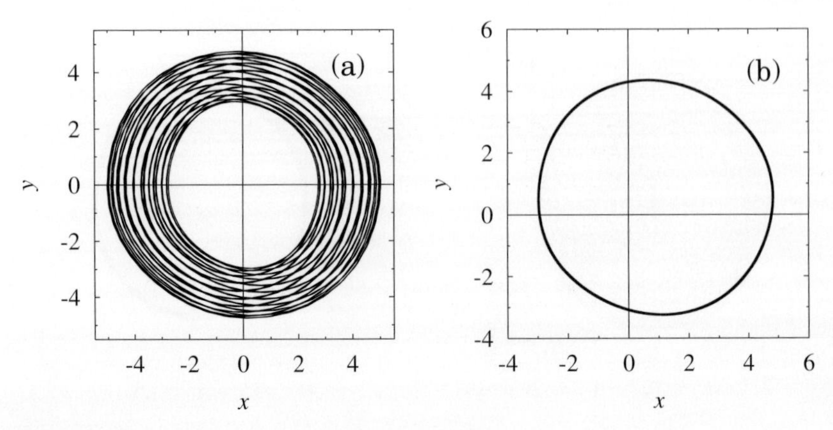

Fig. 3.11. (**a**) Phase portrait and (**b**) Poincaré map of the quasiperiodic orbit of the equation (3.65) for $f = 0.532$ and $\omega = 0.8$

on a two-dimensional torus and in the Poincaré map, points move on an ellipse.

3.7.4 Chaotic Attractor

The attractors such as point attractors, limit cycle attractor and quasiperiodic attractor describe dynamical motions which are asymptotically ($t \to \infty$) stationary, periodic and almost periodic, respectively. These attractors are insensitive to initial conditions, that is, even when the (initial) state of the system is slightly disturbed the underlying trajectory eventually returns to the attractor. (Further details are given in Chap. 5).

Let us consider now an attractor, which is neither a finite set of points nor a closed curve. For example Fig. 3.12a shows the attractor of the Duffing oscillator equation (2.6) for the choice of the parameters $\alpha = 0.5$, $\beta = 1$, $\omega_0^2 = -1$, $\omega = 1$ and $f = 0.8$. It is definitely not a periodic attractor of any of the above types, including the quasiperiodic one, and it is called a *chaotic attractor*. The Poincaré map of the chaotic attractor shown in Fig. 3.12a is given in Fig. 3.12b. It has been drawn using points collected at $2\pi/\omega$ time intervals, that is, at every period of the external periodic force of the system. The geometric structure of the chaotic attractor in the Poincaré map appears as a totally disconnected and uncountable set of points. An important characteristic property of this attractor is that the underlying motion is *highly sensitive to initial conditions*. We will discuss more about chaotic attractors in the next few chapters. In addition, there is another type of attractor termed *strange nonchaotic*. Like the chaotic attractor, it is neither a periodic orbit nor a quasiperiodic orbit. However, it is *not highly sensitive to initial conditions* in contrast to the chaotic attractor. We will also mention about this in Chap. 5.

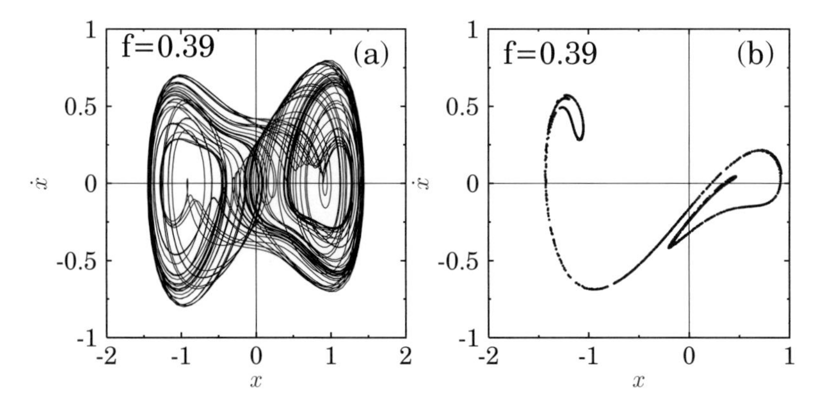

Fig. 3.12. (**a**) Phase portrait and (**b**) Poincaré map of a chaotic attractor of the Duffing oscillator equation (2.5)

Exercise:

10. Numerically integrating the following dynamical systems, obtain the quasiperiodic and chaotic solutions for the specified parametric values. Draw the phase portrait and Poincaré map of these attracting solutions (see also Chap. 5 for such analysis).
    a) van der Pol oscillator:
    $$\ddot{x} - \alpha\left(1 - x^2\right)\dot{x} + x = f\cos\omega t\ .$$
    For $\alpha = 5$, the system exhibits quasiperiodic motion for $f = 1$ and $\omega = 1.5$ and chaotic motion for $f = 5$ and $\omega = 2.463$.
    b) Brusselator equation:
    $$\dot{x} = A + x^2 y - Bx - x + f\cos\omega t\ ,$$
    $$\dot{y} = Bx - x^2 y\ .$$
    For $A = 0.4$ and $B = 1.2$, quasiperiodic motion occurs for $f = 0.005$ and $\omega = 1.5$, while chaotic dynamics occurs for $f = 0.08$ and $\omega = 0.86$.

3.8 Dissipative and Conservative Systems

Dynamical systems can be conservative or nonconservative. Among the former the Hamiltonian systems and among the later dissipative systems constitute the most important classes of them. Such a classification can be based on the following considerations.

Consider a single particle whose position is given by the cartesian coordinates $\boldsymbol{r} = (x, y, z)$. Suppose that the total force acting on the particle can be expressed as

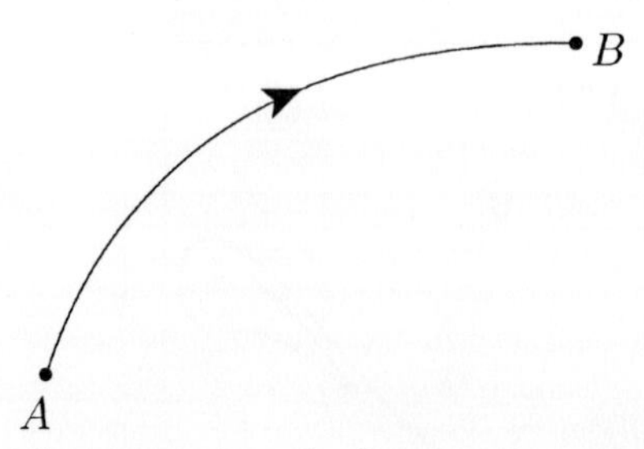

Fig. 3.13. A path of a particle moving from point A to point B

$$\mathcal{F} = -\nabla V \,, \tag{3.66}$$

where the potential energy function V is a single valued function of position only, that is, it is not a function of velocity or time[4]. Such a force is known as a *conservative force*. Now consider the work done by the force $\mathcal{F}$ as the particle moves through an infinitesimal displacement $\mathrm{d}\boldsymbol{r}$:

$$\mathrm{d}W = \mathcal{F} \cdot \mathrm{d}\boldsymbol{r} = -\mathrm{d}V \,. \tag{3.67}$$

The total work W done by the force $\mathcal{F}$ as the particle moves over a certain path between points A and B (Fig. 3.13), is

$$W = \int_A^B \mathcal{F} \cdot \mathrm{d}\boldsymbol{r} = -\int_A^B \mathrm{d}V = V_A - V_B \,, \tag{3.68}$$

which is obviously independent of the path taken. Then the work done over a closed path vanishes. If no other forces do work on the system then the total mechanical energy is conserved and hence the corresponding system is called a *conservative system*. The argument can be extended to an N-particles system such that the condition for a system to be conservative is again the existence of a scalar function $V = V(\boldsymbol{r}_1, \boldsymbol{r}_2, \ldots, \boldsymbol{r}_N)$.

Let us calculate following Ref. [13] the rate of change of a small rectangular volume $\Delta\tau$ in the phase space about the point X_0 in a system described by the equation of motion (see equation (3.1))

$$\frac{\mathrm{d}X}{\mathrm{d}t} = \boldsymbol{F}(X) \,, \tag{3.69}$$

where $X = (x_1, x_2, \ldots, x_n)$ and $\boldsymbol{F} = (F_1, F_2, ..., F_n)$. We write $\Delta\tau$ as

$$\Delta\tau(X_0, t) = \prod_{i=1}^{n} \Delta x_i \,, \tag{3.70a}$$

where

[4] Generalization to include velocity dependent potentials is also possible. For details, see for example, Ref. [1].

$$\Delta x_i(X_0, t) = \frac{\partial x_i(X_0, t)}{\partial x_{i0}} \Delta x_{i0} \tag{3.70b}$$

and $\Delta x_{i0} = \Delta x_i(X_0, 0)$. The rate of change of $\Delta\tau$ can be calculated using (3.70a) as

$$\Lambda(X) = \frac{1}{\Delta\tau}\frac{\partial(\Delta\tau)}{\partial t} = \sum_i \frac{1}{\Delta x_i}\frac{\partial(\Delta x_i)}{\partial t} . \tag{3.71}$$

But from (3.70b) we obtain (after a change of order of differentiations)

$$\frac{\partial \Delta x_i}{\partial t} = \frac{\partial}{\partial x_{i0}}\frac{\partial x_i(X_0, t)}{\partial t} \Delta x_{i0} . \tag{3.72}$$

Using (3.69) and the fact that $\mathrm{d}x_i/\mathrm{d}t = \partial x_i(X_0, t)/\partial t$ in (3.72), $\Lambda(X)$ becomes

$$\Lambda(X) = \sum_i \frac{1}{\Delta x_i}\frac{\partial}{\partial x_{i0}} F_i \Delta x_{i0} . \tag{3.73}$$

Evaluating the above near $t = 0$, the instantaneous rate of change of the volume is

$$\Lambda = \sum_i \frac{\partial F_i}{\partial x_i} = \nabla \cdot \boldsymbol{F} . \tag{3.74}$$

Then the conditions $\nabla \cdot \boldsymbol{F} = 0$, < 0 and > 0 naturally represent *conservative, dissipative* and *volume expanding* dynamical systems, respectively (see Sect. 3.8.1 below). We note that to know whether a given dynamical system is conservative or dissipative we need only the differential equations but not the solutions, that is we need to calculate only $\nabla \cdot \boldsymbol{F}$.

Examples:

1. *Second-Order Conservative System*

$$\ddot{x} + f(x) = 0 . \tag{3.75}$$

With $x = x_1$, $\dot{x}_1 = x_2$ we obtain $\boldsymbol{F} = (x_2, -f(x_1)) = (F_1, F_2)$ and $\nabla \cdot \boldsymbol{F} = \partial F_1/\partial x_1 + \partial F_2/\partial x_2 = \partial x_2/\partial x_1 + \partial f(x_1)/\partial x_2 = 0$ so that all such second-order differential equations are conservative.

2. For the equation

$$\ddot{x} + d\dot{x} + f(x) = 0 , \quad d > 0, \tag{3.76}$$

we have $\boldsymbol{F} = (x_1, -dx_2 - f(x_1))$ and $\nabla \cdot \boldsymbol{F} = -d$, indicating that the phase space area diminishes in time and the system is dissipative.

3. Similarly the Lorenz equations (3.45) is dissipative (verify).

We wish to note that if the set of second-order differential equations describing the given system has only constant coefficients the presence of a term containing the first derivative or some odd powers of their derivatives generally indicates either a dissipative or an energy absorbing character of the system, giving a simple criterion that the system is not conservative. The converse is not true because a lack of such terms does not mean that the system is conservative.

3.8.1 Hamiltonian Systems

In Hamiltonian systems, the phase space volumes remain unchanged under time evolution. As noted in Sect. 3.2, the set of state variables X are identified here with the generalized coordinates q_i and canonically conjugate momenta $p_i, i = 1, 2, ..., n$, such that $X = (x_1, x_2, ..., x_{2n}) \equiv (q_1, q_2, ..., q_n, p_1, p_2, \ldots, p_n)$. The equations of motion of a Hamiltonian system with n degrees of freedom are

$$\dot{q}_i = \frac{\partial H}{\partial p_i}, \quad \dot{p}_i = -\frac{\partial H}{\partial q_i}, \quad i = 1, 2, \ldots, n. \tag{3.77}$$

It is easy to verify that the Hamiltonian flows conserve volume in phase space. Let us calculate $\nabla \cdot \boldsymbol{F}$, using (3.74) and (3.77),

$$\nabla \cdot \boldsymbol{F} = \sum_{k=1}^{2n} \frac{\partial F_k}{\partial x_k} = \sum_{i=1}^{n} \left[\frac{\partial}{\partial q_i} \left(\frac{\partial H}{\partial p_i} \right) + \frac{\partial}{\partial p_i} \left(-\frac{\partial H}{\partial q_i} \right) \right] = 0. \tag{3.78}$$

Thus using the expression (3.74), we can conclude that if we choose an initially ($t = 0$) closed $(N-1)$-dimensional surface S_0 in the N-dimensional phase space, and then evolve each point on the surface S_0 forward in time by using them as initial conditions, then the closed surface S_0 evolves to a closed surface S_t at some later time t, and the N-dimensional volumes $V(0)$ of the region enclosed by S_0 and $V(t)$ of the region enclosed by S_t are the same, that is, $V(t) = V(0)$. Therefore the system is volume preserving in phase space and is *conservative*. Consequently, the nature of trajectories in the phase space will be different in dissipative and conservative systems. In dissipative systems, steady state solutions are attractors. They can be nodes, stars, foci, limit cycles, quasiperiodic and chaotic attractors. Unstable equilibrium points including saddle can also occur in dissipative systems. However, they cannot have center type equilibrium points. In contrast, conservative systems can exhibit concentric periodic, quasiperiodic and chaotic (often called stochastic) solutions but these are not attractors. Further, they can have center and saddle type equilibrium points. Other types of equilibrium points cannot exist in conservative systems (why?).

3.9 Conclusions

In this chapter, we have indicated that the equations of motion of nonlinear dynamical systems are even though difficult in general to solve exactly, one can still make very useful qualitative analysis to start with. Treating the equation of motion of any given dynamical system as a system of n first-order coupled differential equations, one can look for equilibrium solutions or fixed points and investigate their linear stability. Based on the stability property, the equilibrium points can be classified and the motions in their neighbourhood identified. The stable equilibrium points are the point attractors of dynamical systems. Besides these, there exist many other interesting stable attractors associated with dissipative systems. These include limit cycles, quasiperiodic attractors and chaotic attractors. Distinction between dissipative and Hamiltonian systems was also indicated. It now becomes important to understand how transitions between motions corresponding to different types of attractors and solutions occur as one or more of the control parameters are varied. Such a study based on local stability of motion leads to the notion of bifurcations and onset of chaos. We will take up these aspects in the following chapters.

Problems

1. Show that the equilibrium point $(x^*, \dot{x}^*) = (0, 0)$ of the damped, linear oscillator
$$\ddot{x} + 2d\dot{x} + \alpha^2 x = 0 \,, \quad d > 0 \,, \alpha > 0$$
is a stable focus for $d < \alpha$ and a stable node for $d > \alpha$. Verify this from the exact analytical solution.
2. For the linear system
$$\dot{x} = x + \mu y \,,$$
$$\dot{y} = x - y \,,$$
show that
   a) for $\mu > -1$, the origin is a saddle equilibrium point,
   b) for $\mu = -1$, there is a line of equilibrium points with $y = x$ and
   c) for $\mu < -1$, the origin is a center.
3. Show that the undamped quadratic oscillator $\ddot{x} + x - x^2 = 0$ has one elliptic and one saddle equilibrium points. Draw the trajectories in the neighbourhood of the equilibrium points.
4. Determine the two equilibrium points of the quadratic oscillator equation
$$\ddot{x} + d\dot{x} - \alpha x + \beta x^2 = 0 \,, \quad d, \alpha, \beta > 0 \,.$$
Show that the origin is always a saddle point while the other equilibrium point is a stable focus for $d^2 < 4\alpha$, stable node for $d^2 > 4\alpha$, stable star at $d^2 = 4\alpha$ and becomes an elliptic point at $d = 0$.

5. Verify that the undamped anharmonic oscillator equation
$$\ddot{x} - \alpha x + \beta x^3 = 0 \,, \quad \alpha > 0,\ \beta > 0$$
has one saddle and two elliptic equilibrium points. What is the nature of the equilibrium points when $\alpha < 0$ and $\beta < 0$?
6. For the nonlinearly damped oscillator equation
$$\ddot{x} + x + 2\epsilon \left(\frac{\dot{x}^3}{3} - \dot{x} \right) = 0$$
verify that its equilibrium point is origin $((x^*, y^*(= \dot{x}^*)) = (0, 0))$ and it becomes
   a) stable node for $\epsilon < -1$,
   b) stable focus for $-1 < \epsilon < 0$,
   c) unstable focus for $0 < \epsilon < 1$ and
   d) unstable node for $\epsilon > 1$.
7. Show that the equilibrium point $(x^*, y^*(= \dot{x}^*)) = (0, 0)$ of the Toda oscillator
$$\ddot{x} + \mu \dot{x} + \alpha \left(\mathrm{e}^x - 1 \right) = 0$$
is a stable focus for $\alpha > 0$ and a saddle for $\alpha < 0$.
8. Consider the Lotka–Volterra equation
$$\dot{x} = ax - xy \,,$$
$$\dot{y} = xy - by \,, \quad a > 0,\ b > 0 \,.$$
Prove that its equilibrium point $(x^*, y^*) = (0, 0)$ is a saddle while (b, a) is an elliptic point.
9. Obtain the equilibrium points of the system
$$\dot{x} = \mathrm{e}^y - 1 \,,$$
$$\dot{y} = \mathrm{e}^y \cos x \,, \quad x \in [-\pi/2, 3\pi/2] \,.$$
Show that one of its equilibrium points is elliptic type, while the remaining two are saddles.
10. Show that the equilibrium point $(x^*, y^*) = (\alpha, \beta/\alpha)$ of the Brusselator equation
$$\dot{x} = \alpha + x^2 y - \beta x - x \,,$$
$$\dot{y} = \beta x - x^2 y \,, \quad \alpha,\ \beta > 0$$
is a stable node for $\beta < \alpha^2 + 1$ and $\alpha^4 + (1 - \beta)^2 - 2\alpha^2(1 + \beta) > 0$ and undergoes Hopf bifurcation at $\beta = \alpha^2 + 1$.
11. The motion of a current carrying wire in the field of an infinite current-carrying conductor and restrained by linear elastic springs is represented by the equation
$$\ddot{x} + x - \frac{\alpha}{1 - x} = 0 \,, \quad (x < 1) \,.$$
Sketch the potential function of the system. Verify the following.

a) For $\alpha < 0$, both the equilibrium points are elliptic.
b) For $\alpha = 0$, the only equilibrium point of the system is the origin and is elliptic type.
c) For $0 < \alpha < 1/4$, the system has one saddle and one elliptic equilibrium points.
d) For $\alpha > 1/4$, there are no real equilibrium points.

Draw the phase space trajectories for the above cases.

12. The motion of a ring sliding freely on the wire described by the parabola $z = px^2$ which rotates with a constant angular velocity Ω about the z-axis is described by the equation

$$\left(1 + 4p^2x^2\right)\ddot{x} + \alpha x + 4p^2 x\dot{x}^2 = 0 ,$$

where $\alpha = 2gp - \Omega^2$. Show that
a) for $\alpha < 0$, the origin is a saddle,
b) for $\alpha = 0$, all the motions are unbounded and
c) for $\alpha > 0$, the origin is an elliptic point.

Sketch the trajectories in the $(x - \dot{x})$ phase plane for the above three cases.

13. Prove that the equilibrium point $(x^*, y^*) = (0, 0)$ of the system

$$\dot{x} = y + \alpha \sin x ,$$
$$\dot{y} = -x$$

is
a) stable node for $\alpha < -2$,
b) stable focus for $-2 < \alpha < 0$,
c) unstable focus for $0 < \alpha < 2$ and
d) unstable node for $\alpha > 2$.

Also verify that the system undergoes Hopf bifurcation at $\alpha = 0$ and $\alpha = 2$ and a limit cycle attractor exists for $0 < \alpha < 2$.

14. A common continuous model equation used to describe predator-prey interactions between two biological species is the Rosenzweiz–MacArthur equation

$$\dot{x} = \gamma x \left(1 - x - \frac{y}{\alpha + x}\right) ,$$
$$\dot{y} = y \left(\frac{(\mu + 1)x}{\alpha + x} - 1\right) .$$

a) Show that for $\alpha < \mu$, the equilibrium points $(x^*, y^*) = (0, 0)$ and $(1, 0)$ are saddles and $(x^*, y^*) = \left(\alpha/\mu, \alpha(\mu - \alpha)(1 + \mu)/\mu^2\right)$ is either a node or a spiral point.
b) Prove that limit cycle occurs for $H = (\mu - \alpha\mu - 2\alpha)/(2\mu(\mu + 1)) > 0$ [14].

15. Consider the Lotka–Volterra type equation [15]

$$\dot{N}_1 = N_1 \left[g_1(N_1) - aN_1 - N_2 \right] ,$$
$$\dot{N}_2 = N_2 \left[g_2(N_1) - N_1 - bN_2 \right] ,$$

where $g_1(N_1)$ and $g_2(N_2)$ are arbitrary functions of N_1 and N_2 respectively. Let $(N_1, N_2) = (n_1, n_2)$ be an equilibrium point of the above system and define $k_1 = g_1'(n_1) - a$ and $k_2 = g_2'(n_2) - b$.

a) Verify that the eigenvalues of the equilibrium point are

$$\lambda_{1,2} = 0.5 \left(-n_1 k_1 + \frac{n_2}{k_2} \right) \pm \sqrt{\left(n_1 k_1 + \frac{n_2}{k_2} \right)^2 + 4 n_1 n_2} \ .$$

b) Show that the equilibrium point is

i) a stable node if $k_1 < k_2 < 0$,

ii) an unstable node if $k_1 > k_2 > 0$ and

iii) a saddle in the other cases.

16. For the system

$$\dot{x} = ax - xy ,$$
$$\dot{y} = x^2 - y ,$$

i) prove that

a) when $a < 0$, there exists only one equilibrium point $(x^*, y^*) = (0, 0)$ and that it is a stable node and

b) when $a > 0$, there are three equilibrium points, namely, $(x^*, y^*) = (0, 0)$, $(\pm\sqrt{a}\ , a)$ and all are saddles.

ii) What does happen at $a = 0$?

17. Verify that the equilibrium points of the system

$$\dot{x} = x^2 - xy ,$$
$$\dot{y} = x^2 - y$$

are $(x^*, y^*) = (0, 0)$ and $(1, 1)$ with the former being degenerate and the later being an elliptic type.

18. Show that the equilibrium point of the system

$$\dot{x} = -x ,$$
$$\dot{y} = 1 - x^2 - ay$$

is $(x^*, y^*) = (0, 1/a)$ and is a saddle for $a < 0$ and a stable node for $a > 0$.

19. Carry out the fixed point analysis of the system

$$\dot{x} = x^2 - y - 1 ,$$
$$\dot{y} = y(x - a) .$$

i) Verify that

a) the fixed point $(x^*, y^*) = (-1, 0)$ is a stable node for $a > -1$ and a saddle for $a < -1$,

b) the fixed point $(x^*, y^*) = (1, 0)$ is an unstable node for $a < 1$ and a saddle for $a > 1$ and
c) the fixed point $(x^*, y^*) = (a, a^2 - 1)$ is a stable node for $a < -1$, a saddle for $-1 < a < 1$ and an unstable node for $a > 1$.

ii) Draw the trajectories near the fixed points for $a = -1.5$, 0.5 and 1.5.

20. For the system

$$\dot{x} = -x + y^2 ,$$
$$\dot{y} = x^2 - y^3$$

prove the following: Its equilibrium point $(1, 1)$ is a saddle. For the other equilibrium point $(0, 0)$ the eigenvalues are $\lambda_1 = -1$, $\lambda_2 = 0$ and hence it is degenerate and that all points on the y-axis are stable equilibrium points. Sketch the trajectories in the $(x - y)$ phase plane.

21. Consider the pendulum of mass m and length a constrained to oscillate in a plane P rotating with angular velocity Ω about the vertical line. The moment of the centrifugal force acting on the pendulum is $m\Omega^2 a^2 \sin\theta \cos\theta$ and that of gravity is $mga \sin\theta$. The equation of motion of the system is given by [16]

$$I\ddot{\theta} - m\Omega^2 a^2 (\cos\theta - \lambda) \sin\theta = 0 ,$$

where $I = ma^2$ is the moment of inertia and $\lambda = g/(\Omega^2 a)$ is a parameter. Its equilibrium points are $(\theta^*, \dot{\theta}^*) = (0, 0)$, $(\pm\pi, 0)$ and $(\cos^{-1}\lambda, 0)$. Investigate their stability.

22. The roll motions of a ship can be described by the following nonlinearly damped anharmonic oscillator equation

$$\ddot{x} + \left(\omega_0^2 x + \alpha_3 x^3 + \alpha_5 x^5\right) + \left(2\mu_1 \dot{x} + \mu_3 \dot{x}^3\right) = 0 .$$

Show that

a) if $(\alpha_3^2 - 4\omega_0^2 \alpha_5) < 0$ the system has only one real equilibrium point, otherwise there exist three or five equilibrium points depending upon the signs of α_3 and α_5 and
b) for $\omega_0 = 5.278$, $\alpha_3 = -1.402\omega_0^2$ and $\alpha_5 = 0.271\omega_0^2$, the system has five real equilibrium points given by $(x^*, \dot{x}^*) = (\pm 2.0782, 0)$, $(0, 0)$, $(\pm 0.9243, 0)$ and that the first three equilibrium points are elliptic for $\mu_1 = \mu_3 = 0$ and become stable foci for $\mu_1 > 0$, $\mu_3 > 0$, while the remaining two are always saddle points [11].

23. Consider the three-dimensional nonlinear equation

$$\dot{x} = y ,$$
$$\dot{y} = z ,$$
$$\dot{z} = -\eta z - \gamma y - \mu x - K_1 x^2 - K_2 y^2 - K_3 xy - K_4 xz - K_5 x^2 z ,$$

which is used to study a continuous chemical flow reactor for the Belousov–Zhabotinski reaction. Investigate the stability of its two equilibrium points as a function of the parameter μ for fixed values of other parameters [17].

24. The nonlinear dynamics of a solid-state laser driven by an injected sinusoidal field is described by the equation

$$\dot{X} = uX + k\cos\phi ,$$
$$\dot{\phi} = \delta - (k\sin\phi)/X ,$$
$$\dot{u} = B - \left(1 + \sqrt{\sigma u}\right)\left(1 + X^2\right) .$$

a) Show that this system in rectangular coordinates with $x = X\cos\phi$, $y = X\sin\phi$ can be rewritten as

$$\dot{x} = ux - \delta y + k ,$$
$$\dot{y} = uy + \delta x ,$$
$$\dot{z} = B - \left(1 + \sqrt{\sigma u}\right)\left(1 + x^2 + y^2\right) .$$

b) Prove that the equilibrium points of the above equation exist if and only if

$$1 \le B \le B_c = 1 + \frac{k^2}{\delta^2} .$$

c) Determine the equilibrium points and study their stability for $B = 1$ and $B \neq 1$ [18].

25. Show that node type and focus type singular points cannot exist for Hamiltonian systems.

4. Bifurcations and Onset of Chaos in Dissipative Systems

In the earlier chapters, we have seen some basic features associated with nonlinear dynamical systems. Now we want to have a more detailed picture of the types of possible motions admitted by typical nonlinear systems, particularly by dissipative systems. Specifically, we wish to ask the question how does the dynamics change as a control parameter (like the strength of nonlinearity or the strength of the external forcing or its frequency) is smoothly varied. One finds that very interesting features arise in this process: *Bifurcations*, which are sudden qualitative changes in the nature of the motion, occur at critical control parameter values (as noted briefly in the earlier chapter). Often these bifurcations lead to chaotic behaviour of the dynamical system. How and why do these bifurcations occur? What are the *mechanisms* responsible for such bifurcations? How can they be classified? These are some of the important questions which arise naturally. One finds that they often occur in specific ways or *routes* as the control parameter is varied. Typically one type of motion loses stability at a critical value of the parameter as it varies smoothly giving rise to a new type of stable motion. This process can continue further to give rise to newer and newer motions. Thus a possible approach to understand the sudden qualitative changes in the dynamics is to look for the *local stability* properties of the solutions in the neighbourhood of the critical parameter values at which bifurcations occur. However, unfortunately no exact solutions, barring exceptions, are in general available for most of the nonlinear dynamical systems. So what one can do is to prepare a dictionary or a catalogue or a list of the standard types of bifurcations which occur in simple and interesting low dimensional nonlinear systems and then use them as reference systems for identification of the types of bifurcations which occur in the numerical study of various nonlinear oscillators and systems.

With this aim in view, in this chapter, we will first study a few of the simple bifurcations which occur in many low dimensional nonlinear continuous dynamical systems. These will include

(1) the saddle-node,
(2) the pitchfork,
(3) the transcritical and
(4) Hopf bifurcations.

Next, as noted above, in a typical nonlinear system as the parameter changes more than one critical value or bifurcation value can exist giving rise to transitions to different types of motion and ultimately to interesting dynamical situations including chaos. In many nonlinear dissipative dynamical systems chaotic motion is found to set in through the following three predominant *routes*, namely,

(1) Feigenbaum scenario (period doubling route),

(2) Ruelle–Takens–Newhouse scenario (quasiperiodic route) and

(3) Pomeau–Manneville scenario (intermittency route).

Besides, there are several other less prominent routes, such as period adding and Farey sequence bifurcation routes. In order to understand the nature of such routes, it is much simpler to consider dynamical systems in which time varies (or can be considered as varying) in discrete steps. In this case we have discrete dynamical systems described by difference equations or maps. They are quite informative and constitute a field of their own, namely *nonlinear maps*. But they can also faithfully represent continuous dynamical systems in the form of *Poincaré maps*, stroboscopic maps and so on (about which more will be said in the following chapters). Further the study of such maps can throw much light in a transparent way on many new types of bifurcations and routes to chaos. Period doubling bifurcations route is one such example. The other bifurcation routes can also be understood more easily in terms of maps. From these points of view, we will study the dynamics of two important nonlinear maps, namely,

(1) the one-dimensional logistic map and

(2) the two-dimensional Hénon map,

besides mentioning other interesting maps.

After obtaining the basic ideas on bifurcations and onset of chaos through certain routes from these maps, we will extend our studies to continuous time (flow) systems in the next chapter. We strongly suggest the reader to perform his own numerical analysis of the above two maps and on the other systems discussed in the next chapters to reproduce the plots presented in this and subsequent chapters. Even BASIC programs with the relatively poor precision available on a personal computer will be able to illustrate many of the interesting phenomena associated with nonlinear dynamical systems.

4.1 Some Simple Bifurcations

In the previous chapter (Sect. 3.4.3) we have studied the stability of equilibrium points of the force-free Duffing oscillator, equation (3.18). We found that for $d^2 < 8a$ the equilibrium points $(\pm\sqrt{a/b}\ ,0)$ are stable nodes and for $d^2 > 8a$ they are stable foci. Thus, as the parameter d is varied the dynamics of the system changes from one type to another at the critical value $d = \sqrt{8a}$. In general, systems of physical interest typically have one or more parameters. As a parameter is varied, qualitative changes in the behaviour of the system can occur at certain *critical values* of the parameter. Each of such a change is called a *bifurcation* and the corresponding value of the parameter is called the *bifurcation value.* At this value, a vanishingly small perturbation of the system can cause qualitative changes in the solution of the system. In this section we focus upon some of the simple types of bifurcation phenomena in which only equilibrium points and limit cycles are involved and illustrate them by means of simple examples.

4.1.1 Saddle-Node Bifurcation

Consider the dynamical system (for illustrative purpose),

$$\dot{x} = \mu - x^2 , \tag{4.1a}$$

$$\dot{y} = -y , \quad \left(\cdot = \frac{\mathrm{d}}{\mathrm{d}t} \right) \tag{4.1b}$$

where μ is a parameter. The equilibrium points of (4.1) are

$$E_1 : \ (x^*, y^*) = (+\sqrt{\mu}\,, 0) , \tag{4.2a}$$

$$E_2 : \ (x^*, y^*) = (-\sqrt{\mu}\,, 0) . \tag{4.2b}$$

The stability determining eigenvalues for each one of the equilibrium points E_1 and E_2 are given by (refer to our earlier discussions in Sect. 3.4)

$$\lambda_1 = -1 , \quad \lambda_2 = -2x^* . \tag{4.3}$$

Exercise:

1. Verify equation (4.3).

For $\mu < 0$, the system has no real equilibrium points and typical flow near the origin is illustrated in Fig. 4.1a. At $\mu = 0$, the origin $(0, 0)$ becomes a unique equilibrium point and the eigenvalues are $\lambda_{1,2} = -1, 0$. We note that equation (4.1b) is linear in y and independent of x. It can be easily integrated to give the solution, $y(t) = y(0)\exp(-t)$. In the limit $t \to \infty$, $y(t) \to 0$. That is, the y-components of the trajectories of (4.1) asymptotically approach zero. On the other hand, the evolution equation for the x-component, (4.1a), is nonlinear and the solution depends on the value of μ.

For $\mu = 0$, (4.1a) can be readily integrated to give

$$x(t) = \frac{x(0)}{1 + x(0)t} . \tag{4.4}$$

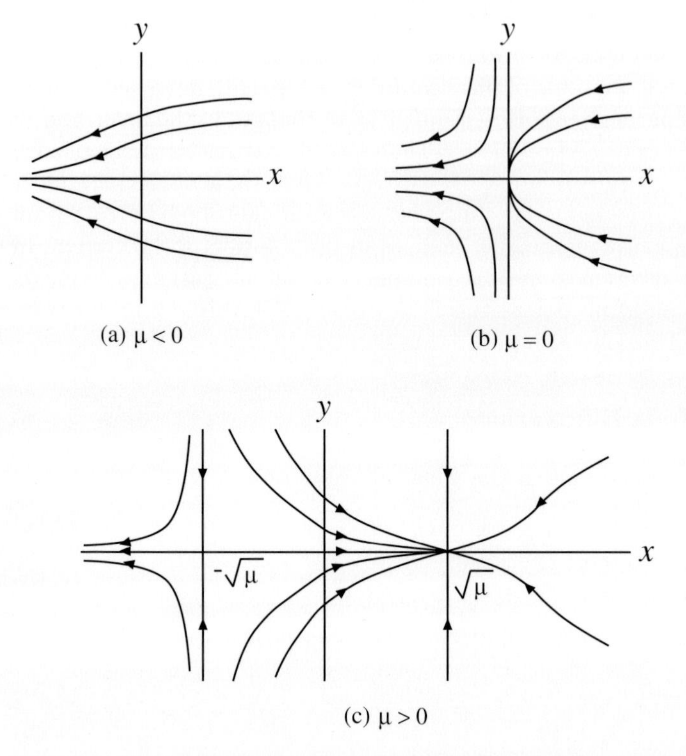

Fig. 4.1. (**a**–**c**) Phase trajectories of the system (4.1) for various μ values illustrating saddle-node bifurcation

Thus for $x(0) > 0$, the trajectories approach the origin, while for $x(0) < 0$, all the trajectories diverge from the origin ($x(t) \to -\infty$ as $t \to \mid 1/x(0) \mid$). That is, in the region $x < 0$ the phase space picture resembles that of trajectories in the vicinity of a saddle while in the region $x > 0$ they appear to be related to a stable node (Fig. 4.1b). This type of an equilibrium point is called a *saddle-node*.

As μ is increased beyond the value zero, the equilibrium point $(0, 0)$ bifurcates into two new equilibrium points E_1 and E_2 and they are given by (4.2). The eigenvalues of the linear stability problem associated with E_1 are -1, $-2\sqrt{\mu}$ and E_2 are -1, $+2\sqrt{\mu}$. The equilibrium point E_2 is thus a saddle which is unstable, while E_1 is a stable node. Figure 4.1c depicts the trajectories near the equilibrium points E_1 and E_2 for $\mu > 0$. As $\mu \to 0$ both the equilibrium points tend to $(0, 0)$ and the eigenvalues tend to 0 and -1. We may then conclude that at the critical value $\mu = 0$ a bifurcation occurs which one might call a *saddle-node bifurcation*. Such a saddle-node bifurcation can occur in systems of arbitrary dimensions as well.

We may then conclude that as the parameter μ in (4.1) varies from negative to positive values, from no solution, suddenly at the critical value $\mu = 0$,

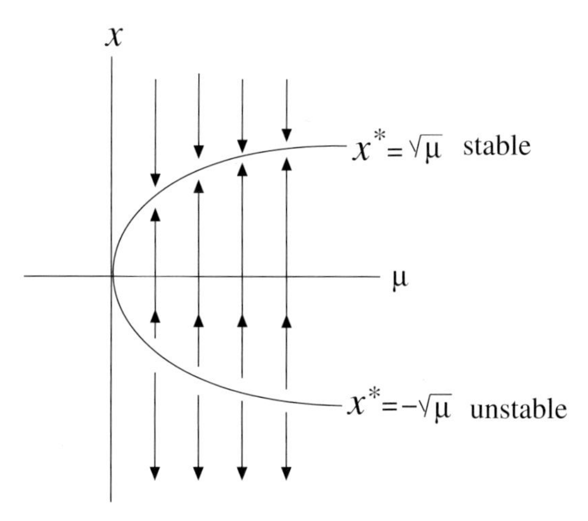

Fig. 4.2. Saddle-node bifurcation diagram. Arrows indicate stable and unstable nature of the equilibrium points

a saddle-node is born. This then bifurcates into a stable node and a saddle for $\mu > 0$. The associated qualitative change can be depicted in the form of a *bifurcation diagram* (x versus μ) as shown in Fig. 4.2. Alternatively, we can say that for $\mu > 0$ the system (4.1) has one stable node and one saddle (unstable) equilibrium points. When μ is decreased from a higher value the two equilibrium points approach each other and form a single equilibrium point at $\mu = 0$ which is of saddle-node type. For $\mu < 0$, the equilibrium point disappears.

Exercises:

2. One can also exactly solve the system (4.1) and verify the above facts. Obtain the solutions

$$x(t) = \sqrt{\mu}\left(\frac{x(0) + \sqrt{\mu}\,\tanh\sqrt{\mu}}{\sqrt{\mu} + x(0)\tanh\sqrt{\mu}\,t}\right) \quad \text{for } \mu > 0 \tag{4.5}$$

   and

$$x(t) = \sqrt{-\mu}\left(\frac{x(0) - \sqrt{-\mu}\,\tan\sqrt{-\mu}\,t}{\sqrt{-\mu} + x(0)\tan\sqrt{-\mu}\,t}\right) \quad \text{for } \mu < 0\,. \tag{4.6}$$

   Of course $y(t) = y(0)\exp(-t)$. Verify that the local stability analysis discussed above does correspond to the nature of the actual solutions.
3. Study the occurrence of saddle-node bifurcation in the system

$$\dot{x} = \mu + x^2\,,$$
$$\dot{y} = -y$$

   using both linear stability analysis and exact solutions. Compare the phase portrait of the above system with that of the system (4.1).

4. For the system

$$\dot{x} = y + v_2 x + x^2,$$
$$\dot{y} = v_1 + x^2$$

find the points in the (v_1, v_2) parameter space at which saddle-node bifurcation occurs.

5. Verify the following for the system

$$\dot{x} = \mu x - x^2 + \alpha\,,$$
$$\dot{y} = -y\,.$$

a) For $\alpha < 0$ there are two real equilibrium points with one being a stable node and the other being a saddle for $|\mu| > 2\sqrt{|\alpha|}$ and that the system undergoes a saddle-node bifurcation at $\mu = \pm 2\sqrt{|\alpha|}$.
b) For $\alpha < 0$, there are no real equilibrium points for $|\mu| < 2\sqrt{|\alpha|}$.

4.1.2 The Pitchfork Bifurcation

Consider now the second dynamical system

$$\dot{x} = \mu x - x^3, \tag{4.7a}$$
$$\dot{y} = -y\,. \tag{4.7b}$$

The equilibrium points of the (4.7) are

$$E_0: \quad (x^*, y^*) = (0, 0)$$
$$E_1: \quad (x^*, y^*) = (-\sqrt{\mu}\,, 0)$$
$$E_2: \quad (x^*, y^*) = (+\sqrt{\mu}\,, 0)$$

Naturally for $\mu < 0$, only $E_0 = (0, 0)$ exists as the other two become complex. For $\mu > 0$, we have three real equilibrium points. For any of these equilibrium points, the stability determining eigenvalues are

$$\lambda_1 = -1\,, \quad \lambda_2 = -3x^{*2} + \mu\,. \tag{4.8}$$

Thus, for $\mu < 0$ the system (4.7) has only one equilibrium point, namely $(0, 0)$, and it is stable. For $\mu = 0$, integrating (4.7a) we get

$$x(t) = \frac{x(0)}{\sqrt{1 + 2x(0)^2 t}}\,. \tag{4.9}$$

Thus, for both positive and negative $x(0)$, the trajectories approach the origin and so the equilibrium point is stable. For $\mu > 0$ the system has two additional equilibrium points. Now the equilibrium point $(0, 0)$ is unstable (saddle) because one of its eigenvalues is positive and the other is negative. For the other two equilibrium points E_1 and E_2, the eigenvalues are $-1, -2\mu$. Both the eigenvalues are negative and hence the equilibrium points are stable nodes. Thus, as μ is increased through $\mu = 0$ the equilibrium point $(x^*, y^*) = (0, 0)$ loses its stability and two new stable equilibrium points E_1 and E_2 emerge. The bifurcation diagram and the phase trajectories near the equilibrium points are depicted in the Figs. 4.3a and 4.4a, respectively. Because of the tuning fork-like appearance of the bifurcation diagram, the above bifurcation is called a *pitchfork bifurcation*. In the above example, the bifurcation is *supercritical* meaning that the new equilibrium points which arise above the critical value $\mu = 0$ are stable. One can verify the above description of the system (4.7) from its analytic solution

$$\begin{aligned} x(t) &= \left[\frac{1}{\mu} + \left(\frac{1}{x(0)^2} - \frac{1}{\mu}\right) \mathrm{e}^{-2\mu t}\right]^{-1/2}, \\ y(t) &= y(0)\,\mathrm{e}^{-t} \quad \text{for } \mu \neq 0 \end{aligned} \tag{4.10a}$$

and

$$\begin{aligned} x(t) &= x(0)\left(1 + 2x(0)^2 t\right)^{-1/2}, \\ y(t) &= y(0)\,\mathrm{e}^{-t} \quad \text{for } \mu = 0\,. \end{aligned} \tag{4.10b}$$

If we consider a slightly different system

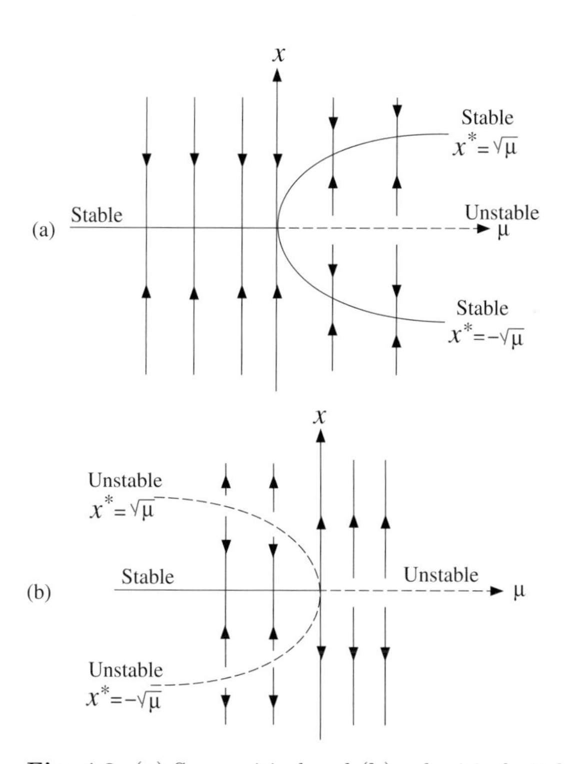

Fig. 4.3. (**a**) Supercritical and (**b**) subcritical pitchfork bifurcation diagrams

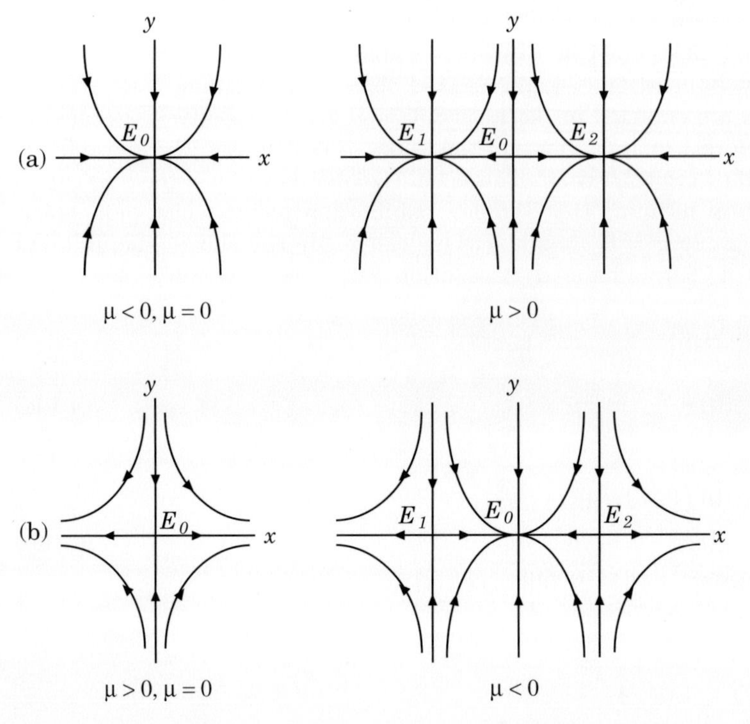

Fig. 4.4. (**a**) Phase trajectories of the system (4.7) illustrating supercritical pitchfork bifurcation. (**b**) Phase trajectories of the system (4.11) illustrating subcritical pitchfork bifurcation

$$\dot{x} = \mu x + x^3 , \tag{4.11a}$$

$$\dot{y} = -y , \tag{4.11b}$$

the situation is analogous to the system (4.7) with the difference that for $\mu > 0$ there is only one equilibrium point (the origin) and it is unstable (verify). At $\mu = 0$, the equilibrium point $(0, 0)$ is still unstable (verify from analytical solution). But for $\mu < 0$, there are three equilibrium points with the origin being stable and the other two being unstable. This is called a *subcritical* pitchfork bifurcation. The corresponding bifurcation diagram and phase trajectories near the equilibrium points are shown in Figs. 4.3b and 4.4b respectively.

Thus in a supercritical pitchfork bifurcation, as a parameter is varied the stable equilibrium point (or even periodic orbit) generally becomes unstable at a critical value and gives birth to two new stable equilibrium points (or periodic orbits). On the other hand, in the subcritical bifurcation, as a control parameter is varied an unstable equilibrium point (or periodic orbit) becomes stable at a critical value where two more equilibrium points (or periodic orbits) are born which are unstable. The pitchfork bifurcation is often found in systems which are invariant under the transformation $\boldsymbol{X} = -\boldsymbol{X}$ (check that this is true for (4.7) and (4.11)).

Exercise:

6. Verify that the following systems undergo pitchfork bifurcation at $\mu = 0$. Sketch the bifurcation diagrams and the phase portraits of the corresponding systems. Compare the results with the systems (4.7) and (4.11).

   a) $\dot{x} = -\mu x - x^3$,
      $\dot{y} = -y$.

   b) $\dot{x} = -\mu x + x^3$,
      $\dot{y} = -y$.

4.1.3 Transcritical Bifurcation

Another example of bifurcations involving only equilibrium points is the transcritical bifurcation where the stability of the equilibrium points are exchanged when the control parameter passes through the bifurcation (point) value. To illustrate this we consider the two-dimensional system

$$\dot{x} = -\mu x + x^2 \,, \tag{4.12a}$$

$$\dot{y} = -y \,. \tag{4.12b}$$

Its equilibrium points are

$$E_0 : \quad (x^*, y^*) = (0, 0) \,,$$

$$E_1 : \quad (x^*, y^*) = (\mu, 0) \,.$$

The eigenvalues are

$$\lambda_1 = -1 \,, \quad \lambda_2 = -\mu + 2x^* \,. \tag{4.13}$$

From the eigenvalues we find that the equilibrium point $(0, 0)$ is unstable for $\mu < 0$ and stable for $\mu > 0$. On the other hand, the equilibrium point $(\mu, 0)$ is stable for $\mu < 0$ and unstable for $\mu > 0$. Thus, for $\mu < 0$ one equilibrium point ($(\mu, 0)$) is stable whereas the other one ((0,0)) is unstable (Fig. 4.5). But for $\mu > 0$ the previously stable one becomes unstable and the unstable one becomes stable (Fig. 4.5). That is, the stability is exchanged (or transferred) at a critical value when the control parameter is varied. For this reason the bifurcation is called a *transcritical* bifurcation or *exchange of stability* bifurcation and is illustrated in the bifurcation diagram, Fig. 4.6.

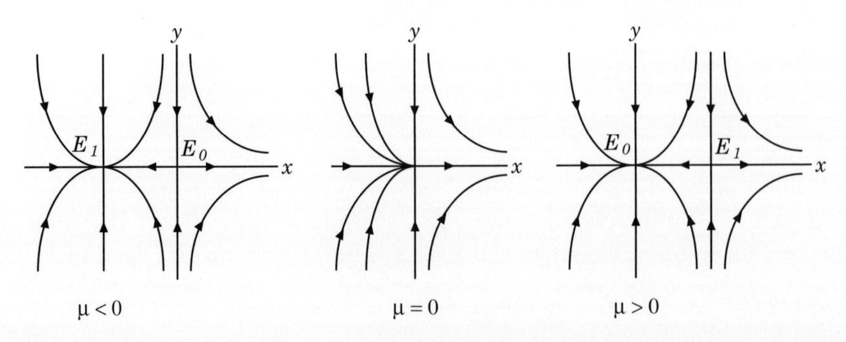

Fig. 4.5. Phase trajectories of the system (4.12) illustrating transcritical bifurcation

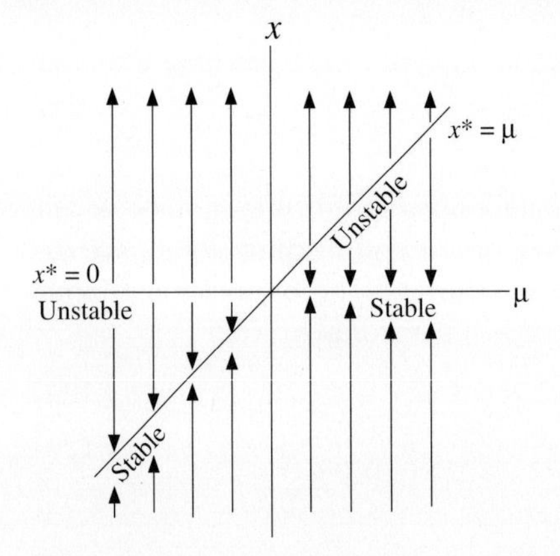

Fig. 4.6. Transcritical bifurcation diagram

Exercises:

7. Like the system (4.1), equation (4.12) can also be solved analytically and its solution is given by

$$x(t) = \mu \frac{A\mathrm{e}^{-\mu t}}{A\mathrm{e}^{-\mu t} - 1}\,, \quad y(t) = y(0)\mathrm{e}^{-t} \tag{4.14a}$$

with

$$A = \frac{x(0)}{x(0) - \mu}\,, \quad \mu \neq 0\,, \tag{4.14b}$$

while for $\mu = 0$,

$$x(t) = \frac{x(0)}{1 - x(0)t}\,, \quad y(t) = y(0)\mathrm{e}^{-t}\,. \tag{4.15}$$

Verify the outcome of linear stability analysis using the above solutions.

8. Discuss the occurrence of transcritical bifurcation in the system

$$\dot{x} = -x\left(x^2 - 2bx - a\right) ,$$
$$\dot{y} = -y .$$

4.1.4 Hopf Bifurcation

Another simple bifurcation is the *Hopf bifurcation*, which we mentioned briefly earlier in Sect. 3.5, where as a control parameter is varied a stable equilibrium point loses its stability at a critical value and gives birth to a limit cycle. Let us consider an example to demonstrate this possibility.

In Sect. 3.5 we have studied the occurrence of limit cycle motion in the van der Pol oscillator equation (3.33). It has an equilibrium point $(0, 0)$. For $-2 < b < 0$, the eigenvalues are complex conjugates with negative real parts and the equilibrium point is a stable focus. For $b = 0$, (3.33) reduces to the equation of motion of a linear harmonic oscillator whose solution is $x(t) = A\cos(t + \phi)$, where A and ϕ are integration constants. For this case the eigenvalues are $\lambda_{\pm} = \pm\mathrm{i}$. The equilibrium point $(0, 0)$ is now of center type and all the solutions are periodic. For $b \in (0, 2)$ the eigenvalues are complex conjugates with positive real parts. The equilibrium point is an unstable focus. Consequently, near the origin the trajectories spiral away from it. However, as shown in Sect. 3.4, in this range of b a limit cycle attractor is found to occur due to a dynamical balance between the positive and negative dampings. That is, the system undergoes a bifurcation at $b = 0$ and a stable limit cycle is developed as b is varied further. This bifurcation is called a *Hopf bifurcation*. It is characterized by a change of the real parts of the pair of complex conjugate eigenvalues associated with an equilibrium point from a negative to positive value while the imaginary part remains greater than zero. It is easy to see that the Hopf bifurcation can occur only in nonlinear systems with dimensions greater than one.

In transcritical and pitchfork bifurcations the real eigenvalues of the least stable equilibrium point increases through zero as a control parameter μ increases (or decreases) through a critical value, say, μ_0 and one or two new stable equilibrium points arise. In contrast, in a Hopf bifurcation, the real parts of a pair of complex conjugate eigenvalues of the least stable equilibrium point increases through zero as μ increases or decreases through μ_0 and a time periodic solution arises.

In (3.33) the bifurcation is *supercritical* because the limit cycle is stable above the bifurcation point. In a *subcritical* Hopf bifurcation the bifurcated limit cycle will be unstable. Figures 4.7 show the phase portraits of supercritical and subcritical Hopf bifurcations. For a two-dimensional system of the form (3.4) the stability of the limit cycles born due to Hopf bifurcation can be determined from the conditions given in the form of Hopf theorem. The theorem can be stated (without proof) as follows. (For fuller details the

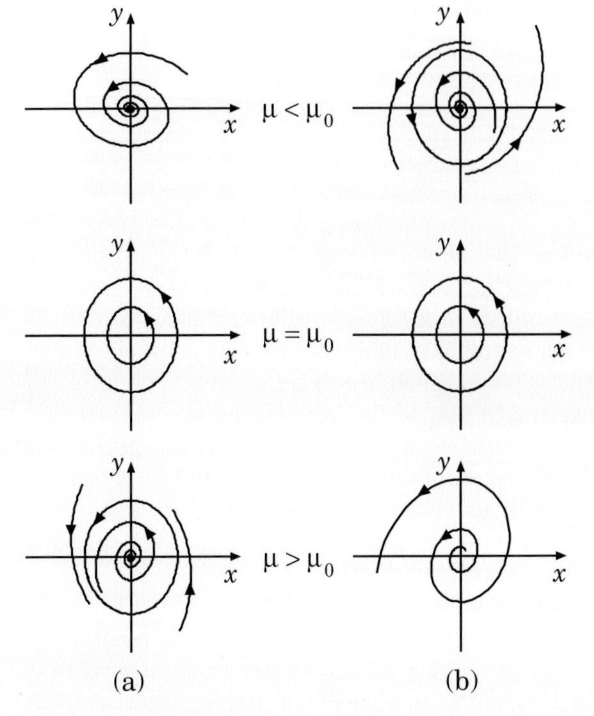

Fig. 4.7. Hopf bifurcations in the phase plane. (**a**) Supercritical Hopf bifurcation. (**b**) Subcritical Hopf bifurcation

readers are referred to more advanced texts such as *Stability, Instability and Chaos* (P. Glendenning, 1994) [1] and *The Hopf Bifurcation and its Application* (J.E. Marsden and M. McCracken, 1976) [2]).

***Theorem**: Suppose that the autonomous system (3.4) has an equilibrium point at* (x_0, y_0)*, and that the associated Jacobian matrix*

$$M = \begin{pmatrix} \partial P/\partial x & \partial P/\partial y \\ \partial Q/\partial x & \partial Q/\partial y \end{pmatrix}\Bigg|_{(x_0, y_0)} \tag{4.16}$$

has a pair of purely imaginary eigenvalues $\lambda(\mu) = \mathrm{i}\omega$ *and* $\lambda^*(\mu) = -\mathrm{i}\omega$ *at* $\mu = \mu_0$*. If*

1. $\dfrac{\mathrm{d}}{\mathrm{d}\mu} Re\lambda(\mu)\Bigg|_{\mu=\mu_0} > 0$ *for some* μ_0*,*
2. $P_{\mu x} + Q_{\mu y} \neq 0$ *and*
3. $a \neq 0$*, where*

$$a = \frac{1}{16}\left[P_{xxx} + Q_{xxy} + P_{xyy}Q_{yyy}\right] + \frac{1}{16\omega}\left[P_{xy}\left(P_{xx} + P_{yy}\right)\right.$$
$$\left. - Q_{xy}\left(Q_{xx} + Q_{yy}\right) - P_{xx}Q_{xx} + P_{yy}Q_{yy}\right] \qquad (4.17)$$

evaluated at (x_0, y_0)*, then a periodic solution occurs for* $\mu < \mu_0$ *if* $a(P_{\mu x} + Q_{\mu y}) > 0$ *or* $\mu > \mu_0$ *if* $a(P_{\mu x} + Q_{\mu y}) < 0$*. The equilibrium point is stable for* $\mu > \mu_0$ *and unstable for* $\mu < \mu_0$ *if* $P_{\mu x} + Q_{\mu y} < 0$*. On the other hand, the equilibrium point is stable for* $\mu < \mu_0$ *and unstable for* $\mu > \mu_0$ *if* $P_{\mu x} + Q_{\mu y} > 0$*. In both the cases the periodic solution is stable if the equilibrium point is unstable while it is unstable if the equilibrium point is stable on the side of* $\mu = \mu_0$ *for which the periodic solution exists.*

The proof of the theorem is based on the so-called center manifold and normal form theorems, see for example Ref. [2]. It is considered to be one of the basic results in the theory of nonlinear oscillations. We illustrate the usefulness of the above theorem by applying it to the van der Pol equation (3.33). Its equilibrium point is $(x_0, y_0) = (0, 0)$. The eigenvalues of the Jacobian matrix M of it are

$$\lambda_{\pm} = \frac{1}{2}\left[b \pm \sqrt{b^2 - 4}\right] . \qquad (4.18)$$

For $-2 < b < 2$ the eigenvalues are complex conjugate and are real for $|b| > 2$. When $b = 0$, they are $\lambda_{\pm} = \pm \mathrm{i}$ and so pure imaginary. For $0 < b < 2$ we find that

(1) $\frac{\mathrm{d}}{\mathrm{d}b}\mathrm{Re}\lambda(b) > 0$

(2) $P_{bx} + Q_{by} = 1$

(3) $a = -b/8$

satisfying the three conditions of the Hopf theorem. Further $a(P_{bx} + Q_{by}) = -b/8 < 0$, thereby implying the existence of a limit cycle solution for $0 < b < 2$. The stability of the limit cycle depends on the sign of the quantity $P_{bx} + Q_{by}$. For the van der Pol equation (3.33), for the equilibrium point $(0, 0)$ this quantity is 1. Therefore, according to the Hopf theorem, for $0 < b < 2$ the origin must be unstable and the limit cycle solution must be stable.

Exercises:

9. Find the conditions for which the equilibrium point $(0, 0)$ of the system

$$\dot{x} = y - x\left(x^2 + y^2 - \alpha\right) ,$$
$$\dot{y} = -x - y\left(x^2 + y^2 - \alpha\right) ,$$

where α is a constant, undergoes a Hopf bifurcation.

10. For the following system (called Bonhoeffer–van der Pol oscillator) determine the interval of the parameter A_0 for which stable limit cycle motion is possible [3]. (Note: One need not explicitly calculate the equilibrium points, rather it is sufficient to use the criterion at which complex conjugate eigenvalues cross the imaginary axis).

$$\dot{x} = x - x^3/3 - y + A_0 ,$$
$$\dot{y} = c(x + a - by) .$$

Here a, b and c are parameters.

11. Discuss the possibility of occurrence of limit cycle in the following systems.

a) Brusselator equations :

$$\dot{x} = a - x - bx + x^2 y ,$$
$$\dot{y} = bx - x^2 y .$$

b) Duffing–van der Pol oscillator:

$$\ddot{x} - b\left(1 - x^2\right)\dot{x} + ax + bx^3 = 0 .$$

11. A predator and prey model equation is

$$\dot{x} = x\left[r\left(1 - \frac{x}{k} - \frac{ay}{b_0 + x}\right)\right],$$
$$\dot{y} = \left(\frac{eax}{b_0 + x} - d\right).$$

Here r, a, b_0, d, k and e are various parameters. Show that the system has a Hopf bifurcation at $b_0 = b_0' = k(ea - d)/(ea + d)$ and the created limit cycle is stable for $b_0 < b_0'$, and further a transcritical bifurcation occurs at $b_0 = b_0'' = k(ea - d)/d$ [4]. Sketch the phase portrait for (a) $b_0 < b_0'$, (b) $b_0' < b_0 < b_0''$ and (c) $b_0 > b_0''$.

12. For the system

$$\dot{x} = y + v_2 x + x^2 ,$$
$$\dot{y} = v_1 + x^2 ,$$

in the (v_1, v_2) parameter space, find the point at which Hopf bifurcation occurs.

So far we have described some simple bifurcations involving equilibrium points and limit cycles only. Further they depend only on one control parameter. They are called *codimension one* bifurcations. One can consider more general bifurcations like codimension two, etc., bifurcations, involving two or more control parameters. They are more complicated to study and possibly less common in occurrence. Nevertheless much work has gone on into the study of such bifurcations. For details we may refer to Ref. [5]. However, even in the case of variation of a single control parameter, there are many other interesting and sophisticated bifurcations such as period doubling, quasiperiodic, intermittent and so on, which can arise in typical nonlinear systems of suitable dimensions (see also Chap. 5 for more details). These bifurcations can often lead to complicated and irregular motions, namely *chaotic motions*. However, in order to appreciate these more exotic bifurcations, it is rather illuminating to investigate a class of dynamical systems called discrete dynamical systems, in which the time variable changes in discrete steps. In the following sections we illustrate the three routes to chaos mentioned at the beginning of the present chapter in some simple discrete dynamical systems.

4.2 Discrete Dynamical Systems

In the previous section we have dealt with dynamical systems in which time is always treated as a continuous (flow) variable. However there is an important subclass of systems, the so-called *discrete dynamical systems*, in which the time variable can be treated as a *discrete* variable rather than a continuous one. This may mean, for example, that it is sufficient or meaningful to measure certain physical variables after finite intervals of time, say an hour, a week, a month, etc. rather than on a continuous basis. Examples include population of an insect species in a forest, radioactive decay, rain fall and temperature in a city. In certain specific scientific contexts, it is natural to represent time as discrete. This is the case in digital electronics, in parts of economics and finance theory, in impulsively driven systems and so on. These systems are often represented by difference equations/ recursion relations/ iterated maps, or simply *maps*. These maps could be of any dimensions, depending on the number of physical variables. For example the rule $x_{n+1} = \cos x_n$ is an example of an *one-dimensional map*, so-called because the points x_n belong to the one-dimensional space of real numbers and n takes only integer values. Generally, for a map $x_{n+1} = f(x_n)$, given an initial value x_0, we compute x_1 and then using x_1 compute x_2, and so on:

$$\begin{aligned} x_1 &= f(x_0) \ , \\ x_2 &= f(x_1) \ , \\ &\dots \\ &\dots \end{aligned}$$

The sequence $x_0, x_1, \dots$ defines the *trajectory* of the dynamical variable x_n, where now the time n is discrete. Similarly one can consider higher dimensional maps.

Further, nonlinear maps are interesting entities to study on their own merit as mathematical block boxes for chaos. Indeed, maps are capable of showing as rich a behaviour as differential equations or even more because the points x_n hop along their orbits rather than flow continuously as shown in Fig. 4.8. Very often they can faithfully mimic continuous systems, in the form

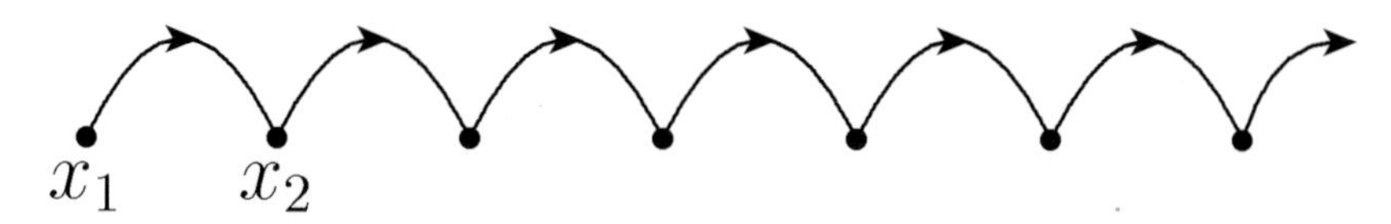

Fig. 4.8. Discrete evolution of the state variable x_n of a map

of Poincaré map, stroboscopic map, etc. as noted in Chap. 3 and also earlier in this chapter. They are also easy and fast to simulate on digital computers where time is inherently discrete. We will study briefly two such maps in the following to understand some of the important bifurcations and chaos phenomena underlying dynamical systems, particularly the period doubling bifurcations route to chaos.

4.2.1 The Logistic Map

Among the different nonlinear map equations, the *logistic map* has played a central role in the development of the theory of chaos. The logistic map has a quadratic nonlinearity and is represented by the equation [6]

$$x_{n+1} = ax_n\,(1 - x_n) = f\,(x_n)\ , \quad n = 0, 1, 2, \dots \tag{4.19}$$

where a is a parameter and we assume that $0 \leq x \leq 1$. This map is a discrete-time analog of the logistic equation for population growth. Here x_n is a dimensionless measure of the population of a given species (such as fly, plankton, etc.) in the nth generation and $a \geq 0$ is the intrinsic growth rate. As shown in Fig. 4.9 the graph of $f(x_n)$ is a parabola with a maximum value $a/4$ at $x = 1/2$. The control parameter a is considered within the range $0 \leq a \leq 4$ so that (4.19) maps the interval $0 \leq x \leq 1$ into itself.

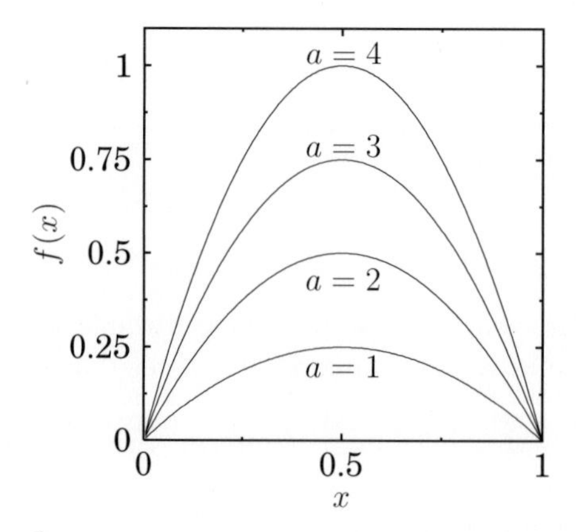

Fig. 4.9. Graph of $f(x)$ for the logistic map

Now we wish to find what this model can tell us about the long time ($n \to \infty$) behaviour of the population fraction x and how that long-term value depends on a, which is the control parameter in the present system.

4.2.2 Equilibrium Points and Their Stability

To start with we calculate the equilibrium points and determine their linear stability similar to the case of differential equations discussed in the previous chapter. Naturally, extending our earlier definition of equilibrium points for a continuous time dynamical system (differential equations) to the discrete one, we can state that if a system is at an equilibrium point x^* at some time, say, n, then it remains at x^* forever. Consequently, for an one-dimensional map the equilibrium points are obtained by writing

$$x_n = x_{n+1} = x^* . \tag{4.20}$$

(Compare this with the definition of equilibrium points, equation (3.5), in Sect. 3.4 for the system of the form given by (3.4)). For the logistic map (4.19), this substitution leads to the following equation

$$x^* = ax^* (1 - x^*)$$

or

$$x^* (1 - a + ax^*) = 0 . \tag{4.21}$$

Solving this equation, we obtain two equilibrium points, namely

$$x_1^* = 0 , \quad x_2^* = \frac{(a-1)}{a} . \tag{4.22}$$

The origin is an equilibrium point for all values of a whereas the equilibrium point $x_2^* = (a-1)/a$ is in the range of allowable x (which is between 0 and 1) only if $a \geq 1$.

What is the stability nature of either of the equilibrium points $x = x^*$? That is if we slightly perturb the state of the system from its equilibrium value, will the asymptotic state approach x^* (for large n) or not? (Again compare with discussions in Chap. 3). To answer this question, for a map $x_{n+1} = f(x_n)$, we consider the difference between the absolute value of x_n and x^* for n arbitrary.

Let

$$| \delta_n | = | x_n - x^* | . \tag{4.23}$$

Then

$$\begin{aligned} | \delta_{n+1} | &= | x_{n+1} - x^* | \\ &= | f(x_n) - x^* | \\ &= | f(x^* + \delta_n) - x^* | \\ &= | x^* + \mathrm{d}f/\mathrm{d}x_n |_{x_n = x^*} \delta_n + \ldots - x^* | \quad \text{(Taylor expansion)} \\ &= | \mathrm{d}f/\mathrm{d}x_n |_{x_n=x^*} | \cdot | \delta_n | , \end{aligned} \tag{4.24}$$

where we have omitted higher orders in $|\delta_n|$ as negligible compared to the leading order term. In the neighbourhood of a stable equilibrium point all the trajectories approach it asymptotically so that $|\delta_{n+1}| \to 0$ as $n \to \infty$. In other words, $|\delta_{n+1}| < |\delta_n|$. Consequently, for stability of the equilibrium point, we require that

$$| \delta_{n+1} | / | \delta_n | < 1 ,$$

so that as n increases x_n can approach x^*. Thus the condition for stability of x^* is

$$|\mathrm{d}f(x)/\mathrm{d}x |_{x=x^*} = | f'(x^*) | < 1 . \tag{4.25}$$

We note that for continuous time dynamical systems, as shown in Sects. 3.4 and 3.6, the condition for stability of an equilibrium point is that the real parts of all the eigenvalues must be less than zero whereas for one-dimensional maps as shown above the stability condition is $|f'(x^*)| < 1$. Applying the criterion (4.25) to the logistic map (4.19), the condition for stability becomes

$$|f'(x^*)| = |a(1 - 2x^*)| < 1 . \tag{4.26}$$

Thus we find that

(1) the equilibrium point $x^* = 0$ is stable for $0 \le a < 1$ and unstable for $a > 1$, since $|f'| = |a|$ is greater than 1 for $a > 1$.
(2) The equilibrium point $x^* = (a-1)/a$ is stable for $1 < a < 3$ (see Fig. 4.10a), and unstable for all other values of a, as $| f' | = | 2 - a |$ is less than 1 for $1 < a < 3$ and greater than 1 for a outside this range.

Further, one may observe that as a increases through 1 the equilibrium point which is stable for $a < 1$ is unstable for $a > 1$, while the equilibrium point which is unstable for $a < 1$ becomes stable for $1 < a < 3$. There is an exchange of stability as a passes through 1. That is, the map (4.19) undergoes a transcritical bifurcation at $a = 1$. At the bifurcation value $a = 1$, the stability determining slope f' for $x^* = 0$ and that of $x^* = (a-1)/a$ are both identically equal to 1. In other words, a transcritical bifurcation occurs when $f' = 1$.

Figure 4.10a shows the numerically iterated values of x_n versus n for $a = 2.8$ with the initial value $x_0 = 0.95$. As n increases, x_n approaches asymptotically the equilibrium point $x^* = (a-1)/a = 0.643$. For the trajectory shown in Fig. 4.10a, roughly, x_0 to x_{15} may be considered as transient. It is a part of the trajectory which is settling to the final (asymptotic) behaviour. Such a transient evolution must be omitted in any calculation.

4.2.3 Stability When the First Derivative Equals to +1 or −1

In the above, in studying the stability of an equilibrium point we have not stated its stability nature when $f' = \pm 1$. When $f'(x^*) = \pm 1$, the equilibrium

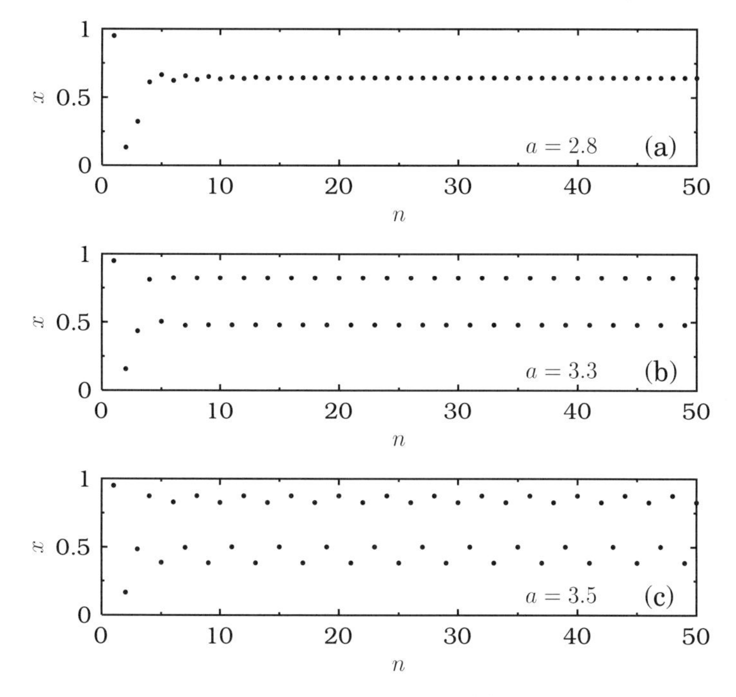

Fig. 4.10. x_n versus n of the logistic map. (**a**) $a = 2.8$, the iterations asymptotically approach the stable equilibrium point $x^* \approx 0.643$. (**b**) $a = 3.3$, the long term behaviour is a period-2 cycle. (**c**) $a = 3.5$, the solution is a period-4 cycle

point can be stable or unstable, that is it is neutral. In this case the stability property depends upon the sign of second and third derivatives $f''(x^*)$ and $f'''(x^*)$. Without going into the details, in the following we give the stability conditions when $f'(x^*) = \pm 1$ (for fuller details, see for example Ref. [7]).

For $f' = 1$,

(1) if $f'' < 0$ then x^* is (semi)stable for $x_0 > x^*$ and unstable for $x_0 < x^*$ and
(2) if $f'' > 0$ then x^* is (semi)stable for $x_0 < x^*$ and unstable for $x_0 > x^*$.

Further if $f'' = 0$, then

(1) x^* is stable for $f''' < 0$, and
(2) x^* is unstable for $f''' > 0$.

In case $f''' = 0$ also, one can state the stability condition in term of $f^{(4)}(x^*) \neq 0$ and so on.

For $f' = -1$, the stability is determined by the sign of the quantity

$$g''' = -2f''' - 3[f'']^2 \,. \tag{4.27}$$

Particularly, if $g''' < 0$ then x^* is stable, and if $g''' > 0$ then it is unstable.

Now we apply the above criteria to the logistic map. For the equilibrium point $x^* = 0$ at $a = 1$ the slope $f' = 1$ and $f'' = -2$. Thus, $x^* = 0$ is stable for $x_0 > 0$ and unstable for $x_0 < 0$. Next, at $a = 3$ the stability determining slope f' of the equilibrium point $x^* = (a-1)/a = 2/3$ is -1 and $g''' = -108$. Thus, x^* is stable at $a = 3$.

4.2.4 Periodic Solutions or Cycles

One can find more general solutions other than the equilibrium points discussed above. The equilibrium points $x^* = 0$ and $(a-1)/a$ are solutions of the map (4.19) which remain unchanged after every iteration. One can find solutions which *repeat* after every two iterations, three iterations, ..., N iterations (N: arbitrary) for different ranges of values of the parameter a. Then we may call the corresponding solutions as 2-periodic, 3-periodic, ..., N-periodic (period-2, period-3, ..., period-N) or 2-cycle, 3-cycle, ..., N-cycle solutions:

$$
\begin{aligned}
2-\text{cycle} &: \quad x_{n+2} = x_n \\
3-\text{cycle} &: \quad x_{n+3} = x_n \\
\ldots \\
N-\text{cycle} &: \quad x_{n+N} = x_n
\end{aligned}
$$

As a consequence, we may call the equilibrium or fixed point solution as the 1-periodic (period-1) or 1-cycle solution.

The above surprising result becomes evident when a is just a bit greater than 3. Figure 4.10b shows the typical trajectory plot of (4.19) for a =3.3. Here the x_n values are oscillating between two values as $n \to \infty$. This type of oscillation, in which x_n repeats after every two iterations, as mentioned above, is the *period-2 cycle*.

A 2-cycle exists if and only if there are two points x_1^* and x_2^* for the given map $x_{n+1} = f(x_n)$ such that $f(x_1^*) = x_2^*$ and $f(x_2^*) = x_1^*$. Equivalently, such a solution must satisfy the relation $f(f(x_1^*)) = x_1^*$, so that we may call the period 2-cycle also as a period-2 equilibrium point. For the logistic map (4.19), the period-2 solutions are obtained from the relations

$$x_2^* = a x_1^* \left(1 - x_1^*\right) , \tag{4.28a}$$

$$x_1^* = a x_2^* \left(1 - x_2^*\right) . \tag{4.28b}$$

The above two equations can be solved as follows to obtain x_1^* and x_2^*. First we subtract (4.28a) from (4.28b) and obtain

$$\left(x_1^* - x_2^*\right) = a\left(x_2^* - x_1^*\right) - a\left(x_2^{*2} - x_1^{*2}\right)$$

or

$$x_1^* + x_2^* = (1+a)/a . \tag{4.29}$$

We now multiply (4.28a) by x_2^* and multiply (4.28b) by x_1^* and subtract to obtain

$$x_2^{*2} - x_1^{*2} = a x_1^* x_2^* \left(x_2^* - x_1^*\right)$$

or

$$x_1^* x_2^* = \left(x_1^* + x_2^*\right)/a = (1+a)/a^2 \ . \tag{4.30}$$

Eliminating x_2^* between (4.29) and (4.30), we have

$$a^2 x_1^{*2} - a(1+a)x_1^* + (1+a) = 0 \ . \tag{4.31}$$

Solving we find

$$x_1^* = \frac{(a+1) \pm \sqrt{(a+1)(a-3)}}{2a} \ , \tag{4.32a}$$

$$x_2^* = \frac{(a+1) \mp \sqrt{(a+1)(a-3)}}{2a} \ . \tag{4.32b}$$

In other words, we obtain the period-2 equilibrium points as

$$x_{1,2}^* = \frac{(a+1) \pm \sqrt{(a+1)(a-3)}}{2a} \ . \tag{4.33}$$

A typical period-2 cycle is shown in Fig. 4.10b.

To obtain the stability condition for a period-2 cycle, we define $|\ \delta_{n+1}\ | = |\ x_{n+1} - x_1^*\ |$ and $|\ \delta_{n+2}\ | = |\ x_{n+2} - x_2^*\ |$. Then

$$\begin{aligned} |\ \delta_{n+2}\ | &= |\ x_{n+2} - x_2^*\ | \\ &= |\ f(x_{n+1}) - x_2^*\ | \\ &= |\ f(x_1^* + \delta_{n+1}) - x_2^*\ | \\ &= |\ f(x_1^*) + f'(x_1^*)\delta_{n+1} - x_2^*\ | \\ &= |\ f'(x_1^*)\delta_{n+1}\ | \ , \end{aligned} \tag{4.34}$$

where we have neglected higher orders in δ_{n+1} in the Taylor series expansion of $f(x_1^* + \delta_{n+1})$ about x_1^*. Similarly, expanding near x_2^* we obtain

$$|\ \delta_{n+1}\ | = |\ f'(x_2^*)\delta_n\ | \ . \tag{4.35}$$

Substituting (4.35) into (4.34), we get

$$|\ \delta_{n+2}\ | = |\ f'(x_1^*) f'(x_2^*)\delta_n\ | \ . \tag{4.36}$$

Now the stability condition $|\ \delta_{n+2}\ | / |\ \delta_n\ | < 1$ becomes

$$|\ f'(x_1^*) f'(x_2^*)\ | < 1 \ . \tag{4.37}$$

For the logistic map, the above stability condition for the period-2 equilibrium point (4.33) becomes $|\ 4 + 2a - a^2\ | < 1$. The interval of a in which this condition is satisfied can be determined as follows. Suppose $s = 4 + 2a - a^2$ is positive. Then we must have $s < 1$ or $a^2 - 2a - 3 > 0$. The solutions of the equation $a^2 - 2a - 3 = 0$ are $a = -1, 3$. That is, $a^2 - 2a - 3$ is > 0 if

$a < -1$ or $a > 3$. On the other hand, if s is negative then we must have $s > -1$ or $a^2 - 2a - 5 < 0$. The solutions of the equation $a^2 - 2a - 5 = 0$ are $a = 1 \pm \sqrt{6}$. From the above, it is clear that the condition $| s |< 1$ is satisfied for $a \in (1 - \sqrt{6}, -1)$ and $a \in (3, 1 + \sqrt{6})$. Since negative values of a are excluded in the present system, we can conclude that the period-2 cycle solution is stable for $3 < a < 1 + \sqrt{6} = 3.449$.

4.2.5 Period Doubling Phenomenon

We note that the period-1 solution $x^* = (a-1)/a$ is unstable for $a > 3$, because the stability determining slope crosses $f' = -1$ at $a = 3$. Similarly, the period-2 solution born at this critical value becomes unstable at $a = 1 + \sqrt{6}$ when its stability determining factor $f'(x_1^*)f'(x_2^*)$ becomes -1. One can proceed further as above and check that a stable period-4 solution is born just after this critical value and again loses its stability when the stability determining factor $f'(x_1^*)f'(x_2^*)f'(x_3^*)f'(x_4^*)$ becomes -1. Figure 4.10c shows a period-4 cycle for $a = 3.5$. Proceeding in this way further, one can check that when the stability determining quantity of a period-k solution becomes -1 a bifurcation occurs giving birth to a stable period-$2k$ solution. Such bifurcations associated with the stability determining quantity crossing the value equal to -1 are called *flip bifurcations*. They are also referred as *period doubling* or *subharmonic bifurcations* for obvious reasons.

Exercises:

13. Verify that the criterion for stability of period-4 solution is

$$\left| f'(x_1^*)f'(x_2^*)f'(x_3^*)f'(x_4^*) \right| < 1 .$$

14. Generalize the above result to show that for a period-N solution the stability criterion is

$$\left| f'(x_1^*)f'(x_2^*) \ldots f'(x_N^*) \right| < 1 .$$

We also note that to obtain the period-4 solution we have to solve a set of four coupled nonlinear algebraic equations. Similarly, to obtain the period-k solution, k-coupled equations have to be solved. In general, three or more coupled nonlinear equations are very difficult to solve analytically or even numerically. For example, simple root-finding methods, such as Newton's method may fail due to the closeness of nearby solutions. However, the map (4.19) itself can be iterated numerically and the long term evolution can be easily studied. The combined analytical and numerical analysis gives the following picture.

(1) Period-1 solution exists in the range $0 < a < 3$, and it loses stability at the critical value of $a = a_1 = 3$, giving birth to (or bifurcating into) a period-2 cycle.

(2) Period-2 solution exists in the range $3 < a < 1 + \sqrt{6}$ (≈ 3.449), and it also loses stability at the critical value of $a = a_2 = 1 + \sqrt{6}$, giving birth to (or bifurcating into) a period-4($= 2^2$) cycle.
(3) Period-2^2 cycle exists for $1 + \sqrt{6} < a < 3.544112$ and this solution bifurcates into a period-2^3($= 8$) cycle solution at $a = a_3 = 3.544112$.
(4) This process proceeds further ad infinitum.

The critical values of a at which successive period doubling bifurcations occur are given in Table 4.1.

Thus, as a is increased a period-2^n equilibrium point or cycle loses its stability and gives birth to a 2^{n+1} cycle at the critical value of the parameter $a = a_n$. This process of bifurcation is called *period doubling phenomenon.* The successive bifurcations occur faster and faster as a is increased. Ultimately, a_n converges to a limiting value a_∞ or $a_c = 3.57$. This convergence is essentially geometric: in the limit of large n, the distance between successive transitions shrinks by a constant factor or universal constant δ, the so-called *Feigenbaum's universal number* or *Feigenbaum's constant* [8],

$$\delta_n = \lim_{n\to\infty} \frac{a_{n+1} - a_n}{a_{n+2} - a_{n+1}} . \tag{4.38}$$

Using the values of a given above we obtain a rough estimate

$$\delta_1 = \frac{3.449489 - 3.0}{3.544112 - 3.449489} = 4.7503144 ,$$
$$\delta_2 = \frac{3.544112 - 3.449489}{3.564445 - 3.544112} = 4.6536665 ,$$
$$\delta_3 = \frac{3.564445 - 3.544112}{3.568809 - 3.564445} = 4.6592576 .$$

The calculated δ values can be shown to approach a constant value $\delta \approx 4.66992\ldots$ [8], which is the value of the Feigenbaum's constant for the present case.

Table 4.1. First few bifurcation values of a

Period of the N-cycle	Bifurcation value of a
2^1	$a_1 = 3.000000$
2^2	$a_2 = 3.449489..$
2^3	$a_3 = 3.544112..$
2^4	$a_4 = 3.564445..$
2^5	$a_5 = 3.568809..$
2^6	$a_6 = 3.569745..$
2^∞	$a_\infty = 3.570000..$

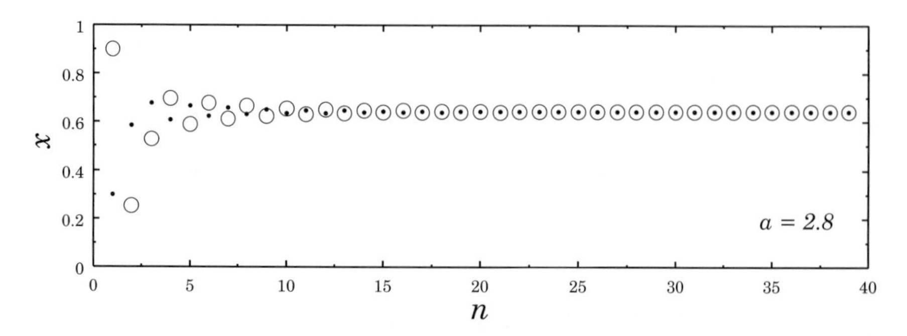

Fig. 4.11. Two trajectories (corresponding to period-1 solution) started with two different initial conditions $x_0 = 0.3$ (dot) and $x_0' = 0.9$ (circle) for $a = 2.8$. After the initial transients, both the trajectories evolve identically, showing insensitiveness to initial conditions

An important characteristic property of the solutions of the logistic map for $a < a_c$ (where only periodic solutions occur) is that they are, in general, *insensitive to initial values* of x provided $0 < x < 1$. Irrespective of the initial value of x the successive iterations asymptotically approach the same periodic solution. As an example, Fig. 4.11 depicts two trajectories started with the initial conditions $x_0 = 0.3$ and $x_0' = 0.9$ for $a = 2.8$. For large n, both x_n (marked by dots) and x_n' (marked by circles) are identical. That is the solution is insensitive to the initial value of x.

4.2.6 Onset of Chaos: Sensitive Dependence on Initial Conditions – Lyapunov Exponent

What is the nature of the sequence $\{x_n\}$ for $a > a_c$? Numerical analysis clearly shows that for values of $a > a_c$, the sequence x_n never settles down to an equilibrium point or to a periodic cycle – instead the long term behaviour is aperiodic, as shown in Fig. 4.12. Such a behaviour is often called *chaotic* since the solution is *highly sensitive to initial conditions*.

To understand the sensitive dependence on initial conditions, we iterate the map with two nearby initial conditions for the same value of $a > a_c$ and study the difference between the two resultant sequences. For example, Fig. 4.13a shows the orbits for two initial conditions with $a = 3.99$. One trajectory (marked by '×') starts from $x_0 = 0.95$, and the second (marked by circles) from $x_0' = 0.95005$. We see that after 8 iterations the trajectories are already separated quite a bit apart displaying the divergence of nearby trajectories. Figure 4.13b shows the distance $S(n)$ (that is, the magnitude of $(x_n' - x_n)$) between the two trajectories. $S(n)$ varies irregularly with n, clearly indicating the sensitive dependence of the dynamics on initial conditions. (We note that for $a = 2.8$, from Fig. 4.11, the distance between the two (period-1) trajectories x_n and x_n' becomes zero in the limit $n \to \infty$). We also

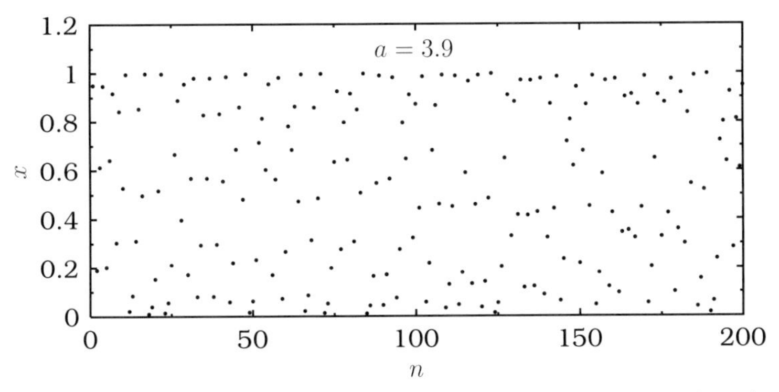

Fig. 4.12. Aperiodic (chaotic) solution (for a single initial condition) of the logistic map for $a = 3.99$

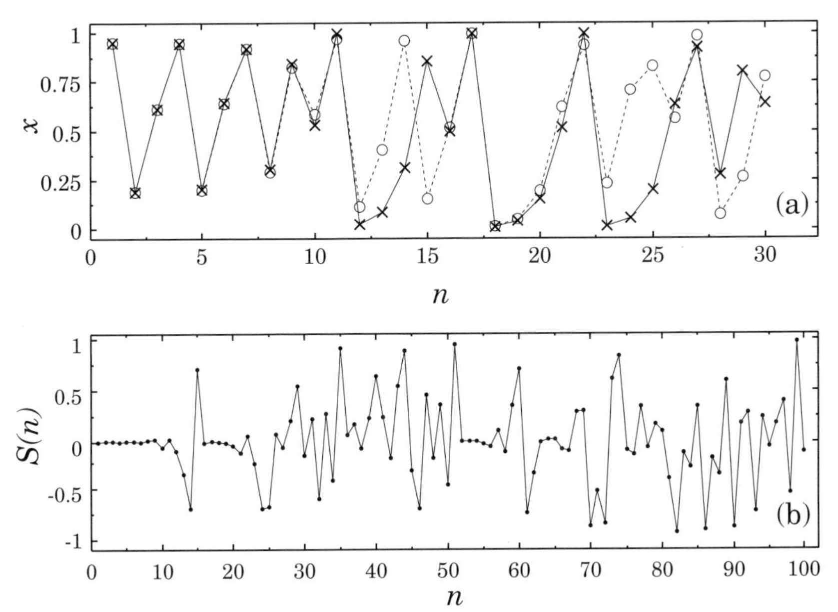

Fig. 4.13. (**a**) Two trajectories started with the initial states $x_0 = 0.95$ (marked as $\times$) and $x_0' = 0.95005$ (circles) for $a = 3.99$. (**b**) Separation distance plot of the two trajectories shown in the subplot (**a**)

observe that the distance $S(n)$ between the two trajectories cannot diverge continuously. If $S(n) \to \infty$ as $n \to \infty$, then at least one of the two trajectories considered must go to ∞ leading to unbounded motion. Since the *chaotic motion* considered is a *bounded solution* (as the range of x is confined to the interval $(0, 1)$), $S(n)$ increases rapidly for a while then decreases and increases and so on. However, on an average the trajectories diverge.

In order to quantify the nature of divergence of the nearby trajectories, we evaluate the distance at time n in terms of the initial ($n = 0$) distance. Denoting $x'_0 = x_0 + \delta_0$, we define the deviation δ_1 at time $n = 1$ as

$$\begin{aligned}\delta_1 &= x'_1 - x_1 \\ &= f(x_0 + \delta_0) - f(x_0) \\ &= f'(x_0)\delta_0 ,\end{aligned} \tag{4.39a}$$

where we have used Taylor series expansion of $f(x_0+\delta_0)$ and neglected higher power terms in δ_0. Next, we write

$$\begin{aligned}\delta_2 &= f(x'_1) - f(x_1) \\ &= \delta_1 f'(x_1) \\ &= \delta_0 f'(x_0) f'(x_1) .\end{aligned} \tag{4.39b}$$

Proceeding further, we can easily see that at time n we have

$$\delta_n = \delta_0 \prod_{i=1}^{n} f'(x_{i-1}) . \tag{4.39c}$$

Let us now define a quantity λ called the *Lyapunov exponent* [9] as

$$\lambda = \frac{1}{n} \ln \left| \frac{\delta_n}{\delta_0} \right| = \frac{1}{n} \sum_{i=1}^{n} \ln |f'(x_{i-1})| , \quad n \to \infty . \tag{4.40}$$

Then we obtain

$$\delta_n = \delta_0 \mathrm{e}^{\lambda n} . \tag{4.41}$$

That is, as time n increases, δ_n diverges exponentially if λ is positive. From a detailed numerical analysis one can check that this is indeed the case of the map (4.19) for most of the values of $a > a_c = a_\infty = 3.57$, where the trajectories are chaotic (see Sect. 4.2.8 below) and show high sensitive dependence on initial conditions. Using (4.41) one can state that in the chaotic regime the trajectories show an exponential divergence of nearby trajectories. For more details on the nature of Lyapunov exponent and the spectrum of it for the logistic map, see Chap. 8, Sect. 8.2. More precisely, any small error δ_0 in the specification of the initial state or round-off error in the numerical computation can get amplified exponentially fast in a finite time interval and can become as large as the system size itself, which is 1 in the present case, and so future prediction becomes inaccurate, although the original map (4.19) itself is purely deterministic. Note that such sensitive dependence cannot occur in a linear system (check! Also refer our discussion in Chap. 2) and it is truly a manifestation of the nonlinearity. And one way to test for chaos is to look for two trajectories that start very close to each other and check for the divergence of nearby trajectories.

Exercises:

15. Show that for the linear map $x_{n+1} = ax_n$, the Lyapunov exponent is negative for bounded solution and so there is no sensitive dependence on initial conditions.
16. Show that for logistic map in the interval $0 \leq a < 3.57$ the Lyapunov exponent is negative or zero. (Hint: Since in this interval the map has period-2^n solutions, $n = 1, 2, \ldots$, use the property of stability determining quantity in (4.40)).

Finally, our discussion above allows us to define what is chaos.

Chaos is the phenomenon of occurrence of bounded nonperiodic evolution in completely deterministic nonlinear dynamical systems with high sensitive dependence on initial conditions.

This is called *deterministic chaos* since the governing equations are deterministic.

Chaos – Butterfly Effect – Weather Predictions

The sensitive dependence of chaotic solution on initial conditions was first observed by E.N. Lorenz (1963) [10] in a system of three coupled first-order ordinary differential equations describing hydrodynamic flow now popularly known as Lorenz system (for more details, see Sect. 5.2). When the notion of sensitive dependence of chaotic solution is applied to atmosphere, which is expectedly nonperiodic, it indicates that prediction of a sufficiently distant future is impossible by any method, unless the present conditions as well as subsequent evolutions are known exactly. In view of the inevitable inaccuracy and incompleteness of weather observations, precise and very long range forecasting would seem to be nonexistent! That is, even a minute perturbation can cause realizable effects in a finite time under chaotic evolution.

To describe the dramatic extreme sensitive dependence of chaotic solution Lorenz coined the term *butterfly effect*. In his own words, it reads as follows [11]: "As small a perturbation as a butterfly fluttering its wings somewhere in the Amazons can in a few days time grow into a tornado in Texas". That is even a minute perturbation can cause realizable effects in a finite time under chaotic evolution.

4.2.7 Bifurcation Diagram

One can summarize the behaviour of the solutions of the logistic map by plotting a *bifurcation diagram*, indicating the nature of the solutions as a function of the control parameter (here 'a'). For a given value of a, one can compute the trajectory from some starting point (after neglecting the transients) and then plot the sequence of allowed values, namely the attracting points for the trajectory, as a function of the parameter a. Figures 4.14a and 4.14b depict the attractor of the map as a function of a. This diagram is plotted numerically by iterating the map (4.19). First, a value of a is chosen and then an initial condition x_0, which lies between $0 < x_0 < 1$, is chosen. Next, the map is iterated for 1000 cycles or so to allow the system to settle down to its eventual asymptotic behaviour. Once the transients have decayed, they are discarded and then the successive orbits, say $x_{1001}, x_{1002}, \dots, x_{1300}$ are plotted above that chosen value of a. Then an adjacent value of a is chosen and by repeating the above process, one eventually sweeps the whole range of a to obtain a complete *bifurcation diagram*. For the map (4.19), it is given in Figs. 4.14.

We may note that for the map (4.19), further increase in the value of $a > 4$ leads to only unbounded motions, that is as $n \to \infty$, $x_n \to \infty$, which is not of interest.

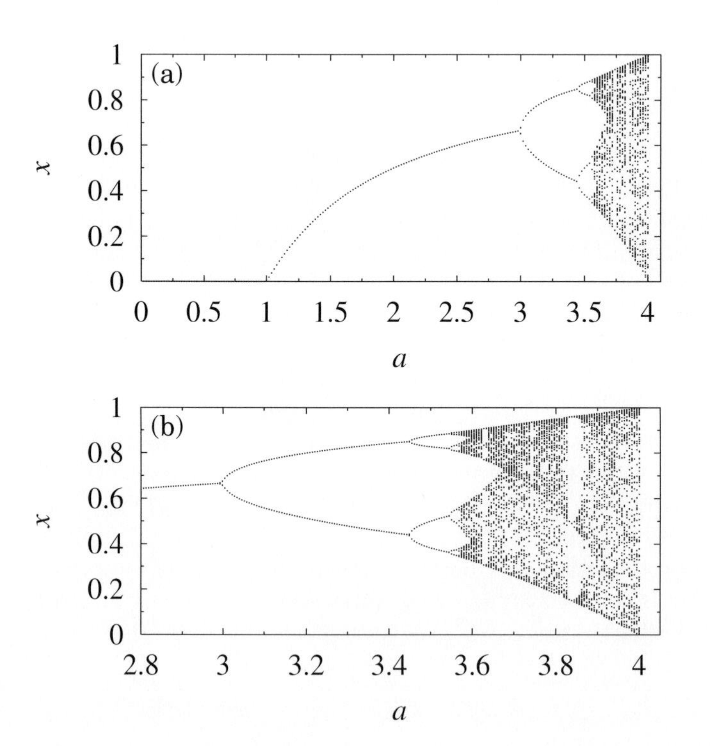

Fig. 4.14. Bifurcation diagrams of the logistic map for (**a**) $a \in (0, 4)$, (**b**) $a \in (2.8, 4.0)$, (**c**) $a \in (3.57, 3.7)$ and (**d**) $a \in (3.82, 3.87)$

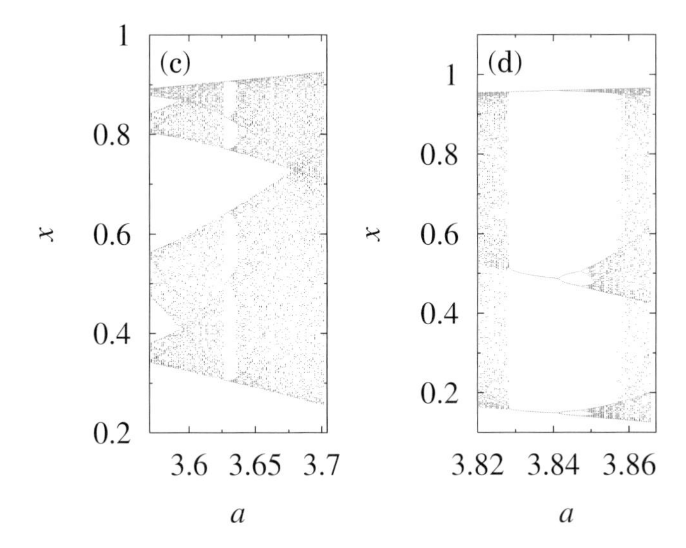

Fig. 4.14. continued

4.2.8 Bifurcation Structure in the Interval $3.57 \leq a \leq 4$

As shown above, for $a < 3.57$, the logistic map exhibits period doubling phenomenon, and chaotic dynamics is first observed at $a \approx 3.57$. For $3.57 \leq a \leq 4$, the dynamics is even more complicated and intricate. This interval of a is not fully occupied by chaotic orbits alone but many fascinating changes in the dynamics take place at different critical values of a. Particularly, the asymptotic motion consists of chaotic orbits interspersed by periodic orbits (*windows*), period doubling of windows and intermittent chaos (which will be described in Sect. 4.4). For example, Fig. 4.14c shows the bifurcation diagram of the map for $a \in (3.57, 3.7)$. For values of a slightly below $a = a_m \approx 3.678$, the chaotic attractor consists of two bands. The points x_n, $x_{n+2}, \ldots$ fall on one band and x_{n+1}, $x_{n+3}, \ldots$ fall on the other band of the attractor. However, the iterated points fall randomly within each of these bands. Moreover, the sizes of both the bands increase smoothly with the increase in the value of the control parameter a and the bands merge together and form a single band chaotic attractor at $a = a_m$. This is known as *band-merging* bifurcation. Generally, for a m-band chaotic attractor, if we take a point in any one of these m bands, the trajectory will come to that band after m iterations.

Next, let us look at the bifurcation diagram 4.14d. We can clearly see many sudden changes in the dynamics as a function of the control parameter a. At $a = a_d \approx 3.828$ a *sudden destruction* of the chaotic attractor occurs. The chaotic attractor is replaced by a period-3 attractor. When the parameter a is increased further this period-3 orbit undergoes a further set of period doubling bifurcations leading to chaotic motion. Another type of bifurcation which is seen in the Fig. 4.14d is the occurrence of *sudden widening* or *sudden increase* in the size of the chaotic attractor at $a = a_w \approx 3.857$. The bifurcations such as sudden widening, sudden destruction and band-merging of chaotic attractors are also termed as *crises* since they represent sudden discontinuous changes in the chaotic attractor as the control parameter is varied. These bifurcations are observed at many values of a in the interval $3.57 \leq a \leq 4$ in the logistic map. The occurrence of intermittency route to chaos is described in Sect. 4.4.

4.2.9 Exact Solution at $a = 4$

Finally, though exact analytic solution of the logistic map for arbitrary values of a is not available, one can easily construct its solution for the special choice $a = 4$. Substituting $x_n = \sin^2 \pi\theta_n$ in (4.19), we obtain

$$\begin{aligned} \sin^2 \pi\theta_{n+1} &= 4\sin^2 \pi\theta_n \cos^2 \pi\theta_n \\ &= \sin^2 2\pi\theta_n \; . \end{aligned} \tag{4.42}$$

That is

$$\begin{aligned} \theta_{n+1} &= 2\theta_n \pmod 1 \\ &= 2^2\theta_{n-1} \\ &= 2^n\theta_0 \pmod 1 \; . \end{aligned} \tag{4.43}$$

Thus, for $a = 4$ the value of x for any arbitrary n and for a given x_0 can be calculated without iterating (4.19). In fact one can argue by expressing the above solution in a binary representation that it does indeed show sensitive dependence on initial conditions and so corresponds to chaotic behaviour (see exercise 17 below).

Exercises:

17. Bring out the sensitive dependence on initial conditions in the map $\theta_{n+1} = 2\theta_n \pmod 1$ for

$$\theta_0 = 0.b_1b_2...b_n = \sum_{k=1}^{\infty} b_k 2^{-k} \; ,$$

where $b_1, b_2, ..., b_n$ are binary numbers (see also Refs. [5,12]).

18. In a one-dimensional map, a period-k equilibrium point with

$$\prod_{i=1}^{k} f'(x_i) = 0$$

is called *maximally stable* or *superstable* or a *supercycle*. For the logistic map, determine the values of the parameter a at which the period-1 and period-2 equilibrium points are superstable. Also show that the value $x^* = 1/2$ always belongs to a supercycle.

4.2.10 Logistic Map: A Geometric Construction of the Dynamics – Cobweb Diagrams

A simple geometric construction of the dynamics of the logistic map can also be developed by a graphical method as detailed below.

In Fig. 4.15, we have plotted $y = f(x)$ for several values of a. Whenever the diagonal line $y = x$ crosses the curve $f(x)$, the map function (4.19) has an equilibrium point since here $x = f(x)$. From Fig. 4.15 it is easy to see that for $a < 1$, the only equilibrium point between 0 and 1 is $x = 0$, but for $a > 1$, there are two equilibrium points in the range of interest.

Using Fig. 4.16, we can see how a trajectory that starts from some value of x different from 0 and 1 approaches an attractor for different values of a. The procedure is the following.

(1) From the starting value of x_0 on the x-axis, draw a line vertically to the f curve. That height is the output x_1.
(2) Then draw a line from the intersecting point parallel to the x-axis to the diagonal ($y = x$) line.
(3) From the intersecting point on the diagonal line, draw a second vertical line to the f curve. That height from the x-axis is the output x_2.
(4) Continuing the procedures 1–3 one obtains a graphical or geometrical representation of the iteration procedure of (4.19) for a given a.

The above geometrical representations for a set of a values are also called *cobweb diagrams* in the literature.

Figure 4.16a shows an iteration sequence, starting from $x = 0.6$, for $a = 0.6$. Successive iterations converge to the origin which is the stable equilibrium

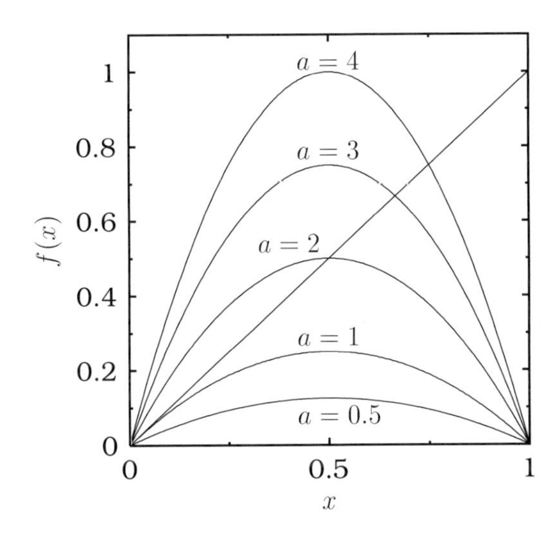

Fig. 4.15. Graph of $f(x)$ of the logistic map for different values of a

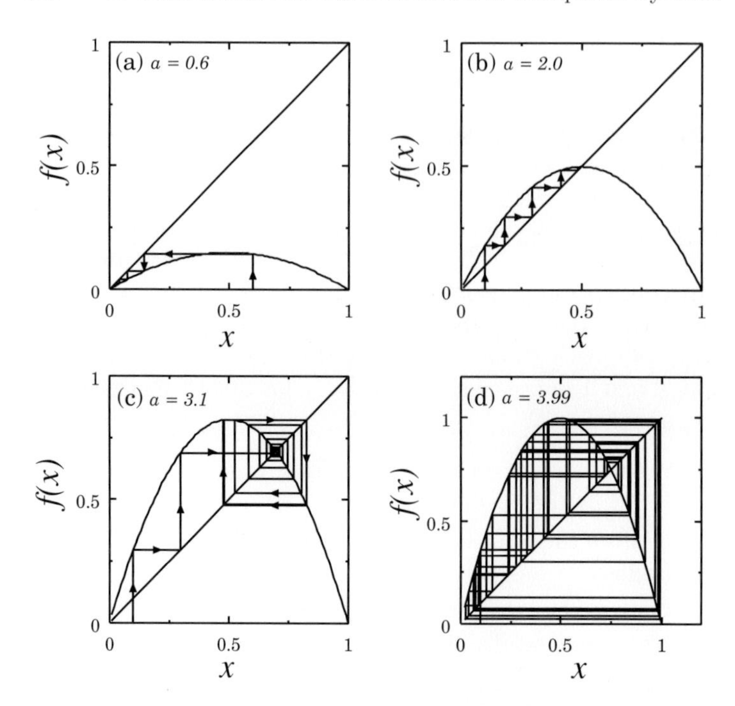

Fig. 4.16. Graphical representations of iterations of the logistic map (cobweb diagrams) for various values of the control parameter a

point of the map for this value of a, as seen in the previous section. What happens when the parameter $a > 1$? In Fig. 4.16b, we have plotted $f(x)$ with $a = 2$ along with the diagonal line $y = x$. If we follow our geometrical trajectory construction method or cobweb diagram, we see that a trajectory starting at $x = 0.1$, for example, now heads for the equilibrium point $x^* = (a-1)/a = 1/2$ for $a = 2$. In fact, any trajectory starting in the range $0 < x < 1$ approaches this attractor.

At this point, one can summarize what the model (4.19) tells us: Given any initial value x_0 lying between 0 and 1, the successive iteration values eventually approach the attracting equilibrium point $x^* = (a-1)/a$ if $1 < a < 3$. For $a > 1$, $x^* = 0$ has become a *repelling equilibrium point*, since trajectories that start near $x = 0$ move away from that value.

For $1 < a < 3$, the attractor of (4.19) consists of a single nonzero equilibrium point $x = (a-1)/a$. For a just greater than 3, the attractor consists of two points, whose values vary as a varies. Figure 4.16c shows the graphical construction of a trajectory that leads to this two-point attractor for $a = 3.1$.

Similarly for any other value of a, the cobweb diagram can be given. In the chaotic region, one notes that the trajectory keeps on winding indefinitely, showing the aperiodic/chaotic nature as shown in Fig. 4.16d for $a = 3.99$ (the first 50 iterations alone are shown).

4.3 Strange Attractor in the Hénon Map

So far we have considered the dynamics of one-dimensional nonlinear maps. In particular the logistic map is a *noninvertible* one, as every x_{n+1} is associated with two x_n values in general (see also Sect. 5.4.2). In order to consider the properties of *invertible* maps, one will have to consider higher dimensional maps. It will be then interesting to consider bifurcations and chaos phenomena in such maps as well.

Like continuous time dynamical systems, maps can also be conservative or dissipative. If an N-dimensional map $\boldsymbol{X}_{n+1} = \boldsymbol{F}(\boldsymbol{X}_n)$, where $\boldsymbol{X} = (x_1, x_2, \ldots, x_N)$ and $\boldsymbol{F} = (F_1, F_2, \ldots, F_N)$, preserves the N-dimensional phase space volume on each iteration then it is called a *conservative* or *volume preserving map*. A map is volume preserving, if the magnitude of its Jacobian matrix

$$J(\boldsymbol{X}) = \det\left[\frac{\partial \boldsymbol{F}(\boldsymbol{X})}{\partial \boldsymbol{X}}\right] \equiv \begin{vmatrix} \partial F_1/\partial x_1 & \partial F_1/\partial x_2 & \ldots & \partial F_1/\partial x_N \\ \partial F_2/\partial x_1 & \partial F_2/\partial x_2 & \cdots & \partial F_2/\partial x_N \\ \cdots & \cdots & \cdots & \cdots \\ \partial F_N/\partial x_1 & \partial F_N/\partial x_2 & \cdots & \partial F_N/\partial x_N \end{vmatrix} \tag{4.44}$$

is 1. On the other hand, if $| J(\boldsymbol{X}) | < 1$ in some regions, then we say that the map is dissipative. $|J(\boldsymbol{X})| > 1$ represents volume expanding character of the map.

In this section we consider a two-dimensional map, namely the celebrated Hénon map [13], and illustrate the characteristic property of the chaotic attractor associated with it, namely, the self-similar structure. A two-dimensional map is described by a set of two first-order difference equations. The Hénon map is

$$x_{n+1} = 1 - a x_n^2 + y_n \, , \tag{4.45a}$$

$$y_{n+1} = b x_n \, , \quad a, b > 0 \, . \tag{4.45b}$$

It is in this two-dimensional map the so-called self-similar property of chaotic attractor was first observed by Hénon in the year 1976 [13], which will be illustrated later in this section. The Jacobian matrix of (4.45) is given by

$$\det M = \begin{vmatrix} \partial x_{n+1}/\partial x_n & \partial x_{n+1}/\partial y_n \\ \partial y_{n+1}/\partial x_n & \partial y_{n+1}/\partial y_n \end{vmatrix} = \begin{vmatrix} -2a x_n & 1 \\ b & 0 \end{vmatrix} = -b \, . \tag{4.46}$$

Thus, the cases $b < 1$, $b = 1$ and $b > 1$ correspond, respectively, to dissipative, conservative and area expanding nature of the map (4.45). For x_0 very large, the quadratic term in (4.45) makes $| x_{n+1} | \to \infty$. However, for (x_0, y_0) within some finite area near the origin, the solution converges toward an attractor. For $b = 0$, the map (4.45) can be transformed into the logistic map (4.19) by suitably redefining the variable x and the parameter a (verify).

4.3.1 The Period Doubling Phenomenon

The period-1 equilibrium points of (4.45) are (which by definition are given by $(x_{n+1} = x_n = x^*,\ y_{n+1} = y_n = y^*)$)

$$(x^*, y^*) = \left(\frac{1}{2a} \left[(b-1) \pm \sqrt{(b-1)^2 + 4a} \right],\ bx^* \right) . \tag{4.47}$$

Similar to the two-dimensional differential equation (3.4), the type of stability of an equilibrium point of a two-dimensional map as well as an N-dimensional map can be determined by the sign and magnitudes of the eigenvalues of the Jacobian matrix M of the given map. An equilibrium point is stable if the absolute value of the real parts of all the eigenvalues are less than 1, that is, $\mid \mathrm{Re}\lambda_i \mid < 1$. If at least one of the eigenvalues has $\mid \mathrm{Re}\lambda_i \mid > 1$ then the equilibrium point is unstable.

For the Hénon map the stability determining eigenvalues associated with the equilibrium points are obtained from the condition

$$\det|M - \lambda I| = \begin{vmatrix} -2ax^* - \lambda & 1 \\ b & -\lambda \end{vmatrix} = 0 . \tag{4.48}$$

Thus, the eigenvalues become

$$\lambda_{\pm} = -ax^* \pm \sqrt{a^2 x^{*2} + b} . \tag{4.49}$$

Substituting the values of x^* in (4.49), we find that (x^*_-, y^*_-) is always unstable, while (x^*_+, y^*_+) is stable for

$$a < \frac{3(1-b)^2}{4} . \tag{4.50}$$

One can proceed to calculate period-k, $k = 2, 3, \ldots$ fixed points and their stability. Such an analysis is a time consuming one and it involves solving up of a number of coupled nonlinear equations. Alternatively, one can numerically iterate the Hénon map, say for a fixed value of b, and study the change in the dynamics by varying the parameter a. We fix the value of b at 0.3. The map is now dissipative. When a is increased from zero, the map exhibits period doubling bifurcations. At $a = a_1 = 0.3675$, the period-1 equilibrium point $(x^*_{n+1} = x_n,\ y^*_{n+1} = y_n)$ becomes unstable and gives a birth to a period-2 solution. The period-2 equilibrium $(x^*_{n+2} = x_n,\ y^*_{n+2} = y_n)$ point is stable for $a_1 < a < a_2 = 0.905$. Further increase in a leads to successive bifurcations which accumulate at $a = a_c \approx 1.06$, where the onset of chaos observed. Figure 4.17 depicts the bifurcation phenomenon in the Hénon map. Figure 4.18a shows the chaotic attractor in x_n versus y_n phase space for $a = 1.4$.

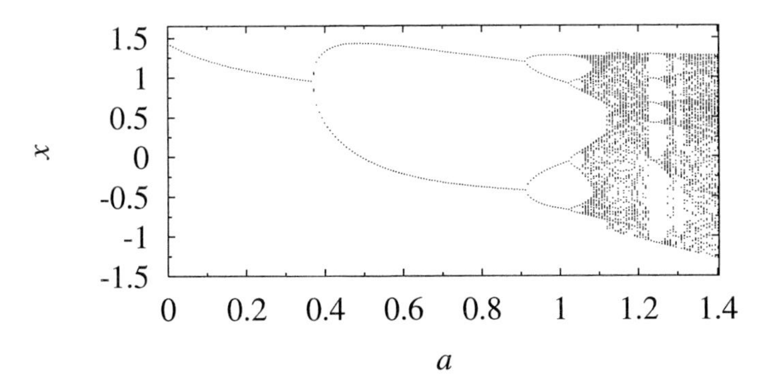

Fig. 4.17. Bifurcation diagram of the Hénon map (4.45) for $b = 0.3$

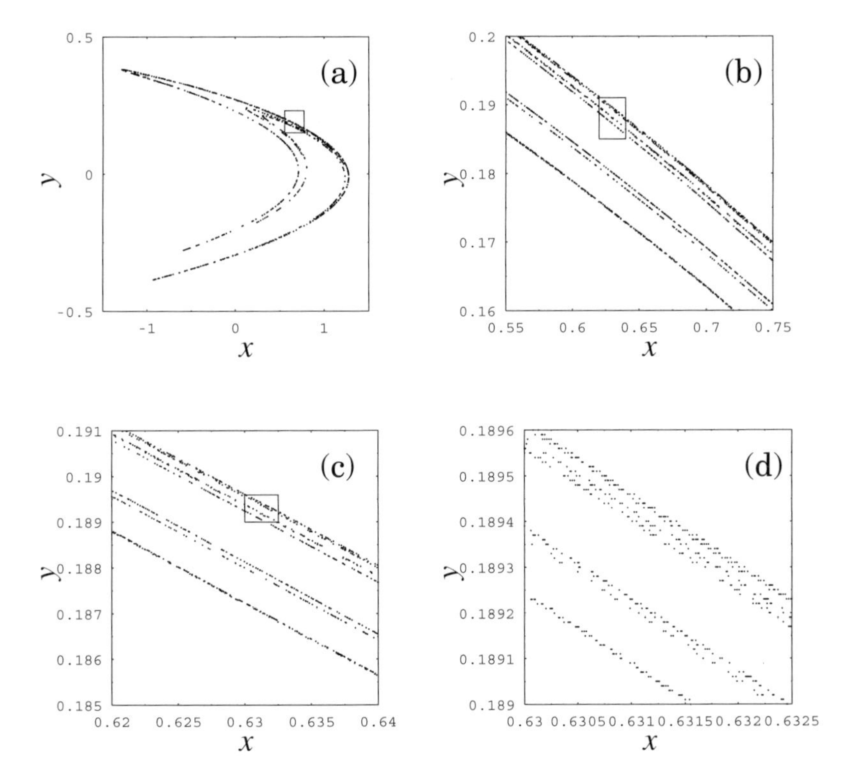

Fig. 4.18. (**a**) Chaotic attractor of the Hénon map for $a = 1.4$, $b = 0.3$. (**b**) Magnification of a small rectangular region of the subplot (**a**). (**c–d**) Second and third stages of magnifications of a small region of the attractor

---

Exercise:

19. Verify the stability properties of the equilibrium points $(x^*_\pm, y^*_\pm)$.

4.3.2 Self-Similar Structure

A characteristic feature of a typical chaotic attractor in dissipative systems is its *self-similar structure*. This can be clearly understood from the chaotic attractor of the Hénon map with $b < 1$. Figure 4.18a is obtained from 3000 iterations of the map (4.45) with $b = 0.3$. When the area within the small rectangle in Fig. 4.18a is magnified, 10^5 iterations yield a more detailed view of the attractor and it is shown in Fig. 4.18b. Six thin layers can be seen in Fig. 4.18b. When the rectangle in Fig. 4.18b, containing only three layers, is magnified, 5×10^5 iterations yield still more details of the structure across the leaves, as shown in Fig. 4.18c. A final magnification, shown in Fig. 4.18d for 10^7 iterations, shows again the six layered structure. Figures 4.18b–d, obtained by repeated magnifications, all look identical with six layers though only parts containing three layers alone are magnified. That is the attractor has structure inside of structure inside of structure and so on. The only difference between these figures is the change in the scale. That is, the many-leaved structure of the attractor repeats itself on finer and finer scales. This scale invariance or similarity across the layer is known as *self-similarity structure* of the attractor. All chaotic attractors of dissipative systems possess the self-similar structure. (Note: The chaotic attractor of the logistic map also possesses self-similar structure). An attractor with a self-similar structure is called a *strange attractor*, a term coined by David Ruelle [14,15]. For more details see Chap. 8 (Sect. 8.5).

The self-similarity of a chaotic attractor arises due to simple stretching and folding operations occurring in the phase space. Because attractors have finite size, two orbits on a chaotic attractor cannot diverge exponentially forever but must be confined to a finite region of the phase space. Consequently, the orbit must fold over onto itself. Although orbits diverge and follow different paths, they must pass close to one another again and again. This combination of stretching and folding is similar to shuffling of a deck of cards. In this case the deck is first expanded so that the cards are slightly separated and is later folded on itself. The orbits on a chaotic attractor are shuffled by this process which is the cause of the randomness of the chaotic orbits. The process of stretching and folding happens repeatedly, creating folds within folds infinite times and leads to an infinitely-layered structure. Thus, a chaotic attractor has a much more complicated structure than a predictable attractor such as a point, a limit cycle or a quasiperiodic one. The self-similar property of a chaotic attractor can be characterized by a noninteger dimension called *fractal dimension* [9], which will be discussed in some detail in Chap. 8 and appendix E.

The bounded orbits of the dissipative Hénon map as well as of any other dissipative system asymptotically approach an attractor. Particularly the geometric structure of a chaotic attractor in the phase space is strange, that is, it possesses self-similar structure. In contrast, in conservative Hénon map as well as in any other conservative system, because of volume preserving nature, the bounded orbits such as period-k equilibrium point, periodic orbit and chaotic orbit never attract nearby trajectories. That is, conservative systems do not have attractors. We will study more details about periodic and chaotic orbits of conservative systems in Chap. 7.

4.4 Other Routes to Chaos

So far we have studied the occurrence of chaos through the period doubling bifurcations in the logistic and Hénon maps. One often calls this process as *period doubling route* to chaos through the mechanism of period doubling bifurcations. It is also called *Feigenbaum scenario* for chaos. There exist many other routes also by which chaos can occur in nonlinear dynamical systems. These include quasiperiodic [16] and intermittency [17–19] routes. The quasiperiodic route to chaos was identified by Ruelle, Newhouse and Takens [16] and it is alternatively called the *Ruelle, Newhouse and Takens scenario*. Similarly, the intermittency route to chaos is also denoted as *Pomeau–Manneville scenario* [17–19]. In this section, we will describe briefly the salient features of these routes to chaos. We do not consider more exotic routes such as period adding, Farey sequences, etc. in this book as they are beyond the scope of our present study.

4.4.1 Quasiperiodic Route to Chaos

In the quasiperiodic or Ruelle–Takens–Newhouse scenario, one can identify the following sequence of bifurcations when a control parameter is changed. The system is initially in a stationary state and becomes unstable and undergoes Hopf bifurcation after a change of the control parameter. At the Hopf bifurcation, a limit cycle is generated around the equilibrium point the stability of which depends on the control parameter. When the parameter is varied further, the system undergoes one more Hopf bifurcation so that a two frequency periodic orbit occurs. If the two frequencies of this oscillation, ω_1 and ω_2, are not commensurate (that is, ω_1/ω_2 is not a rational number), then the observed motion is not periodic but *quasiperiodic* (see also Sect. 3.7). This quasiperiodic orbit then bifurcates into a chaotic motion as the control parameter varies further. Figure 4.19 depicts a schematic representation of the quasiperiodic route to chaos.

Quasiperiodic route to chaos can occur in maps as well as in continuous dynamical systems, under suitable circumstances. As a typical example, we may consider the dynamics of the so-called dissipative circle map [9],

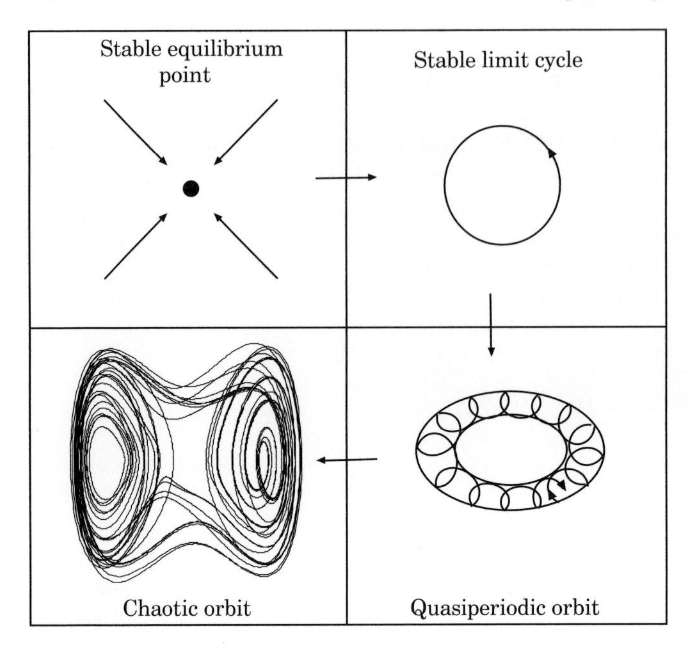

Fig. 4.19. Schematic representation of quasiperiodic route to chaos

$$x_{n+1} = x_n + \Omega - \frac{K}{2\pi} \sin 2\pi x_n + b y_n \,, \quad (\text{mod } 1) \tag{4.51a}$$

$$y_{n+1} = b y_n - \frac{K}{2\pi} \sin 2\pi x_n \,. \tag{4.51b}$$

The above two-dimensional map describes the motion of a periodically kicked, damped rotator. Here x_n is the angle of the kicked rotator at time n and y_n is a variable proportional to the angular velocity. One can study the behaviour of the map by varying the control parameter K for $\Omega = 0.612$ and $b = 0.5$. Breakup of the torus (representing quasiperiodic motion) is observed when the control parameter K is increased from 0.814 to 1.2. Figures 4.20a and 4.20c show the quasiperiodic attractor for $K = 0.8$ in the x_n versus y_n plane and $u(= (1+4y)\cos 2\pi x)$ versus $v(= (1+4y)\sin 2\pi x)$ plane, respectively. In the map (4.51), discontinuity occurs only along the x-direction whenever x crosses 1 and y_n's are always less than 1. Consequently, in the (x_n, y_n) plane, the points do not appear to move on a closed path. However, the torus is clearly seen in the new coordinates (u_n, v_n) defined above. The points now appear to be uniform on the torus. As K is increased breakup of the torus occurs leading to chaotic motion. Figures 4.20b and 4.20d show the chaotic attractor for $K = 1.2$ in the (x_n, y_n) and (u_n, v_n) planes, respectively, where the points are nonuniform. An example of a continuous system, exhibiting quasiperiodic route to chaos, will be presented in the next chapter.

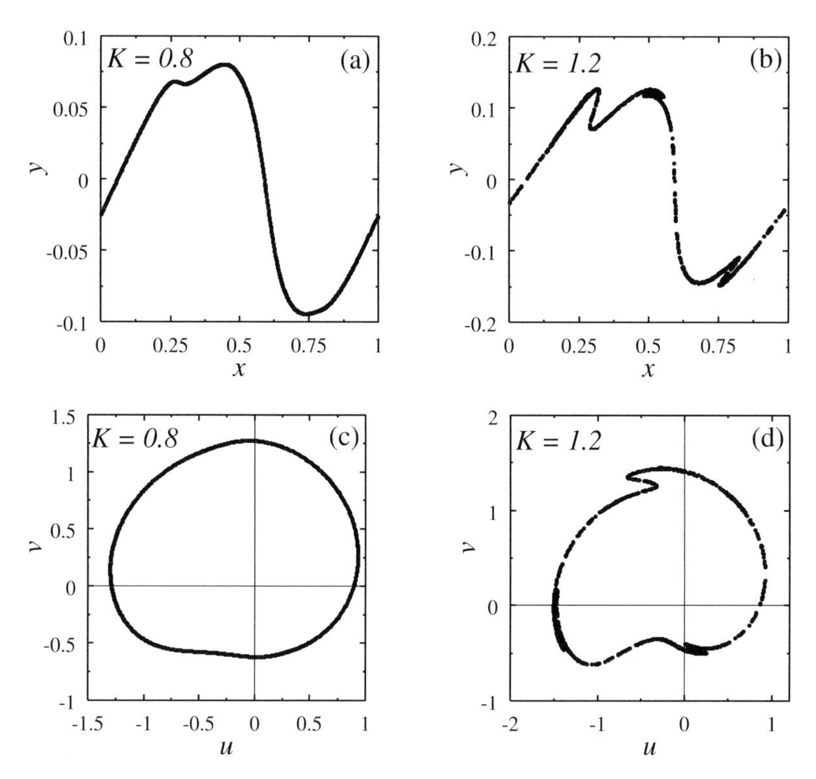

Fig. 4.20. (**a**) and (**c**) Quasiperiodic attractor and (**b**) and (**d**) chaotic attractor of the map (4.51) for $K = 0.8$ and $K = 1.2$ respectively. The values of the other parameters are $\Omega = 0.612$ and $b = 0.5$. The variables u and v are given by $u = (1 + 4y)\cos 2\pi x$ and $v = (1 + 4y)\sin 2\pi x$

4.4.2 Intermittency Route to Chaos

There is yet another important route to chaos called Pomeau–Manneville scenario or intermittency route which proceeds through the appearance of intermittent dynamical behaviour. When a motion alternates randomly between long regular or laminar phases and relatively short irregular bursts, it is said that the motion is *intermittent* or that there is an *intermittency*. In the intermittent route to chaos, initially the time series consists of regular laminar motion interrupted by irregular bursts. The laminar motions between two successive bursts have different durations which are randomly distributed over the time intervals. As a control parameter is increased, the length of the laminar region decreases and the bursts become very frequent so that at a critical value of the parameter the laminar phases disappear altogether and the motion becomes fully chaotic. In other words, for a control parameter, say, p less than p_c, the attractor is a periodic orbit. For p slightly larger than p_c, there are long laminar phases during which the orbit appears to be periodic and closely resembles the orbit for $p < p_c$. In addition, there are short chaotic bursts. As p increases sufficiently above p_c, the motion becomes fully chaotic without laminar phase. This intermittent behaviour was first found by Pomeau and Manneville [17] in 1979 in the Lorenz equations.

In the intermittency transition, one has a periodic orbit which is replaced by the chaotic orbit as p passes through p_c. This implies that the stable periodic orbit either becomes unstable or is destroyed as p increases through p_c. This can occur through one of the three bifurcations, namely saddle-node, Hopf and inverse-period doubling, and the corresponding intermittency transitions are called *type-I, type-II* and *type-III*, respectively. Since type-I intermittency is found to occur more frequently, we describe its occurrence here. For other types we refer to Ref. [9]. In addition, several other exotic types of intermittencies, including multi-intermittency [20], crisis-induced intermittency [21] and on-off intermittency [22], exist. Interested reader may refer to the relevant references.

4.4.3 Type-I Intermittency

First, we describe the mechanism by which this intermittency route to chaos occurs. For illustrative purpose, let us consider an one-dimensional map of the form $x_{n+1} = f(x_n, p)$, where p is a control parameter. Type-I intermittency corresponds to the saddle-node bifurcation in which stable and unstable fixed points coalesce (inverse to what was described in Sect. 4.1.1, see Figs. 4.1 also) and disappear, when the function $f(x)$ is tangent to the bisection $y = x$, as depicted in Fig. 4.21.

Near the fixed point x_c at which the stability determining slope $f' = 1$ and the critical value of the parameter $p = p_c$, the map f can be expanded about x_c as

$$\begin{aligned} f(x,p) &= f[x_c + (x-x_c), p_c + (p-p_c)] \\ &= f(x_c,p_c) + (x-x_c) f_x(x_c,p_c) + (p-p_c) f_p(x_c,p_c) \\ &\quad + \frac{1}{2}(x-x_c)^2 f_{xx}(x_c,p_c)\,, \quad f_x = \frac{\partial f}{\partial x}\,, \quad f_p = \frac{\partial f}{\partial p}\,, \end{aligned} \tag{4.52}$$

where the other higher order terms are neglected. The above equation can be rewritten as

$$f(x) = x_c + (x-x_c) + \alpha_c (x-x_c)^2 + \beta_c (p-p_c)\,, \tag{4.53}$$

where we have used $f_x(x_c) = 1$, $\alpha_c = f_{xx}(x_c)/2$ and $\beta_c = f_p$. Substituting $u = (x - x_c)/\beta_c$, $\epsilon = p - p_c$ and $\alpha = \alpha_c/\beta_c > 0$ in the map $x_{n+1} = f(x_n, p)$ and using the relation (4.53) we obtain

$$u_{n+1} = u_n + \alpha u_n^2 + \epsilon\,. \tag{4.54}$$

The fixed points of the map (4.54) are $u_\pm^* = \pm\sqrt{-\epsilon/\alpha}$ and they exist only if $\epsilon < 0$. Since $\mathrm{d}u_{n+1}/\mathrm{d}u_n = 1 + 2\alpha u_n$, the fixed point u_+^* is unstable, while u_-^*

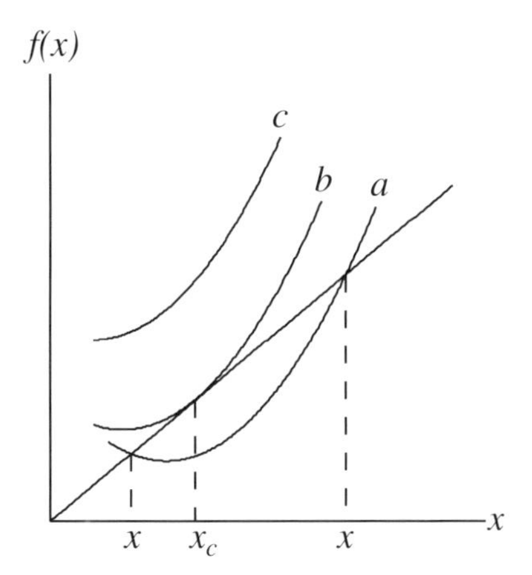

Fig. 4.21. Graph of a function $f(x)$ near the saddle-node bifurcation. Curves a, b and c represent $f(x)$ just before the bifurcation, at the bifurcation and just after the bifurcation respectively

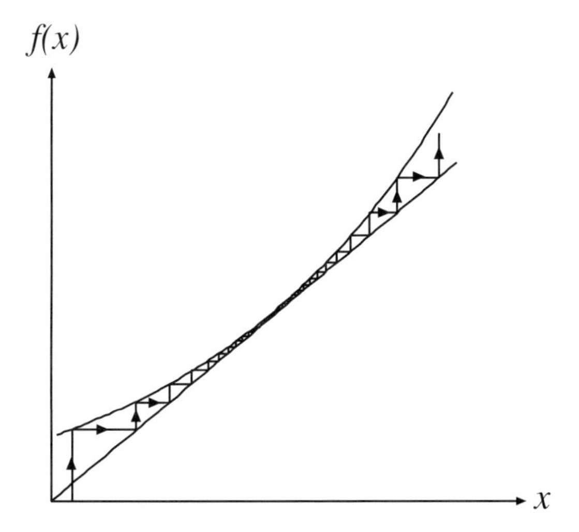

Fig. 4.22. Typical slow passage of iterations through a constricted zone

is stable. Now if $\epsilon > 0$, that is, $p > p_c$ the fixed points disappear, however, there remains a kind of memory of the fixed point x_c in the form of a channel between $f(x)$ and the bisector $y = x$. As shown in Fig. 4.22, it may take many iterations before x_n gets out of the channel and during this process its value is nearly constant, which results in almost regular oscillations. Once it leaves this region it will move chaotically until another portion of the map reinjects it into the constricted zone.

We further illustrate the above phenomenon in the logistic map (4.19). For values a a slightly larger than $a_c \approx 1 + \sqrt{8} \approx 3.828427\ldots$, the map has a period-3 attractor as shown in the bifurcation diagram Fig. 4.23.

At $a = a_c$, intermittent dynamics is developed. As a is decreased from a_c, the plots given by Figs. 4.24a, 4.24b and 4.24c are obtained. We see that the bursts become more frequent as a decreases. For $a = 3.82841$, the values of x constituting the laminar phase are $x = (0.161, 0.5145, 0.956)$. Figure 4.25 shows a blowup of x_n versus x_{n+3} plot near $x = 0.161$. The straight line is the $y = x$ line. The circled dots represent a sequence of third iterates. The slow passage of the iterates, indicated by staircase path as x_n approaches $x_c = 0.161$, is clearly evident. After passing through the constricted region shown in the Fig. 4.25, the iterates move widely and exhibit chaotic motion determined by the specific form of the map (4.19). The orbit is then reinjected into the constricted region when by chance the chaotic orbit lands in the neighbourhood of 0.161 or a similar one near 0.5145 or 0.956. This produces the regular laminar-like regions separated by bursts shown in Figs. 4.24a and 4.24b.

4.4.4 Standard Bifurcations in Maps

The four simple bifurcations discussed in Sect. 4.1 can also be analysed in terms of simple maps. For example, Table 4.2 summarizes the nature of bifurcations and bifurcation values in some maps. It is left as an exercise to the reader to verify these and also to draw the bifurcation diagram in each of the examples.

A summary of the various commonly occurring bifurcations in flows and maps are given later in Tables 5.2 and 5.3 of Chap. 5. Having studied chaotic dynamics in some important maps, now one may ask: Given a map, without iterating it, can one determine whether it will exhibit chaos or not? Though

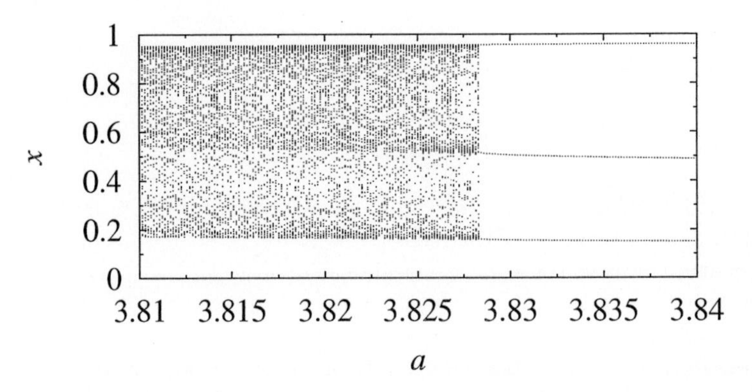

Fig. 4.23. Bifurcation diagram of the logistic map for $a \in (3.81, 3.84)$. Period-3 attractor can be seen for $a > a_c \approx 3.828427\ldots$

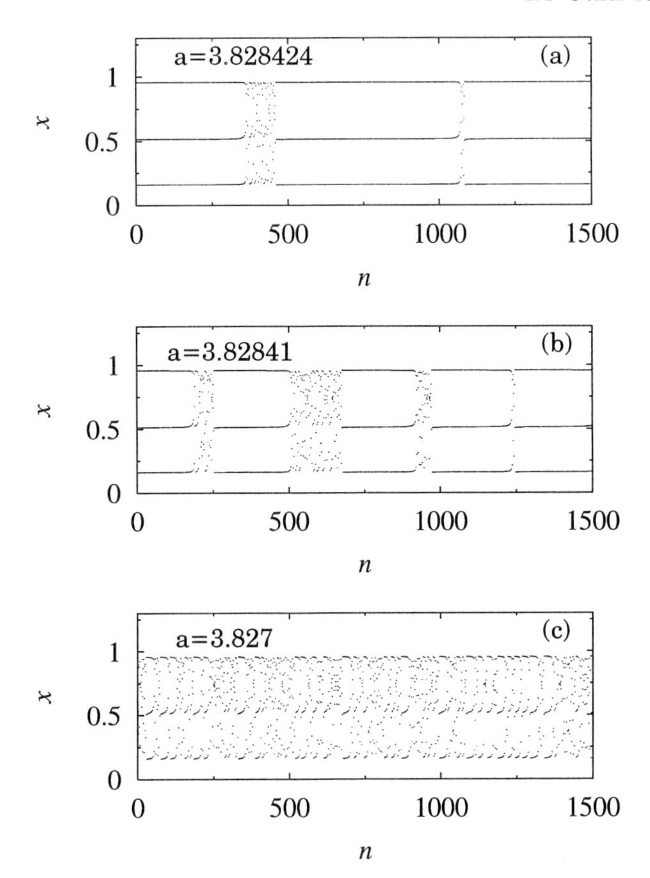

Fig. 4.24. x_n versus n of the logistic map for three values of a showing (**a**–**b**) intermittent chaos and (**c**) fully developed chaos

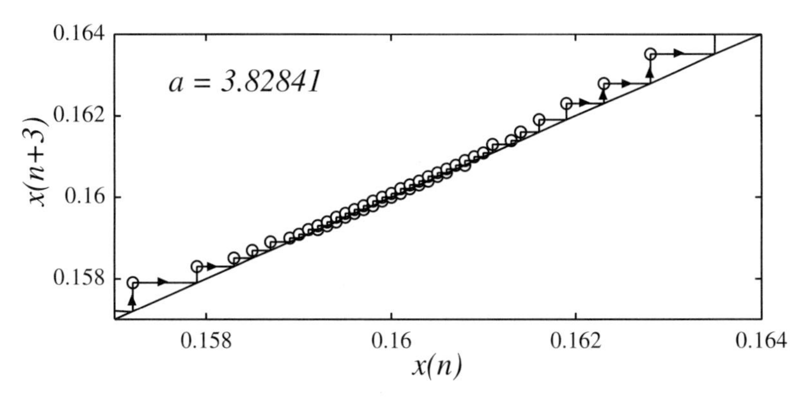

Fig. 4.25. Third iterates of the logistic map near $x_c \approx 0.161$ forming a laminar phase for $a = 3.82841$

Table 4.2. Some simple bifurcations in maps

No.	Map	Nature of bifurcation	Bifurcation value
1.	$x_{n+1} = a - x_n^2$	saddle-node	$a = -1/4$
2.	$x_{n+1} = ax_n(1 - x_n)$	transcritical	$a = 1$
3.	$x_{n+1} = x_n^3 - ax_n$	pitchfork (subcritical)	$a = -1$
4.	$x_{n+1} = -x_n^3 - ax_n$	pitchfork (supercritical)	$a = -1$
5.	$x_{n+1} = y_n$ $y_{n+1} = ay_n(1 - ax_n)$	Hopf (supercritical)	$a = 2$

the complete answer to this problem is still not known, it is possible to provide certain necessary conditions for the occurrence of chaos in maps and also in continuous dynamical systems. We will address this question also in the next chapter.

Problems

1. For the logarithmic map

$$x_{n+1} = \ln(a|x_n|) , \quad a > 0$$

verify that
   a) saddle-node bifurcation occurs at $x^* = 1$, $a = \mathrm{e}$ and
   b) chaos occurs for $1/\mathrm{e} < a < \mathrm{e}$ [23].
2. Show that the period-1 fixed points of the exponential map

$$x_{n+1} = x_n \exp[A(1 - x_n)] , \quad A > 0$$

are $x^* = 0$ and 1 and the former is always unstable while the later is stable for $A < 2$. Find the value of A at which $x^* = 1$ is superstable. Also verify that $x = 1/A$ is always a member of the superstable period-n cycle [24].
3. Compute period-1 fixed points and their stability determining quantity for the so-called Bellows map [24]

$$x_{n+1} = rx_n / \left(1 + x_n^b\right) .$$

   a) Verify that the fixed point $x^* = 0$ is stable for $r < 1$, while $x^* = (r-1)^{1/b}$ exists for $r > 1$ and is always unstable.
   b) Show that $x = (r-1)^{1/b}$ is always a member of the superstable period-n cycle.
   c) Investigate the occurrence of period doubling and intermittency routes to chaos by varying r from 0 to 5 for $b = 6$.

4. Using the bifurcation diagram, investigate the occurrence of period doubling, band-merging, sudden widening and intermittency bifurcations in the cubic map
$$x_{n+1} = x_n^3 - a x_n^2 - b$$
for $a = 2.0625$, by varying b in the interval $[-1.72, 0.2]$ [25].
5. Study the dynamics of the one-dimensional map [26]
$$x_{n+1} = x_n^2 \exp(A - x_n)$$
by varying the parameter A from 0 to 3 with the initial condition $x_0 > 1$. What does happen if $x_0 < 1$?
6. Numerically show that the map
$$x_{n+1} = Q - A x_n + x_n^3$$
exhibits finite period doubling bifurcations for $A = 1.7$ and $Q \in [-1.5, 1.5]$ and period doubling route to chaos and reverse period doubling for $A = 1.8$ and $Q \in [-1.5, 1.5]$ [27,28].
7. Consider the following nonpolynomial one-dimensional map
$$x_{n+1} = Q + \frac{A x_n}{1 + x_n^2} \, .$$
For $A = 6$ the map has three fixed points.
   a) Draw a plot of Q versus x^* for $Q \in [-2.5, 2.5]$.
   b) Identify the stable and unstable fixed points.
   c) Numerically verify that when Q is varied from -5 to $+5$, finite period doubling bifurcations occur for $A = 9$ and 11.5 [27].
8. Numerically verify that the circle map
$$\theta_{n+1} = \theta_n + \phi + \mu \sin(p\, \theta_n) \pmod{2\pi}$$
exhibits
   a) intermittency route to chaos when μ is varied from 1.65 to 2 for $p = 1$, $\phi = 6 + 1.17\mu$ and
   b) period doubling route to chaos when μ is varied from 2.5 to 4.5 for $p = 1$, $\phi = 6 + 0.8\mu$ [29].
9. The following notions are of considerable importance in chaotic dynamics:
   a) *frequency-locking*
   b) *devil's staircase.*
   Prepare a write up on these aspects [30].
10. Numerically investigate the dynamics of the power-law map
$$x_{n+1} = \begin{cases} 1 - r|x_n|^z \, , & \text{if } x_n < 0 \\ 1 - r x_n^2 \, , & \text{if } x_n \geq 0 \end{cases}$$
as a function of the parameter r in the interval for $r \in [0, 2]$, for $z = 0.4$ and 0.6 [31].

11. A one-dimensional map which is used to describe the dynamical behaviour of intramolecular processes and isomerization is [32]

$$x_{n+1} = r x_n \left(\omega^3 - 2\omega x_n^2 + x_n^4\right) .$$

Fix the parameter $\omega = 0.8$ and study various routes to chaos by considering r as the control parameter.

12. The quadratic map

$$x_{n+1} = p - x_n^2$$

exhibits intermittent switching between three bands of a chaotic attractor for a range of values of p just above $p_c = 1.79$. Draw x_{3n} for $p = 1.79$, 1.79033, 1.7904 and 1.791 and observe the intermittent dynamics. Calculate the mean life time τ on a chaotic band as a function of $(p - p_c)$ for $0 < (p - p_c) < 0.1$. Assuming that τ scales as $\tau \sim C(p - p_c)^{-\gamma}$, estimate C and γ [33].

13. An one-dimensional map exhibiting a variety of dynamics is the sine-square map

$$X_{n+1} = A \sin^2 (X_n - B) ,$$

which describes the dynamics of a liquid crystal hybrid optical bistable device. For $B = 1.2$, when A is varied from 0, period doubling bifurcations occur and these accumulate at $A = A_\infty = 1.92163$. Intermittency route to chaos can be observed when A is decreased from 2.305. Band merging and sudden widening of chaotic attractor occur at $A = 2.05$ and 2.3882 respectively [34]. Verify the above dynamics by numerically iterating the map.

14. Investigate the stability of fixed points of the complex map [35]

$$z_{n+1} = \left(z_n + \frac{\epsilon}{2}\right) \exp\left[\, \mathrm{i}/|z_n + (\epsilon/2)|^2\right] + \frac{\epsilon}{2} .$$

15. Both analytically and numerically show that the period-2 fixed point of the Hénon map (4.45) with $b = 0.3$ is a stable node for $0.3675 < a < 0.49$, a stable focus for $0.49 < a < 0.79$ and again a stable node for $0.79 < a < 0.9125$. Study the variation of the eigenvalues λ_1 and λ_2 in the $(\lambda_{\mathrm{real}} - \lambda_{\mathrm{imag}})$ plane for $a \in [0.3675, 0.9125]$.

16. Consider the predator-prey model map

$$x_{n+1} = a x_n (1 - x_n - y_n) ,$$
$$y_{n+1} = b x_n y_n .$$

a) Show that its fixed point $(x^*, y^*) = (1/b,\ (ab - b - a)/ab)$ is unstable for $b > 2a/(a - 1)$ and undergoes Hopf bifurcation at $b = 2a/(a - 1)$.
b) Choose $b = 1$. Verify that the map undergoes transcritical bifurcations at $a = 1$ and 2.
c) What happens when a is increased from 2?

17. For the coupled logistic maps

$$x_{n+1} = 1 - (a + \delta)x_n^2 ,$$
$$y_{n+1} = 1 - (a - \delta)y_n^2 + 2b(x_n - y_n) ,$$

obtain the expressions for stability determining eigenvalues of period-n fixed point. Then analytically determine the conditions on a and b for $\delta = 0.01$ for stability of period-1 and period-2 cycles. In the $(a-b)$ parameters space sketch the regions in which period-1 and period-2 attractors occur. Verify the analytical predictions by numerically iterating the map. Also investigate the occurrence of period doubling route to chaos.

18. For the Burgers map

$$x_{n+1} = (1 - \gamma)x_n - y_n^2 ,$$
$$y_{n+1} = (1 + \mu + x_n)y_n ,$$

verify both analytically and numerically that it
   a) has one fixed point at the origin for $0 \leq \gamma \leq 2$ and $-2 \leq \mu < 0$,
   b) undergoes pitchfork bifurcation at $\mu = 0$ for $\gamma = 1$ and
   c) undergoes Hopf bifurcation at $\mu = 1/2$ for $\gamma < 4$ [36].
   d) Using the bifurcation diagram and x_n versus y_n plot, study the dynamics as a function of μ in the interval $[0, 1]$ for $\gamma = 1$.

19. The dissipative standard map is given by

$$y_{n+1} = Bx_n + (K/2\pi)\sin 2\pi x_n ,$$
$$x_{n+1} = x_n + y_{n+1} .$$

Fix the parameter b at 0.3.
   a) Show that the period-1 fixed point of the map is a stable node for $0 < K < 0.2$, a stable focus for $0.2 < K < 2.4$, again a stable node for $2.4 < K < 2.6$ and an unstable node for $K > 2.4$.
   b) Numerically study the dynamics for $K > 2.6$.

20. A nonperiodic attractor with noninteger dimension and negative maximal Lyapunov exponent is termed as strange nonchaotic attractor. For the quasiperiodically forced circle map

$$\phi_{n+1} = \phi_n + 2\pi K + V\sin\phi_n + C\cos\theta_n , \quad (\text{mod } 2\pi)$$
$$\theta_{n+1} = \theta_n + 2\pi\omega \quad (\text{mod } 2\pi)$$

with $\omega = (\sqrt{5} - 1)/2$, $C = 0.6$ and $K = 0.4$ verify that the map exhibits three-frequency quasiperiodic attractor $\longrightarrow$ strange nonchaotic attractor $\longrightarrow$ chaotic attractor transition when V is varied from 0 to 1.1 [37].

21. Determine the stability of the four fixed points of the map [38]

$$x_{n+1} = bz_n ,$$
$$y_{n+1} = b\left(1 + y_n - az_n^2\right) ,$$
$$z_{n+1} = 1 + x_n - \left(a/b^2\right) y_{n+1}^2 .$$

5. Chaos in Dissipative Nonlinear Oscillators and Criteria for Chaos

In the previous chapter we have discussed various basic bifurcations, and the chaotic dynamics underlying the logistic and Hénon maps. The maps are represented by difference equations, where the time variable varies in discrete steps. In the present chapter we shall study the bifurcations phenomena and chaotic solutions of continuous time (flow) dynamical systems described by ordinary differential equations, by making use of our earlier understanding of bifurcations and chaos. To start with, we will study the occurrence of period doubling phenomenon and chaotic dynamics in two typical nonlinear dissipative systems, namely,

(1) the Duffing oscillator and

(2) the Lorenz equations,

which are often used as prototypes for the study of nonlinear dynamics, and then briefly discuss the behaviour of a few other ubiquitous nonlinear oscillators. The underlying phenomena are qualitatively the same as discussed in the case of maps in the last chapter. We then discuss the conditions necessary for chaos to occur in dynamical systems, both for continuous and discrete systems.

In the past three decades or so, a large number of analytical, numerical and experimental studies have been carried out on different nonlinear oscillator systems with an effort to understand the various features associated with the occurrence of chaotic behaviour. As in the case of maps, one essentially tries to vary one or more of the control parameters in the system so that the parameter ranges for which periodic and chaotic behaviours occur can be identified. Again, as in the previous chapter, one finds that in many nonlinear dissipative dynamical systems chaotic motion sets in through the three predominant routes, namely, period doubling, quasiperiodic and intermittency bifurcations[1]. These routes to chaos are also universal in the sense that many different systems exhibit the same pattern. We will now present some details of these aspects.

---

[1] As noted in the earlier chapters, there are many other not so predominant bifurcations present in dissipative chaotic dynamical systems. These include period adding, period bubbling, torus doubling, etc.

5.1 Bifurcation Scenario in Duffing Oscillator

As noted earlier in Chap. 2, the Duffing oscillator equation is a ubiquitous nonlinear differential equation, which makes its presence in many physical, engineering and even biological problems. Originally the model was introduced by the Dutch physicist Duffing in 1918 [1] to describe the hardening spring effect observed in many mechanical problems and is represented by

$$\ddot{x} + \alpha\dot{x} + \omega_0^2 x + \beta x^3 = f \sin \omega t \;, \quad \alpha > 0 \;. \tag{5.1}$$

Here α is the damping constant, ω_0 is the natural frequency and β is the stiffness constant which plays the role of the nonlinear parameter. Equation (5.1) can be also thought of as the equation of motion for a particle of unit mass in the potential well [2,3]

$$V(x) = \frac{1}{2}\omega_0^2 x^2 + \frac{\beta}{4}x^4 \;, \tag{5.2}$$

subjected to a viscous drag force of strength α and driven by an external periodic signal of period $T = 2\pi/\omega$ and strength f. We can distinguish three types of potential wells of physical relevance here:

(1) $\omega_0^2 > 0$, $\beta > 0$: A single-well with a potential minimum at the equilibrium point $x = 0$ (see Fig. 5.1a).
(2) $\omega_0^2 < 0$, $\beta > 0$: A double-well with potential minima at $x = \pm\sqrt{(\mid \omega_0^2 \mid /\beta)}$ and a local maximum at $x = 0$ (see Fig. 5.1b).
(3) $\omega_0^2 > 0$, $\beta < 0$: A double-hump potential well with a local minimum at $x = 0$ and maxima at $x = \pm\sqrt{(\omega_0^2/ \mid \beta \mid)}$ (Fig. 5.1c).

Each one of the above three cases has become a classical central model to describe inherently nonlinear phenomena. In Chap. 2 we have shown the occurrence of nonlinear resonances and jump phenomenon in the system (5.1) for weak nonlinearity ($\beta \ll 1$). Such behaviours can be analysed using various approximation methods. For sufficiently large values of β the system (5.1) exhibits a rich variety of regular (periodic) and complex (chaotic) motions which can coexist or exist in neighbouring parametric regimes of say f and ω. As this equation cannot be solved analytically, we will have to carry out essentially numerical analysis.

In the following, we will consider the more interesting case of the dynamics of the double-well Duffing oscillator (5.1) with $\omega_0^2 < 0$ and $\beta > 0$. The other two cases, namely the single-well and double-hump, can be analysed similarly, see for example Ref. [3]. A mechanical model [2] having double-well potential is depicted in Fig. 5.2.

Consider the case in which a steel beam is clamped to a rigid framework. In the absence of any external force, the elastic forces of the beam keep it straight. When magnets are placed on either side of the beam as shown in Fig. 5.2, their attractive forces overcome the elastic forces and

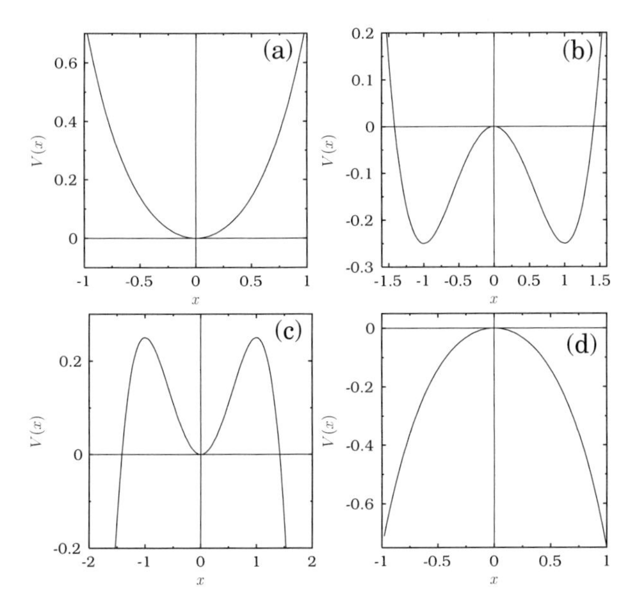

Fig. 5.1. Shape of the potential function, (5.2), for (**a**) $\omega_0^2 > 0$, $\beta > 0$ (single-well potential), (**b**) $\omega_0^2 < 0$, $\beta > 0$ (double-well potential), (**c**) $\omega_0^2 > 0$, $\beta < 0$ (double-hump potential) and (**d**) $\omega_0^2 < 0$, $\beta < 0$ (inverted single-well potential), not of physical relevance

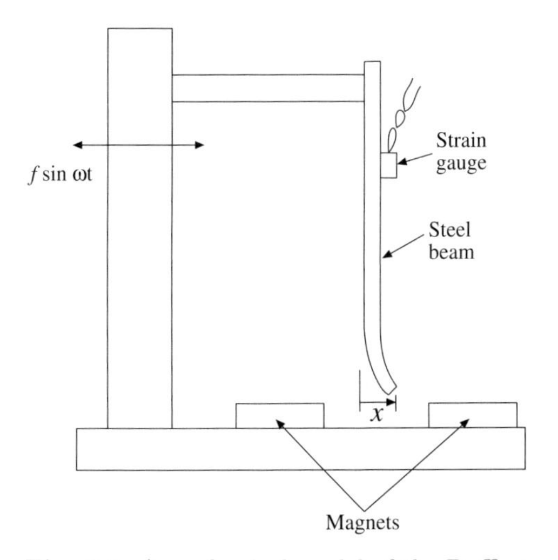

Fig. 5.2. A mechanical model of the Duffing equation (5.1)

the tip of the beam is now close to one of the magnets as shown in the Fig. 5.2. These two positions can be considered as stable equilibrium points of the system. The central position is unstable. The simplest model for such a potential is the double-well potential shown in Fig. 5.1b. Now from the Newton's second law of motion, a model for the horizontal displacement of the beam is given by $\ddot{x} = -\mathrm{d}V/\mathrm{d}x$. Adding the dissipation due to friction, viscous damping from the surrounding air and magnetic damping by a linear velocity dependent term, we obtain (5.1) with $f = 0$. Suppose the apparatus is shaken sinusoidally using an electromagnetic vibration generator, which can be represented by the addition of a forcing term $f \sin \omega t$, one obtains the full equation (5.1).

One can easily understand through a simple scaling argument ($x \to x' = \omega_0 x/\sqrt{\beta}$, $t \to t' = t\omega_0$, $\alpha \to \alpha' = \alpha/\omega_0$, $\omega \to \omega' = \omega/\omega_0$, $f \to f' = \sqrt{\beta}/\omega_0^3$) that there are only three free parameters in (5.1), which we can take for convenience to be α, f and ω. However, it is quite complicated and intricate to study the dynamics in a three parameter (α, f, ω) space or even in a two parameter space such as ($\alpha - f$) or ($f - \omega$) or ($\alpha - \omega$) subspaces numerically (for details see, for example, Ref. [3]). However, some ideas about the intricacies of the system can be obtained even by varying just one of the above three parameters while keeping all the other parameters in (5.1) fixed.

As an example, let us study the dynamics of (5.1) corresponding to a double-well[2] by fixing the values of the parameters ω_0^2, β, α and ω, thereby scanning the drive amplitude f upward from 0. The restoring force parameters are fixed as $\omega_0^2 = -1$ and $\beta = 1$. The other two control parameters are fixed at the values $\alpha = 0.5$ and $\omega = 1$. From the nature of the results of the numerical investigations obtained by solving (5.1) for the above parameters, using the standard *fourth-order Runge–Kutta integration algorithm* (for details, see appendix C), we infer the following picture. The results are substantiated by

(1) trajectory plot in the ($x - y(= \dot{x})$) phase space and
(2) bifurcation diagram.

For more sophisticated characterizations such as Lyapunov exponents, Fourier spectrum, and so on we refer to Chap. 8.

The phenomena of resonances and hysteresis discussed in Chap. 2 can be easily identified if we fix the parameter f and vary ω suitably. This is left as an exercise to the reader.

5.1.1 Period Doubling Route to Chaos

To start with, we observe that for $f = 0$ the trajectory approaches asymptotically the stable equilibrium point, which is now a focus (the left equilib-

[2] One can study the dynamics for single-well ($\omega_0^2 > 0$, $\beta > 0$) and double-hump ($\omega_0^2 > 0$, $\beta < 0$) Duffing oscillators also and identify similar bifurcations and chaos phenomena (see also below). For more details see (for example Ref. [3]).

rium point $(x_1^*, y_1^*) = (-1, 0)$ in the present study), as shown in Fig. 5.3a, corresponding to purely damped oscillations. One can also start with a *different* initial condition so as to reach the other stable equilibrium point $((x_2^*, y_2^*) = (1, 0))$ in the right well asymptotically. As f is increased from 0, a stable period-$T(= 2\pi/\omega)$ limit cycle about the left equilibrium point $(x_1^*, y_1^*) = (-1, 0)$ occurs, which persists up to $f = 0.3437$. Figure 5.3b depicts such a limit cycle for $f = 0.33$. One can easily convince oneself by comparing with the discussion on Hopf bifurcations in Sect. 4.1.4 that the present bifurcation is also of the same type. As f is increased to a critical value 0.3437, the phase trajectory bifurcates into a new limit cycle of period-$2T$, again about (x_1^*, y_1^*), as shown in Fig. 5.3c for $f = 0.35$. The system then undergoes further period doubling bifurcations as the control parameter f is smoothly varied. For example, at $f = 0.355$, the period-$2T$ limit cycle becomes unstable and gives birth to a period-$4T$ solution. Figure 5.3d shows the period-$4T$ orbit for $f = 0.357$. This cascade of bifurcations continues further as $8T$, $16T$,..., solutions and accumulates at $f = f_c \approx 0.3589$, beyond which

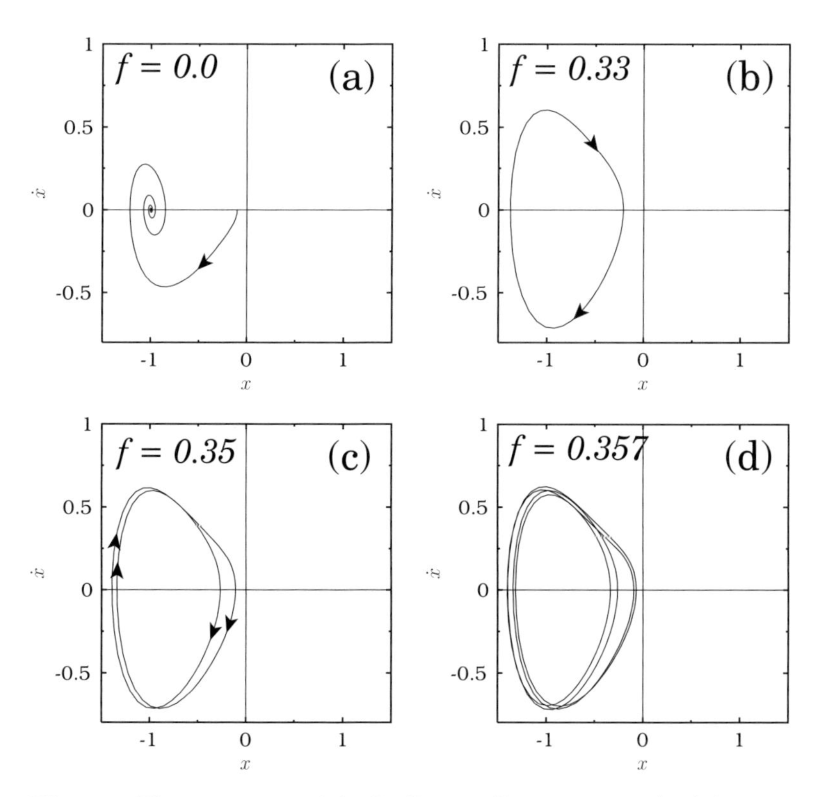

Fig. 5.3. Phase portraits of the Duffing oscillator equation (5.1) for various values of forcing amplitude f. The other parameters are kept at $\alpha = 0.5$, $\omega_0^2 = -1$, $\beta = 1$ and $\omega = 1$

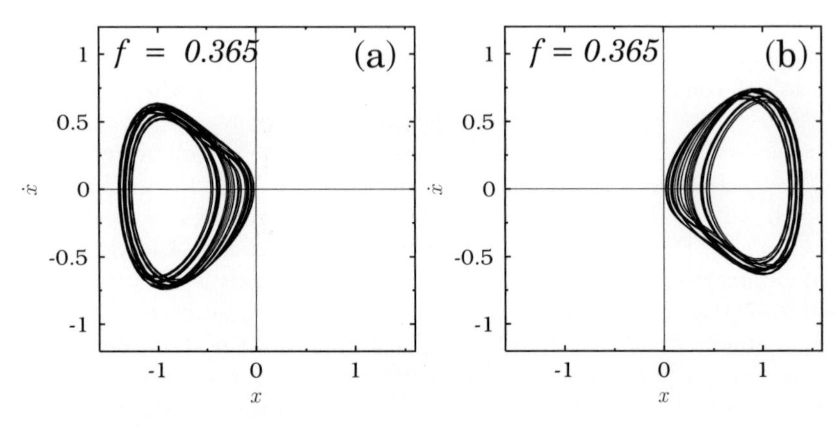

Fig. 5.4. One-band chaotic attractors confined to (**a**) the left well and (**b**) the right well for $f = 0.365$

chaotic motion starts occurring. Figure 5.4a shows a typical one-band (left well) chaotic attractor for $f = 0.365$. The chaotic attractor which coexists in the right well is depicted in Fig. 5.4b, for comparison, which is obtained for a different initial condition.

The above results can be put in a nutshell in the form of a bifurcation diagram (Fig. 5.5) and also qualitatively as in Table 5.1. Figure 5.5 illustrates the behaviour of the left well attractor as a function of the forcing amplitude. (Similar behaviour occurs for the right well attractor also). In the bifurcation diagram, for each value of f the values of x collected at $2\pi/\omega$ (the period of the driving force) time intervals are plotted. The bifurcation dia-

Table 5.1. Summary of bifurcation phenomena of the double-well Duffing oscillator, equation (5.1), with $\alpha = 0.5$, $\omega_0^2 = -1$, $\beta = 1$ and $\omega = 1$

Value of f	Nature of solution
$f = 0$	Damped oscillation to the stable spiral equilibrium $x_1^* = -1, y_1^* = 0$
$0 < f < 0.3437$	Period-T oscillation
$0.3437 \le f < 0.355$	Period-$2T$ oscillation
$0.355 \le f < 0.3577$	Period-$4T$ oscillation
$0.3577 \le f < 0.3589$	Further period doublings
$0.3589 \le f < 0.3833$	One-band chaos
$0.3833 \le f \le 0.42$	Double-band chaos
$f > 0.42$	Chaos, periodic windows, period doublings of windows, reverse period doublings, etc.

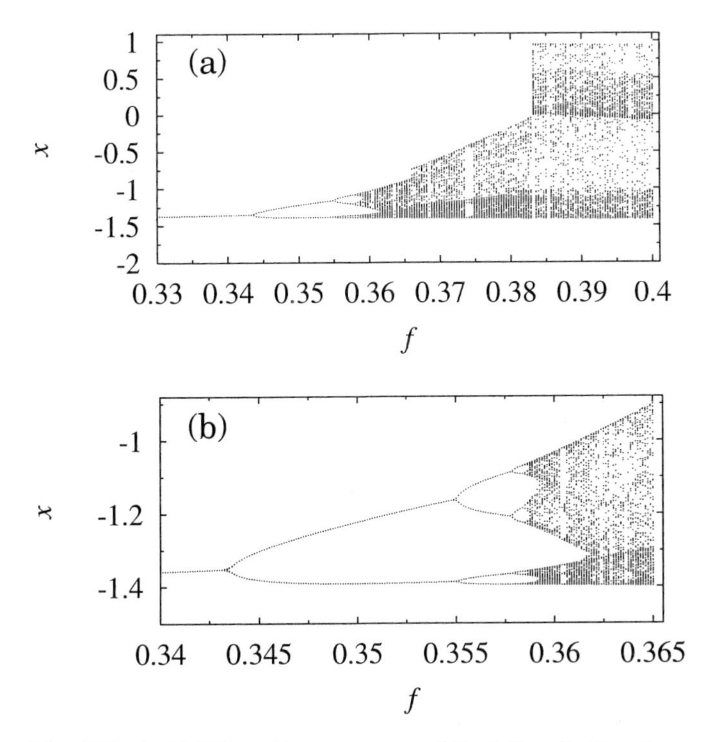

Fig. 5.5. (**a,b**) Bifurcation sequence of the left well attractor

gram clearly depicts the associated period doubling phenomenon leading to chaos as the control parameter f is varied. (This may be compared with the bifurcation diagram of the logistic map discussed in Sect. 4.2.5). When the parameter f is further increased from 0.365, the single well chaotic attractor bifurcates to a cross-well (double-band) chaotic attractor where the trajectory jumps between left and right wells. Figure 5.6a shows the phase portrait of the double-band chaotic attractor, for $f = 0.42$. The Poincaré map (see Sect. 3.7.3) of the chaotic attractor shown in Fig. 5.6a is given in Fig. 5.6b. It has been drawn using the points collected at $2\pi/\omega$ time intervals (same as what we have done to obtain the bifurcation diagram) of the system. The geometric structure of the chaotic attractor in the Poincaré map appears as a totally disconnected and uncountable set of points. One can check that the extreme sensitive dependence on initial conditions discussed in the previous chapter is present here also in the chaotic region. The nature of chaotic and periodic attractors can also be confirmed by characterising them in terms of Lyapunov exponents, power spectra, autocorrelations, dimensions, etc. These are discussed later in Chap. 8.

One might ask what happens beyond $f = 0.42$. Typically, one finds that chaotic orbits persist for a range of f values, interspersed by periodic window

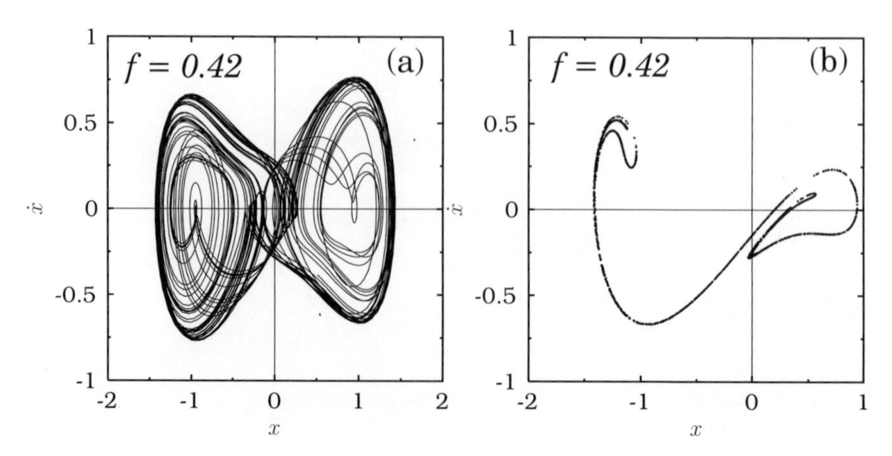

Fig. 5.6. (**a**) Phase portrait and (**b**) Poincaré map of the double-band chaotic attractor for $f = 0.42$

regions, where interesting new kinds of bifurcations and routes to chaos occur in a very sophisticated way (details of which are beyond the scope of this book, however, some details are given in the following). One can not only change the parameter f but also ω, for example, in (5.1). Then a very complex picture of bifurcations phenomena, present even in such a simple nonlinear system as the Duffing oscillator equation (5.1), can be realized. A summary of the basic bifurcations is given in Table 5.1. For further details, see for example, Ref. [3].

Exercises:

1. Draw the Poincaré plot for the single band chaotic attractor of Fig. 5.4.
2. Show the sensitive dependence of the chaotic attractors corresponding to the single-band (Fig. 5.4) and double-band (Fig. 5.6), by plotting $x(t)$ versus t for two nearby initial conditions.
3. Investigate the periodic window region, for $f > 0.42$.
4. Investigate the bifurcations phenomena in the Duffing oscillator equation (5.1) for (i) the single-well, $\omega_0^2 > 0$, $\beta > 0$ and (ii) double-hump, $\omega_0^2 > 0$, $\beta < 0$, potentials (see also Ref. [3]).

5.1.2 Intermittency Transition

Next, we show the occurrence of intermittency transition to chaos in the Duffing oscillator equation (5.1), which is quite similar to the case of the logistic map discussed in Sect. 4.4.3. Figure 5.7 shows the bifurcation phenomenon for $f \in (0.538, 0.5445)$. We see that just above $f = f_c \approx 0.54215$ there is a stable period-$3T$ orbit, while just below f_c there is chaos. Figure 5.8 shows

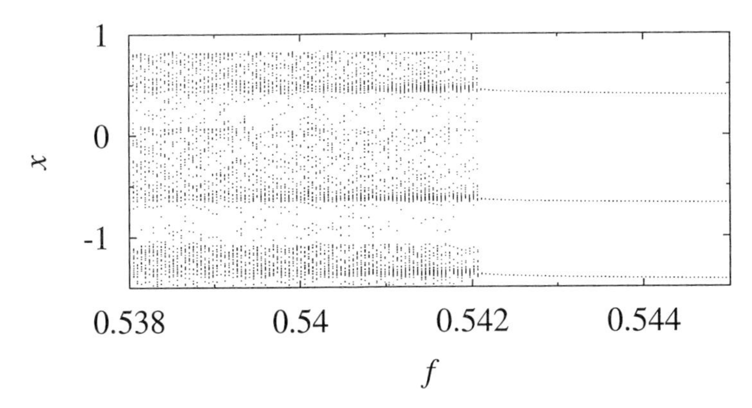

Fig. 5.7. Bifurcation diagram of the Duffing equation (5.1) in the intermittency region (see also Fig. 5.5)

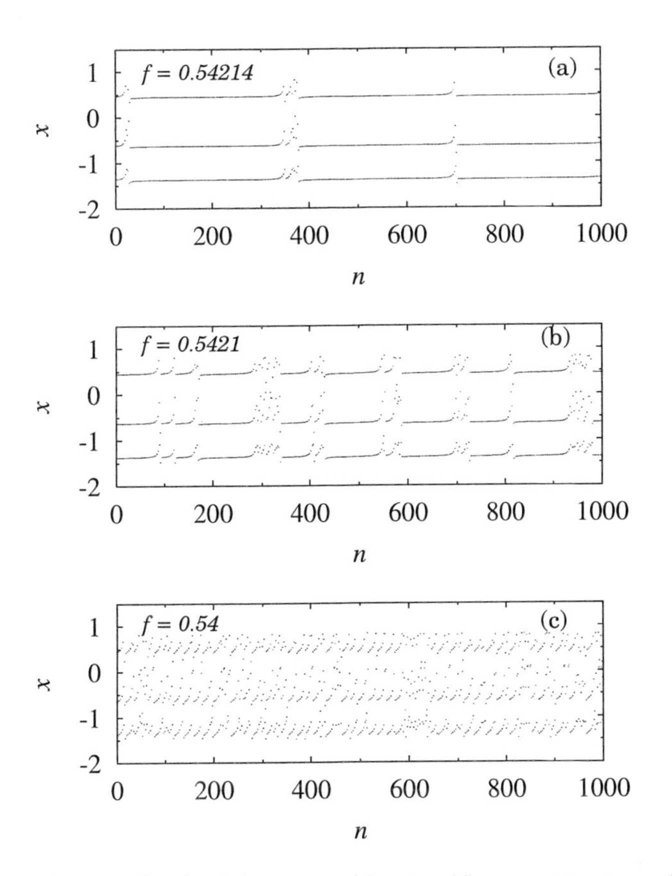

Fig. 5.8. **(a–c)** $x(n)$ versus n (that is $x(t)$ versus t in steps of $2\pi/\omega$) of the Duffing equation (5.1) illustrating intermittency route to chaos

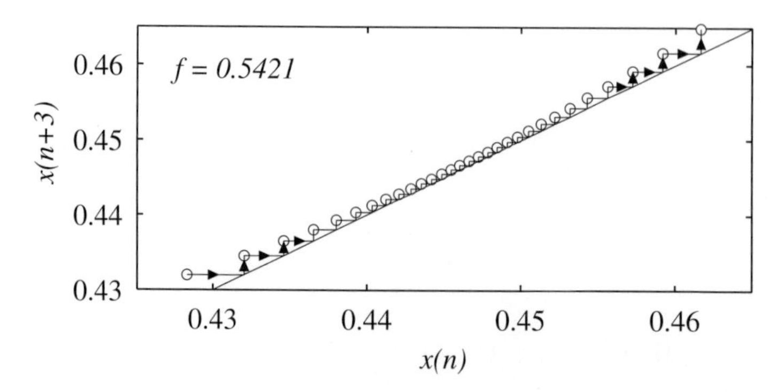

Fig. 5.9. $x(n+3)$ versus $x(n)$ near $x = 0.446$ showing the slow passage of x thereby forming a laminar phase

x versus t (in steps of $2\pi/\omega$) for three values of f. Intermittent dynamics is clearly seen in Figs. 5.8a and 5.8b. Figure 5.9 illustrates the orbit in the vicinity of $x = 0.446$ in terms of the $x(n)$ versus $x(n+3)$ plot, where n is the time t in steps of $2\pi/\omega$.

The straight line is the $x(n + 3) = x(n)$ line. The circles represent a sequence of third iterates. The slow passage of the iterates indicated by a staircase path as $x(n)$ approaches $x = 0.446$ is clearly seen. After passing through the constricted region shown in Fig. 5.9, the iterates move widely and exhibit chaotic motion until the orbit is reinjected again into the constricted region. Thus, as the amplitude of the external force is decreased through f_c we have an intermittency transition from a period-$3T$ attractor to a fully developed chaotic attractor. We can compare the Figs. 5.7–5.9 of the present case with Figs. 4.23–4.25 of the logistic map to convince ourselves that the transition is indeed type-I intermittency.

As shown in the case of the logistic map in the previous chapter and in the Duffing oscillator equation above, it is clear that in the intermittency route the periodic orbit is replaced by chaos as a control parameter passes through the critical value. This implies that the stable periodic orbit either becomes unstable or is destroyed as it passes through the critical value. The orbit is not replaced by another stable periodic orbit, as is the case in period doubling bifurcation.

5.1.3 Quasiperiodic Route to Chaos

Quasiperiodic route to chaos is observed when (5.1) with $\omega_0^2 > 0$ (that is single-well) is modified by replacing the linear damping term with a nonlinear damping term of the form $(1-x^2)\dot{x}$. The Duffing oscillator equation with this damping term and with a periodic driving force $f \cos \omega t$ can be written as

$$\ddot{x} - p\left(1 - x^2\right)\dot{x} + \omega_0^2 x + \beta x^3 = f \cos \omega t \,. \tag{5.3}$$

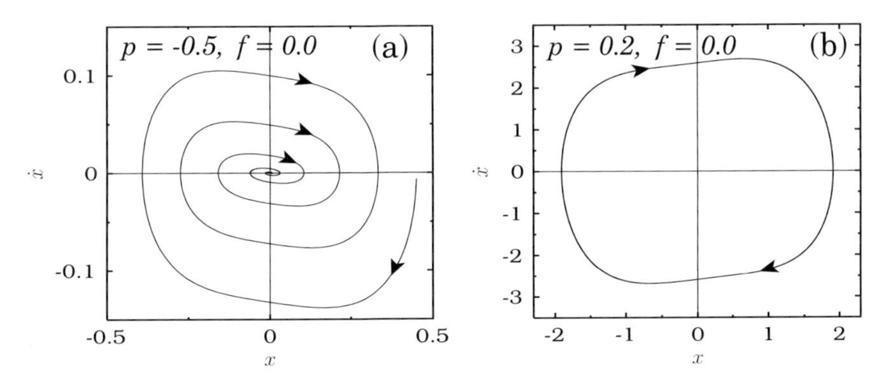

Fig. 5.10. Phase portraits of the (5.3) with $f = 0.0$, $\omega_0^2 = 0.011$ $\beta = 1$ and for (**a**) $p = -0.05$ and (**b**) $p = 0.2$

Equation (5.3) is also called the Duffing–van der Pol oscillator equation [3]. We note that when $\beta = 0$ (5.3) becomes the forced van der Pol oscillator equation (5.9) considered later in Sect. 5.3. When $f = 0$ and $\beta = 1$, (5.3) has only one real equilibrium point at $(x^*, y^*) = (0, 0)$. Its stability determining eigenvalues are (Sect. 3.4.2) $\lambda_\pm = (p \pm \sqrt{p^2 - 4\omega_0^2}\,)/2$. For $p < -2\sqrt{\omega_0^2}$, both the eigenvalues are less than zero and the equilibrium point is a stable node. The eigenvalues become complex conjugate with negative real parts for $-2\sqrt{\omega_0^2} < p < 0$ and correspondingly the equilibrium point becomes a stable focus. For example, Fig. 5.10a shows a typical trajectory converging to the equilibrium point $(0, 0)$ along a spiral path for $f = 0$, $\omega_0^2 = 0.011$ and $p = -0.05$. At $p = 0$, $\lambda_\pm = \pm \mathrm{i}\,\omega_0$, that is, the real part vanishes. Thus the system undergoes a Hopf bifurcation at $p = 0$. For $0 < p < 2\sqrt{\omega_0^2}$, the real parts of both the complex conjugate eigenvalues are positive and so the equilibrium point becomes unstable. However, in this interval of p a stable limit cycle does exist due to the Hopf bifurcation at $p = 0$. Figure 5.10b shows such a limit cycle for $f = 0.0$, $\omega_0^2 = 0.011$ and $p = 0.2$. In the presence of the forcing term $f \cos \omega t$ in (5.3), a second Hopf bifurcation occurs, which gives rise to a stable quasiperiodic solution. For example, Figs. 5.11a and 5.12a show the phase portrait and Poincaré map, respectively, of the quasiperiodic attractor of (5.3) for $\omega_0^2 = 0.011$, $p = 0.2$, $\beta = 1$, $f = 1$ and $\omega = 0.92$. As ω is increased further, a breakup of the torus occurs leading to chaotic motion. This is illustrated in Figs. 5.11b and 5.12b for $\omega = 0.94$.

5.1.4 Strange Nonchaotic Attractors (SNAs)

Another interesting type of attractor observed in many nonlinear oscillator systems and maps is the strange nonchaotic attractor (SNA). Such attractors are not chaotic but aperiodic and display strange geometric properties. They are generic in the quasiperiodically forced systems and are typically

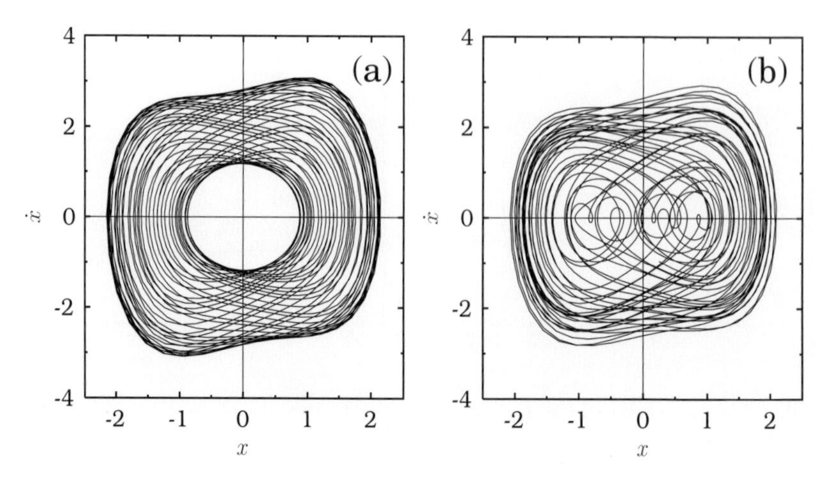

Fig. 5.11. Phase portraits of (**a**) the quasiperiodic attractor for $\omega = 0.92$ and (**b**) the chaotic attractor for $\omega = 0.94$ of the (5.3). The other parameters are fixed at $p = 0.2$, $\omega_0^2 = 0.011$, $\beta = 1$ and $f = 1$

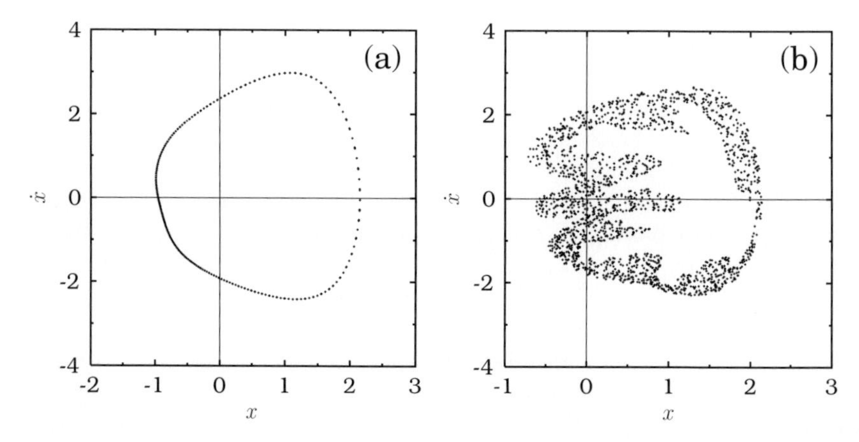

Fig. 5.12. Poincaré maps of the attractors shown in Fig. 5.11: (**a**) quasiperiodic attractor and (**b**) chaotic attractor

found in the neighbourhood of related periodic or quasiperiodic or strange chaotic attractors in parameter space. In a sense they represent dynamics which is intermediate between quasiperiodic and chaotic: there is no sensitive dependence on initial conditions, similar to motion on regular (periodic or quasiperiodic) attractors, but the motion is aperiodic similar to the dynamics of chaotic attractors [4].

As a simple example, SNAs are found to occur in the parametrically driven Duffing oscillator equation [5]

$$\ddot{x} + \alpha\dot{x} + \omega_0^2(1 + \eta(f\cos\omega t + \cos\Omega t))x + \beta x^3 = 0 \,. \tag{5.4}$$

When $\eta = 0$ (5.4) is the unforced, damped Duffing oscillator equation. We fix $\beta = 1$, $\omega_0^2 = -1$, $\eta = 0.3$, $\omega = 1$, $\Omega = (\sqrt{5}+1)/2$, $f = 0.47$ and study the dynamics of (5.4) by varying the damping parameter α. For values of α just below $\alpha_c = 0.088689$ a two-frequency quasiperiodic orbit exists. Strange nonchaotic attractor occurs in the region $0.088689 < \alpha < 0.088962$. On further increase in the value of α to 0.088963 we find the emergence of a chaotic attractor. There exist many routes to chaos via SNAs. These include torus breakdown, gradual fractalization, intermittency (type-I and type-III), torus collision and quasiperiodic routes [4].

Exercises:

5. Study the dynamics of (5.4) for α values in the interval $[0, 0.3]$ using bifurcation diagram, phase portrait and Poincaré map. Also verify the nonchaotic character of SNA by calculating the distance between two nearby trajectories [5].
6. A simple map exhibiting SNA is the quasiperiodically forced cubic map

$$x_{n+1} = Q\cos(2\pi y_n) - Ax_n + x_n^3 ,$$
$$y_{n+1} = y_n + \omega \pmod 1 .$$

Fix the parameter ω at $(\sqrt{5}-1)/2$. Study the transition from (i) torus to SNA by varying A from 1.8865 to 1.8875 for $Q = 0.65$ and (ii) two-torus to SNA by varying A from 2.16 to 2.168 for $Q = 0.1$ [6].

The above discussed routes to chaos are found in numerous other physical, chemical and biological systems also. Chaotic oscillation is observed in such wide ranging systems as in a simple pendulum placed in a periodically varying magnetic field, CO_2 laser, neural networks, electronic circuits with nonlinear elements (see the following chapter), certain chemical reactions, neural membranes, heart beats and so on (see also Chap. 1, Sect. 1.4). EEG recordings of human epileptic seizure shows the existence of chaotic behaviour. It is not difficult to create noisy, chaotic like acoustic effects in a trumpet or a wind instrument. Bifurcations phenomena and occurrence of chaos in some more important dynamical systems are presented in the next two sections.

5.2 Lorenz Equations

One of the first models which was shown to exhibit chaotic behaviour in numerical simulation was the fluid convection model introduced in 1963 by E.N. Lorenz [7] in his studies on atmospheric weather (briefly discussed earlier in Sect. 3.6.1):

$$\dot{x} = \sigma(y - x) , \tag{5.5a}$$

$$\dot{y} = rx - y - xz , \tag{5.5b}$$

$$\dot{z} = -bz + xy . \tag{5.5c}$$

The model corresponds to the following situation. Consider the fluid dynamical problem in which a fluid slab of finite thickness is heated from below and a fixed temperature difference is maintained between the top cold surface and the bottom hot surface. The fluid motion is described by the well known Navier–Stokes equation, which is a set of complicated nonlinear partial differential equations. When the motion is assumed to be two-dimensional, the flow can be characterized by two variables, the stream function ξ for the motion and the departure θ of the temperature from that of the state of no convection. The set of partial differential equations for the flow can then be transformed to a set of ordinary differential equations by expanding ξ and θ in double Fourier series in the spatial variables with functions of t alone occurring in the Fourier coefficients. Equations (5.5) are then obtained by truncating the series with the first three Fourier coefficients only [7].

In the above model equations (5.5), $x(t)$ represents the amplitude of the convection motion and $y(t)$ and $z(t)$ measure the horizontal and vertical temperature variations, respectively. Here σ represents the Prandtl number which is a ratio of kinematic viscosity to thermal conductivity, r is called the Rayleigh number and is proportional to the temperature difference between the upper and lower surfaces of the fluid; and b is a geometric factor. Consequently, all the three parameters are taken to be positive. The general form of these equations also serves as a simple model for complex dynamics in certain laser devices and many other physical problems.

5.2.1 Period Doubling Bifurcations and Chaos

The set of Lorenz equations (5.5) has a natural symmetry under the change of variables $(x, y, z) \to (-x, -y, z)$. This symmetry persists for all values of the parameters. The z-axis, $x = y = 0$, is invariant: All trajectories which start on the z-axis remain on it and tend towards the origin $(0, 0, 0)$. Let us check whether the set of Lorenz equations is dissipative or conservative. Referring to Sect. 3.8, identifying (5.5) with $\mathrm{d}\boldsymbol{X}/\mathrm{d}t = \boldsymbol{F}(\boldsymbol{X})$, (3.69), we find that

$$\begin{aligned} \nabla \cdot \boldsymbol{F} &= \frac{\partial}{\partial x}\sigma(y - x) + \frac{\partial}{\partial y}(rx - y - xz) + \frac{\partial}{\partial z}(-bz + xy) \\ &= -(\sigma + b + 1) < 0 . \end{aligned} \tag{5.6}$$

The system is thus dissipative (see our definition in Sect. 3.8). Consequently a volume element, V, in the phase space is contracted by the flow into a volume element $V \exp\left[-(\sigma + b + 1)t\right]$ after a time t. Depending upon the initial conditions, the system admits different attractors.

Exercise:

7. Show that the Duffing oscillator equation (5.1) is dissipative. What is the nature of the Duffing–van der Pol equation (5.3)?

In Sect. 3.6.1 we have carried out a linear stability analysis of the equilibrium point solutions of the Lorenz equations. The main properties of (5.5) turn out to be the following:

(1) For $0 < r < 1$, there is only one stable equilibrium point which is the origin.

(2) For $1 < r < 1.346$, two new stable nodes are born and the origin becomes a saddle. In other words, as r passes through the value 1, the preceding solution becomes unstable and two new stable equilibrium points are created. That is, the system undergoes a pitchfork bifurcation at $r = 1$.

(3) For $1.346 < r < 24.74$, the equilibrium points which are stable nodes previously now become stable spirals.

Now we present further properties of the Lorenz equations based on numerical analysis of (5.5).

An interesting observation is that chaotic motion is found to coexist with the stable equilibrium points for r values in the interval 24.06–24.74. In other words, each attractor has its own basin of attraction or region of initial conditions (see also our previous discussion in Chap. 3, Sect. 3.6.1). We illustrate the coexistence of three attractors for $r = 24.1$. Figure 5.13a shows the convergence of the trajectory to the equilibrium point $X_+ = (7.85, 7.85, 23.1)$. For illustrative purpose, only the evolution of the x-component is shown. For clarity the values of the x-component collected at every three units of time are plotted. The initial condition used is $(1.0, 7.85, 23.1)$. Figure 5.13b depicts the convergence of the trajectory to the other equilibrium point $X_- = (-7.85, -7.85, 23.1)$, where now the initial condition is $(-1, -7.85, 23.1)$. The chaotic attractor obtained for the initial condition $(25, 18, 120)$ is shown in Fig. 5.14. The orbit neither converges to the equilibrium point X_+ nor to the equilibrium point X_-, but it moves around each of the two equilibrium points forever. In general, a trajectory, starting from an arbitrary initial condition, ends up in one of these three attractors. As noted earlier, the stable equilibrium points become unstable for $r > 24.74$.

(4) For $r > 24.74$, the system exhibits a very complicated dynamics including chaos, reverse period doubling bifurcations, periodic windows, etc.

(5) One can also check the presence of reverse period doubling phenomenon in the Lorenz equations. Let us study the dynamics by decreasing the value of r from 101. Figure 5.15a shows the attractor for $r = 100.795$.

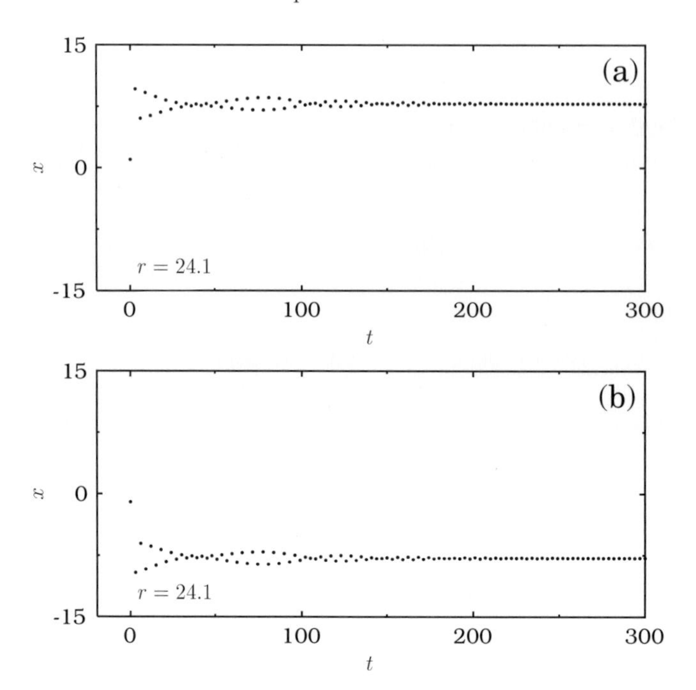

Fig. 5.13. Convergence of the trajectories of the Lorenz equations (5.5) to the equilibrium points (**a**) X_+ for the initial condition $(1.0, 7.85.23.1)$ and (**b**) X_- for the initial condition $(-1.0, -7.85, 23.1)$ for $r = 24.1$, $b = 8/3$ and $\sigma = 10$. For illustrative purpose, only the x-component, collected at every three units of time, is shown

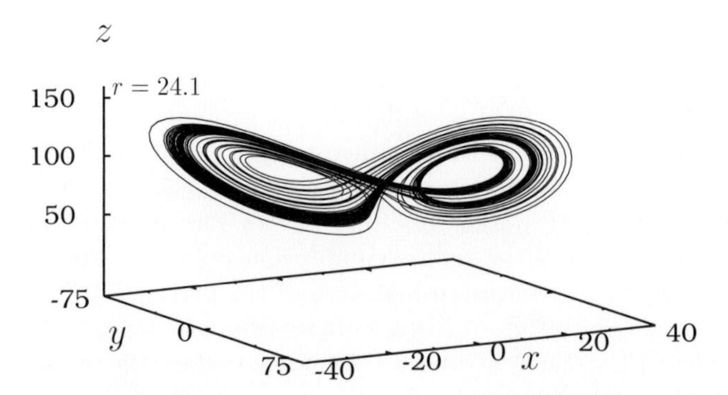

Fig. 5.14. Chaotic attractor of the Lorenz equations (5.5) for $r = 24.1$, $b = 8/3$ and $\sigma = 10$. The initial condition is $(x, y, z) = (25, 18, 120)$. This attractor coexists (that is, for the same parameter values of r, b and σ) with the two stable equilibrium points $X_\pm(x^*, y^*, z^*) = (\pm 7.85, \pm 7.85, 23.1)$, see Fig. 5.13

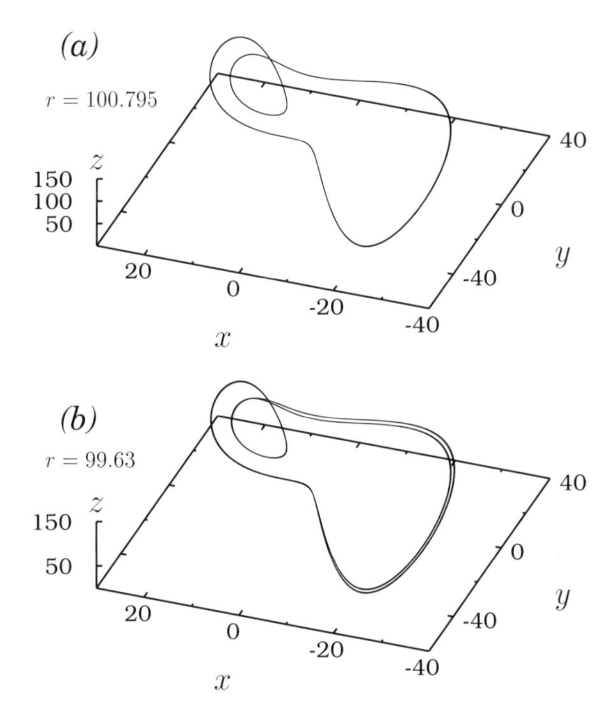

Fig. 5.15. Phase portraits of (**a**) period-T (for $r = 100.795$) and (**b**) period-$2T$ (for $r = 99.63$) attractors of the Lorenz equations (5.5), indicating reverse period doubling phenomenon

It is a periodic orbit and its period is $T \approx 1.1$. We call it period-T or period-1 limit cycle attractor. At $r = r_1 = 99.96$ this orbit becomes unstable and a period-$2T$ orbit is born. Figure 5.15b shows the period-$2T$ orbit for $r = 99.63$. As the parameter r is further decreased bifurcations into period-$4T$, $8T$,... orbits occur. This period doubling phenomenon (in the reverse direction of the control parameter) accumulates at $r_c \approx 99.524$. Thus the interval $99.524 < r < 100.795$ is a period doubling region in the reverse sense. (When the parameter r is increased from r_c we observe transition from chaos to period-T orbit through reverse period doubling phenomenon). For r values above and below this region complicated sequence of bifurcations occur which can be studied numerically using the phase portrait and identifying the nature of the attractors. For more quantitative characterization of attractors, see Chap. 8.

The bifurcation diagram of the Lorenz equations (5.5) can be drawn as follows. In the Duffing oscillator equation (5.1), we collected the values of the state variable x at every period of the driving force $f \sin \omega t$ for a set of values of f and plotted them in the f versus x coordinate system. We cannot use the same procedure for Lorenz equations (5.5), since they have no explicit periodic time dependent term. However, we can define a suitable *surface of section*, which will be described in detail in Chap. 7, and consider the points of intersections of the trajectory on it. For example, let us consider the $y-z$ plane with $x=0$ as the surface of section in the full (x,y,z) phase space. For a given value of the parameter r and the initial conditions $(x(0),y(0),z(0))$, we integrate the Lorenz equations (5.5) and collect the values of the state variables y and z whenever $x=0$. We leave the first 500 points as transients and collect the next 300 points. Then we plot the obtained values of y (or z) as a function of the parameter r. Figure 5.16 shows the bifurcation diagram in the interval $r \in (99.4, 100.1)$. This figure clearly illustrates the occurrence of period doubling bifurcations leading to chaos, when the control parameter r is decreased from the value 101.

Figure 5.17 shows the phase portrait of the chaotic orbit for $r=99$. For this value of r the system has three equilibrium points, $X_{\pm} = (x^*, y^*, z^*) = (\pm 16.166, \pm 16.166, 98)$, $X_0 = (0,0,0)$, and all of these are unstable. The trajectories make a number of spirals around X_+ and X_-. The two bands of the attractor appears as wings of a butterfly. In order to check whether the attractor is chaotic or not, we test the sensitive dependence of the orbit on the initial conditions. In fact, Lorenz was led to the discovery [7] of sensitivity with respect to small changes in initial conditions by paying careful attention to the results of an attempted numerical integration of his model equations (5.5). Lorenz equations (5.5) were integrated numerically with two sets of initial conditions, $X(0) = (x,y,z) = (10,1,100)$ and $X'(0) = (10,1,100.01)$. Defining the distance $S(t)$ between the two trajectories $X(t)$ and $X'(t)$ (see also Chap. 8 for more details) as

$$S(t) = \sqrt{(x(t)-x'(t))^2 + (y(t)-y'(t))^2 + (z(t)-z'(t))^2}\ , \qquad (5.7)$$

$S(t)$ can be plotted as in Fig. 5.18. It neither decays to zero nor diverges but varies irregularly with time. That is, the motion is highly sensitive to

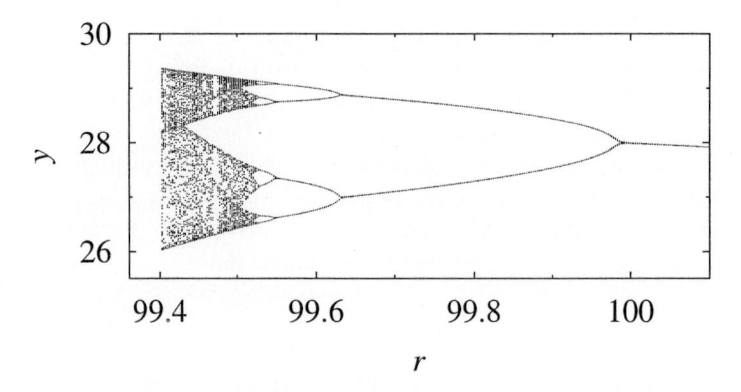

Fig. 5.16. Bifurcation diagram of the Lorenz equations (5.5)

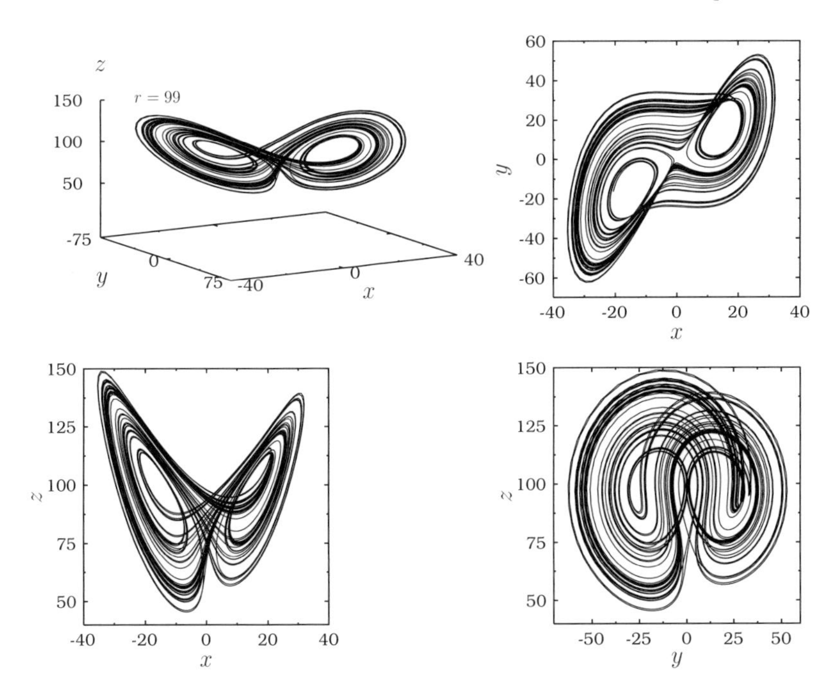

Fig. 5.17. Phase portraits of the chaotic attractor of the Lorenz equations (5.5) for $r = 99$, $b = 8/3$ and $\sigma = 10$, with the initial condition $(X(0) = (-20, -15, 113))$

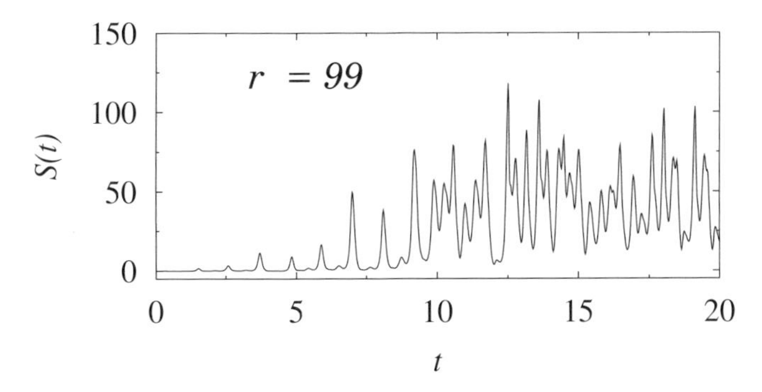

Fig. 5.18. $S(t)$ versus t for the two trajectories of the Lorenz equations (5.5) started with the nearby initial states $X(0) = (10, 1, 100)$ and $X'(0) = (10, 1, 100.01)$ for $r = 99$, $b = 8/3$ and $\sigma = 10$

infinitesimal perturbation or uncertainty in the initial conditions. This is exactly similar to the highly sensitive dependence of the dynamics on initial conditions during evolution in the case of the logistic map as discussed in Sect. 4.2.6. Thus the prediction of sufficiently distant future is impossible by any method unless the present conditions are known exactly and that no errors are introduced in any subsequent measurements.

5.3 Some Other Ubiquitous Chaotic Oscillators

In this section, we wish to give a brief sketch of the basic features associated with some of the other important nonlinear oscillators. Among these, the van der Pol equation and the damped, driven pendulum have played historically a paradigmic role in the development of the field of chaotic dynamics [2,8–10]. On the other hand, the Rössler system [11] has been one of the pioneering models studied to understand the nature of the chaotic attractor, while the Morse system [12] played a crucial role in understanding the interatomic potential and for fitting the vibrational spectra of diatomic molecules. We will merely summarize the basic features associated with the dynamics of these systems here and for more details the readers may refer to the relevant literature.

5.3.1 Driven van der Pol Oscillator

As a mathematical model to describe the oscillations in a vacuum tube circuit, van der Pol [8] introduced the second-order nonlinear differential equation with a linear restoring force and a nonlinear damping of the form (see Sect. 3.5 also)

$$\ddot{x} - \epsilon\left(1 - x^2\right)\dot{x} + x = 0\,. \tag{5.8}$$

It also describes self-excited oscillations in several problems in electronics, physics, biology, neurophysics and many other disciplines. Equation (5.8) admits limit cycle oscillations for small ϵ in the region $(x, \dot{x}) = (-2, 2)$, refer Sect. 3.5.

When driven by external periodic forcing, (5.8) exhibits interesting bifurcation structures. A detailed investigation of the driven van der Pol oscillator

$$\ddot{x} - \epsilon\left(1 - x^2\right)\dot{x} + x = f\cos\omega t \tag{5.9}$$

has brought out the existence of various mode-locking[3] responses and period doubling cascades [9]. The results (for the case $\epsilon = 5$) can be summarized as in Table 5.2 and Figs. 5.19–5.22.

5.3.2 Damped, Driven Pendulum

The pendulum that is damped as well as driven by an oscillatory torque has been widely investigated, because the corresponding equation of motion also models many other physical phenomena [10], example: the time dependence

[3] By mode-locking we mean that if a system exhibits a natural periodic stable limit cycle of frequency ω_1 and is then perturbed by an external periodic force of frequency ω_2, the resultant periodic oscillation of the combined system is the mode-locked oscillation.

Table 5.2. Dynamics of the driven van der Pol oscillator for some ranges of f and ω

Value of f	Region of ω	Type of orbits	Figures
1.0	$(0, 1.5)$	mode-locking and quasiperiodicity	5.19 and 5.20
2.5	$(0, 6)$	mode-lockings, large period oscillations and quasiperiodicity	5.21
5.0	$(2.424, 2.502)$	periodic windows, period doubling bifurcations and chaos	5.22

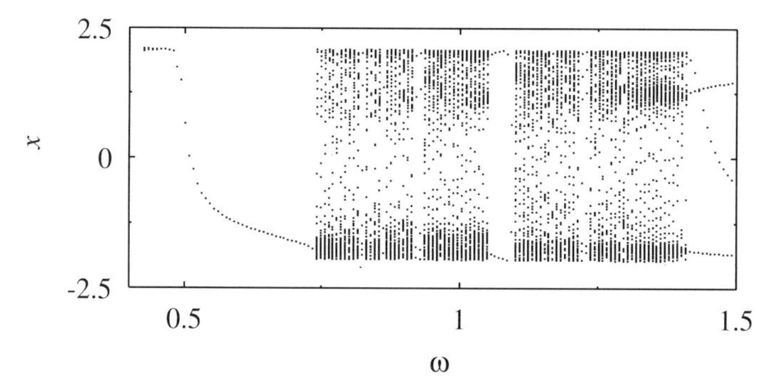

Fig. 5.19. Bifurcation diagram for $f = 1.0$ and $\epsilon = 5$ of the driven van der Pol oscillator equation (5.9) (Ref. [9]), showing the presence of mode-locking and quasiperiodic responses

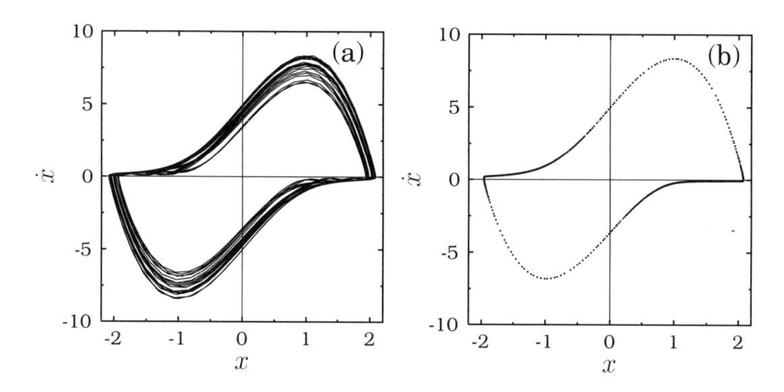

Fig. 5.20. (**a**) Phase portrait and (**b**) Poincaré map of an attractor of the driven van der Pol equation (5.9), lying on an invariant torus (quasiperiodic motion) in the phase space for $\epsilon = 5.0$, $f = 1.0$ and $\omega = 1$ (Ref. [9])

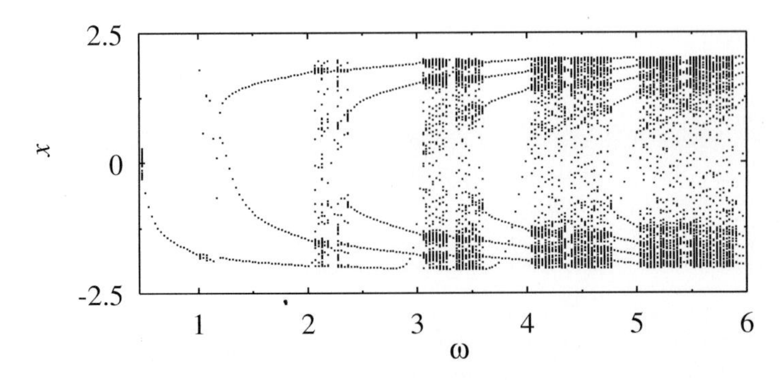

Fig. 5.21. Bifurcation diagram for $f = 2.5$ of the driven van der Pol equation (5.9), showing periodic windows and quasiperiodic orbits

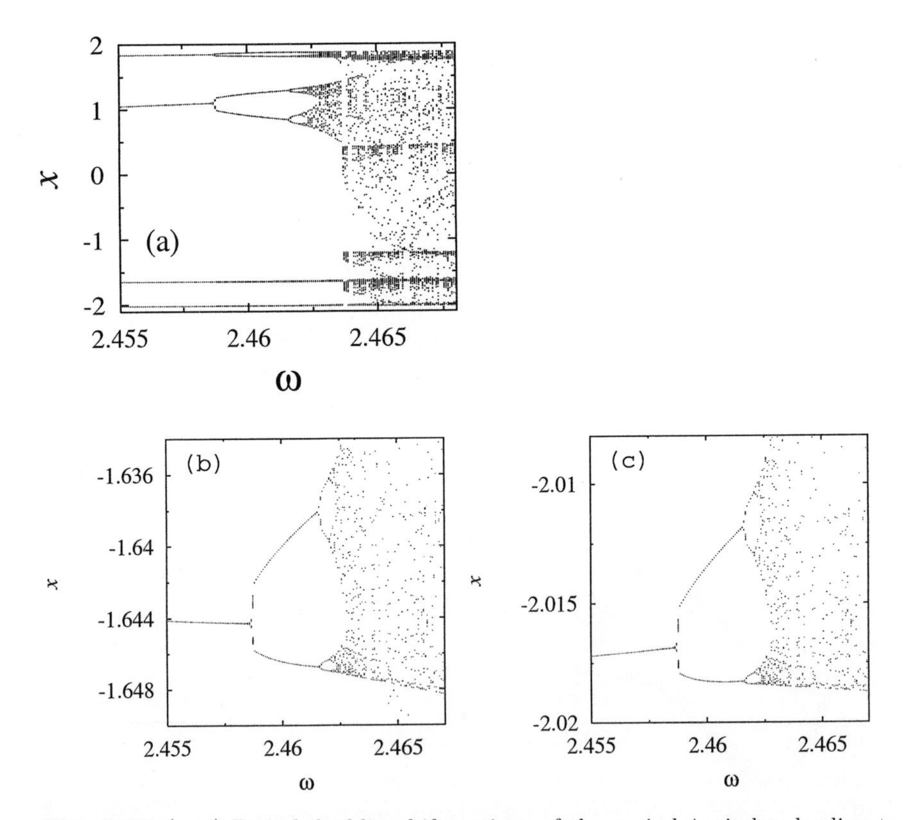

Fig. 5.22. (**a–c**) Period doubling bifurcations of the period-4 window leading to chaos in the driven van der Pol oscillator (equation (5.9))

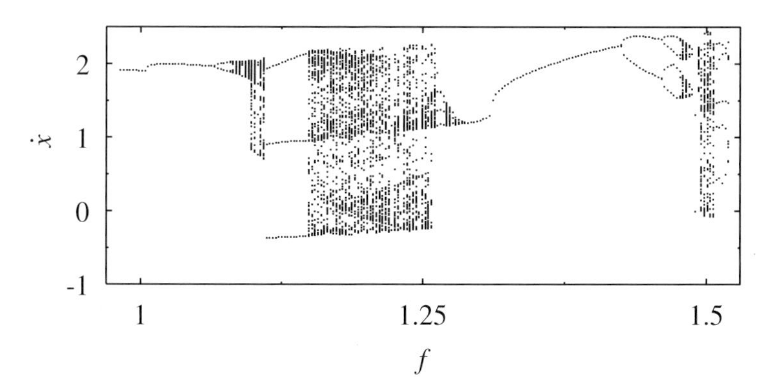

Fig. 5.23. Bifurcation diagram of the damped driven pendulum (equation (5.10)) in the $\dot{x} - f$ plane for $\alpha = 0.5$, $\omega = 2/3$ [10]

of the quantum phase difference of a Josephson junction. Further, this model has been simulated with an analog circuit to explore extensively the dynamic behaviour. The differential equation for the damped, driven pendulum is given by

$$\ddot{x} + \alpha\dot{x} + \sin x = f \cos \omega t \ , \tag{5.10}$$

where α, f and ω measure the strength of damping, forcing amplitude and forcing frequency, respectively. If $\alpha > 0$ and $f = 0$, due to damping, almost all the orbits approach the origin. If f is included, and varied, then the system admits the familiar period doubling bifurcation sequences to chaos. A typical bifurcation diagram in the $(\dot{x} - f)$ space for $\alpha = 0.5$, $\omega = 2/3$ is shown in Fig. 5.23.

5.3.3 Morse Oscillator

The equation of motion of the damped and driven Morse oscillator [12] is represented in terms of dimensionless variables as

$$\ddot{x} + \alpha\dot{x} + \beta \mathrm{e}^{-x}\left(1 - \mathrm{e}^{-x}\right) = f \cos \omega t \ . \tag{5.11}$$

This model has been widely used and we indicate below some of the contexts:

(1) as a model for infrared multiphoton excitation,

(2) for the problem of laser isotope separation,

(3) for explanation of the anomalous gains observed in the stimulated Raman emission and

(4) for dissociation of van der Waals complexes.

A typical study of the system (5.11) by fixing the parameters as $\alpha = 0.8$, $\beta = 8$ and $f = 3.5$ exhibits chaotic phenomenon through Feigenbaum's period

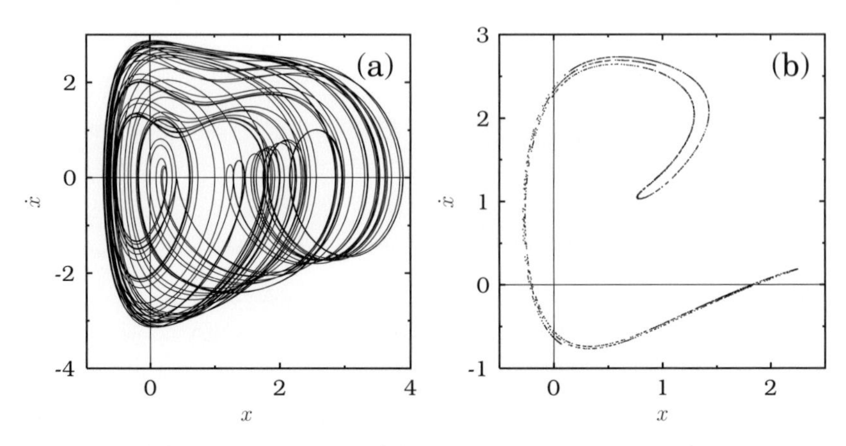

Fig. 5.24. (**a**) Phase portrait and (**b**) Poincaré map of the Morse oscillator equation (5.11) for $\alpha = 0.8$, $\beta = 8$, $f = 3.5$ and $\omega = 2$

doubling route when the angular frequency ω is varied. Figure 5.24 shows the phase portrait and Poincaré map of the chaotic attractor for $\omega = 2$.

5.3.4 Rössler Equations

A simple model motivated by the dynamics of chemical reactions in a stirred tank is the following set of equations proposed by Rössler [11]:

$$\dot{x} = -(y + z) , \tag{5.12a}$$

$$\dot{y} = x + ay , \tag{5.12b}$$

$$\dot{z} = b + z(x - c) . \tag{5.12c}$$

If $x < c$, $z(t)$ settles down near to the value $b/(c-x)$ (and rapidly, if $(c-x)$ is large), whereas if $x > c$, $z(t)$ increases exponentially. The system often studied is the case $a = b = 1/5$. Period-1, 2 and 4 motions can be found to occur for $c = 2.6$, 3.5 and 4.1 respectively. Chaotic motions can be found for $c > 4.23$. Figure 5.25 shows one such Rössler chaotic attractor for $c = 5.7$.

Exercises:

8. Analyse the various bifurcations and routes to chaos in the driven van der Pol oscillator.
9. Investigate the various bifurcations with reference to Fig. 5.23.
10. Deduce the bifurcation diagram for the Morse oscillator (5.11) as a function of ω and discuss the dynamics.
11. Draw the bifurcation diagram for the Rössler equations (5.12) as a function of the parameter c and analyse the underlying bifurcations and transition to chaos.

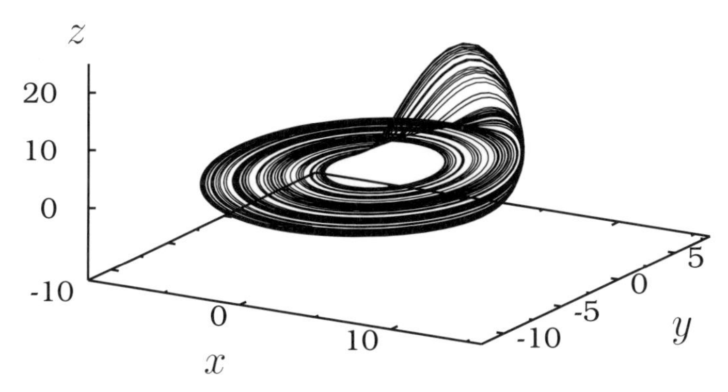

Fig. 5.25. Phase portrait in the $x-y-z$ plane of the Rössler equations (5.12) for $a=b=1/5$ and $c=5.7$

12. Investigate the highly sensitive dependence of the chaotic attractors of the various above oscillators on initial conditions.

5.4 Necessary Conditions for Occurrence of Chaos

After studying the existence of chaos in discrete dynamical systems such as the logistic and Hénon maps and in continuous time dynamical systems like the Duffing oscillator and Lorenz equations, one may ask, what are the necessary and sufficient conditions for the occurrence of chaos in dynamical systems? In this section, we first point out the types of attractor that may occur in the first, second and higher order autonomous continuous dynamical systems and establish the necessary conditions for chaos to occur. Then we give the conditions for chaos to occur in maps. However the question of sufficiency has not been answered at the present time satisfactorily in either of the cases.

5.4.1 Continuous Time Dynamical Systems (Differential Equations)

A linear dynamical system with constant coefficients is exactly solvable and the bounded motion must be regular. Similar conclusion can also drawn for nonautonomous systems (refer to Chap. 2). Thus, linear systems preclude chaotic motion.

Next, let us consider the situation in nonlinear systems. Suppose the system is of first-order (autonomous differential equation) only, then the motion will take place on a one-dimensional phase space, a phase line. The noncrossing of trajectories severely restricts the type of motion on the phase line. In this case trajectories starting from various initial conditions at time $t = t_0$ either approach an equilibrium point or go to $\pm\infty$ as $t \to \infty$. Thus first-order nonlinear systems preclude chaotic motion.

If the system is of second-order, then in addition to point attractors, it may also have a limit cycle attractor as discussed in Chap. 3. Further, according to the Poincaré-Bendixson theorem (see Sect. 3.5.1), a bounded trajectory in the phase space can either approach an equilibrium point or a limit cycle as $t \to \infty$. Thus second-order nonlinear autonomous systems preclude chaotic dynamics.

On the other hand, second-order nonlinear nonautonomous systems or higher order ($n > 2$) nonlinear autonomous systems can have, besides point attractors and limit cycle attractors, quasiperiodic, strange nonchaotic and chaotic attractors also. For example, we have seen in detail that the Duffing oscillator equation (5.1) and the Lorenz equations (5.5) exhibit chaotic motion. Thus, the necessary conditions for a system of the form (3.1) to exhibit chaotic motion are that

(1) it must have a nonlinear element and

(2) the order or dimension of it must be ≥ 3.

We note that these are only the necessary conditions for chaos to occur in a dynamical system. Sufficient conditions on the form of (3.1) for the occurrence of chaos, as mentioned above, are not known at present. For example, for a given three-dimensional nonlinear system can one identify the parametric range for which regular motion exists and the range for which chaotic motion exists? Some progress in this direction has been made by looking for the singularity structure of the solution in the complex time plane. For details see Chap. 10.

5.4.2 Discrete Time Systems (Maps)

Now we consider the required conditions for chaos in maps. Linear maps, like linear continuous time systems, cannot show chaotic behaviour.

In nonlinear maps one must distinguish between *invertible* and *noninvertible* maps. A map M is said to be invertible if, given X_n, we can obtain $X_{n+1} = F(X_n)$ uniquely and so also, given X_{n+1}, we can evaluate

$$X_n = F^{-1}(X_{n+1}) \tag{5.13}$$

uniquely, where F^{-1} is the inverse of F.

A. *Invertible Maps*

Let us first consider one-dimensional invertible maps

$$x_{n+1} = f(x_n) \ . \tag{5.14}$$

If a map is invertible then $f(x)$ must be a monotonically increasing or decreasing function of x. If it decreases for a while and then increases for a while and so on, then as shown in Fig. 5.26, for some values of $f(x)$, there exists more than one value of x and hence the map is noninvertible. Now let us find the possible attractors in one-dimensional invertible maps. Obviously, period-1 fixed point (stable or unstable) can occur in an invertible map.

Let us investigate further the possibility of a period-2 orbit (again stable or unstable). Suppose $f(x)$ is a monotonically increasing function of x. Let (x_1^*, x_2^*) be a period-2 solution. For simplicity, we drop the superscript '*' in the following discussion. We arrange the members of the orbit in the order $x_1 < x_2$. Since $f(x)$ is monotonically increasing function of x we must have

$$f(x_1) < f(x_2) \tag{5.15}$$

because $x_1 < x_2$. Since $f(x_1) = x_2$ and $f(x_2) = x_1$ for a period-2 orbit, the relation (5.15) becomes

$$x_2 < x_1 \quad \text{or} \quad x_1 > x_2 \,, \tag{5.16}$$

which is contradictory to our assumption $x_1 < x_2$. Therefore, a period-2 solution is not possible. For a period-$n(> 2)$ orbit ϕ, consisting of n real numbers $(x_1, x_2, \ldots, x_n)$, they can be ordered according to their magnitudes as

$$x_{k_1} < x_{k_2} < x_{k_3} < \ldots < x_{k_n} \,, \tag{5.17}$$

where $k_l \in (1, 2, \ldots, n)$. If f is an increasing function then

$$\begin{aligned} f(x_{k_1}) &< f(x_{k_2}) < f(x_{k_3}) < \ldots < f(x_{k_n}) \\ &= x_{k_1+1} < x_{k_2+1} < x_{k_3+1} < \ldots < x_{k_n+1} \,. \end{aligned} \tag{5.18}$$

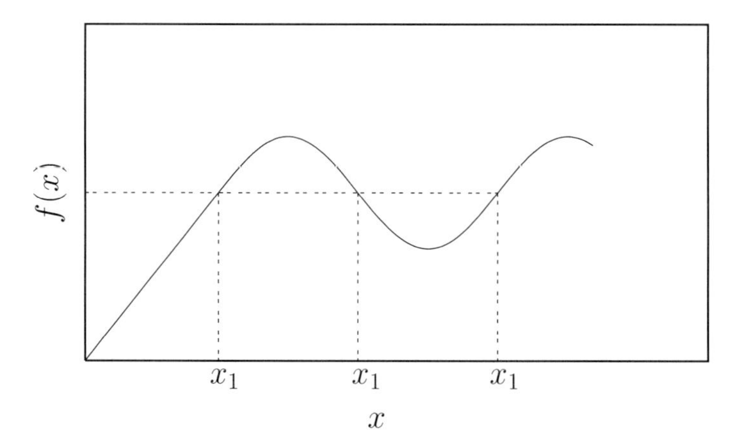

Fig. 5.26. Graph of a map function $f(x)$ whose inverse has multivaluedness in x. The map is obviously noninvertible

The index $k_l + 1 = 1$ if $k_l = n$ otherwise $k_l + 1 = k_{l+1}$. Suppose (5.17) is of the form

$$x_2 < x_1 < x_3 < \ldots < x_n \, , \tag{5.19}$$

then the relation (5.18) takes the form

$$\begin{aligned} f(x_2) &< f(x_1) < f(x_3) < \ldots < f(x_n) \\ &= x_3 < x_2 < x_4 < \ldots < x_1 \, , \end{aligned} \tag{5.20}$$

which is different from (5.19). Similarly one can easily check that for any form of (5.17), the relation (5.18) is not satisfied.

Next, let us assume that $\phi = (x_1, x_2, \ldots)$ is a nonperiodic orbit, obtained from an arbitrary initial condition x_1, which does not converge to a period-1 orbit (which is the only possible periodic orbit as shown above) in the limit $n \to \infty$. Then we have

$$f(x_1) < f(x_2) < f(x_3) < \ldots = x_2 < x_3 < x_4 < \ldots \, . \tag{5.21}$$

That is, the magnitude of the successive iterations increase and consequently diverge to ∞. Remember that chaotic orbit is a bounded solution. Therefore nonperiodic orbit ϕ existing within a finite interval of x is also not possible. Thus, period-1 equilibrium point is the only possible orbit in invertible and monotonically increasing maps. This precludes the possibility of chaos.

Exercise:

13. Verify that for the invertible and monotonically increasing map $x_{n+1} = \lambda \exp(x_n)$, $\lambda > 0$ the only possible attractor is period-1 equilibrium point.

Next, we consider one-dimensional invertible maps with monotonically decreasing function of x. In this case for a period-2 equilibrium point, with the arrangement $x_1 < x_2$, we must have

$$f(x_1) > f(x_2) \, . \tag{5.22}$$

Replacing $f(x_1)$ by x_2 and $f(x_2)$ by x_1, we obtain

$$x_2 > x_1 \quad \text{or} \quad x_1 < x_2 \, . \tag{5.23}$$

Therefore period-2 solution is possible. For a period-$n(\geq 3)$ orbit, the (5.18) becomes

$$\begin{aligned} f(x_{k_1}) &> f(x_{k_2}) > f(x_{k_3}) > \ldots > f(x_{k_n}) \\ &= x_{k_1+1} > x_{k_2+1} > x_{k_3+1} > \ldots > x_{k_n+1} \, . \end{aligned} \tag{5.24}$$

This is not possible for any n greater than 2. Let us verify this for $n = 3$. Now, suppose $x_1 < x_2 < x_3$ then the relation (5.24) is

$$f(x_1) > f(x_2) > f(x_3) = x_2 > x_3 > x_1 \, , \tag{5.25}$$

which is not true. Similarly, for any arrangement of the members of ϕ for $n = 3$ and as well as for $n > 3$, the relation (5.24) is not satisfied. Therefore, period-1 and 2 solutions are alone possible. Further, if $\phi = (x_1, x_2, \ldots)$ is nonperiodic, then we must have

$$f(x_1) > f(x_2) > f(x_3) > \ldots = x_2 > x_3 > x_4 > \ldots . \tag{5.26}$$

which implies that ϕ cannot be a chaotic orbit. Thus, chaos cannot occur in one-dimensional invertible maps with monotonically decreasing function of x also.

Exercise:

14. Show that attractors other than period-1 and period-2 cannot occur in the invertible map $x_{n+1} = \lambda \exp(-x_n)$, $\lambda > 0$, based on the above treatment. Verify it numerically.

Next, we consider two-dimensional maps. The Hénon map (4.45) is invertible, since we can also rewrite it as

$$x_n = y_{n+1}/b , \tag{5.27a}$$

$$y_n = x_{n+1} - 1 + \frac{a}{b^2} y_{n+1}^2 . \tag{5.27b}$$

As we have seen in Sect. 4.4, the invertible two-dimensional Hénon map is a chaotic dynamical system, exhibiting various types of bifurcations and chaos. Therefore, we can conclude that invertible two-dimensional maps can exhibit chaos.

B. *Noninvertible Maps*

What is the situation in noninvertible one-dimensional maps? Let us consider the logistic map (4.19). This map is not invertible, because for a given x_{n+1}, there are two values of x_n from which it could have come. In the previous chapter, we have already seen the occurrence of chaos in the logistic map. Thus chaos may occur in one-dimensional noninvertible maps.

Chaos is found in certain two-dimensional noninvertible maps also. For example the dissipative circle map (4.51) exhibits chaotic motion for $\Omega = 0.612, b = 0.5$ and $K = 1.2$ as shown in Sect. 4.4.1.

Thus, we can conclude that if the nonlinear map is invertible then there can be no chaos unless the dimension is greater than one. If the nonlinear map is noninvertible, chaos is possible even in one-dimension.

5.5 Computational Chaos, Shadowing and All That

In this and earlier chapters we have shown the occurrence of regular and chaotic dynamics in certain prototypical nonlinear systems. Regular motion is not highly sensitive to initial conditions whereas chaotic motion is highly sensitive even for small perturbations. The physical consequence to real systems is the *butterfly effect* mentioned in Sect. 4.2.6. What is the consequence in numerical analysis? Since round-off errors are unavoidable in any computer calculation and every numerical integration schemes of ordinary differential equations have errors in their own algorithm one would naturally ask: (i) Is computer generated chaos an artifact of finite-precision arithmetic calculation? (ii) Does the computer generated trajectory in the case of chaotic evolution truthfully depict the actual system trajectory? (iii) If the computer generated trajectory is not the true trajectory of a chaotic system then why should one carry out numerical study at all and how acceptable are the conclusions based on such studies?

It is true that because of highly sensitive dependence on infinitesimal deviations in the initial conditions the numerical trajectory $\boldsymbol{X}(t)$ obtained with the initial condition $\boldsymbol{X}(0) = \boldsymbol{X}_1(0)$ may not be a true trajectory corresponding to the specified initial condition $\boldsymbol{X}_1(0)$. However, it is conceivable that the presently generated trajectory is the true trajectory of another physically realizable initial condition $\boldsymbol{X}_1'(0)$.

In order to understand the effect of computer calculations on chaotic systems, for example, Dalling and Goggin [13] iterated the logistic map (4.19) for $a = 3.99$ (chaotic region) and $a = 3.1$ (regular region) with various precisions upto 408 decimal digits. The iterates of various precisions were compared with each other to determine the number of iterations required before the Mth digit of one answer disagreed by one unit with the Mth digit of a more precise answer. The rate at which initially close orbits in the chaotic case separate is found to be independent of the precision used in the calculation. For a nonchaotic $a = 3.1$, the answers agreed to the limits of the precision with which they were calculated. Moreover, the error introduced by rounding grows as the square root of the number of arithmetic operations [14]. In contrast, in chaotic systems errors grow exponentially fast whatever be the precision. Essentially, rounding is the source of the difference between calculations of different precision. Because of its slow growth rate round-off error is not the source of high sensitivity of a chaotic system.

For low-dimensional chaotic systems computational techniques [15,16] have been proposed to obtain, and verify the existence of, a true trajectory of the system that closely follows the computer generated (pseudo)trajectory. The process of obtaining such a true trajectory is called *shadowing* of trajectory. In a shadowing process a computer generated trajectory is continuously deformed into a true trajectory in such a way that the errors at each trajectory points are decreased monotonically to zero. So, one may conclude that in general computer generated trajectory can be treated as a trajectory close to the true trajectory. For further details, see Refs. [15,16].

5.6 Conclusions

In the previous and present chapters we discussed some simple and complex bifurcations in certain typical nonlinear dynamical systems. The various bifurcation diagrams clearly show that these systems exhibit qualitative (and quantitative) changes at critical values of the control parameters as they are varied smoothly. These correspond to the different bifurcations exhibited by the systems. The question is how to understand the existence of such bifurcations and mechanisms by which they arise. One possible approach is to look for the local stability properties of solutions in the neighbourhood of the critical parameter values at which bifurcations occur. However, very few nonlinear dynamical systems have exact analytical solutions. Since, generally, most of the nonlinear dynamical systems have no exact solutions, one can prepare a dictionary of bifurcations that occur in simple and interesting low dimensional nonlinear systems, and then use them as reference systems for the identification of the bifurcations from the numerical analysis of the various nonlinear systems. Based on our discussions in the present and previous chapters, we summarize in Tables 5.3 and 5.4 the salient features of some important simple and complex bifurcations, respectively, for appropriate ranges of parameter values.

Finally in the following set of problems, we introduce a number of physically interesting nonlinear systems which exhibit chaos. More details about them can be obtained from the indicated references.

Problems

1. The quasiperiodically driven Ueda's equation is given by
$$\ddot{x} - \mu\left(1 - x^2\right)\dot{x} + x^3 = f\cos\omega_1 t + f\cos\omega_2 t\ .$$
For $\mu = 0.2$, $\omega_1 = 1$ and $\omega_2 = (\sqrt{5} - 1)/2$, when the forcing strength is decreased from 5 to 2, the equation exhibits two-frequency quasiperiodicity $\rightarrow$ strange nonchaos $\rightarrow$ chaos $\rightarrow$ strange nonchaos $\rightarrow$ chaos transition [17]. Study this transition dynamics using bifurcation diagram and Poincaré map.
2. Consider the problem of motion of a particle of mass m sliding freely on a wire described by the parabola $z = \sqrt{\lambda/2}\ x^2$, which rotates with a constant angular velocity $[\Omega = (-\omega_0^2 + g\sqrt{\lambda}\ )^{1/2}]$ about its z-axis. A physical realization of this model is that of a man riding a motor bike on a rotating parabolic well in a circus. This model is represented by the damped and driven velocity dependent nonlinear oscillator equation
$$\left(1 + \lambda x^2\right)\ddot{x} + \lambda x\dot{x}^2 + \omega_0^2 x + \alpha\dot{x} = f\cos\omega t\ .$$
Fix the values of the parameters as $\omega_0^2 = 0.25$, $\lambda = 0.5$, $\alpha = 0.1$, $\omega = 1$. Study the bifurcation phenomenon as a function of f in the interval $[0, 4]$ [18].

Table 5.3. Some simple bifurcations

Type of bifurcation	Model equations	Illustration
Flows		
1. Saddle-node	$\dot{x} = \mu - x^2$	Fig. 4.2
	$\dot{y} = -y$	
2. Pitchfork		
(a) Supercritical	$\dot{x} = \mu x - x^3$	Fig. 4.3a
	$\dot{y} = -y$	
(b) Subcritical	$\dot{x} = \mu x + x^3$	Fig. 4.3b
	$\dot{y} = -y$	
3. Transcritical	$\dot{x} = -\mu x + x^2$	Fig. 4.6
	$\dot{y} = -y$	
4. Hopf		
(a) Supercritical	$\ddot{x} - b(1 - x^2)\dot{x} + x = 0$	Fig. 4.7a
(b) Subcritical	$\ddot{x} + b(1 - x^2)\dot{x} + x = 0$	Fig. 4.7b
Maps		
1. Saddle-node	$x_{n+1} = a - x_n^2$	----
2. Pitchfork		
(a) Supercritical	$x_{n+1} = -x_n^3 - ax_n$	----
(b) Subcritical	$x_{n+1} = x_n^3 - ax_n$	----
3. Transcritical	$x_{n+1} = ax_n(1 - x_n)$	----
4. Hopf		
Supercritical	$x_{n+1} = y_n$	----
	$y_{n+1} = ay_n(1 - x_n)$	

3. Consider the Duffing–van der Pol oscillator

$$\ddot{x} - p\left(1 - x^2\right)\dot{x} + \omega_0^2 x + \beta x^3 = f \sin \omega t \,.$$

Choose the parameters as $p = f = 5$, $\omega_0^2 = 1$ and $\omega = 2.466$. When β is varied from 0 to 0.02, a variety of attractors such as symmetrical periodic, asymmetrical periodic, symmetrical chaotic and asymmetrical chaotic occurs [19]. Identify their regions of β using phase portrait.

4. For the Duffing oscillator equation

$$\ddot{x} + \dot{x} - 10x + 100x^3 = f \sin 3.5t \,,$$

verify that for f values just below $f_c \approx 0.849$ two chaotic attractors coexist one confined to the well in $x < 0$ and one confined to the well in $x > 0$ and intermittent switching between them occurs for f values just above f_c. Define the quantity τ as the average time between successive

Table 5.4. Standard complex bifurcations

Type of bifurcation	Model equations	Illustration
Maps		
1. Period doubling route to chaos	$x_{n+1} = ax_n(1 - x_n)$	Fig. 4.14
2. Quasiperiodic route to chaos	$x_{n+1} = x_n + \Omega - (K/2\pi)\sin 2\pi x_n + by_n, \pmod 1$ $y_{n+1} = by_n - (K/2\pi)\sin 2\pi x_n$	Fig. 4.20
3. Intermittency route to chaos	$x_{n+1} = ax_n(1 - x_n)$	Fig. 4.24
Flows		
1. Period doubling route to chaos	$\ddot{x} + \alpha\dot{x} + \omega_0^2 x + \beta x^3 = f\sin\omega t$	Fig. 5.5
2. Quasiperiodic route to chaos	$\ddot{x} - p(1 - x^2)\dot{x} + \omega_0^2 x + \beta x^3 = f\cos\omega t$	Figs. 5.11 and 5.12
3. Intermittency route to chaos	$\ddot{x} + \alpha\dot{x} + \omega_0^2 x + \beta x^3 = f\sin\omega t$	Fig. 5.8

switching. Compute τ for f values in the interval $[0.849, 0.850]$ and show that τ scales as $\sim (f - f_c)^{-0.70}$ [20].

5. The nerve impulses in a neuronal membrane is often described by the Bonhoeffer–van der Pol oscillator equation

$$\dot{x} = x - x^3/3 - y + A_1\cos\omega t\ ,$$
$$\dot{y} = c(x + a - by)\ ,$$

where x and y represent voltage across the neural membrane and quantity of refractoriness, respectively. Choose the values of the parameters as $a = 0.7$, $b = 0.8$, $c = 0.1$ and $\omega = 1$. The system shows period doubling phenomenon for $f \in [0, 0.7182]$, onset of chaos at $f = 0.7182$, type-I intermittency dynamics for $f \in [0.78, 0.78144]$, $[1.092, 1.094]$ and $[1.162, 1.164]$, band merging crisis at $f = 1.28653$, sudden widening of a chaotic attractor at $f = 0.747486$ and reverse period doubling phenomenon for $f \in (1.29, 2)$ [21–23]. Numerically verify the above dynamics.

6. The Brusselator model equation describing hypothetical three-molecular chemical reaction with autocatalytic step under far-from-equilibrium conditions is

$$\dot{x} = A + x^2 y - Bx - x + f\cos\omega t\ ,$$
$$\dot{y} = Bx - x^2 y\ .$$

For $A = 0.4$ and $B = 1.2$, investigate the occurrence of periodic, quasiperiodic and chaotic motions in $f - \omega$ parameter space with $f \in [0, 0.4]$ and $\omega \in [0, 4]$ [24,25].

7. The dynamical equation for the round-trip phase shifts induced by the laser-induced molecular reorientation and laser heating, respectively, in a Fabry–Perot interferometer system is governed by the equation

$$\dot{x} = \frac{1}{\tau_1}\left[-x + \frac{A(1 + f\sin\omega t)}{1 + F\sin^2\{(P + x + y)/2\}}\right],$$

$$\dot{y} = \frac{1}{\tau_2}\left[-y + \frac{A(1 + f\sin\omega t)}{1 + F\sin^2\{(P + x + y)/2\}}\right].$$

For $\tau_1 = 125$, $\tau_2 = 1$, $A = 6.37$, $B = -0.98$, $P = 4$, $F = 30$, the system exhibits a rich variety of dynamics in the $(f - \omega)$ parameter space. Numerically verify the occurrence of

a) period doubling route to chaos of a period-$9T (T = 2\pi/\omega)$ attractor for $f = 0.2$ and $\omega \in [0.6544, 0.6568]$,
b) quasiperiodic route to chaos for $f = 0.068$, $\omega \in [0.2673, 0.2774]$,
c) intermittency route to chaos for $\omega = 0.646$, $f \in [0.159, 0.161]$ and
d) draw the bifurcation diagram for $f = 0.2$, $\omega \in [0.3, 1.5]$, $[0.58, 0.68]$ and analyse the observed bifurcation phenomena [26,27].

8. The Helmholtz–Duffing oscillator equation, containing both quadratic and cubic nonlinearities, is given by

$$\ddot{x} + \mu\dot{x} + x + c_2x^2 + c_3x^3 = P\cos\Omega t .$$

Fix the parameters at $c_2 = 35.952$, $c_3 = 534.43$, $P = 0.4$ and $\mu = 0.1$. Investigate the bifurcation phenomena in the interval $\Omega \in [0, 5]$ [28].

9. Consider the following anharmonic oscillator equation

$$\dddot{x} + \ddot{x} + s\dot{x} - rx + x^2 = 0 .$$

Show that the system has two equilibrium points of which $(x^*, \dot{x}^*, \ddot{x}^*) = (0, 0, 0)$ is always unstable, while $(r, 0, 0)$ undergoes a Hopf bifurcation at $r = s$. Verify that finite period doubling bifurcation occurs when r is increased from 0.75 to 0.95 for $s = 0.5$ and 0.8 and r from 2.3 to 2.8 for $s = 1.5$. Also verify the occurrence of period doubling route to chaos by varying r from 3 to 3.8 for $s = 2$ [29].

10. The governing rate equations of a three-dimensional autocatalator is

$$\dot{x} = \left(\frac{1}{1+k} + \epsilon\mu\right)(k + z) - x - xy^2 ,$$

$$\epsilon\dot{y} = x + xy^2 - y ,$$

$$\dot{z} = y - z .$$

Investigate the dynamics of the above system for $k = 2.5$, $\epsilon = 0.013$ and varying the value of μ from 0 [30]. Also carry out a fixed point analysis.

11. The dynamics of an NMR laser with modulated parameters is given by

$$\begin{aligned}\dot{x} &= \sigma y - \sigma x/(1 + A\cos\omega t)\ ,\\ \dot{y} &= -y(1 + ay) + rx - xz\ ,\\ \dot{z} &= -bz + xy\ .\end{aligned}$$

For $\sigma = 4.875$, $a = 0.2621$, $r = 1.807$, $b = 0.0002$ investigate the dynamics in the (A, ω) parameter space [31].

12. Consider the Lorenz type equations

$$\begin{aligned}\dot{x} &= -0.4x + y + 10yz\ ,\\ \dot{y} &= -x - 0.4y + 5xz\ ,\\ \dot{z} &= \alpha z - 5xy\ .\end{aligned}$$

Show that
  a) the system is dissipative for $\alpha < 0.8$, conservative for $\alpha = 0.8$ and volume expanding for $\alpha > 0.8$.
  b) It has four equilibrium points for $0 < \alpha < 0.8$ and all are saddles.
  c) It has two limit cycle attractors for $\alpha = 0.205$.

Also study the period doubling bifurcations of the coexisting limit cycle attractors by varying α from 0.205 [32].

13. Verify that the FitzHugh–Nagumo equations

$$\begin{aligned}\dot{V} &= W\ ,\\ \dot{W} &= -V + V^3/3 + R - uW\ ,\\ \dot{R} &= -(c/u)(V + a - bR)\end{aligned}$$

undergoes period doubling bifurcation leading to chaos, when the parameter u is decreased from 1.08 for $a = 0.6$, $b = 0.5$, $c = 0.1$. Draw the bifurcation diagram using the values of V intersecting the plane $W = 0$ [33].

14. A high frequency oscillator, namely, a typical nuclear spin generator which generates and controls the oscillations of the motion of a nuclear magnetization vector in a magnetic field is described by the coupled nonlinear differential equations

$$\begin{aligned}\dot{x} &= -\beta x + y\ ,\\ \dot{y} &= -x - \beta y(1 - kz)\ ,\\ \dot{z} &= \beta\alpha(1 - z) - \beta k y^2\ ,\end{aligned}$$

where x, y, z are the components of the nuclear magnetization vector and β, α, $k > 0$.
  a) Show that the system has three real equilibrium points for $k > (1 + \beta^2)/\beta^2$ and one for $k < (1 + \beta^2)/\beta^2$. Investigate the stability of the equilibrium points.
  b) Obtain the condition for Hopf bifurcation.

c) For $\alpha = 0.15$, $\beta = 0.75$ study the occurrence of various routes to chaos by varying the parameter k [34].

15. A three-dimensional system possessing five coexisting attractors is

$$\dot{x} = \alpha(y - x) - \alpha c_1(x + d_1)(c_2(x + d_2)^2 - 1) ,$$
$$\dot{y} = x - y + z ,$$
$$\dot{z} = -\beta y .$$

Fix the parameters at $\alpha = 7$, $\beta = 10.5$, $c_1 = 1.39$, $c_2 = 0.1$, $d_1 = 0.02$ and $d_2 = 0.115$. Numerically verify that it has two stable equilibrium points given by $(x^*, y^*, z^*) = (1.5866, 0, -1.5866)$, $(-1.765, 0, 1.765)$, two strange attractors enclosing each of the above equilibrium points and a stable limit cycle enclosing all the above four attractors [35]. Draw the phase portrait of all the attractors in (x, y, z) and (x, y) phase planes.

16. A semiclassical model for homogeneously broadened two-mode laser system is given by

$$\dot{I} = -2I - 4\delta\rho + rRnI ,$$
$$\dot{r} = -2r - 2\delta\rho + RnI ,$$
$$\dot{\rho} = -2\rho + 2\delta r + 2Rn\rho ,$$
$$\dot{n} = -G(n - 1 + nI) ,$$

where I is the dynamical variable proportional to the total field intensity, r and ρ are proportional to real and imaginary components of the interference term for two modes respectively and n is the normalised mean population inversion. The parameter R is proportional to pumping, δ is the intermode frequency spacing of a hollow cavity and G is proportional to inversion relaxation.

a) Obtain the three real equilibrium points of the system and study their stability dynamics.

b) Verify that for $G = 0.3$, $R = 1.2$ the system exhibits stable stationary solution for $\delta < 0.759$, chaos for $0.759 < \delta < 0.806$, coexistence of chaos and steady state solution for $0.806 < \delta < 0.821$, irregular intensity pulsations for $0.821 < \delta < 0.829$, onset of intermittency at 0.829 and a stable limit cycle for $0.829 < \delta < 1$ [36].

6. Chaos in Nonlinear Electronic Circuits

Even though numerical analysis can help to bring out a detailed picture of the dynamics of a nonlinear system such as the Duffing oscillator (equation (5.1)), it requires much computer power and enormous time to scan the entire parameters space, particularly, if more than one control parameters are involved, in order to understand the rich variety of bifurcations and chaotic phenomena. In this connection, *analog simulation* studies of nonlinear oscillators through appropriate electronic circuits are often helpful in a dramatic way for a quick scan of the parameters space.

From another point of view, in recent times a variety of nonlinear electronic circuits consisting of either real nonlinear physical devices such as nonlinear diodes, capacitors, inductors and resistors or devices constructed with ingenious piecewise-linear circuit elements have been utilized as veritable block boxes to study complex dynamics. These circuits are easy to build, easy to analyse and easy to model. With the great advances in nonlinear circuit theory over the past thirty years or so it is now possible to understand highly complex nonlinear behaviours with simple models and minor extensions of linear circuit theory. In this chapter, we will introduce nonlinear circuit elements, discuss bifurcations and chaotic dynamics in a few simple nonlinear electronic circuits and also illustrate how dynamical systems can be analog simulated to study their behaviour.

6.1 Linear and Nonlinear Circuit Elements

In this section, we shall briefly discuss the properties of various circuit elements, namely resistors, capacitors and inductors. They can be linear or nonlinear depending upon their characteristic curves, namely the v-i (voltage-current), v-q (voltage-charge) and i-ϕ (current-magnetic flux) curves.

(a) *Resistors*

A two-terminal resistor (Fig. 6.1a) whose current $i(t)$ and voltage $v(t)$ falls on some fixed (characteristic)curve in the v-i plane, represented by the equation $f_R(v,i) = 0$ at any time t is called a *time-invariant resistor*. If the measured

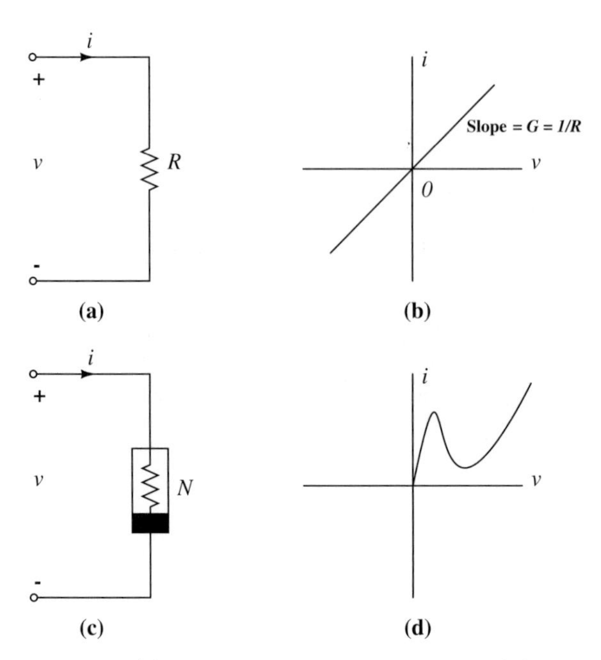

Fig. 6.1. (**a**) A two-terminal linear resistor, (**b**) its v-i characteristic curve, (**c**) a nonlinear resistor and (**d**) its v-i characteristic curve

v-i characteristic curve is a straight line passing through the origin as shown in Fig. 6.1b, then the resistor is said to be *linear* and it satisfies the Ohm's law

$$v(t) = Ri(t) \,. \tag{6.1}$$

If a resistor is characterized by a v-i curve other than a straight line, it is called a *nonlinear resistor*. The characteristic curve of a typical nonlinear resistor (Fig. 6.1c) is given in Fig. 6.1d.

(b) *Capacitors*

A two-terminal circuit element (Fig. 6.2a) whose charge $q(t)$ and voltage $v(t)$ falls on some fixed (characteristic)curve in the q-v plane, represented by the equation $f_C(q,v) = 0$ at any time t, is called a *time-invariant capacitor*. If the measured q-v characteristic curve is a straight line passing through the origin as shown in Fig. 6.2b, then the capacitor is said to be *linear* and it satisfies the current-voltage relation

$$i = C\frac{\mathrm{d}v}{\mathrm{d}t} \,. \tag{6.2}$$

In all other cases, the capacitor is *nonlinear*. The characteristic curve of a typical nonlinear capacitor (Fig. 6.2c) is given in Fig. 6.2d.

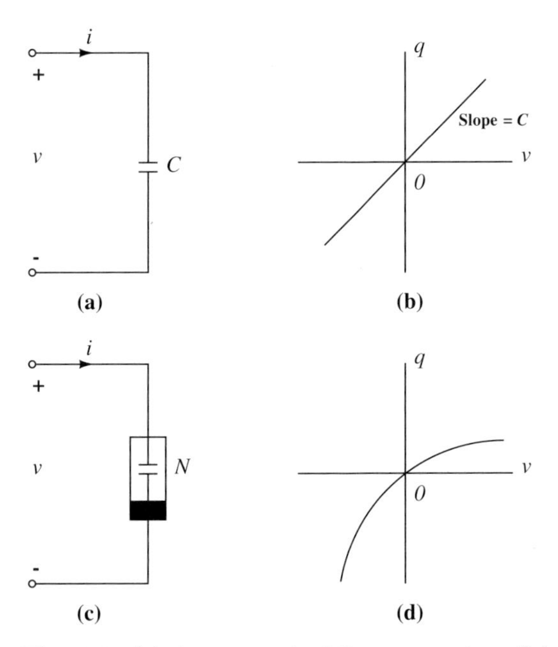

Fig. 6.2. (**a**) A two-terminal linear capacitor, (**b**) its v-q characteristic curve, (**c**) a nonlinear capacitor and (**d**) its v-q characteristic curve

(c) *Inductors*

A two-terminal circuit element (Fig. 6.3a) whose flux $\phi(t)$ and current $i(t)$ falls on some fixed (characteristic)curve in the ϕ-i, plane represented by the equation $f_L(\phi, i) = 0$ at any time t, is called a *time-invariant inductor*. If the measured ϕ-i characteristic curve is a straight line passing through the origin as shown in Fig. 6.3b, then the inductor is said to be *linear* and it satisfies the voltage-current relation

$$v = L\frac{\mathrm{d}i}{\mathrm{d}t} . \tag{6.3}$$

In all other cases, the inductor is *nonlinear*. The characteristic curve of a nonlinear inductor (Fig. 6.3c) is given in Fig. 6.3d.

6.2 Linear Circuits: The Resonant RLC Circuit

As noted above, linear circuits are those which contain linear elements (linear resistors, capacitors and inductors) only. Consider the LC circuit shown in Fig. 6.4a. By applying Kirchhoff's laws to this circuit, the state equations can be written as

$$L\frac{\mathrm{d}i_L}{\mathrm{d}t} = -v , \quad C\frac{\mathrm{d}v}{\mathrm{d}t} = i_L \tag{6.4}$$

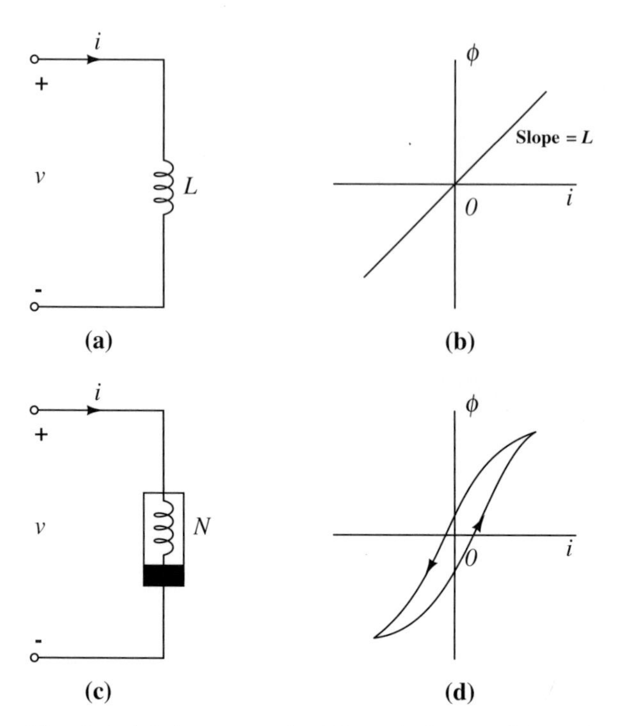

Fig. 6.3. (**a**) A two-terminal linear inductor, (**b**) its ϕ-i characteristic curve, (**c**) a nonlinear inductor and (**d**) its ϕ-i characteristic curve

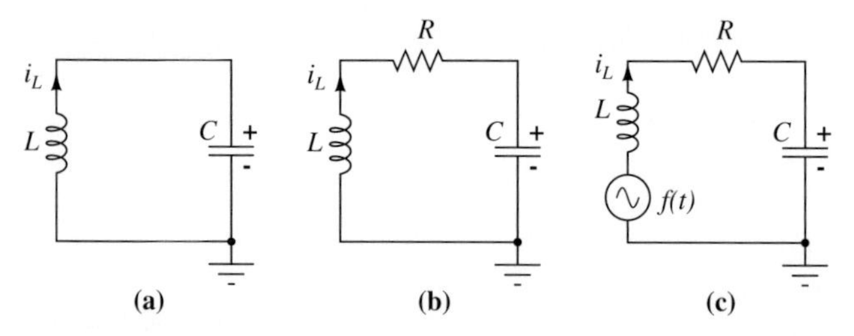

Fig. 6.4. (**a**) A linear LC circuit. (**b**) A linear RLC circuit. (**c**) A linear RLC circuit driven by a time-dependent voltage source $f(t) = F \sin \omega t$

with $i_L(0) = i_{L_0}$ and $v(0) = v_0$. Here $C > 0$ and $L > 0$. The above equation can be rewritten as

$$\frac{\mathrm{d}^2 v}{\mathrm{d}t^2} + \frac{1}{LC} v = 0 . \tag{6.5}$$

Equation (6.5) corresponds to the equation of motion of a linear harmonic oscillator and solving it, the capacitor voltage v can be written as (see Sect. 2.1.1)

$$v(t) = v_0 \cos\left(\sqrt{1/LC}\, t\right) , \tag{6.6}$$

where $v(0) = v_0$ and $\dot{v}(0) = 0$. The energy that was initially stored in the capacitor and inductor cannot be dissipated, but simply has to oscillate back and forth between these two elements. In the absence of damping, this sinusoidal oscillation continues indefinitely; the circuit then mimics a harmonic oscillator.

Next, we consider the circuit (Fig. 6.4b) with a linear resistor R included between the inductor L and the capacitor C. The state equations are given by

$$L\frac{\mathrm{d}i_L}{\mathrm{d}t} = -v - Ri_L\ , \quad C\frac{\mathrm{d}v}{\mathrm{d}t} = i_L\ . \tag{6.7}$$

It can obviously be reexpressed as

$$\frac{\mathrm{d}^2 v}{\mathrm{d}t^2} + \frac{R}{L}\frac{\mathrm{d}v}{\mathrm{d}t} + \frac{1}{LC}v = 0\ . \tag{6.8}$$

Equation (6.8) is nothing but the damped but unforced linear oscillator equation discussed in Sect. 2.1.2. The form of the solution depends on whether R is large or small. When R is less than $2\sqrt{L/C}$, the solution has the form

$$v(t) = \frac{v_0}{\beta\sqrt{LC}}\exp\left(-\frac{R}{2L}t\right)\cos(\beta t - \delta)\ , \tag{6.9a}$$

where

$$\beta = \sqrt{\frac{1}{LC} - \left(\frac{R}{2L}\right)^2}\ , \quad \delta = \tan^{-1}\left(\frac{R}{2L\beta}\right) \tag{6.9b}$$

and $v(0) = v_0$ and $\dot{v}(0) = 0$. When R is greater than $2\sqrt{L/C}$, the solution is

$$v(t) = \mathrm{e}^{(-R/2L)t}\left[A_1\mathrm{e}^{\beta t} + A_2\mathrm{e}^{-\beta t}\right]\ , \tag{6.9c}$$

where now

$$\beta = \sqrt{\left(\frac{R}{2L}\right)^2 - \frac{1}{LC}} \tag{6.9d}$$

and A_1, A_2 are integration constants. If R is positive, the resistor is said to be *dissipative*. The energy initially stored in the capacitor and inductor is dissipated away as heat in the resistor as the magnetic and electric fields collapse, and v and i_L approach zero in the form of exponentially decaying signals.

Finally, let us consider the circuit of Fig. 6.4c. It consists of a linear resistor, a linear inductor, a linear capacitor and a time-dependent voltage source $f(t) = F\sin\omega t$. Applying Kirchhoff's voltage law to the circuit of Fig. 6.4c, we obtain

$$C\frac{\mathrm{d}v}{\mathrm{d}t} = i_L\ , \quad L\frac{\mathrm{d}i_L}{\mathrm{d}t} + Ri_L + v = F\sin\omega t\ , \tag{6.10}$$

with $i_L(0) = i_{L_0}$ and $v(0) = v_0$. Substituting (6.2) into (6.10) and simplifying, we obtain the second-order linear inhomogeneous ordinary differential equation with constant coefficients:

$$\frac{\mathrm{d}^2 v}{\mathrm{d}t^2} + \frac{R}{L}\frac{\mathrm{d}v}{\mathrm{d}t} + \frac{1}{LC}v = \frac{F}{LC}\sin\omega t\,. \tag{6.11}$$

Equation (6.11) is identical in form with the resonant damped forced linear oscillator equation (2.1) with the identification, $v \to x$, $(R/L) \to \alpha$, $(1/LC) \to \omega_0^2$, $(F/LC) \to f$.

Now using the solution (2.4) for (6.11), one can easily write down the solution to the voltage v across the capacitor as

$$v(t) = \frac{v_0}{\beta\sqrt{LC}}\exp\left(-\frac{R}{2L}t\right)\cos(\beta t - \delta) + F_p\cos(\omega t - \gamma)\,, \tag{6.12a}$$

where β and δ are given by (6.9b) and

$$F_p = \frac{F/LC}{[\{(1/LC)^2 - \omega^2\}^2 + (R\omega/L)^2]}\,,$$

$$\gamma = \tan^{-1}\left[\frac{L\left(\omega^2 - (1/LC)^2\right)}{R\omega}\right]. \tag{6.12b}$$

Obviously, the asymptotic orbit of the RLC circuit driven by a sinusoidal force is a closed orbit (ellipse) in the $(v - i_L)$ plane corresponding to the sinusoidal steady state of the circuit. All the waveforms v and i_L converge to a periodic waveform with a frequency that resonates with the input signal of frequency ω. The actual experimental circuit result in the form of a phase portrait in the $(v - i_L)$ plane in a CRO is shown in Fig. 6.5 and it mimics exactly the phase portrait of the damped and driven linear harmonic oscillator, Fig. 2.3. This is the most regular asymptotic behaviour which one can expect from a linear circuit with a sinusoidal source.

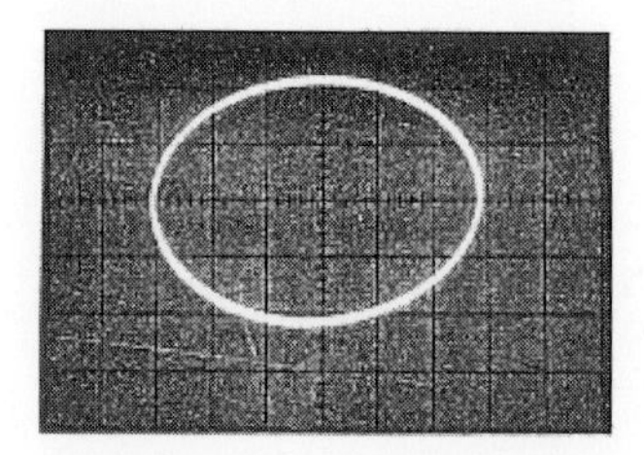

Fig. 6.5. Phase portrait of the asymptotic solution of the experimental circuit of Fig. 6.4c for $R = 500\Omega$, $L = 18$mH, $C = 10$nF, $F = 0.1$V and frequency $\omega = 8890$Hz

---

Exercise:

1. We can generalize the RLC circuit (Fig. 6.4b) further by replacing the linear capacitor C by two linear capacitors C_1 and C_2 connected as in

Fig. 6.6 below. Applying Kirchhoff's laws to this linear circuit, construct the equation of motion for the variables v_1, v_2 and i_L, which are the voltages across the capacitors C_1 and C_2 and the current i_L through the inductor L, respectively. Find its analytical solution and study its dynamics.

6.3 Nonlinear Circuits

A circuit is said to be *nonlinear* if it contains at least one nonlinear circuit element like a nonlinear resistor, a nonlinear capacitor or a nonlinear inductor. Another basic inventory in nonlinear circuit analysis is the use of piecewise-linear circuit elements designed ingeniously for specific needs, whose characteristic curves are piecewise-linear. These include piecewise-linear resistors, capacitors and inductors.

Of all the possible nonlinear circuit elements nonlinear resistors are easy to build and model. In this connection, *Chua's diode* is a simple nonlinear resistor with piecewise-linear characteristic and is widely used by circuit theorists [1,2] and electronic engineers [3]. In the following we shall discuss some simple circuits which contain this nonlinear resistor along with additional linear circuit elements and investigate the underlying dynamics.

6.3.1 Chua's Diode: Autonomous Case

One can generalize the modified RLC circuit (Fig. 6.6) into a nonlinear circuit (Chua's circuit), by placing the piecewise-linear (and so a nonlinear) resistor N, namely the *Chua's diode*, parallel to the capacitor C_1, as shown in Fig. 6.7.

The most interesting part of the typical (v-i) characteristic curve of the Chua's diode consists of the odd-symmetric three segment piecewise-linear form as shown in Fig. 6.8.

Its most important qualitative features are its piecewise nature and the *negative resistance* behaviour as the slope is negative between the two break points. In this figure $G_a = m_1$ and $G_b = m_0$ are the inner and outer slopes respectively, and the breakpoints are located at $v_R = -B_P$ and $v_R = B_P$. The mathematical representation of the characteristic curve [1,2] is given by

$$\begin{aligned} i_R &= h(v_R) \\ &= m_0 v_R + 0.5(m_1 - m_0)[\,|\,v_R + B_P\,| - |\,v_R - B_P\,|\,]\,, \end{aligned} \tag{6.13a}$$

or

$$h(v_R) = \begin{cases} m_0 v_R + (m_0 - m_1)B_P & \text{if } v_R < -B_P \\ m_1 v_R & \text{if } -B_P \le v_R \le B_P \\ m_0 v_R + (m_1 - m_0)B_P & \text{if } v_R > B_P\,, \end{cases} \tag{6.13b}$$

where $B_P > 0$.

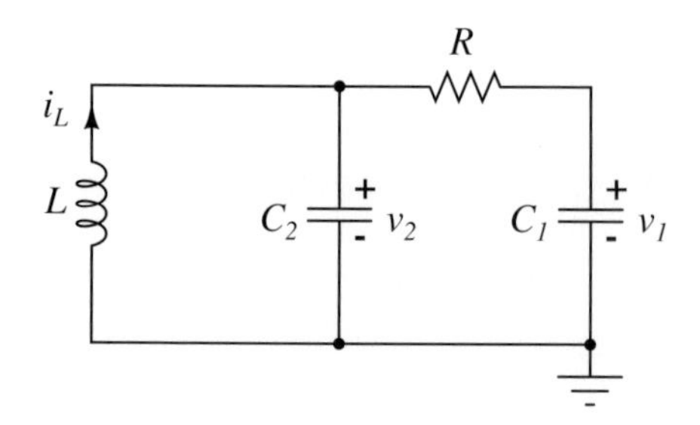

Fig. 6.6. A linear circuit with one inductor L, one resistor R and two capacitors C_1 and C_2

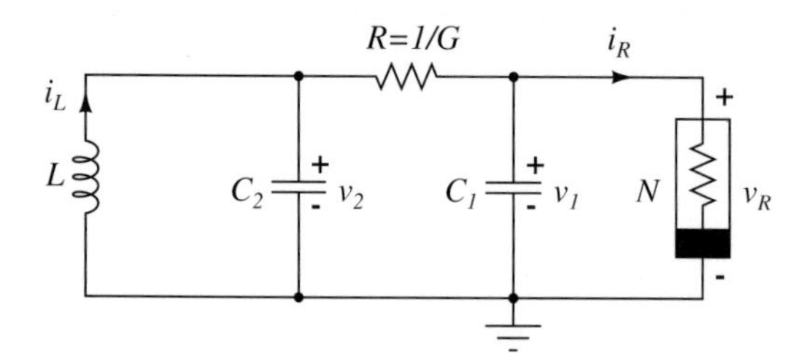

Fig. 6.7. A modified LCR circuit (Chua's circuit) with Chua's diode as the nonlinear element

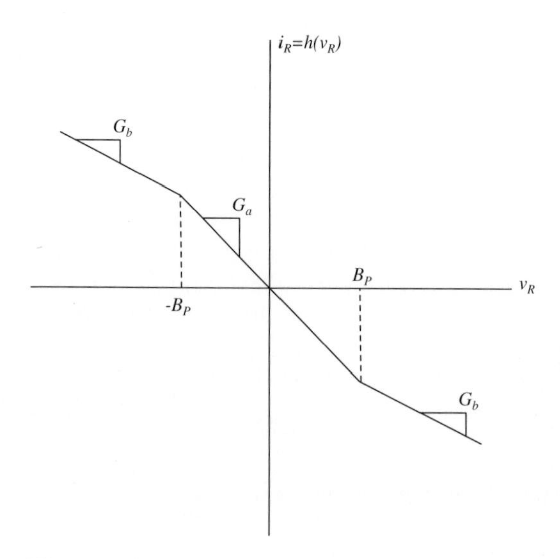

Fig. 6.8. Nonlinear element function $h(v_R)$ as given in (6.13)

6.3.2 A Simple Practical Implementation of Chua's Diode

In this section, we describe a simple practical implementation [4] of Chua's diode (N) using two operational amplifiers (op-amps) and six resistors. Figure 6.9 shows such an implementation. The Chua's diode circuit is then constructed using analog devices **AD712** (dual **BiFET**) op-amps, two 9 volts batteries and six linear resistors. The op-amp circuit consisting of the op-amps A_1, A_2 and the linear resistors R_1–R_6 functions as a nonlinear resistor N with a driving point characteristic as shown in Fig. 6.10. The typical values of the resistances R_1–R_6 are 220Ω, 220Ω, 2.2KΩ, 22KΩ, 22KΩ and 3.3KΩ respectively. The use of two 9V batteries to power the op-amps gives the voltages $V_+ = 9$V and $V_- = -9$V.

The slopes of the characteristic curve are given by $G_a = m_1 = -R_2/(R_1R_3) - R_5/(R_4R_6) = -0.758$mA/V and $G_b = m_0 = -R_2/(R_1R_3) + (1/R_4) = -0.409$mA/V. The breakpoints are determined by the saturation voltages E_{sat} of the op-amps. They are calculated as $\bar{B}_P = E_{\text{sat}}R_3/(R_2 + R_3) \approx 7.61$V and $B_P = E_{\text{sat}}R_6/(R_5 + R_6) \approx 1.08$V.

The v_R-i_R characteristic curve (Fig. 6.10) of the op-amp based Chua's diode differs from the desired piecewise-linear characteristics shown in Fig. 6.8 in that as noted above in Sect. 6.3.1 the latter has only three segments while the former has five segments, the outer two of which have positive slopes $G_c = m_2 = (1/R_2) = (1/220)$mA/V. Every physical resistor is eventually passive, meaning simply that for a large enough voltage across its terminals, the instantaneous power $P(t)(= v(t)i(t))$ consumed by a real resistor is positive. For large enough $\mid v \mid$ or $\mid i \mid$, therefore, the v_R-i_R characteristics must lie only in the first and third quadrants of the v-i plane. Hence, the v_R-i_R characteristic of a real Chua's diode must include at least two outer segments with positive slopes in the first and third quadrants (Fig. 6.8). From a practical point of view, as long as the voltages and currents in a given nonlinear circuit with Chua's diode are restricted to the negative resistance region of the characteristics, the above mentioned outer segments will not affect the circuit's dynamical behaviour. This is what we will assume in the present study.

6.3.3 Bifurcations and Chaos

By fixing the circuit parameters at $C_1 = 10$nF, $C_2 = 100$nF and $L = 18$mH (the values used by Chua's group in their experimental study) and by reducing the variable resistor (R) from 2KΩ downwards to zero, the Chua's circuit (Fig. 6.7) is readily seen to exhibit a sequence of bifurcations ranging from an equilibrium point through a Hopf bifurcation and period doubling sequence to a spiral-Chua (chaotic) attractor and a double-scroll Chua (chaotic) attractor, as illustrated in Fig. 6.11. In the experimental study, a two-dimensional projection of the attractor is obtained by connecting v_1 and v_2 to the X and Y channels, respectively, of an oscilloscope. This figure depicts the projection

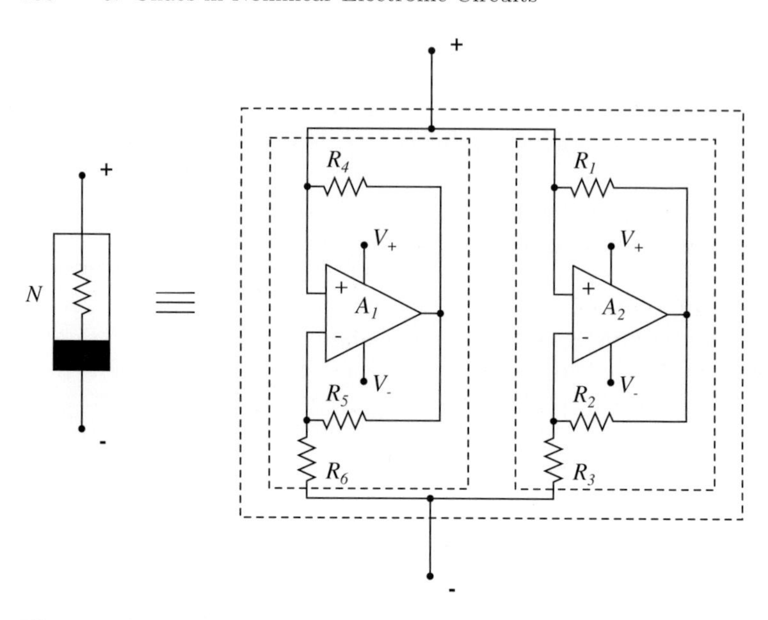

Fig. 6.9. Schematic diagram of the nonlinear resistor N of Figs. 6.7 and 6.8

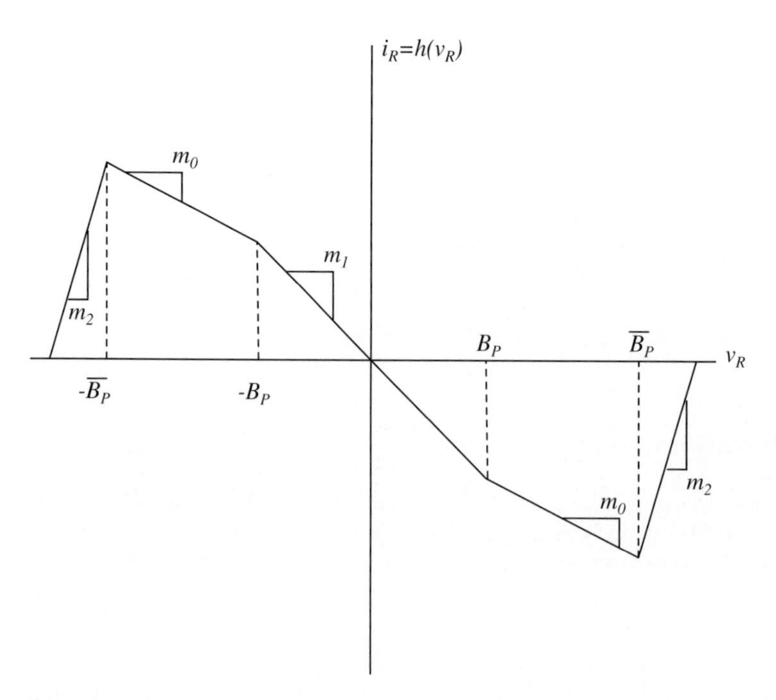

Fig. 6.10. Typical characteristic curve of a Chua's diode (see Fig. 6.9)

of v_1 versus v_2 phase portrait. Inserting a small current sensing resistor R_s in series with the inductor L, the current i_L can also be measured.

For $R = 2\text{K}\Omega$, the circuit has a stable equilibrium point as shown in Fig. 6.11a. When R is decreased, at a critical value the system undergoes Hopf bifurcation and a stable limit cycle is created and the equilibrium point becomes unstable. This is shown in Fig. 6.11b. Figures 6.11c–6.11e illustrate the period doubling phenomenon culminating in chaos. For $R = 1.79\text{K}\Omega$ one-band chaos is found to occur. Figure 6.11f depicts a double-scroll (two paper-roll like) attractor for $R = 1.74\text{K}\Omega$, where the trajectory spirals about, say, the left equilibrium point for a while and then moves towards right equilibrium point and makes several spirals about it and so on. For $R = 1.4\text{K}\Omega$ a large limit cycle corresponding to the outer segments of the (v-i) characteristic is observed.

An alternative way to view the bifurcation sequence is by adjusting C_1 or C_2 or L. For example, by fixing the values of L, C_2 and R at 18mH, 100nF and $1.8\text{K}\Omega$, respectively and varying C_1 from 12nF to 6nF, the full range of bifurcations from equilibrium point through Hopf bifurcation, period doubling sequence to spiral Chua attractor and double-scroll Chua attractor can be obtained. Some of the important features associated with Chua's circuit are the following.

(1) Chua's circuit exhibits a number of different scenarios or routes towards the onset of chaos. These include transition to chaos through period doubling cascade [3], through breakdown of invariant torus [2] and through intermittency (Chaps .4 and 5).

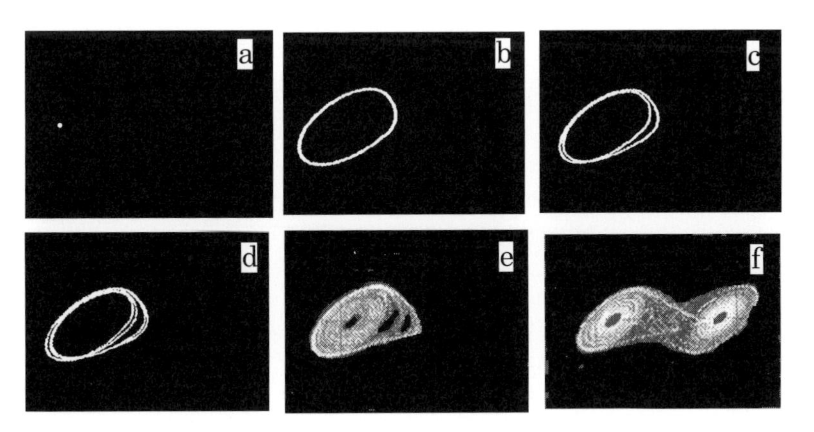

Fig. 6.11. Experimental results in the $(v_1 - v_2)$ plane of the Chua's circuit given in Fig. 6.7. (**a**) Stable equilibrium point for $R = 2\text{K}\Omega$. (**b**) Period-T limit cycle ($R = 1.88\text{K}\Omega$). (**c**) Period-$2T$ limit cycle ($R = 1.85\text{K}\Omega$). (**d**) Period-$4T$ limit cycle ($R = 1.84\text{K}\Omega$). (**e**) One-band chaos ($R = 1.79\text{K}\Omega$). (**f**) Double-scroll chaotic attractor ($R = 1.74\text{K}\Omega$). Other values of the circuit parameters are fixed at $C_1 = 10\text{nF}$, $C_2 = 100\text{nF}$ and $L = 18\text{mH}$

(2) Chua's circuit exhibits a typical chaotic attractor called the *double-scroll Chua's attractor*. It appears at the conjunction of a pair of nonsymmetric spiral attractors. Three unstable equilibrium states are visible in this attractor, which indicates that the double-scroll Chua's attractor is *multistructural*.

(3) As regards their mathematical nature, the attractors which occur in Chua's circuit are quite complicated.

(4) By applying Kirchhoff's laws to the various branches of circuit of Fig. 6.7 we obtain the state equations for the Chua's circuit as

$$C_1\dot{v}_1 = (1/R)\,(v_2 - v_1) - h\,(v_1)\;, \tag{6.14a}$$

$$C_2\dot{v}_2 = (1/R)\,(v_1 - v_2) + i_L\;, \tag{6.14b}$$

$$L\dot{i}_L = -v_2\;, \quad \left(\cdot = \frac{\mathrm{d}}{\mathrm{d}t}\right) \tag{6.14c}$$

where $h(v_1)$ is the mathematical representation of Chua's diode characteristic curve which is given by (6.13). In dimensionless form, (6.14) can be rewritten as

$$\dot{x} = \alpha(y - x - h(x))\;, \tag{6.15a}$$

$$\dot{y} = x - y + z\;, \tag{6.15b}$$

$$\dot{z} = -\beta y\;, \tag{6.15c}$$

where $h(x) = bx + 0.5(a-b)[\mid x+1 \mid - \mid x-1 \mid]$, and $v_1 = xB_P$, $v_2 = yB_P$, $i_L = B_P G z$, $t = C_2\tau/G$, $G = 1/R$, $a = RG_a$, $b = RG_b$, $\alpha = C_2/C_1$, $\beta = C_2R^2/L$. We wish to add that when the set of equations (6.15) is solved numerically we get exactly the same results obtained from the actual circuit itself.

The above discussions clearly show that the kind of nonlinear circuit investigated above can be considered as a prototypical nonlinear dynamical system on its own merit. In the next section, we will consider an even more simple but nonautonomous circuit and study its dynamics through (i) the actual electronic circuit and (ii) by solving the corresponding equation of motion numerically. We show that both the approaches give the same results for a set of values of the parameters of the circuit.

Exercises:

2. Show that (6.15) follow from (6.14) by the change of variables indicated below (6.15).
3. Numerically integrate (6.15) and verify the experimental results on the routes to chaos for different parameter settings.

6.4 Chaotic Dynamics of the Simplest Dissipative Nonautonomous Circuit: Murali–Lakshmanan–Chua (MLC) Circuit

In this section we wish to point out that apart from higher order nonautonomous circuits, even a much simpler second-order nonautonomous dissipative nonlinear circuit consisting of the Chua's diode as the only nonlinear element, as suggested by Murali, Lakshmanan and Chua [5,6], can exhibit a rich variety of bifurcations and chaos phenomena.

6.4.1 Experimental Realization

The actual circuit realization of the simple nonautonomous Murali–Lakshmanan–Chua (MLC) circuit [5], shown in Fig. 6.12, is as follows. It contains a capacitor, an inductor, a linear resistor, an external periodic forcing, and only one nonlinear element, namely the Chua's diode (N). In order to measure the inductor current i_L in the experiment, a small current sensing resistor R_s can be used. By applying Kirchhoff's laws to this circuit, the governing equations for the voltage v across the capacitor C and the current i_L through the inductor L are represented by the following set of two first-order nonautonomous differential equations:

$$C\frac{\mathrm{d}v}{\mathrm{d}t} = i_L - h(v)\,, \tag{6.16a}$$

$$L\frac{\mathrm{d}i_L}{\mathrm{d}t} = -R\,i_L - R_s i_L - v + F\sin\omega t\,, \tag{6.16b}$$

where $h(v)$ is the piecewise-linear function defined by (6.13). F is the amplitude and ω is the angular frequency of the external periodic force. The parameters of the circuit elements are fixed at $C = 10\mathrm{nF}$, $L = 18\mathrm{mH}$, $R = 1340\Omega$, $R_s = 20\Omega$, and the frequency $\nu(= \omega/2\pi)$ of the external forcing source is 8890Hz. First, let us find the equilibrium points of (6.16) with $F = 0$ (no external forcing) and study their stability. Next, we construct the explicit analytical solution with $F \neq 0$. Then we study the nonlinear dynamics of the MLC circuit.

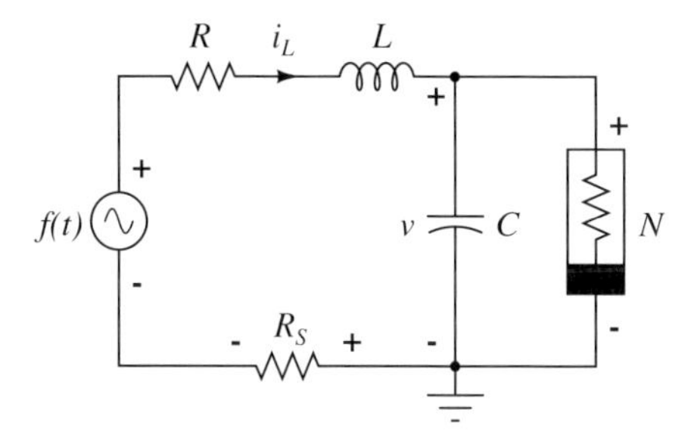

Fig. 6.12. Schematic diagram of the simple nonautonomous MLC circuit

6.4.2 Stability Analysis

The actual values of the slopes and break-point voltage of the Chua's diode are fixed as $m_1 = -0.7\text{ms}$, $m_0 = -0.41\text{ms}$, and $B_P = 1\text{V}$. Rescaling (6.16) with $v = xB_p$, $i_L = yGB_P$, $G = 1/R$, $\omega' = \omega C/G$ and $t = t'C/G$ and then dropping the primes in t and ω, the following set of normalized equations are obtained:

$$\dot{x} = y - h(x) = F_1 \ , \tag{6.17a}$$

$$\dot{y} = -\beta(1+\nu)y - \beta x + f \sin \omega t = F_2 \ , \tag{6.17b}$$

where $\beta = C/(LG^2)$, $\nu = GR_s$ and $f = F\beta/B_P$. Obviously $h(x)$ is represented by

$$h(x) = \begin{cases} bx + a - b & x \geq 1 \\ ax & | x | \leq 1 \\ bx - a + b & x \leq -1 \ . \end{cases} \tag{6.17c}$$

Here $a = m_1/G$, $b = m_0/G$. Now the dynamics of (6.17) depends on the parameters ν, β, a, b, ω and f. For the experimental parameter values considered in the previous section, we obtain $\beta = 1$, $\nu = 0.015$, $a = -1.02$, $b = -0.55$ and $\omega = 0.75$.

A. *Equilibrium Points*

Because of the piecewise-linear nature of $h(x)$ in (6.17), it is straightforward to see that when $f = 0$ there are three equilibrium points in the following three regions:

Region	Equilibrium point	
$D_1 = \{(x,y) \mid x \geq 1\}$	$P_+ = (-k_1, -k_2)$	(6.18a)
$D_0 = \{(x,y) \mid \mid x \mid \leq 1\}$	$P_0 = (0,0)$	(6.18b)
$D_{-1} = \{(x,y) \mid x \leq -1\}$	$P_- = (k_1, k_2)$	(6.18c)

Here $k_1 = \sigma(a-b)/(\beta + \sigma b)$, $k_2 = \beta(b-a)/(\beta + \sigma b)$ and $\sigma = \beta(1+\nu)$.

B. *Linear Stability*

In each of the three regions D_1, D_0 and D_{-1}, (6.17) is linear. The stability determining matrix for (6.17) takes the form (see (3.8))

$$A = \begin{pmatrix} \partial F_1/\partial x & \partial F_1/\partial y \\ \partial F_2/\partial x & \partial F_2/\partial y \end{pmatrix} \Bigg|_{(x = x^*, \ y = y^*)}$$

$$= \begin{pmatrix} -h'(x^*) & 1 \\ -\beta & -\sigma \end{pmatrix} . \tag{6.19}$$

Then the stability property of each of the equilibrium points P_0, P_+ and P_- are as follows.

(i) *Stability of P_0*

$P_0 = (0,0)$ is the equilibrium point lying in the region $|x| < 1$. In this region, from (6.17c), we find that $h'(x^*) = a$. Thus the stability matrix becomes

$$A_0 = \begin{pmatrix} -a & 1 \\ -\beta & -\sigma \end{pmatrix} \tag{6.20}$$

so that the eigenvalues are

$$\lambda_{1,2} = \frac{1}{2}\left[-(a+\sigma) \pm \sqrt{(a-\sigma)^2 - 4\beta} \right] . \tag{6.21}$$

When the numerical values of the parameters $a = -1.02$, $\beta = 1$ and $\sigma = 1.015$ are used, they become $\lambda_1 = 0.1904$ and $\lambda_2 = -0.1854$. So P_0 is a saddle and it corresponds to a hyperbolic (unstable) equilibrium point.

(ii) *Stability of P_+*

From (6.17c) it is clear that in the interval $x \geq 1$, $h'(x^*) = b$. Then from (6.19) the stability determining eigenvalues are obtained as

$$\lambda_{1,2} = \frac{1}{2}\left[-(b+\sigma) \pm \sqrt{(b-\sigma)^2 - 4\beta} \right] . \tag{6.22}$$

For the parametric values chosen as above, the eigenvalues are $\lambda_{1,2} = -0.2325 \pm \mathrm{i}\ 0.623$, which are complex conjugates with negative real parts. Thus the equilibrium point P_+ is a stable focus.

(iii) *Stability of P_-*

In the interval $x \leq -1$, $h'(x^*) = b$ is identical to its value in the range $x \geq 1$. Therefore the stability of the equilibrium point P_- is exactly same as that of P_+. Thus, P_- is also a stable focus.

Naturally these equilibrium points can be observed depending upon the initial conditions $x(0)$ and $y(0)$ in (6.17) when $f = 0$. As the forcing signal is included ($f > 0$), stable limit cycles are created about both the stable equilibrium points. As f is increased further the system exhibits period doubling bifurcations, starting from the period-1 limit cycle, and finally transition to chaos as discussed below in Sect. 6.4.4.

6.4.3 Explicit Analytical Solutions

Actually (6.17) can be explicitly integrated in terms of elementary functions in each of the three regions D_0, D_1 and D_{-1} and matched across the boundaries to obtain the full solution as shown below.

It is quite easy to see that in each one of the regions D_0, D_{-1}, D_1, (6.17) can be represented as a single second-order inhomogeneous differential equation for the variable $y(t)$ (by using one of the equations (6.17) into the other):

$$\ddot{y} + (\beta + \beta\nu + \mu)\dot{y} + (\beta + \mu\beta\nu + \beta\mu)y$$
$$= \Delta + \mu f \sin \omega t + f\omega \cos \omega t \, , \qquad (6.23a)$$

where

$$\mu = a \, , \quad \Delta = 0 \text{ in region } D_0 \, , \qquad (6.23b)$$
$$\mu = b \, , \quad \Delta = \pm\beta(a - b) \text{ in region } D_{\pm 1} \, . \qquad (6.23c)$$

The general solution of (6.23) can be written as

$$y(t) = C^1_{0,\pm} \exp(\alpha_1 t) + C^2_{0,\pm} \exp(\alpha_2 t) + E_1 + E_2 \sin \omega t$$
$$+ E_3 \cos \omega t \, , \qquad (6.24a)$$

where $C^1_{0,\pm}$ and $C^2_{0,\pm}$ are integration constants in the appropriate regions D_0, $D_\pm$, and

$$\alpha_1 = (-A + \sqrt{A^2 - 4B}\,)/2 \, , \quad \alpha_2 = (-A - \sqrt{A^2 - 4B}\,)/2 \, ,$$
$$E_1 = 0 \text{ in region } D_0 \text{ and } E_1 = \Delta/B \text{ in region } D_{\pm 1} \, ,$$
$$E_2 = \left[(f\omega^2(A - \mu) + \mu f B)/(A^2\omega^2 + (B - \omega^2)^2\right] \, ,$$
$$E_3 = f\omega \left[(B - \omega^2 - \mu A)/(A^2\omega^2 + (B - \omega^2)^2\right] \, ,$$
$$A = \beta + \beta\nu + \mu \text{ and } B = \beta + \mu\beta\nu + \beta\mu \, . \qquad (6.24b)$$

Knowing $y(t)$, $x(t)$ can be obtained from (6.17) as

$$x(t) = (1/\beta)\{-\dot{y} - \beta y(1 + \nu) + f \sin \omega t\}$$
$$= (1/\beta)\{-\dot{y} - \sigma y + f \sin \omega t\}$$
$$= (1/\beta)\{-C^1_{0,\pm}(\alpha_1 + \sigma)\mathrm{e}^{\alpha_1 t} - C^2_{0,\pm}(\alpha_2 + \sigma)\mathrm{e}^{\alpha_2 t}$$
$$-(E_2\omega + E_3\sigma) \cos \omega t + (f - E_2\sigma + E_3\omega) \sin \omega t - E_1\sigma\} \, . \qquad (6.25)$$

Thus, if we start with an initial condition in D_0 the arbitrary constants C^1_0 and C^2_0 in (6.24) get fixed. Then $x(t)$ evolves as given in (6.25) upto either $t = T_1$, when $x(T_1) = 1$ and $\dot{x}(T_1) > 0$, or $t = T'_1$, when $x(T'_1) = -1$ and $\dot{x}(T_1) < 0$. Knowing whether $T_1 > T'_1$ or $T_1 < T'_1$ we can determine the next region of interest ($D_{\pm 1}$), and the arbitrary constants of the solutions of that region can be fixed by matching the solutions. This procedure can be continued for each successive crossing. In this way explicit solutions can be obtained in each of the regions D_0, $D_{\pm 1}$. However, it is clear that sensitive dependence on initial conditions is introduced in each of these crossings at appropriate parameter regimes during the inversion procedure of finding T_1, T'_1, T_2, T'_2, ... etc. from the solutions, and thereby leading to chaos.

6.4.4 Experimental and Numerical Studies

In order to study some basic aspects of the dynamics of this circuit, we can use one of the parameters, say the amplitude F of the forcing signal, as the control parameter. By increasing the amplitude F from zero, the circuit of Fig. 6.12 is found to exhibit experimentally a sequence of bifurcations. Starting from an equilibrium point, the solution bifurcates through Hopf bifurcation to a limit cycle, and then by period doubling sequence to one-band attractor, double-band attractor, periodic windows, etc. These are illustrated in Fig. 6.13. The left side figures depict phase portrait while the right side ones depict the corresponding wave form $v(t)$. One division in horizontal and vertical axes correspond to 0.5V and 5mV respectively. For small values of F, period-T limit cycle attractor occurs. This is shown in Fig. 6.13a for $F = 0.0365V_{\mathrm{rms}}$. This limit cycle is stable for $0 < F < 0.071V_{\mathrm{rms}}$. At $F = 0.071V_{\mathrm{rms}}$, the period-1 limit cycle becomes unstable and a period-$2T$ orbit is born and is stable for $0.071V_{\mathrm{rms}} \leq F < 0.089V_{\mathrm{rms}}$. Figure 6.13b shows period-$2T$ orbit for $F = 0.0549V_{\mathrm{rms}}$. As the parameter F is further varied period-$4T$, $8T$, ... orbits are found to occur leading to the onset of chaos at $F = 0.093V_{\mathrm{rms}}$. Figure 6.13d shows a typical one-band chaotic attractor for $F = 0.0723V_{\mathrm{rms}}$. At a critical value of F the one-band attractor bifurcates into a double-band attractor (Fig. 6.13e). The detailed dynamical behaviour of the MLC circuit is summarized in Table 6.1.

The dynamics of the circuit (6.12) can also be studied by solving the associated equation of motion (6.17) numerically. Using the standard fourth-order Runge–Kutta integration method, (6.17) is analysed with the rescaled circuit parameters of Fig. 6.12, $\beta = 1$, $\nu = 0.015$, $a = -1.02$, $b = -0.55$, $\omega = 0.75$, and with f as the control parameter.

Figure 6.14 depicts the phase portraits for different values of f corresponding to Fig. 6.13. From Figs. 6.13 and 6.14, we observe that the actual experimental circuit and the associated equation of motion give identical results.

What we have indicated here is only the basic features of the above nonlinear circuits. More intricate aspects of the underlying chaotic dynamics can

Table 6.1. Summary of bifurcation phenomena of MLC circuit

Amplitude(F)(V_{rms})	Nature of solution
$0 < F < 0.05$	Period-T limit cycle
$0.05 \leq F < 0.063$	Period-$2T$ limit cycle
$0.063 \leq F < 0.0658$	Period-$4T$ limit cycle
$0.0658 \leq F < 0.1343$	Chaos
$0.1343 \leq F < 0.246$	Period-$3T$ window
$0.246 \leq F < 0.353$	Chaos
$0.353 \leq F < 0.442$	Period-$3T$ window
$F > 0.442$	Period-T limit cycle

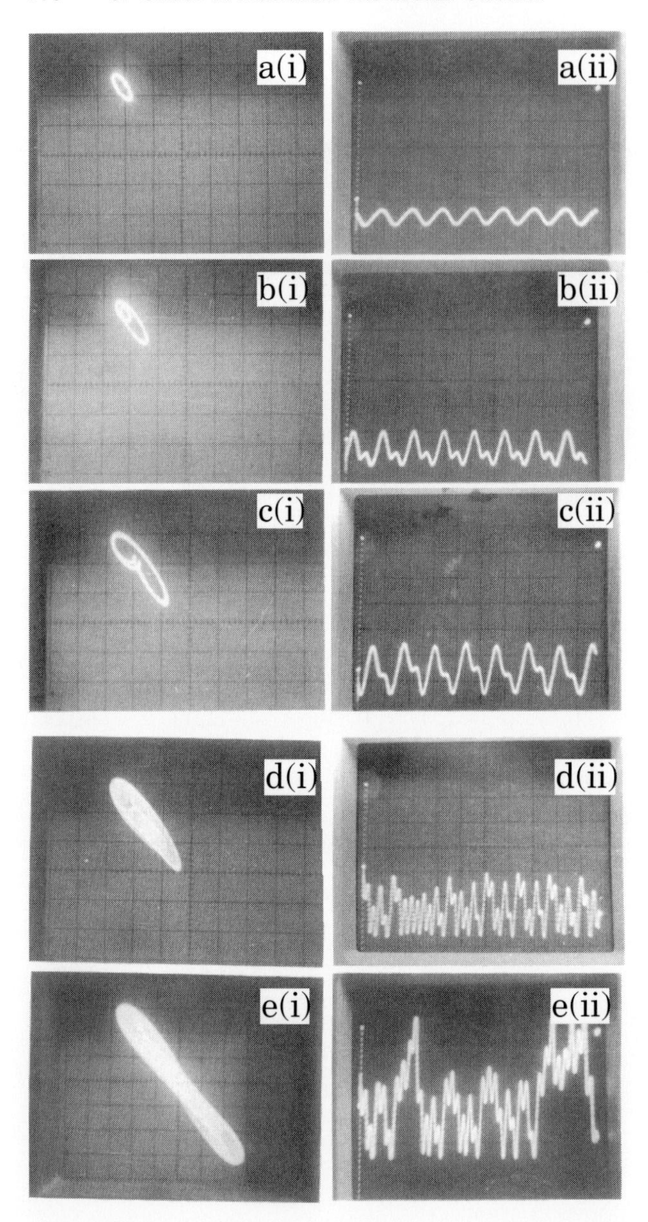

Fig. 6.13. The MLC circuit – Experimental results: Period doubling bifurcation sequences. (i) Horizontal axis v : 0.5V/div, vertical axis ($v_s = R_s i_L$): 5mV/div; (ii) wave form $v(t)$. (**a**) $F = 0.0365V_{\rm rms}$; Period-1. (**b**) $F = 0.0549V_{\rm rms}$; Period-2. (**c**) $F = 0.064V_{\rm rms}$; Period-4. (**d**) $F = 0.0723V_{\rm rms}$; one-band chaos. (**e**) $F = 0.107V_{\rm rms}$; Double-band chaos. (**i**) Phase portrait v versus v_s; (**ii**) wave form $v(t)$. The other parameters of the circuit elements are fixed at $C = 10$nF, $L = 18$mH, $R = 1340\Omega$, $R_s = 20\Omega$ and the frequency $\nu(=\omega/2\pi)$ of the external forcing source is 8890Hz

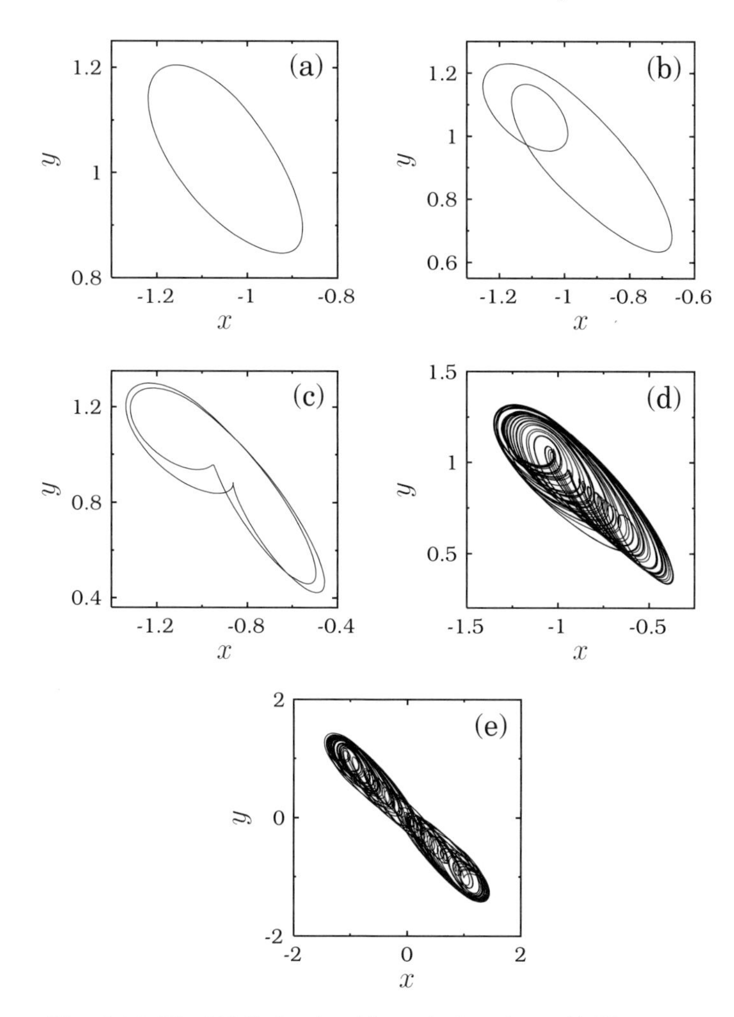

Fig. 6.14. The MLC circuit – Numerical analysis: (i) Phase portraits in the $(x-y)$ plane of (6.17) for the parameters $\beta = 1.0$, $\nu = 0.015$, $a = -1.02$, $b = -0.55$ and $\omega = 0.75$. (**a**) $f = 0.065(F = 0.046V_{\text{rms}})$, Period-1 limit cycle; (**b**) $f = 0.08(F = 0.0565V_{\text{rms}})$, Period-2 limit cycle; (**c**) $f = 0.091(F = 0.06435V_{\text{rms}})$, Period-4 limit cycle; (**d**) $f = 0.1(F = 0.0707V_{\text{rms}})$, One-band chaos; (**e**) $f = 0.15(F = 0.106V_{\text{rms}})$, double-band chaos; (Here f is the calculated value of the forcing amplitude using the relation $F = fB_P/\beta$, see (6.17)). These figures may be compared with the experimental results given in Figs. 6.13

be studied as discussed in the previous chapter, which we refrain from doing so here. For more details, see Ref. [6].

6.5 Analog Circuit Simulations

Apart from the construction of ingenious nonlinear circuits to study chaotic dynamics, one can also utilize suitable analog simulation circuits to mimic the dynamics of typical nonlinear systems, modelled by nonlinear differential equations. Such circuits are constructed using suitable op-amp modules and multipliers. Usually op-amps are used as integrators, differentiators, sign-changers, adders, and so on. These operations are generally obtained using individually an op-amp with resistor feedback or an op-amp with capacitor feedback. These op-amp mathematical modules are better choices because of the following advantages:

(1) The high input impedance allows one to add many input signals.
(2) The low output impedance allows one to connect these devices to several others without changing the characteristic time of the single modules (especially integrators and differentiators).

Because of the great development of the electronic devices today, it turns out to be easy and inexpensive to perform product and division operations using multiplier and divider integrated circuit chips. A typical multiplier/divider chip is the AD532 or AD534 made by the Analog Devices. Because of the differential nature of the inputs, we are in a position to assemble *minimum component devices* using op-amp modules and multipliers so as to simulate nonlinear differential equations. In this section we describe the analog simulation circuit study of the Duffing oscillator equation (5.1) with the double-well potential case.

The electronic analog simulator of (5.1) can be easily constructed by the conventional op amps and four-quadrant multipliers. Replacing now $x(t)$ and $\dot{x}(t)(= \mathrm{d}x/\mathrm{d}t)$ by

$$x(t) = \int \dot{x}\mathrm{d}t\,, \quad \dot{x}(t) = \int \ddot{x}\mathrm{d}t\,, \tag{6.26}$$

the equivalent schematic simulation circuit [6] is shown in Fig. 6.15. In this figure IC_1 and IC_2 represent two integrators. IC_3 is an invertor (sign-changer) and M_1 and M_2 are four-quadrant multipliers. The operational amplifiers used in IC_1, IC_2 and IC_3 are op-amp μA741C and the two multipliers M_1 and M_2 are Analog Devices multipliers AD532 or AD534. In the circuit, the resistors and capacitors are fixed as $R_1 = R_3 = R_4 = R_5 = R_6 = 100\mathrm{K}\Omega$, $R_E = R_2 = R_7 = 10\mathrm{K}\Omega$, $C_1 = C_2 = 0.01\mu F$. The multipliers produce output $(-XY/10)$ for inputs X and Y. By applying Kirchhoff's current law at the junctions A and B of Fig. 6.15, we obtain the following equation:

$$R_4R_5C_1C_2\frac{\mathrm{d}^2x}{\mathrm{d}t^2} = -R_5C_2P_1\frac{\mathrm{d}x}{\mathrm{d}t} + \frac{R_4P_2}{R_3}x - \frac{R_4P_3}{10R_2}x^3 + \frac{R_4}{R_1}f\sin\omega' t\,. \tag{6.27}$$

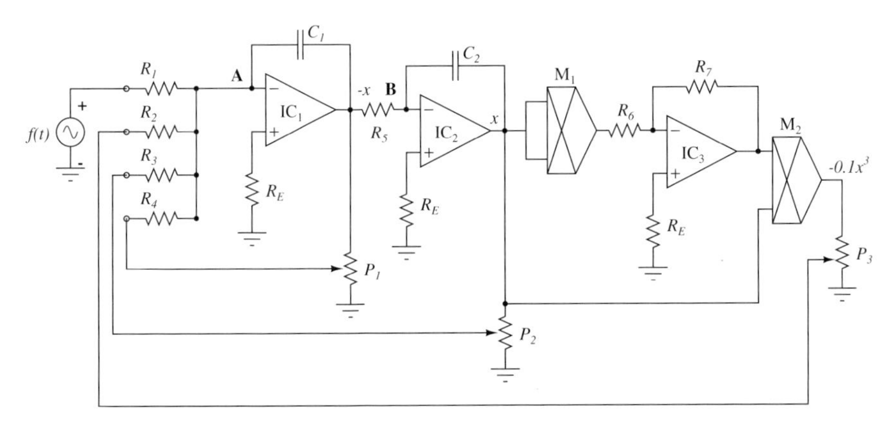

Fig. 6.15. Electronic analog simulation circuit of the Duffing oscillator equation (5.1) with a double-well potential

With the rescaling $t \to R_1 C_1 t'$ and redefinition of the parameters $t' = t$, $P_1 = \alpha$, $P_2 = \omega_0^2$, $P_3 = \beta$ and $R_1 C_1 \omega' = \omega$, along with the above choice of circuit elements, (6.27) can be seen to be identical to the Duffing oscillator equation (5.1). Here, the quantities P_1, P_2, P_3 and the frequency of the external signal ($= \omega'/2\pi$) can be changed. Changes in these parameters correspond to changes in α, ω_0^2, β and ω in (5.1).

The frequency of the external sinusoidal signal is fixed at 160Hz so that the corresponding redefined frequency $\omega = R_1 C_1 \omega' = 100\text{K}\Omega \times 0.01\mu\text{F} \times (2\pi \times 160\text{Hz}) \approx 1.0$. Choosing $\omega_0^2 = -1$, $\beta = 1$ and $\alpha = 0.5$, the circuit shown in Fig. 6.15 will then give immediately the waveforms x and $\dot{x}$ and they can be traced by an oscilloscope to observe the trajectory plot in the $(x - t)$ plane and the phase-portrait in the $(x - \dot{x})$ plane. The dynamical behaviour of (5.1) is easily investigated by scanning the driving amplitude f upwards from 0. The initial conditions are chosen as $x(0) = -0.2$ and $\dot{x}(0) = 0.01$. From the nature of the phase portrait in the $(x - \dot{x})$ plane obtained from both *numerical* (Figs. 5.3–5.6 discussed earlier in Chap. 5) and *experimental* (Fig. 6.16) investigations, the following results are obtained.

For $f = 0$, the system (5.1) asymptotically approaches a stable equilibrium point (focus) (see Figs. 5.3a and 6.16a). As f is increased slightly, a stable period$-T(= 2\pi/\omega)$ limit cycle occurs as shown in Figs. 5.3b and 6.16b for $f = 0.33$, which is an one-dimensional curve closing by itself after one forcing cycle. As f is increased to a critical value 0.35, the phase trajectory bifurcates into a period$-2T$ limit cycle as shown in Figs. 5.3c and 6.16c. The system then undergoes further period doubling bifurcations as shown in Figs. 5.3d and 6.16, when f is smoothly increased further. This cascade of bifurcations accumulates at f_c and the sequence converges geometrically at a rate given by Feigenbaum. From the above results, the calculated ratio turns out to be $\delta = 4.77674$, which is in fair agreement with the Feigenbaum ratio

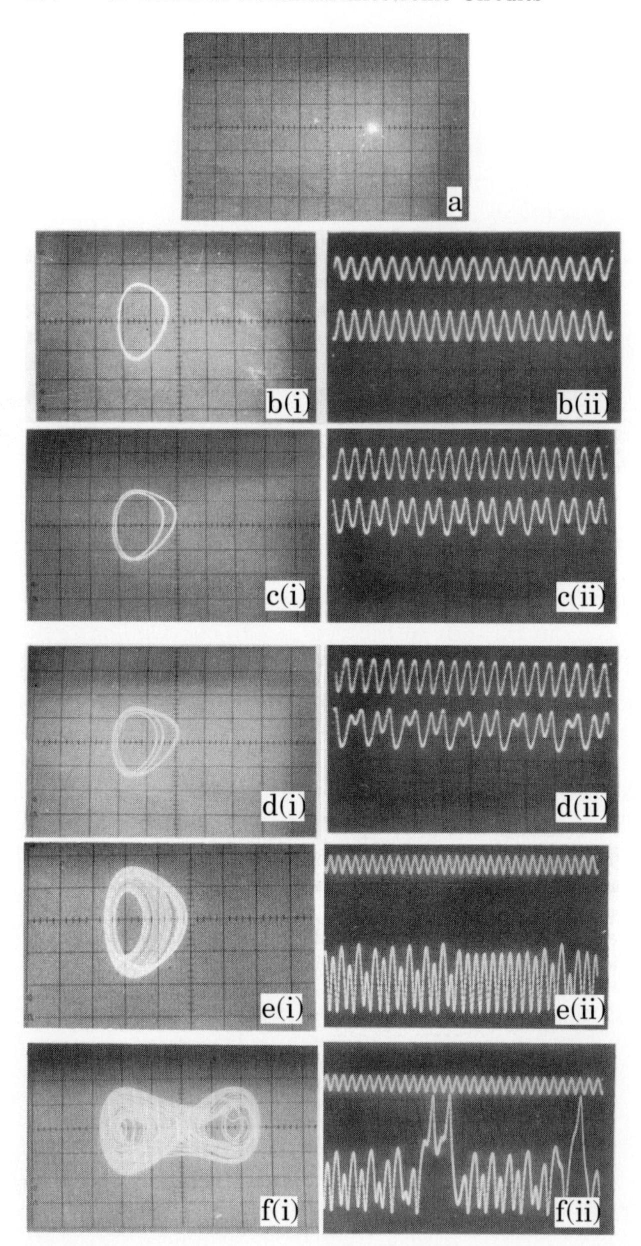

Fig. 6.16. Experimental analog circuit simulation result for the Duffing oscillator (5.1). (**a**) Phase portrait denoting the equilibrium point ($x^* < 0$) for $f = 0$. (**b**) (**i**) Period-$T(= 2\pi/\omega)$ limit cycle for $f = 0.33$. (**c**) (**i**) Period-$2T$ limit cycle for $f = 0.35$. (**d**) (**i**) Period-$4T$ limit cycle for $f = 0.357$. (**e**) (**i**) One-band chaos for $f = 0.365$. (**f**) (**i**) Double-band chaos for $f = 0.42$. In the subplots (**ii**) lower traces depict wave forms $x(t)$ of (**i**) and upper traces depict external periodic signal. These results may be compared with the numerical results given in Figs. 5.3–5.6 in Sect. 5.2

of 4.6672.... As discussed in Chap. 4, this convergence rate δ tells us how quickly successive bifurcations occur as the control or bifurcation parameter f is varied. Beyond certain critical value of f, chaotic orbits occur. Figures 5.4 and 6.16e show a typical one-band chaotic attractor for $f = 0.365$. When f is increased a little above 0.365, the orbit begins to migrate into the right valley of the phase plane and exhibits a hopping state as shown in Figs. 5.6 and 6.16f.

6.6 Some Other Useful Nonlinear Circuits

During the past few years, chaotic signals have been shown to be very useful in a large number of applications, including secure communications, signal encryption, medical field, neural networks and even music generation. For some of these applications, see Chap. 9 later on. Due to the increase in such chaotic signal applications, an increasing demand on electronic chaotic signal generators that are simple, easy to construct, easy to tune and easy to operate in different frequency bands has emerged. It is also generally required that the generated signal has a large voltage swing and persists for a wide and continuous range of parameter values and that periodic as well as chaotic waveforms can be sustained. In this section we describe some more of the simple nonlinear circuits exhibiting periodic and chaotic dynamics. They are good exercises both for numerical and experimental studies.

6.6.1 RL Diode Circuit

An RL diode circuit [7] is shown in Fig. 6.17. The circuit equation is

$$L\frac{\mathrm{d}^2Q}{\mathrm{d}t^2} + R\frac{\mathrm{d}Q}{\mathrm{d}t} + \nu(Q) = A\cos\Omega t\,, \tag{6.28a}$$

where $\nu(Q)$ is the voltage-charge characteristic relation of the diode D. For MP60 diode the measured voltage-charge relation can be approximated as

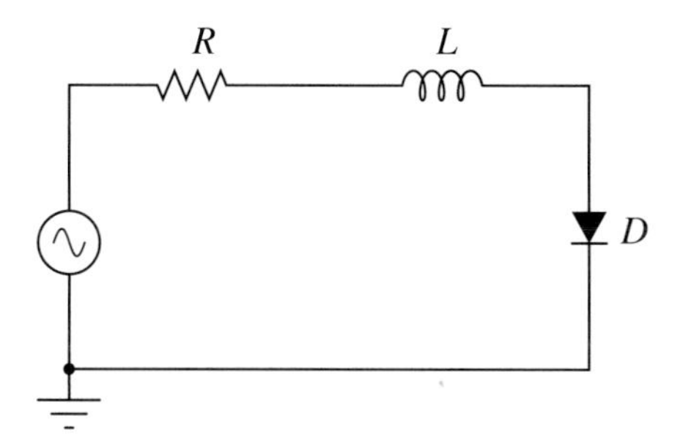

Fig. 6.17. The RL diode nonlinear circuit

$$\nu(Q) = aQ + bQ^2 , \tag{6.28b}$$

where a and b are constants. It can be determined by quadratic curve fitting to the experimentally measured voltage-charge data. One can study the various nonlinear phenomenon of the circuit given in Fig. 6.17 experimentally and also solve numerically the underlying dynamical equation (6.28) and compare the details. This is left as an exercise.

6.6.2 Hunt's Nonlinear Oscillator

Figure 6.18 shows the circuit [8,9] where the transistor Q_1 acts as the diode to ground and provides a current to charge C_1, while Q_1 is conducting. After the conduction, a portion of the cycle C_1 discharges through Q_2 causing it to conduct for an additional portion of the cycle. The larger the current that passes through Q_1, the longer Q_2 remains conducting. The nonlinear capacitors associated with the transistor junctions are negligible compared with the 1nF series capacitor. Consequently, the major cause of the period doubling and chaotic behaviour is the reverse-recovery-time effect. The circuit is driven sinusoidally by a voltage $A \cos \Omega t$. The transistors Q_1 and Q_2 are varactor diodes-2N3393 (or any other equivalent element).

6.6.3 p-n Junction Diode Oscillator

Figure 6.19 shows the circuit diagram with the p-n junction diode (IN4004) as a nonlinear element [10]. Typical values of the circuit elements are $L = 2$mH, $L_1 = 3.2$mH, $C_1 = 330$pF, $R_1 = 12.2\Omega$, $R = 51.2\Omega$ and $\omega = 45$KHz. Phase-locking, quasiperiodic and various types of intermittency dynamics, period doubling bifurcations leading to chaos can be seen by varying the amplitude V_0 of the drive signal.

6.6.4 Modified Chua Circuit

Figure 6.20 depicts the circuit diagram which also exhibits a double-scroll

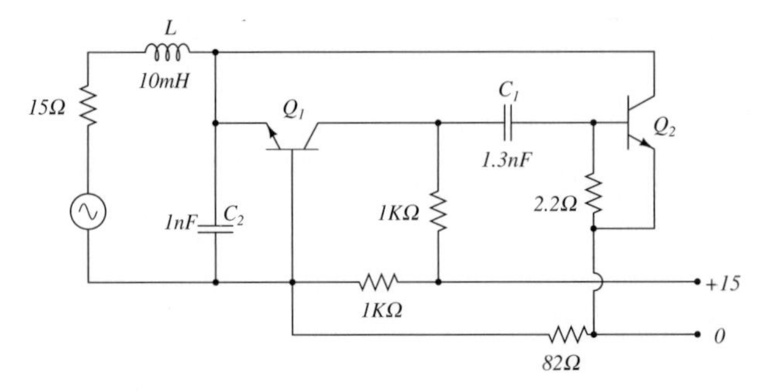

Fig. 6.18. Hunt's nonlinear oscillator circuit

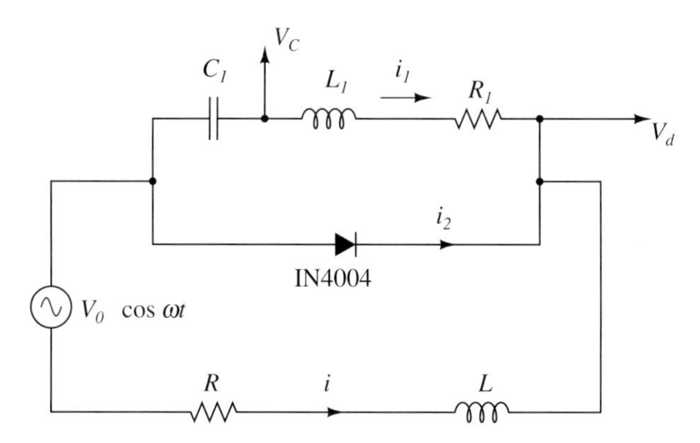

Fig. 6.19. A circuit with p-n junction diode as nonlinear capacitance

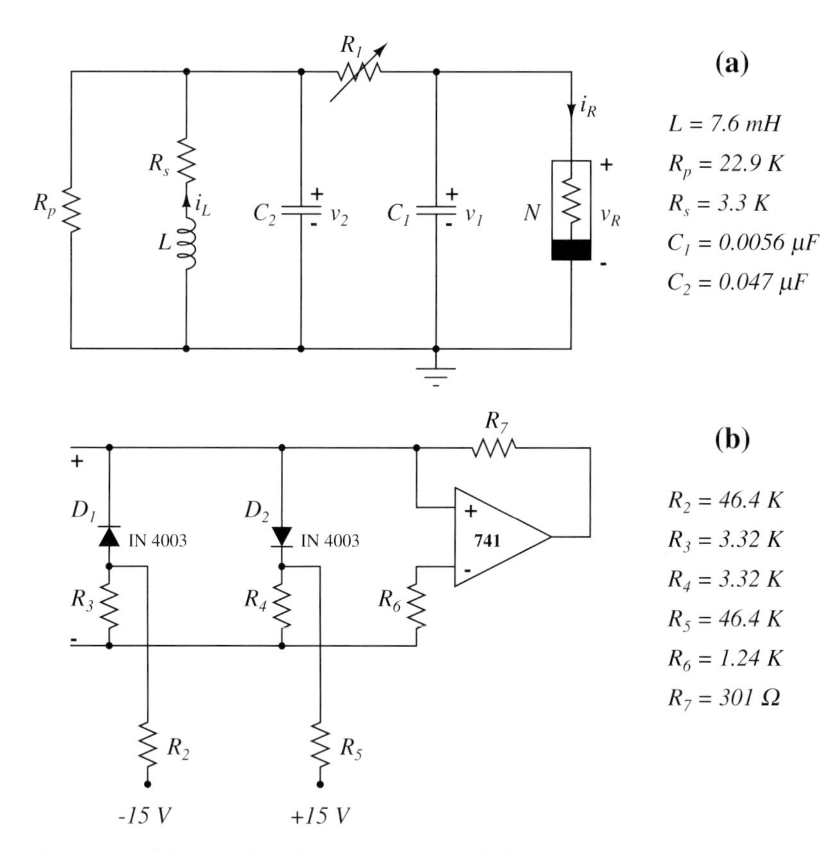

Fig. 6.20. (**a**) Modified Chua circuit and (**b**) circuit realization of the nonlinear element

attractor [11]. It is a modified form of the Chua's circuit (Fig. 6.6). In the circuit (6.20) the inductor is implemented as an active inductor composed of resistors, capacitors and an amplifier. The two resistors R_s and R_p have been included to provide an accurate model of the equivalent series and parallel resistance of the active inductor. The nonlinear element is a piecewise linear negative conductance defined by the function (6.13). This function can be implemented either using the circuit shown in Fig. 6.8 or the circuit depicted in Fig. 6.20b. The equations of motion for the circuit (6.20) can be obtained by applying Kirchhoff's laws. The reader can compare the obtained equations with (6.14). When R_1 is decreased from 1.6KΩ period doubling cascade leading to one-band chaos, double-band chaos and double scroll attractor occurs.

6.6.5 Colpitt's Oscillator

Figure 6.21 shows a simple piecewise-linear oscillatory circuit exhibiting chaotic dynamics, namely, the Colpitt's oscillator [12]. The circuit consists of a single bipolar junction transistor (BJT) Q. It is biased in its active region by means of V_{EE}, R_E and V_{CC}. The feedback network consists of an inductor L with a series resistance R_L and a capacitive divider composed of C_1 and C_2. Typical values of circuit elements are $R_L = 35\Omega$, $L = 98.5\mu\text{H}$, $C_1 = C_2 = 54\text{nF}$, $V_{CC} = 5\text{V}$, $V_{EE} = -5\text{V}$, $R_{EE} = 400\Omega$ and the BJT is type 2N2222. If we assume that the BJT acts as a purely resistive element, then the circuit may be described by the following set of three autonomous differential equations:

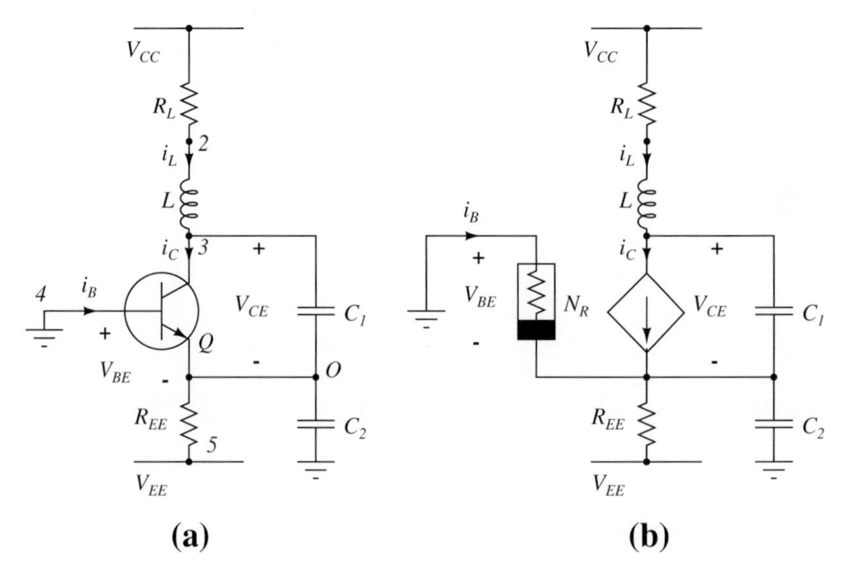

Fig. 6.21. (**a**) BJT Colpitt's oscillator and (**b**) its equivalent circuit

$$C_1\frac{\mathrm{d}V_{CE}}{\mathrm{d}t} = I_L - I_C\ , \tag{6.29a}$$

$$C_2\frac{\mathrm{d}V_{BE}}{\mathrm{d}t} = -\frac{V_{EE}+V_{BE}}{R_{EE}}I_L - I_B\ , \tag{6.29b}$$

$$L\frac{\mathrm{d}I_L}{\mathrm{d}t} = V_{CC} - V_{CE} + V_{BE} - I_L R_L\ . \tag{6.29c}$$

Experimentally one observes that the BJT operates in just two regions: forward active and cut off. Consequently, the transistor can be modelled as a two-segment piecewise-linear voltage-controlled resistor N_R and a linear current-controlled current source as shown in the Fig. 6.21b. Thus

$$I_B = \begin{cases} 0 & \text{if } V_{BE} \le V_{TH} \\ (V_{BE} - V_{TH})/R_{ON} & \text{if } V_{BE} > V_{TH} \end{cases} \tag{6.30a}$$

$$I_C = \beta_F I_B\ , \tag{6.30b}$$

$$\tag{6.30c}$$

where V_{TH} is the threshold voltage (≈ 0.75V), R_{ON} is the small signal on-resistance of the base-emitter junction and β_F is the forward current gain of the device.

6.7 Nonlinear Circuits as Dynamical Systems

The above analysis should convince the readers that nonlinear electronic circuits may be treated as interesting dynamical systems on their own merit or as analog simulations of existing dynamical systems. In either way, nonlinear electronic circuits can play a very useful role in understanding nonlinear dynamical systems. More importantly they can be used as testing grounds from applications point of view of the various concepts emerging in chaotic dynamics (see for example, Ref. [6]). Some of them are indicated in later chapters.

Problems

1. Write down the state equations for the linear circuits shown below in Fig. 6.22. Also construct the exact solutions of the systems.
2. Assume that the voltage-current characteristic of the nonlinear element N in the Duffing–van der Pol circuit shown in Fig. 6.23 is $I(V) = aV + bV^3$ ($a < 0$, $b > 0$) [13]. Construct a nonlinear resistor for the above I-V relation. Show that the equation of motion of the circuit after suitable scaling becomes

$$\dot{x} = -\mu\left(x^3 - \alpha x - y\right)\ ,$$
$$\dot{y} = x - y - z\ ,$$
$$\dot{z} = \beta y\ .$$

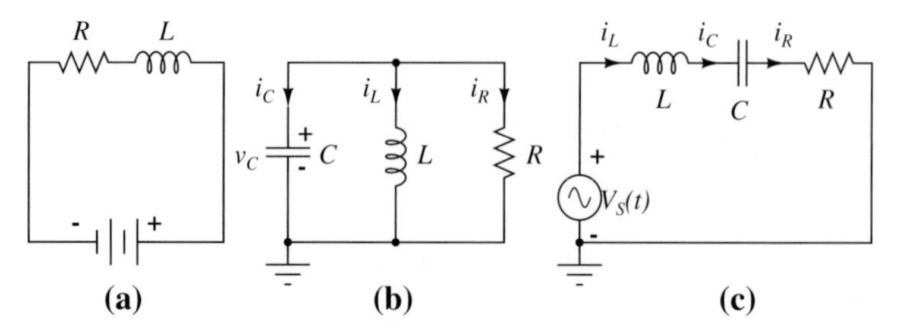

Fig. 6.22. Some linear circuits

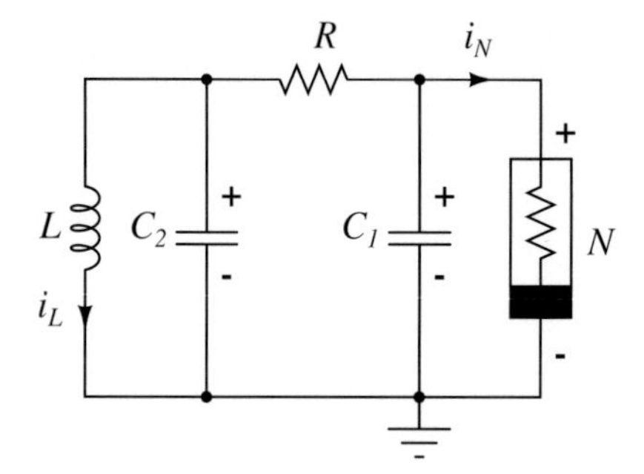

Fig. 6.23. A circuit realization of the Duffing–van der Pol oscillator

Investigate the dynamics of the circuit both experimentally and numerically for $\mu = 100$, $\alpha = 0.35$ and varying β from 200 to 1000. (See also Ref. [6]).

3. Show that the mathematical model describing the dynamics of Matsumoto–Chua–Kabayashi circuit [14] given in (Fig. 6.24) can be written as (after using Kirchhoff's laws)

$$\begin{aligned}
\dot{x} &= y - N(x, z) , \\
\dot{y} &= x + by , \\
\epsilon \dot{z} &= -w + N(x, z) , \\
\mu \dot{w} &= z ,
\end{aligned}$$

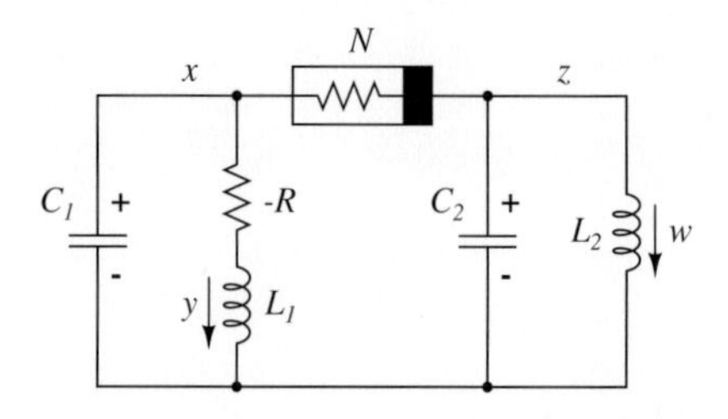

Fig. 6.24. Matsumoto–Chua–Kobayashi circuit

where N is a nonlinear resistor. For the nonlinear function

$$N(x, z) = \begin{cases} 0 & \text{if } x - z \leq 1 \\ c(x - z - 1) & \text{if } x - z > 1 \end{cases}$$

construct a nonlinear resistor circuit. Study the dynamics of the circuit as well as the model equation for $c = 3$, $\epsilon = 0.1$ and $\mu = 0.66$ as a function of b in the interval $[0, 0.5]$.

4. The dynamics of the circuit shown in Fig. 6.25 is given by

$$C\dot{v} = i - f(v) ,$$
$$L\dot{i} = -v + R_1 i + E \cos \omega t .$$

Rewrite the above equation of motion in dimensionless form. Fix the parameters in the circuit as $C = 9.81$nF, $L = 128$mH, $R_1 = 10\Omega$, $R_2 = 268\Omega$, $R_3 = 1\text{K}\Omega$, $\omega = 2\pi \times 4500$ rad/sec. Draw the $v - i$ characteristic of the nonlinear resistor of the circuit. Assuming $i(v) = f(v) = av + bv^3$, find a and b from the experimental data. Then investigate the dynamics by varying the parameter E [15]. Compare the experimental result with dynamics of the corresponding equation of motion of the circuit.

5. Figure 6.26 depicts an oscillator circuit consisting of current feedback op-amplifier (CFOA) with two resistors and two capacitors [16]. Show that a condition for oscillation is $C_1 = 2C_2$, $R_2 = 2R_1$ and the frequency of oscillation is $\omega_0 = 1/(R_2 C_2)$.

6. Figure 6.27 shows a nonlinear circuit [16] with junction field effect transistor (JFET) as a nonlinear element and current feedback op-amplifier (CFOA) as the active building block. In this circuit, the gate and drain terminals of JFET are shorted. For AD844 CFOA biased with ± 9V supply and a JFET of the type J2N4338, draw the current flowing in the JFET versus its drain to source voltage for $R = 900\Omega$ and $C_1 = C_2 = C_3 = C = 0.01\mu$F. Verify that the JFET characteristic is nonlinear. Investigate the dynamics of the circuit (use $V_{C_1} - V_{C_2}$ trajectory for analysis) in the R-C parameter space.

7. Consider the following pendulum equation which has been used to represent the radio-frequency-driven superconducting quantum interference device (SQUID) with inertia and finite damping

$$\ddot{x} + k\dot{x} + x + \frac{\beta}{2\pi} \sin 2\pi x = q_1 \sin \omega_1 t + q_2 \sin \omega_2 t .$$

Construct an analog simulator for the above equation. Fix the parameters at $k = 2$, $\beta = 2$, $\omega_1/\omega_2 = (1 + \sqrt{5}\,)/2$ (golden mean) and $q_1 = 2.768$. Investigate the dynamics by varying q_2 from 0 [17]. Compare the result by numerically integrating the equation of motion.

8. Construct an analog circuit for the Duffing–van der Pol oscillator equation

$$\ddot{x} + p\left(x^2 - 1\right)\dot{x} + \omega_0^2 x - \beta x^3 = f \sin \omega t .$$

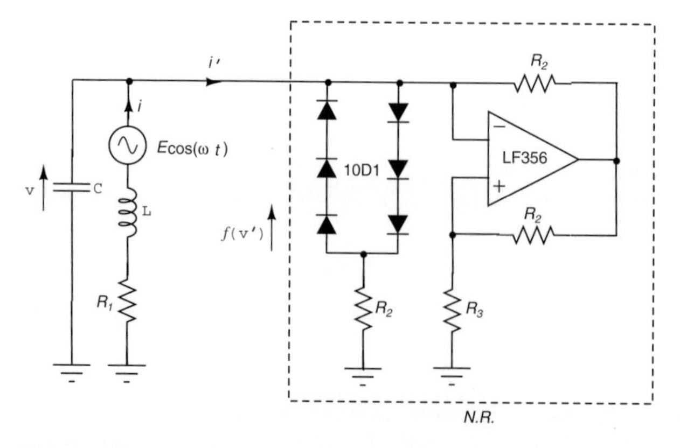

Fig. 6.25. A nonlinear circuit

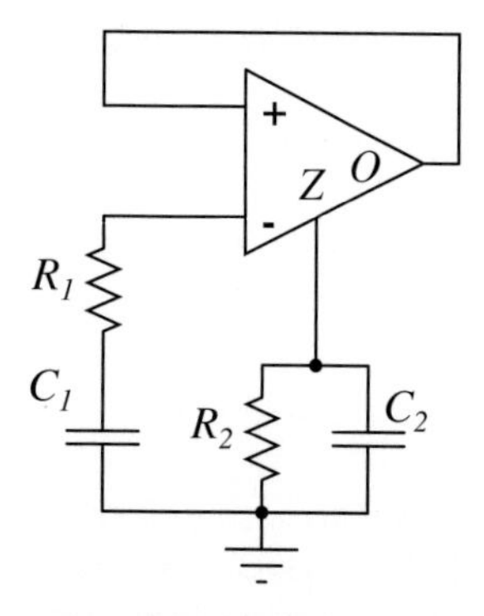

Fig. 6.26. An oscillator circuit with current feedback op-amplifier

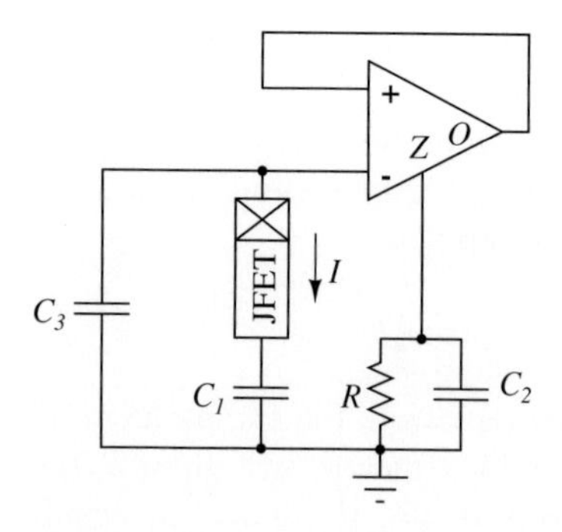

Fig. 6.27. A nonlinear circuit with JFET as a nonlinear element

Study the dynamics of the system for $p = 0.1$, $\omega_0^2 = 1$, $\beta = 0.1$ and $\omega = 1.57$ by varying the control parameter f [18]. Compare the result by numerically integrating the equation of motion.

9. Modify the Chua's circuit shown in Fig. 6.7 by adding an external periodic forcing $f(t) = f' \sin \omega t$ in series with the inductor. Show that the equation of motion of the circuit after suitable scaling becomes

$$\dot{x} = \alpha(y - x - h(x)) ,$$
$$\dot{y} = x - y + z ,$$
$$\dot{z} = -\beta y + f \sin \Omega t ,$$

where $h(x) = bx + 0.5(a - b)[|x + 1| - |x - 1|]$. Study the dynamics of the circuit both experimentally and numerically for $\alpha = 7.0$, $\beta = 14.87$, $\Omega = 3$, $a = -1.27$ and $b = -0.68$ and varying f from 0 to 1.5 [19].

10. The MLC circuit shown in Fig. 6.12 can be modified by placing the inductor L in parallel with the capacitor C. Write down the state equation for this circuit. Carry out the following analysis [20].
    a) Identify the fixed points and find their linear stability of the undriven system ($f(t) = 0$).
    b) In the presence of the external periodic signal find the explicit analytical solution of the circuit equation in the three different regions of the characteristic curve. Obtain the various dynamical phenomena associated with the system by making use of this solution.
    c) Investigate the dynamics of the circuit both experimentally and numerically for $L = 445$mH, $C = 10$nF, $R = 1475\Omega$, $\omega = 2\pi \times 1115$Hz and varying f from 0 to 500mv.
11. Construct a suitable analog simulation circuit to study the dynamics of DVP oscillator equation given in problem 2.
12. Study the dynamics of the Lorenz equations (5.5) by means of analog simulation circuit [21].

7. Chaos in Conservative Systems

In the previous chapters, we have discussed in some detail the nonlinear phenomena associated with dissipative dynamical systems. Typically the long time bounded motions of such systems are described by attractors. A bounded trajectory starting from an arbitrary initial condition approaches an attractor asymptotically (in the limit $t \to \infty$). The attractor may be an equilibrium point or a periodic orbit or even a chaotic orbit. Further, a dissipative nonlinear system may admit one or more attractors as shown in the case of the Duffing oscillator and Lorenz equations in Chap. 5, depending upon the values of the system parameters and initial conditions. When a control parameter is varied qualitative changes in the dynamics, namely bifurcations, occur at certain critical values of the control parameters.

We now ask: What is the situation in conservative (energy/volume preserving) nonlinear systems? As we have mentioned in Sect. 3.8, the phase space volume is conserved and so the trajectory cannot reach an equilibrium point asymptotically. This implies that nodes and spirals are not possible in conservative systems (Verify). However, elliptic (center) and hyperbolic (saddle) equilibrium points can and do occur. In fact, attracting solutions are not possible at all in conservative systems (Verify). Then, what are the possible orbits in conservative systems? How does chaos occur here? What are the possible routes to chaos? These are some of the questions we wish to consider in this chapter.

In the numerical study of many conservative systems (continuous systems as well as maps), one finds the following general picture: Initially for a certain range of a control parameter the entire accessible phase space region is filled with regular orbits on *invariant tori*. Then, above a certain critical value of the control parameter part of the phase space is filled with invariant tori and the rest is occupied by *chaotic trajectories*, often called *stochastic trajectories* . The volume occupied by the invariant tori decreases abruptly with an increase in the parameter value and the system exhibits a variety of orbits depending upon the initial conditions. A seminal numerical example, illustrating the occurrence of different kinds of orbits and the transition from regular to stochastic motion, was identified by Hénon and Heiles in 1964 [1] to simulate the behaviour of a three body problem such as the motion of an asteroid around the sun, perturbed by the motion of Jupiter. Another simple system exhibiting both regular and chaotic orbits in phase space is the undamped (conservative), but periodically driven, Duffing oscillator, which is often treated as a perturbed Hamiltonian system. The model equations of Hénon–Heiles and conservative Duffing oscillator are nonlinear ordinary differential equations, where time is continuous.

Alternatively, as noted earlier in Chap. 4, discrete time systems or maps play a crucial role in understanding the various nonlinear phenomena. The map which is often used as a prototype model to study the dynamics of conservative systems is the *standard map*. In the present chapter we illustrate many of the concepts associated with nonlinear conservative systems with reference to three specific systems, namely,

(1) the Hénon–Heiles system,

(2) the conservative Duffing oscillator equation under periodic forcing and

(3) the standard map.

Earlier in Sect. 5.4, we have shown that transition from regular to chaotic motion occurs only in dynamical systems with (phase space)dimension greater than two. When the dimension is greater than three (or even equal to three in some cases), it will be difficult to visualize the orbits in the phase space and to find the characteristic features of them. Further, a nonlinear Hamiltonian system with one degree of freedom cannot admit chaos as its phase space dimension is only two. So the next possibility is a two degrees of freedom Hamiltonian system, which is certainly a candidate for chaos (where phase space dimension is four). Otherwise, a nonautonomous perturbation of the one degree of freedom Hamiltonian system can also exhibit chaos. More often, the dynamics of higher dimensional systems is investigated on a section of the phase space called *Poincaré surface of section* briefly mentioned in Chap. 3. In this chapter, we will illustrate the dynamics of the Hénon–Heiles system on such a surface of section (Sect. 7.3). Therefore, first we describe the basic concepts of Poincaré surface of section (Sect. 7.1) and then the possible orbits with reference to such a surface of section (Sect. 7.2). We also indicate in Sect. 7.4 how a periodic perturbation to a one-dimensional anharmonic oscillator can show chaos. The above results are confirmed further by discussing the dynamics of the standard map in Sect. 7.5. A brief introduction to the celebrated Kolmogorov–Arnold–Moser theorem is also given towards the end of the chapter (Sect. 7.6).

7.1 Poincaré Cross Section or Surface of Section

Earlier in Sect. 3.7.3, we have introduced the notion of Poincaré map for periodically driven dynamical systems. In these systems, for a clear visualization of the geometry of complicated attractors, one can collect the values of the state variables at every period of the driving force. In this way, from a continuous-time trajectory we can obtain a discrete sequence of data. What is the analogous situation in dynamical systems not driven by a periodic force? What is the scheme to be followed to obtain an appropriate discrete set of state values for autonomous systems such as the Lorenz equations or the Hénon–Heiles system? We will consider briefly these questions in this section.

For autonomous dynamical systems, we can visualize the evolution on a suitably defined surface of section of the full phase space. For illustrative purpose, let us consider a three-dimensional dynamical system with state variables (x_1, x_2, x_3). As discussed in Chap. 3, a state in the phase space is represented by the instantaneous values of $(x_1(t), x_2(t), x_3(t))$. Suppose we define a plane S with $x_3 = c$, a constant (often chosen to be zero). The coordinates in this plane will be x_1 and x_2. In other words, only the intersections of a trajectory with $x_3(t) = c$ are considered. Infinitely many other suitable planes (for example, with different values of c) can also be constructed. A typical trajectory cuts the plane S at the points P_1, P_2, ... (Fig. 7.1) so that any one of them is determined by the preceding one. This plane is the *Poincaré cross section* or *surface of section* (SOS) and one can define the Poincaré map as the corresponding mapping of the plane onto itself,

$$P_k \longrightarrow P_{k+1} = F(P_k) \ . \tag{7.1}$$

Essentially, a continuous-time trajectory in the phase space is now replaced by a set of points $P_1, P_2, ...$ on the SOS. The time durations between the successive intersections are not necessarily equal. When the given system is driven by a periodic force, say $f \cos \omega t$, then as described in Sect. 3.7.3, the SOS is obtained by means of a stroboscopic illumination, that is, picking out the points at $t_n = n2\pi/\omega + t_0$, where n is an integer and $2\pi/\omega$ is the period of

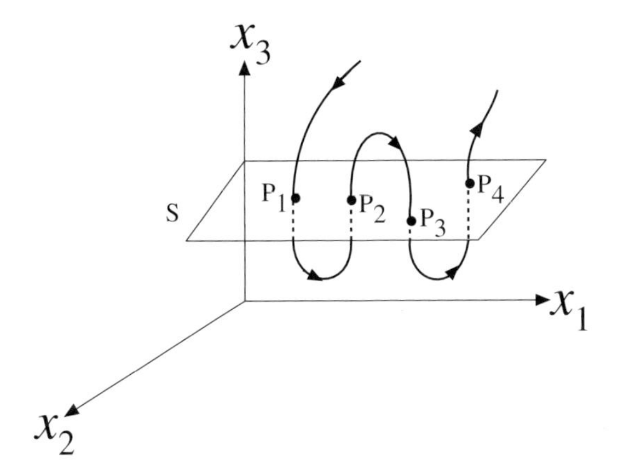

Fig. 7.1. Intersections of a trajectory on a Poincaré surface of section (SOS)

the driving force, and t_0 is arbitrary. In this case the time durations between successive points are equal to the period of the driving force.

Example:

For illustrative purpose, let us consider the following system of three coupled first-order nonlinear ordinary differential equations introduced to model dynamic conflict with mediation, for example, conflict between townspeople, university students and police (also see the problem 12 of Chap. 5)

$$\dot{x} = -0.4x + y + 10yz \,, \tag{7.2a}$$

$$\dot{y} = -x - 0.4y + 5xz \,, \tag{7.2b}$$

$$\dot{z} = \alpha z - 5xy \,. \tag{7.2c}$$

Figure 7.2a depicts a chaotic orbit of it in the $(x - y - z)$ space. Suppose we choose the SOS as the $y - z$ plane with $x = 0$. From Fig. 7.2a, we note that the trajectory passes through this plane several times as it evolves. We collect the values of y and z whenever the orbit crosses this plane (that is, $x(t) = 0$) from the negative x-axis. Figure 7.2b shows the plot of these data in the $y - z$ plane.

The above ideas can be extended to conservative Hamiltonian systems as well. For example, if we consider two degrees of freedom Hamiltonian systems, the dimension of the phase space is four. However, since the energy is conserved, one can fix the value of the energy and consider the associated energy 'surface', which is three-dimensional. As in the three-dimensional autonomous case, one can then associate suitable SOS.

It is interesting to note that in spite of a reduction of the dimension of the phase space the SOS retains many of the properties of the original flow. If the motion is conservative in an n-dimensional phase space, the same is true on the SOS. In other words, if the flow preserves volume in the phase

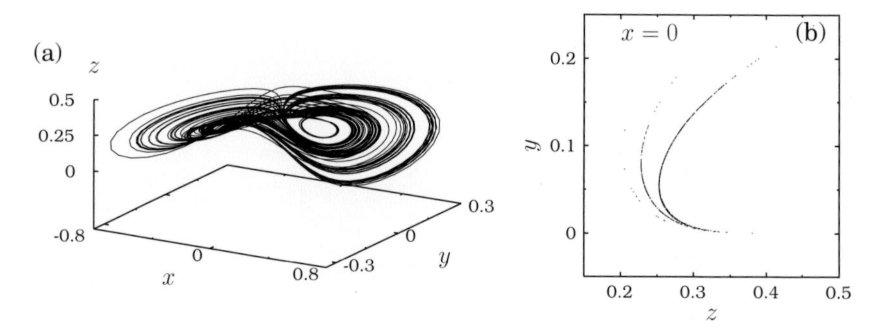

Fig. 7.2. Chaotic attractor (**a**) in the x-y-z phase space and (**b**) on the surface of section of (7.2) for $\alpha = 0.175$

space then the sequence of points which are intersections of the flow on the two-dimensional SOS also preserves area. If the system is dissipative, then there is both contraction of volume in the phase space and of area in the SOS. Further, a periodic orbit in the phase space of a system generates finite number of repeating points on the SOS (Fig. 7.3a). That is, for all of the intersections P_j of the orbit with the SOS, there is a number K such that

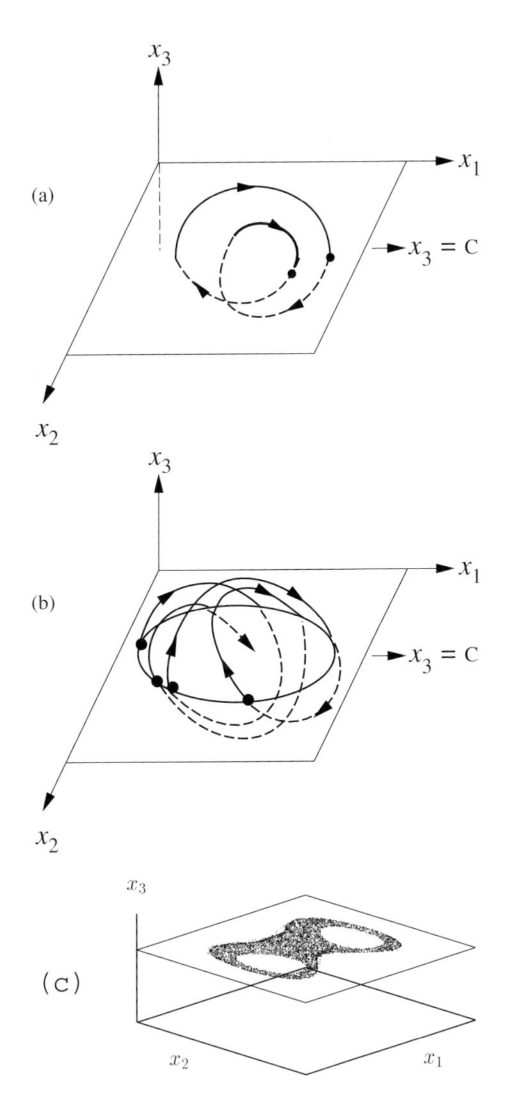

Fig. 7.3. Intersections of (**a**) a periodic orbit, (**b**) a quasiperiodic orbit and (**c**) a chaotic orbit in a conservative system by a SOS. Note that in (**c**) we have suppressed the orbits and included only the intersections on the SOS

$$F^K(P_j) = P_j \; . \tag{7.3}$$

Thus, a periodic cycle is a fixed point of some powers of the Poincaré map. For a quasiperiodic motion the orbit in the phase space is unclosed and hence the points on the SOS will not be periodic. However, they fill up a smooth curve on the SOS (Fig. 7.3b). For a chaotic evolution, the points on the SOS appear to bounce around haphazardly as shown in Fig. 7.2b for (7.2) or Fig. 7.3c for conservative systems.

Further, in conservative systems the chaotic orbit also preserves phase space area on the SOS whereas it is not so in dissipative systems. A consequence of this is that the basin of attraction of a chaotic orbit in conservative systems is identical to the region occupied by it. In contrast, in dissipative systems the orbit is a subset of its basin of attraction. That is, in dissipative chaotic systems there are regions in the phase space which are not parts of the chaotic attractor but the trajectories starting from these regions are attracted asymptotically to the chaotic attractor. This important property is characterized by the dimension of the attractor (for details see Sect. 8.5). The dimension of the chaotic orbit of conservative systems will be integer while that of dissipative systems will be noninteger.

Exercises:

1. Show that attractors are not possible at all in conservative systems.
2. Obtain the SOS plot for the Lorenz system corresponding to (5.5), in the $x = 0$ plane, for different types of orbits.
3. Obtain the stroboscopic plot of the Duffing oscillator equation (5.1) for various orbits and compare with the SOS plots of the Lorenz system.

7.2 Possible Orbits in Conservative Systems

Let us consider a nonlinear conservative system with N degrees of freedom and described by the Hamiltonian

$$H = H_0\left(q_i, p_i\right) , \quad i = 1, 2, \ldots, N \; . \tag{7.4}$$

As we have discussed in Sect. 3.7.1, if one is able to find a suitable generating function $F(q_i, J_i)$, then the Hamiltonian can be expressed purely in terms of the new momenta J_i, which are now the action variables, as

$$H = H_0\left(J_i\right) , \quad i = 1, 2, \ldots, N \; . \tag{7.5}$$

Actually such a possibility arises when the N degrees of freedom system with the Hamiltonian H_0 admits N linearly independent, involutive integrals of motion including the energy integral $E = H_0$. In this case the system is known to be integrable or completely integrable in the Liouville sense and the motion is confined to the N-dimensional integral surface called *invariant torus*. Further discussions on such integrable systems will be given later in Chaps. 10, 13–14.

It is simple to see that the corresponding canonical equations of motion (see also Sect. 13.4 for more details) can be trivially integrated to yield

$$J_i = \text{constant} , \tag{7.6a}$$

$$\theta_i = \left(\frac{\partial H_0}{\partial J_i}\right) t + \text{constant} \equiv \nu_i t + \text{constant} , \quad i = 1, 2, \ldots, N \tag{7.6b}$$

so that the dynamics can be expressed in toroidal representation as discussed earlier and in Sect. 3.7.1. For example, for two degrees of freedom J_1 and J_2 (actions) represent the minor and major radii of a torus and θ_1 and θ_2 represent the corresponding angles. What happens if this system is perturbed by additional forces?

For illustrative purpose, let us consider an autonomous system with two degrees of freedom given by the Hamiltonian H,

$$H(J_1, J_2, \theta_1, \theta_2) = H_0(J_1, J_2) + \epsilon H_1(J_1, J_2, \theta_1, \theta_2) , \quad \epsilon \ll 1 , \tag{7.7}$$

where J_i's and θ_i's ($i = 1, 2$) are the action and angle variables, respectively, of the unperturbed system ($\epsilon = 0$). The perturbation parameter ϵ is assumed to be small, H_0 is a function of the actions alone and H_1 is a periodic function of the angle variables (θ's). As mentioned in the beginning of this chapter, because of volume preservation of any small volume element in the phase space, conservative systems do not have attractors. However, they can exhibit different types of orbits depending upon the values of the control parameters and initial conditions. Typically, the existence of the various types of orbits for given parameter values and initial conditions can be identified by detailed numerical analysis. However, very meaningful information can be deduced through appropriate perturbation analysis of the Hamiltonian (7.7), for example through canonical perturbation theory (as indicated in appendix B). Or more profitably one can make a judicious combined numerical and analytical effort to understand the dynamics underlying systems of the type (7.7). The basic conclusions are as follows. For more details, see Refs. [2–6]. The above types of studies reveal that conservative systems admit the following broad categories of motion. (1) Regular motion corresponding to periodic and quasiperiodic orbits and (2) irregular motion consisting principally of chaotic trajectories. The details are as follows.

7.2.1 Regular Trajectories

Regular trajectories are generally either periodic orbits with definite periodicity or quasiperiodic (almost periodic) orbits. For a system with two degrees of freedom, these orbits are described in general by specifying the trajectory in a four-dimensional phase space with coordinates $(J_1, J_2, \theta_1, \theta_2)$. However, for $\epsilon = 0$, the system (7.7) possesses two single-valued constants of motion, namely, the action variables J_1 and J_2. Consequently, the motion takes place in a two-dimensional subspace of the phase space labelled by the action variables. This subspace is usually a two-dimensional torus. Points on a given torus are labelled by the angle variables θ_1 and θ_2. For the two degrees of freedom systems, as shown in Sect. 3.7.1, the tori are two-dimensional surfaces, similar to the surface of a doughnut. The trajectory of the system winds around the torus. The motion around the major axis of the torus is periodic in θ_1 with period $2\pi/\omega_1$.

A natural way to study the motion on the torus is to look at the dynamics on the SOS of the torus, for example, the section with $\theta_1 = c \pmod{2\pi}$, where c is a constant. The successive intersecting points θ_{20}, θ_{21},... will be

$$\theta_{2m} = \theta_{20} + m\alpha \,, \quad \pmod{2\pi} \quad m = 0, 1, 2, ... \tag{7.8a}$$

where $\alpha = 2\pi\omega_2/\omega_1$ and

$$\omega_i = \partial H / \partial J_i \,, \quad i = 1, 2 \,. \tag{7.8b}$$

Now the typical orbit on the torus depends on whether the frequency ratio ω_2/ω_1 is rational or irrational.

A. *Rational ω_2/ω_1: Primary Resonance*

If the frequency ratio ω_2/ω_1 is a rational number l/k, then ω_1 and ω_2 satisfy the (resonance)condition, see (3.61),

$$k\omega_2 - l\omega_1 = 0 \,. \tag{7.9}$$

Then ω_1 and ω_2 can be written as integral multiples of some frequency ω_0. In a time interval $2\pi/\omega_0$ both the angle variables θ_1 and θ_2 increase by integral multiples of 2π and the system returns to its original state as shown in Fig. 7.4. The motion is then periodic. Now on the SOS we have

$$\theta_{2m} = \theta_{20} + 2\pi m \frac{l}{k} \pmod{2\pi} \,. \tag{7.10}$$

This implies that when $m = k$ the point $\theta_{2m} = \theta_{20}$. Thus, on the SOS k points repeat. This is shown in Fig. 7.5a for $k = 5$, $l = 2$, where the SOS is defined by $\theta_1 = c$. More precisely, the trajectory is

(1) periodic in θ_1 and θ_2,

(2) a closed curve on the torus and

(3) a set of repeating points on the SOS.

These k intersecting points can be regarded as fixed points of the motion on the SOS (refer Sect. 7.1). This kind of motion is referred to as belonging to a *primary resonance* since it is a closed periodic trajectory.

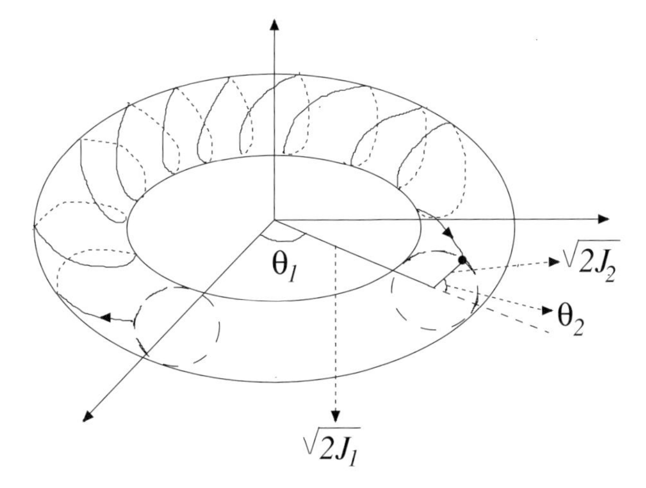

Fig. 7.4. A periodic orbit on a two-dimensional torus

B. *Rational ω_2/ω_1: Primary Islands and Secondary Resonance*

What is the nature of a trajectory in the neighbourhood of the primary resonance of Fig. 7.5a when the perturbation in (7.7) is switched on ($\epsilon \neq 0$) for ω_2/ω_1 rational? We know that in a conservative system the possible fixed points are elliptic and saddle (hyperbolic). Near the elliptic point trajectories encircle it and do not approach it as $t \to \infty$. Consequently, on the SOS points fall on a closed curve in the neighbourhood of the primary resonance. This is depicted in Fig. 7.5b, where five smooth curves encircle the five fixed points. These curves are known as *primary islands* or simply *islands*. In the case of the perturbed two degrees of freedom system (7.4), the motion takes place on island tori. On the SOS, the sequence of intersecting points on the primary islands surrounding the fixed points will be definite periodic, contain repeating set of points. Figure 7.5c shows the SOS for a closed periodic trajectory which winds three times about the $k = 5$, $l = 2$ primary resonance in fifteen circuits of θ_1 which is an example discussed in Ref. [2]. This is a *secondary resonance*. Essentially, the primary resonances give rise to primary islands which give rise to secondary resonances and their islands and so on. If the intersecting points are not periodic then they densely fillup the invariant curves.

C. *Irrational ω_2/ω_1: Quasiperiodic Orbit*

When l/k is irrational then $\theta_{2m} \neq \theta_{20}$ for all m and the points on the SOS are nonrepeating and the corresponding orbit on the torus never closes. Since the flow is on the torus the points on the SOS of the torus fall on a well defined smooth closed curve called *invariant curve*. This is shown in Fig. 7.5d.

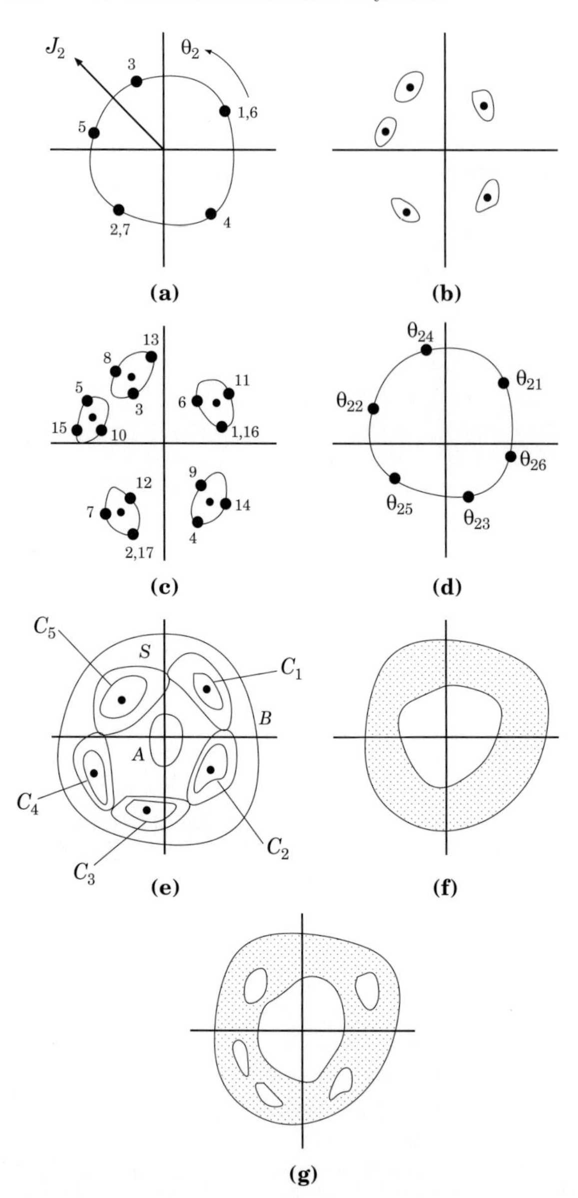

Fig. 7.5. Intersection of a trajectory lying on the surface of section defined by θ_1 =constant. (**a**) Primary resonance $k = 5$, $l = 2$. (**b**) Primary islands around the $k = 5$, $l = 2$ fixed points. (**c**) Secondary resonance with three steps around the $k = 5$, $l = 2$ primary resonance. (**d**) Invariant curve for irrational ω_2/ω_1. (**e**) Coexistence of various regular orbits. (**f**) Stochastic region between two invariant curves. (**g**) Stochasticity layer bounded by both primary and secondary invariant curves. Adapted from Ref. [2]

On the SOS, the various regular orbits described above may coexist and can be obtained for different initial conditions. Figure 7.5e depicts such a situation where one can see invariant curves (marked as A and B), primary resonance (five fixed points) and primary islands ($C_1 - C_5$). The curve S in this figure separating invariant curves from islands is known as a *separatrix*.

7.2.2 Irregular Trajectories

As noted above primary islands consists of elliptic fixed points. Between two neighbouring elliptic points there exists a saddle or hyperbolic point. Figure 7.5e shows the chain of 5 alternating elliptic and hyperbolic points. Neighbouring hyperbolic points are smoothly joined by separatrices. Typical example of such a scenario is the phase plane of the pendulum and one can show that in the neighbourhood of resonances the phase space is pendulum-like for a weakly perturbed Hamiltonian of type (7.7) [2,6].

Now what happens in the neighbourhood of the hyperbolic points and separatrices? Hyperbolic points play key role for the origin of complicated irregular motion in conservative systems. As we move away from the elliptic points the period of the island oscillation generally increases. The period is infinite on the separatrix trajectory. Near the separatrix naturally the period of island oscillation takes a very large value, that is, the frequency ω of island oscillation approaches zero.

Let us consider the separatrix joining two nearest neighbouring hyperbolic points. The portion of the separatrix which enters a hyperbolic point is called a stable manifold. The orbit which emanates from the hyperbolic point is called an unstable manifold. For an integrable system ($\epsilon = 0$), (7.7), the stable and unstable manifolds are smoothly joined. When $\epsilon \neq 0$ the manifolds do not join. Suppose the manifolds of two different hyperbolic points intersect transversely at a point called a *heteroclinic point*. Or the stable and unstable manifolds of the same hyperbolic point meet transversely at the so-called *homoclinic point*. Then it has been established that one intersecting point implies infinitely many [2–4]. These points are more and more closely spaced as the hyperbolic points are approached. Since the phase space volume of a conservative system has to be preserved the oscillations must grow in amplitude as the hyperbolic points are approached.

Further, near a saddle as seen through the linear stability analysis in Sect. 3.4, the trajectories diverge exponentially which is a source of chaotic motion. Consequently the orbits in the neighbourhood of separatrices are nonperiodic, chaotic and fill densely the volume available to them in the phase space. For small perturbation this region is thin and is referred as a *resonance layer* [2–6]. Such resonance layers are separated by invariant curves (in the SOS). Motion from one layer to another layer is forbidden by the invariant curves. When the perturbation is increased the invariant curves separating the resonance layers are destroyed. This results in a decrease in the volume occupied by invariant curves (tori) and increase in the volume occupied by resonance layers. When all the invariant curves are destroyed the resonance layers merge or overlap. The Kolmogorov–Arnold–Moser theorem, to be mentioned in Sect. 7.6 gives a solid basis for the existence of these results.

Figure 7.5f shows an annular layer of chaotically filled points by a single trajectory lying between two invariant curves. Occurrence of both chaotic orbits and islands is also possible between two invariant curves. This is illustrated in Fig. 7.5g. A clear demonstration of the various above mentioned orbits and transitions will be given in Sects. 7.4–7.7 by analysing the dynamics of the Hénon–Heiles, perturbed quartic oscillator and standard map systems.

7.2.3 Canonical Perturbation Theory: Overlapping Resonances and Chaos

A qualitative understanding of the occurrence of the above type of regular and irregular trajectories and in particular how stochastic motion sets in near the separatrices can be obtained by developing a suitable canonical perturbation theory for the Hamiltonian (7.7) in powers of the small parameter ϵ as discussed in appendix B, and identifying overlap of resonances at higher and higher orders as the main cause for the onset of Hamiltonian chaos. The conclusions can then be corroborated by appropriate numerical analysis. These ideas can be illustrated with typical examples, for example through the model systems studied by Walker and Ford [7]. For more details, see for example Ref. [5].

Example 1: A (2, 2) Resonance System

Consider the Hamiltonian

$$H = H_0\,(J_1, J_2) + \epsilon J_1 J_2 \cos\,(2\theta_1 - 2\theta_2) = E\ , \tag{7.11}$$

where the unperturbed Hamiltonian is

$$H_0\,(J_1, J_2) = J_1 + J_2 - J_1^2 - 3J_1J_2 + J_2^2\ . \tag{7.12}$$

Defining the canonical transformation

$$\mathcal{J}_1 = J_1 + J_2 = I', \quad \mathcal{J}_2 = J_2, \quad \phi_1 = \theta_2, \quad \phi_2 = \theta_2 - \theta_1\ , \tag{7.13}$$

the Hamiltonian (7.11) becomes

$$\mathcal{H} = \mathcal{J}_1 - \mathcal{J}_1^2 + 3\mathcal{J}_2^2 + \epsilon\mathcal{J}_2\,(\mathcal{J}_1 - \mathcal{J}_2)\cos 2\phi_2 = E\ . \tag{7.14}$$

Writing down the corresponding the canonical equations of motion for the dynamical variables $(\mathcal{J}_1, \mathcal{J}_2, \phi_1, \phi_2)$, one can find the fixed points to be $\phi_1 = n\pi/2$, $\mathcal{J}_2 = \mathcal{J}_{\prime}$, where $\mathcal{J}_0$ is a solution of the equation $-I'+6\mathcal{J}_0+\epsilon\cos(n\pi)(I'-2\mathcal{J}_0) = 0$. By carrying out a linear stability analysis (see Chap. 3), one can show that for n even the fixed points are hyperbolic and for n odd, they are elliptic in nature. These fixed points are separated by separatrices. The nature of the trajectories in the neighbourhood of the fixed points inside and outside the separatrices consisting of islands and invariant curves can be identified by computing the level curves through the canonical perturbation theory and confirmed by numerical analysis. The analysis shows that in the region inside and in the immediate neighbourhood outside the separatrix, which one might call as the nonlinear resonance zone, large changes in the action, $\mathcal{J}_2$, occur signalling a large exchange of energy between the modes of the system [5].

Example 2: A $(2,3)$ *Resonance System*

In a similar manner, considering the Hamiltonian

$$H = H_0\left(J_1, J_2\right) + \epsilon J_1 J_2^{3/2} \cos\left(2\theta_1 - 3\theta_2\right) = E \,, \tag{7.15}$$

where $H_0(J_1, J_2)$ is given by (7.12) in the previous example, one can show that (7.15) is integrable, exhibiting hyperbolic and elliptic fixed points and a $(2,3)$ resonance zone (see Refs. [5,7]).

Example 3: A Mixed $(2,2)$ *and* $(2,3)$ *Resonance System*

It can be established that any Hamiltonian system containing two or more resonance terms are not integrable because the second integral of motion does not exist in these cases. Here as the perturbing parameter increases, one can show that overlap of resonance regions can occur leading to a transition to chaos. Considering, for instance, the example of Walker and Ford [7] for a two resonances Hamiltonian system,

$$\begin{aligned} H &= H_0\left(J_1, J_2\right) + \epsilon_1 J_1 J_2 \cos\left(2\theta_1 - 2\theta_2\right) + \epsilon_2 J_1 J_2^{3/2} \cos\left(2\theta_1 - 3\theta_2\right) \\ &= E \,, \end{aligned} \tag{7.16}$$

one can show that as the energy increases, the resonance regions overlap and chaos occurs.

Exercises:

4. Show that for systems (7.11) and (7.15), the second integral of motion can be found (see Ref. [5] and also the appendix B).
5. Write down the equation of motion associated with the Hamiltonian system (7.14) and identify the fixed points. Carry out a linear stability analysis to investigate their nature.
6. Carry out a canonical perturbation theory analysis for system (7.11) by treating the term proportional to ϵ as weak. Bring out how the problem of small denominator arises and investigate the region of validity of the analysis. Draw the level curves.

7. Investigate the dynamics of (7.11) on suitable SOS by numerical analysis of the equation of motion. Compare the perturbation analysis results with the numerical results.
8. Repeat the exercises 5–7 to systems (7.15) and (7.16) [5,7].

With this brief survey, we are now in a position to turn our attention to concrete systems to understand clearly the chaotic dynamics underlying conservative systems. To start with we will consider the celebrated Hénon–Heiles systems.

7.3 Hénon–Heiles System

Let us consider the motion of a particle of unit mass in a plane corresponding to the two-dimensional potential $V(x, y)$. The Newton's equations of motion are given by

$$\ddot{x} = -\frac{\partial V}{\partial x}, \quad \ddot{y} = -\frac{\partial V}{\partial y}. \tag{7.17}$$

To illustrate the nature of the potential $V(x, y)$, we can treat x and y as horizontal coordinates and V as the vertical coordinate in a three-dimensional Euclidean space. Then $V(x, y)$ represents a surface and (7.17) describe the motion of a particle on this surface. Obviously, the Hamiltonian corresponding to (7.17) is

$$H = \frac{1}{2}\left(p_x^2 + p_y^2\right) + V(x, y), \tag{7.18}$$

where $p_x = \dot{x}$ and $p_y = \dot{y}$. We consider the form of V investigated by Hénon and Heiles, namely

$$V(x, y) = \frac{1}{2}\left(x^2 + y^2 + 2x^2 y - \frac{2}{3}y^3\right), \tag{7.19}$$

which is a simple model potential for the motion of a star in a cylindrically symmetric and gravitationally smooth galactic potential [1]. It can also provide a simple model for simulation of a three-atom solid and a vibrating triatomic molecule [3]. Figure 7.6a depicts the form of the potential. The equipotential lines $V =$ constant in the (x, y) plane are shown in Fig. 7.6b. Near the origin, the quadratic terms of (7.19) dominate and the equipotentials are approximately circular; this corresponds to small values of V where the last two terms in (7.19) can be neglected. Away from the origin, the curves are distorted, where a three-fold symmetry can be observed. For $V = 1/6$, the equipotential is an equilateral triangle. Finally, for $V > 1/6$, the equipotential lines are open. That is, the well varies from a harmonic potential at small x and y to triangular equipotential lines on the edges of the potential as illustrated in Fig. 7.6b.

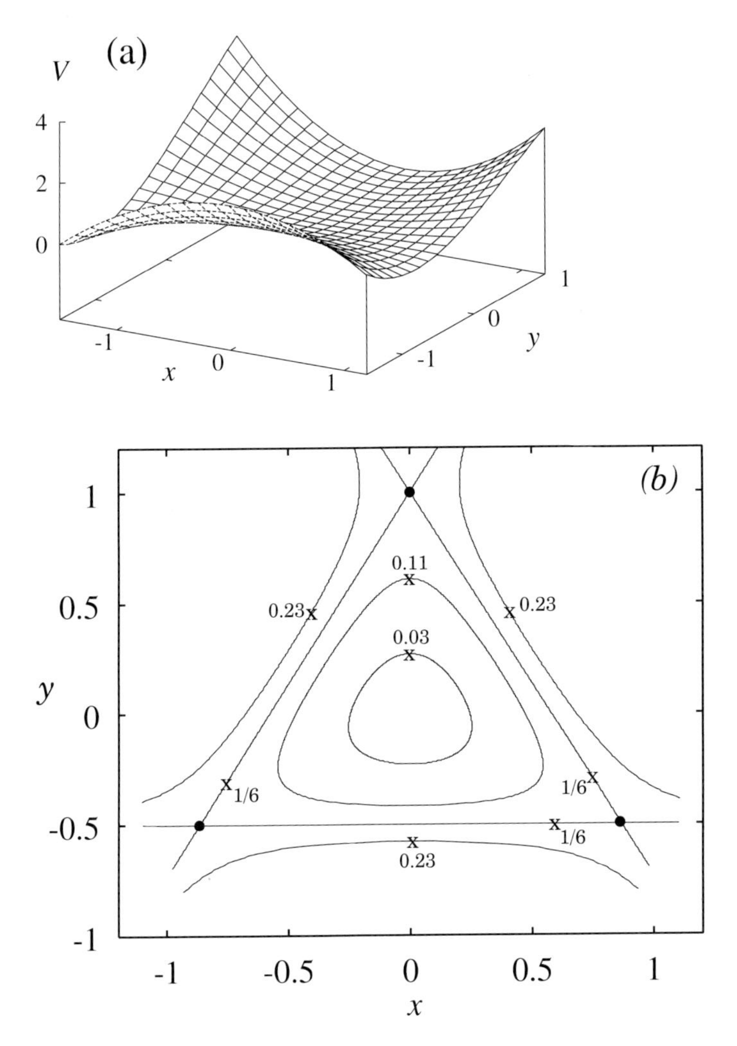

Fig. 7.6. (**a**) The potential well of the Hénon–Heiles Hamiltonian, (7.19). (**b**) The curves of the constant potential V of the Hénon–Heiles Hamiltonian, (7.19). The values of V are marked near the curves

---

Exercise:

9. Verify the above properties of the Hénon–Heiles potential. Express the Hamiltonian (7.18) in terms of the action and angle variables of the linear oscillators.

Now we can identify some more properties of the Hénon–Heiles system straight-forwardly. Since the total energy of the systems is given by $E = H$, we have from (7.18) that

$$V(x, y) \leq E \,. \tag{7.20}$$

Consequently, the trajectory is constrained to remain in that part of the $(x - y)$ plane where this inequality is satisfied. Thus, if the trajectory is started inside the equipotential line $V(x, y) = E$ in Fig. 7.6b it must lie entirely within that line.

The velocity is also constrained. From the expression (7.18) for the Hamiltonian, we have

$$\frac{1}{2}\left(p_x^2 + p_y^2\right) \leq E \,, \tag{7.21}$$

since V is positive in the region represented by Fig. 7.6. So the trajectory must lie inside a finite volume of the phase space. Because of this property, we shall restrict our analysis to values of the energy in the range $0 \leq E \leq 1/6$ and to the inside of the triangle of Fig. 7.6b.

For $V > 1/6$, the equipotential lines open, and the trajectories can no longer be guaranteed to be confined to a finite region of the (x, y) plane and most trajectories escape to infinity.

To start with let us find the equilibrium points of the system and determine their stability property. Then we will study the more complicated dynamics of the system by numerically integrating the equations of motion.

7.3.1 Equilibrium Points

For the Hénon–Heiles potential, $V(x, y)$, given by (7.19), the Hamilton's canonical equations of motion can be written as

$$\dot{x} = p_x \,, \tag{7.22a}$$

$$\dot{p}_x = -(x + 2xy) \,, \tag{7.22b}$$

$$\dot{y} = p_y \,, \tag{7.22c}$$

$$\dot{p}_y = -\left(y + x^2 - y^2\right) \,. \tag{7.22d}$$

The equilibrium points of (7.22) are obtained by substituting $\dot{x} = \dot{p}_x = \dot{y} = \dot{p}_y = 0$, which then gives

$$x^* \left(1 + 2y^*\right) = 0 \,, \quad y^* + x^{*2} - y^{*2} = 0 \,. \tag{7.23}$$

Solving the above set of equations, we obtain the following four equilibrium points $X^* = (x^*, p_x^*, y^*, p_y^*)$:

$$X_1^* = (0, 0, 0, 0) \,, \tag{7.24a}$$

$$X_2^* = (0, 0, 1, 0) \,, \tag{7.24b}$$

$$X_{3,4}^* = \left(\pm\sqrt{3/4}\,, 0, -1/2, 0\right) \,. \tag{7.24c}$$

The stability determining eigenvalues (refer Sect. 3.6) are obtained by solving the characteristic equation

$$\begin{vmatrix} -\lambda & 1 & 0 & 0 \\ -1-2y^* & -\lambda & -2x^* & 0 \\ 0 & 0 & -\lambda & 1 \\ -2x^* & 0 & -1+2y^* & -\lambda \end{vmatrix} = 0\,. \tag{7.25}$$

Expanding the above determinant we get

$$\lambda^4 + 2\lambda^2 + 1 - 4\left(x^{*2} + y^{*2}\right) = 0\,. \tag{7.26}$$

In a conservative system only center (elliptic) and saddle (hyperbolic) equilibrium points occur (see Sect. 3.8.1). Thus, if all the eigenvalues are pure imaginary then the equilibrium point is a center, otherwise it is a saddle. For the equilibrium point X_1^*, the eigenvalue equation (7.26) becomes

$$\lambda^4 + 2\lambda^2 + 1 = 0\,. \tag{7.27}$$

The roots of the above equation are $\lambda = \pm\,\mathrm{i}$. All the eigenvalues are pure imaginary and thus the origin is an elliptic equilibrium point. The trajectories in its neighbourhood move around it, without ever reaching it. For the other three equilibrium points X_2^*, X_3^* and X_4^*, the eigenvalues are $\lambda_{1,2} = \pm 1$ and $\lambda_{3,4} = \pm\sqrt{3}\,\mathrm{i}$. (Verify). Thus, the equilibrium points are hyperbolic (unstable). Trajectories starting from the neighbourhood of these equilibrium points generally diverge away from them.

7.3.2 Poincaré Surface of Section of the System

The phase space of the system (7.22) is of four dimensions, consisting of the coordinates x, $\dot{x}(=p_x)$, y, $\dot{y}(=p_y)$. For a clear understanding of the nature of the dynamics, we investigate the solution of (7.22) on a suitable SOS. For this purpose, in this section, we describe how to construct the SOS of the system and obtain the points on it.

To start with let us fix the energy E (that is we are on the given *energy shell*), so that one of the coordinates, say $\dot{x}$, gets fixed. Suppose now, we choose a SOS with $x = 0$. The coordinates in the SOS now will be $(y, \dot{y})$. Now let us verify whether two numbers corresponding to these variables define a starting point for a trajectory. From (7.18) and (7.19), we have (with $x = 0$)

$$\dot{x} = p_x = \pm\left(2E - \dot{y}^2 - y^2 + \frac{2}{3}y^3\right)^{1/2}\,. \tag{7.28}$$

The $\pm$ sign implies that one point of the SOS with $x = 0$ corresponds to two possible values of $\dot{x}$ or two possible trajectories. This ambiguity can be eliminated by redefining the SOS as

$$x = 0 \ \text{ and } \ \dot{x} \geq 0\,. \tag{7.29}$$

In other words, only the intersections of a trajectory with $x = 0$ in the positive direction are considered. Of course, one can also define the SOS as $x = 0$ and $\dot{x} \leq 0$. In fact, infinitely many other suitable SOS can also be constructed (How?). In our numerical analysis we will choose the SOS given by (7.29).

At this stage one should note that the position of equilibrium points and their stability nature on the SOS, (7.29), will be different from those obtained from the previous subsection for the full system. This is because the equilibrium points X^* given by (7.24) are steady state solutions in the four-dimensional phase space with $\dot{x} = \dot{p}_x = \dot{y} = \dot{p}_y = 0$, whereas the SOS under consideration has the property $x = 0, \dot{x} \geq 0$ and y, p_y are arbitrary. The equilibrium points of the SOS, generally, cannot be determined analytically.

The previously studied cases of the Duffing oscillator equation (5.1) and the Lorenz equations (5.5) in Chap. 5 contained a number of system (control)parameters. Varying one of them and fixing the others at constant values, we studied the bifurcation phenomenon. Now in (7.22) apparently there is no such free parameter. However, we can treat the energy E of the system as a control parameter and investigate the dynamics by varying it. Further, for a fixed value of E, we can generate infinite number of initial conditions. For example, from (7.18) and (7.19) we may write

$$p_x = \pm \left[2E - \left(x^2 + y^2 + 2x^2 y - \frac{2}{3} y^3 \right) - p_y^2 \right]^{1/2} . \qquad (7.30)$$

For different values of $x(0)$, $y(0)$, $\dot{y}(0)$ we can have the same fixed energy value provided $p_x(0)$ is chosen according to (7.30). However, trajectories started with different initial conditions do not asymptotically approach an attractor because of the conservative nature of the phase space volume. The characteristic nature of the orbits obtained for various initial conditions may or may not be identical. Thus, it is important to study the dynamics of the system for different initial conditions for a given energy. Then we can study the qualitative changes in the dynamics on the SOS by varying the energy as well as the initial conditions.

7.3.3 Numerical Results

First, we describe how to obtain points on a suitably chosen SOS. The orbits on the SOS can be obtained as follows:

(1) Fix the value of E.

(2) Choose an initial value $y(0)$, $p_y(0)$ and set $x(0) = 0$. Determine $p_x(0)$ from (7.30).

(3) Integrate the equations of motion (7.22) using a numerical integration scheme (for example, fourth-order Runge–Kutta method as indicated in appendix D) and collect $y(t)$ and $p_y(t)$ whenever $x(t) = 0$, $p_x(t) \geq 0$. Ensure that the obtained values correspond to the chosen value of E to

within the desired accuracy. Collect large number of, say, 2000 or more, data points.

(4) Plot the data in the $(y - p_y)$ phase plane.

(5) Repeat the steps 2–4 for different values of $y(0)$ and $p_y(0)$.

The resultant phase plot displays the overall dynamics on the SOS.

A Technical Point:

How does one find the values of $y(t)$ and $p_y(t)$ which exactly correspond to $x(t) = 0$ and $p_x(t) \geq 0$? Since (7.22) is integrated numerically with a fixed time step Δt, the requirement that $x(t) = 0$ is hardly met. Generally, in the crossing region, one gets $x(t) < 0$ and $x(t + \Delta t) > 0$ or $x(t) > 0$ and $x(t + \Delta t) < 0$. In these cases $x(t) = 0$ occurs in between the times t and $t + \Delta t$. This technical difficulty can be easily overcome. What one has to do now is that whenever the signs of $x(t)$ and $x(t + \Delta t)$ are different use an interpolation scheme to determine the time t' at which $x \approx 0$. Next, calculate p_x, y, and p_y at $t = t'$ employing again the interpolation scheme using their values at t and $t + \Delta t$. Collect the data $y(t')$ and $p_y(t')$ if and only if $p_x(t')$ is ≥ 0. Then continue the integration to get the next intersection point.

We now present the numerical results on (7.22).

(1) We can start the analysis with low values of energy, say, $E = 1/12$. Figure 7.7 shows the phase portrait in the $(x - p_x)$ and $(y - p_y)$ phase planes obtained from a single initial condition given by

$$
\begin{aligned}
& x(0) = 0, \quad y(0) = 0.45, \\
& p_y(0) = 0, \quad p_x(0) = \left[2E - y(0)^2 + \frac{2}{3} y(0)^3 \right]^{1/2} .
\end{aligned}
\tag{7.31}
$$

The trajectory is clearly quasiperiodic. Figure 7.8 shows a sequence of points in the Poincaré SOS corresponding to Fig. 7.7. Successive points have been numbered. The points are seen to lie on a curve, the so-called *invariant curve* discussed in the previous section. In order to show this, a curve passing through the points is drawn in Fig. 7.8. Successive points appear to move regularly on the curve but not with definite periodicity, that is they are not a set of repeating points. To characterize the underlying motion we can define a quantity called *rotation number*, ν. It is analogous to the number of winds of a wire of unit length around a coil. In the present case, it is the number of complete traces on the invariant curve made by the orbit per intersection on the SOS. ν is rational if and only if the orbit is periodic and is irrational if the orbit is quasiperiodic. Therefore, ν can be used to distinguish periodic and quasiperiodic orbits. We can obtain an estimate for the rotation number ν of the quasiperiodic orbit shown in Fig. 7.8. When we go from points 1 to

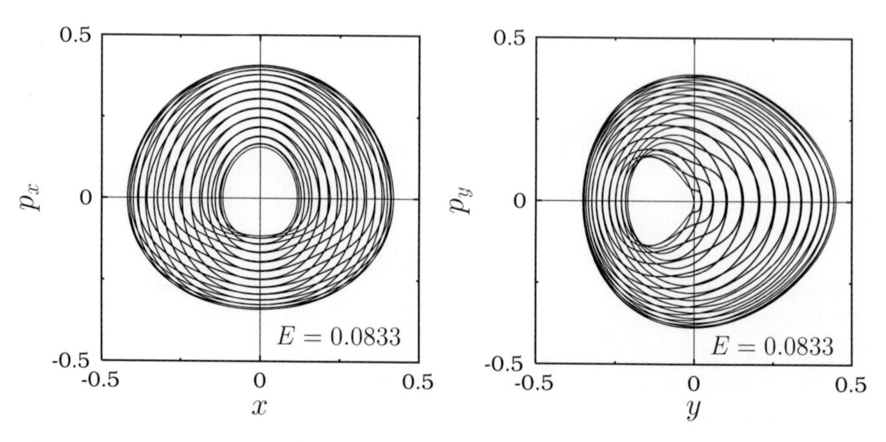

Fig. 7.7. A quasiperiodic orbit of the Hénon–Heiles equation (7.22) for $E = 1/12$ with the initial conditions given by (7.31)

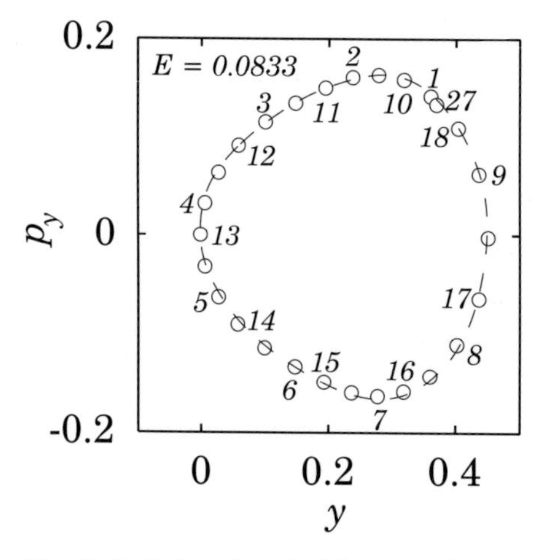

Fig. 7.8. Points (marked by circles) on the surface of section defined by the (7.29) corresponding to the quasiperiodic orbit in Fig. 7.7. Successive intersecting points are numbered sequentially

9 (anticlockwise) we do not complete a revolution, therefore the period T is > 8 or $\nu < 1/8$; but 10 lies after 1, therefore $T < 9$ or $\nu > 1/9$; after two more revolutions, we reach 27 which lies slightly before 1, therefore $\nu < 3/26$, and so on; the value of ν can be progressively refined as more and more points are computed.

(2) What will happen if we change the initial conditions keeping the energy value still at $E = 1/12 = 0.0833...$? Figure 7.9a depicts the result obtained for few different initial conditions. There are four elliptic and three hyperbolic equilibrium points. In each one of the cases, the sequence of points lie on

well defined curves which are sections of tori. (Recall our discussions on the motion on tori in Sect. 7.2). Thus, the motions computed are all regular. The outer curve marks the boundary of the accessible region in the $(y - p_y)$ plane. This is defined by the condition that p_x given by (7.28) must be real, that is

$$p_y^2 + y^2 - \frac{2}{3}y^3 < 2E \,. \tag{7.32}$$

In the neighbourhood of the points e in Fig. 7.9a the orbits are deformed ellipses implying that the points e are elliptic equilibrium points. On the other hand, trajectories near the points S indicate that they are hyperbolic equilibrium points.

(3) Similarly, for values of the energy $E < 1/12$ also, one may obtain plots similar to Fig. 7.9a. One finds that there is no qualitative difference with the result for $E = 1/12$.

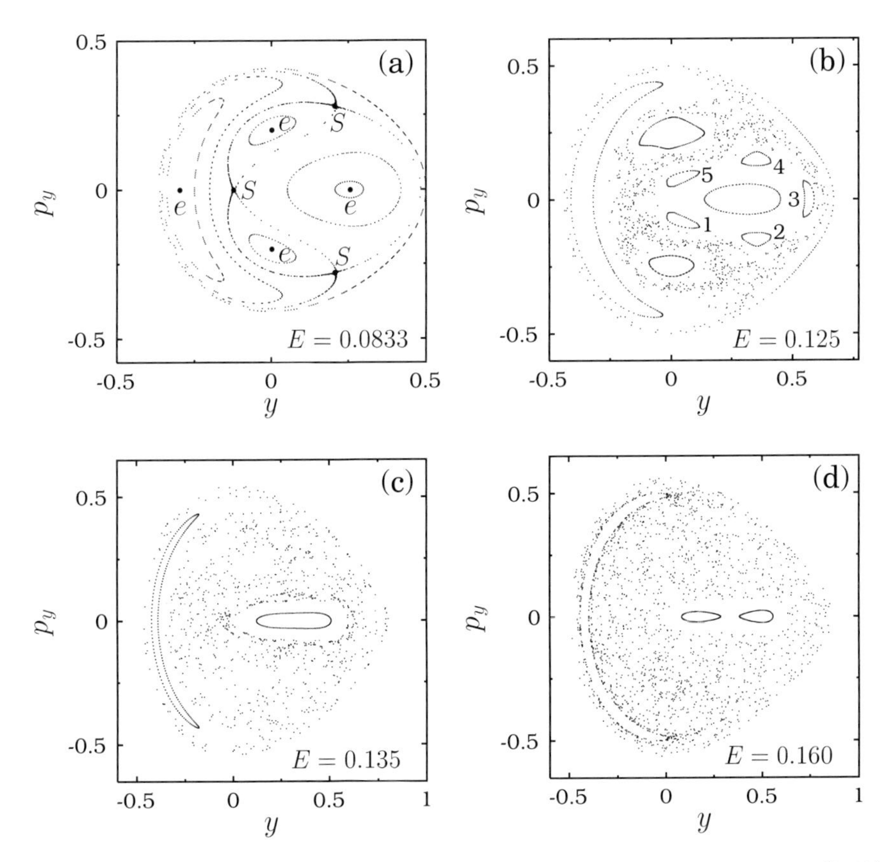

Fig. 7.9. (**a–d**) Surface of sections of trajectories of Hénon–Heiles equation (7.22) for various values of E (indicated in the figures)

(4) When the energy, E, is increased to higher values, we get fascinating and surprising results. For example, Fig. 7.9b depicts the orbits for $E = 0.125$. Three types of orbits are clearly seen.

(i) Simple invariant curves occur about the elliptic equilibrium points as with the lower energy values.

(ii) There are some orbits that do not seem to lie on invariant curves, that is, the corresponding tori have been destroyed. A multiple loop orbit represented by the chain of five small islands (marked by the numbers $1-5$ in Fig. 7.9b), similar to Fig. 7.5b, in which the points jump from loop to loop is seen. This orbit is obtained with a single initial condition only. The trajectory returns to its initial neighbourhood on the $(y - p_y)$ plane after intersecting that plane five times.

(iii) For some other suitable initial conditions the points fall on a two-dimensional region irregularly. Figure 7.10 shows the phase portrait of the *chaotic orbit* in the $(x - p_x)$ and $(y - p_y)$ planes. This orbit can be compared with the quasiperiodic orbit of Fig. 7.7. In the latter case motion on the torus is clearly visible. In contrast, the orbit of Fig. 7.10 irregularly and densely fills up the region in the phase space (see also the scattered points in the SOS in Fig. 7.9b). The orbit is clearly nonperiodic and also not quasiperiodic. This orbit is actually chaotic and the region which it occupies is called a *chaotic region*. In fact, sensitivity with respect to small changes in initial conditions can be verified for the chaotic region of the present system as in the case of dissipative systems discussed in Sect. 4.2.6.

Thus, the system for a fixed energy value exhibits both quasiperiodic and chaotic motions, depending upon the initial conditions.

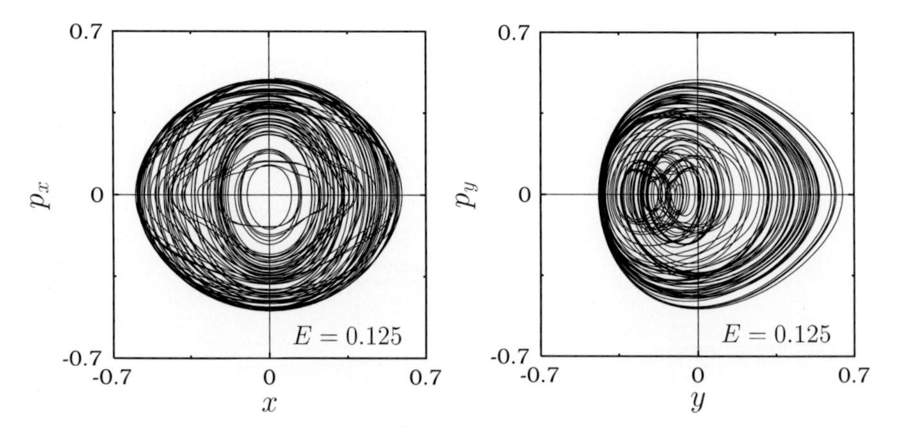

Fig. 7.10. A chaotic orbit of the Hénon–Heiles equation (7.22) for $E = 0.125$

(5) Figures 7.9c and 7.9d show the orbits for $E = 0.135$ and 0.16, respectively. For certain initial conditions, we still find that some sequence of points lie on curves. In Fig. 7.9c ($E = 0.135$) we find small invariant curves. As the energy is increased further, these curves are broken into smaller islands, which one calls them second-order islands, as shown in Fig. 7.9d for $E = 0.16$. Summarizing, the Figs. 7.9a–d suggest the following:

(i) For low energies practically the whole accessible region in the SOS is covered by invariant curves, corresponding to quasiperiodic motion on tori.

(ii) For a range of relatively higher energy values both quasiperiodic and chaotic orbits coexist.

(iii) As the energy E increases, the number of elliptic points increases. Correspondingly, the number of hyperbolic points also increases. Our ability to see these points graphically is limited by finite precision arithmetic and finite resolution of computer graphics.

(iv) As the energy E increases further the region occupied by the invariant curves decreases abruptly while the chaotic region increases dramatically.

We thus realize that the Poincaré SOS gives a good visualization prediction about the dynamics of the system. Particularly, one can qualitatively understand the various regions of phase space. One can carry out a similar analysis for other interesting Hamiltonian systems, such as coupled quartic anharmonic oscillators and pendula, classical form of the hydrogen atom in an external magnetic field and so on. Some of these systems are indicated in the Problems given at the end of this chapter. In all these systems, the qualitative features remain the same as above.

Exercise:

10. Following exercise 9, develop a canonical perturbation theory to the present case of equal frequencies. Extend the analysis to eighth-order and compare the results of perturbation theory with the numerical results of Figs. 7.9 [8].

7.4 Periodically Driven Undamped Duffing Oscillator

In Sect. 5.1, we have studied the dynamics of a periodically driven and damped Duffing oscillator, equation (5.1). Particularly, we had shown the occurrence of period doubling and intermittency routes to chaos. In this section, we consider the undamped, but periodically driven Duffing oscillator, and illustrate the occurrence of regular and chaotic motions, by treating it as a (time dependent)Hamiltonian system with an integrable unperturbed part and a perturbing periodic potential in the angle variable of the form (7.4) for a single degree of freedom system.

To start with, consider the system (5.1) without damping ($\alpha = 0$) and external forcing ($f = 0$). Now (5.1) becomes

$$\dot{x} = y \,, \tag{7.33a}$$
$$\dot{y} = -\omega_0^2 x - \beta x^3 \,. \tag{7.33b}$$

Its exact solution (see appendix A) is (see also (2.7))

$$x(t) = A \mathrm{cn}\left(\bar{\omega}\left(t - t_0\right), \, k\right) \,, \tag{7.34a}$$

where A and t_0 are integration constants and k, $\bar{\omega}$ are given by

$$k^2 = \beta A^2/2\left(\omega_0^2 + \beta A^2\right), \quad \bar{\omega}^2 = \omega_0^2 + \beta A^2 \,. \tag{7.34b}$$

What are the equilibrium points and their stability property of (7.33)? (See also Sect. 3.4.2). The system (7.33) has three equilibrium points and they are given by

$$(x^*, y^*) = (0,0), \ \left(\pm\sqrt{-\omega_0^2/\beta}\,, 0\right) \,. \tag{7.35}$$

For our further analysis we fix the parameters at $\omega_0^2 = -1$ and $\beta = 1$. For this choice the potential of the system is of double-well nature (Fig. 5.1). The associated equilibrium points are now given by

$$(x^*, y^*) = (0,0), \ (\pm 1, 0) \,. \tag{7.36}$$

From the stability determining eigenvalues

$$\lambda_{\pm} = \pm\sqrt{1 - 3x^{*2}} \tag{7.37}$$

we find that the equilibrium point $(0,0)$ is of saddle type while the other two are of elliptic type. Figure 7.11a shows some of the orbits in the $(x - y)$ phase plane. In the neighbourhood of each of the elliptic equilibrium points the trajectories (marked as A) move around it and never approach it. These orbits are periodic. The orbits C encloses all the three equilibrium points and is rotation modulated by oscillation. These are again periodic orbits. The orbit B connects the saddle point $(0,0)$ to itself and is completely lying in the region $x \geq 0$. The orbit B' is also of type B, however, lying in the region $x \leq 0$. The orbits B and B' are called *separatrices* as they separate two types of periodic orbits.

What happens to these orbits when the system is driven by an external periodic force? In the presence of the external force $f \sin \omega t$ the equation of motion (7.33) becomes

$$\dot{x} = y \,, \tag{7.38a}$$
$$\dot{y} = -\omega_0^2 x - \beta x^3 + f \sin \omega t \,. \tag{7.38b}$$

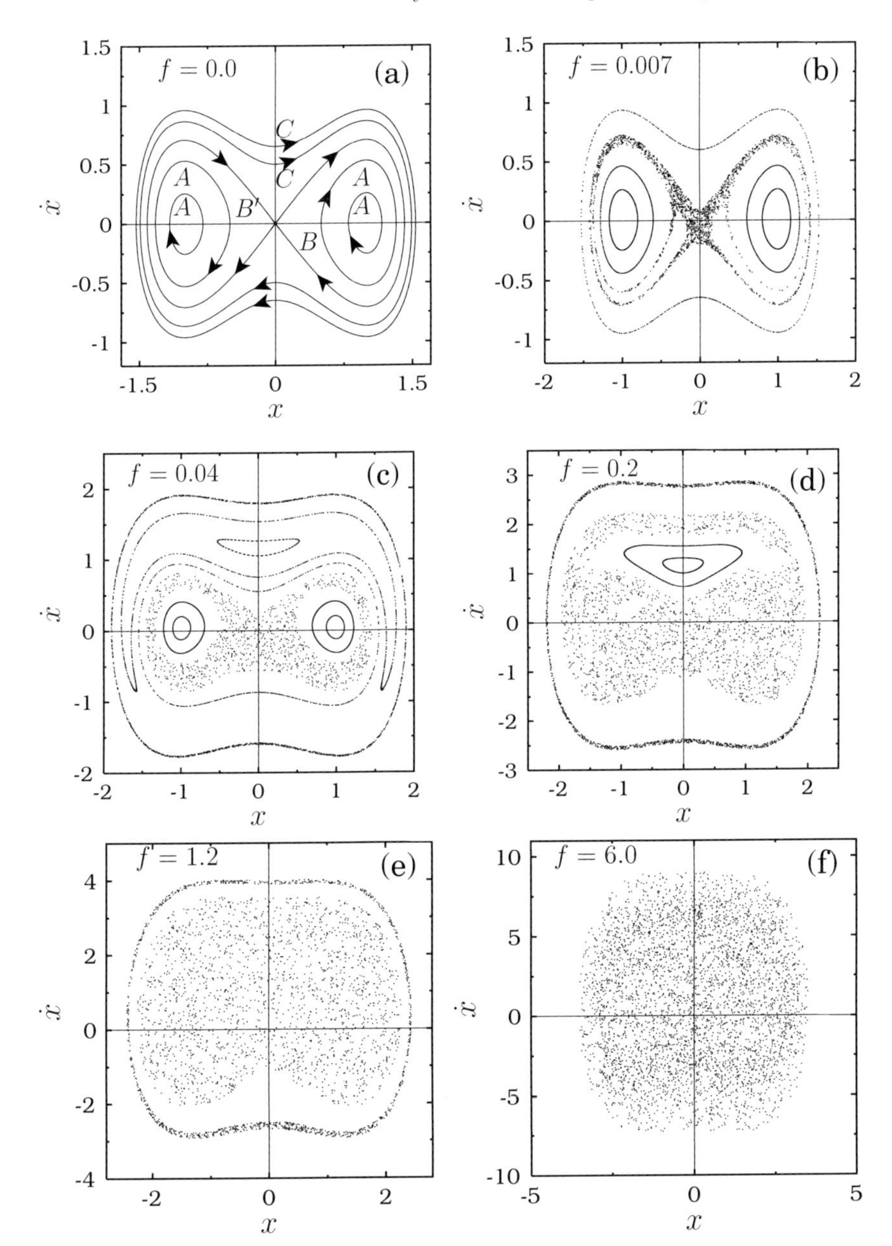

Fig. 7.11. (**a**) Orbits of undamped and force free Duffing oscillator equation (7.33). (**b–f**) Orbits of periodically driven undamped Duffing oscillator equation (7.38) in the Poincaré map for $f = 0.007$(**b**), $f = 0.04$(**c**), $f = 0.2$(**d**), $f = 1.2$(**e**) and $f = 6$(**f**). The other parameters have been fixed at $\omega_0^2 = -1$, $\beta = 1$ and $\omega = 1$

We study the influence of the external force by fixing ω at 1 and varying f from a small value. Since (7.38) is nonautonomous, the stability of equilibrium points cannot be studied by linear stability analysis. Therefore, we perform a numerical investigation. Figures 7.11b–f show the change in the dynamics in the Poincaré map as a function of the parameter f as well as the initial conditions. In these figures each orbit corresponds to different initial conditions and x, y values are collected at every period of the driving force. For small values of f the two equilibrium points $(\pm 1, 0)$ in the Poincaré map are still stable. In the $(x - y)$ phase plane the closed periodic orbits (A and C) found for $f = 0$ now become quasiperiodic orbits. Consequently in the Poincaré map the points fall on invariant curves. This is shown in Fig. 7.11b for $f = 0.007$. Further when $f \neq 0$ the separatrices are destroyed by the perturbation and for small values of f a thin chaotic layer is formed (Fig. 7.11b) in the vicinity of separatrices. In each of the two potential wells inside the layer there are invariant curves encompassing the elliptic equilibrium point. Figure 7.11c shows orbits for $f = 0.04$. The area of chaotic region is increased. Further, we find the occurrence of an additional elliptic point around which closed invariant curves are seen. For $f = 0.2$ (Fig. 7.11d) invariant curves are found in a small region inside the chaotic region. There is almost no invariant curves around the equilibrium points $(\pm 1, 0)$. As the parameter f is further increased more and more invariant curves are destroyed and the area of the chaotic region increases. This is shown in Figs. 7.11e and 7.11f for $f = 1.2$ and 6, respectively.

Exercises:

11. Write down the time-dependent Hamiltonian corresponding to (7.38). For the $\epsilon = 0$ case obtain a suitable canonical transformation to action and angle variables, and express the Hamiltonian in terms of them. Develop a canonical perturbation theory to investigate the overlapping of resonances (see also Refs. [5,6]).
12. Investigate the transition from regular to chaotic dynamics in the undamped and periodically driven Duffing oscillator equation (7.33) corresponding to the double-hump potential ($\omega_0^2 = 1$ and $\beta = -4$) and single-well potential ($\omega_0^2 = 0.3$ and $\beta = 1$).

7.5 The Standard Map

The numerical study of continuous-time Hamiltonian systems with phase space dimension greater than two is quite time consuming because in order to get a point on a suitably chosen SOS one has to numerically integrate the equations of motion over a long time. A rather simple but effective set of systems to understand and appreciate the general features of Hamiltonian systems is the so-called *area-preserving maps*. They can be easily iterated numerically and they exhibit essentially the same features that typical conservative continuous-time dynamical systems exhibit. In this section we consider one such example of an area-preserving map and illustrate some of the major features observed in the Hénon–Heiles and the periodically perturbed conservative Duffing oscillator systems.

One of the most studied area-preserving maps is the *standard map*, introduced originally by Taylor and by Chirikov in 1979 [9]. The standard map is the discrete time analogue of the equation of motion of the vertical pendulum, in which the continuous gravitational force is replaced by a sequence of impulses at equal time intervals. The state of the system at an integer time n is represented by the coordinates θ and I, where θ is the angle of rotation and I is the conjugate (angular) momentum. The map is then given by

$$I_{n+1} = I_n + K\sin\theta_n \equiv f\,, \quad (\text{mod } 2\pi) \tag{7.39a}$$

$$\theta_{n+1} = \theta_n + I_{n+1} \equiv g\,, \quad (\text{mod } 2\pi) \tag{7.39b}$$

where K is a positive parameter which determines the dynamics of the map and mod 2π corresponds to modulo 2π. Its Jacobian J is given by

$$\begin{aligned} J &= \begin{vmatrix} \partial f/\partial I_n & \partial f/\partial\theta_n \\ \partial g/\partial I_n & \partial g/\partial\theta_n \end{vmatrix} \\ &= \begin{vmatrix} 1 & K\cos\theta_n \\ 1 & 1+K\cos\theta_n \end{vmatrix} = 1\,. \end{aligned} \tag{7.40}$$

Since $J = 1$, the map is conservative or area-preserving.

For $K = 0$, the standard map becomes

$$I_{n+1} = I_n\,, \quad (\text{mod } 2\pi) \tag{7.41a}$$

$$\theta_{n+1} = \theta_n + I_{n+1}\,. \quad (\text{mod } 2\pi) \tag{7.41b}$$

Its solution is $\theta_n = (\theta_0 + nI_0)$ mod 2π and $I_n = I_0$. If $I_0 = 2\pi N/M$, where N and M are integers, then

$$\theta_n = \theta_0 + n\frac{N}{M}2\pi \quad (\text{mod } 2\pi) \tag{7.42}$$

and $\theta_M = \theta_0$. Consequently, the points on the torus repeat after M iterates. Therefore, the orbit on the torus is periodic with period M. When $I_0 \neq 2\pi N/M$, then $\theta_n \neq \theta_0$ for any n. Now the iterated points are nonperiodic and densely fill the line $I = I_0$. That is, the orbit on the torus never closes back itself and is essentially quasiperiodic. Thus, for $K = 0$, the motion is either definite periodic or quasiperiodic depending upon the initial value I_0. (See Sect. 7.5.2 below for further discussions). Further, each value of θ_0 issues a distinct periodic orbit with period M.

7.5.1 Linear Stability and Invariant Curves

The above analysis for $K = 0$ does not hold good for $K \neq 0$. In the following we investigate the nature of the dynamics for $K \neq 0$. To begin with we will consider explicitly the period-1 and period-2 fixed points, and their stability, and then consider the onset of chaos.

A. Period-1 Fixed Points

Many of the regular features of the map (7.39) can be realized analytically by finding its fixed points and determining their stability property. First let us find the period-1 fixed points and study their stability. Naturally, for the map (7.39), the period-1 fixed points are obtained by solving the equations

$$I^* = I^* + K \sin\theta^* , \quad (\text{mod } 2\pi) \tag{7.43a}$$

$$\theta^* = \theta^* + I^* . \quad (\text{mod } 2\pi) \tag{7.43b}$$

The resultant solutions are $\theta^* = 0$, or π and $I^* = 0$. Therefore, the fixed points are $(\theta^*, I^*) = (0, 0)$ and $(\pi, 0)$.

A.1 Linear Stability of a Fixed Point of a Two-Dimensional Map

For a conservative two-dimensional map the stability of a fixed point can be stated in terms of the trace of its Jacobian matrix without writing the eigenvalue equation. For a two-dimensional map of the form

$$x_{n+1} = f(x_n, y_n) , \tag{7.44a}$$

$$y_{n+1} = g(x_n, y_n) , \tag{7.44b}$$

the Jacobian matrix is

$$J = \begin{pmatrix} \partial f/\partial x_n & \partial f/\partial y_n \\ \partial g/\partial x_n & \partial g/\partial y_n \end{pmatrix} = \begin{pmatrix} a & b \\ c & d \end{pmatrix} . \tag{7.45}$$

If the map is conservative then $\det J = 1$, as noted earlier in Sect. 4.3. This implies that $ad - bc = 1$. Now the stability determining eigenvalues of J are obtained from

$$\lambda^2 - \lambda(a + d) + ad - bc = 0$$

or

$$\lambda^2 - \lambda \text{Tr}J + 1 = 0 . \tag{7.46}$$

The solutions for λ are

$$\lambda_{1,2} = \frac{1}{2}\left[\text{Tr}J \pm \sqrt{(\text{Tr}J)^2 - 4}\,\right] . \tag{7.47}$$

It is easy to see that $\lambda_1\lambda_2 = 1$ and $\lambda_1 + \lambda_2 = \text{Tr}J$. For two-dimensional maps, a fixed point is stable if the absolute values of the real parts of both the eigenvalues are less than 1; otherwise it is unstable. Thus, for the two-dimensional area-preserving maps stability of the fixed point can be determined from the value of $\text{Tr}J$. We have the following two cases.

(i) λ_1 *and* λ_2 *are complex conjugates*

From (7.47), we note that if $|\text{Tr}J| < 2$ then the eigenvalues are complex conjugates. Since $\lambda_1\lambda_2 = 1$, the eigenvalues must be of the form

$$\lambda_{1,2} = \mathrm{e}^{\pm \mathrm{i}\alpha} \equiv \cos\alpha \pm \mathrm{i}\sin\alpha. \tag{7.48}$$

The absolute values of the real parts of both the eigenvalues are now always less than 1. Therefore, the perturbation does not grow with time. The fixed point is stable.

(ii) λ_1 *and* λ_2 *are real*

Suppose $|\text{Tr}J| > 2$. Then both the eigenvalues λ_1 and λ_2 are real and of the form $|\lambda_{1,2}| = \exp(\pm\alpha)$. Further, $\lambda_2 = 1/\lambda_1$, since $\lambda_1\lambda_2 = 1$. Therefore, the absolute values of one of the eigenvalues is always greater than 1 and the other is always less than 1. The solutions are now growing and the fixed point is unstable. From the above it is clear that the fixed point of an area-preserving map is stable if $|\text{Tr}J| < 2$ and is unstable for $|\text{Tr}J| > 2$. The bifurcation occurs at $|\text{Tr}J| = 2$.

In dissipative maps, if a fixed point is stable then in the neighbourhood of it trajectories asymptotically approach it. In contrast to this, in conservative maps, because of the area-preserving property, in the neighbourhood of a stable fixed point (that is, $|\text{Tr}J| < 2$) trajectories starting from different initial conditions form closed orbits and the fixed point is of center type (elliptic). For the second case, that is $|\text{Tr}J| > 2$, the fixed point is hyperbolic or saddle. The case with $|\text{Tr}J| = 2$, which gives $\lambda_1 = \lambda_2 = \pm 1$, is a special case and it corresponds to a bifurcation or change in the stability, that is the exact transition between the stable and unstable cases.

A.2 Stability of Period-1 Fixed Points

For the standard map, the Jacobian matrix is

$$J = \begin{pmatrix} 1 & K\cos\theta^* \\ 1 & 1 + K\cos\theta^* \end{pmatrix} \tag{7.49}$$

and its trace is

$$\text{Tr}J = 2 + K\cos\theta^* \,. \tag{7.50}$$

Then, we have the following behaviour of period-1 fixed points:

(1) For the fixed point $(0, 0)$, the stability condition becomes

$$|\text{Tr}J| < 2 \Longrightarrow |2 + K| < 2 \,. \tag{7.51}$$

Thus the fixed point $(0, 0)$ is unstable for $K > 0$.

(2) For the fixed point $(\pi, 0)$, the above stability condition becomes $|2 - K| < 2$. Thus, it is stable for $0 < K < 4$. Therefore, for $K < 4$, in the neighbourhood of $(\pi, 0)$ the iteration of the map produces closed orbits for a set of initial conditions. For $K > 4$, the fixed point $(\pi, 0)$ is unstable.

Next, we look for period-2 fixed points.

B. Period-2 Fixed Points

The period-2 fixed points [2] can be obtained by solving the set of equations

$$I_2 = I_1 + K \sin\theta_1 \, , \tag{7.52a}$$
$$\theta_2 = \theta_1 + I_2 - 2\pi m_1 \, , \tag{7.52b}$$
$$I_1 = I_2 + K \sin\theta_2 \, , \tag{7.52c}$$
$$\theta_1 = \theta_2 + I_1 - 2\pi m_2 \, , \tag{7.52d}$$

where m_1 and m_2 are integers. Adding (7.52a) and (7.52c), we get $\sin\theta_2 = -\sin\theta_1$. Therefore, we have two possibilities: (i) $\theta_2 = -\theta_1$ or (ii) $\theta_2 = \theta_1 - \pi$, with $0 < \theta_1 \leq \pi$. We will consider both the cases separately.

Case (i) $\theta_2 = -\theta_1$

From (7.52d), using $\theta_2 = -\theta_1$, we obtain

$$2\theta_1 - I_1 + 2\pi m_2 = 0 \, . \tag{7.53}$$

Next, in the above equation, substituting I_1 from (7.44c) and then using (7.52b) for I_2 we get

$$3\theta_1 - K \sin\theta_2 - \theta_2 - 2\pi\,(m_1 - m_2) = 0 \, . \tag{7.54}$$

Replacing θ_2 by $-\theta_1$ we have

$$4\theta_1 + K \sin\theta_1 - 2\pi p = 0 \, , \tag{7.55}$$

where $p = m_1 - m_2$. Equation (7.55) determines θ_1. When $p = 1$, we get the primary family of fixed points and when $p = 2$ the secondary family is obtained. I_1 and I_2 are then obtained from (7.52d) and (7.52b) as

$$I_1 = 2\theta_1 + 2\pi m_2 \, , \tag{7.56a}$$
$$I_2 = -2\theta_1 + 2\pi m_1 \, . \tag{7.56b}$$

For $K \ll 1$, one can neglect $K \sin\theta_1$ in (7.55) and obtain $\theta_{1,2} = \pm\pi/2$ for $p = 1$. From (7.56), we find that $I_{1,2} = 2\pi(m_2 + 1/2)$ for $p = 1$. Thus, for $\theta_2 = -\theta_1$, the primary period-2 fixed point is

$$(\theta_1^*, I_1^*) = (\pi/2, 2\pi\,(m_2 + 1/2)) \, , \tag{7.57a}$$
$$(\theta_2^*, I_2^*) = (-\pi/2, 2\pi\,(m_2 + 1/2)) \, . \tag{7.57b}$$

Substituting $m_2 = 0, \pm 1, ...$ and taking modulo 2π for both I and θ, we obtain

$$(\theta_1^*, I_1^*) = (\pi/2, \pi) \, , \quad (\theta_2^*, I_2^*) = (3\pi/2, \pi) \, . \tag{7.58}$$

Further, for any nonzero value of K, from (7.55) one can show that θ_1 must be $\leq \pi/2$ (Verify).

Case (ii): $\theta_2 = \theta_1 - \pi$

For $\theta_2 = \theta_1 - \pi$, from (7.52), the equation for θ_1 is obtained as

$$K \sin \theta_1 + 2\pi(1 - p) = 0 \, . \tag{7.59}$$

I_1 and I_2, obtained from (7.52c) and (7.52d), are given by

$$I_1 = 2\pi \, (m_2 + 1/2) \, , \tag{7.60a}$$

$$I_2 = 2\pi \, (m_2 + p - 1/2) \, . \tag{7.60b}$$

For $p = 1$, from (7.59–60), we obtain the primary period-2 fixed points as (after taking mod 2π)

$$(\theta_1^*, I_1^*) = (\pi, \pi) \, , \quad (\theta_2^*, I_2^*) = (0, \pi) \, . \tag{7.61}$$

C. Linear Stability of Period-2 Fixed Points

In analogy with one-dimensional maps (see Sect. 4.2.4), here, for the stability of the period-2 fixed points given by (7.58) and (7.61) of the two-dimensional standard map, we consider the trace of the product matrix

$$\begin{aligned} A &= J\,(\theta_1^*, I_1^*) \, J\,(\theta_2^*, I_2^*) \\ &= \begin{bmatrix} 1 & K\cos\theta_2 \\ 1 & 1 + K\cos\theta_2 \end{bmatrix} \begin{bmatrix} 1 & K\cos\theta_1 \\ 1 & 1 + K\cos\theta_1 \end{bmatrix} \, . \end{aligned} \tag{7.62}$$

The trace of the above matrix can be easily seen to be

$$\begin{aligned} \mathrm{Tr}A &= 1 + K\cos\theta_2 + K\cos\theta_1 + (1 + K\cos\theta_1)\,(1 + K\cos\theta_2) \\ &= 2 + 2K\,(\cos\theta_1 + \cos\theta_2) + K^2 \cos\theta_1 \cos\theta_2 \, . \end{aligned} \tag{7.63}$$

As noted earlier in this section, a fixed point is stable if $|\mathrm{Tr}A| < 2$. This leads to the condition

$$-4 < 2K\,(\cos\theta_1 + \cos\theta_2) + K^2 \cos\theta_1 \cos\theta_2 < 0 \, . \tag{7.64}$$

For the case (i) (that is $\theta_2 = -\theta_1$), substituting the values of θ_1^* and θ_2^* from (7.58), the above stability condition becomes

$$-4 < K\cos\theta_1 < 0 \, . \tag{7.65}$$

Since $\theta_1 \leq \pi/2$ (as noted earlier), $K\cos\theta_1$ is always positive and hence the above stability condition is not satisfied for any positive value of K. The period-2 fixed point given by (7.58) is thus always unstable (saddle). Next, for the case (ii) ($\theta_2 = \theta_1 - \pi$), the stability condition (7.64) becomes

$$K^2 \cos^2 \theta_1 < 4 \,. \tag{7.66}$$

Substituting $\theta_1^* = \pi$, the above stability condition becomes

$$K^2 < 4 \,. \tag{7.67}$$

That is, the period-2 fixed point given by (7.61) is stable (elliptic) for $K < 2$ and unstable (saddle) for $K > 2$.

To summarize, the period-1 fixed point $(\theta^*, I^*) = (0, 0)$ is unstable for $K > 0$ while $(\theta^*, I^*) = (\pi, 0)$ is stable for $0 < K < 4$. The primary period-2 cycle given by (7.58) is always unstable. The other primary period-2 cycle ((7.61)) is stable for $0 < K < 2$ and unstable for $K > 2$. That is, for $0 < K < 2$ both the period-1 fixed point and the period-2 cycle ((7.61)) are stable. They coexist with different basins of attraction.

A secondary fixed point with $p = 2$ and $\theta_2 = -\theta_1$ in (7.55) appears at $K = 4$ and it becomes stable for $4 < K < 2\pi$. On the other hand, at $K = 2\pi$ another secondary fixed point with $p = 2$ and $\theta_2 = \theta_1 - \pi$ is born and it is stable for $6.28 < K < 6.59$. For more details about secondary fixed points, the reader may refer to Ref. [9].

The above linear stability analysis gives typical nature of solutions only in the neighbourhood of the primary fixed points. For the standard map, a detailed numerical analysis can be easily performed to show that chaotic solutions coexist along with stable fixed points. To understand the detailed dynamics of the standard map we now discuss the numerical results.

7.5.2 Numerical Analysis: Regular and Chaotic Motions

Equation (7.39) for the standard mapping, with I and θ taken mod 2π, can be iterated easily for several thousands of steps, thus examining numerically the $(\theta - I)$ phase plane for various values of K.

A. Completely Regular Dynamics

For $K = 0$, as noted earlier the orbits are either nonperiodic and densely fillup the line $I = I_0$, when $I_0/2\pi$ is irrational, or periodic and constitute a finite set of points for rational values of $I_0/2\pi$ (Fig. 7.12). For sufficiently small values of K, some of the orbits found for $K = 0$ may continue to be present for $K \neq 0$ also, while the others may be destroyed.

Figure 7.13a shows the phase plane for a relatively small value of K, namely, 0.5. The ten orbits marked as $1, 2, ..., 10$ correspond to ten distinct initial conditions. The orbits $1, 2, 3$ and 7 are closed thereby indicating area-preserving nature of the map. There are many orbits such as 4-6, 8-10 running roughly horizontally from $\theta = 0$ to $\theta = 2\pi$. These orbits are those that originate from the nonresonant orbit of the map (7.39) with $K = 0$ (identifiable from Fig. 7.12) and have survived the perturbation.

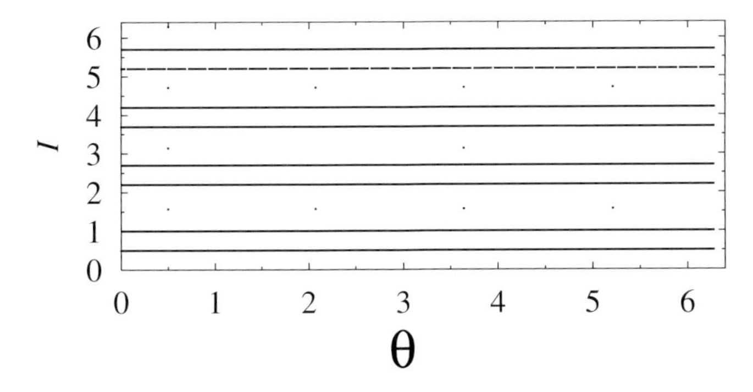

Fig. 7.12. Orbits of the standard map (7.41) ($K = 0$). There are eight orbits running horizontally and densely filling the θ axis with constant I. These are nonresonant (nonperiodic) orbits with irrational $I_0/(2\pi)$. Three resonant (definite periodic) orbits corresponding to rational $I_0/(2\pi)$ such as $I_0 = \pi/2$ (period-4), π (period-2) and $3\pi/2$ (period-4) with $\theta_0 = 0.5$ are clearly seen

Previously, in the Sect. 7.5.1, we obtained period-1 and period-2 fixed points. These fixed points are marked in the Fig. 7.13a. E_1 and S_1 are period-1 fixed points. E_2=(E_{21}, E_{22}) and S_2=(S_{21}, S_{22}) are primary period-2 fixed points. For $K = 0.5$, E_1 and E_2 are stable. As a result, there are closed orbits in the neighbourhood of these points. The orbits 1–3 move around the period-1 fixed point E_1. The orbit 7 is obtained by a single initial condition. The points fall on two smooth closed curves encircling the period-2 cycle. The fixed points S_1 and S_2 are unstable and orbits generally diverge from them.

B. Complicated Dynamics

When the parameter K is increased further, more of the deformed surviving nonresonant orbits are destroyed. This is shown in Fig. 7.13b, for $K = 1$, where only very few nonresonant orbits can be seen. For $K = 1$, the fixed points E_1 and E_2 are still stable and hence there are closed orbits in the neighbourhood of them. There are other orbits such as islands (marked as I) and chaos (marked as C). We observe chaotic regions with interspersed island chains. Each island breaks into chains of still smaller islands as K is continuously increased from a small value. Such a hierarchy of orbits suggests that new details emerge continuously as one examines the system on a finer and finer scale. For example, let us look at Fig. 7.13b. There is a chain of two islands surrounding the two points of the period-2 elliptic cycle $(\pi, \pi) \rightleftharpoons (0, \pi)$. The positions of the intermediate two points of the hyperbolic cycle are also well seen. Now we enlarge the vicinity (marked by a rectangle in the Fig. 7.13b) of the hyperbolic point with approximate coordinates $(\theta, I) = (0 - 2, 2 - 2.4)$. We now obtain Fig. 7.14a. Figure 7.14b is a magnification of a small region of Fig. 7.14a. More and more detailed structures can be seen

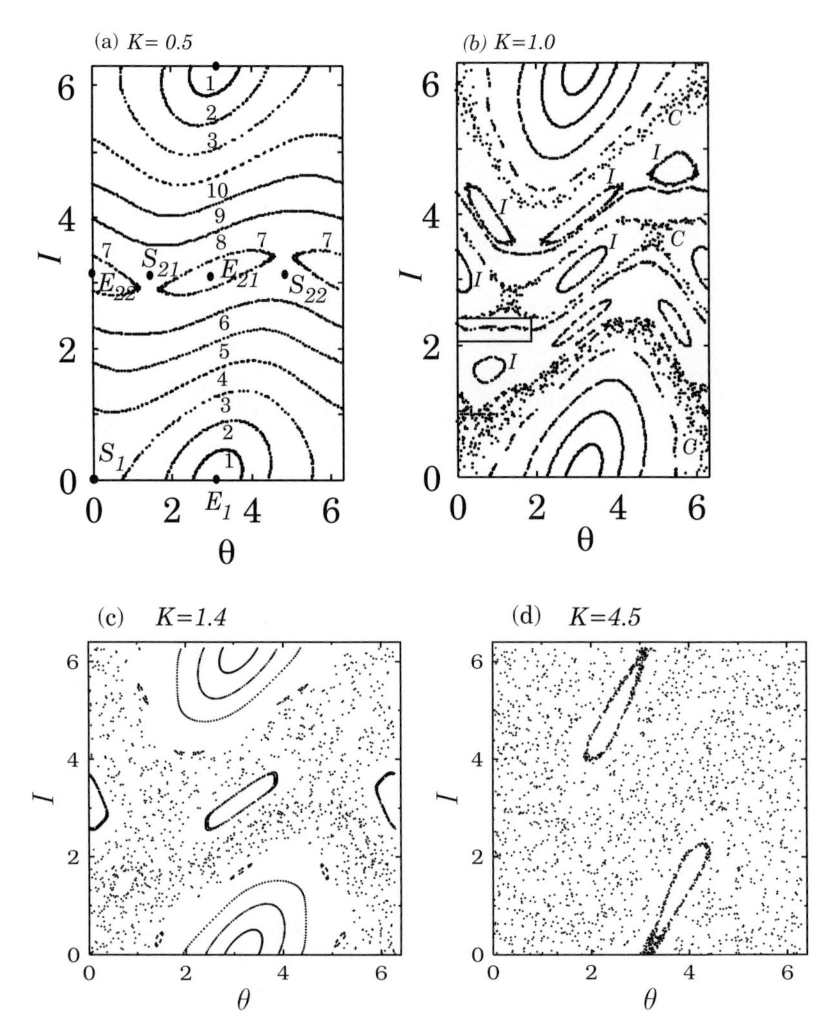

Fig. 7.13. Phase plane of the standard map (7.31) for various values of K. (**a**) $K = 0.5$, (**b**) $K = 1$, (**c**) $K = 1.4$ and (**d**) $K = 4.5$

to emerge in these figures which cannot be seen in the Fig. 7.13b. That is, the real situation is more complex than the Fig. 7.13b.

As the parameter K is further increased, many of the island chains disappear and the chaotic region enlarges. This is clear from Fig. 7.13c ($K = 1.4$), where we note that closed orbits occur only in a small region of the phase space. In the remaining region, the motion is mostly chaotic. The iterated points are randomly distributed over a wide range of phase space. For $2 < K < 4$, the period-2 cycle (E_{21}, E_{22}) is unstable while the period-1 fixed point $(\pi, 0)$ (E_1) is stable. Consequently, in the (θ, I) space, no closed or-

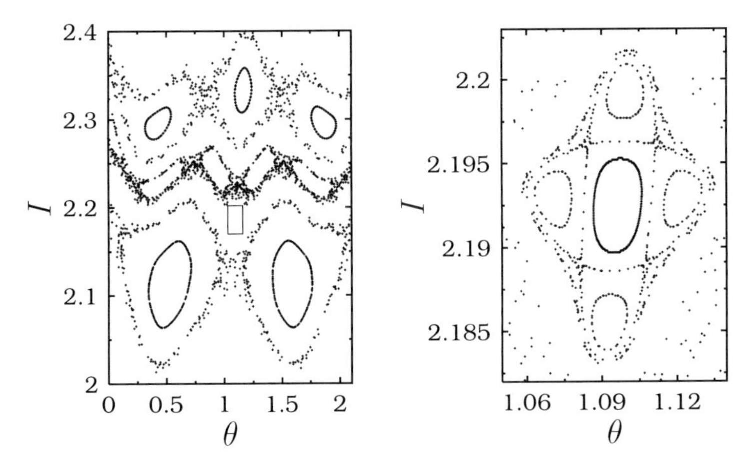

Fig. 7.14. Successive magnifications of a small portion of the Fig. 7.13b

bits occur around the period-2 cycle. Closed orbits around the period-1 fixed point (E_1) occur only in a small region of the phase space. Most of the region is occupied by chaotic solution.

At $K = 4$, period doubling bifurcation occurs and is typically followed by an infinite period doubling cascade as the value of K increases. In such a cascade, the period-q elliptic orbit destablizes and becomes hyperbolic with the simultaneous appearance of a period-$2q$ elliptic orbit, which then period doubles to produce periodic orbits of period 2^2q elliptic orbit and so on. At some $K = K_\infty$, the period doubling phenomenon terminates. This is a Hamiltonian version of the period doubling cascade phenomenon. Figure 7.13d corresponds to $K = 4.5$, where only a very small region is occupied by periodic orbits and a single chaotic orbit densely fills the remaining region.

An important characteristic property of two-dimensional area-preserving maps is that the area bounded by two invariant curves is itself invariant. That is, solution curves (regular or chaotic) obtained for arbitrary initial conditions lying between two invariant curves lie within the same invariant curves and they cannot diffuse through them. In other words, the region between any two invariant curves must map into itself. This implies that in area-preserving maps many chaotic regions separated by invariant curves may coexist in phase space. This can be clearly seen in the Fig. 7.13b. When a control parameter is varied the invariant curves are destroyed and the various chaotic regions merge together.

We note that in dissipative systems like the Duffing oscillator, equation (5.1), and Lorenz equations (5.3), the trajectories near a stable period-n orbit asymptotically approach it and after the transient evolution the motion is a purely period-n solution. In contrast to this, in Hamiltonian systems, because of volume preserving nature, the trajectories near a stable period-n

cycle move around the cycle as shown in the Fig. 7.13a for standard map and in the Fig. 7.9a for Hénon–Heiles equation.

We have seen earlier in Chaps. 4 and 5 that in dissipative systems in the period doubling regions the ratio of the distance between the two successive bifurcations geometrically converges to the value 4.6692... Period doubling bifurcations have been studied in certain area-preserving maps and also in continuous-time dynamical systems. In all these systems, the Feigenbaum ratio is found to be $\delta_F = 8.7210...$(For more details see Refs. [2,10]).

7.6 Kolmogorov–Arnold–Moser Theorem

A path breaking development in the understanding of nonlinear dynamical systems of Hamiltonian type was provided by a theorem outlined originally by Kolmogorov in 1954, which was later on proved by Arnold and Moser in early 1960s [11–13]. It is the well known KAM theorem. It provides a starting point for an explanation for the transition from regular or quasiperiodic to chaotic type motion in Hamiltonian systems. Here we briefly present the details of the theory without any proof or rigor and restrict our attention to systems with two degrees of freedom only. For more details, see for example Refs. [2,4,6,11–13].

Let us consider a system of two-coupled oscillators whose equations of motion can be completely integrated to give exact analytical solution. As pointed out in Sect. 7.2, we can introduce action-angle variables (J_i, θ_i) so that the Hamiltonian H has the form $H_0(J_1, J_2)$, a function of the action variables alone. The motion then occurs on invariant tori in phase space (as discussed in Sect. 7.2). A particular torus can be labelled by the action variables J_i and points on the torus can be labelled by the angle variables θ_i, $i = 1, 2$. For example, this is the case in the standard map (7.39) for $K = 0$. For each set of initial values of (I, θ), I_n remains as I_0 whereas θ_n runs over the interval $(0, 2\pi)$.

Now we modify the Hamiltonian H slightly by adding a small perturbation $\epsilon H_1(J_1, J_2, \theta_1, \theta_2)$. The new Hamiltonian is

$$H = H_0 (J_1, J_2) + \epsilon H_1 (J_1, J_2, \theta_1, \theta_2) \ , \tag{7.68}$$

where $\epsilon \ll 1$. For $\epsilon = 0$, the frequencies associated with the unperturbed motion are defined by

$$\omega_i = \partial H_0 / \partial J_i, \quad i = 1, 2 \ . \tag{7.69}$$

If $\det|\partial^2 H_0 / \partial J_i \partial J_j| \neq 0$, then the frequencies ω_2 and ω_1 are linearly independent or incommensurate. Now, in the presence of a perturbation $\epsilon \neq 0$, we ask: What is the effect of ϵH_1 on the unperturbed orbits with ω_1/ω_2 irrational? What happens to the definite periodic orbits with the frequency ratio ω_1/ω_2 rational? These can be answered by the KAM theorem.

KAM Theorem:

If the unperturbed Hamiltonian system is nondegenerate, that is, the frequencies ω_i *given by (7.69) are functionally independent so that the Jacobian of the frequencies is nonzero,* $|\partial\omega_i/\partial J_j| \neq 0$*, then those tori with sufficiently irrational* ω_1/ω_2 *with*

$$\left|\frac{\omega_1}{\omega_2} - \frac{m}{s}\right| > \frac{K(\epsilon)}{s^{2.5}}, \quad (K(\epsilon \to 0) \to 0) \tag{7.70}$$

where m *and* s *are mutually prime integers, will not disappear, but only undergo a slight deformation.*

The function $K(\epsilon)$ is unspecified but goes to zero with ϵ. For large enough ϵ the perturbation destroys all tori. The system phase trajectory in regions of the destroyed tori is quite complicated and will be randomly filling the destroyed region. We have observed the same picture in the cases of the Hénon–Heiles equation (7.22) and the standard map (7.39). In the Hénon-Heiles equation for low energies the SOS has closed orbits about elliptic fixed points. Similarly, the standard map with $K = 0$ is integrable and the phase space is filled with definite periodic and quasiperiodic orbits. For $K \ll 1$, for example, $K = 0.5$ (Fig. 7.13a), there are orbits running roughly horizontally from $\theta = 0$ to $\theta = 2\pi$. These are the orbits originated from the nonresonant orbits of the map (7.39) with $K = 0$ and only their shape is deformed by the nonlinear perturbation.

The proof of the KAM theorem is highly nontrivial [11–13]. The theorem concentrates on proving the existence of individual tori in weakly perturbed Hamiltonian systems which satisfy certain conditions as stated above. The proof essentially contains two basic aspects: (1) A superconvergent perturbation theory. (2) A number theoretic analysis that determines how irrational the frequencies ω_1/ω_2 should be so that a particular torus is not destroyed. The KAM theorem does not concern itself about the nature of the 'rational tori' under perturbation, which is in some sense get destroyed, for example, through overlapping of resonances as discussed in Sect. 7.2. Obviously, here lies the origin of chaos and so the importance of KAM theorem.

7.7 Conclusions

In this chapter we have shown that the dynamics of nonlinear conservative systems can exhibit very many distinguishing features compared to nonlinear dissipative systems: There are no strange attractors and chaotic orbits can fill the entire accessible phase space region. In particular for integrable and near integrable systems, the motion essentially takes place on the surface of tori corresponding to periodic or quasiperiodic orbits. This can be best demonstrated by going over to action and angle variables through appropriate canonical transformations. What happens when additional perturbation is applied or the strength of perturbation is increased? Concentrating on the structure of the orbits with reference to a suitably chosen Poincaré surface of section, which essentially corresponds to a mapping of successive points of intersections of the orbits with the SOS, in particular with reference to two degrees of freedom systems it has been pointed out that when the nature of unperturbed frequencies ω_1/ω_2 is rational primary resonances lead to elliptic fixed points which are encircled by island tori. These in turn give rise to secondary resonances which in turn give rise to secondary islands and so on. When the ratio of frequencies is irrational, invariant curves corresponding to quasiperiodic orbits are found which separate the islands. In addition when the ratio is rational, in the neighbourhood of hyperbolic fixed points resonance layers consisting of chaotic motion start appearing, whose size increases with increase in perturbation/energy. There can also be coexistence of islands and chaotic orbits depending on the initial conditions and other parameters. Thus an extremely complicated picture of dynamics arises, which is illustrated by means of Hénon–Heiles system, perturbed double-well quartic oscillator and standard map.

In our considerations so far we have not discussed in detail how one may characterize and quantify chaotic motion either in dissipative systems or in conservative systems except for pointing out their highly sensitive dependence on initial conditions. In the next chapter, we will introduce several characteristic quantities to distinguish between chaotic motion and regular motions, which can help to corroborate and confirm all our discussions more concretely.

Problems

1. ***Two-Coupled Anharmonic Oscillators***

   The Hamiltonian of a system of two coupled anharmonic oscillators is given by

   $$H = \frac{1}{2}\left(p_x^2 + p_y^2\right) + \frac{1}{2}\left(x^2 + y^2\right) + \lambda\left(x^4 + y^4 + 2Cx^2y^2\right) ,$$

   where x, y, p_x, p_y represent the displacements and momenta of the two oscillators, C represents coupling parameter and λ is a scaling parameter. The above Hamiltonian has arisen in problems associated with molecular dynamics, bistable lasers and other topics.

   a) Write down the equations of motion of the system.
   b) The equation of motion associated with it has exact analytical solutions for three particular values of the coupling parameter C, namely, $C = 0$, 1 and 3 and λ arbitrary [14]. That is, for these three values of C the behaviour of the system is always regular and never chaotic, irrespective of the values of energy. Verify.
   c) What is the nature of the dynamics for other values of C?

Deng and Hioe [15] studied the dynamics of the system for the initial conditions $x(0) = 5$, $p_x(0) = 0$, $y(0) = 10$ and $p_y(0) = 0$. They found chaotic motion for $C \leq -0.21$, regular motion for $-0.21 < C < 5.2$ and again chaotic motion for $C \geq 5.2$. Thus the motion transits from chaos to regular and again to chaos as C is continuously varied from $-\infty$ to $+\infty$. Verify the above results numerically, by drawing the trajectory plots and SOS plots. Also study the dynamics for various initial conditions.

2. ***Two-Coupled Nonlinear Oscillators***

The Hamiltonian of an unforced system of two coupled nonlinear oscillators is

$$H = \frac{1}{2}\left(p_x^2 + p_y^2\right) + \frac{1}{2}\left(\omega_x^2 x^2 + \omega_y^2 y^2\right) + \lambda x\left(y^2 + \eta x^2\right) ,$$

where (x, p_x, ω_x) and (y, p_y, ω_y) denote the coordinate, momentum and zeroth-order frequency, respectively. For the parametric choices $\omega_x = 1.3$, $\omega_y = 0.7$, $\lambda = 0.1$ and $\eta = -1$ the equation of motion associated with the above Hamiltonian is found to show transition from regular to chaos behaviour when its energy is varied [16]. Write the equation of motion for the above Hamiltonian and determine the stability of the equilibrium points. Study the dynamics as a function of the energy in the interval $[0.0 - 6.5]$ in the SOS (y, p_y) with $x = 0$, $p_x > 0$.

3. ***Chaos in Hydrogen Atom***

A real atomic system which is experimentally found to show various chaotic features is the hydrogen atom in a uniform magnetic field. If we consider two identical atoms (neutral) separated by a distance d (which is large in comparison with the radii of the atoms), then the atoms induce dipole moments on each other which will cause an attractive interaction between the atoms. This is called van der Waals interaction. The Hamiltonian of the problem can be reduced under certain assumptions to the form

$$H = \frac{p^2}{2} - \frac{1}{r} + \nu\left(x^2 + y^2 + \beta^2 z^2\right) , \quad \nu = \frac{-1}{16d^3} .$$

In cylindrical coordinate system, after suitable rescaling of the variables and parameters, the above Hamiltonian becomes

$$H = 2 = \frac{1}{2}\left(p_u^2 + p_v^2\right) - E\left(u^2 + v^2\right) + A\left(u^6 + v^6\right) + B\left(u^4 v^2 + u^2 v^4\right).$$

Its equation of motion is

$$\ddot{u} - 2Eu + 6Au^5 + 4Bu^3 v^2 + 2Buv^4 = 0 ,$$
$$\ddot{v} - 2Ev + 6Av^5 + 4Bu^2 v^3 + 2Bu^4 v = 0 .$$

Fix the value of E as -1×10^{-8} and A as $1/6$. For these choices the system is found to be integrable for $B = 0$, 0.5 and 2.5. That is, the system shows only regular behaviour at these values of B. When B is increased from -0.1, chaos $\longrightarrow$ regular ($B = 0$) $\longrightarrow$ chaos $\longrightarrow$ regular ($B = 0.5$) $\longrightarrow$ chaos $\longrightarrow$ regular ($B = 2.5$) $\longrightarrow$ chaos transition is observed [17]. (For a review on hydrogen atom problem, see for example Ref. [18]). Numerically investigate the above transitions and explore the dynamics in the (u, p_u) SOS with $v = 0$, $p_v > 0$.

4. ***Abelian–Higgs Model***

   The Abelian–Higgs model is considered as a relativistic field theory of type-II superconductors. In a simplified version, the Hamiltonian of the Abelian–Higgs model is given by

   $$H = \frac{1}{2}\left(\dot{q}_1^2 + \dot{q}_2^2\right) + \frac{1}{2}q_1^2 q_2^2 + \frac{P}{4}\left(q_2^2 - \frac{1}{P}\right)^2 ,$$

   where q_1 and q_2 describe time variation of vector and scalar fields in type-II superconductors.

   a) Write the potential function and equation of motion for the Abelian–Higgs model.
   b) Draw the potential curves for $P = 1$.
   c) Describe the nature of phase space trajectories for E (energy) < 0.25 and $E > 0.25$. In the $q_2 - \dot{q}_2$ SOS with $q_1 = 0$ study the occurrence of chaos for $P = 1$ and by varying the parameter E (energy) from a small value [19].

5. ***Atom-Molecule Collision System***

   The Hamiltonian of an atom-molecule collision system [16] is

   $$H = \frac{1}{2}\left(p_1^2 + p_2^2\right) + \frac{1}{2}\left(a_1 q_1^2 + a_2 q_2^2\right) + \lambda q_1 q_2^2 .$$

   For $a_1 = 1.6$, $a_2 = 0.9$, $\lambda = -0.08$ investigate the dynamics in a suitable SOS in the energy interval $[0, 25]$.

6. ***Two Coupled Rotators***

   The Hamiltonian of a system of two coupled rotators is given by [20]

   $$H = \frac{1}{2}\left(p_1^2 + p_2^2\right) - \epsilon \cos\left(q_2 - q_1\right) .$$

   Study the occurrence of regular and chaotic motion by varying the energy for a fixed value of ϵ.

7. ***Billiards Problem***

   The Hamiltonian of the billiards problem [21] is given by

   $$H = \frac{1}{2}\left(p_x^2 + p_y^2\right) + \frac{1}{2}\left(x^2 + \lambda y^2\right)^k = 1 .$$

a) Draw the orbits in the $x-y$, $x-p_x$ and $y-p_y$ planes for $\lambda = 2$ and for six values of k namely, $k = 1, 2, 3, 10, 20$ and 40. Use the initial conditions as $x = 1.2$, $y = 0$, $p_x = -0.2$ and $p_y = 0.721110255093$. Write a short note on the observed dynamics.
b) Study the dynamics in the $y-p_y$ SOS with $x = 0$ for $\lambda = 53/\pi$ and by varying the parameter k in the interval $[1, 8]$.

8. ***Double Pendulum***

The Hamiltonian of a double pendulum consisting of two equal point masses m, one suspended from a fixed support by a weightless rod of length L and the second suspended by the first by a similar rod is given by [22]

$$H = \frac{p_1^2 + p_2^2 - 2p_1p_2\cos(q_1 - q_2)}{2mL^2\left(1 + \sin^2(q_1 - q_2)\right)} + mgL\,(3 - 2\cos q_1 - \cos q_2)\ ,$$

where p_1 and p_2 are canonical momenta conjugate to the coordinates q_1 and q_2 respectively. Study the dynamics of the system in the SOS $p_1 - \dot{p}_1$ with $p_2 = 0$ and $\dot{p}_2 > 0$ as a function of energy by solving the equation of motion numerically.

9. ***Nonlinear Propagation of Electromagnetic Waves in a Collisionless Plasma***

The following Hamiltonian occurs in the study of nonlinear propagation of relativistically intense electromagnetic waves in a collisionless plasma [23]

$$H = \frac{1}{2}\left(\dot{X}^2 + \dot{Z}^2\right) + |\beta - \beta_b|\sqrt{\beta^2 - 1 + X^2 + Z^2 + Z(1 - \beta\beta_b)}\ ,$$

where X and Z are the variables proportional to the x and z components of the momentum of the wave respectively and β_b is the equilibrium velocity of the plasma and $\beta = u/c$ where u is the velocity of the frame and c is velocity of light. Write down the equations of motion for the above Hamiltonian and
a) investigate the stability of equilibrium points and
b) study the dynamics in the Z, $\dot{Z}$ SOS with $X = 0$, $\dot{X} > 0$ for $\beta = 1.1$, $\beta_b = 0.7$ as a function of energy.

10. ***Coupled Bloch Oscillators Equations***

The system of coupled Bloch oscillators equations representing electronic excitations of molecular dimers is

$$\begin{aligned}
\dot{x} &= -\gamma q y\ ,\\
\dot{y} &= 2Vz + \gamma q x\ ,\\
\dot{z} &= -2Vy\ ,\\
\dot{q} &= p\ ,\\
\dot{p} &= -\omega^2 q - \gamma z
\end{aligned}$$

with the total energy E given by

$$E = \frac{1}{2}\left(p^2 + \omega^2 q^2\right) + \frac{\gamma}{\sqrt{2}}\, zq - Vx \,.$$

Here γ, V and ω are parameters.

a) Investigate the stability of equilibrium points.

b) Study the occurrence of regular and chaotic dynamics in the SOS $q-p$ for $V = 10$, $\omega = 2.435$, $\gamma = 20$ and varying E in the interval $[-15, 15]$ [24].

11. ***Undamped and Periodically Driven Pendulum***

The equation of motion of an undamped and periodically driven pendulum is given by

$$\ddot{x} + \sin x = f \sin(kx - \omega t) \,.$$

When $f = 0$, typical orbits in the $(x - \dot{x})$ phase plane are depicted in Fig. 3.3. Fix the values of the parameters at $\omega = 4$ and $k = 75$. When $f \neq 0$ the separatrices are destroyed and chaotic motion is developed in its neighbourhood. In the Poincaré map inside the chaotic layer there are invariant tori enclosing the elliptic equilibrium point. As the control parameter f is increased the dynamics becomes complex and more and more invariant curves are destroyed [25]. Analyse the behaviour of the system by varying the parameter f.

12. ***Undamped and Periodically Driven Torsional Pendulum***

The equation of motion of an undamped torsional pendulum is [26]

$$\ddot{x} + kx - \alpha \sin x = f \sin t \,.$$

a) Show that the unforced system ($f = 0$) has only one equilibrium point for $k \geq \alpha$ and multiple equilibria for $k < \alpha$. For $\alpha = 0.5$ find the equilibrium points and obtain their stability.

b) Investigate the occurrence of chaos in the Poincaré map by varying the parameter f for $\alpha = 0.5$ and $k = 0.6$.

13. ***Chaos in a Dielectric Liquid System***

Consider the problem of a dielectric liquid placed between two electrodes. Injection of ions is assumed to occur from one of the electrodes when a voltage difference is applied. Experimentally the ion trajectory was shown to be chaotic [27]. The flow equations are

$$\dot{x} = -2xy \,,$$
$$\dot{y} = -A(1 + \epsilon \sin \Omega t) + y^2 + x^2/2 \,,$$

where x and y are cartesian coordinates and A is a parameter related to the fluid velocity in the cell.

a) Verify that for $\epsilon = 0$ and $A > 0$ the system has two saddles at $(0, \pm\sqrt{A}\,)$ and two centers at $(\pm\sqrt{2A}\,, 0)$.

b) For $A = 0.01$ investigate the dynamics of the system in the Poincaré map by varying ϵ for fixed values of Ω.

14. ***Chaotic Dynamics of a Satellite***

The equation of motion of an arbitrarily shaped satellite in an elliptic orbit about a point is given by

$$(1 + \epsilon \cos t)\ddot{\Psi} - 2\epsilon \sin t(1 + \dot{\Psi}) + 3K \sin \Psi \cos \Psi = 0 ,$$

where Ψ, ϵ and K are the libration angle in the orbital plane as measured from the local vertical, orbital eccentricity and inertia moment ratio respectively. Investigate the occurrence of regular and chaotic dynamics in the Poincaré map as a function of initial conditions for $K = 0.05$ and thereby varying the parameter ϵ from 0 to 0.3 [28].

15. ***Area-Preserving Hénon Map [10]***

The conservative Hénon map is given by

$$x_{n+1} = 1 - ax_n^2 + y_n ,$$
$$y_{n+1} = x_n .$$

Show that period-1, period-2 and period-4 elliptic fixed points occur in the interval $0 < a < 3$, $3 < a < 4$ and $4 < a < 4.120452497..$ respectively (see Ref. [5]). Also investigate the occurrence of chaos.

16. ***Whisker Map [4]***

The whisker map

$$w_{n+1} = w_n + \epsilon\pi\omega_0 \operatorname{sech}\left(\omega_0\pi/\left(2\sqrt{2}\right)\right) \sin\phi_n ,$$
$$\phi_{n+1} = \phi_n + \left(\omega_0/\sqrt{2}\right) \ln\left(64/w_{n+1}\right)$$

is a map of the energy change and phase change of a trajectory in the neighbourhood of the separatrix for each period of the motion for a periodically driven undamped Duffing oscillator equation [4]. Study the dynamics of the map as a function of the parameter ϵ for a fixed value of ω_0.

17. ***The Web Map***

The web map describing a periodically kicked oscillator is given by

$$u_{n+1} = v_n ,$$
$$v_{n+1} = -u_n - k \sin v_n .$$

Investigate the stability of fixed points of the map. Numerically study the dynamics as a function of the parameter k [29].

18. ***An Area-Preserving Map***

The map

$$\theta_{i+1} = \theta_i + \alpha + \sin^{-1} s_i \ ,$$
$$s_{i+1} = s_i + 2\nu \sin(2\theta_{i+1}) \ ,$$

is introduced to describe the ground state properties of some discotic liquid crystals [30]. Show that the map is area-preserving as long as $|s_i| < 1$ and investigate the dynamics for a fixed value of α and varying ν from 0.

19. ***Spin Map***

The following map arises in the study of spin patterns in a field perturbed Heisenberg chain [31],

$$X_{i+1} = X_i - r\sin(\theta_i - \pi) \ , \quad -1 \leq X_i \leq 1$$
$$\theta_{i+1} = \theta_i + \sin^{-1} X_{i+1}, \quad (\text{mod } 2\pi) \ .$$

a) Show that the map is area-preserving.
b) Verify that the map with $r > 0$ has an elliptic fixed point at $(X^*, \theta^*) = (0, \pi)$ and a hyperbolic fixed point at $(X^*, \theta^*) = (0, 0)$ (mod 2π).
c) Study the dynamics of the map by varying r from a small value.

20. ***Particle Kicked by a Potential***

The motion of a particle which is periodically kicked by the potential $V = (V_0/\beta)/(1+x^2)^{\beta/2}$ is described by the two-dimensional map [32]

$$x_{n+1} = x_n + p_{n+1} \ ,$$
$$p_{n+1} = p_n - V_0 x_n \left(1 + x_n^2\right)^{-(\beta+2)/2} \ ,$$

where x and y represent position and momentum of the particle, and V_0 and β are the parameters controlling the depth and asymptotic behaviour of the well, respectively. Show that the map has a stable fixed point at the origin for $0 < V_0 < 4$ and period doubling cascade occurs for $V_0 > 4$.

21. ***A Generalized Standard Map***

Consider the map

$$J_{n+1} = J_n + (K/2\pi)\sin(2\pi R\theta_n) \ ,$$
$$\theta_{n+1} = \theta_n + J_{n+1} \quad (\text{mod } 1) \ ,$$

where K and R are parameters. When $R = 1$, the map reduces to the standard map (7.39) with suitable redefinition of the variables. Fixing K at some low value, study the dynamics as a function of the parameter R. In particular show that the tori of the $R = 1$ case get destroyed even for slight changes in R. Compare the dynamics for $R = 0.95$, 1.0 and 1.05 [33].

8. Characterization of Regular and Chaotic Motions

We have pointed out in the previous chapters that many important nonlinear dynamical systems in physics, chemistry and biology clearly display different types of regular and chaotic behaviours, depending upon factors such as the strength of control parameters, initial conditions, nature of external forcings, and so on. Thus in order to identify *definitively* whether a given motion of a typical dynamical system is, for example, periodic or quasiperiodic or chaotic, one would like to have specific quantitative measures in addition to the various qualitative features such as the geometric structures of attractors, stability features and so on. Moreover we have seen earlier that the motion of typical nonlinear systems undergo characteristic qualitative changes as certain control parameters smoothly change, namely bifurcations. These are again identified by means of the changes of the system attractors or phase space structures and stability properties. However, since deterministic chaos is associated with random behaviour arising from sensitive dependence on initial conditions, it is quite natural to expect quantitative criteria to distinguish between chaotic and regular motions should be based on statistical measures. Indeed there are many such measures available in the literature for this purpose and some of the most prominent of them are

(1) Lyapunov exponents,
(2) power spectrum,
(3) correlation function and
(4) dimension.

These measures are capable of distinguishing different degrees of complexity of attractors and motions. In this chapter we will explain briefly the classification and characterization of attractors in terms of these statistical measures. We will also discuss the procedures to estimate these quantities.

8.1 Lyapunov Exponents

While studying chaotic motions, earlier we have met with the crucial notion of sensitive dependence on initial conditions and seen simple examples of neighbouring orbits which separate exponentially (see for example, Sect. 4.2.6).

This sensitive dependence on initial conditions can be characterized using Lyapunov exponents [1,2].

The Lyapunov exponents describe essentially the rate of divergence or convergence of nearby trajectories onto the attractor in different directions in the phase space. Let us now define the Lyapunov exponents for the trajectory $\boldsymbol{X}(t) = (x_1(t), x_2(t), ..., x_n(t))$ of the system (3.1). We consider two trajectories (Fig. 8.1) in an n-dimensional phase space starting from two nearby initial conditions $\boldsymbol{X}_0$ and $\boldsymbol{X}'_0 = \boldsymbol{X}_0 + \delta\boldsymbol{X}_0$. They evolve with time yielding the vectors $\boldsymbol{X}(t)$ and $\boldsymbol{X}(t) + \delta\boldsymbol{X}(t)$, respectively, with the Euclidean norm

$$d(\boldsymbol{X}_0, t) = ||\delta\boldsymbol{X}(\boldsymbol{X}_0, t)|| \equiv \sqrt{\delta x_1^2 + \delta x_2^2 + ... + \delta x_n^2} \,. \tag{8.1}$$

$d(\boldsymbol{X}_0, t)$ is simply a measure of the distance between two trajectories $\boldsymbol{X}(t)$ and $\boldsymbol{X}'(t)$. The time evolution of $\delta\boldsymbol{X}$ is found by linearizing (3.1) to obtain

$$\delta\dot{\boldsymbol{X}} = M(\boldsymbol{X}(t)) \,.\, \delta\boldsymbol{X} \,, \tag{8.2}$$

where $M = \partial\boldsymbol{F}/\partial\boldsymbol{X}|_{\boldsymbol{X}=\boldsymbol{X}_0}$ is the Jacobian matrix of $\boldsymbol{F}$. Then the mean rate of divergence of two close trajectories, defined as the *Lyapunov exponent*, is given by (see also our earlier discussions in Sect. 4.2.6)

$$\lambda(\boldsymbol{X}_0, \delta\boldsymbol{X}) = \lim_{t\to\infty} \frac{1}{t} \log\left(\frac{d(\boldsymbol{X}_0, t)}{d(\boldsymbol{X}_0, 0)}\right) \,. \tag{8.3}$$

Furthermore, there are n-orthonormal vectors $\boldsymbol{e}_i$ of $\delta\boldsymbol{X}$, $i = 1, 2, ..., n$, (see also Sect. 3.6, (3.40)) such that

$$\delta\dot{\boldsymbol{e}}_i = M(\boldsymbol{X}_0)\,\boldsymbol{e}_i \,, \quad M = \mathrm{diag}(\lambda_1, \lambda_2, ..., \lambda_n) \,. \tag{8.4}$$

That is, there are n-Lyapunov exponents given by

$$\lambda_i(\boldsymbol{X}_0) = \lambda_i(\boldsymbol{X}_0, \boldsymbol{e}_i) \,, \quad i = 1, 2, ..., n \,. \tag{8.5}$$

These can be ordered as $\lambda_1 \geq \lambda_2 \geq ... \geq \lambda_n$. From (8.3) and (8.5) we may write

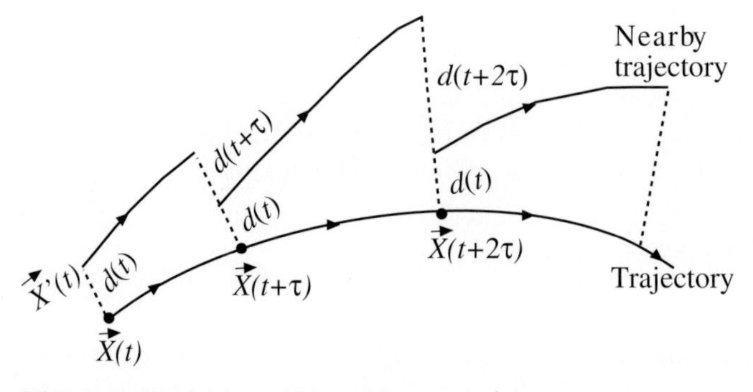

Fig. 8.1. Evolution of two nearby trajectories: Numerical calculation of the Lyapunov exponent (τ is a finite time interval). For details, see Sect. 8.2.2

$$d_i\left(\boldsymbol{X}_0, t\right) \approx d_i\left(\boldsymbol{X}_0, 0\right) \mathrm{e}^{\lambda_i t} , \quad i = 1, 2, ..., n . \tag{8.6}$$

To identify whether the motion is periodic or chaotic it is sufficient to consider the largest nonzero Lyapunov exponent λ_m. The following cases can be easily identified.

(i) $\lambda_m < 0$: *Equilibrium points/periodic attractors*

As t increases, $d(\boldsymbol{X}_0, t)$ (or simply $d(t)$) decreases from $d(0)$ and for $t \to \infty$, $|d(t)| \to 0$. This is the case for a stable equilibrium point or a stable periodic solution where two nearby trajectories converge towards the same attractor in the limit $t \to \infty$. There is no divergence or irregular variation of any infinitesimal perturbation.

(ii) $\lambda_m > 0$: *Chaotic attractor*

As t increases, $|d(t)|$ grows exponentially fast implying sensitive dependence on the initial perturbation and so on the initial condition. Thus, a positive Lyapunov exponent is the essence of deterministic chaos. The basic idea is that small errors $\approx |d(0)|$ in the initial conditions or round off errors in numerical computation are magnified exponentially fast, so that future state of the system cannot be predicted with the desired accuracy.

The Lyapunov exponents can also be interpreted in terms of information theory as giving the rate of loss of information about a location of the initial point $\boldsymbol{X}(0)$. In general, λ's can only be found by numerical computation but in some simple cases they can be evaluated analytically (see next section).

Some of the general properties of the Lyapunov exponents associated with dynamical systems are as follows:

(1) Since a volume element in the neighbourhood of a stable fixed point is contracted in all directions, all the Lyapunov exponents associated with it are negative.

(2) For a limit cycle attractor a volume element is neither contracted or elongated along the direction of the orbit and therefore one of the Lyapunov exponents is always zero and the others are negative.

(3) For a periodic orbit with a single frequency ω, one of the Lyapunov exponents is zero as stated in (2). Similarly, for a quasiperiodic orbit with two frequencies ω_1 and ω_2 with ω_1/ω_2 being irrational two of the Lyapunov exponents are zero and the others are negative. Generally, for a quasiperiodic orbit with k-linearly independent frequencies, k-Lyapunov exponents are zero and the others are negative. That is, for a stable k-torus, $\lambda_1 = \lambda_2 = ... = \lambda_k = 0$ and $\lambda_i < 0$ for $i = k+1, ..., N$.

(4) For chaotic behaviour at least one of the Lyapunov exponents must be positive so that neighbouring trajectories diverge. When more than one of the Lyapunov exponents are positive, then the motion is referred as *hyperchaos* [3].

(5) The exponential rate of growth of distance between nearby trajectories is given by the maximal Lyapunov exponent. On the other hand, the rate of change of an N-dimensional volume element is given by the sum of all the Lyapunov exponents. Earlier in Sect. 3.8 we have shown that the rate of change of the initial phase space volume is given by $\nabla \cdot \boldsymbol{F}$. Thus, if the divergence of the force field, $\nabla \cdot \boldsymbol{F}$, is constant then

$$\sum_{i=1}^{N} \lambda_i = \log |\nabla \cdot \boldsymbol{F}| \,. \tag{8.7}$$

(6) For a Hamiltonian system with N degrees of freedom the dimension of the phase space is $2N$ and hence there are $2N$ independent Lyapunov exponents. If the system has N functionally independent (and involutive) constants of motion then all the $2N$ Lyapunov exponents become zero. In other cases, since energy is always one of the constants of motion, at least two of the Lyapunov exponents should be zero. Further, the Lyapunov exponents of conservative systems are symmetric about zero so that the sum of all the Lyapunov exponents is zero implying that the overall phase space volume is conserved. This means that for a two degrees of freedom Hamiltonian system with a periodic or quasiperiodic motion all the four Lyapunov exponents are zero. On the other hand, for a chaotic orbit we have $\lambda_1 > 0$, $\lambda_2 = \lambda_3 = 0$, and $\lambda_4 = -\lambda_1$. Generalizing this, one can easily show that for a Hamiltonian system with N degrees of freedom, the relation

$$\lambda_i + \lambda_{2N+1-i} = 0 \tag{8.8}$$

holds for all $0 < i < N+1$, where the phase space dimension is $2N$.

(7) For a *purely random signal*, predictability is not possible even for a short time. In fact, if $\boldsymbol{X}(t)$ is a random solution at time t then predictability is lost at the next moment itself and hence even λ_1 is ∞.

8.2 Numerical Computation of Lyapunov Exponents

The Lyapunov exponents associated with a given attractor can be computed by linearizing the equation of motion about the corresponding orbit. In order to formulate a useful method for the numerical computation of Lyapunov exponents we begin with the case of a one-dimensional map. Then we apply the method to differential equations.

8.2.1 One-Dimensional Map

For a one-dimensional map of the form $x_{n+1} = f(x_n)$, the evolution of an infinitesimal initial perturbation δx is given by (refer Sect. 4.2)

$$\delta x_{n+1} = \delta x_1 \frac{\mathrm{d}f^n(x_1)}{\mathrm{d}x_1} , \tag{8.9}$$

where $f^n(x_1) = f(f...f(x_1))$. Using the chain rule for derivative, namely,

$$\frac{\mathrm{d}}{\mathrm{d}x_1} f^n(x_1) = \prod_{i=1}^{n} \frac{\mathrm{d}f(x_i)}{\mathrm{d}x_i} , \tag{8.10}$$

(8.9) can be rewritten as

$$\begin{aligned} |\delta x_{n+1}| &= |\delta x_1| \prod_i \frac{\mathrm{d}f(x_i)}{\mathrm{d}x_i} \\ &= |\delta x_1| \exp\left[\sum_i \log\left|\frac{\mathrm{d}f(x_i)}{\mathrm{d}x_i}\right|\right] \\ &= |\delta x_1| \, \mathrm{e}^{\lambda n} , \end{aligned}$$

where

$$\lambda = \frac{1}{n} \sum_i \log\left|\frac{\mathrm{d}f(x_i)}{\mathrm{d}x_i}\right| . \tag{8.11}$$

Equation (8.11) can be used to calculate λ by taking the limit $n \to \infty$.

For the logistic map, (4.19), λ is given by

$$\lambda = \frac{1}{n} \sum_{i=1}^{n} \log |a - 2ax_i| . \tag{8.12}$$

For $a = 3.99$, Fig. 8.2 depicts the variation of λ with n. We note that λ approaches a constant value for sufficiently large n. Therefore in any numerical computation of λ to assure the convergence of λ one should choose a sufficiently large value of n. Figure 8.3 shows the bifurcation diagram and the corresponding numerically computed Lyapunov exponent for $a \in (2.8, 4.0)$, where n is chosen as 10^4. It is seen that λ is negative for period-n solutions and positive for chaotic solutions. At the bifurcation value λ is found to be nearly zero.

8.2.2 Computation of Lyapunov Exponents for Continuous Time Dynamical Systems

Now we will indicate how to find the Lyapunov exponents associated with a continuous system of the form (3.1). To find all the n-Lyapunov exponents of such a n-dimensional continuous system, a reference trajectory is created by integrating the nonlinear equations of motion (3.1). Simultaneously the linearized equations of motion (3.37) or (8.2) are integrated for n-different initial conditions defining an arbitrarily oriented frame of n-orthonormal vectors $(\boldsymbol{\Delta X}_1, \boldsymbol{\Delta X}_2, ..., \boldsymbol{\Delta X}_n)$. There are two technical problems [3] in evaluating

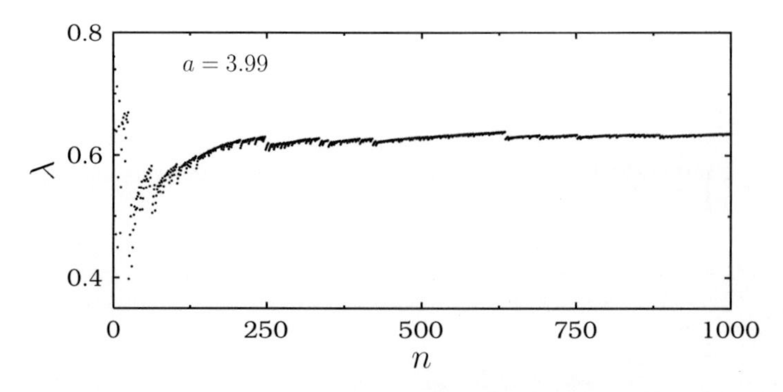

Fig. 8.2. Variation of λ as a function of n for the logistic map (4.19) for $a = 3.99$, showing convergence of it to a constant value in the limit $n \to \infty$

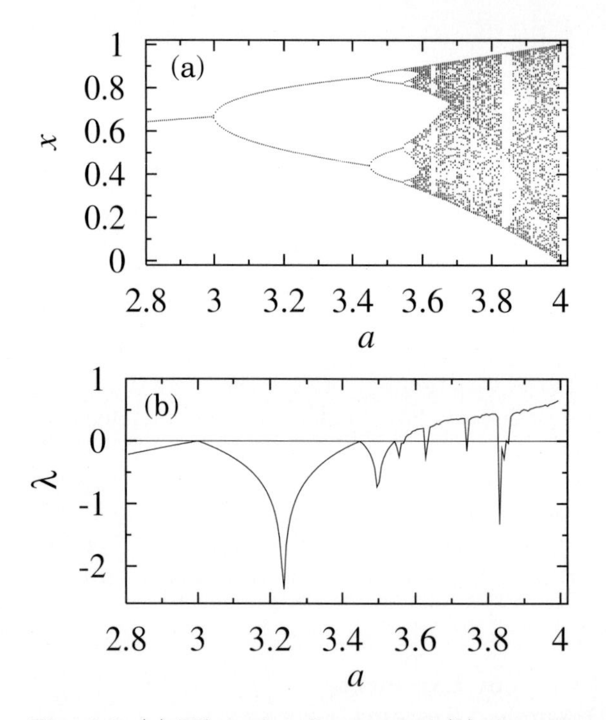

Fig. 8.3. (**a**) Bifurcation diagram and (**b**) the corresponding Lyapunov exponent of the logistic map, (4.19), in the range $a \in (2.8, 4.0)$

the Lyapunov exponents directly using (8.3), namely, the variational equations have at least one exponentially diverging solution leading to a storage problem in the computer memory. Further, the orthonormal vectors evolve in time and tend to fall along the local direction of most rapid growth. Due to the finite precision of computer calculations the collapse toward a common direction causes the tangent space orientation of all the axis vectors to become indistinguishable. Both the problems can be overcome by a repeated use of what is known as Gram-Schmidt reorthonormalization (GSR) procedure [4] which is well known in the theory of linear vector spaces. We apply GSR after τ time steps which orthonormalize the evolved vectors to give a new set $\{\boldsymbol{u}_1, \boldsymbol{u}_2, ..., \boldsymbol{u}_n\}$:

$$\boldsymbol{v}_1 = \boldsymbol{\Delta X}_1 \,, \tag{8.13a}$$

$$\boldsymbol{u}_1 = \boldsymbol{v}_1/||\boldsymbol{v}_1|| \,, \tag{8.13b}$$

$$\boldsymbol{v}_i = \boldsymbol{\Delta X}_i - \sum_{j=1}^{i-1} \langle \boldsymbol{\Delta X}_i \,, \boldsymbol{u}_j \rangle \, \boldsymbol{u}_j \,, \quad i = 2, 3, ..., n \tag{8.13c}$$

$$\boldsymbol{u}_i = \boldsymbol{v}_i/||\boldsymbol{v}_i|| \,, \tag{8.13d}$$

where $\langle , \rangle$ denotes inner product. In this way the rate of growth of evolved vectors can be updated by the repeated use of GSR. Then, after the N-th stage, for N large enough, the one-dimensional Lyapunov exponents are given by

$$\lambda_i = \frac{1}{N\tau} \sum_{k=1}^{N} \log ||\boldsymbol{v}_i^{(k)}|| \,. \tag{8.14}$$

For a given dynamical system, τ and N are chosen appropriately so that the convergence of Lyapunov exponents is assured. A fortran code algorithm implementing the above scheme can be found in Ref. [3].

As an example, we now calculate the Lyapunov exponents for the Duffing oscillator, equation (5.1), by employing the above procedure. The values of the parameters are fixed at $\alpha = 0.5$, $\beta = 1$, $\omega_0^2 = -1.0$, $\omega = 1.0$. τ and N in (8.13) are chosen as $(2\pi/\omega)/10$ and 10^4, respectively. Figure 8.4 shows the bifurcation diagram and the calculated maximal Lyapunov exponent.

We note further that for two-dimensional and higher dimensional maps also all the Lyapunov exponents can be computed using the method described above for continuous time dynamical systems. For example, Fig. 8.5 depicts the bifurcation diagram and the corresponding maximal Lyapunov exponent for the Hénon map (4.45) with $b = 0.3$ and $\tau = 500$.

Exercises:

1. For the following maps
   a) Obtain the expression for the one-dimensional Lyapunov exponent.
   b) Draw the bifurcation diagram and compute the Lyapunov exponent for the specified range of values of the control parameter r.
   c) Characterize regular and chaotic dynamics and bifurcations using bifurcation diagram and the Lyapunov exponent.

   (i) $x_{n+1} = x_n \exp\left[r\left(1 - x_n\right)\right]$, $r \in (0, 4)$ [5].
   (ii) $x_{n+1} = rx_n/\left(1 + x_n^6\right)$, $r \in (1, 5)$ [6].

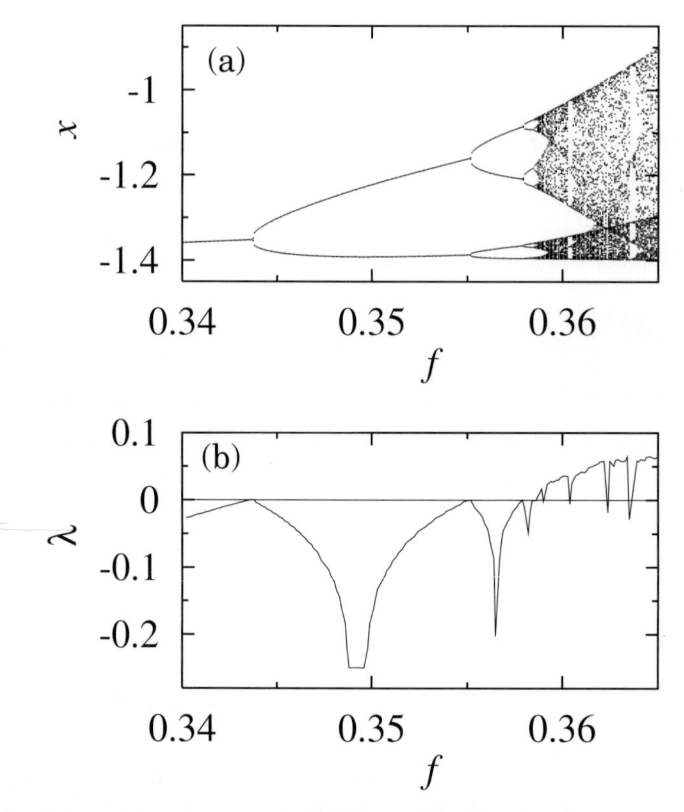

Fig. 8.4. (**a**) Bifurcation diagram and (**b**) the corresponding maximal Lyapunov exponent for the Duffing oscillator, equation (5.1)

(iii) $x_{n+1} = x\,(a + b/\,(1 + x^r))$, $a = 0.55$, $b = 3.45$, $r \in (0, 100)$ [7].

(iv) $x_{n+1} = \begin{cases} 1 - r|x_n|^a & \text{if } x_n < 0 \\ 1 - rx_n^2 & \text{if } x_n \geq 0, \end{cases}$

where $a = 0.4$, $r \in (0.71, 1.07)$ and $a = 0.8$, $r \in (0.7, 1.06)$ [8].

2. For the following two-dimensional maps obtain the formulas for Lyapunov exponents. Study the dynamics of these maps in the specified range of values of the control parameter r using Lyapunov exponents, bifurcation diagram and x_n versus x_{n+1} plot.

(a) ***Modulated Logistic Map [9]***

$$x_{n+1} = 4a_n x_n\,(1 - x_n) \ ,$$
$$a_{n+1} = 4ra_n\,(1 - a_n) \ , \quad r \in (0, 1) \ .$$

(b) ***Quasiperiodically Excited Circle Map [10]***

$$x_{n+1} = x_n + 2\pi K + r \sin x_n + C \cos y_n \ , \quad (\text{mod } 2\pi)$$
$$y_{n+1} = y_n + 2\pi\omega \ , \quad (\text{mod } 2\pi)$$

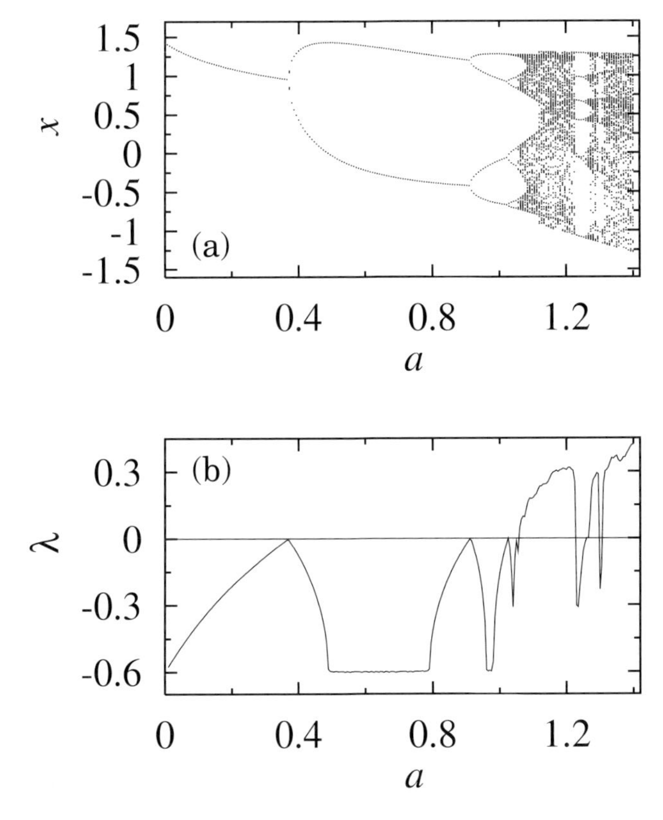

Fig. 8.5. (**a**) Bifurcation diagram and (**b**) the corresponding maximal Lyapunov exponent for the Hénon map (4.45) with $b = 0.3$

where $\omega = (\sqrt{5} - 1)/2$, $K = 0.3$, $C = 0.6$ and $r \in (0, 1.2)$.

(c) ***Quasiperiodically Excited Logistic Map [11]***

$$x_{n+1} = ax_n (1 - x_n) + r \sin 2\pi y_n \,, \quad (\text{mod } 1)$$
$$y_{n+1} = y_n + \omega \,, \quad (\text{mod } 2\pi)$$

where $\omega = (\sqrt{5} - 1)/2$, $a = 2.8$ and $r \in (0, 0.25)$.

3. What is the Lyapunov exponent of the following tent map?

$$x_{n+1} = 2\mu \begin{cases} x_n & \text{if } \; 0 \le x < 0.5 \\ 1 - x_n & \text{if } \; 0.5 \le x \le 1.0. \end{cases}$$

4. Show that the formula for the Lyapunov exponent of the logarithmic map $x_{n+1} = \ln(a|x_n|)$ is

$$\lambda = \ln a - \lim_{n \to \infty} \frac{1}{n} \sum_{i=1}^{n} x_i \,.$$

5. Explain the reason for the dip observed about $a = 1 + \sqrt{5} \approx 3.23607$ in the Lyapunov exponent diagram (8.3) of the logistic map.
6. In Fig. 8.5 the maximal Lyapunov exponent of the Hénon map is constant for $a \in (0.49, 0.79)$. Investigate the reason for this. [Hint: Compute the eigenvalues of the period-2 cycle and analyse its variation].
7. Can the Lyapunov exponent take a constant value in an interval of the control parameter in the case of the logistic map? Why?
8. Analytically prove that the one-dimensional Lyapunov exponent of the exponential map

$$x_{n+1} = x_n \exp\left(A\left(1 - x_n\right)\right)$$

becomes $-\infty$ at $A = 1$.
9. Write the variational equations for the
   a) Lorenz equations (5.5),
   b) van der Pol equation (5.9),
   c) pendulum equation (5.10) and
   d) MLC circuit equation (6.17).

   Develop fortran code algorithms to compute all the Lyapunov exponents for the attractors of these systems. Draw the maximal Lyapunov exponent plot corresponding to the bifurcation diagrams (5.16), (5.22), (5.23) and (6.14).
10. Write down the variational equations for the equations of problem 7 in Chap. 5. Compute the Lyapunov exponents corresponding to the bifurcation diagram in the interval $f \in (0, 0.6)$ for $\omega = 0.646$.
11. Compute all the Lyapunov exponents for the problem 4 in Chap. 6. Identify hyperchaotic region where more than one Lyapunov exponents become positive.
12. For the logistic map (4.19) calculate the Lyapunov exponent as a function of a control parameter in the neighbourhood of onset of chaos, sudden widening and intermittency and band-merging bifurcations. Verify that it varies linearly near band-merging and abruptly in the form $\lambda \propto (a - a_0)^\alpha$, where a_0 is the bifurcation point and α is the scaling exponent, for other bifurcations [12].
13. The local expansion rate of an one-dimensional map $x_{n+1} = f(x_n)$ is defined as $\Lambda(x_n) = \ln|\mathrm{d}f(x_n)/\mathrm{d}x_n|$. Consider the variance

$$\langle [S_n(x) - n\lambda]^2 \rangle = \frac{1}{N} \sum_{m=1}^{N} [S_n(x_m) - n\lambda]^2 \ ,$$

where

$$S_n(x_1) = \sum_{t=1}^{n} \Lambda(x_t) \ , \quad n = 1, 2, ...$$

and λ is the Lyapunov exponent. Calculate the variance as a function of n. Verify that

a) at the onset of chaos the plot of variance versus $\log_2 n$ exhibits self-similar pattern and
b) for a chaotic orbit the variance grows with n for large n [13].

For further developments on developing improved methods of finding Lyapunov exponents, difficulties involved in the existing methods and ways of overcoming them, analyse the references [14–17].

8.3 Power Spectrum

A simple way to check for periodicity and chaos of a dynamical system is to consider the time series of a dynamical variable and to compute its Fourier transform. We briefly explain below the details.

8.3.1 The Power Spectrum and Dynamical Motion

Considering the Fourier series expansion of a function $x(t)$, we can write

$$x(t) = \sum_{k=-\infty}^{\infty} C_k \mathrm{e}^{\mathrm{i}kt} , \tag{8.15a}$$

where each the Fourier coefficients C_k is calculated by the integral

$$C_k = \frac{1}{2\pi} \int_0^{2\pi} x(t) \mathrm{e}^{-\mathrm{i}kt} \mathrm{d}t . \tag{8.15b}$$

The Fourier coefficients C_k of $x(t)$ can be used to understand the periodic and aperiodic character of the function. For example, let $x(t)$ be a 2π-periodic function. If we write

$$C_k = |C_k| \mathrm{e}^{\mathrm{i}\theta_k} , \tag{8.16}$$

then the Fourier series of $x(t)$ becomes

$$x(t) \approx 2 \sum_{k=0}^{\infty} |C_k| \cos(\theta_k + kt) . \tag{8.17}$$

That is, $x(t)$ is expressed as a sum of simple harmonic oscillations. The kth mode

$$x_k = 2|C_k| \cos(\theta_k + kt) \tag{8.18}$$

has amplitude $2|C_k|$, frequency $k/2\pi$, angular frequency k, period $2\pi/k$ and phase angle θ_k. Here $|C_k|$ measures the extent to which a simple harmonic motion of angular frequency k is present in the total motion. The sequence $|C_0|^2$, $|C_1|^2$, ..., is then called the *power spectrum* $P(k)$ of $x(t)$, that is,

$$P(k) = |C_k|^2 , \quad k = 0, 1, ..., \infty . \tag{8.19}$$

For a continuous-time dynamical system described by a set of differential equations, the numerical data are based on a discretization of time, while the measurements are also usually made at regular intervals of time. Therefore, it is not possible to calculate the Fourier coefficients given by (8.15b) exactly, because the integral cannot be evaluated in closed form. However, numerical integration is possible. Therefore, we concentrate on a time series $x(t_m), m = 1, 2, ...,$ where $t_m = m\Delta t$ with Δt being the time duration with which the data are collected. The composite trapezoidal rule for numerical integration of a function $g(t)$ in the interval 0 to 2π is given by

$$\int_0^{2\pi} g(t)\mathrm{d}t \approx \frac{2\pi}{N} \sum_{m=0}^{N-1} g\left(\frac{2\pi m}{N}\right) , \tag{8.20}$$

where $g(t)$ is 2π-periodic. Then

$$C_k = \frac{1}{N} \sum_{m=0}^{N-1} x(2\pi m/N) \exp(-\mathrm{i}2\pi km/N) . \tag{8.21}$$

Now a periodic orbit $\{x_1^p, x_2^p, ..., x_q^p\}$ with period $q = 2^p$ can be represented by its Fourier expansion

$$x_k^p = \sum_j C_j^p \exp(\mathrm{i}\Omega_j k) , \tag{8.22}$$

where the frequency $\Omega_j = 2\pi j/2^p$, $j = 0, 1, 2, ..., 2^{p-1}$. We note that the complex Fourier transform $F(\Omega)$ of $\exp(\mathrm{i}\omega k)$ is $2\pi\delta(\Omega - \omega)$. Consequently, the Fourier transform of the above periodic orbit is

$$\overline{x}(\Omega) = 2\pi \sum_j C_j^p \delta(\Omega - \omega_j) . \tag{8.23}$$

Thus the power spectrum $P(\Omega) = |\overline{x}(\Omega)|^2$ will consist of delta functions at the frequencies ω_j with the corresponding amplitudes given by $|C_j^p|^2$.

We can now characterize the power spectrum associated with the various attractors.

(1) The power spectrum of a periodic motion with frequency ω_1 has Dirac-delta peaks at ω_1 and its various harmonics. The number of components of the spectrum will double after each period doubling, $p \to p + 1$.

(2) A quasiperiodic motion with m rationally independent frequencies ω_1, ω_2, ..., ω_m has Dirac delta peaks at all linear combinations of the basic frequencies with integer coefficients.

(3) On the other hand, the spectrum of a fixed point solution contains a single peak at zero frequency.

(4) The spectrum of a chaotic solution is quite different from that of the periodic and quasiperiodic solutions. It has continuous and broad-band nature. In addition to the broad-band component, it is quite common for

chaotic spectra to contain spikes indicating the predominant frequencies of the solutions.

Equation (8.21) is usually taken only as an approximation to C_k with $|k| \leq N/2$. From (8.21) we find that the evaluation of any particular C_k requires N multiplications and N additions. Therefore the calculation of $N/2$ number of C's would take N^2 operations. Thus, for 10^3 sample points we would need 10^6 operations which is a serious time-consuming computation. For efficient calculation of power spectrum it is convenient to use the *fast Fourier transform* (FFT) [18] algorithm to evaluate $\overline{x}(\Omega_k)$; in this case N is taken to be an integral power of 2, typically, 2048 or 4096. The advantage of FFT is that the number of computer operations, that is, the number of multiplications and additions are reduced considerably. It takes advantage of certain symmetry properties in the trigonometric functions at their points of evaluation in order to achieve an increase in speed over more conventional methods. If N is the number of data points in the time series then the FFT requires about $N \log_2 N$ operations only compared to N^2 operations in conventional methods.

To obtain an accurate FFT, a "cosine bell" function or some other "windowing" function can be applied to the sampled time series before the FFT is applied. This eliminates spurious frequency components associated with sharp edges in the time series.

For an N-point time series the discrete FFT is

$$C_k = \sum_{j=0}^{N-1} x_j \omega_j \exp(\mathrm{i}2\pi jk/N) \,, \quad k = 0, 1, ..., N-1 \tag{8.24}$$

where ω_j is the "Welch window" function given by

$$\omega_j = 1 - \left(\frac{2j - N}{N} \right)^2 . \tag{8.25}$$

Then the power spectrum is defined at $(N/2) + 1$ frequencies as

$$P(\Omega_0) = |C_0|^2 / W_{ss} \,, \tag{8.26a}$$

$$P(\Omega_k) = [\, |C_k|^2 + |C_{N-k}|^2 \,] / W_{ss} \,, \quad k = 1, 2, ..., N/2 - 1 \,, \tag{8.26b}$$

$$P(\Omega_c) = |C_{N/2}|^2 / W_{ss} \,, \tag{8.26c}$$

where W_{ss} stands for "window squared and summed",

$$W_{ss} = N \sum_{j=0}^{N} \omega_j^2 \tag{8.26d}$$

and

$$\Omega_k = \frac{k}{N \Delta t} = \frac{2\Omega_c k}{N} \,, \quad k = 0, 1, 2, ..., N/2 \,. \tag{8.26e}$$

For fuller details, see for example, Ref. [18].

In Sect. 5.1, we have studied the period doubling phenomenon leading to chaos in the Duffing oscillator equation (5.1). We now calculate the power spectrum for the equilibrium point, periodic and chaotic solutions of the Duffing equation. A set of 2^{12} data collected at a time interval $(2\pi/\omega)/10$ are used. For $f = 0$, the attractor in the left well is an equilibrium point. Figure 8.6a shows the log of power density of the solution. The power density has a peak at zero frequency and it decays exponentially with the frequency Ω. As noted above for a periodic orbit with frequency Ω the power spectrum must have δ-peaks at $\Omega, 2\Omega, ...$, that is at $1/T, 2/T, ...$ Figure 8.6b depicts the power spectrum corresponding to period-T orbit ($f = 0.33$). In this figure peaks at $1/T, 2/T, 3/T, 4/T, 5/T$ are clearly visible. The peak at $6/T$ is not visible in the scale used. Figure 8.6c corresponds to a period-$2T$ solution ($f = 0.35$). The doubling of the number of peaks in this figure indicates the period doubling phenomenon. The power spectrum of a chaotic solution ($f = 0.365$) is shown in Fig. 8.6d where the continuous, broad-band and noisy spectrum supports the chaotic nature of the solution. The sharp peaks in the continuous spectrum indicates the presence of large number of unstable periodic orbits embedded in the chaotic attractor (see Sect. 9.4 for some more details).

Next, we consider the MLC circuit equation with quasiperiodic forcing,

$$\dot{x} = y - h(x) \,, \tag{8.27a}$$

$$\dot{y} = -\beta(1+\nu)y - \beta x + f_1 \sin \omega_1 t + f_2 \sin \omega_2 t \,, \tag{8.27b}$$

where $h(x) = bx + 0.5(a-b)[\,|x+1| - |x-1|\,]$. For the parameters values $\beta = 1$, $\nu = 0.015$, $a = -1.02$, $b = -0.55$, $\omega_1 = 1$, $\omega_2 = (\sqrt{5} - 1)/2$, $f_1 = 0.08$ and $f_2 = 0.05$, the system has a quasiperiodic attractor and is shown in Fig. 8.7a. The numerically computed power spectrum is depicted in Fig. 8.7b. Peaks at the fundamental frequencies Ω_1, Ω_2 and some of their linear combinations are clearly seen.

For the nature of power spectrum of some simple and complicated functions such as beats, rectangular pulse and δ-function the reader can refer to Ref. [19] and also the exercises given below.

Exercises:

14. For the periodic functions $x(t) = \sin t$ and $\sin^2 t$ numerically compute the FFT and verify that their power spectra have peaks at the fundamental frequencies $1/2\pi$ and $1/\pi$, respectively.
15. Consider the beats consisting of two partial waves

    $$x(t) = \cos 2\pi f_1 t + \cos 2\pi f_2 t \,.$$

    The power spectrum of the signal $x(t)$ will have peaks at the frequencies f_1 and f_2. Verify this for $f_1 = 3$ and $f_2 = 4$.
16. A function with three harmonics is the amplitude modulation given by

    $$x(t) = (1 + \sin 2\pi f_M t) \sin 2\pi f_C t \,.$$

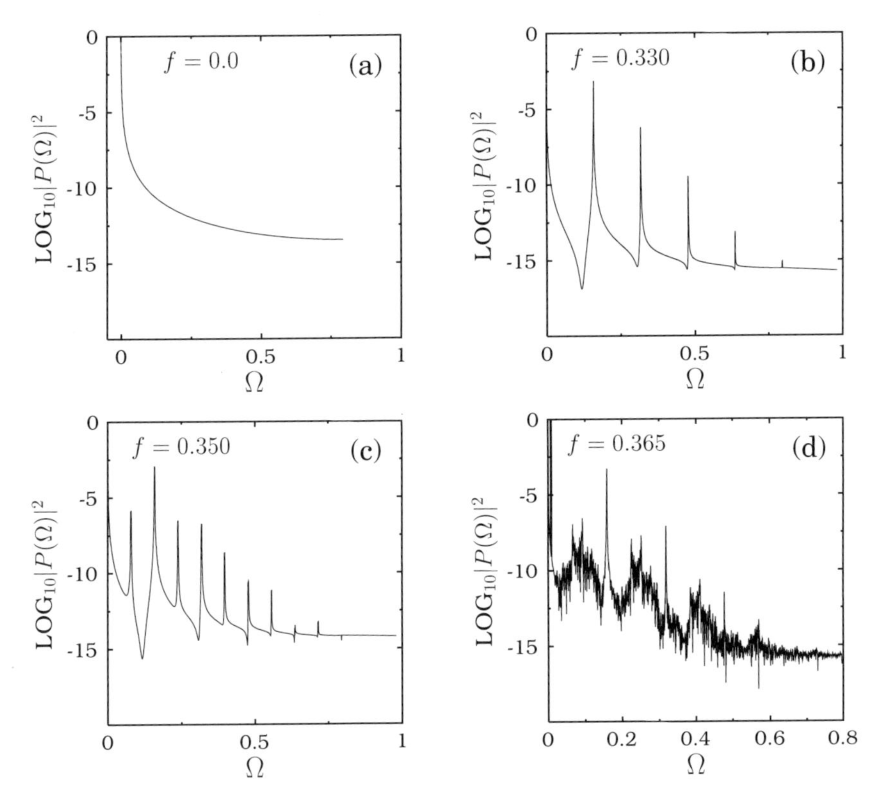

Fig. 8.6. Power spectrum of (**a**) fixed point solution, (**b**) period-T solution, (**c**) period-$2T$ solution and (**d**) chaotic solution of the Duffing oscillator equation (5.1). The parameters values are $\alpha = 0.5$, $\omega_0^2 = -1$, $\beta = 1$ and $\omega = 1$

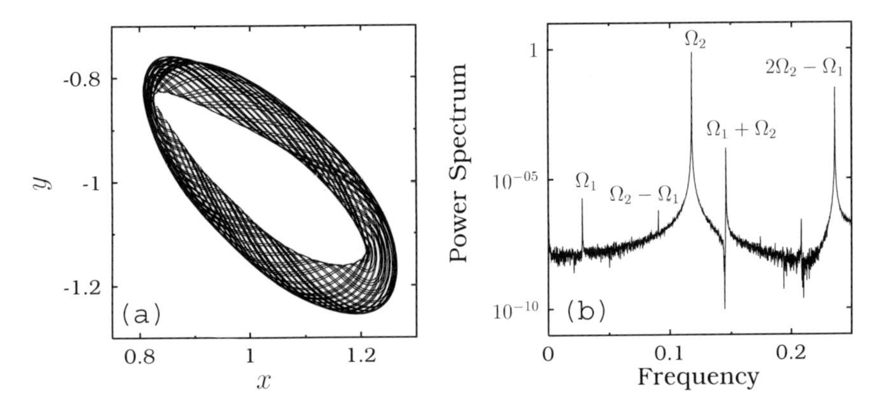

Fig. 8.7. (**a**) Quasiperiodic attractor and (**b**) the corresponding power spectrum of the MLC equation with quasiperiodic forcing, equation (8.27)

Numerically compute the power spectrum of $x(t)$ and show that it has peaks at the frequencies f_C, $f_C - f_M$ and $f_C + f_M$ for suitable choice of f_M and f_C.

17. For the sequence of square wave

$$x(t) = \begin{cases} 1 & \text{if } n2\pi < t < (2n+1)\pi \\ 0 & \text{if } (2n+1)\pi < t < 2(n+1)\pi \,, \end{cases}$$

where $n = 0, 1, 2, \ldots$ numerically compute the power spectrum and show that it has peaks at odd-integer multiples of its fundamental frequency $1/2\pi$. Explain the observed spectrum.

18. Compute the FFT of quasiperiodic attractors shown in Figs. 4.20a, 5.11a and 5.20 and identify the fundamental frequencies and some of their linear combinations.
19. From the obtained power spectrum of the previous problem, numerically show that for quasiperiodic attractors the number of peaks $N(\sigma)$ in the power spectrum exceeding a threshold amplitude σ scale as $N(\sigma) \propto \log(1/\sigma)$ [20]. (Hint: Compute $\log |P(\Omega)/\text{max}.P(\Omega)|^2$ and then calculate $N(\sigma)$ for a range of values of σ. Draw a graph between $N(\sigma)$ and $\log(1/\sigma)$. The data will fall roughly on a straight line).
20. Consider the quasiperiodically driven pendulum equation

$$\ddot{x} + d\dot{x} + \sin x = K + f\,(\cos\omega_1 t + \cos\omega_2 t)$$

with $\omega_1 = (\sqrt{5} - 1)/2$, $\omega_2 = 1$, $d = 3$, $f = 0.55$. For $K = 1.33$ the system exhibits strange nonchaotic attractor. Obtain its power spectrum. Verify that the number of peaks $N(\sigma)$ in the power spectrum exceeding a threshold amplitude σ scale as $N(\sigma) \propto \sigma^{-\alpha}$, $1 < \alpha < 2$ [20].

21. The equation of motion for the damped, driven pendulum is

$$\ddot{\theta} + \frac{1}{Q}\dot{\theta} + \sin\theta = g\,\cos\omega_d t \,.$$

The system shows type-I intermittency for $Q = 2$, $\omega_d = 2/3$, $g = 1.491908$, intermittent switching between two bands of the chaotic attractor (the so-called crisis-induced intermittency) for $g = 1.49546$ and fully developed chaos for $g = 1.5$. Compute the power spectrum for these three distinct chaotic orbits and draw the plot of the power spectrum versus frequency Ω in log-log scale. Verify that for intermittent dynamics the power spectrum has $1/\Omega$ dependence while for fully developed chaos it has $1/\Omega^2$ dependence [21].

8.4 Autocorrelation

Another important statistical tool to analyse a given time series is the set of autocorrelation coefficients, which measure the correlation between observations at different distances (times) apart. These functions most clearly exhibit the loss of information along the trajectory.

Given N observations $x_1, x_2, ..., x_N$, on a discrete time lattice we can perform $(N-1)$ pairs of (nearest neighbour) observations, namely, (x_1, x_2), (x_2, x_3), ..., (x_{N-1}, x_N). Regarding the first observation in each pair as one variable, the second observation as a second variable, the correlation between x_t and x_{t+1} is given by

$$C_1 = \frac{\sum_{t=1}^{N-1} (x_t - \bar{x})(x_{t+1} - \bar{x})}{\sum_{t=1}^{N} (x_t - \bar{x})^2} , \tag{8.28a}$$

where

$$\bar{x} = \frac{1}{N} \sum_{t=1}^{N} x_t . \tag{8.28b}$$

In a similar manner we can find the correlation between observations a distance k apart, which is given by

$$C_k = \frac{\sum_{t=1}^{N-k} (x_t - \bar{x})(x_{t+k} - \bar{x})}{\sum_{t=1}^{N} (x_t - \bar{x})^2} , \tag{8.29}$$

where $\bar{x}$ is again given by (8.28b). This is called the *autocorrelation coefficient* at a time lag k. In practice, the autocorrelation coefficients are usually calculated by computing the time series of auto-covariance coefficients C_k, which are defined as

$$C_k = \frac{1}{N-k} \sum_{t=1}^{N-k} (x_t - \bar{x})(x_{t+k} - \bar{x}) . \tag{8.30}$$

We then compute

$$C(k) = \frac{C_k}{C_0} \tag{8.31}$$

for $k = 1, 2, ..., m$, where $m \leq N/2$. The correlation function $C(k)$ measures how much x_j differs from its mean value. In other words, the function $C(k)$ gives a measure of how much $x_j - \bar{x}$, which is the difference between the signal and its mean value, keeps a memory of its value after an interval k of time. If the signal changes only slightly over k steps, almost all the terms in the sum will have the same sign and the correlation will be high. Then $C(k)$ will be large. If, on the other hand, if the signal changes erratically, the summands will cancel each other and $C(k)$ will be small. Thus if $C(k)$ is large, the x_t's change little after k steps, being thus still predictable after a time lag k. On the other hand, if it is small or zero, x_{t+k} is not close to x_t on the average and prediction is not possible. If a time series is completely random, then for large N, $C(k) = 0$ for all nonzero values of k.

Extending the above arguments, one can easily define the correlation function for a continuous signal $x(t)$ as

$$C(k) = \lim_{T \to \infty} \frac{1}{T} \int_0^T x(t) x(t+k) \mathrm{d}t . \tag{8.32}$$

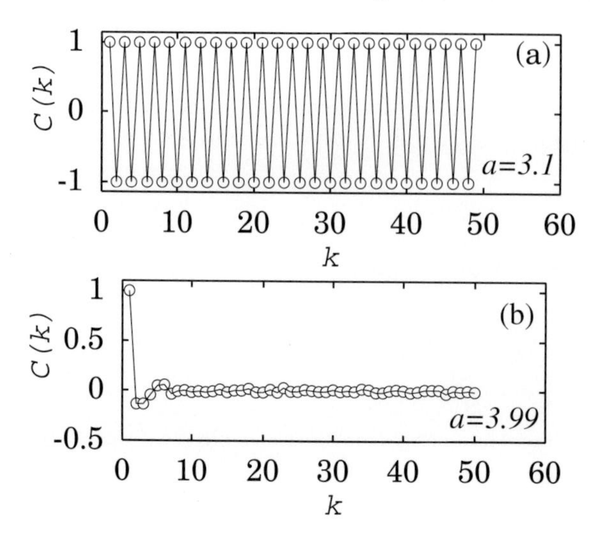

Fig. 8.8. Correlation function plot for the solution of the logistic map for (**a**) $a = 3.1$ (period-2 solution) and (**b**) $a = 3.99$ (chaotic solution)

For a periodic signal $C(k)$ becomes constant or oscillates about a mean value. For example, $C(k)$ for the signal $x(t) = \cos t$ is

$$\begin{aligned} C(k) &= \lim_{T \to \infty} \frac{1}{T} \int_0^T \cos t \ \cos(t+k)\mathrm{d}t \\ &= (\cos k)/2 \end{aligned} \tag{8.33}$$

so that $C(k)$ does not decay to zero as a consequence of the regularity of the signal. For irregular and chaotic signals $C(k)$ decays to zero. Thus, the decay of $C(k)$ can be used to characterize chaos.

As an example, Fig. 8.8a shows the calculated correlation function $C(k)$ for the period-2 solution of the logistic map (4.19) with $a = 3.1$. $C(k)$ oscillates periodically and no loss of information occurs. For $a = 3.99$ the solution is chaotic and the corresponding $C(k)$ shown in Fig. 8.8b decays to zero, revealing the fact that information is lost.

Exercise:

22. Compute the correlation function for period-1, 2, 4 and chaotic attractors of Hénon map (4.45) and van der Pol equation (5.9). Develop a short note on the characterization of period doubling route to chaos by correlation function in the above equations.

8.5 Dimension

Another important notion in dynamics is the *dimension* of the attractor (in phase space). After all we know that familiar geometric objects like a point, a curve, a square or a cube have dimensions zero, one, two and three, respectively, which are all integers. Then one can easily check that the dimension of a point attractor is zero, a periodic attractor is one and a two-frequency quasiperiodic attractor is two and so on. Then what is the dimension of a chaotic attractor? To describe the geometric and probabilistic features of the chaotic attractor several different dimensions can be used. These include the fractal dimension or capacity or Hausdorff dimension (D_F), information dimension (D_I) and the correlation dimension (D_C) [1]. These dimensions are all smaller in value than the number of degrees of freedom.

(i) *Fractal or Hausdorff Dimension* (D_F)

Let us denote the N points of a long-time trajectory by $\{\boldsymbol{X}_i\} = \{\boldsymbol{X}(t + i\tau)\}$, $i = 1, 2, ..., N$, where τ is an arbitrary but fixed time increment. For a dynamical system, $\boldsymbol{X}_i$'s can be a set of points of an attractor on a suitable Poincaré surface of section or Poincaré map. Then the fractal dimension [1] is defined through the expression

$$D_F = \lim_{\epsilon \to 0} \frac{\log M(\epsilon)}{\log(1/\epsilon)} , \tag{8.34}$$

where $M(\epsilon)$ is the number of d-dimensional cubes of size ϵ needed to cover the entire set. For small ϵ, we may write

$$M(\epsilon) \sim K\epsilon^{-D_F}; \quad K = \text{constant} . \tag{8.35}$$

Note that the above definition is a natural generalization of the geometric notion of dimension. In fact, for a point, $M(\epsilon) = 1$ and D_F is zero. For a line segment $M(\epsilon) = 1/\epsilon$ and therefore $D_F = 1$. For a two-dimensional plane $M(\epsilon) = (1/\epsilon)^2$ and hence $D_F = 2$. Similarly for a cube $D_F = 3$. In these examples, the dimension is an integer. Since the chaotic attractor of a dissipative dynamical system is neither a finite number of points nor fill up a d-dimensional phase space, obviously, its dimension must be noninteger. However, the chaotic orbit of a conservative system preserves phase space volume and so its dimension is integer. There are several popular examples available which illustrate vividly the concept of fractal dimension: Cantor set, Koch curve, Julia set and so on. For some further details see appendix D.

The algorithm (8.34) is called *box-counting* because one counts the number of boxes $M(\epsilon)$, that covers the set, for boxes of size ϵ. Though the algorithm (8.34) appears simple, it turns out that it is a time consuming process so that the task of computing D_F becomes very difficult whenever the phase space dimension is greater than two.

(ii) *Correlation Dimension* (D_C)

Another, practically useful measure is the spatial correlation dimension D_C, which can be computed using an algorithm suggested by Grassberger and Procaccia [22]. The correlation dimension is defined by

$$D_C = \lim_{r \to 0} \frac{\log C(r)}{\log r} , \tag{8.36a}$$

where $C(r)$, the correlation function, is given by

$$C(r) = \lim_{N \to \infty} \frac{1}{N^2} \sum_{i=1}^{N} \sum_{j=1}^{N} H\left(r - |\boldsymbol{X}_i - \boldsymbol{X}_j|\right) . \tag{8.36b}$$

Here r is the radius of spheres with their centres as $\boldsymbol{X}_i$ and $H(\theta) = 1$ for $\theta > 0$, $H(\theta) = 0$ for $\theta < 0$. $H(\theta)$ is also called the Heaviside step function. The special values of correlation dimension, $D_C = 0$ and $D_C = 1$ are obviously associated with point and limit cycle attractors, respectively, while a chaotic attractor is characterized by a noninteger dimension.

The correlation dimension is usually estimated with a large number of data points, say, 10^4 or more, for several values of r. From (8.36a) we note that

$$\log C(r) = D_C \log r \quad \text{as} \quad r \to 0 \tag{8.37}$$

so that D_C is the slope of the $\log C(r)$ versus $\log r$ plot. In practice, we make a least-square fit of the $\log C(r)$ and $\log(r)$ values to satisfy the linear relation

$$y(= \log C(r)) = a\, x(= \log r) + b , \tag{8.38}$$

where a and b are given by

$$a = \frac{K \sum x_i y_i - \sum x_i \,.\, \sum y_i}{K \sum x_i^2 - (\sum x_i)^2} , \tag{8.39}$$

$$b = \frac{\sum y_i \,.\, \sum x_i^2 - \sum x_i \,.\, \sum x_i y_i}{\sum x_i^2 - (\sum x_i)^2} , \tag{8.40}$$

where $\sum$ represents sum over $i = 1, 2, ..., K$ with K being the number of r values used. Hence the D_C, the slope of the log–log plot, that is $\mathrm{d}y/\mathrm{d}x$ in (8.38), is determined from the relation $D_C = a$. As an illustration, we employ the above procedure to calculate the correlation dimension of the strange attractor of the Hénon map (4.45) for $a = 1.4$ and $b = 0.3$. Figure 8.9 shows the $\log_{10} C(r)$ versus $\log_{10}(r)$ plot, where $N = 2 \times 10^4$. Circles represent numerical results and continuous line is the best straight line fit. The numerical data points nicely fall on a straight line implying the power-law relation. The calculated D_C, which is the slope of the straight line in Fig. 8.9, is 1.26.

The other types of dimension are less frequently used in the literature. For more details we refer the reader to [1].

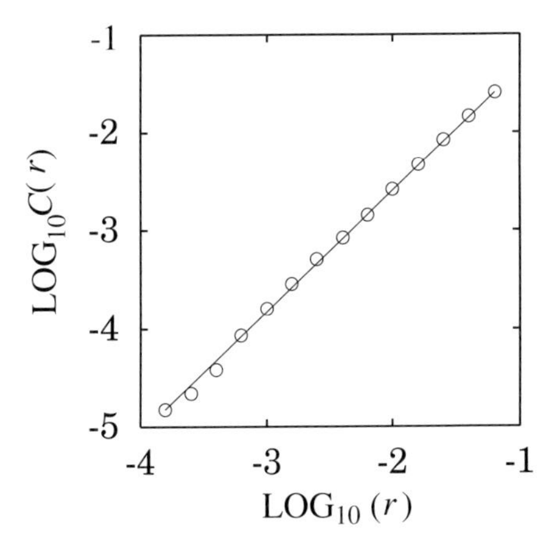

Fig. 8.9. $\log_{10} C(r)$ versus $\log_{10} r$ plot for the chaotic attractor of the Hénon map (4.45) for $a = 1.4$ and $b = 0.3$. Circles represent numerical data and continuous line is the best straight line fit to the data

---

Exercises:

23. Numerically calculated correlation dimension of chaotic attractors of some nonlinear systems are given in Table 8.1. Develop a computer program to compute the correlation dimension and verify the data in the table. In all the systems 10^4 data points are used.
24. Consider the Duffing oscillator equation

$$\ddot{x} + \gamma\dot{x} - \frac{1}{2}\left(x - x^3\right) = f\cos\omega t$$

with $f = 0.16$ and $\omega = 0.8333$. Compute the correlation dimension D_C with $0.01 \leq r \leq 0.158$ of the chaotic attractor of the Duffing equation as a function of the damping parameter γ in the interval $[0.05, 0.23]$ in steps of 0.01 and show that D_C increases with increase in γ [23].

8.6 Criteria for Chaotic Motion

From the above discussion, one can now introduce the following possible quantitative criteria for identification of chaotic motion in dissipative/conservative nonlinear systems:

(1) at least one positive Lyapunov exponent
(2) power spectrum with continuous, broad-band and noisy component
(3) decaying autocorrelation function.

Table 8.1. Correlation dimension of chaotic attractors of some nonlinear systems

No.	System	Figure/ Parameters value	Range of r	Value of D_C
1.	Logistic map (4.19)	$a = 3.7$	$10^{-3} - 0.005$	0.987
2.	Hénon map (4.45)	Fig. 4.18	$10^{-4} - 10^{-1}$	1.26
3.	Duffing oscillator, equation (5.1)	Fig. 5.6	$10^{-5} - 10^{-2}$	1.07
4.	Duffing–van der Pol oscillator, equation (5.3)	Fig. 5.11b	$10^{-3} - 0.3$	1.89
5.	Morse oscillator, equation (5.11)	Fig. 5.24	$10^{-5} - 10^{-2}$	1.16
6.	Dipsy model, equation (7.2)	Fig. 7.2b	$10^{-3} - 10^{-1}$	1.094
7.	Lorenz equations (5.5) (Points are collected after every 0.2 time interval)	$\sigma = 10$, $b = 8/3$ and $r = 99$	$0.1 - 0.25$	1.75

(4) In dissipative systems, in addition to the above characteristic properties the chaotic attractor possesses noninteger dimension (fractal). The dimension of the chaotic orbit of conservative systems is integer because of volume preservation property of the phase space.

In addition to the above four quantities one can also use the so-called Kolmogorov entropy (K) [1] to characterize periodic and chaotic motions. $K = 0$ corresponds to regular motion while for chaos $0 < K < \infty$. On the other hand, inhomogeneity of chaotic attractors can be characterized by generalized dimensions D_q and $f(\alpha)$ spectrum [1] (some details are given in appendix D).

For clarity, the characteristic properties of different types of attractors are illustrated in Table 8.2. From the table we can also note a connection between the dimension of the attractor (in the dissipative case) and Lyapunov exponents. If in an N-dimensional phase space all the Lyapunov exponents are negative, the dimension is 0; if one is zero and the others are negative, the dimension is 1; if $K(< N)$ Lyapunov exponents are zero and the remaining are negative, it is K; if at least one is positive, it is noninteger. The above results can be stated in a general relation, given by Kaplan and Yorke [24],

$$D = j + \sum_{i=1}^{j} \frac{\lambda_i}{|\lambda_{j+1}|} , \tag{8.41}$$

Table 8.2. Characteristic properties of different types of attractors/motions

Attracting set/motion		*Characteristic properties*			
Steady state	**Poincaré map**	*Dimension*	*Lyapunov exponents*	*Power spectrum*	*Correlation function*
Equilibrium point	——	0	$\lambda_i < 0,\ i = 1, 2, ..., n$	exponentially decaying spectrum with no peaks	0
Periodic orbit	finite number of points	1	$\lambda_1 = 0,\ \lambda_i < 0,$ $i = 2, 3, ..., n$	discrete spectrum with sharp peaks at basic frequency ω and at its harmonics	oscillates
Quasiperiodic (two periodic)	one or more closed curves	2	$\lambda_1 = \lambda_2 = 0,\quad \lambda_i < 0,$ $i = 3, ..., n$	discrete spectrum with peaks at basic frequencies and at their linear combinations	oscillates
Quasiperiodic (K-periodic)	one or more $(K-1)$ tori	K	$\lambda_i = 0,\ i = 1, 2, ..., K$ $\lambda_j < 0,\ j = K+1, ..., n$	discrete spectrum with peaks at basic frequencies and at their linear combinations	oscillates
Chaotic:					
Dissipative systems	self-similar structure	noninteger	at least one positive λ	continuous broad-band noisy spectrum	decays to zero
Chaotic:					
Conservative systems	space filling points	integer	at least one positive λ	continuous broad-band noisy spectrum	decays to zero

where the exponents are labelled as $\lambda_1 \geq \lambda \geq ... \geq \lambda_N$ and j is the largest integer for which the sum of the first j exponents is positive. This conjecture has been shown to be valid in the cases to which it has been applied, however, a proof (or a disproof) of its validity is still lacking.

Problems

1. Consider the various important dissipative systems discussed in Chaps. 3–6, including exercises and Problems. For each of the system, compute/verify the various entries in Table 8.2.
2. Carry out the above tasks for conservative systems discussed in Chap. 7.

9. Further Developments in Chaotic Dynamics

In the earlier chapters we have illustrated various transitions from regular to chaotic motions in certain simple nonlinear systems and identified suitable characterizations of the associated attractors using certain statistical measures. At this point one may ask several further interesting questions. Some of them are as follows.

(1) Given a set of experimental data in the form of a time series for a dynamical variable, how can one identify regular and chaotic dynamics? How do we quantify them?

(2) Often chaos might be harmful to the physical system. Is it possible to eliminate the presence of chaos by minimal perturbation of the system?

(3) What are the possible effects of external noise on nonlinear systems?

(4) Can one direct a chaotic system to follow another identical chaotic system started with a different initial condition?

(5) Can we speak of chaos in quantum mechanical systems?

During the past one decade or so a great deal of interest has been focussed on to address the above mentioned and other related problems. Many interesting and fascinating results have been obtained, which have enlarged the scope of nonlinear and chaotic dynamics considerably. Some of the topics under which recent investigations are being carried out include

(1) Time series analysis
(2) Stochastic resonance
(3) Chaotic scattering
(4) Controlling of chaos
(5) Synchronizing chaotic systems
(6) Quantum chaos

and so on.

In this chapter we wish to give a brief account of the salient features of the above topics so as to give a flavour of these ideas. For more details the readers are referred to more specialized works (see for example Refs. [1–5]).

9.1 Time Series Analysis

Let us assume that for an experimental system a scalar time series $s(t)$, $t' \leq t \leq t''$, is measured at regular intervals of time for a single physical variable and is denoted mathematically by

$$s(t) = h(\boldsymbol{x}) + \eta(t) \ , \quad \boldsymbol{x} = (x_1, x_2, ..., x_n) \ , \tag{9.1}$$

where h is the measurement function (a single scalar quantity which is a function of the set of variables $\boldsymbol{x}$) and η is the noise that is invariably associated with any measurement process. As an example, one may consider the measurement of current through a nonlinear resistor in an electrical circuit or amplitude of oscillation of a buckled beam or temperature difference between the lower and upper surface of a fluid flow. Now we ask: Can we reconstruct the as yet unknown dynamical equations for the experimental system, given only a finite scalar time series $s(t)$? With the presently available techniques, reconstruction of the full dynamics from $s(t)$ appears impossible. However, many useful informations about the underlying dynamical process can be obtained by a careful analysis of the time series $s(t)$. Particularly, it is possible to calculate the largest Lyapunov exponent to identify whether the given time series is regular or chaotic.

To start with, from the time series $s_i = s(t_i)$, $i = 1, 2, ..., N$ with $\Delta t = t_{i+1} - t_i$, construct a vector $\boldsymbol{S}_i$ out of the time-delayed copies of s:

$$\boldsymbol{S}_i = \{s_i, s_{i+\tau}, s_{i+2\tau}, ..., s_{i+(m-1)\tau}\} \ , \tag{9.2}$$

where τ is an arbitrary but fixed value. $\boldsymbol{S}$ provides the coordinates of a point in an n-dimensional space and these coordinates will evolve as t is varied by keeping τ fixed. τ is called the *delay time* and m is the *embedding dimension* ($m \leq n$). It has been shown that the trajectory of $\boldsymbol{S}$ in its n-dimensional space can be simply related to the trajectory of the original system in its full phase space. In the following we briefly describe the estimation of τ, m and then the calculation of the largest Lyapunov exponent.

9.1.1 Estimation of Time-Delay and Embedding Dimension

To make an estimate of optimal delay, one can calculate the autocorrelation function

$$C(\tau) = \langle\langle s_i s_{i+\tau} \rangle\rangle \tag{9.3}$$

of the time series, where $\langle\langle , \rangle\rangle$ denotes time averaging after subtracting out the mean from the time series. We determine the position, say τ^* at which $C(\tau)$ becomes zero. The idea is that $C(\tau^*) = 0$ implies that s_i and $s_{i+\tau^*}$ become statistically independent and so could be considered as the components of a vector with τ^* as the delay time.

For calculating the embedding dimension one often employs the Grassberger–Procaccia algorithm described earlier in Sect. 8.5. Essentially, one calculates

$$C^m(r) = \frac{1}{N'^2} \sum_{i=1}^{N'} \sum_{j=1}^{N'} H\left(r - \left|\boldsymbol{S}_i^{(m)} - \boldsymbol{S}_j^{(m)}\right|\right) , \tag{9.4}$$

where $\boldsymbol{S}_i^{(m)}$ is the m-dimensional τ^* delay vector, N' denotes the total number of delay vectors that can be constructed from the time series of length N, $N > N'$, and is given by $N' = N - (m-1)\tau^*$ and H is the Heaviside step function. A plot of $\log C^m(r)$ versus $\log r$ is made. If M is the appropriate embedding dimension one would find that for $m \geq M$ the plot would consist of parallel lines in the scaling regime of r, and the slope would yield the correlation dimension D_C.

9.1.2 Largest Lyapunov Exponent

To calculate the largest Lyapunov exponent of the time series, one first finds the Euclidean norm

$$d_{i,j}(0) = |\boldsymbol{S}_i - \boldsymbol{S}_j| \tag{9.5a}$$

and

$$d_{i,j}(k) = |\boldsymbol{S}_{i+k} - \boldsymbol{S}_{j+k}| \ . \tag{9.5b}$$

Here k is an appropriately chosen time over which the dynamics evolves and one considers only those vectors $\boldsymbol{S}_i$ and $\boldsymbol{S}_j$ with $|j-i| > (m-1)\tau^*$. Then, if we plot

$$r = \log\left(d_{i,j}(k)/d_{i,j}(0)\right) \tag{9.6}$$

against $\log d_{i,j}(0)$ for all possible vectors, it gives a local exponential divergence plot. Next, one calculates the average value of r, $\langle r \rangle$, as a function of k. The points fall on a straight line passing through the origin. The slope of the line gives a reasonable estimate of the largest Lyapunov exponent. A positive Lyapunov exponent implies that the time series is unambiguously chaotic in nature.

For more details on time series analysis the readers may refer to Refs. [1,6]. A detailed analysis for a time series obtained from a set of four-coupled first order differential equations has been presented in Ref. [7].

Problems:

1. Consider the time series $x(t)$ consisting of 2000 points collected at every time interval $\Delta t = \pi/25$ of the Rössler equation (5.12a), where $a = 0.15$, $b = 0.2$, $c = 10$. For $k = 9$ draw the local divergence plot as a function of (m, τ), where m is the embedding dimension and τ is the delay time.

Show that the minimum embedding dimension is $m = 3$ and optimal delay time is $\tau^* = 8$. Also calculate $< r >$ as a function of k for $\tau^* = 8$ and $m = 3$ and show that the maximal Lyapunov exponent is ≈ 0.067 [6].

2. For the time series of Hénon map (4.45) with $a = 1.4$ and $b = 0.3$ consisting of 2000 points, show that the minimal embedding dimension is $m = 2$, optimal delay time is $\tau^* = 1$ and the maximal Lyapunov exponent is $\lambda = 0.421$ [6].
3. Verify that for the time series consisting of 3000 points of the Lorenz equations (5.5), collected at every time interval $\Delta t = 0.03$ with $\sigma = 10$, $b = 8/3$ and $r = 45.92$, the minimal embedding dimension is 3, optimal delay time is 3 and the maximal Lyapunov exponent is 1.48 [6].

9.2 Stochastic Resonance

Another topic of recent interest in chaotic dynamics is *stochastic resonance* [2], a phenomenon in which noise enhances the response of a nonlinear system to a weak periodic or nonperiodic input signal. The word resonance (as we have seen in Chap. 2) is usually applied in physics to phenomenon in which a dynamical system having periodic oscillations at some frequencies ω_i, when subjected to a periodic forcing of frequency nearer to one of the ω_i's, shows a marked response. Stochastic resonance is a term given to an effect that generally occur in multistable (that is, coexistence of more than one stable state) nonlinear systems driven by noise and a weak periodic or aperiodic signal. In this phenomenon the information flow through the system in the form of periodic signal is enhanced by the noise. For every frequency of the modulation, the information flow is optimum for a specific noise intensity.

A prototype system [8] exhibiting stochastic resonance is the double-well *overdamped* version of the double-well Duffing oscillator equation (5.1) with a noise term $\eta(t)$ added to the right hand side of the equation:

$$\dot{x} = \omega_0^2 x - \beta x^3 + f \cos \omega t + \sqrt{D}\, \eta(t) \,. \tag{9.7}$$

The noise term is often assumed to be zero-mean Gaussian distributed white noise such that $\langle \eta(t) \rangle = 0$ and $\langle \eta(t)\eta(t+\tau) \rangle = \delta(\tau)$ and D is the amplitude (variance) of the noise. The amplitude of the periodic force in equation (9.7) is assumed to be small so that in the absence of noise this force is unable to move a particle from one well to the other. When noise is added, for noise intensity D less than a critical value D_C the particle continues to remain in one of the two wells as depicted in Fig. 9.1a for $\omega_0^2 = 32$, $\beta = 1$, $f = 8$, $\omega = 0.19$ and $D = 75$Hz. For $D \geq D_C$ the particle makes transitions over the barrier at random times and resides in one or the other well for a random length of time and then switches to another well and so on. This is shown in Fig. 9.1b for $D = 250$Hz. We say that a particle has resided in a potential

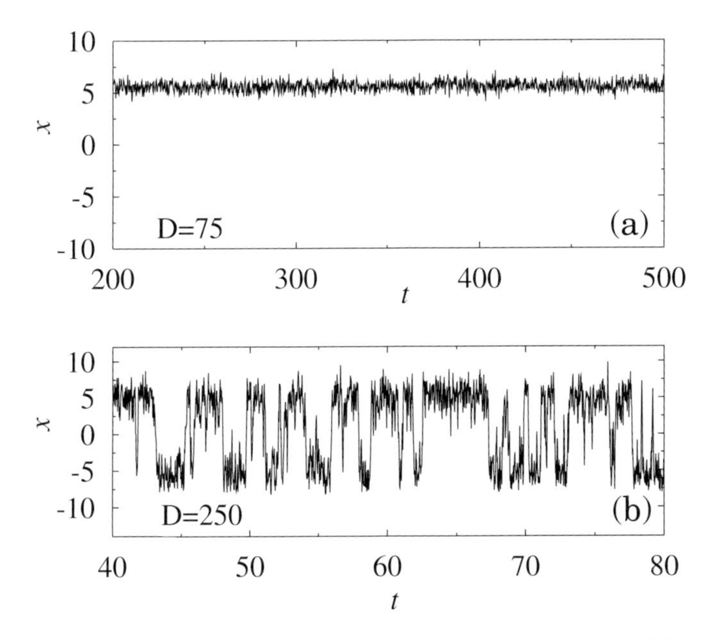

Fig. 9.1. (**a**) $x(t)$ versus t for noise intensity $D = 75$Hz. $x(t)$ remains in the right well. Noise is added at every 0.005 time step. (**b**) $x(t)$ versus t for $D = 250$Hz showing transition from one potential well to another at random times

well only if it stayed in that potential for a time interval greater than T', where T' is a preassumed constant.

One can introduce a quantity called switching rate defined as the number of times the particle jumped from one well to another (after spending a time greater than T') per unit time. Obviously, for $D < D_C$ the switching rate is zero. At $D = D_C$ the switching rate is nonzero and one expects that it will vary with D. For large noise intensity, that is $D \gg D_C$, the switching between the two potential wells is erratic. That is, the particle moves from one potential well to another before spending a time $T > T'$ in a potential well. Consequently, the switching rate becomes zero. Thus, for vanishingly small noise intensity, $D \to 0$, the switching rate approaches zero. For sufficiently large noise also, again the switching rate approaches zero. What happens in between these limits? Between these two limits, initially the switching rate is small which then increases with D. Interestingly, it reaches a maximum value at a particular noise intensity D_m and then decreases with further increase in D. That is, the response of the nonlinear system to the weak periodic signal is optimized by the presence of a particular intensity of noise. This phenomenon is called *stochastic resonance*.

Stochastic resonance can be characterized by power spectrum, autocorrelation function of an appropriate state variable and the probability density of the residence time in a potential well. The most frequently used one is the

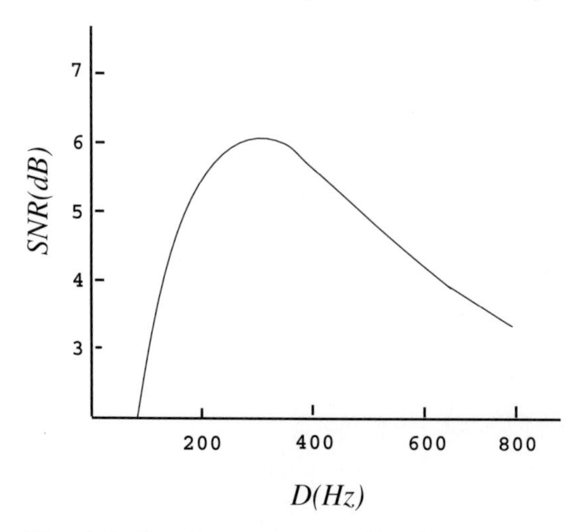

Fig. 9.2. Signal-to-noise ratio (SNR) as a function of D for (9.7) (from Ref. [8])

power spectrum. From the power spectrum one calculates signal-to-noise ratio (SNR). It is the ratio of the strength of the signal peak to the mean amplitude of the background noise at the input signal frequency. Stochastic resonance is demonstrated by observing that the SNR increases from zero and passes through a maximum with noise intensity. For the double-well overdamped Duffing oscillator equation (9.7) the result [8] is given in Fig. 9.2. Stochastic resonance has been observed in many experimental systems [2], including electronic circuits, sensory neurons, optical systems, magnetic systems, mechanical systems, ring lasers and tunnel diode. It has been speculated that stochastic resonance can be used for improving signal detection and enhancing information transmission.

Problems:

1. Study the stochastic resonance phenomenon associated with (9.7) using probability density of residence times [8]. Construct an analog simulator of the above equation and then experimentally investigate the stochastic resonance.
2. An electronic circuit experiment on stochastic resonance was done by Fauve and Heslot [9] using a Schmitt trigger. The dynamics of it is represented by a two-state model. The circuit diagram and the input-output voltage characteristic of this device are depicted in Fig. 9.3. For $\epsilon = 0$, in the presence of noise the trigger randomly switches between the two states $+V_1$ and $-V_1$. When the periodic signal is also added, the measured power spectrum of the output shows a sequence of delta-like peaks. The SNR can be measured in the usual way. Construct the Schmitt trigger circuit and observe stochastic resonance calculating SNR.

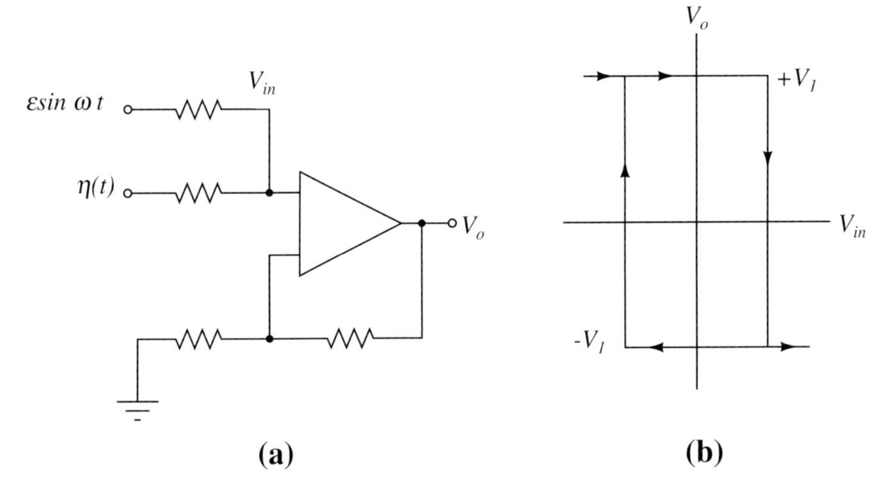

Fig. 9.3. (**a**) Circuit diagram of a Schmitt trigger driven by noise and a periodic function. (**b**) The transfer characteristic of the Schmitt trigger (see problem 2 below)

3. Stochastic resonance has been observed [10] in the periodically driven Lorenz equations

$$\begin{aligned} \dot{x} &= \sigma(y - x) , \\ \dot{y} &= rx - y - xz + A \cos \Omega t , \\ \dot{z} &= -bz + xy . \end{aligned}$$

Fix the parameters at $\sigma = 10$, $b = 8/3$, $A = 30$ and $\Omega = 1$. Verify that at $\omega \approx \Omega$ there is a marked peak in the power spectrum for $r \in [15, 50]$. Plot $|P(\Omega)|^2$ as a function of r and show that a sudden jump in $|P(\Omega)|^2$ occurs at $r = 24.74$.

4. The magnetic flux through a periodically driven SQUID (superconducting quantum interference device) loop is governed by the equation

$$\ddot{q} + \Gamma \dot{q} + B\left(q - q_{\mathrm{dc}}\right) + \sin q = \eta(t) + A \cos \Omega t ,$$

where $\eta(t)$ is a Gaussian noise with mean zero. Investigate the occurrence of stochastic resonance using SNR for $A = 0.005$, $\Omega = 0.39$, $B = 0.1$, $q_{\mathrm{dc}} = -1$ by varying the intensity of noise [11].

5. The stochastic resonance phenomenon has been observed in the Chua circuit equation driven by a periodic force and noise:

$$\begin{aligned} \dot{x} &= \alpha[y - h(x)] , \\ \dot{y} &= x - y + z , \\ \dot{z} &= -\beta y + \eta(t) + A(1 + m \cos \Omega t) \cos \omega_0 t , \end{aligned}$$

where $h(x) = bx + 0.5(a-b)[\, |x+1| - |x-1|\,]$, and $\eta(t)$ is a Gaussian noise with mean zero. Fix the parameters at $a = -1/7$, $b = 2/7$, $\beta = 14.286$,

$A = 0.1$, $\omega_0 = 0.5652$, $\Omega = 0.1256$, $m = 0.5$ and $\alpha = 8.55$. Compute SNR (use power spectral density at ω_0) for the noise intensity $D \in [0, 0.02]$ and observe the occurrence of stochastic resonance. It can also occur in the system in the absence of noise. For $D = 0$, varying α from 8.5 to 9.1 study the stochastic resonance phenomenon [12].

9.3 Chaotic Scattering

The study of scattering of particles in a given potential is quite fundamental to physics. We are all familiar with scattering of particles in a Coulomb potential and the resultant Rutherford scattering formula [13]. However, scattering in noncentrosymmetric potentials is less well known. Particularly, it has become clear in recent time that in a large number of classical scattering problems one sees that scattering exhibits regular or chaotic nature [3].

A typical scattering experiment consists of sending a large number of incident particles with different initial conditions into some interaction region. Typically, the particles move toward the scattering region from outside it, interact with the scatterer or scattering potential and then leave the scattering region as shown in the Fig. 9.4. Now we ask how does the motion far from the scatterer after scattering depends on the motion of the incident particles? In order to answer the above question, one often defines an input variable b (the impact parameter) and an output parameter ϕ (the scattering angle) as shown in the Fig. 9.4. Then the character of the functional dependence of ϕ on b is investigated.

Figure 9.5 depicts the variation of ϕ as a function of b for the potential [3]

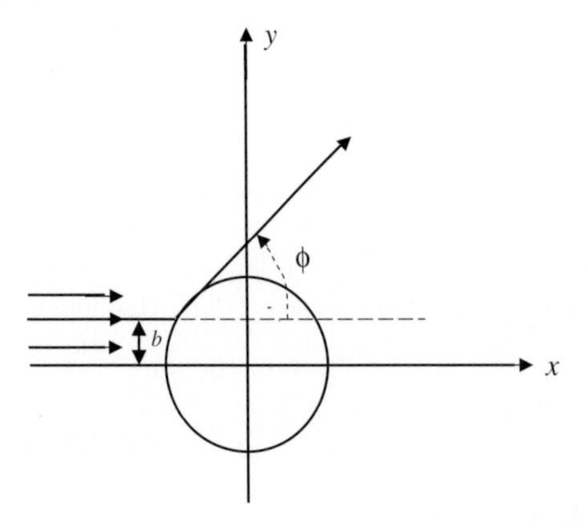

Fig. 9.4. Schematic illustration of a scattering problem in two-dimensional space

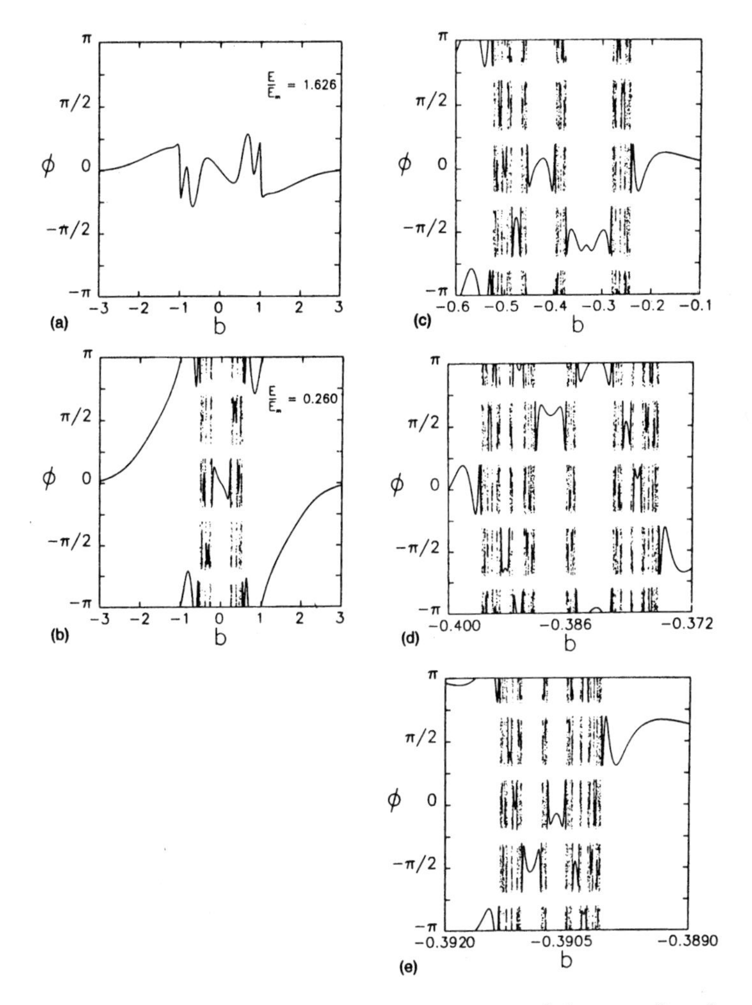

Fig. 9.5. ϕ versus b for the potential $V(x, y) = x^2y^2 \exp(-x^2 - y^2)$. (**a**) For a case where $E > E_m$. (**b**) For a case with $E < E_m$. (**c**–**e**) Successive blowups of a small region in b for the case $E < E_m$ shown in (**b**) (from Ref. [3])

$$V(x, y) = x^2y^2 \exp\left(-x^2 - y^2\right) . \tag{9.8}$$

This potential consists of four hills whose peaks are located at $(x, y) = (\pm 1, \pm 1)$. For $E/E_m = 1.626$, where E_m is the maximum value of energy at the hill peaks, the relation between ϕ and b is a simple smooth curve (Fig. 9.5a). For $E/E_m = 0.26$ (Fig. 9.5b) there are some regions in which ϕ varies too rapidly with an infinitesimal increase in b. This is because before leaving the scattering region, during finite time scale the particle exhibits chaotic motion. Because of the sensitive dependence of chaotic motion on initial conditions, a small change in b leads to a large change in the output variable ϕ. Magnifications of small region of Fig. 9.5b shown in Figs. 9.5c–e fail to resolve the function. That is, there is a fractal set of b values for the scattering angle ϕ. This implies that it will be very difficult to obtain accurate values of the scattering angle if there is a small uncertainty in the specification of b. Generally, a scattering process is said to be chaotic if the scattering functions are sensitive to changes in the initial conditions. In particular, an arbitrary small change in the initial condition (for example, the value of the impact parameter) produces a large change in the exit properties.

Chaotic scattering occurs in a variety of situation of practical interest including satellite encounters in celestial mechanics, molecular dynamics, microwaves, chemical reactions, collisions between nuclei, magnetic dipole and electron collisions with a He^+ ion and so on.

Problems:

1. ***Two-Coupled Anharmonic Oscillators [14]***

   The Hamiltonian of a system of two-coupled anharmonic oscillators is

   $$H = \frac{1}{2}\left(p_x^2 + p_y^2\right) + \frac{1}{2}\left(x^2 + y^2\right) - \frac{1}{4}\left(x^4 + y^4\right) + \epsilon x^2 y^2 .$$

   a) Draw the potential function for $\epsilon = 0.05$ in the range $x, y \in [-1.6, 1.6]$.
   b) Choose the initial conditions as $x = 1.8$, $y = b$, $p_x = -\sqrt{2E}$ and $p_y = 0$ so that b is the impact parameter. Study the occurrence of chaotic scattering by numerically measuring the angle ϕ between the outgoing particle velocity and x-axis as a function of b for $E = 0.45$, $\epsilon = 0.1$.

2. ***The Electron Scattering on the H_2^+ Ion [15]***

   The electron scattering on the H_2^+ ion is described by the Hamiltonian

   $$H = \frac{1}{2}\left(p_x^2 + p_y^2\right) + V(x,y) , \quad V(x,y) = -\alpha y^2 \exp\left[-\beta\left(x^2 + y^2\right)\right] .$$

   a) Plot the equipotential curves for $\alpha = -1$, $\beta = 1$.
   b) Show that under the change of variables
   $$p_x = \sqrt{2E - 2V(x,y)}\ \cos\phi ,$$
   $$p_y = \sqrt{2E - 2V(x,y)}\ \sin\phi$$
   the equation of motion of the system reduces from four coupled first-order equations to a set of three coupled first-order equations with the variables x, y and ϕ.

   c) Choose the initial conditions as $(x, p_x, y, p_y) = (-5,\ \sqrt{2E},\ b,\ 0)$ so that b is the impact parameter. Verify that for $b \in [0, 4]$ chaotic scattering does not occur for $E = 0.125$ and it occurs for $E = 0.005$.

Also characterize the chaotic scattering using escape time defined as the time duration spent by the trajectory in the scattering region.

3. ***Two-Coupled Morse System [16]***
The Hamiltonian of a two-coupled Morse system is given by

$$H = \frac{1}{2}\left(p_x^2 + p_y^2\right) + \left(\mathrm{e}^{-x} - 1\right)^2 + \left(\mathrm{e}^{-y} - 1\right)^2 - \epsilon p_x p_y \ ,$$

where x, y and p_x, p_y are coordinates and conjugate momenta respectively. Investigate the occurrence of chaotic scattering for fixed values of ϵ and energy.

4. ***Scattering in a Two-Dimensional Map [17]***
Consider the two-dimensional map of the problem 16 in Chap. 7. Becker and Eckelt have shown that chaotic scattering in this map can be characterized using the energies of the incoming and outgoing trajectories. Define $E = p^2/2$ and denote E_{in} and E_{out} as energies of the in-asymptote and out-asymptote of the scattering orbit $x(t)$. Characterize the chaotic scattering in the map for $V_0 = 8$, $\beta = 4$ by calculating E_{out} for $E_{\mathrm{in}} \in [0.75, 1]$.

5. ***Chaotic Scattering in a Complex Map [18]***
Another map exhibiting chaotic scattering is

$$z_{n+1} = \left(z_n + \frac{\epsilon}{2}\right) \exp\left(\mathrm{i}/\left|z_n + \frac{\epsilon}{2}\right|^2\right) + \frac{\epsilon}{2} \ ,$$

where z is complex. Transform the above one-dimensional complex map into a two-dimensional real map by the substitution $z = x + \mathrm{i}y$. Then consider the evolution of the initial conditions $(x, y) = (-x_0, b)$, where $x_0 \gg 1$. Iterate the map until it crosses the line x_0. In general (x, y) will never land exactly on the line $x = x_0$. However by a linear interpolation one may obtain $y(x_0)$ and denote it as b'. Treating b as an impact parameter investigate chaotic scattering by calculating b' as a function of b in the interval $[0, 7.5]$ for $\epsilon = 0.39$. Also characterize the chaotic scattering process using the residence time R_t needed for a particle starting at $(-x_0, b)$ to reach (x_0, b').

9.4 Controlling of Chaos

Although chaos is ubiquitous, widespread and observed in numerous physical, chemical and biological systems, in many situations chaos is an unwanted phenomenon. For example, increasing drag in flow systems, erratic fibrillations of heart beating, complicated population dynamics of species, noisy type oscillations in electronic circuits, erratic vibrations of mechanical systems are some situations where chaos is harmful. Thus, one may wish to avoid or control chaos whenever it is harmful with minimal efforts and without altering the underlying system significantly [4].

9.4.1 Controlling and Controlling Algorithms

In a chaotic system the state values are nonperiodic; however, they lie in a finite region of the phase space (namely, the chaotic attractor). Consequently, during the evolution, the state of the system often comes closer to the already visited regions. Interestingly, the system stays for a short time in the neighbourhood of an *unstable periodic orbit* (UPO) (that is unstable period-1, $2, ..., n$ equilibrium points or limit cycles as the case may be) and then deviate from it, evolve randomly and then again come closer to the same UPO or another UPO and so on. Generally, a chaotic orbit contains a large number of UPOs. A chosen UPO can be stabilized by changing the value of the parameter of the system slightly or by adding an appropriate weak perturbation in the form of a feedback to the system. In this way, the long time behaviour of the system can be converted from chaotic motion to a desired periodic motion. This is called *controlling of chaos*. In an alternative approach, chaos can be eliminated by creating new periodic orbits by adding an appropriate small constant or weak periodic perturbation and it may be called *suppression of chaos*. In the nonlinear dynamics literature, controlling of chaos and suppression of chaos are often used synonymously for *conversion of chaos to regular motion*. Whatever be the method used, the important thing is that modification introduced should be minimal so that the system does not get altered drastically.

Let us consider a general n-dimensional nonlinear system as in Chap. 3,

$$\dot{\boldsymbol{X}} = \frac{\mathrm{d}\boldsymbol{X}}{\mathrm{d}t} = F(\boldsymbol{X}; p; t) , \tag{9.9}$$

where $\boldsymbol{X} = (x_1, x_2, ..., x_n)^{\mathrm{T}}$ represents the n state variables and p is a control parameter such that the system is executing chaotic motion. Recall the various examples discussed in Chaps. 3–8. In recent years several chaos control algorithms have been proposed and applied to both theoretical model equations and experimental systems. In the existing control algorithms either the parameter p is slightly perturbed suitably or an appropriate additional perturbation $\boldsymbol{C}(X, t) = (c_1, c_2, ..., c_n)^{\mathrm{T}}$ is added to the right hand side of (9.9). In practice a c_i is added to one of the suitable components of the system. Many of the existing chaos control methods can be broadly classified into (i) feedback and (ii) nonfeedback methods. Feedback methods essentially make use of the intrinsic properties of chaotic systems, including their high sensitivity to initial conditions, to stabilize orbits embedded in the chaotic attractor. Nonfeedback methods modify the underlying chaotic systems weakly so that stable periodic solutions appear. For more details see for example, Refs. [4,19–33].

Table 9.1 summarizes the typical forms of the controller and the most important characteristic properties of the various control algorithms and their notable advantages and disadvantages. Thus, there are surprisingly a large number of practically implementable control algorithms available for controlling chaos. The applicability and efficacy of these algorithms have been studied in many chaotic systems. Typical systems in which control of chaos has been achieved include a chaotic diode resonator circuit, different types of laser systems, electrochemical cell, magnetoelastic ribbon, rat brain, and various nonlinear electronic circuits, to name a few. By manipulating a chaotically behaving oscillator by a small control, the oscillator output can be made to carry information [4]. Interested reader may refer to, for example, Refs. [4,19–33].

9.4.2 Stabilization of UPO

To illustrate the concept of chaos control, we now demonstrate the stabilization of an UPO in the logistic map (4.19) by employing the method proposed by Ott, Grebogi, Yorke (OGY) [21] which has now been applied to many dynamical systems. As an example, let us consider the logistic map in the form

$$x_{n+1} = 1 - Ax_n^2 = F(x_n) \; . \tag{9.10}$$

It can be obtained from the standard form (4.19) by the change of variables

$$x' = 2(2x-1)/(a-2) \, , \quad A = a(a-2)/4 \tag{9.11}$$

and then dropping the prime. Suppose for a given value of A, the map (9.10) exhibits chaotic solution. By definition, for this parametric value, there exists a large number of periodic orbits which are all unstable (UPO). Recall the discussion on period doubling route to chaos in Chap. 4. Now suppose we want to stabilise unstable period-1 cycle or the fixed point x^*. The basic idea of the OGY approach is as follows. Define the small deviation

$$\delta x_{n+1} = \frac{\partial F}{\partial x_n} \cdot \delta x_n + \frac{\partial F}{\partial A} \cdot \delta A_n \, , \tag{9.12}$$

where the derivatives are evaluated at $x_n = x^*$ and $A_n = A$. The OGY formula is then obtained by requiring $\delta x_{n+1} \to 0$ for a specific small perturbation in the parameter A, namely δA_n. This results in

$$\delta A_n = -\left(\frac{\partial F}{\partial x_n} \cdot \delta x_n \right) \Big/ \left(\frac{\partial F}{\partial A} \right) \; . \tag{9.13}$$

For the logistic map we obtain

$$\delta A_n = \frac{-2A\,(x_n - x^*)}{x^*} \; . \tag{9.14}$$

Table 9.1. This table gives the form of control introduced in various control schemes. The general form of the dynamical system is $\dot{\boldsymbol{X}} = F(\boldsymbol{X}, p) + \boldsymbol{C}(t)$, where $\boldsymbol{C}(t)$ is the perturbation introduced to control chaos. Here, $\boldsymbol{X}_F(p)$: location of UPO, λ_u: unstable eigenvalue of the desired UPO, $\boldsymbol{f}_u$: unstable eigenvector, $\delta\boldsymbol{X}_n = \boldsymbol{X}_{n+1} - \boldsymbol{X}_F$, T: period of the desired UPO, $\boldsymbol{g}$: chosen goal dynamics, $S(t)$: switching function, for example, $S(t) = 0$ for $t \leq 0$ and $S(t) = 1$ for $t > 0$, $\delta(t)$: Dirac-delta function, ϵ: amplitude of the control and A: constant matrix whose eigenvalues all have negative real part

No.	Method	Controller	Special features
	Feedback methods		
1.	Adaptive control algorithm (Huberman and Lumer [23,24])	$\boldsymbol{C}(t) = 0$, $\dot{p}(t) = \epsilon G(\boldsymbol{X}(t) - \boldsymbol{X}_F(t))$	UPO corresponding to other values of p can be stabilized. System dynamics for a range of values of p must be known.
2.	Ott–Grebogi–Yorke method [21]	$\boldsymbol{C}(t) = 0$, $p_n = p_0 + \delta p_n$, $\delta p_n = \alpha/\beta$, $\alpha = \lambda_u \boldsymbol{f}_u \cdot \delta\boldsymbol{X}_n$, $\beta = (\lambda_u - 1)\boldsymbol{f}_u \cdot \boldsymbol{v}$, $\boldsymbol{v} = \partial\boldsymbol{X}_F(p)/\partial p$	Perturbation is predetermined and knowledge of the evolution equation is not necessary.
3.	Singer–Wang–Bau feedback method [26]	$\boldsymbol{C}(t) = \epsilon\,\mathrm{sgn}(\boldsymbol{X}(t) - \boldsymbol{X}_F(t))$	Evolution equation of the system is not required. Stable control is possible only for a certain range of values of ϵ.
4.	Pyragas feedback method [29]	$\boldsymbol{C}(t) = \epsilon(\boldsymbol{X}(t) - \boldsymbol{X}(t - nT))$	Evolution equation of the system and the location of the UPO are not required. The method stabilizes any one of the UPO of period-nT.
5.	Method of Chen and Dong [30]	$\boldsymbol{C}(t) = \epsilon(\boldsymbol{X}(t) - \boldsymbol{X}_F(t))$	Evolution equation of the system is not required. Stable control to a desired UPO occurs for certain range of values of ϵ.

Table 9.1. Continued

No.	Method	Controller	Special features
	Nonfeedback methods		
6.	Weak periodic parametric perturbation (Lima and Pettini [25])	$\boldsymbol{C}(t) = 0,\ p(t) = p(1 + \eta \cos \Omega t)$	Additional dynamical equation is not required. Suppression of chaos can be observed only for a certain range of values of η and Ω.
7.	Addition of weak periodic force (Braiman and Goldhirsch [27])	$C_j(t) = \eta \cos \Omega t,\ C_i = 0,\ i \neq j$	Easy to implement in mechanical and electrical circuit systems. Suppression of chaos occurs for a certain range of values of η and Ω.
8.	Weak periodic Dirac delta force (Rajasekar, Murali and Lakshmanan [31])	$C_j(t) = \alpha\delta(t \bmod \tau)$, $C_i = 0,\ i \neq j$	Constant amplitude perturbation is added to the system only at discrete times. System must be studied in (α, τ) space in order to choose suitable values of α and τ to eliminate chaos.
9.	Constant bias (Parthasarathy and Sinha [32])	$C_j(t) = C_0$ =constant, $C_i = 0,\ i \neq j$	Perturbation is constant and easy to implement in dynamical systems. Before implementation, the system dynamics must be known for various values of C_0.
10.	Entrainment open-loop control (Hubler and Luscher [22])	$\boldsymbol{C}(t) = [\boldsymbol{g} - \boldsymbol{F}(\boldsymbol{g}, p)]S(t)$	Any arbitrary goal dynamics can be stabilized. Evolution equation is required and a particular solution of the system cannot be entrained.
11.	Migration control (Jackson [28])	$\boldsymbol{C}(t) = [\boldsymbol{g} - \boldsymbol{F}(\boldsymbol{g}, p)]S(t)$ $+(\mathrm{d}\boldsymbol{F}/\mathrm{d}\boldsymbol{g} + \boldsymbol{A})(\boldsymbol{X} - \boldsymbol{g})$	Perturbation vanishes once the goal dynamics is achieved. Evolution equation is necessary.

The main effect of the OGY perturbation (9.14) is to locally alter the map F in the neighbourhood of x^* so that the perturbed map has zero slope at x^*. Recall that the necessary condition for a fixed point of an one-dimensional map to be stable is $|F'(x^*)| < 1$. For $A = 2$, the map exhibits chaotic behaviour and the value of x^* embedded in the chaotic orbit is found to be 0.5. Figure 9.6 shows the stabilization of this unstable fixed point with $x_0 = -0.665$. The control is switched on when $|x_n - x^*| \leq \delta = 0.05$. Once the control is switched on the parameter A changes according to (9.14) such that successive iterated values approach the desired x^*. Here the stabilization of x^* is always guaranteed because in deriving the formula we set $\delta x_n \to 0$. Stabilization of higher periodic orbits is also possible. For example, for $A = 1.6$, there is a period-4 unstable fixed point given by $(-0.424, 0.712, 0.189, 0.943)$. Stabilization of this orbit is illustrated in Fig. 9.7 where $x_0 = 0.65$.

For a review on various chaos control algorithms the readers may refer to Refs. [4,19,20].

Problems:

1. The logistic map with the addition of the controller $\epsilon(x_n - \bar{x})$ can be written as

$$x_{n+1} = 1 - Ax_n^2 + \epsilon\,(x_n - \bar{x}) ,$$

where $\bar{x}$ is the desired unstable periodic orbit and ϵ is the stiffness of the controller. Prove that stabilization of an unstable period-1 orbit is possible only for $\epsilon \in [-1 + 2A\bar{x}, 1 + 2A\bar{x}]$. Iterating the map study the stabilization of the unstable orbit $\bar{x} = 0.5375$ for $A = 1.6$.

2. Obtain the condition on the control parameter ϵ for stabilization of unstable period-1 orbit in the logistic map (9.10) by the Pyragas feedback method.

3. The logistic map with adaptive control implemented to the parameter A is written as

$$x_{n+1} = 1 - A_n x_n^2 ,$$
$$A_{n+1} = A_n + \epsilon\,(x_n - \bar{x}) ,$$

where $\bar{x}$ is the desired unstable fixed point. Applying linear stability analysis determine the range of values of ϵ for which chaos can be suppressed by stabilizing the unstable fixed point $\bar{x}$. Study the suppression of chaos for $A = 1.6$ by stabilizing the fixed point $\bar{x}$ corresponding to $A = 0.6$.

4. The logistic map with parametric feedback controller can be written as

$$x_{n+1} = 1 - (A + \delta A_n)\,x_n^2 , \quad \delta A_n = \epsilon\,(\bar{x} - x_n)$$

where $\bar{x}$ is the desired unstable fixed point. For $A = 1.6$ and $\epsilon = 0$ the map exhibits chaotic motion and a period-1 UPO is $\bar{x} = 0.5375$. Applying linear stability analysis show that chaos can be controlled by stabilizing the fixed point $\bar{x}$ for ϵ in the interval

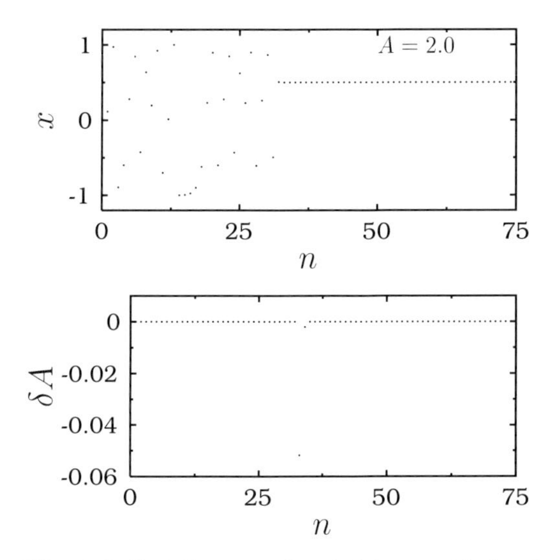

Fig. 9.6. Stabilization of period-1 unstable fixed point of the map (9.10) for $A = 2$

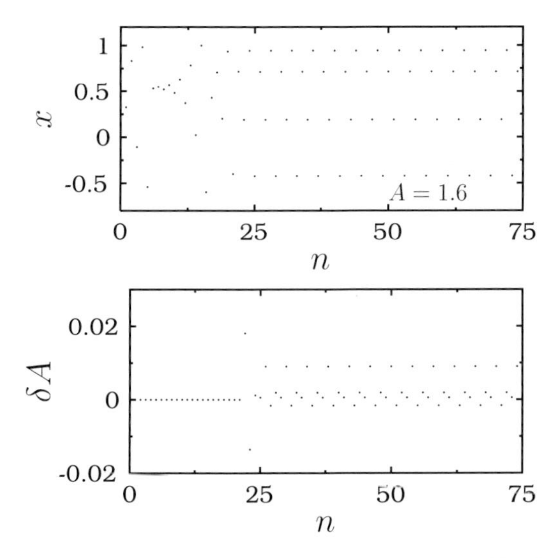

Fig. 9.7. Stabilization of a period-4 unstable fixed point of the map (9.10) for $A = 1.6$

$$\epsilon \in \left[\frac{2A\bar{x} - 1}{\bar{x}^2}, \frac{1 + 2A\bar{x}}{\bar{x}^2}\right] .$$

Numerically verify the above condition.

5. A period-1 fixed point of the Hénon map

$$x_{n+1} = a - x_n^2 + by_n ,$$

$$y_{n+1} = x_n$$

is

$$(x^*, y^*) = \frac{1}{2}\left[b - 1 \pm \sqrt{(b-1)^2 + 4a}\,\right] .$$

For $a = 1.4$ and $b = 0.3$ the map exhibits chaotic dynamics and the fixed point (x^*, y^*) is unstable. Treating a as an adjustable parameter obtain the OGY formula for the Hénon map. Verify the formula [34].

6. The Brusselator equation (see Chap. 16 for more details) with adaptive control implemented to the parameter β is given by

$$\dot{x} = \alpha + x^2 y - \beta x - x ,$$
$$\dot{y} = \beta x - x^2 y ,$$
$$\dot{\beta} = \epsilon(\bar{y} - y) ,$$

where $(\bar{x}, \bar{y}) = (\alpha, \beta/\alpha)$ is a desired unstable fixed point. The system has a limit cycle attractor for $\alpha = 1$, $\beta = 3$ and $\epsilon = 0$. Show that transition from limit cycle solution to the fixed point solution $(1, 1)$ corresponding to the parametric values $\alpha = 1$ and $\beta = 1$ can be achieved only for $\epsilon = 1/2$.

7. The FitzHugh–Nagumo oscillator equation with the controller $C(t) = -\epsilon(V - \bar{V})$ can be written as

$$\dot{V} = W ,$$
$$\dot{W} = -V - V^3/3 + R - uW - A_0 - \epsilon(V - \bar{V}) ,$$
$$\dot{R} = -(c/u)(V + a - bR) ,$$

where $\bar{V}$ is the value of the V-component of the desired unstable fixed point of the system for $\epsilon = 0$. For $a = 0.5$, $b = 0.5$, $c = 0.1$, $u = 0.7$, $A_0 = 2.2$ and $\epsilon = 0$ the system exhibits chaotic motion. An unstable fixed point of the system is $(\bar{V}, \bar{W}, \bar{R}) = (0.931, 0.0, 2.862)$. Study the stabilization of the above unstable fixed point both by numerically and by linear stability analysis [35].

8. The Duffing oscillator equation with a parametric perturbation of the cubic term is given by

$$\ddot{x} + d\dot{x} - x + \beta(1 + \eta\cos\Omega t)x^3 = \nu\cos\omega t .$$

For $\eta = \Omega = 0$, the equation shows chaotic behaviour for $d = 0.154$, $\beta = 4$, $\omega = 1.1$ and $\nu = 0.088$. Investigate suppression of chaos in (η, Ω) parameter space [25].

9. For the logistic map (9.10) and the Bonhoeffer–van der Pol oscillator equation

$$\dot{x} = x - x^3/3 - y + A_1\cos\omega t ,$$
$$\dot{y} = c(x + a - by) ,$$

carry out a comparative study of the various feedback and nonfeedback methods with special emphasis on

a) variation of correction signal with time,
b) region of stable control and
c) dependence of recovery time, R_T (which is defined as $t_0' - t_0$, where t_0 and t_0' are the times at which control is initiated and after which the distance between the desired and actual solutions is always less than a preassumed value, respectively) on the stiffness parameter of the control signal.

For some details, for example see Refs. [31,36–38].

9.5 Synchronization of Chaos

A consequence of extremely high sensitive dependence on initial conditions is that two identical but independently evolving chaotic systems can never synchronize with one another, that is the corresponding state variables of the two systems cannot evolve identically, as any infinitesimal deviations in the starting conditions or in the system specification (such as slight parameter mismatches) can lead to exponentially diverging trajectories making synchronization impossible. Contrast this with the case of linear systems (and also regular motion of nonlinear systems), where the evolution of two identical systems can be very naturally synchronized. In this connection, the recent suggestion of Pecora and Carroll [39,40] that it is possible to synchronize even chaotic systems by introducing appropriate coupling between them has changed our outlook on chaotic systems, synchronization and controlling of chaos, paving ways for new and exciting technological applications: spread spectrum secure communications of analog and digital signals (see also Chap. 16 for further discussions). In this section we bring out briefly these exciting developments.

9.5.1 Chaos in the DVP Oscillator

In order to appreciate the concept of chaos synchronization, let us consider for illustrative purpose a typical chaotic dynamical system, namely the Duffing–van der Pol (DVP) oscillator, especially from a circuit theoretic point of view keeping in mind the ensuing signal transmission applications (though other dynamical systems can be equally well considered). A typical circuit realization of the DVP oscillator equation (6.30) is given in Fig. 6.24. It consists of the linear circuit elements L (inductance), C_1, C_2 (capacitors) and R (linear resistor) and a nonlinear resistor (N) whose current-voltage characteristics can be represented as $i_N = aV_N + bV_N^3$ $(a < 0, b > 0)$. Typically, the nonlinear resistor in Fig. 6.24 can be constructed using a set of diodes and operational amplifiers, or it can be approximated by the Chua's diode corresponding to piecewise linearity. To see chaos, we can numerically analyse (6.30). A numerical simulation of (6.30) with fixed values of $\mu = 100$ and $\alpha = 0.35$

exhibits period doubling bifurcations leading to chaos as the parameter β is decreased from a large value. One observes period-1, period-2, period-4 limit cycles, one-band and double-band chaos respectively at $\beta = 800$, 750, 710, 600 and 300. In this simulation, the initial conditions were chosen as $x(0) = 0.5$, $y(0) = 0.1$ and $z(0) = 0.1$. Naturally, in the chaotic regime, the motions from two nearby initial conditions diverge exponentially until they become completely uncorrelated.

9.5.2 Synchronization of Chaos in the DVP Oscillator

Driving a nonlinear system with a periodic signal is a common feature in nonlinear dynamics. However, the idea of using a chaotic signal to drive a nonlinear system and so for maintaining synchronization in phase and amplitude among the respective state variables of the driving and driven systems is rather new. For example, two identical chaotic systems started at nearly the same initial conditions have by definition trajectories which quickly become completely uncorrelated, even though each system traces the same strange attractor in phase space for the same set of parametric values. It was the idea of Pecora and Carroll [39,40] that a subsystem of a chaotic system can be synchronized with a separate but identical chaotic system under certain conditions. The idea is to treat one of the chaotic systems as a drive (master) system and then other as a response (slave) system so that the first chaotic system drives the second one to induce synchronization so that asymptotically the response system variables faithfully follow the drive system variables. It has also been demonstrated recently that such a synchronization can be achieved through a kind of one-way coupling between the two chaotic systems [41] as in the case of transmission of signals: transmitter $\rightarrow$ receiver.

According to the original scheme of Pecora and Carroll, one can identify those variables which have negative Lyapunov exponents (corresponding to contraction of phase space trajectories in these directions) and retain only those parts in the response system, while the chaotic signal of the remaining drive part (corresponding to positive Lyapunov exponent) is fed directly into the response system as shown schematically in Fig. 9.8a. As a consequence for treating the DVP system (6.30) the variable x is identified as the chaotic signal with positive Lyapunov exponent, while y and z variables correspond to negative Lyapunov exponents. Then the response system (denoted by prime variables) becomes

$$\dot{x}' = x \,, \tag{9.15a}$$

$$\dot{y}' = x - y' - z' \,, \tag{9.15b}$$

$$\dot{z}' = \beta y' \,, \tag{9.15c}$$

to obtain full synchronization $y' \rightarrow y$, $z' \rightarrow z$ asymptotically. Thus one can generate from (9.15) the variables y' and z' which are identical to y and z of (6.30) under the influence of the single driving variable x. One can generalize

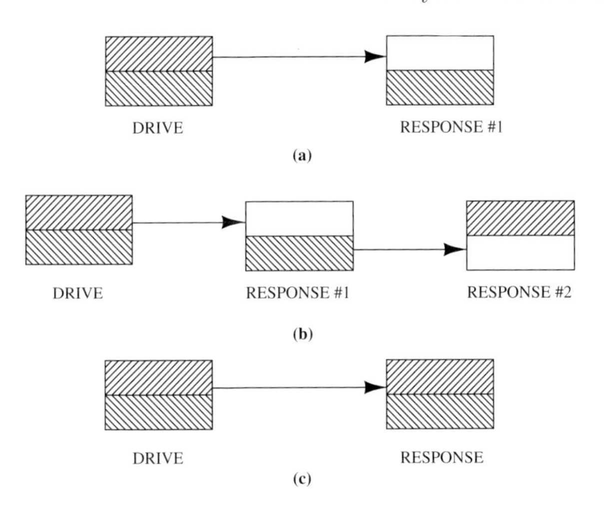

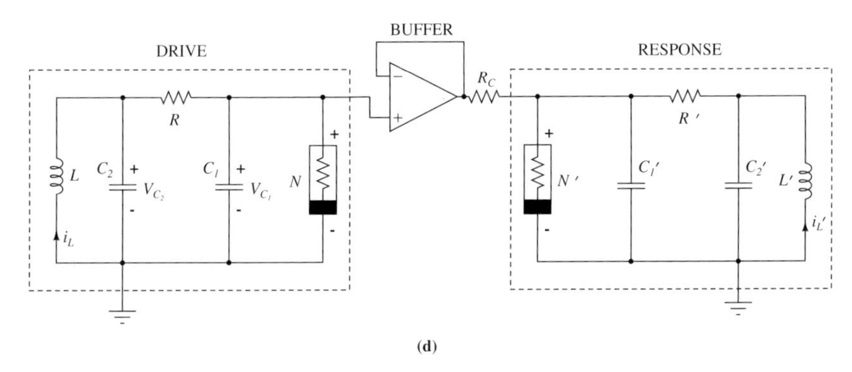

Fig. 9.8. Schematic of (**a**) drive-response scenario of chaos synchronization, (**b**) cascading synchronization systems and (**c**) two identical chaotic oscillators with one-way driving. (**d**) Circuit realization of two unicoupled DVP oscillators

further and develop a cascading procedure to regenerate all the variables identically as shown the Fig. 9.8b, so that (9.15) are replaced by

Response 1:

$$\dot{y}' = x - y' - z' , \tag{9.16a}$$

$$\dot{z}' = \beta y' , \tag{9.16b}$$

Response 2:

$$\dot{x}'' = -\nu \left[(x'')^3 - \alpha x'' - y' \right] , \tag{9.16c}$$

so as to obtain the full synchronization $x'' \to x$, $y' \to y$, $z' \to z$ asymptotically.

Recently it has been pointed out that one can obtain complete synchronization even for two identical systems by an appropriate one-way coupling

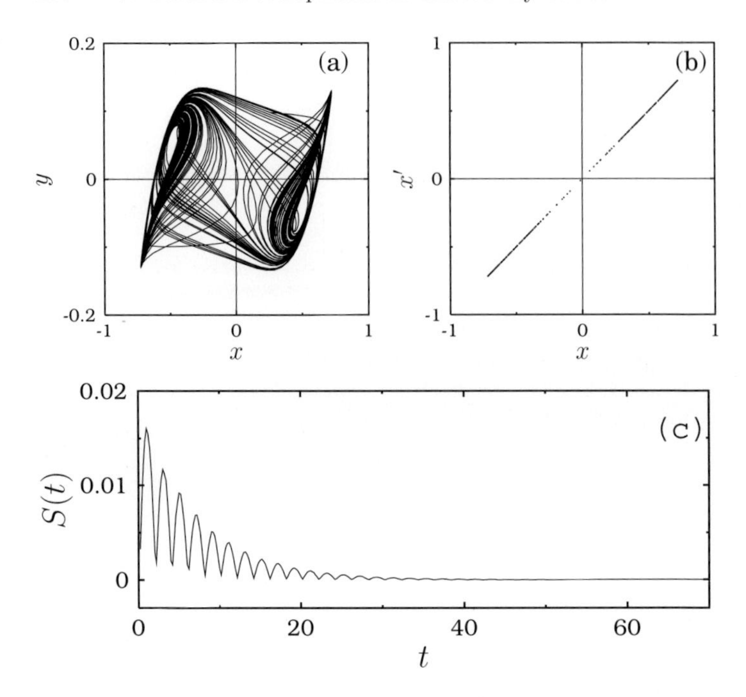

Fig. 9.9. Synchronization of chaos of (9.17). (**a**) Phase portrait in the $(x-y)$ plane. (**b**) Plot in $(x-x')$ plane. (**c**) $S(t) = \sqrt{(x-x')^2 + (y-y'^2) + (z-z')^2}$ versus t plot

(Fig. 9.8c). For example, for the DVP oscillator the drive system is just (6.30), while the response system (Fig. 9.8d) is

$$\dot{x}' = \nu\left[(x')^3 - \alpha x' - y'\right] + \nu\epsilon\,(x - x') \ , \tag{9.17a}$$

$$\dot{y}' = x' - y' - z' \ , \tag{9.17b}$$

$$\dot{z}' = \beta y' \ . \tag{9.17c}$$

Here $\epsilon = (R'/R_c)$ is the coupling parameter. From the numerical simulation of (9.17), the synchronized chaotic behaviour between x and x' variables for $\epsilon = 1.0$, $\alpha = 0.35$, $\nu = 100$ and $\beta = 300$ is shown in Fig. 9.9. The initial conditions have been fixed as $x(0) = 0.1$, $y(0) = 0.1$, $z(0) = 0.2$, $x'(0) = 0.15$, $y'(0) = 0.2$, $z'(0) = 0.3$.

9.5.3 Chaotic Signal Masking and Transmission of Analog Signals

We now focus on the use of synchronized chaotic signals as vehicles for effective transmission of analog signals in the context of secure communications. We illustrate the idea again with the use of the chaotic DVP oscillator. The point is that when a chaotic signal is transmitted it cannot be synchronized in general by a second person unless one has the full information about the transmitting (chaotic nonlinear) system to couple with it appropriately. Thus, the information bearing signal $s(t)$ which is to be transmitted is masked by the noise-like chaotic signal by adding it at the transmitter and at the receiver the masking is removed.

The basic idea is to use the received signal to regenerate the information-bearing signal by subtracting the masking chaotic signal (received separately through synchronization) to obtain $s(t)$. This task is feasible with the synchronizing receiver system since the ability to synchronize is robust, that is, it is not highly sensitive to perturbation in the drive signal. It is assumed that for masking, the power level of $s(t)$ is significantly lower than that of the chaotic drive signal $x(t)$. Then one can exploit the robustness of synchronization using $r(t) = x(t) + s(t)$ as the synchronizing drive signal at the receiver. If the receiver or response has synchronized with $r(t)$ as the drive signal, then $x_r(t) \approx x(t)$ and consequently $s(t)$ is recovered as $s'(t) = r(t) - x_r(t)$. Figure 9.10 illustrates this approach schematically.

The above procedure can be demonstrated with the aid of the DVP oscillator system (6.30) easily through numerical simulation by using the signal $x(t)$ of the drive system as the masking signal for the information signal $s(t)$. The response system is now modified as

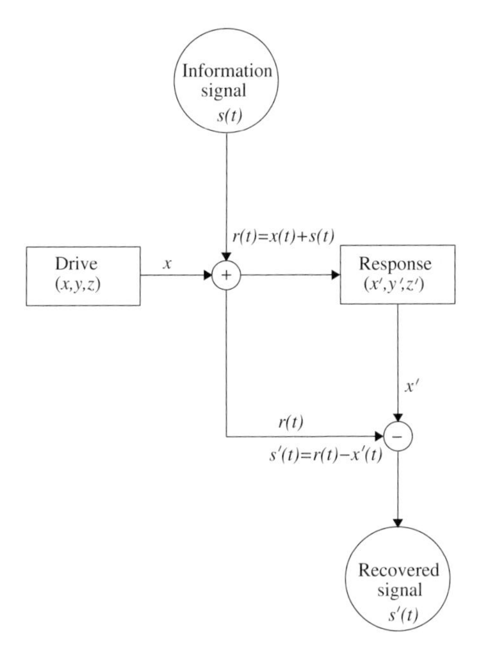

Fig. 9.10. Schematic diagram of chaos based 'signal masking' technique for analog signal transmission

$$\dot{x}' = \nu\left[(x')^3 - \alpha x' - y'\right] + \nu\epsilon\left[r(t) - x'\right] , \tag{9.18a}$$

$$\dot{y}' = x' - y' - z' , \tag{9.18b}$$

$$\dot{z}' = \beta y' , \quad r(t) = s(t) + x(t) . \tag{9.18c}$$

The information signal is recovered as

$$s(t) = r(t) - x'(t) = x(t) + s(t) - x'(t) \approx s'(t) . \tag{9.19}$$

9.5.4 Chaotic Digital Signal Transmission

It is not only the analog signals which can by securely transmitted through chaos synchronization. Binary valued bit signals can be equally well transmitted by this method. Here the idea is to essentially modulate a component parameter associated with the transmitter using the information bearing digital waveform and accordingly transmit the chaotic signal. At the receiver, the coefficient of modulation will produce a synchronization error between the received drive signal and receiver's regenerated drive signal, with an error signal amplitude that depends on the modulation. Using the synchronization error the modulation can be detected.

Now we explain the method for the DVP oscillator, which is illustrated schematically in Fig. 9.11. Here the coefficient β in (6.30) is modulated by the information w88aveform $s(t)$, which is a binary coded signal.

The information is carried over the channel by the chaotic signal $x(t)$ which serves as the driving input to the receivers #1 and #2. At the receivers the modulation is detected by forming the difference between $x(t)$

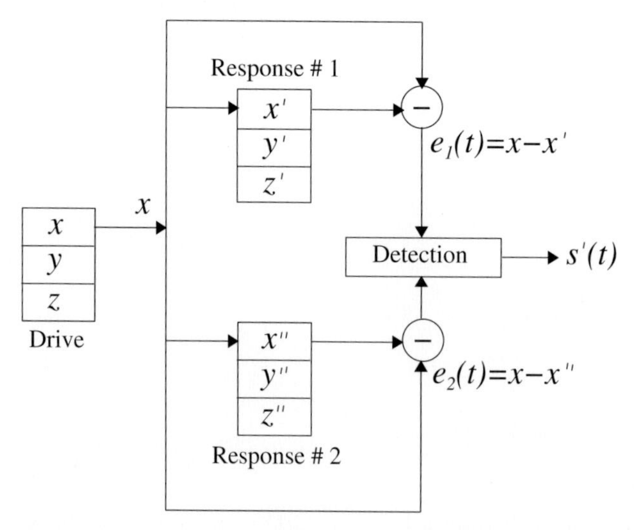

Fig. 9.11. Schematic of chaos based digital signal transmission technique

and the reproduced signal $x'(t)$ of #1 and $x''(t)$ of #2. Then the synchronization error $e_1(t) = x(t) - x'(t)$ will be relatively large in amplitude during the time period when '1' is transmitted and small in amplitude during '0' transmission. Thus the synchronization receiver can be recognized as a form of matched filter for the chaotic transmitted signal $x(t)$. From the error signal $e_1(t)$ or $e_2(t)$ after suitable filtering and thresholding, the transmitted binary information signal can be recovered. A sample illustration is provided in Fig. 9.12. In recent times, much progress has been made in the understanding of various synchronization aspects of chaos, including compounding chaotic signals, generalized concepts of synchronization, lag synchronization, anticipating synchronization, phase synchronization and so on. For some of

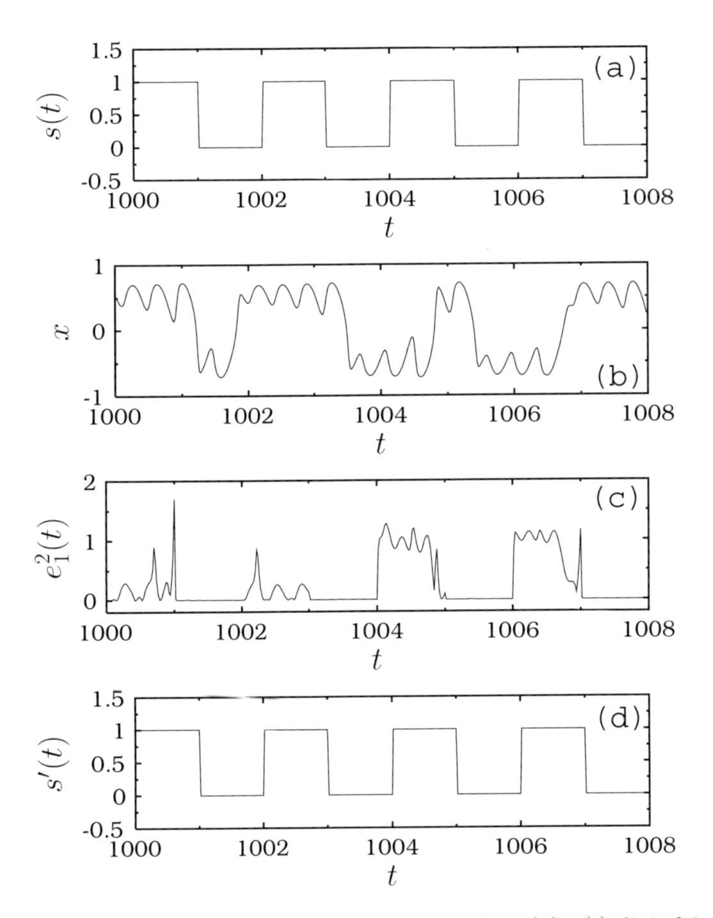

Fig. 9.12. Secure digital signal transmission. (**a**) $s(t)$-digital information signal. (**b**) Actual transmitted signal. (**c**) Error signal $e_1^2(t)$. (**d**) Recovered information signal $s'(t)$

these ideas and rigorous and unifying definitions of synchronization of chaos, see for example Refs. [42,43].

Problems:

1. Investigate how chaos synchronization can be effected in
   a) Lorenz system, equations (5.5),
   b) Chua's circuit, equation (6.15) and
   c) MLC circuit equation (6.17)

   by (i) Pecora–Carroll method and (ii) one-way coupling method.
2. Work out a scheme in which a sinusoidal signal can be transmitted through chaos synchronization in the Duffing oscillator, equation (5.1). Extract the signal at the receiver end (see for example, Ref. [4]).

9.6 Quantum Chaos

So far we have considered the various dynamical features associated with nonlinear systems, which are essentially classical and so macroscopic in nature. For microscopic systems a natural question arises: What is the quantum effect on classical chaos, particularly in the case of Hamiltonian systems?

Search for quantum manifestations of classical chaos in the practical sense, which goes by the terminology *quantum chaos* or *quantum chaology*, has recently attracted considerable interest [5]. For example, one might look for possible fingerprints of chaos in the eigenvalue spectrum, wave function patterns and so on. Considerable investigations have been carried out along these lines in the literature. For details, see for example Refs. [5,44–51].

In recent times it has been found that highly excited Rydberg atoms and molecules (which are effectively one-electron systems) under various external fields are veritable gold mines for exploring the quantum aspects of chaos [44,45]. These systems are particularly appealing as they are not merely mathematical models but important physical systems which can be realized in the laboratory. Particular examples are the hydrogen atom in external magnetic fields, crossed electric and magnetic fields, van der Waals force, periodic microwave radiation and so on. In this section we wish to bring out the salient features of these developments and point out the future promises very briefly.

9.6.1 Quantum Signatures of Chaos

The main question is given the potential, the underlying Schrödinger wave equation is essentially linear. Then how and where does the effect of chaos will be felt in quantum mechanics. For this purpose, several signatures of chaos have been identified as fingerprints of quantum manifestations of classical chaos. These include energy level statistics and wave function behaviour. Some details are given below.

A. Energy Levels and Universality

Bounded quantum systems are characterized by discrete eigenvalues and normalizable eigenfunctions. The macroscopic chaos (namely classical chaos) is essentially due to the long-time behaviour of trajectories in phase space. The Heisenberg's uncertainty relation, namely $\Delta E \Delta T \geq \hbar$ implies that long-time behaviour should necessarily be associated with short energy intervals. Thus, one possible approach to understand the implications of chaos in quantum systems is to look for short-range correlations between energy levels (spacings). Indeed, by examining a large class of quantal systems such as billiards of various types, coupled anharmonic oscillators, atomic and molecular systems, it has been realized that there exists generically a universality [5] in the spacing distribution of the quantum version of classically integrable as well as chaotic systems. For regular systems, nearest neighbour spacings follow a Poisson distribution, while chaotic systems follow either one of the three universality classes (depending upon the underlying symmetry and the value of the angular momentum (spin)) given in Table 9.2. For chaotic systems the three universality classes correspond to *Gaussian Orthogonal Ensemble (GOE)* or *Wigner* statistics, *Gaussian Unitary Ensemble (GUE)* statistics and *Gaussian Symplectic Ensemble (GSE)* statistics, similar to the ones

Table 9.2. Various universality classes of level statistics

Classical dynamics	**Symmetry**	**Quantum level statistics**	**Examples**
Regular (integrable)	--	Poisson	Integrable billiards, (un)coupled oscillators, etc.
Near-integrable	--	Intermediate (Izrailev, Berry–Robnik, Brody)	Near-integrable cases of coupled oscillators
Chaotic	Time reversal invariance	GOE (Wigner)	Hydrogen atom in strong magnetic field, hyperbolic systems, anisotropic Kepler system
	Broken time reversal invariance	GUE	Aharanov–Bohm billiard
	Time reversal invariance + (1/2) odd integer total angular momentum	GSE	Neutrino billiard

which occur in random matrix theory of nuclear physics developed by Wigner, Dyson, Mehta and others during 1960s. Finally, the near integrable and intermediate cases are found to satisfy either the Brody or Berry–Robnik or Izrailev distribution [46].

B. Energy Level Dynamics

In fact, one can rigorously analyse the scheme of transition from Poisson statistics to the Gaussian random matrix ensemble statistics (corresponding to transition from regular to chaotic motion in classical dynamics) in the framework of 'level dynamics'. Considering a generic Hamiltonian $H = H_0 + \lambda V$, where λ is typically the nonintegrability parameter which is now taken as a 'time' variable, one can deduce [47] the 'equations of motion' for the nondegenerate discrete energy levels and eigenfunctions under the λ-flow from the Schrödinger eigenvalue problem $H|n(\lambda)\rangle = x_n(\lambda)|n(\lambda)\rangle$:

$$\frac{\mathrm{d}x_n}{\mathrm{d}t} = \langle n|V|n\rangle = p_n \, , \quad (t = \lambda) \tag{9.20a}$$

$$\frac{\mathrm{d}p_n}{\mathrm{d}t} = 2\sum_{m\neq n} V_{nm} \cdot V_{mn} \left(x_n - x_m\right)^{-1} \, , \tag{9.20b}$$

$$\frac{\mathrm{d}|n\rangle}{\mathrm{d}t} = \sum_{m\neq n} |m\rangle V_{mn} \left(x_n - x_m\right)^{-1} \, , \tag{9.20c}$$

$$\frac{\mathrm{d}\langle n|}{\mathrm{d}t} = \sum_{m\neq n} \langle m| V_{nm} \left(x_n - x_m\right)^{-1} \, , \quad V_{nm} = \langle n|V|m\rangle \, . \tag{9.20d}$$

Equation (9.20) can be shown to be equivalent [47] to a completely integrable N-particle Calogero–Moser system endowed with an additional internal complex vector space structure. The solutions of this system determine the possible energy spectrum and wave function patterns at an arbitrary flow value of λ, the nonintegrability parameter. Further, one can use these solutions to describe the origin of successive avoided crossings observed in the energy level structure of bounded systems. Then by making use of a grand canonical ensemble one can derive the universal spacing distributions as in the case of random matrix theory [5].

C. Wave Function Behaviour: Scarring and Localization

The phase trajectories of classically integrable and near integrable Hamiltonian systems have been shown in Chap. 7 to lie on energy shells corresponding to invariant tori, the so-called KAM tori. Correspondingly, in the semiclassical limit, the Einstein–Brillouin–Keller (EBK) quantization rules hold good and the quantal eigenstates are known to be localized on the invariant tori. When the classical system is chaotic, the phase trajectories traverse the accessible phase space ergodically. Typically, the chaotic trajectories consist of infinite number of unstable periodic orbits. Surprisingly, the corresponding quantum eigenfunctions in generic examples have also been found to have structures with high amplitudes along short, unstable periodic orbits. Heller (see for example, Ref. [48]) described them as having *scars* of classical periodic orbits and provided theoretical arguments for the constructive quantum interference responsible for their appearance. The enhanced probability density of the scarred wave functions is ascribed to the focal points of the periodic orbit. The scars play an important role in interpreting experimental results, for example modulation of intensity.

A second, but related, aspect of quantal wave functions of classically chaotic system is *localization* [44–46]. By considering a periodically kicked rotator, it has been shown that while in the classical case the energy diffuses as a function of time, while in the quantal case, after a finite time, the energy (diffusion) displays a bounded fluctuation about a finite mean value. By analogy with Anderson localization in solid-state physics, one can argue that the quantum-mechanical suppression of diffusion is essentially due to the localization of the wave function arising from the interference effects.

9.6.2 Rydberg Atoms and Quantum Chaos

Atomic systems are considered to be the most suitable testing grounds for understanding the various ideas on quantum chaos discussed above. One of the reasons behind this is that the underlying systems are amenable to classical, semiclassical and quantum treatment. They are also experimentally realizable in the laboratory. Moreover, one can make a systematic comparison of experimental and theoretical (analytical and numerical) results.

We know that hydrogen atom is the simplest atomic system containing only one electron. Its classical analogue is the Kepler problem of a planet moving around a relatively stationary star under the action of the gravitational force. Both the quantum and classical problems are solvable and comparisons can be made. In the presence of a physically realizable and meaningful external field, the classical analogue of the hydrogen atom (the perturbed Kepler problem) shows a rich variety of nonlinear phenomena and is a typical system exhibiting Hamiltonian chaos [49]. The corresponding quantum system represents the so-called highly excited Rydberg atoms (effective one-electron atoms, where the outermost electron is excited to a higher energy level so that the remaining electrons with the nuclear core form an effective nuclear charge) in external fields. Thus, the quantum manifestations of classical chaos can be inferred from the study of such real atomic systems in external fields.

A. Quadratic Zeeman Effect

To appreciate that the Rydberg atoms in external fields are indeed laboratory-realizable systems, let us consider for a moment the problem of hydrogen atom kept in a constant magnetic field along the z-direction. The associated Hamiltonian can be written as

$$H = \left(\frac{(\boldsymbol{p} - e\boldsymbol{A}/c)^2}{2m} \right) - \left(\frac{e^2}{r} \right) , \tag{9.21}$$

where the magnetic field is derived from the vector potential $\boldsymbol{A}$ as $\boldsymbol{B} = (\boldsymbol{\nabla} \times \boldsymbol{A}) = (0, 0, B)$, $\boldsymbol{A} = (\boldsymbol{B} \times \boldsymbol{r})/2$. Here m is the reduced mass, $\boldsymbol{p}$ the momentum and e the electronic charge. Then one can rewrite the Hamiltonian (9.21) as

$$H = \frac{p^2}{2} - \left(\frac{e^2}{r} \right) + \omega L_z + \frac{1}{2} m\omega^2 (x^2 + y^2) , \tag{9.22}$$

where L_z is the z-component of the angular momentum ($= xp_y - yp_x$) and ω is half the cyclotron frequency,

$$\omega = \frac{1}{2}\omega_c = \frac{eB}{2mc} . \tag{9.23}$$

Thus, when the magnetic field strength is small, the last term on the right hand side of (9.22), the so-called quadratic Zeeman term, can be neglected. In order that this quadratic term to have an appreciable effect, it should be of the order of the Coulomb interaction. An estimate of the required field strength can be obtained as follows.

Comparing the Rydberg energy (that is, ground-state energy $R = -me^4/2\hbar^2$) with the (two-dimensional) oscillator energy $\hbar^2\omega$ gives a critical field strength of

$$B = B_0 = \frac{m^2 e^3 c}{\hbar^3} \approx 2.35 \times 10^9 G = 2.35 \times 10^5 \mathrm{T} . \quad (\mathrm{T}: \ \mathrm{Tesla}) \tag{9.24}$$

Thus, one requires an astronomical field strength of 2.35×10^5T in order to realize the effect of the quadratic Zeeman term, which in the classical case leads to chaos. However, one can quickly realize that with the recent advent of tunable lasers, atoms can be excited to higher levels, even up to the order of the principle quantum number $n \approx 50$ to 100, so that there is a dramatic reduction of the field strength required in the laboratory, as given in Table 9.3. It is clear that for $n = 50$ state, just a 2T field, comparable to that of the Coulomb field, is enough to realize the effect of chaos.

In fact, a large body of investigations has been performed both analytically (theoretical and numerical) and experimentally to realize the quantum effects of chaos in the quadratic Zeeman problem. We summarize the details below.

The motion of the electron is regular in a weak magnetic field and chaotic on a large scale in a strong magnetic field. In the intermediate region there is a smooth transition. As far as the energy spectrum of the corresponding quantum system is concerned, in the weak field regime the nearest-neighbour spacing distribution follows a Poisson distribution, whereas in the strong field regime they follow the GOE statistics. In the intermediate regime, where

Table 9.3. Variation of certain physical entities of Rydberg states as a function of the principle quantum number n

Physical entity	**Scaling relation**	$n = 1$	$n = 50$
Size	n^2	1 A°	0.5 μm
Energy	$1/n^2$	13.6 eV	5 meV
Electron velocity	$1/n$	2×10^6 m/sec	4×10^4 m/sec
Critical electric field	$1/n^4$	10^9 V/cm	100 V/cm
Critical magnetic field	$1/n^3$	2×10^5 T	2 T

both regular and chaotic trajectories coexist, it is described by intermediate statistics.

In the chaotic region, near the ionization limit, Garton and Tomkins observed a series of broad unresolved resonances called quasi-Landau resonances equally spaced by 1.5 cyclotron frequency. This phenomenon shows that the long-range correlations exist in the spectrum which can be analysed using the so-called Gutzwiller's trace formula which involves the summations over all classical periodic orbits [5]. The experimentally observed spectra are shown to have an impressive agreement with these classical periodic orbits.

By making use of Husimi distribution in phase space (which represents the overlap of the eigenstate with a coherent state), the quantum localization near the classical invariant torus is identified for weak field limit whereas in the strong field limit no such localization is observed. Thus, there is a qualitative difference between regular and chaotic eigenstates of the quantum systems.

B. Hydrogen Atom in Other External Fields/Interactions

It is not only the problem of hydrogen atom in a constant magnetic field that is of interest in quantum chaos. There are a host of other interesting problems in which the hydrogen atom interacts with external fields which are also of considerable interest both from theoretical and experimental points of view. They are enumerated in Table 9.4.

9.6.3 Hydrogen Atom in a Generalized van der Waals Interaction

If we consider two identical atoms (neutral) separated by a distance d (which is large in comparison with the radius of the atoms), then the atoms induce dipole moments on each other, which will cause an attractive interaction between them. This is the so-called *van der Waals interaction* or *London interaction* or *induced dipole-dipole interaction*. It is possible in this case to

Table 9.4. Various physically meaningful perturbations of hydrogen atom: $H = H_0 + H_1$; $H_0 = (p^2/2) - (1/r)$

Physical problem	**Typical form of H_1**	**Properties**
Stark effect	Fz	Separable, no chaos
Magnetic field (constant)	$\omega L_z + (1/2)m\omega^2(x^2+y^2)$	Nonintegrable, order to large chaos and smooth transitions
Oscillating electric field	$Fz \cdot \cos\omega t$	Order to chaos depending on F and ω values (nonintegrable)
Crossed electric and magnetic field (constant)	$\gamma(x^2+y^2)+Fz$	Depending on γ and F value either regular or chaotic behaviour (nonintegrable)
Spherical quadratic Zeeman problem	$\gamma(x^2+y^2+z^2)$	Separable, no chaos (integrable)
Instantaneous van der Waals field	$\gamma(x^2+y^2+2z^2)$	Near-integrable, order to small scale chaos transition
Wave guide	$\gamma(x^2+y^2+\alpha z^2)$, $\alpha = [2, 8/3]$	Near-integrable, order to small scale chaos transitions

show that the interaction between the hydrogen atom and a nearby metal surface (called the instantaneous van der Waals interaction) can be described by the Hamiltonian

$$H = \frac{p^2}{2} - \left(\frac{1}{r}\right) + \gamma_I \left(x^2 + y^2 + 2z^2\right) , \quad \gamma_I = -\frac{1}{16d^3} . \tag{9.25}$$

By generalizing the above equation (9.25), one can introduce the so-called generalized van der Waals interaction with the Hamiltonian

$$H = \frac{p^2}{2} - \left(\frac{1}{r}\right) + \gamma \left(x^2 + y^2 + \beta^2 z^2\right) , \tag{9.26}$$

which encompasses many of the physically interesting systems discussed above for appropriate choices and range of the parameters γ and β. Interestingly, this system has also a close analogy with the Paul trap problem of ions in precision atomic spectroscopy [50]. In a Paul trap, an rf electric field is applied between the end plates and a ring-shaped electrode at the center so that the fields create hyperbolic potentials in which the motion of ions is harmonic to a first order. At the center of the trap, the ion inhabits virtually a field-free region. The Hamiltonian of such a Paul trap is essentially given by

$$H = \frac{p^2}{2} + \left(\frac{1}{r}\right) + \frac{1}{2}\left(x^2 + y^2 + \beta^2 z^2\right) , \tag{9.27}$$

where β represents the deformation of the secular oscillator. One may note that (9.27) is analogous to (9.26) except for a change in sign of the Coulombic term.

The Hamiltonian (9.26) is not an exactly solvable problem for arbitrary values of β. By using cylindrical and semiparabolic coordinates one can show that the z-component of the angular momentum is a constant of the motion so that the system becomes effectively a problem of two coupled sixth-power anharmonic oscillators. Then all the recently developed techniques to deal with chaotic, nonintegrable Hamiltonian systems can be used to study the classical and semiclassical aspects. Detailed analysis [49] shows that as the parameter β increases for arbitrary γ, there is a

chaos → *order* → *chaos* → *order* → *chaos* → *order* → *chaos*

type of transition, with exactly integrable behaviour at $\beta = 1/2$, 1 and 2 due to the existence of certain nontrivial dynamical symmetries of the system. The corresponding quantum problem can be analysed by solving the Schrödinger equation numerically in terms of the SO(4, 2) group generators using a scaled set of normalized Coulombic wave functions. In this way a large number of converged eigenvalues can be obtained and the level statistics can be analysed. Again as the parameter β increases, one obtains a

GOE → *Poisson* → *Brody* → *Poisson* → *Brody* → *Poisson* → *GOE*

type of transition corresponding to classical dynamics, thereby confirming the classical and quantum connections. Some representative details are given in Fig. 9.13. Similar studies can also be performed for wave function dynamics.

9.6.4 Outlook

It is not only the study of Rydberg atoms that is of interest in quantum chaos. Recent investigations show that Rydberg molecules are also of paramount importance in view of the nonuniversality properties of the correlations between intensities and spacings and the classical chaotic auto-ionization mechanism proposed for the experimentally observed ionization process. In a related development, for the He atom for collinear configuration with both electrons on the same side of the nucleus, one finds regular behaviour, while for electrons on different sides of the nucleus, fully developed chaotic behaviour occurs, whereas the motion on the Wannier ridge shows a mixed behaviour.

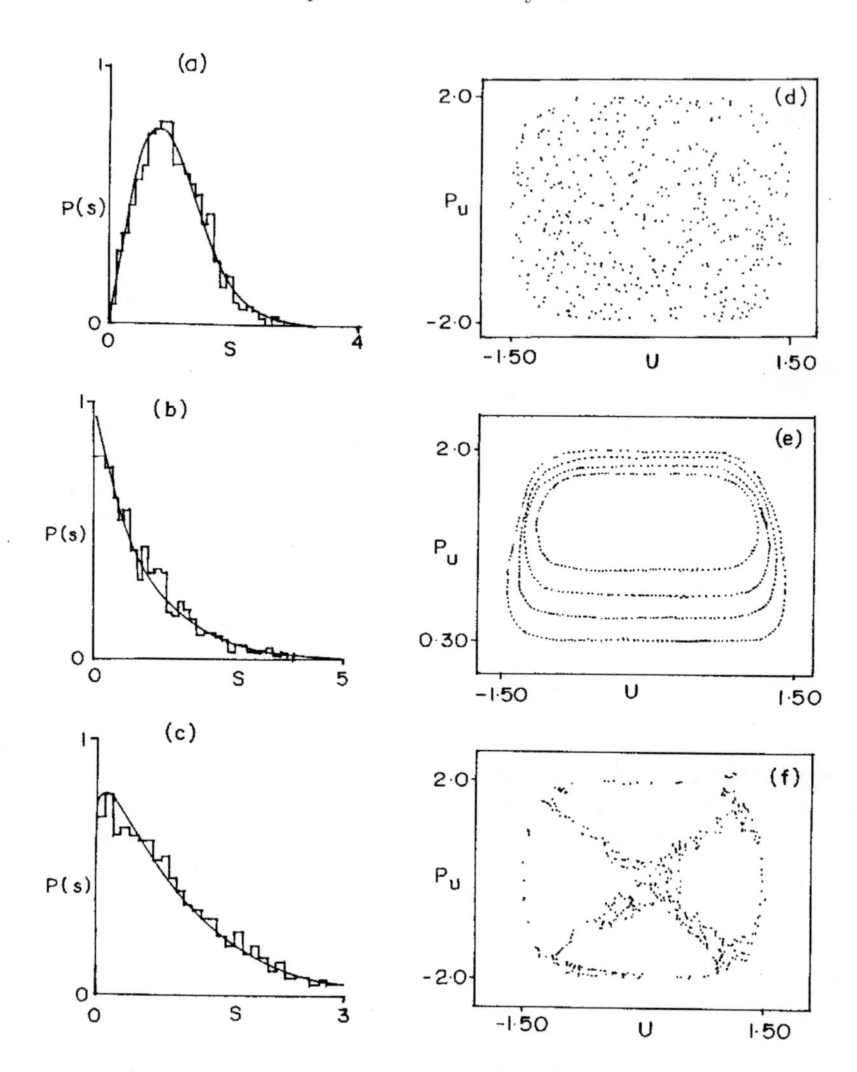

Fig. 9.13. (**a–c**) Nearest neighbour spacing distribution of the hydrogen atom in a generalised van der Waals potential problem for $\beta = 1/4$, $1/2$ and $\sqrt{0.4}$. (**d–f**) Poincaré SOS of Hamiltonian (9.27) (using the corresponding oscillator counterpart Hamiltonian) for $\beta = 1/4$, $1/2$ and $\sqrt{0.4}$. For details see Ref. [50]

Interestingly, all these three regions have been quantized semiclassically for the first time. Moreover, there seems to exist a close agreement with pure quantum-mechanical and experimental results. Also, at present many interesting classical, semiclassical and quantal investigations are ongoing along this direction.

In the case of mesoscopic systems (whose size $\approx 10^{-9}$m), by looking at the dispersion of energy levels to external perturbations (like magnetic field), the so-called generalized conductance was recently introduced. Indeed remarkably, the conductivity properties of these mesoscopic systems share analogous universality properties exhibited by quantum chaotic systems. These findings have brought much practical importance in the field of quantum chaos [5].

Problems:

1. Study the quantum manifestations of classical chaos in the case of two coupled quartic oscillators given by the potential
$$V = \frac{1}{2}kx^2 + \frac{1}{2}ky^2 + \frac{1}{4}\lambda x^2y^2 \ .$$
2. Deduce (9.20) for the 'motion' of eigenvalues and eigenfunctions, starting from time-independent Schrödinger equations [47].

9.7 Conclusions

In this chapter, we have indicated that the concept of chaos in dynamical systems is giving rise to very many novel concepts, ideas and applications. We have touched only a few of them. These in turn have great ramifications in very many areas of physics, chemistry, biology and so on [52]. One might say that these developments are revolutionizing our understanding of natural phenomena. Many more challenging problems remain to be understood: the problem of coupled dynamical systems and arrays, turbulence, spatiotemporal patterns and chaos and so on, to name a few. One might say that what has been understood is only a minuscule when compared to the large number of other complicated nonlinear dynamical phenomena which remain yet to be explored.

In the meantime, we will turn our attention to another important and interesting class of nonlinear dynamical systems, namely the so-called integrable systems. They may be either finite dimensional or infinite dimensional, described by ordinary or partial differential (or even difference) equations, respectively. These systems have beautiful underlying mathematical structures and physical properties. Though they may be considered as measure zero in number (that is relatively small in number), still the list of them is impressive and increasing. In the next few chapters, we will concentrate on these aspects.

10. Finite Dimensional Integrable Nonlinear Dynamical Systems

We have seen in the earlier chapters that dynamical equations of typical nonlinear systems are not amenable to exact solutions in terms of known functions, except for isolated cases like the equations of motion of the undamped and unforced Duffing oscillator, the pendulum, the Kepler particle and so on. Generally speaking, finite dimensional nonlinear systems of the type (3.1), for $n > 2$, are not explicitly solvable, and they are often chaotic depending upon the values of the control parameters. Integrating the equations of motion completely, obtaining analytic solutions and finding acceptable constants of motion/integrals of motion/invariants for such nonlinear systems seem to be rare. Yet they do exist in isolated, but a large number of cases; one says that they are measure zero systems among the totality of nonlinear systems.

From a physical point of view, the existence of such *integrable* nonlinear dynamical systems often means the existence of very regular motion (note that regular motion is not precluded in nonintegrable systems), and from a mathematical point of view they imply the existence of beautiful analytic and geometric structures (as will be seen in this and next few chapters). One can learn a lot about nonlinear dynamical systems in general from such examples, as analytic results are much easier to use, to interpret and to generalize. They can be further utilized to develop suitable approximation schemes, including perturbation theories, in order to deal with nonintegrable systems [1,2], by treating the integrable cases as basic zeroth order exact results.

With this point of view, we will consider in this chapter several examples of integrable nonlinear dynamical systems with finite degrees of freedom (Hamiltonian as well as dissipative) as well as integrable mappings. We will also discuss the meaning and possible criteria of integrable systems and methods to identify them, including Painlevé analysis, symmetry analysis and direct methods. Actually a much larger and important class of integrable systems occur in the case of infinite degrees of freedom (particularly in the case of continuous systems, namely soliton systems). They will be treated in some detail in the next few Chaps. 11–14.

10.1 What is Integrability?

Two obvious fundamental questions arise when we talk of integrable systems. They are: (1) What does one mean by integrability of a dynamical system and (2) when does it occur for a given system? The answer to the former question is somewhat vague as the concept of integrability is itself in a sense not well defined and there seems to be no unique definition for it as yet [3,4]. The answer to the latter is even more difficult, as no well defined criteria seem to exist to identify integrable systems in a rigorous sense. Even then one can attempt to give useful definitions and criteria for integrability based on a practical knowledge.

From a qualitative point of view, integrability can be considered as a mathematical property that can be successfully used to obtain more predictive power and quantitative informations to understand the dynamics of the system locally as well as globally. Recent investigations (for reviews see for example, Refs. [4–7]), which are in a sense revival of the efforts of the mathematicians and physicists of the past, show that the integrability nature of dynamical systems can be methodologically investigated using at least the following two broad notions (among many other possibilities).

(i) *Integrability in the Complex Time Plane: Painlevé Property*

The first one uses essentially the literal meaning: integrable–integrated with sufficient number of integration constants; nonintegrable–proven not to be integrable (in terms of *meromorphic*[1] functions). This loose definition of integrability can be related to the existence of single-valued, meromorphic solutions, a concept originally advocated by Fuchs, Kovalevskaya, Painlevé and others towards the second half of the nineteenth century and the beginning of the twentieth century for differential equations. Such a definition then leads to the notion of integrability in the complex time plane, which has now been refined to some extent and is generally called the *Painlevé property* [4–7].

(ii) *Complete Integrability and Liouville Integrability*

The second notion is concerned with the existence of *sufficient* number of integrals of motion. In particular, complete integrability may mean that these integrals exist in sufficient number. For a system of N first-order autonomous ordinary differential equations of the form (3.1), 'sufficient' may mean $(N-1)$ time-independent integrals (whereupon the system can be reduced to a single quadrature) or N time-dependent ones (in which case the solutions can be obtained by solving an algebraic problem). In the case of Hamiltonian

[1] A function of the complex variable z is said to be meromorphic in a domain D if it is analytic in D except for poles, that is, all its singularities in D are poles.

systems with N degrees of freedom, the above requirement essentially degenerates into the existence of N involutive first integrals of motion (including the Hamiltonian), which are functionally independent (see also Sect. 10.7.1 below and recall our discussions earlier in Sec. 7.2 on conservative systems). Note that even though the Hamilton's equations of motion are $2N$ first-order differential equations, existence of N above type of integrals imply integrability, because the remaining N integrals in the form of 'angles' are ensured in principle through Hamilton–Jacobi procedure. (See also Sect. 13.4.1-D in Chap. 13 for some more details on this aspect). This is the Liouville integrability.

Another interesting aspect of the above concept of integrability via existence of global integrals of motion is the connection with the existence of symmetries and invariance under continuous transformations of the dynamical system exemplified by the Noether's theorem [7].

In fact the combination of the above mentioned two notions has met with remarkable success in recent years in predicting integrable cases of nonlinear dynamical systems. We will indicate some of the details in the following sections as applicable to finite dimensional systems, including maps. Some remarks regarding continuous (infinite-dimensional) systems (pde's) will be included towards the end of Chaps. 13, 14 and in appendix J.

10.2 The Notion of Integrability

As pointed out in the previous section, the notions of integrability and complete integrability are not yet uniquely or conclusively defined. The search for practical definitions can be well appreciated with the aid of simple examples, which we present in the following (see for more details, for example Ref. [4]).

(a) *One-Dimensional Hamiltonian Systems*

One-dimensional Hamiltonian systems defined by

$$H = E = \frac{p^2}{2m} + V(q) \,, \tag{10.1}$$

where the energy E is a constant of motion, can always be considered to be integrable. This is because, making use of the equations of motion

$$\dot{q} = \frac{p}{m} \,, \quad \dot{p} = -\frac{\mathrm{d}V}{\mathrm{d}q} \,, \tag{10.2}$$

(10.1) can be integrated once to reduce it to a quadrature,

$$t - t_0 = \int_{q_0}^{q} \frac{\mathrm{d}q'}{\sqrt{E - V(q')}} \,. \tag{10.3}$$

Here t_0 can be thought of as the second constant. On evaluation of the integral on the right hand side of (10.3), the solution $q(t)$ at an arbitrary time t can be obtained, in terms of the initial value $q(0) = q_0$ (which is infact related to E) at $t = t_0$.

(b) *A Simplified Rikitake System*

The so-called Rikitake two-disk dynamo model, which has been proposed for the description of the time variation of the earth's magnetic field (see for example, Ref. [4]), has the following form in its simplified version,

$$\dot{x} = yz \,, \quad \dot{y} = -xz \,, \quad \dot{z} = -xy \,. \tag{10.4}$$

Using the expressions for the derivatives of the Jacobian elliptic functions (see appendix A), one can immediately write down the explicit solutions to (10.4) as

$$x = A\,\mathrm{sn}\,[p\,(t - t_0)]\,, \quad y = A\,\mathrm{cn}\,[p\,(t - t_0)]\,, \quad z = p\,\mathrm{dn}\,[p\,(t - t_0)], \tag{10.5}$$

where the square of the modulus $k^2 = A^2/p^2$. Here A, p and t_0 are the three constants (integrals) of the system. In other words, (10.4) has been integrated to obtain the solutions (10.5) giving local as well as global informations on the system.

Exercise:

1. Integrate explicitly (10.4) to obtain the solutions (10.5).

(c) *Systems with Central Potentials*

The well known two body Kepler problem is considered to be a typical example of an integrable system [8]; in fact it is a *superintegrable* system. Indeed one knows that the motion of two particles under the influence of a mutual central force with the potential, $V(|\boldsymbol{r}_1 - \boldsymbol{r}_2|)$, depending only on the distance between the particles, can always be reduced to that of an effective one particle motion for the relative coordinates $\boldsymbol{r} = (\boldsymbol{r}_1 - \boldsymbol{r}_2)$, about the center of mass motion. The reduced Hamiltonian

$$H = E = \frac{1}{2\mu}p^2 + V(r)\,, \quad r = \sqrt{x^2 + y^2 + z^2}\,, \tag{10.6}$$

where the reduced mass $\mu = m_1 m_2/(m_1 + m_2)$ and the associated momentum $\boldsymbol{p} = \mu(\boldsymbol{p}_1/m_1 - \boldsymbol{p}_2/m_2)$, has three (involutive) conserved quantities, namely the total energy E, the square of the total angular momentum L^2 (where $\boldsymbol{L} = \boldsymbol{r} \times \boldsymbol{p}$) and, say, the z-component of the angular momentum, L_z. Using these integrals, the equation of motion can be easily reduced to quadratures, as is well known in classical mechanics [8].

However, in the case of the gravitational potential $V = -k/r$, there exists an additional independent integral invariant, namely, the Runge–Lenz vector

$$\boldsymbol{A} = \frac{1}{\mu}(\boldsymbol{p} \times \boldsymbol{L}) - k\frac{\boldsymbol{r}}{r} \tag{10.7}$$

and so the system is considered to be superintegrable. This superintegrability is also argued to be the main reason for the Kepler problem to be separable in different coordinate systems.

The above examples clearly illustrate the fact that integrable systems admit one or more of the following properties:

(1) The equation of motion is exactly solvable in terms of 'elementary' functions, which are analytic, so that the nature of the system may be known at every instant of time as well as in a global sense.

(2) The system possesses sufficient number of integrals of motion[2] so that global properties such as conservation of energy, momentum, etc. may be inferred.

(3) The integrals of motion may be used to reduce the equation of motion to quadratures so that the equation of motion can be integrated exactly.

Exercises:

2. Show that the Euler's equations describing the motion of a torque free rigid body given by

$$I_1 \frac{d\omega_1}{dt} = (I_2 - I_3)\,\omega_2\omega_3 \,,$$
$$I_2 \frac{d\omega_2}{dt} = (I_3 - I_1)\,\omega_3\omega_1 \,,$$
$$I_3 \frac{d\omega_3}{dt} = (I_1 - I_2)\,\omega_1\omega_2$$

can be solved in terms of Jacobian elliptic functions. Find the independent integrals of motion.

3. Solve the equation of motion of the isotropic harmonic oscillator acted upon by the force $\boldsymbol{F} = -k\boldsymbol{r}$, $k > 0$, by separation of variables in spherical polar coordinates. What are the involutive integrals of motion? Investigate whether it is a 'superintegrable' system.

---

[2] For a Hamiltonian system with N degrees of freedom this may be mean N functionally independent involutive integrals of motion in the Poisson bracket [8] sense. Some more details are given later in Chap. 13 (Sect. 13.4.1).

10.3 Complete Integrability – Complex Analytic Integrability

Given a dynamical system, how can one decide whether it is integrable or not in the above sense? In the following sections, we will use certain criteria based on the so-called 'analytic integrability' [4] so that the system admits single-valued, analytic (meromorphic) solutions. As a consequence, one can hope to ensure the existence of globally single-valued analytic integrals of motion and introduce the notion of 'complete integrability' as the requirement of sufficient number of independent integrals of motion as pointed out earlier in Sect. 10.2.

Example 1: Three-Dimensional Lotka–Volterra System [4]

Consider the system

$$\dot{x} = x(Cy + z)\,, \quad \dot{y} = y(x + z)\,, \quad \dot{z} = z(x + y)\,. \tag{10.8}$$

It admits the two independent integrals of motion

$$I = (x - Cy)\left(1 - \frac{z}{y}\right) \tag{10.9}$$

and

$$\Omega = \frac{xz}{y}\left(1 - \frac{y}{z}\right)^{C+1}\,. \tag{10.10}$$

Note that in the case C integer, the integrals are analytic and satisfy the analytic complete integrability property.

Example 2: A System of Two-Coupled Quartic Oscillators

The system

$$\ddot{x} + 4x^3 + 12xy^2 = 0\,, \tag{10.11a}$$

$$\ddot{y} + 32y^3 + 12x^2y = 0 \tag{10.11b}$$

possesses the Hamiltonian

$$H = \frac{1}{2}\left(p_x^2 + p_y^2\right) + x^4 + 6x^2y^2 + 8y^4\,. \tag{10.12}$$

Then the quantity

$$I = p_x^4 + 4x^2\left(x^2 + 6y^2\right)p_x^2 - 16x^3yp_xp_y + 4x^4p_y^2 + 4x^4\left(x^4 + 4x^2y^2 + 4y^4\right) \tag{10.13}$$

is the second functionally independent analytic, involutive, integral of motion, thereby showing the complete integrability of the system (10.11). In fact, the system is also separable and solvable in terms of elliptic functions (see Sect. 10.3.3 below), which are in fact single-valued and analytic (meromorphic) functions, satisfying the requirement of analytic integrability.

Exercises:

4. Verify that I and Ω given by (10.9) and (10.10) are integrals of (10.8).
5. Show that the second integral (10.13) is involutive with respect to the Hamiltonian (10.12).

10.3.1 Real Time and Complex Time Behaviours

One can even go one step further and argue that locally every differential equation is integrable in a trivial sense. Consider the system (3.1) in the form

$$\dot{x}_i(t) = f_i\left(t, x_1, ..., x_n\right) , \quad i = 1, 2, ..., n \tag{10.14}$$

subject to the initial condition $x_i(t_0) = c_i$. One can take locally the initial conditions as the constants of motion of the system. Then the general solution (valid locally) to (10.14) can be written as

$$x_i = F_i\left(t, c_1, ..., c_n\right), \quad i = 1, 2, ..., n. \tag{10.15}$$

Inverting (10.15), which is equivalent in principle to integrating backwards in time from t to t_0, we can write the n constants as

$$c_i = I_i\left(t, x_1, x_2, ..., x_n\right) , \quad i = 1, 2, ..., n. \tag{10.16}$$

But the crux of the problem is that this inversion may not be unique or single-valued in general, leading to multivaluedness of the solution in the complex time plane. This is indeed a source of nonintegrability.

One may raise the philosophical question as to why should one worry about the behaviour of the solutions in the complex time, while physical time is real? The answer is not obvious and straightforward; however, recent investigations make one thing clear: the real time behaviour of the solution of a dynamical system is heavily influenced by its behaviour in the complex time plane as pointed out in the next section on Painlevé singularity structure analysis. To illustrate possible differences between real time and complex time behaviours, Martin Kruskal (see for example, Ref. [4]) considers the following ordinary differential equation (ode) in the complex plane:

$$\frac{\mathrm{d}x}{\mathrm{d}t} = \frac{\alpha}{t-a} + \frac{\beta}{t-b} + \frac{\gamma}{t-c} , \tag{10.17}$$

where α, β, γ, a, b and c are constants. On integrating, we obtain

$$I = x - \alpha \log(t-a) - \beta \log(t-b) - \gamma \log(t-c) , \tag{10.18}$$

where I is the integration constant. If one considers complex time, then since the logarithm of a complex number is defined only upto an integral multiple of $2\mathrm{i}\pi$, the value of I from (10.18) is determined only upto an additive term $2\mathrm{i}\pi(m\alpha + n\beta + s\gamma)$, where m, n, s are integers. When any one or two of the quantities α, β and γ are zero, one can define I or x in a unique way within a two or one-dimensional lattice, respectively. However, if all of them are nonzero and linearly independent over the integers, there is a strong indeterminacy in I and so x or vice versa. Then one should consider the system as not compatible with integrability.

10.3.2 Partial Integrability and Constrained Integrability

When the dynamical system admits integrals which are less in number than the sufficient ones, we may say that the system is partially integrable. In this sense all Hamiltonian systems which are not integrable, that is those systems which do not possess sufficient number of involutive integrals of motion, fall within this class as at least the energy is conserved. A typical example is the classical Hamiltonian of hydrogen atom in external uniform magnetic field along the z-direction,

$$H = \frac{p^2}{2m} - \frac{e^2}{r} + \alpha \left(x^2 + y^2\right) , \tag{10.19}$$

where α is a constant. Here $E(= H)$ and L_z are the only constants of motion; however one requires at least three involutive integrals of motion for complete integrability. So it is partially integrable, but not fully integrable. In fact, it is a chaotic system (see for example, Chap. 7, problem 3).

Similarly, consider the Lorenz system, (3.45). For $b = 2\sigma$, it has one time-dependent integral

$$x^2 - 2\sigma z = C\mathrm{e}^{-2\sigma t} , \tag{10.20}$$

which again leads to partial integrability.

Finally, one can also identify another type of incomplete integrability, called *constrained integrability*. For example, if a Hamiltonian system is integrable only for a fixed value of energy, say $H = E = 0$, it corresponds to constrained integrability. For more details, see for example [4].

10.3.3 Integrability and Separability

As we have seen above, one of the most conclusive proofs of integrability is the explicit construction of a sufficient number of independent integrals of motion. Then making use of these integrals one can reduce the problem to quadratures and obtain explicit solutions. For Hamiltonian systems, this can be done in principle by constructing the angle variables, while treating the N involutive integrals as 'action' variables. However, in many cases, the problem turns out to be *s*eparable, either for the equation of motion or for the underlying Hamilton–Jacobi equation, sometimes through nontrivial transformations or variables. As examples, one can cite the central potential problems (including the Kepler problem) which are separable in terms of spherical polar coordinates. We illustrate the separability aspects of some other nontrivial examples.

Example 1: Coupled Quartic Oscillators

The equations of motion

$$\ddot{x} + 2\left(A + 2\alpha x^2 + \delta y^2\right) x = 0\ , \tag{10.21a}$$

$$\ddot{y} + 2\left(B + 2\beta y^2 + \delta x^2\right) y = 0 \tag{10.21b}$$

corresponds to the Hamiltonian

$$H = \frac{p_x^2}{2} + \frac{p_y^2}{2} + Ax^2 + By^2 + \alpha x^4 + \beta y^4 + \delta x^2 y^2\ . \tag{10.22}$$

There are four specific choices of parameters [7] in (10.21) for which second integrals of motion exist so that the system becomes integrable in each of these cases. These four cases along with their second integrals of motion are given below.

Case (i): $A = B$, $\alpha = \beta$, $\delta = 6\alpha$

$$I = p_x p_y + 2\left[A + 2\alpha\left(x^2 + y^2\right)\right] xy\ . \tag{10.23}$$

Case (ii): $\alpha = \beta$, $\delta = 2\alpha$

$$I = (xp_y - yp_x)^2 + \frac{2}{\alpha}(B - A)\left[\frac{1}{2}p_x^2 + Ax^2 + \alpha\left(x^2 + y^2\right)x^2\right]\ . \tag{10.24}$$

Case (iii): $A = 4B$, $\alpha = 16\beta$, $\delta = 12\beta$

$$I = (yp_x - xp_y)\, p_y + 2\left(B + 4\beta x^2 + 2\beta y^2\right) xy^2\ . \tag{10.25}$$

Case (iv): $A = 4B$, $\alpha = 8\beta$, $\delta = 6\beta$

$$\begin{aligned} I = {} & p_y^4 + 4y^2\left(B + 6\beta x^2 + \beta y^2\right) p_y^2 - 16\beta x y^3 p_x p_y + 4\beta y^4 p_x^2 \\ & + 4B^2 y^4 + 4\beta\left[2B + \beta\left(2x^2 + y^2\right)\right]\left(2x^2 + y^2\right) y^4\ . \end{aligned} \tag{10.26}$$

It has been found that the first three cases are also separable under coordinate transformations, in addition to being integrable. The fourth case is also separable but after a complicated transformation involving momenta also [4]. Brief details are given below.

Case (i) :

Under the linear transformation

$$u = x + y\,, \quad v = x - y \tag{10.27}$$

the equation of motion (10.21) for $A = B$, $\alpha = \beta$ and $\delta = 6\alpha$ becomes separable as

$$\ddot{u} + 2Au + 4\alpha u^3 = 0\,, \tag{10.28a}$$

$$\ddot{v} + 2Av + 4\alpha v^3 = 0\,. \tag{10.28b}$$

Each of the above equations correspond to the independent cubic anharmonic oscillators solvable in terms of Jacobian elliptic functions (see appendix A for details) as shown in Sect. 2.2.1.

Case (ii) :

For the case $\alpha = \beta$, $\delta = 2\alpha$, when $A = B$, if we transform the cartesian coordinates x and y to polar coordinates $x = r\cos\theta$, $y = r\sin\theta$, the Hamiltonian (10.22) is independent of θ, $H = \left(p_r^2 + 2Ar^2 + 2\alpha r^4 + I_2^2/r^2\right)/2$, where the angular momentum $I_2 = (xp_y - yp_x)$ is an integral of motion. The resultant equation of motion can be again solved in terms of elliptic functions.

However, when $A \neq B$, the system is not separable in polar coordinates. In terms of the elliptic coordinates

$$x = \frac{\epsilon\eta}{c}\,, \quad y = \frac{1}{c}\left[\left(\epsilon^2 - c^2\right)\left(c^2 - \eta^2\right)\right]^{1/2}\,, \quad c^2\alpha = B - A \tag{10.29}$$

the Hamiltonian becomes

$$H = \frac{1}{2}\frac{1}{\epsilon^2 - \eta^2}\left[\frac{1}{\epsilon^2 - c^2}p_\epsilon^2 + \frac{1}{c^2 - \eta^2}p_\eta^2 + 2\alpha\left(\epsilon^6 - \eta^6\right)\right.$$
$$\left. + 2(2A + B)\left(\epsilon^4 - \eta^4\right) - 2c^2A\left(\epsilon^2 - \eta^2\right)\right]\,. \tag{10.30}$$

The corresponding Hamilton–Jacobi equation becomes

$$\frac{1}{2}\,\frac{1}{\epsilon^2 - \eta^2}\left[\frac{1}{\epsilon^2 - c^2}S_\epsilon^2 + \frac{1}{c^2 - \eta^2}S_\eta^2 + 2\alpha\left(\epsilon^6 - \eta^6\right) + 2(2A + B)\left(\epsilon^4 - \eta^4\right)\right.$$
$$\left. - 2c^2A\left(\epsilon^2 - \eta^2\right)\right] = E\,, \quad c^2\alpha = B - A\,, \tag{10.31}$$

where E is the energy and subscript denotes partial derivative. Obviously (10.31) is separable.

Case (iii) :

Similarly, for the parametric restrictions $A = 4B$, $\alpha = 16\beta$, $\delta = 12\beta$, if we transform the cartesian coordinates to parabolic coordinates, $x = (\epsilon^2 - \eta^2)/2$, $y = \epsilon\eta$, the resulting Hamiltonian becomes

$$H = \frac{1}{\epsilon^2 + \eta^2} \left[\frac{1}{2} \left(p_\epsilon^2 + p_\eta^2 \right) + B \left(\epsilon^6 + \eta^6 \right) + \beta \left(\epsilon^{10} + \eta^{10} \right) \right] , \tag{10.32}$$

and hence the associated Hamilton–Jacobi equation is separable.

Case (iv) :

Finally, we note that for the parametric choice $\alpha = 8\beta$, $\delta = 6\beta$ and $A = 4B$, the equation of motion becomes separable under a more general transformation involving both the coordinates and momenta (see for example, Ref. [4])

$$u = \frac{p_y^2 + c}{x^2} + 2y^2 + 4x^2 , \tag{10.33a}$$

$$v = \frac{p_y^2 - c}{y^2} + 2y^2 + 4x^2 , \tag{10.33b}$$

where c is the square-root of the second constant of motion. Then the equation of motion separates and one can reduce it to quadratures. For example, for $A = 0$,

$$\dot{u}^2 = 2u^3 - 8u(2h + c) , \quad (h : \text{total} \quad \text{energy}) \tag{10.34a}$$

$$\dot{v}^2 = 2v^3 - 8v(2h - c) . \tag{10.34b}$$

Naturally (10.34) can be integrated in terms of Jacobian elliptic functions. Knowing u and v, the variables x and y may be also expressed explicitly in terms of elliptic functions. Equations (10.34) can be generalized to the case $A \neq 0$ also.

10.4 How to Detect Integrability: Painlevé Analysis

Given the equation of motion of a dynamical system, how can one determine whether it is integrable or not? As noted earlier, one way to look at this problem is to analyse the singularity structure (meromorphicity property) of the solution in the complex time plane, a property commonly referred to as *Painlevé property* [4–7,9]. In order to develop this notion, we have to first introduce the notion of *fixed and movable singularities* admitted by ordinary differential equations. A brief idea of them is given in the following. (For more details, see for example Ref. [7]).

10.4.1 Classification of Singular Points

In the analysis of ordinary differential equations singular points play a crucial role. So it is very pertinent to look at their basic properties.

A. Fixed and Movable Singularities

Let us consider odes in the complex plane of the independent variable. The general solution of odes may cease to be analytic at certain points, called *singularities*. Poles, branch points (both algebraic and logarithmic) and essential singularities are all such points. Since the solutions are functions of the constants of integration, these singular points may depend on the constants also (ultimately on the initial conditions of the problem). In turn, these singular points may be placed anywhere in the complex plane and in such a case they may be called *movable singular points*. On the other hand, if the singularities do not depend upon the integration constants, they are *fixed singular points*. Among the singular points, the branch points and essential singularities are usually referred to as *critical points*.

B. Linear Ordinary Differential Equations and Fixed Singularities

Consider an nth order linear ode,

$$\frac{\mathrm{d}^n w}{\mathrm{d}z^n} + p_1(z)\frac{\mathrm{d}^{n-1} w}{\mathrm{d}z^{n-1}} + \ \ldots\ + p_n(z)w = 0\,, \tag{10.35}$$

where the coefficient functions p_i, $i = 1, 2, ..., n$, are analytic at the ordinary point $z_0 \in C_1$ and also that the functions $(z - z_1)^i p_i(z)$ are analytic at the regular singular point z_1. Obviously, (10.35) admits n linearly independent analytic solutions w_i, $i = 1, 2, ..., n$, in the neighbourhood of z_0 and z_1 and therefore the general solution is given by

$$w(z) = \sum_{i=1}^{n} c_i w_i(z)\,, \tag{10.36}$$

where c_i's are integration constants. The singularities of the solution of (10.35) must be located at the singularities of the coefficients, p_i, $i = 1, 2, ..., n$, which are all fixed. Thus, one may conclude that the locations of the singularities admitted by linear odes are fixed and independent of integration constants.

C. Nonlinear Ordinary Differential Equations and Movable Singularities

In contrast to the linear case, the solution of nonlinear odes can admit movable singularities as well as fixed singular points [9]. We may illustrate this nature with the following examples:

$$\frac{\mathrm{d}w}{\mathrm{d}z} + w^2 = 0 , \tag{10.37a}$$

$$\frac{\mathrm{d}w}{\mathrm{d}z} - \mathrm{e}^w = 0 , \tag{10.37b}$$

$$\frac{\mathrm{d}w}{\mathrm{d}z} + w(\log w)^2 = 0 , \tag{10.37c}$$

$$z\frac{\mathrm{d}^2w}{\mathrm{d}z^2} - w^2 = 0 . \tag{10.37d}$$

The general solution of (10.37a) is $w(z) = (z - z_0)^{-1}$, where z_0 is the integration constant, which also locates the singularity, namely a movable pole. In the case of (10.37b) the general solution takes the form $w(z) = \log(z - z_0)$ and hence admits a movable logarithmic branch point. It can be straightforwardly checked that the general solution of (10.37c) is $w(z) = \exp(z - z_0)^{-1}$ and therefore it has a movable essential singularity. On the other hand, (10.37a) has the solution $w(z) = 2/z$ with a local Laurent series expansion $(2/z)\times [z_0^2(z - z_0)^{-2} + (24/5)z_0(z - z_0)^{-1} + (39/25) + ...]$, where z_0 is the integration constant, and so $z = 0$ is a fixed singularity independent of the integration constant z_0.

As noted earlier, if the critical points are independent of the initial conditions, they are fixed. Then the fundamental existence theorem guarantees that the general solution of an nth-order ode is completely and uniquely specified by the knowledge of the values which $w(z)$ and its $n - 1$ derivatives assume at the noncritical point z_0. On the other hand, if a movable critical point is admitted, then there will be considerable difficulties in the problem of analytic continuation of solutions; the latter requires a specific and unique path, but the movable critical points prevent one from choosing such paths due to the multiple number of choices, and so the uniqueness theorem may not hold. This in turn may lead to the nonintegrable nature of the corresponding odes. (Refer to (10.17) in this connection).

Possibly, motivated by the above considerations and by the development of the theory of complex variables, a problem of intense interest arose in the theory of differential equations in the second half of nineteenth century. It was concerned with the classification of odes according to the singularities, movable critical or not, admitted by the solutions.

10.4.2 Historical Development of the Painlevé Approach and Integrability of Ordinary Differential Equations

A. First-Order Nonlinear Ordinary Differential Equations and Kovalevskaya's Rigid-Body Problem

By considering a first-order ode of the form

$$\frac{\mathrm{d}w}{\mathrm{d}z} = F(w, z) , \tag{10.38}$$

where F is rational in w and analytic in z, Fuchs (in the year 1884) concluded that out of all the forms of F in (10.38), the only equation which is free from movable critical points is the generalized Riccati equation,

$$\frac{\mathrm{d}w}{\mathrm{d}z} = p_0(z) + p_1(z)w + p_2(z)w^2 , \qquad (10.39)$$

where the p_i, $i = 0, 1, 2$, are analytic in z. The fact that (10.39) is free from movable critical points also leads to the conclusion that if w_1, w_2 and w_3 are three particular solutions, the general solution may be obtained by a nonlinear superposition principle,

$$(w - w_1) / (w - w_2) = A\,(w_3 - w_1) / (w_3 - w_2) , \qquad (10.40)$$

where A is a constant.

Possibly, familiar with the above results and the then current Jacobi's work on elliptic functions, Sophya Kovalevskaya (in the year 1889) was led to analyse the problem of heavy rigid body under the influence of gravity in terms of the singularity structure exhibited by the general solution. The equation of motion is of the form

$$\frac{\mathrm{d}\boldsymbol{I}}{\mathrm{d}t} = \boldsymbol{I} \wedge \boldsymbol{\Omega} + mg\boldsymbol{r}_0 \wedge \boldsymbol{e} , \quad \frac{\mathrm{d}\boldsymbol{e}}{\mathrm{d}t} = \boldsymbol{e} \wedge \boldsymbol{\Omega} , \qquad (10.41)$$

where the angular velocity $\boldsymbol{\Omega}$ and the angular momentum $\boldsymbol{I}$ are given by

$$\boldsymbol{\Omega} = \sum_{i=1}^{3} \Omega_i \boldsymbol{e}_i , \quad \boldsymbol{I} = A\Omega_1\boldsymbol{e}_1 + B\Omega_2\boldsymbol{e}_2 + C\Omega_3\boldsymbol{e}_3 , \qquad (10.42)$$

with respect to a moving trihedral of unit vectors $\boldsymbol{e}_i$, $i = 1, 2, 3$, fixed on the body. Then the vertical unit vector $\boldsymbol{e}$ and the center of mass $\boldsymbol{r}_0$ are given by

$$\boldsymbol{e} = \alpha\boldsymbol{e}_1 + \beta\boldsymbol{e}_2 + \gamma\boldsymbol{e}_3 , \quad \boldsymbol{r}_0 = x_0\boldsymbol{e}_1 + y_0\boldsymbol{e}_2 + z_0\boldsymbol{e}_3 , \qquad (10.43)$$

where α, β and γ refer to the direction cosines which define the orientations of the heavy rigid body.

Equations (10.41) can be rewritten in component form as

$$A\frac{\mathrm{d}\Omega_1}{\mathrm{d}t} = (B - C)\Omega_2\Omega_3 - \beta z_0 + \gamma y_0 , \quad \frac{\mathrm{d}\alpha}{\mathrm{d}t} = \beta\Omega_3 - \gamma\Omega_2 , \qquad (10.44a)$$

$$B\frac{\mathrm{d}\Omega_2}{\mathrm{d}t} = (C - A)\Omega_3\Omega_1 - \gamma x_0 + \alpha z_0 , \quad \frac{\mathrm{d}\beta}{\mathrm{d}t} = \gamma\Omega_1 - \alpha\Omega_3 , \qquad (10.44b)$$

$$C\frac{\mathrm{d}\Omega_3}{\mathrm{d}t} = (A - B)\Omega_1\Omega_2 - \alpha y_0 + \beta x_0 , \quad \frac{\mathrm{d}\gamma}{\mathrm{d}t} = \alpha\Omega_2 - \beta\Omega_1 . \qquad (10.44c)$$

Before Kovalevskaya's investigations two specific cases, due to Euler (1750) and Lagrange (1788), respectively,

$$x_0 = y_0 = z_0 = 0; \quad x_0 = y_0 = 0 \;\; (z_0 > 0, A = B) \qquad (10.45)$$

(besides the straightforward case $A = B = C$), were known to be integrable and the associated solutions were given in terms of meromorphic (Jacobian elliptic) functions. The connection between integrability and meromorphism in the above integrable cases had presumably prompted Kovalevskaya to find all the parametric choices of (10.44) for which the general solution is meromorphic. Kovalevskaya approached the problem by demanding that the solutions of (10.44) be single-valued, meromorphic functions in order that the system might be integrable, and by expressing each of the six functions as Laurent series in the neighbourhood of a movable singular point.

After a detailed investigation, Kovalevskaya concluded that apart from the two existing nontrivial possibilities, found already by Euler and Lagrange, there is one more parametric choice with

$$y_0 = z_0 = 0 \,, \quad A = B = 2C \,, \tag{10.46}$$

for which the general solution admits the required number of arbitrary constants and is meromorphic. Kovalevskaya went on to prove the integrability for the above parametric choice by constructing a sufficient number (four) of integrals of motion as

$$I_1 = C\left[2\left(\Omega_1^2 + \Omega_2^2\right) + \Omega_3^2\right] - (\alpha/C)x_0 \,, \tag{10.47a}$$

$$I_2 = 2\left(\alpha\Omega_1 + \beta\Omega_2\right) + \gamma\Omega_3 \,, \tag{10.47b}$$

$$I_3 = \alpha^2 + \beta^2 + \gamma^2 \,, \tag{10.47c}$$

$$I_4 = \left[\Omega_1^2 - \Omega_2^2 + (\alpha/C)x_0\right]^2 + \left[2\Omega_1\Omega_2 + (\beta/C)x_0\right]^2 \,, \tag{10.47d}$$

with respect to the Poisson brackets

$$\{\Omega_i, \Omega_j\} = -\epsilon_{ijk}\Omega_k \,, \quad \{\Omega_i, e_j\} = -\epsilon_{ijk}e_k \,, \quad \{e_i, e_j\} = 0 \,, \tag{10.48}$$

where ϵ_{ijk}, $i, j, k = 1, 2, 3$, is the usual Levi–Civita tensor. For more details, see for example Refs. [4,7,9].

In fact, Kovalevskaya went on to integrate (10.44) explicitly for the choice (10.46) and obtained the general solution in terms of hyperelliptic functions. Thus, Kovalevskaya's discovery of a new integrable case for (10.44) provided strong evidence for the close connection between meromorphic solutions and the integrability of nonlinear dynamical systems.

B. Second-Order Nonlinear Ordinary Differential Equations and Painlevé Transcendental Equations

The next important development was at the turn of the twentieth century due to the efforts of the French mathematician-politician Paul Painlevé and his coworkers. By considering the second-order odes having the form

$$\frac{\mathrm{d}^2 w}{\mathrm{d}z^2} = F(z, w, \mathrm{d}w/\mathrm{d}z) \,, \tag{10.49}$$

where F is rational in w, algebraic in $\mathrm{d}w/\mathrm{d}z$, locally analytic in z, it was found that there were only fifty canonical equations possessing the property of having no movable critical points. Among them, forty four can be integrated in terms of elementary functions, including the elliptic functions. The remaining six, which are now referred to as Painlevé transcendental equations, needed special treatment and introduction of new transcendental functions. Their explicit forms are as follows:

$$P_I \quad : \quad w'' = 6w^2 + z \,, \tag{10.50}$$

$$P_{II} \quad : \quad w'' = 2w^3 + zw + \alpha \,, \tag{10.51}$$

$$P_{III} \quad : \quad w'' = \frac{1}{w}w'^2 - \frac{1}{z}\left(w' - \alpha w^2 - \beta\right) + \gamma w^3 + \frac{\delta}{w} \,, \tag{10.52}$$

$$P_{IV} \quad : \quad w'' = \frac{1}{2w}w'^2 + \frac{3}{2}w^3 + 4zw^2 + 2\left(z^2 - \alpha\right)w + \frac{\beta}{w} \,, \tag{10.53}$$

$$\begin{aligned} P_V \quad : \quad w'' = {} & \left(\frac{1}{2w} + \frac{1}{w-1}\right)w'^2 - \frac{1}{z}w' + \frac{(w-1)^2}{z^2}\left(\alpha z + \frac{\beta}{w}\right) \\ & + \gamma\frac{w}{z} + \delta w\frac{w+1}{w-1} \,, \end{aligned} \tag{10.54}$$

$$\begin{aligned} P_{VI} \quad : \quad w'' = {} & \frac{1}{2}\left(\frac{1}{w} + \frac{1}{w-1} + \frac{1}{w-z}\right)w'^2 \\ & - \left(\frac{1}{z} + \frac{1}{z-1} + \frac{1}{w-z}\right)w' \\ & + w\frac{(w-1)(w-z)}{z^2(z-1)^2}\left(\alpha + \beta\frac{z}{w^2}\right. \\ & \left. + \gamma\frac{z-1}{(w-1)^2} + \delta z\frac{z-1}{(w-z)^2}\right) \,, \end{aligned} \tag{10.55}$$

where prime denotes differentiation with respect to z and α, β, γ and δ are constants.

In some sense (10.55) for P_{VI} may be considered as the most general Painlevé transcendental equation which by repeated limiting and coalescence procedures can be related to the other Painlevé equations. The treatise of Ince [9] contains greater details of Painlevé transcendental equations.

Finally, attempts have also been made by Bureau and Chazy to classify third- and higher-order odes in the same way as was done for the second-order odes at the turn of the previous century, though the analysis remains rather incomplete.

C. Painlevé (P) Property of Ordinary Differential Equations: Strong and Weak P-Properties

The classical work of Kovalevskaya, Painlevé and his team and others did not draw much attention for several years, probably due to the advent of quantum mechanics in the late 1920's and the ensued interest in linear odes.

However, the situation has changed dramatically after the discovery of solitons in the Korteweg–de Vries equation by Zabusky and Kruskal in the mid 1960's. (See Chaps. 11–14 for further details). It was shown that the soliton possessing evolution equations are completely integrable through the so-called inverse scattering transform (IST) method (discussed in some details later in Chaps. 13 and 14). Furthermore, it was realized that a deep connection exists between the soliton possessing nonlinear evolution equations solvable by the IST method and odes of P-type (possessing the P-property). Ablowitz, Ramani and Segur (ARS) (see Ref. [6]) conjectured that every ode obtained by an exact reduction of a soliton possessing partial differential equation (pde) solvable by IST is of P-type. This has been verified for a large class of nonlinear pdes by Lakshmanan and Kaliappan [10] using the underlying Lie point symmetries. The term *strong* P-property is now used in the ARS sense meaning that the solution of an ode (strong P-type) in the neighbourhood of a singularity z_0 can be expressed as a Laurent expansion with the leading term proportional to

$$\tau = (z - z_0)^{-p}, \quad z \to z_0 \tag{10.56}$$

where p is a positive integer.

The fact that odes under suitable transformations can exhibit P-property, for example P_{VI}, has also given rise to the definition of weak P-property. Ramani, Grammaticos and Dorizzi have suggested that the existing P-property (ARS) can be generalised to include the so-called *weak* P-property [4,6] by which it is meant that the general solution in the neighbourhood of a movable singularity z_0 can be expressed as an expansion in powers of

$$\tau = (z - z_0)^{-1/n}, \tag{10.57}$$

where n must be "natural" and depends purely on the dominant behaviour of the singularity as well as the nature of the nonlinearity. Indeed, a number of integrable Hamiltonian systems having weak P-property have been identified during the past few years. We will give some more examples for the strong and weak P-properties in the following.

10.4.3 Painlevé Method of Singular Point Analysis for Ordinary Differential Equations

For an ode to be of P-type, it is necessary that its solution possesses no movable branch points, either algebraic or logarithmic. We do not consider in our following analysis the presence of essential singularities, whose treatment appears to be much more complicated and the theory is probably not complete to locate them. In the following, we describe the ARS algorithm, which provides a systematic way to investigate the presence of movable critical points of branch-point type, and to determine whether the given ode is of P-type or not.

To be specific, we consider an nth order ode,

$$\frac{\mathrm{d}^n w}{\mathrm{d}z^n} = F\left(z; w, \mathrm{d}w/\mathrm{d}z, ..., \mathrm{d}^{n-1}w/\mathrm{d}z^{n-1}\right) , \tag{10.58}$$

or equivalently n first-order equations

$$\frac{\mathrm{d}w_i}{\mathrm{d}z} = F_i\left(z; w_1, w_2, ..., w_n\right) , \quad i = 1, 2, ..., n. \tag{10.59}$$

Here F and the F_i's are analytic in z and rational in their other arguments. We expand the solution of (10.58) or (10.59) as a Laurent series in the neighbourhood of a movable singular point z_0 (see (10.61) below). Then the ARS algorithm essentially consists of the following three steps.

(1) Determination of leading-order behaviour of the Laurent series in the neighbourhood of the movable singular point z_0;
(2) determination of 'resonances', that is, the powers at which arbitrary constants of the solution of (10.58) (or (10.59)) can enter into the Laurent series expansion (the term 'resonance' used here should not be confused with the one we used earlier in the study of oscillations); and
(3) verification that a sufficient number of arbitrary constants exist without the introduction of movable critical points.

At the end of the above three steps one will be in a position to check the necessary conditions for the existence of a P-type solution and integrability of (10.58). Whether these conditions are *sufficient* is a further intricate problem: One has to show that the associated Laurent series does converge so that the general solution defines a well defined function. In typical dynamical problems this is assumed to be so, as the actual analysis becomes too complicated, or the existence is proved by other means. We now give briefly the details of the above three steps.

A. Leading-Order Behaviour

Considering the nth order ode (10.58), we assume that the dominant behaviour of $w(z)$ in a sufficiently small neighbourhood of the arbitrary movable singularity z_0 is algebraic, that is,

$$w(z) \approx a_j \left(z - z_0\right)^{q_j} \quad \text{as} \quad z \to z_0 , \tag{10.60}$$

where (a_j, q_j) are constants to be determined from (10.58) and $\mathrm{Re}q_j < 0$. Each allowed pair (a_j, q_j) is said to belong to the jth branch or possibility of the solution. Note that, in principle, one can also allow for the possibility of logarithmic leading-order terms and positive powers corresponding to Taylor series, but for simplicity we do not consider them here. Using (10.60) into (10.58), one can see that for certain values of q_j two or more terms may balance each other, while the remaining terms can be ignored as $z \to z_0$. The balancing terms are referred to as leading-order terms. Knowing the q_j's, the a_j's can be easily evaluated. From the leading-order behaviour, we may observe the following.

(1) If at least one of the q_j is an irrational or complex number, then we can conclude, at this stage, that the solution is severely multivalued and so (10.58) is non-P-type leading to nonintegrability.

(2) If all the q_j's are negative integers, then (10.60) may represent the first term of a Laurent series, for each q_j valid in a deleted neighbourhood of z_0. This can then be an indication of the strong P-property and further analysis is necessary to confirm this.

(3) If any of the q_j is not an integer, but a rational number, then from the dominant behaviour (10.60) of $w(z)$ near z_0, we find that the solution will have a movable algebraic branch point. This may then possibly be associated with weak P-property.

In either of the latter two cases, the solution $w(z)$ may be shown to take the form

$$w(z) = (z - z_0)^{q_j} \sum_{m=0}^{\infty} a_{j,m} (z - z_0)^m , \quad 0 < |z - z_0| < R . \tag{10.61}$$

Let us illustrate the leading-order behaviour terminology with some typical examples.

Example 1:

$$w''' + ww'' - 2w^3 + w^2 + \mu w = 0 , \tag{10.62}$$

where the prime means differentiation and μ is a parameter. Assuming the dominant behaviour to be $w(z) \approx a_0 \tau^q$, $\tau = (z - z_0) \to 0$, with z_0 an arbitrary movable singularity, (10.62) becomes

$$a_0 q(q-1)(q-2)\tau^{q-3} + a_0^2 q(q-1)\tau^{2q-2} - 2a_0^3 \tau^{3q} = 0 . \tag{10.63}$$

Then there are two possible choices:

(1) $q = -1$, $a_0 = 3$, the leading-order terms are w''' and ww''.

(2) $q = -2$, $a_0 = 3$, the leading-order terms are ww'' and $-2w^3$.

Example 2:

$$w'' + \sin w = 0 . \tag{10.64}$$

Since the nonlinearity of (10.64) is nonalgebraic, it is convenient to transform (10.64) into a different form

$$vv'' - v'^2 + \frac{1}{2}\left(v^3 - v\right) = 0 , \quad v = \mathrm{e}^{\mathrm{i}w} . \tag{10.65}$$

From the leading-order analysis, that is $v(z) \approx a_0 \tau^q$, $\tau = (z - z_0) \to 0$, we find $q = -2$, $a_0 = -4$.

B. Resonances

In the Laurent series solution (10.61), z_0 is an arbitrary constant which is also the position of the singularity. In addition, if $n-1$ of the coefficients a_j in (10.61) are also arbitrary, there are totally n constants of integration of the ode (10.58) and so (10.61) is the general solution in the deleted neighbourhood of z_0. The powers at which these arbitrary constants enter inside the summation in (10.61) are called *resonances* of the Laurent series.

In order to find the resonance values, for each leading-order pair (q_j, a_j) or branch, one constructs a simplified equation that retains only the leading terms of (10.58). Substituting (hereafter we omit suffix j in a and q for simplicity)

$$w(z) = a\tau^q + \beta\tau^{q+r}, \quad \tau = (z - z_0) \to 0 \tag{10.66}$$

into the simplified equation, to leading-order in β one can obtain

$$Q(r)\beta\tau^{\widehat{q}} = 0\,, \quad \widehat{q} \geq q + r - n\,, \tag{10.67}$$

where n is the order of the ode (10.58). Here if the highest derivative of (10.58) is a leading-order term, then $Q(r)$ is a polynomial of order n. If not, $\widehat{q} > q + r - n$. Then the roots of $Q(r)$, given by

$$Q(r) = 0\,, \tag{10.68}$$

determine the resonance values. We may note the following characteristics at this stage.

(1) One root of (10.68) is always -1, representing the arbitrariness of z_0. This may be inferred in the following way: by perturbing the complex variable $\tau = z - z_0$ in (10.60) by $\tau + \epsilon$, $\epsilon \ll 1$, and expanding it in terms of ϵ, one can find that the first term appears at $\tau^{q_i - 1}$.

(2) Suppose that the constant a_j in (10.60) is arbitrary, then one of the roots is at $r = 0$.

(3) Roots with Re $r < 0$ can be ignored because they violate the leading-order hypothesis. The occurrence of such resonances indicates that the associated Laurent series-expansion solution is a singular one. However, such a case need to be analysed with care.

(4) Any root with Re $r > 0$, but not a real integer, indicates a movable branch point at $z = z_0$, in general. The associated solutions, in general, are of non-P type.

(5) Any root with Re r which is a positive rational number $r = p/q$, with q as in the denominator of dominant behaviour, indicates in general a movable algebraic branch point and this may then be associated with the weak P-property.

(6) If at least one of the roots of (10.68) is an irrational or complex number, then we can conclude that the given ode (10.58) is of non-P-type.

(7) If for every possible set of values (q, a) in (10.60), excluding -1 and possibly 0, all the remaining roots are positive real integers, then there are no algebraic branch points.

For the Laurent series expansion (10.61) to be the general solution of the ode (10.58), $Q(r)$ must have $n-1$ non-negative distinct roots of real rational numbers including integers. If for every allowed (q, a), $Q(r)$ has fewer than $n-1$ such roots, then none of the local solutions is general.

Let us illustrate all these, again with reference to the two examples cited in the previous subsection.

Example 1:

Substituting (10.66) into (10.62), we obtain for the branch $q = -1$, $a_0 = 3$,

$$\big[(q+r)(q+r-1)(q+r-2)\tau^{q+r-3} \\ +a_0(q+r)(q+r-1)\tau^{2q+r-2} + a_0 q(q-1)\tau^{2q+r-2}\big]\,\beta = 0\,, \qquad (10.69)$$

which gives

$$(r+1)\left(r^2 - 4r + 6\right) = 0 \quad \text{or} \quad r = -1, 2 \pm \mathrm{i}\sqrt{2}\,. \qquad (10.70)$$

On the other hand for the second branch $q = -2$, $a_0 = 3$, we find

$$r = -1, 6\,. \qquad (10.71)$$

Thus the first branch leads to complex resonances, thereby showing the presence of movable critical points. So (10.62) is not of P-type.

Example 2:

For (10.65), we obtain the resonances at

$$r = -1, 2\,. \qquad (10.72)$$

C. Evaluation of Arbitrary Constants

The final step in the ARS algorithm consists of verifying whether or not in the Laurent series expansion (10.61) a sufficient number of arbitrary constants exist (without the introduction of movable critical points) at the resonance values. For a given leading-order set (q, a_0) or branch, let $r_1 \le r_2 \le \cdots \le r_s$ denote the positive rational roots of $Q(r) = 0$, $s \le n-1$. So we substitute

$$w(z) = a_0\tau^q + \sum_{k=1}^{\infty} a_k \tau^{q+k} \qquad (10.73)$$

into the full ode (10.58). Requiring now that the coefficients of τ^{q+k-n} shall vanish, we have

$$Q(k)a_k - R_k\left(z_0; a_0, a_1, ..., a_{k-1}\right) = 0 \;. \tag{10.74}$$

For $k \le r_1$, (10.74) determines the coefficients a_k. At $k = r_1$, $Q(r_1) = 0$. Then, there are two possibilities.

(1) If $R_{r_1} = 0$, a_{r_1} is an arbitrary constant and we can proceed to find the next coefficient. The procedure can be continued until the Laurent series solution possesses n arbitrary constants so that the ode (10.58) is of P-type.
(2) If $R_{r_1} \neq 0$, then (10.74) cannot be satisfied identically and a_{r_1} is not an arbitrary constant and there is no solution of the form (10.61).

In order to capture the arbitrariness of a_{r_1}, we introduce logarithmic terms as follows:

$$w(z) = a_0\tau^q + \sum_{k=1}^{r_1-1} a_k\tau^{q+k} + \left(a_{r_1} + b_{r_1}\log\tau\right)\tau^{q+r_1} + \;... \tag{10.75}$$

Now the coefficient of $\tau^{q+r_1-n}\log\tau$ is

$$Q(r_1)b_{r_1} = 0 \;, \tag{10.76}$$

but b_{r_1} is determined by demanding that the coefficient of τ^{q+r_1-1} vanishes; a_{r_1} is arbitrary. Suppose the resulting coefficient a_{r_1} is not arbitrary, then again we have to introduce more singular terms like $\log(\log\tau)$ in (10.61) and repeat the procedure until the coefficient a_{r_1} becomes arbitrary. In this case, (10.61) signals the presence of a movable logarithmic branch point which shows that the ode (10.58) is of non-P-type.

Having analysed the nature of the Laurent series (10.61), it is necessary now to verify whether the series converges and if so what its radius of convergence is. However, due to the highly nonlinear nature of the recurrence relations between the various coefficients involved, the analysis is too difficult to carry out in general. Usually, instead of proving the convergence of the Laurent series, one takes an alternate route by establishing the existence of sufficient number of integrals of motion.

We now illustrate the last step on our examples considered earlier.

Example 1:

As noted earlier, (10.62) is not of P-type, as it admits complex resonances indicating the presence of movable critical point.

Example 2:

One can check that an arbitrary constant exists for $r = 2$ in the Laurent series solution for (10.65) and the associated integral of motion is

$$\Phi\left(v, v'\right) = v'^2 + v^3 + v - kv^2 \;, \quad k = \text{constant} \;, \tag{10.77}$$

so that (10.65) is completely integrable.

D. Singularity Structure Analysis in the Non-Painlevé Cases

The above type of Painlevé singularity structure analysis isolates the cases of dynamical systems whose equations of motion are free from movable critical points and thereby helps to identify integrable systems. Now the natural question arises as to what will be the nature of singularities exhibited by nonintegrable and chaotic systems corresponding to non-Painlevé cases, for example the Lorenz system, Duffing oscillator, forced van der Pol oscillator, Hénon–Heiles system, etc. discussed in the earlier chapters. An independent study of the singularity structures in the complex time plane of these systems shows that in many of these nonintegrable and chaotic cases the solutions exhibit an immensely complicated, multiarmed, infinite-sheeted structure of singularities, in contrast to the regular and ordered patterns of singularities (like the lattice array of complex poles of a Jacobian elliptic function) in the case of integrable and regular dynamical systems. However, this is an intricate and complex subject of study and we will not consider this aspect in this book (however, see problems 6–9 given at the end of the chapter).

Figure 10.1 illustrates the main points associated with the singularity structure analysis of equations of motion of nonlinear dynamical systems.

10.5 Painlevé Analysis and Integrability of Two-Coupled Nonlinear Oscillators

The Painlevé analysis described in the previous section can be profitably used to investigate the integrability property of nonlinear dynamical systems. In particular, it can be used to isolate and identify integrable parametric choices both in the Hamiltonian and dissipative systems. To illustrate this idea, we consider in the following a typical nonlinear Hamiltonian system. Similar procedure can be applied to any other dynamical system.

10.5.1 Quartic Anharmonic Oscillators

Let us now consider the problem of two-coupled quartic nonlinear oscillators described by the equations of motion (10.21). This system is widely used as a model in lattice dynamics, condensed matter theory, field theory, astrophysics, etc. For clarity we will describe the three steps of the ARS approach separately in the following subsections.

A. Leading-Order Behaviour

Considering (10.21), we assume that the leading-order behaviour of $x(t)$ and $y(t)$ in a sufficiently small neighbourhood of the movable singularity t_0 is

$$x(t) \approx a_0 \tau^p , \quad y(t) \approx b_0 \tau^q , \quad \tau = t - t_0 \to 0 . \tag{10.78}$$

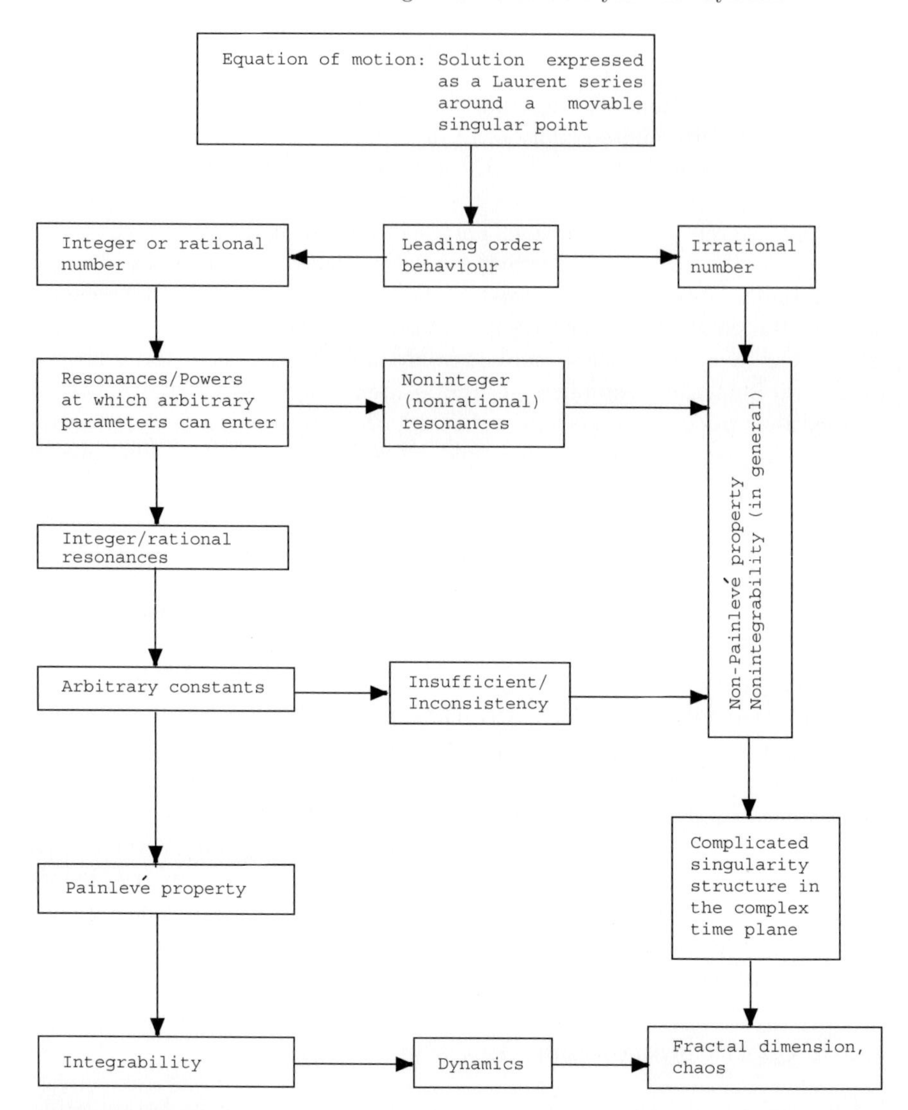

Fig. 10.1. Singularity structure analysis of equations of motion of nonlinear dynamical systems

To determine p, q, a_0, and b_0, we use (10.78) in (10.21) and obtain a pair of leading-order equations

$$a_0 p(p-1)\tau^{p-2} + 4\alpha a_0^3 \tau^{3p} + 2\delta a_0 b_0^2 \tau^{p+2q} = 0 \,, \tag{10.79a}$$

$$b_0 q(q-1)\tau^{q-2} + 4\beta b_0^3 \tau^{3q} + 2\delta a_0^2 b_0 \tau^{2p+q} = 0 \,. \tag{10.79b}$$

From (10.79), we identify the following two distinct sets of possibilities.

Case 1: $p = q = -1$

$$a_0^2 = (\delta - 2\beta)\Delta_1 \ , \quad b_0^2 = (\delta - 2\alpha)\Delta_1 \ , \quad \Delta_1 = \left(4\alpha\beta - \delta^2\right)^{-1} \ . \tag{10.80}$$

Case 2a: $p = -1$, $q = \frac{1}{2} + \frac{1}{2}\left[1 + 4\delta/\alpha\right]^{1/2}$

$$a_0^2 = -\frac{1}{2\alpha} \ , \quad b_0^2 = \text{arbitrary} \ . \tag{10.81a}$$

Case 2b: $p = -1$, $q = \frac{1}{2} - \frac{1}{2}\left[1 + 4\delta/\alpha\right]^{1/2}$

$$a_0^2 = -\frac{1}{2\alpha} \ , b_0^2 = \text{arbitrary} \ . \tag{10.81b}$$

Due to the complete symmetry between x and y variables in (10.21), we do not treat the other possibility $q = -1$, $p > -1$ as distinct from (10.81) in the following.

B. Resonances

For finding the resonances, we substitute

$$x(t) \approx a_0\tau^p + \Omega_1\tau^{p+r} \ , \quad y(t) \approx b_0\tau^q + \Omega_2\tau^{q+r} \ , \quad \tau \to 0 \ , \tag{10.82}$$

into (10.21). Retaining only the leading-order terms, we obtain a system of linear algebraic equations,

$$M_2(r)\Omega = 0 \ , \quad \Omega = (\Omega_1, \Omega_2) \ , \tag{10.83}$$

where $M_2(r)$ is a 2×2 matrix dependent on r. In order to have a nontrivial set of solutions (Ω_1, Ω_2) we require that

$$\det M_2(r) = 0 \ . \tag{10.84}$$

For case 1, the form of $M_2(r)$ is

$$M_2(r) = \begin{bmatrix} r^2 - 3r + 8\alpha a_0^2 & 4\delta a_0 b_0 \\ 4\delta a_0 b_0 & r^2 - 3r + 8\beta b_0^2 \end{bmatrix} \ , \tag{10.85}$$

so that using (10.80), (10.84) becomes

$$\left(r^2 - 3r - 4\right)\left(r^2 - 3r + X_0\right) = 0 \ , \quad X_0 = 4\left[1 + 2\left(\alpha a_0^2 + \beta b_0^2\right)\right] \ . \tag{10.86}$$

Thus for case 1, the resonances occur at

$$r = -1, \ \ r = 4, \ \ r = \frac{3}{2} \pm \frac{1}{2}\left(9 - 4X_0\right)^{1/2} \ . \tag{10.87}$$

For the leading orders (10.81) of case 2, the expression for $M_2(r)$ degenerates to

$$M_2(r) = \text{diag}\left[r^2 - 3r + 8\alpha a_0^2, \ r(r + 2q - 1)\right] \ . \tag{10.88}$$

Furthermore, for P-property, all the resonances must be non-negative rationals, including integers and thus the following sets of resonances are isolated along with specific parametric restrictions.

Case 1(i):

$$r = -1, 1, 2, 4, \quad \delta = 2\left[\alpha + \beta \pm \left(\alpha^2 + \beta^2 - \alpha\beta\right)^{1/2}\right] . \tag{10.89}$$

Case 1(ii):

$$r = -1, 0, 3, 4, \quad \alpha = \beta, \quad \delta = 2\alpha . \tag{10.90}$$

Case 2b:

$$r = -1, 0, 2, 4, \quad 3\alpha = 4\delta . \tag{10.91}$$

Note that case 2a is omitted, since the resonance values contradict the leading-order singularity nature, $q \geq 1/2$. Obviously, the resonance at $r = -1$ in (10.91) is associated with the arbitrariness of t_0.

C. Evaluation of Arbitrary Constants

Introducing now the series expansions

$$x(t) = \sum_{k=0}^{4} a_k \tau^{p+k} , \quad y(t) = \sum_{k=0}^{4} b_k \tau^{q+k} , \tag{10.92}$$

into the full equations (10.21), one can obtain a pair of recursion relations. For example, for case 1, they are

$$(j-1)\ (j-2)a_j + 2Aa_{j-2} + 4\alpha \sum_l \sum_m a_{j-l-m} a_l a_m$$
$$+2\delta \sum_l \sum_m b_{j-l-m} a_l b_m = 0 , \tag{10.93a}$$
$$(j-1)\ (j-2)b_j + 2Bb_{j-2} + 4\beta \sum_l \sum_m b_{j-l-m} b_l b_m$$
$$+2\delta \sum_l \sum_m a_{j-l-m} a_l b_m = 0 , \quad 0 \leq l,\ m \leq j \leq 4. \tag{10.93b}$$

Solving equations (10.93) successively, one can obtain the various a_k's and b_k's explicitly. For example, for case 1(i) we have the following system of algebraic equations:

$$2\alpha a_0^2 + \delta b_0^2 = -1 , \tag{10.94a}$$
$$\delta a_0^2 + 2\beta b_0^2 = -1 \tag{10.94b}$$

for $j = 0$,

$$\left(4\alpha a_0^2 - 1\right) a_1 + 2\delta a_0 b_0 b_1 = 0 , \tag{10.95a}$$
$$2\delta a_0 b_0 a_1 + \left(4\beta b_0^2 - 1\right) b_1 = 0 \tag{10.95b}$$

for $j = 1$,

$$Aa_0 + 6\alpha a_0 a_1^2 + 2\delta b_0 a_1 b_1 + \delta a_0 b_1^2 + \left(4\alpha a_0^2 - 1\right) a_2 + 2\delta a_0 b_0 b_2 = 0, \tag{10.96a}$$

$$Bb_0 + \delta b_0 a_1^2 + 2\delta a_0 a_1 b_1 + 6\beta b_0 b_1^2 + (4\beta b_0^2 - 1)b_2 + 2\delta a_0 b_0 a_2 = 0 \quad (10.96b)$$

for $j = 2$, and similar equations for $j = 3, 4$. Now one can easily check that the coefficients a_0 and b_0 satisfy (10.80). Similarly from (10.95), we find that a_1 or b_1 becomes arbitrary iff

$$\alpha = \beta, \quad \delta = 2\alpha,\ 6\alpha. \quad (10.97)$$

However, from (10.80) we infer that for the values $\alpha = \beta$, $\delta = 2\alpha$ either a_0 or b_0 becomes arbitrary, which is not indicated by the resonance values in (10.89), and hence this choice is omitted.

Proceeding further, it is straightforward to check the arbitrariness of a_2 or b_2 which requires an additional constraint on the parameters, that is, $A = B$. In a similar manner, from the equations for $j = 3, 4$, we can uniquely determine a_3 and b_3, while either one of the coefficients a_4 or b_4 is arbitrary without any further restrictions on the parameters and thus leading to a full four-parameter branch of the solution. We call such branches the *main branches* (MB) of the solutions $x(t)$ and $y(t)$.

For the P-property to hold, it is necessary to verify that the remaining or *subsidiary branches* (SB) of the solution are also free from movable critical points (within the strong and weak P-concepts). We do verify explicitly that this is indeed the case for the above choice, namely

$$\alpha = \beta, \quad \delta = 6\alpha, \quad A = B. \quad (10.98)$$

One can proceed in a similar manner for the cases 1(ii) and 2 and obtain all the possible P-cases, which are indicated in Table 10.1. It also presents the leading orders, resonance values, parametric restrictions and associated integrals of motion. Methods to obtain the integrals of motion will be discussed in the following sections.

Exercise:

6. Carry out the Painlevé analysis for the cases 1(ii) and 2(b) given by (10.90) and (10.91), respectively, and identify the parametric restrictions for which the Painlevé property is satisfied (see Ref. [7] for details).

The Painelvé analysis illustrated above for the case of coupled anharmonic oscillators can be extended to any other finite dimensional nonlinear dynamical system either of Hamiltonian or non-Hamiltonian type. Many such systems have been investigated successfully [4–7], confirming that P-analysis is a useful tool to investigate integrability.

10.6 Symmetries and Integrability

Symmetries, invariance, conservation laws and integrals of motion are aspects of dynamical systems which are closely interrelated [8]. In particular,

Table 10.1. P-cases of the system of two-coupled quartic anharmonic oscillators (M is the number of arbitrary constants)

Case	Parametric restriction	MB/ SB	Leading order p	q	Resonances r	M	Integrals of motion $I_1(=H)$, I_2
1(i)	$A=B$, $\alpha=\beta$, $\delta=6\alpha$	MB	-1	-1	$-1,1,2,4$	4	$I_1=\frac{1}{2}(p_x^2+p_y^2)+\alpha(x^4+y^4+6x^2y^2)$
		SB	-1	-1	$-5,-1,0,4$	3	$+A(x^2+y^2)$
							$I_2=p_xp_y+2\left[A+2\alpha(x^2+y^2)\right]xy$
1(ii)	$\alpha=\beta$, $\delta=2\alpha$	MB	-1	-1	$-1,0,3,4$	4	$I_1=\frac{1}{2}(p_x^2+p_y^2)+Ax^2+By^2+\alpha(x^2+y^2)^2$
		SB	-1	2	$-3,-1,0,4$	3	$I_2=(xp_x-yp_x)^2+\frac{2}{\alpha}(B-A)\left[\frac{1}{2}p_x^2+Ax^2\right.$
							$\left.+\alpha(x^2+y^2)x^2\right]$
2b(i)	$A=4B$, $\alpha=16\beta$, $\delta=12\beta$	MB	-1	$-\frac{1}{2}$	$-1,0,2,4$	4	$I_1=\frac{1}{2}(p_x^2+p_y^2)+B(4x^2+y^2)$
		SB-1	-1	-1	$-2,-1,4,5$	3	$+\beta(16x^4+y^4+12x^2y^2)$
		SB-2	-1	$\frac{3}{2}$	$-2,-1,0,4$	3	$I_2=(yp_x-xp_y)p_y$
							$+2(B+4\beta x^2+2\beta y^2)xy^2$
2b(ii)	$A=4B$, $\alpha=8\beta$, $\delta=6\beta$	MB	-1	$-\frac{1}{2}$	$-1,0,2,4$	4	$I_1=\frac{1}{2}(p_x^2+p_y^2)+B(4x^2+y^2)$
		SB-1	-1	-1	$-2,-1,4,8$	3	$+\beta(8x^4+y^4+6x^2y^2)$
		SB-2	-1	$\frac{3}{2}$	$-2,-1,0,4$	3	$I_2=p_y^4+4y^2(B+6\beta x^2+\beta y^2)p_y^2$
							$-16\beta xy^3p_xp_y+4\beta y^4p_x^2+4B^2y^4$
							$+4\beta\left[2B+\beta(2x^2+y^2)\right](2x^2+y^2)y^4$

Noether's theorem[3] establishes an important connection between invariance and existence of integrals of motion for Lagrangian systems. Such connections between dynamical systems and symmetries can be very profitably used to isolate integrable cases and their integrals. In this section, we will briefly introduce this approach.

Sophus Lie advocated almost a century ago the study of differential equations through invariance analysis under one-parameter continuous transformation groups associated with symmetry properties as a means of analysing them. This method has in recent times been further developed by Ovsjannikov, Bluman and Cole and others [11,12]. Conventionally the theory of one-parameter Lie groups of continuous transformations is applied to dynamical systems, particularly to Lagrangian systems, by assuming the infinitesimal generators of the group to be functions of the independent and dependent variables alone and not their derivatives. However, this approach has a rather limited range of applications as it deals with point symmetries alone. The limitations in the above approach can be circumvented by including the velocity terms in the argument of the infinitesimals. Such transformations operate on trajectories in space-time and are called *dynamical symmetries*, if they preserve the equations of motions of the given system. The study of the invariance of differential equations under derivative-dependent transformations has drawn wide attention in recent times. Several Lagrangian systems have been identified to possess dynamical symmetries by many authors. In the following, we present a brief outline of Lie's extended theory of one-parameter continuous transformations applicable to two-dimensional Lagrangian systems and apply the method to the nonlinear oscillators of our interest. The method can in principle be extended to more general situations without much difficulty.

10.6.1 Invariance Conditions, Determination of Infinitesimals and First Integrals of Motion

Let us consider a two-dimensional (two degrees of freedom) Lagrangian system related to the Hamiltonian $H = \frac{1}{2}(p_x^2 + p_y^2) + V(x,y)$. Its form is

$$L = \frac{1}{2}\left(\dot{x}^2 + \dot{y}^2\right) - V(x,y)\,, \tag{10.99}$$

where the dot means derivative with respect to time. The Euler–Lagrange equations of motion are

$$\ddot{x} = \partial L/\partial x = \alpha_1(x,y)\,, \quad \ddot{y} = \partial L/\partial y = \alpha_2(x,y)\,. \tag{10.100}$$

[3] Noether's theorem established by Emmy Noether (1918) states that for a given Lagrangian L, if the action integral is invariant under a continuous one parameter group of transformations, there exists a conserved quantity. For details see for example Ref. [8].

A. Invariance Conditions

For (10.100) to be invariant under the action of the infinitesimal transformations of a one-parameter (ϵ) continuous Lie group,

$$x \to X = x + \epsilon\eta_1(t,x,y,\dot{x},\dot{y}) + O\left(\epsilon^2\right) , \tag{10.101a}$$

$$y \to Y = y + \epsilon\eta_2(t,x,y,\dot{x},\dot{y}) + O\left(\epsilon^2\right) , \tag{10.101b}$$

$$t \to T = t + \epsilon\xi(t,x,y,\dot{x},\dot{y}) + O\left(\epsilon^2\right), \quad \epsilon \ll 1, \tag{10.101c}$$

we require the following invariance conditions to be satisfied [7]:

$$\ddot{\eta}_1 - \dot{x}\ddot{\xi} - 2\dot{\xi}\alpha_1 = E\left(\alpha_1\right), \quad \ddot{\eta}_2 - \dot{y}\ddot{\xi} - 2\dot{\xi}\alpha_2 = E\left(\alpha_2\right) , \tag{10.102}$$

where the infinitesimal operator E is given by

$$E = \xi\frac{\partial}{\partial t} + \eta_1\frac{\partial}{\partial x} + \eta_2\frac{\partial}{\partial y} + \left(\dot{\eta}_1 - \dot{\xi}\dot{x}\right)\frac{\partial}{\partial \dot{x}} + \left(\dot{\eta}_2 - \dot{\xi}\dot{y}\right)\frac{\partial}{\partial \dot{y}} . \tag{10.103a}$$

Here the quantities $\dot{\eta}_1$, $\dot{\eta}_2$, $\dot{\xi}$, $\ddot{\eta}_1$, $\ddot{\eta}_2$ and $\ddot{\xi}_2$ are given by

$$\dot{\eta}_1 = \eta_{1t} + \eta_{1x}\dot{x} + \eta_{1y}\dot{y} + \alpha_1\eta_{1\dot{x}} + \alpha_2\eta_{1\dot{y}} , \tag{10.103b}$$

$$\begin{aligned}\ddot{\eta}_1 = {} & \eta_{1tt} + \left(\eta_{1x} + 2\eta_{1\dot{x}t}\right)\alpha_1 + \left(\eta_{1y} + 2\eta_{1\dot{y}t}\right)\alpha_2 + \eta_{1\dot{x}\dot{x}}\alpha_1^2 + \eta_{1\dot{y}\dot{y}}\alpha_2^2 \\ & + 2\eta_{1\dot{x}\dot{y}}\alpha_1\alpha_2 + \left[2\eta_{1xt} + 2\alpha_1\eta_{1x\dot{x}} + 2\alpha_2\eta_{1x\dot{y}} + \alpha_{1x}\eta_{1\dot{x}} + \alpha_{2x}\eta_{1\dot{y}}\right]\dot{x} \\ & + \left[2\eta_{1yt} + 2\alpha_1\eta_{1\dot{x}y} + 2\alpha_2\eta_{1y\dot{y}} + \alpha_{1y}\eta_{1\dot{x}} + \alpha_{2y}\eta_{1\dot{y}}\right]\dot{y} \\ & + \eta_{1xx}\dot{x}^2 + \eta_{1yy}\dot{y}^2 + 2\eta_{1xy}\dot{x}\dot{y}\end{aligned} \tag{10.103c}$$

and so on.

B. Determination of Infinitesimals

From (10.102), it is clear that the invariance conditions form an incomplete system in η_1, η_2 and ξ. This suggests that one will have to assume specific forms for η_1, η_2 and ξ in order to solve (10.102) consistently. A trivial choice is

$$\eta_1 = \eta_2 = 0, \quad \xi = C = \text{constant} , \tag{10.104}$$

and so $E = C\partial/\partial t$, from which we may infer (see below) that the Hamiltonian H is an integral of motion. To determine the existence of other nontrivial infinitesimal symmetries, we may assume for example η_1, η_2 and ξ to be polynomials in velocities $\dot{x}$ and $\dot{y}$ and then find the t, x and y dependence consistently.

As an example, let us first consider the linear form

$$\begin{aligned}&\xi = a_1 + a_2\dot{x} + a_3\dot{y} , \quad \eta_1 = b_1 + b_2\dot{x} + b_3\dot{y} , \\ &\eta_2 = c_1 + c_2\dot{x} + c_3\dot{y} ,\end{aligned} \tag{10.105}$$

where the a_i, b_i and c_i, $i = 1,2,3$ are functions of t, x, y only. Making use of (10.105) in (10.102) and equating the various coefficients of $\dot{x}^m\dot{y}^n$, $m, n = 1,2,3,4$ to zero, we obtain a system of overdetermined linear pdes,

$$a_{2xx} = 0, \;\; 2a_{2xy} + a_{3xx} = 0, \;\; a_{2yy} + 2a_{3xy} = 0, \;\; a_{3yy} = 0 \; , \tag{10.106}$$

$$b_{2xx} - (a_{1xx} + 2a_{2xt}) = 0, \tag{10.107a}$$
$$(2b_{2xy} + b_{3xx}) - 2(a_{1xy} + a_{2yt} + a_{3xt}) = 0, \tag{10.107b}$$
$$b_{2yy} + 2b_{3xy} - a_{1yy} - 2a_{3yt} = 0, b_{3yy} = 0 \; , \tag{10.107c}$$

$$b_{1xx} + 2b_{2xt} - 2a_{1xt} - a_2\alpha_{1x} - 5\alpha_1 a_{2x} - a_{2tt}$$
$$-2\alpha_2 a_{3x} - \alpha_2 a_{2y} = 0 \; , \tag{10.108a}$$
$$2(b_{1xy} + b_{2yt} + b_{3xt}) - (2a_{1yt} + a_2\alpha_{1y} + 4\alpha_1 a_{2y})$$
$$-2(a_3\alpha_{2y} + a_{3tt} + 3\alpha_2 a_{3y} + 3\alpha_1 a_{3x}) = 0 \; , \tag{10.108b}$$
$$b_{1yy} + 2b_{3yt} + 2\alpha_1 a_{3y} = 0 \; , \tag{10.108c}$$

$$2b_{1xt} + b_2\alpha_{1x} + 3\alpha_1 b_{2x} + b_3\alpha_{2x} + b_{2tt} + 2\alpha_2 b_{3x} + \alpha_2 b_{2y}$$
$$-(a_{1tt} + 4\alpha_1 a_{2t} + \alpha_2 a_{1y} + 2\alpha_2 a_{3t} + 3\alpha_1 a_{1x})$$
$$-(b_2\alpha_{1x} + c_2\alpha_{1y}) = 0 \; , \tag{10.109a}$$
$$2b_{1yt} + b_2\alpha_{1y} + 2\alpha_1 b_{2y} + b_3\alpha_{2y} + b_{3tt} + 3\alpha_2 b_{3y} + \alpha_1 b_{3x}$$
$$-2\alpha_1(a_{1y} + a_{3t}) - (b_3\alpha_{1x} + c_3\alpha_{1y}) = 0 \; , \tag{10.109b}$$

$$b_{1tt} + 2\alpha_1 b_{2t} + 2\alpha_2 b_{3t} + \alpha_1 b_{1x} + \alpha_2 b_{1y} - 2\alpha_1(a_{1t} + \alpha_1 a_2 + \alpha_2 a_3)$$
$$-(b_1\alpha_{1x} + c_1\alpha_{1y}) = 0 \; , \tag{10.110}$$

$$c_{2xx} = 0 \; , \tag{10.111a}$$
$$c_{2yy} + 2c_{3xy} - 2(a_{1xy} + a_{2yt} + a_{3xt}) = 0 \; , \tag{10.111b}$$
$$2c_{2xy} + c_{3xx} - (a_{1xx} + 2a_{2xt}) = 0 \; , \tag{10.111c}$$
$$c_{3yy} - (a_{1yy} + 2a_{3yt}) = 0 \; , \tag{10.111d}$$

$$c_{1xx} + 2c_{2xt} - 2\alpha_2 a_{2x} = 0 \; , \tag{10.112a}$$
$$2(c_{1xy} + c_{2yt} + c_{3xt}) - (2a_{1xt} + a_2\alpha_{1x} + 3\alpha_1 a_{2x} + a_3\alpha_{2x})$$
$$+(a_{2tt} + 4\alpha_2 a_{3x} + 3\alpha_2 a_{2y}) = 0 \; , \tag{10.112b}$$
$$c_{1yy} + 2c_{3yt} - (2a_{1yt} + a_2\alpha_{1y} + 2\alpha_1 a_{2y} + a_3\alpha_{2y})$$
$$+(a_{3tt} + 5\alpha_2 a_{3y} + \alpha_1 a_{3x}) = 0 \; , \tag{10.112c}$$

$$2c_{1xt} + c_2\alpha_{1x} + 3\alpha_1 c_{2x} + c_3\alpha_{2x} + c_{2tt} + 2\alpha_2 c_{3x} + \alpha_2 c_{2y}$$
$$-2\alpha_2(a_{1x} + a_{2t}) - (b_2\alpha_{2x} + c_2\alpha_{2y}) = 0 \; , \tag{10.113a}$$
$$2c_{1yt} + c_2\alpha_{1y} + 2\alpha_1 c_{2y} + c_3\alpha_{2y} + c_{3tt} + 3\alpha_2 c_{3y} + \alpha_1 c_{3x}$$
$$-(a_{1tt} + 2\alpha_1 a_{2t} + 4\alpha_2 a_{3t} + \alpha_1 a_{3x} + 3\alpha_2 a_{1y})$$
$$-(b_3\alpha_{2x} + c_3\alpha_{2y}) = 0 \; , \tag{10.113b}$$

$$c_{1tt} + 2\alpha_1 c_{2t} + 2\alpha_2 c_{3t} + \alpha_1 c_{1x} + \alpha_2 c_{1y}$$
$$-2\alpha_2(a_{1t} + \alpha_1 a_2 + \alpha_2 a_3) - (b_1\alpha_{2x} + c_1\alpha_{2y}) = 0 \; , \tag{10.114}$$

where subscripts denote partial derivatives. By successively solving the determining equations (10.106)–(10.114) together with the equations of motion (10.100), we can find the forms of η_1, η_2 and ξ explicitly.

In a similar way one can start from a cubic form for ξ, η_1 and η_2 in the velocities $\dot{x}$ and $\dot{y}$ as

$$\xi = \sum_{i,j=0}^{3} a_{ij}\dot{x}^i\dot{y}^j , \quad \eta_1 = \sum_{i,j=0}^{3} b_{ij}\dot{x}^i\dot{y}^j , \quad \eta_2 = \sum_{i,j=0}^{3} c_{ij}\dot{x}^i\dot{y}^j , \tag{10.115}$$

where the a_{ij}, b_{ij} and c_{ij} are functions of t, x, y only and obtain determining equations. Solving them consistently, one can again obtain the generalised symmetries. For details see Ref. [7].

C. Integrals of Motion

In order to find the explicit form of integrals of motion using the dynamical symmetries derived earlier, one can make use of the Noether's theorem. It can be shown that given the infinitesimal symmetries η_1, η_2, and ξ and the Lagrangian L, the integral of motion, if it exists, may be written as

$$I(t,x,y,\dot{x},\dot{y}) = (\xi\dot{x} - \eta_1)\frac{\partial L}{\partial \dot{x}} + (\xi\dot{y} - \eta_2)\frac{\partial L}{\partial \dot{y}} - \xi L + f , \tag{10.116a}$$

where f is a function of x, y only and is to be determined from the equation

$$E(L) + \dot{\xi} L = \dot{f} . \tag{10.116b}$$

As mentioned earlier for (10.104), one can easily check that $I = H$, the Hamiltonian of the system, is indeed an integral of motion. The above procedure can be applied to any Lagrangian system; for example, it can be applied to the case of two-coupled anharmonic oscillators discussed in the previous section and the integrals of motion given in Table 10.1 can be deduced (see exercises below). We now illustrate the method to obtain the integrable cases of a Hénon–Heiles like system.

10.6.2 Application – The Hénon–Heiles System

The Lagrangian of a generalised Hénon–Heiles system (recall the form in Chap. 7) is

$$L = \frac{1}{2}\left(\dot{x}^2 + \dot{y}^2\right) - \frac{1}{2}\left(Ax^2 + By^2\right) - \left(\alpha x^2 y - \frac{1}{3}\beta y^3\right) , \tag{10.117}$$

where A, B, α and β are parameters. From the equations of motion,

$$\ddot{x} + Ax + 2\alpha xy = 0 , \quad \ddot{y} + By + \alpha x^2 - \beta y^2 = 0 , \tag{10.118}$$

we have

$$\alpha_1 = -(Ax + 2\alpha xy) , \quad \alpha_2 = -\left(By + \alpha x^2 - \beta y^2\right) . \tag{10.119}$$

Now analysing (10.106), we obtain

$$a_2 = a_{20}y^2 + a_{21}xy + a_{22}x + a_{23}y + a_{24} , \tag{10.120a}$$

$$a_3 = -a_{21}x^2 - a_{20}xy + a_{31}x + a_{32}y + a_{33} , \tag{10.120b}$$

where the coefficients a_{ij} are functions of t only. Also, from (10.107) and (10.111), we find

$$b_2 - a_1 = \dot{a}_{21}x^2y + \dot{a}_{20}xy^2 + \dot{a}_{22}x^2 + b_{20}xy + \frac{1}{2}b_{21}y^2 + b_{22}x + b_{23}y + b_{24} , \tag{10.121a}$$

$$b_3 = -\dot{a}_{21}x^3 - \dot{a}_{20}x^2y + \frac{1}{2}b_{30}x^2 + b_{31}xy + b_{32}x + b_{33}y + b_{34} , \tag{10.121b}$$

$$c_3 - a_1 = -\dot{a}_{21}x^2y - \dot{a}_{20}xy^2 + \dot{a}_{32}y^2 + \frac{1}{2}c_{30}x^2 + c_{31}xy + c_{32}x + c_{33}y + c_{34} , \tag{10.122a}$$

$$c_2 = \dot{a}_{20}y^3 + \dot{a}_{21}xy^2 + \frac{1}{2}c_{20}y^2 + c_{21}xy + c_{22}x + c_{23}y + c_{24} , \tag{10.122b}$$

$$2b_{20} + b_{32} = 2(\dot{a}_{23} + \dot{a}_{31}) = c_{20} + 2c_{31} , \tag{10.123a}$$

$$b_{21} + 2b_{30} = 2\dot{a}_{32} , \quad c_{30} + 2c_{21} = 2\dot{a}_{22} , \tag{10.123b}$$

where the coefficients a_{ij}, b_{ij} and c_{ij} are functions of t alone. Similarly, from (10.108c) and (10.112a) we get

$$b_1 = \left(A + \frac{2}{3}\alpha y\right)(a_{20}x - a_{32})\,xy^3 + \left(\ddot{a}_{20}x - \dot{b}_{30}\right)xy^2 - \dot{b}_{31}y^2 + B_{10}y + B_{11} , \tag{10.124}$$

$$c_1 = -\tfrac{1}{6}\,\alpha\,(a_{21}y + a_{22})\,x^4 - [B\,(a_{21} + a_{22}) + \dot{c}_{21}]\,x^2y - [\beta\,(a_{22} + \ddot{a}_{21}y) + a_{22}]\,x^2y^2 - \dot{c}_{22}x^2 - C_{10}x - C_{11} , \tag{10.125}$$

where B_{10}, B_{11}, and C_{10}, C_{11} are functions of t, x and t, y, respectively, and the remaining coefficients are functions of t alone. Making use of the above coefficient values (10.121)–(10.125) in the remaining equations of (10.108)–(10.110) and (10.112)–(10.114), one can explicitly find the values of the a_i, b_i and c_i. However, for practical purposes, particularly for constructing the time-independent integral of motion, it is sufficient to consider that the functions b_i and c_i are independent of time while the a_i's vanish. As a result, one can straightforwardly check that the consistency condition holds only for the following three parametric choices (excluding the trivial choice $\alpha = \beta = 0$). For example, for the parametric choice

$$A = B, \quad \alpha = -\beta , \tag{10.126a}$$

we have

$$a = 0, \quad b_1 = b_2 = c_1 = c_3 = 0, \quad b_3, c_2 = \text{constant} , \tag{10.126b}$$

and so the infinitesimal symmetries become

$$\xi = 0, \quad \eta_1 = k\dot{y}, \quad \eta_2 = k\dot{x}, \quad k = \text{constant} . \tag{10.126c}$$

The other two parametric restrictions are

$$A, B \text{ arbitrary}, \quad 6\alpha = -\beta; \quad 16A = B, \quad 16\alpha = -\beta . \tag{10.127}$$

The associated forms of the infinitesimals ξ, η_1 and η_2 are displayed in Table 10.2.

Earlier, we have pointed out that in order to construct the integrals of motion associated with the infinitesimal symmetries derived above, it is necessary to solve (10.116b), which can be written in the present case as

$$\eta_1 \frac{\partial L}{\partial x} + \eta_2 \frac{\partial L}{\partial y} + \left(\dot{\eta}_1 - \dot{x}\dot{\xi} \right) \dot{y} + \dot{\xi} L = \dot{x} \frac{\partial f}{\partial x} + \dot{y} \frac{\partial f}{\partial y} . \tag{10.128}$$

Solving (10.128), one obtains the explicit form of f. It turns out that effectively in all the integrable cases the explicit form of f can be chosen without much difficulty. For example, for the set of infinitesimal symmetries $\xi = 0$, $\eta_1 = k\dot{y}$, $\eta_2 = k\dot{x}$ together with the parametric values $A = B$, $\alpha = -\beta$, (10.128) gives

$$\frac{\partial f}{\partial y} = 2\ddot{x} = -2(A + 2\alpha y)x , \quad \frac{\partial f}{\partial x} = 2\ddot{y} = -\left(Ay + \alpha x^2 + \alpha y^2 \right) \tag{10.129a}$$

and so

$$f = -2 \left(Axy + \frac{1}{3}\alpha x^3 + \alpha x y^2 \right) . \tag{10.129b}$$

As a result we obtain the second integral of motion as

$$I = \dot{x}\dot{y} + \left(Ay + \frac{1}{3}\alpha x^2 + \alpha y^2 \right) x . \tag{10.129c}$$

The other integrable cases of Hénon–Heiles and coupled quartic oscillators can also be similarly worked out (table 10.2). For further details, see Ref. [7].

Exercises:

7. Using the form of the Lagrangian L given in (10.99), prove that $\mathrm{d}I/\mathrm{d}t = 0$, where I is given by (10.116).
8. Work out the symmetries and integrals of motion associated with the parametric choices (10.127).
9. Identify the four parametric choices associated with the coupled quartic oscillator system (10.21) and obtain the associated symmetries and integrals of motion and verify that they agree with the results given in Sect. 10.3.3 (see also ref. [7]).

Table 10.2. Infinitesimal symmetries and integrals of motion of the generalized Hénon–Heiles system (10.117). Here $L_{xy} = x\dot{y} - y\dot{x}$, $\rho^2 = x^2 + y^2$

Cases	Parametric restriction	Infinitesimal symmetries			Second integral of motion I_2
		ξ	η_1	η_2	
1	$A = B$, $\alpha = -\beta$	0	$k\dot{y}$	$k\dot{x}$	$\dot{x}\dot{y} + Axy + \frac{1}{3}\alpha x^3 + \alpha xy^2$
2	A, B arbitrary, $6\alpha = -\beta$	0	$4\alpha(x\dot{y} - 2\dot{x}y)$ $+2(4A - B)\dot{x}$	$4\alpha x\dot{x}$	$4L_{xy}\dot{x} + (4Ay + \alpha x^2 + 4\alpha y^2)x^2$ $+\frac{4A-B}{\alpha}\left(\dot{x}^2 + Ax^2\right)$
3	$16A = B$, $16\alpha = -\beta$	0	$4\dot{x}^3 + 4(A + 2\alpha y)x^2\dot{x}$ $-\frac{4}{3}\alpha x^3\dot{x}$	$-\frac{4}{3}\alpha x^3\dot{x}$	$\dot{x}^4 + 2(A + 2\alpha y)x^2\dot{x}^2 - \frac{4}{3}\alpha x^3\dot{x}\dot{y}$ $+A^2x^4 - \frac{4}{3}\alpha(A + \alpha y)x^4y - \frac{2}{9}\alpha^2x^6$

10.7 A Direct Method of Finding Integrals of Motion

After identifying the parametric choices for which a given dynamical system is integrable, say by Painlevé analysis, how does one can go about to obtain the sufficient number of integrals of motion? As pointed out in the previous section, one can look for the associated symmetries and using them one can deduce the integrals of motion, for example using the Noether's theorem. Apart from this one can also make a direct approach to deduce the integrals of motion, particularly for integrable Hamiltonian systems. We demonstrate this procedure as developed in Whittaker's treatise [13] for two degrees of freedom systems. The method can be generalized to higher order systems as well. See for example the book of Chandrasekhar [14] for the analysis of three degrees of freedom systems [7].

In general, a given integral of motion is of the form $I = I(x, y, p_x, p_y)$. Let us illustrate the procedure to a specific case by restricting ourselves to integrals which are fourth degree polynomial in momenta p_x, p_y (for convenience), for two degrees of freedom systems. In such a situation the most general form of the second integral of motion can be written as

$$\begin{aligned} I = f_1 p_x^4 &+ f_2 p_x^3 p_y + f_3 p_x^2 p_y^2 + f_4 p_x p_y^3 + f_5 p_y^4 \\ &+ f_6 p_x^2 + f_7 p_x p_y + f_8 p_y^2 + f_9 \,, \end{aligned} \tag{10.130}$$

where the f_i, $i = 1, 2, ..., 9$, are functions of (x, y) alone. To obtain the f_i's, we demand that the Poisson bracket $[I, H]_{PB}$, where H is the Hamiltonian, vanishes. Equating now the coefficients of each monomial of the momenta $p_x^m p_y^n$, $m, n = 1, 2, ...5$ in the resultant expression separately to zero, we obtain a system of overdetermined linear pdes. The first set of these pdes is

$$\begin{array}{lll} f_{1x} = 0, & f_{1y} + f_{2x} = 0, & f_{2y} + f_{3x} = 0 \\ f_{3y} + f_{4x} = 0, & f_{4y} + f_{5x} = 0, & f_{5y} = 0 \,, \end{array} \tag{10.131}$$

where subscripts denote partial derivatives. By successively solving (10.131) we find

$$f_1 = \epsilon_0 y^4 + \epsilon_1 y^3 + \epsilon_2 y^2 + \epsilon_3 y + \epsilon_4 \,, \tag{10.132a}$$

$$f_2 = \left(4\epsilon_0 y^3 + 3\epsilon_1 y^2 + 2\epsilon_2 y + \epsilon_3\right) x + \eta_0 y^3 + \eta_1 y^2 + \eta_2 y + \eta_3 \,, \tag{10.132b}$$

$$\begin{aligned} f_3 = &\left(6\epsilon_0 y^2 + 3\epsilon_1 y + \epsilon_2\right) x^2 - \left(3\eta_0 y^2 + 2\eta_1 y + \eta_2\right) x + \sigma_0 y^2 \\ &+ \sigma_1 y + \sigma_2 \,, \end{aligned} \tag{10.132c}$$

$$\begin{aligned} f_4 = &-\left(4\epsilon_0 y + \epsilon_1\right) x^3 + \left(3\eta_0 y + \eta_1\right) x^2 - \left(2\sigma_0 y + \sigma_1\right) x \\ &+ \Theta_0 y + \Theta_1 \,, \end{aligned} \tag{10.132d}$$

$$f_5 = \epsilon_0 x^4 - \eta_0 x^3 + \sigma_0 x^2 - \Theta_0 x + \Theta_2 \,, \tag{10.132e}$$

where the ϵ_i, η_i, σ_i and Θ_i are integration constants.

The next set of pdes is

$$\begin{array}{ll} 4 f_1 \ddot{x} + f_2 \ddot{y} + f_{6x} = 0 \,, & 3 f_2 \ddot{x} + 2 f_3 \ddot{y} + f_{6y} + f_{7x} = 0 \,, \\ 2 f_3 \ddot{x} + 3 f_4 \ddot{y} + f_{7y} + f_{8x} = 0 \,, & f_4 \ddot{x} + 4 f_5 \ddot{y} + f_{8y} = 0, \end{array} \tag{10.133}$$

where the dot stands for differentiation with respect to t. From (10.133) we obtain the compatibility condition

$$\begin{aligned}(f_4\ddot{x} + 4f_5\ddot{y})_{xxx} - 2\,(f_3\ddot{x} + 3f_4\ddot{y})_{xxy} + (3f_2\ddot{x} + 2f_3\ddot{y})_{xyy} \\ - (4f_1\ddot{x} + f_2\ddot{y})_{yyy} = 0\,. \end{aligned} \tag{10.134}$$

The final set of pdes is

$$2f_6\ddot{x} + f_7\ddot{y} + f_{9x} = 0\,, \quad f_7\ddot{x} + 2f_8\ddot{y} + f_{9y} = 0\,, \tag{10.135}$$

and so the second compatibility condition becomes

$$(2f_6\ddot{x} + f_7\ddot{y})_y - (f_7\ddot{x} + 2f_8\ddot{y})_x = 0\,. \tag{10.136}$$

Therefore, it is clear that an integral of motion which is quartic in momenta exists whenever the equations of motion satisfy (10.134) and (10.136).

Following the above procedure, we find that for the two-coupled quartic oscillators (10.21) the compatibility condition holds only for the four parametric choices possessing the P-property and the explicit forms of the first (Hamiltonian) and second integrals of motion are obtained in the form given earlier. Similarly for the generalized Hénon–Heiles system (10.117), the integrals for the three integrable cases can be given. The method can be applied to any other two degrees of freedom Hamiltonian systems as well. Extension to higher dimensional systems may also be similarly considered, see for example Refs. [7,14].

Exercises:

10. Use the direct method to identify all the integrable cases of the two-coupled quartic oscillators problem and to deduce the associated second integrals of motion.
11. Using the direct method, obtain the second integral of motion for all the three integrable choices of the generalized Hénon–Heiles system.

10.8 Integrable Systems with Degrees of Freedom Greater Than Two

Just as we have identified nontrivial integrable systems with two degrees of freedom, we can extend our analysis to higher degrees of freedom as well, except that the analysis becomes more involved. We only indicate a few typical examples of such integrable systems, without detailed discussions. For more details see for example Ref. [7].

A. Three Degrees of Freedom of Hamiltonian Systems

Example: A System of Three-Coupled Quartic Oscillators

$$L = \frac{1}{2}\left(\dot{x}^2 + \dot{y}^2 + \dot{z}^2\right) - \left(Ax^2 + By^2 + Cz^2 + \alpha x^4 + \beta y^4 + \gamma z^4 + \delta x^2 y^2 + \epsilon y^2 z^2 + \omega x^2 z^2\right) . \tag{10.137}$$

There are eight nontrivial and distinct parametric choices for which the associated equation of motion is integrable (for details see Ref. [7]). As examples, we cite two of them.

Case 1: $\alpha = \beta = \gamma$, $\delta = \epsilon = \omega = 2\alpha$, $A = B = C$

$$I_1 = \frac{1}{2}\left(\dot{x}^2 + \dot{y}^2 + \dot{z}^2\right) + A\left(x^2 + y^2 + z^2\right) + \alpha\left(x^2 + y^2 + z^2\right)^2 , \tag{10.138a}$$

$$I_2 = L_{xy}^2 + L_{yz}^2 + L_{zx}^2 , \tag{10.138b}$$

$$I_3 = L_{xy} , \quad L_{xy} = (x\dot{y} - y\dot{x}) , \quad L_{yz} = (y\dot{z} - \dot{y}z) , \quad L_{zx} = (z\dot{x} - \dot{z}x) . \tag{10.138c}$$

Note that the case $A \neq B \neq C$ is also integrable [7].

Case 2: $\alpha = \beta = \gamma$, $\delta = 2\alpha$, $\epsilon = \omega = 6\alpha$, $A = B = C$

The three integrals of motion:

$$I_1 = \frac{1}{2}\left(\dot{x}^2 + \dot{y}^2 + \dot{z}^2\right) + A\left(x^2 + y^2 + z^2\right) + \alpha\left(x^2 + y^2 + z^2\right)^2 + 4\alpha\left(x^2 + y^2\right) z^2 , \tag{10.139a}$$

$$I_2 = (x\dot{y} - y\dot{x})^2 , \tag{10.139b}$$

$$I_3 = (x\dot{x} + y\dot{y})\,\dot{z} + (x\dot{y} - y\dot{x})^2\left[\left(x^2 + y^2\right)^{-1}\dot{z}^2 + 8\alpha z^2\right] + 2\left\{\left[A + 2\alpha\left(x^2 + y^2 + z^2\right)\right]\left(x^2 + y^2\right) z\right\}^2\left(x^2 + y^2\right)^{-1} . \tag{10.139c}$$

B. N-Degrees of Freedom of Hamiltonian Systems

Example: A System of N-Coupled Quartic Oscillators

$$L = \frac{1}{2}\sum_{i=1}^{N} \dot{x}_i^2 - \sum_{i=1}^{N} A_i x_i^2 - \sum_{i=1}^{N} \alpha_i x_i^4 - \frac{1}{2}\sum_{i,j,i\neq j} \beta_{ij} x_i^2 x_j^2 . \tag{10.140a}$$

For this system there are at least $(4N-4)$ nontrivial and distinct parametric choices for which the system is integrable. (Again see Ref. [7] for details). One simple example is

$$\alpha_i = \alpha_1, \quad \beta_{ij} = 2\alpha_1, \quad A_i = A_1, \quad i,j = 1,2,...,N, \quad i \neq j. \tag{10.140b}$$

Then the N involutive integrals of motion are

$$I_1 = \frac{1}{2}\sum \dot{x}_i^2 + A_1 r^2 + \alpha_1 r^4 \,, \quad r^2 = \sum_{i=1}^{N} x_i^2 \tag{10.140c}$$

and

$$I_j = \sum_{k=1}^{N} L_{jk}^2 \,, \quad L_{jk} = (x_j \dot{x}_k - \dot{x}_j x_k) \,, \quad j = 2, ..., N, \;\; j \neq k. \tag{10.140d}$$

10.9 Integrable Discrete Systems

So far in this chapter we have considered integrable dynamical systems, which evolve continuously in time. However, we have seen in the earlier chapters that discrete dynamical systems, such as the logistic map or Hénon's map are also equally important. The question is then, do we have integrable nonlinear systems which evolve discretely in time as well? Indeed integrable discrete systems or maps do exist in rather large numbers, which are very often discrete analogs of integrable continuous systems. Search for such integrable discrete systems form an active area of research in recent times [4]. In this section, we will give only some rudimentary details of such systems in order to give a feel of the nature of these systems. More details can be obtained for example in Ref. [4].

1. Integrable Discrete Riccati Equation

As a simple example of integrable discrete system, one can identify an integrable discrete version of the particular Riccati equation

$$\dot{x} = \alpha x^2 + \beta x + \gamma \,, \tag{10.141}$$

where α, β and γ are constants. Note that (10.141) is linearizable through the transformation $x = P/Q$. (Example: $Q = \psi$, $P = \psi_x$). The discretization is done by replacing $\dot{x}$ by $(x_{n+1} - x_n)/\Delta t$, where Δt is the fixed step size, and x^2 by $x_n x_{n+1}$. Then, (10.141) can be written in the form of a discrete map

$$x_{n+1} = \frac{b x_n + c}{1 + a x_n} \,, \tag{10.142}$$

where a, b and c are constant parameters. This is linearizable again by the transformation $x = P/Q$. Note that any other replacement of x^2 such as x_n^2, x_{n+1}^2, etc. lead in general to nonintegrable mappings.

2. Integrable Elliptic Function Maps [4]

The work of R.J. Baxter on exactly solvable statistical models has shown that elliptic functions obey addition relations which can be interpreted as discrete equations. As a result the two-point correspondence

$$\alpha x_{n+1}^2 x_n^2 + \beta x_{n+1} x_n \left(x_{n+1} + x_n\right) + \gamma \left(x_{n+1}^2 + x_n^2\right) \\ + \epsilon x_{n+1} x_n + \zeta \left(x_{n+1} + x_n\right) + \mu = 0 , \tag{10.143}$$

where $\alpha, \beta, \gamma, \epsilon, \zeta$ and μ are constants, can be parameterized in terms of elliptic functions. In other words, the x_n's are given in terms of elliptic functions over an one-dimensional mesh of points. From a different point of view, one can show that the three-point mapping

$$x_{n+1} = \frac{f_1(x_n) - x_{n-1} f_2(x_n)}{f_2(x_n) - x_{n-1} f_3(x_n)} , \tag{10.144}$$

where f_i's are specific functions of x_n, has an integral of the form (10.143) which is independent of n. Their specific forms are as follows:

$$f_1 (x_n) = u_1 u_2 - u_3 u_4 , \tag{10.145a}$$

$$f_2 (x_n) = u_5 u_4 - u_1 u_6 , \tag{10.145b}$$

$$f_3 (x_n) = u_3 u_6 - u_5 u_2 . \tag{10.145c}$$

where

$$\begin{aligned} u_1 &= \left(\gamma_0 x_n^2 + \lambda_0 x_n + \mu_0\right) , & u_2 &= \left(\beta_1 x_n^2 + \epsilon_1 x_n + \lambda_1\right) , \\ u_3 &= \left(\beta_0 x_n^2 + \epsilon_0 x_n + \lambda_0\right) , & u_4 &= \left(\gamma_1 x_n^2 + \lambda_1 x_n + \mu_1\right) , \\ u_5 &= \left(\alpha_0 x_n^2 + \beta_0 x_n + \gamma_0\right) , & u_6 &= \left(\alpha_1 x_n^2 + \beta_1 x_n + \gamma_1\right) , \end{aligned} \tag{10.145d}$$

α_i, β_i, γ_i, λ_i, μ_i, $i = 0, 1$ are arbitrary parameters. The associated integral is

$$I_n = \frac{u_5 x_{n-1}^2 + u_3 x_{n-1} + u_1}{u_6 x_{n-1}^2 + u_2 x_{n-1} + u_4} . \tag{10.146}$$

Finally one may note that by considering the quantities f_i's as functions of x_n as well as of n, that is deautonomising (10.144), for suitable choices one can obtain the so-called discrete Painlevé equations [4].

Example: The McMillan Map

With $f_3 = 0$, $f_2 = 1$, $f_1 = 2\mu x_n/(1 - x_n^2)$, the map (10.144) degenerates into the so-called McMillan map,

$$x_{n+1} + x_{n-1} = \frac{2\mu x_n}{1 - x_n^2} . \tag{10.147}$$

The solution to (10.147) can be written as

$$x_n = x_0 \mathrm{cn}(\Omega n, k) , \quad k = x_0 \frac{\mathrm{dn}\Omega}{\mathrm{sn}\Omega} , \quad \mu = \frac{\mathrm{cn}\Omega}{\mathrm{dn}^2\Omega} = \text{constant} . \tag{10.148}$$

It has an integral of motion of the form (10.146). Also the mapping (10.144) or (10.147) are time-discretized versions of suitable Hamiltonian systems or symplectic ones. Since they possess integrals of motion, it ensures their complete integrability in the sense of Liouville.

3. Integrable Trigonometric Function Maps

One can also consider integrable discrete maps possessing trigonometric solutions. For example the simple map

$$x_{n+1}x_{n-1} = x_n^2 - \omega_0^2 \tag{10.149}$$

possesses exact solutions of the form

$$x_n = A\sin(n\omega T + \delta) \tag{10.150a}$$

where A and δ are constants and

$$\sin^2 \omega T = \frac{\omega_0^2}{A^2}\,. \tag{10.150b}$$

Equation (10.149) can be considered as the integrable discrete version of the system (2.28) in the $\lambda \to \infty$ limit. The associated integral is given by

$$I = \frac{(x_{n+1} + x_{n-1})}{x_n} = 2\cos\omega T\,. \tag{10.151}$$

Exercises:

12. Show that the Riccati equation (10.141) can be transformed into a linear second-order differential equation by the transformation $x = P/Q$.
13. Show that the map (10.142) is linearizable through suitable transformation $x = P/Q$.
14. Prove that the form (10.148) is indeed the solution of the McMillan map.
15. Show that (10.150) satisfies the map (10.149).

10.10 Integrable Dynamical Systems on Discrete Lattices

So far we have considered integrable finite dimensional/degrees of freedom (continuous and discrete) systems and analysed them using various techniques. Now there also exists many interesting classes of integrable dynamical systems on discrete lattices in one spatial dimension (and also sometimes in higher spatial dimensions). The most famous of them is the so-called Toda lattice, originally introduced by M. Toda in the year 1967. It consists of an one-dimensional array of equal mass points connected by nearest neighbour couplings and having nonlinear forces of exponential type. Typically, its form reads as (see Chap. 14 and appendix I for details)

$$\ddot{x}_n = \exp\left[-(x_n - x_{n-1})\right] - \exp\left[-(x_{n+1} - x_n)\right] , \quad n = 0, 1, 2, ... \quad (10.152)$$

Using addition theorems for elliptic functions, Toda had demonstrated that the lattice (10.150) admits travelling wave type solutions represented by Jacobian elliptic functions. Most interestingly, a limiting form of such a wave solution is the localized pulse-like form, called the solitary wave. In fact this solitary wave has further remarkable properties under collision, leading to the concepts of soliton in nonlinear discrete lattices and integrable discrete lattices. We will discuss further details of these aspects in the following chapters, see especially Chap. 14. and appendix I.

Another interesting and important integrable discrete nonlinear lattice is the so-called Ablowitz–Ladik lattice. The equation of motion associated with it is given by [15]

$$\mathrm{i}\dot{q}_n = (q_{n+1} + q_{n-1} - 2q_n) + q_n^* q_n (q_{n+1} + q_{n-1}) , \quad (10.153)$$

where $q_n(t)$ is a complex dynamical variable. Some of the properties of this system will be discussed in appendix I. Similarly, yet another interesting integrable nonlinear lattice system which has found considerable application in recent times is the Calogero–Moser system (see for example Ref. [16]). The equation of motion of this system reads

$$\ddot{x}_i = 2g^2 \sum_{i(\neq k)} \frac{1}{(x_i - x_k)^3} - \omega^2 x_i, \quad i = 0, 1, 2, ... \quad (10.154)$$

Again some details of the system are given in appendix I. There are several other such integrable nonlinear lattices [15,16] known in the recent literature, which are considered to be discrete integrable versions of the so-called soliton systems (treated later in Chaps. 11–14). For more details the relevant literature may be referred.

10.11 Conclusion

We have seen that there exist typical integrable nonlinear dynamical systems which show completely regular properties in contrast to the chaotic systems investigated earlier. Though they are relatively few, they can be systematically identified by suitable methods through appropriate criteria such as complex integrability (Painlevé property), Liouville integrability (existence of sufficient integrals of motion or symmetries) and so on. Such integrable systems exist even in the case of nonlinear discrete dynamical systems. These and other related results have led to intensive investigations on the structure and properties of integrable dynamical systems and a separate branch namely, *integrable systems* has developed in the field of nonlinear dynamics.

In particular, associated with a class of integrable systems the notion of 'soliton' has been developed with great ramifications in the mathematical study and physical applications of nonlinear systems. We will take up some of these aspects in the next few chapters.

Problems

1. Find the integral (Hamiltonian) for each of the following equations of motion:
   (a) ***Particle on a rotating parabola***
   $$\left(1+\lambda x^2\right)\ddot{x}+\lambda x\dot{x}^2+\omega_0^2 x=0 \, .$$
   (b) ***A nonpolynomial oscillator***
   $$\left(1+\lambda x^2\right)\ddot{x}-\lambda x\dot{x}^2+\omega_0^2 x=0 \, .$$
   Also obtain the corresponding solutions.
2. Prove that (a) the Kepler problem and (b) the isotropic three-dimensional oscillator are integrable in the Liouville sense by identifying the required number of involutive integrals of motion.
3. Investigate the Painlevé property of the equation
   $$v\ddot{v}-3\dot{v}^2+v^4=0$$
   and show that it satisfies the P-property. Obtain the first integral and the explicit solution.
4. Show that the simplified Rikitake system (10.4) satisfies the Painlevé property and that the leading order singularity is a pole of order 1.
5. Show that the Euler's equations describing the torque free motion of a rigid body given in Sect. 10.2 satisfy the Painlevé property.
6. Investigate the Painlevé property of the Duffing oscillator
   $$\ddot{x}+p\dot{x}+\omega_0^2 x+\beta x^3=f\cos\omega t$$
   and show that the P-property holds good only for the choices (i) $f=0$, $p=0$ and (ii) $f=0$, $p=\pm(3/\sqrt{2}\,)\omega_0$ [17]. Solutions for these cases are discussed in Chap. 2 and appendix A.
7. Show that the equation of motion of the driven pendulum
   $$\ddot{x}+\alpha\dot{x}+\omega_0^2\sin x=\gamma\cos\omega t$$
   admits the Painlevé property only for the case $\gamma=0$, $\alpha=0$ [18]. (Hint: Use the transformation $y=\exp ix$ and then carry out the P-analysis).
8. Show that the oscillator equation
   $$\ddot{x}+\dot{x}+\frac{1}{2}x^2-a=0\, ,$$

where a is a parameter satisfies the P-property only for the value $a = 18/625$. Also show that for this choice, the equation can be transformed to the form

$$\mathrm{d}^2w/\mathrm{d}z^2 + 6w^2 = 0\ ,$$

where

$$x(t) = \mathrm{e}^{-27/5}w(z) + (6/25),\ \ t = \frac{-5\mathrm{i}}{\sqrt{12}}\mathrm{e}^{-z/5}\ .$$

Then obtain the general solution for this value of a.

9. Carry out the Painlevé analysis for the Kovalevskaya rigid body problem modelled by (10.44) and verify that only for the parametric choices given by (10.45) and (10.46) the P-property holds [19].
10. Verify that the integrals of motion (10.47) are independent in the sense of Poisson brackets (10.48).
11. Carry out a Painlevé analysis of the equation

$$w'' = 6w^2 + f(z)$$

and show that the P-property holds only for the case $f(z) = az + b$, where a and b are constants. Relate the resultant equation to the Painlevé I transcendental equation.

12. Verify that the P_{II} equation given by (10.56) and P_{III} equation given by (10.57) satisfy the P-property.
13. Carry out the Painlevé analysis for the generalized Hénon–Heiles system described by the equation of motion (10.118) and verify that the P-property holds only for the three integrable cases given by the parametric restrictions (10.126a) and (10.127).
14. Consider the system of three-coupled quartic anharmonic oscillator given by the Hamiltonian

$$H = \frac{p_x^2}{2} + \frac{p_y^2}{2} + \frac{p_z^2}{2} + Ax^2 + By^2 + Cz^2 + \alpha x^4 + \beta y^4 + \gamma z^4 \\ + \delta x^2y^2 + \epsilon y^2z^2 + \omega x^2z^2\ .$$

Write down the equation of motion and carry out a Painlevé analysis. Show that there exists a set of eight distinct parametric choices for which the P-property holds [7].

15. (a) Show that the free particle equation of motion $\ddot{x} = 0$ admits an eight parameter symmetry group of point transformations.
(b) Show that the linear harmonic oscillator equation admits an eight parameter symmetry group of point transformations.
16. Obtain the conditions on parameters and the associated second integrals of motion for the perturbed Kepler problem corresponding to the Lagrangian

$$L = \frac{1}{2}\left(\dot{x}^2 + \dot{y}^2\right) - \left[-\frac{g}{\sqrt{x^2+y^2}} + ax^M + by^N\right]$$

using Lie symmetry analysis [20]. Here g, a, b, M and N are constants.

17. Show that the system described by the inverse square potential

$$L = \frac{1}{2}\left(\dot{x}^2 + \dot{y}^2\right) - \left[Ax^2 + \dot{B}y^2 + \left(x^2 + y^2\right)^2 + \frac{C}{x^2} + \frac{D}{y^2}\right]$$

is integrable for any real values of the parameters A, B, C and D [20].

18. By applying P-analysis to the system

$$H = \frac{1}{2}\left(p_u^2 + p_v^2\right) + \frac{1}{2}\left(u^2 + v^2\right) + A\left(u^6 + v^6\right) + B\left(u^4v^2 + u^2v^4\right),$$

which occurs in the study of dynamical symmetries of perturbed hydrogen atom [21], show that it is integrable at least for the three case (i) $B = 0$, (ii) $B = 3A$, and (iii) $B = 15A$. Using symmetry analysis, find the corresponding second integrals of motion.

19. Using the direct method, find the second integral of motion for the perturbed Kepler problem given in problem 17.
20. Using the direct method, deduce the second integral of motion for the inverse square potential given in the problem 18 above.
21. For all the three integrable cases of the Hamiltonian given in problem 17, find the corresponding second integral of motion using the direct method.
22. Considering the map

$$x_{n+1} = \frac{a_n x_n + b_n}{x_n + c_n},$$

linearize it through the transformation $x_n = (y_{n+1}/y_n) - c_n$ and solve.

23. Show that the mapping

$$cx_n^2\left(x_{n+1} + x_{n-1}\right) + \alpha - x_n = 0 \quad (c, \alpha : \text{constants})$$

has an integral

$$I = cx_n^2x_{n+1}^2 - x_nx_{n+1} + \alpha\left(x_n + x_{n+1}\right).$$

24. Show that the Harper map [22]

$$x_{n+1} = f\left(x_n, \phi_n\right) = -\frac{1}{\left[x_n - E + 2\epsilon\cos 2\pi\phi_n\right]},$$

with rigid rotation dynamics $\phi_n = n\omega + p_0$, can be linearized into the Harper equation

$$\psi_{n+1} + \psi_{n-1} + 2\epsilon\cos\left[2\pi\left(n\omega + \phi_0\right)\right]\psi_n = E\psi_n$$

through the transformation $x_n = \psi_{n-1}/\psi_n$. Show that for $\epsilon > 1$, the two Lyapunov exponents are 0 and $-\left[\ln\epsilon\right]^{-1}$, so that only regular motion exists [23].

25. Show that the Hamiltonian which gives rise to the Toda equation (10.150) is

$$H = \frac{1}{2}\sum_n p_x^2 + \sum_n \phi\left(x_{n+1} - x_n\right),$$

where the nonlinear force

$$\phi = \exp\left[-\left(x_{n+1} - x_n\right)\right] - \left(x_{n+1} - x_n\right) .$$

26. Show that the equation of motion for the Calogero–Moser system (10.152) follows from the Lagrangian

$$L = \frac{1}{2}\left[\sum_i p_i^2 + g^2 \sum_{i(\neq j)} \frac{1}{(q_i - q_j)^2} - \omega^2 \sum_i q_i^2\right] .$$

27. The system

$$\dot{x} = x^2 + 3xy ,$$
$$\dot{y} = y^2 + 3xy$$

under the change of variables $u = x^2 + y^2$, $v = x + y$ takes the form

$$\dot{u} = 2v^3 ,$$
$$\dot{v} = -2u + 3v^2 .$$

Find whether both the above systems posses P-property. Also find the symmetries associated with the above systems and the integral of motion through symmetry analysis.

28. Show that the system

$$\dot{u} = vw ,$$
$$\dot{v} = uw ,$$
$$\dot{w} = uv$$

possesses the Painlevé property. Show that it can also be rewritten as

$$u\dddot{u} - u\ddot{u} - 4\dot{u}u^3 = 0 .$$

Now carry out a Lie symmetry analysis to determine the associated infinitesimals and hence the invariants via reduction of order.

29. Find for what values of λ, the system

$$\dot{u} = (2u + \lambda v)w ,$$
$$\dot{v} = (-\lambda u + 2v)w ,$$
$$\dot{w} = u^2 + v^2 + w^2$$

passes the Painlevé test. Find the corresponding integrals.

30. Show that the system

$$\dot{u} = (v - w)u - vx ,$$
$$\dot{v} = [bu + (b - 2)w]v ,$$
$$\dot{w} = -(u - w)w - vx ,$$
$$\dot{x} = -[(b - 1)u + (b - 3)w]x$$

does not possess P-property for any value of b.

11. Linear and Nonlinear Dispersive Waves

So far, in the previous chapters, we have considered nonlinear systems with finite degrees of freedom. The dynamics of these systems are governed by nonlinear ordinary differential equations (and maps). However, very often, the variation of a particular physical property may depend in a continuous fashion on both space and time variables. Typical examples include the vibrations of a string, propagation of electromagnetic waves, waves on the surface of water and so on. Again such systems may be classified into linear and nonlinear ones. In these cases, the dynamics of the underlying systems are governed by *partial differential equations* and they are often considered to be infinite dimensional (in phase space) or as infinite degrees of freedom continuous systems. We are all familiar with many of the linear systems described by linear partial differential equations like the sound equation, heat equation, Maxwell equations, Laplace equation, Schrödinger equation and so on. Naturally one would expect many interesting and novel phenomena to occur in nonlinear continuous systems also in analogy with the case of finite dimensional systems.

Among the nonlinear continuous systems, we will concentrate on the special case of the so-called *nonlinear dispersive wave systems* in the present and next few chapters while we will relegate the study of nonlinear diffusive systems to Chap. 15. The main reason is that a certain class of these nonlinear dispersive systems exhibits an exciting type of wave solution of finite energy, having remarkable stability properties, called *soliton*. Physically the solitons behave like stable particles, exhibiting in general elastic collision property on collision with other solitons in one spatial dimension. They occur in a wide variety of physical systems and lead to the identification of *completely integrable, infinite dimensional, nonlinear dynamical systems*. As the solitons are basically solitary waves, we will study such waves also in some detail in this chapter. As a prelude to such a study we will start our discussion with an introduction to the properties of linear dispersive waves.

11.1 Linear Waves

We are all familiar with different kinds of waves even in our day to day life. Human beings are always fascinated by them. But what is a wave? One can define it to be the result of some physical event that sends tremors or disturbances through a medium. Such tremors carry energy away from the point of disturbance [1].

Examples:

(1) When a stone is thrown into a still pond, the waves radiate in a circular pattern: the energy given by the stone is transferred into the motion on the surface of the pond. This motion sets in waves that travel in the form of periodic undulations with crests and troughs but overall in the form of a wave packet.

(2) When a girl plays (on) a veena, the notes we hear, like all sounds, are the result of vibrations pulsating through the air.

(3) When a radio station broadcasts its signals, the electromagnetic energy from its transmitter radiates outward in an identical fashion.

These are all examples of linear dispersive waves. In each of the cases, the waves disperse and diminish over distance. The water settles, the melody fades and the signal weakens. Why is it so? One way to understand them is to note that they are all described by linear partial differential equations which admit propagating wave solutions associated with dispersion relations connecting frequency and wave number. Coupled with the linear superposition principle, then it is easy to see that these linear dispersive systems cannot sustain waves of permanence. In order to appreciate these aspects, we will first consider the wave propagation in nondispersive linear media and then proceed to the case of dispersive media.

11.2 Linear Nondispersive Wave Propagation

Let us consider the problem of one-dimensional longitudinal vibrations of small amplitude in an elastic rod. It can be studied in the simplest approximation by treating the elastic rod as a discrete system of mass points, each of mass m, connected by *harmonic force springs* of strength $\widehat{k}$ (Fig. 11.1). When the equilibrium distance 'a' between neighbouring points approaches zero, the corresponding equation of motion for the displacements is [2]

$$\frac{m}{a}\ddot{\eta}_i = \widehat{k}a\frac{\eta_{i+1}-\eta_i}{a^2} - \widehat{k}a\frac{\eta_i-\eta_{i-1}}{a^2}\,, \quad i = 1, 2, \ldots \tag{11.1}$$

In the continuous limit, $a \to 0$, it takes the form of a linear partial differential equation, namely, the so-called *(linear)wave equation* for the displacement $u(x,t) = \eta(x = na, t)$,

$$\frac{1}{c^2}\frac{\partial^2 u}{\partial t^2} - \frac{\partial^2 u}{\partial x^2} = 0\,, \quad c^2 = \frac{\widehat{k}a^2}{m}\,. \tag{11.2}$$

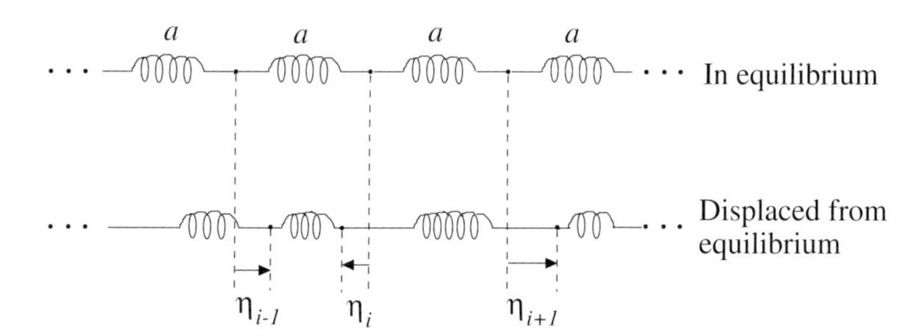

Fig. 11.1. A discrete system of equal mass points connected by springs, as an approximation to a continuous elastic rod

Similarly, the transverse motion of a stretched elastic string can also be described by a wave equation of the above form.

Equation (11.2) admits progressing waves of the form

$$u = A\cos(\omega t - kx + \delta) \tag{11.3a}$$

with

$$\omega = \pm ck\ , \qquad A\ , \delta = \text{ constants}\ , \tag{11.3b}$$

which travel with a *wave* or *phase velocity* $v_p = \pm c$. Even more generally, (11.2) admits solutions of the form

$$u(x,t) = f(x - ct) + g(x + ct)\ , \tag{11.4}$$

where f and g are arbitrary functions, subject to appropriate initial and boundary conditions. Physically, this corresponds to an arbitrary disturbance which is a combination of two wave profiles propagating with a velocity c to the left or right, but *without change of form* (as shown in Fig. 11.2), just as in the case of plucking the string of a veena. It is a *nondispersive* wave propagation.

Exercises:

1. Obtain the general solution of the wave equation (11.2), subject to the initial condition $u(x,0) = \text{sech}x$, $\partial u(x,0)/\partial t = \tanh x$.
2. Express Maxwell's equations in vacuum as wave equations for the vector and scalar potentials. Write down the corresponding wave solutions.

11.3 Linear Dispersive Wave Propagation

In the above, the rod is assumed to be perfectly elastic, the string is considered to be perfectly flexible and so on. The real strings and rods are stiff, tend to

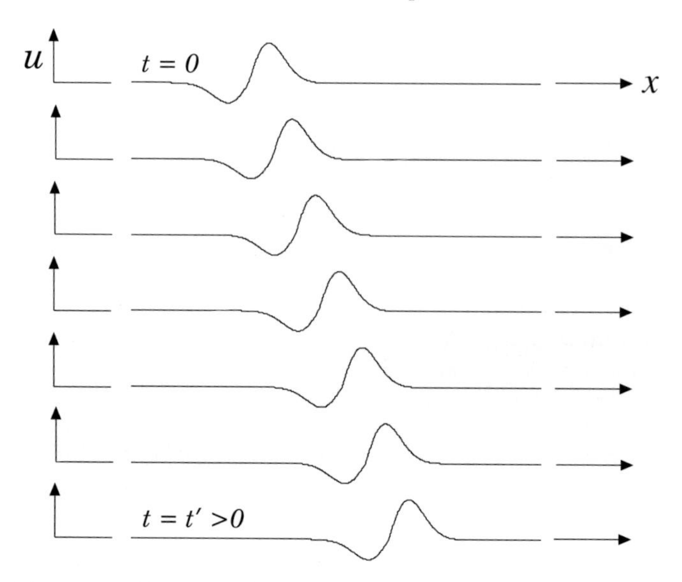

Fig. 11.2. Dispersionless wave propagation. Note the preservation of shape of the wave profile during propagation

straighten and restore. The inclusion of such effects necessarily modifies the wave equation into forms such as

$$\frac{1}{c^2}\frac{\partial^2 u}{\partial t^2} - \frac{\partial^2 u}{\partial x^2} + m^2 u = 0\,, \tag{11.5a}$$

$$\frac{1}{c^2}\frac{\partial^2 u}{\partial t^2} - \frac{\partial^2 u}{\partial x^2} + \alpha^2 \frac{\partial^4 u}{\partial x^4} = 0 \tag{11.5b}$$

and so on. The sinusoidal travelling wave solutions of the form (11.3a) now exist only if

$$\omega^2 = c^2\left(k^2 + m^2\right)\,, \quad \omega = \pm\, c\sqrt{k^2 + m^2} \tag{11.6a}$$

for (11.5a) and

$$\omega^2 = c^2 k^2\left(1 + \alpha^2 k^2\right)\,, \quad \omega = \pm\, ck\sqrt{1 + \alpha^2 k^2} \tag{11.6b}$$

for (11.5b). Then, the wave translates with a phase velocity of the form

$$v_p = \pm\, c\frac{\sqrt{k^2 + m^2}}{k} \tag{11.7a}$$

in the case of (11.5a) and

$$v_p = \pm\, c\,\sqrt{1 + \alpha^2 k^2} \tag{11.7b}$$

in the case of (11.5b) (see Fig. 11.3). In both the above examples, the wave or phase velocity depends on the wave number/wavelength, which is the hallmark of *dispersion*. Relations of the form (11.6a) and (11.6b) connecting the

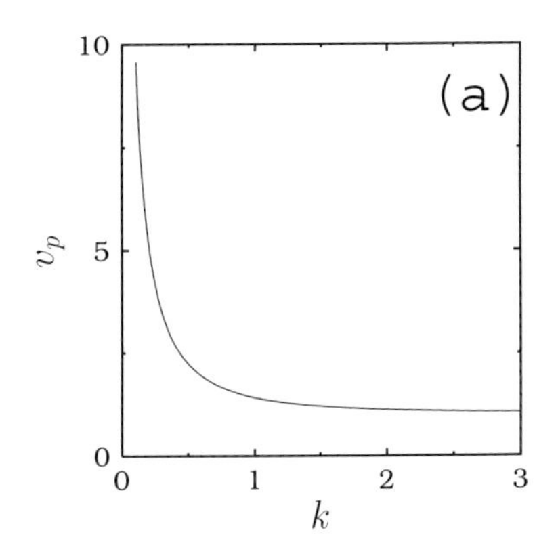

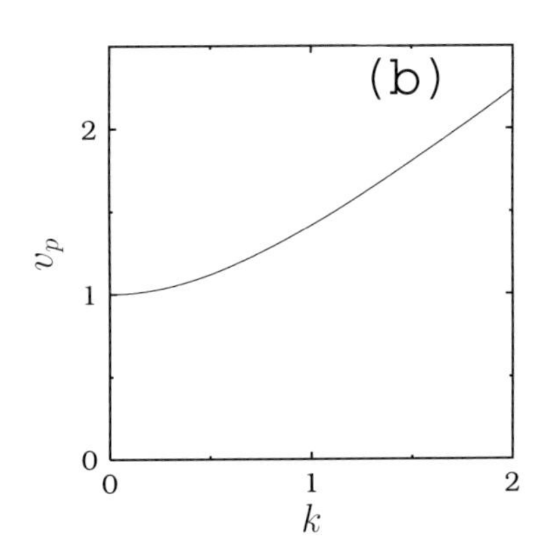

Fig. 11.3. Behaviour of wave or phase velocity as a function of wave number (**a**) $v_p = c\sqrt{k^2+m^2}/k$ and (**b**) $v_p = c\sqrt{1+\alpha^2 k^2}$. We have chosen $c = 1$, $m = 1$ and $\alpha = 1$

angular frequency ω and wave number k are called *dispersion relations*. From these relations it is clear that wave packets which are linear combinations of waves of different wave numbers cannot be propagated as wave packets without change of shape (see below). The wave group will get dispersed and die down in due course. To realize this aspect more specifically, let us consider the general solution of the initial value problem of any linear dispersive wave equation of the form (11.5).

Exercise:

3. Under certain circumstances the unidirectional wave propagation in shallow water surfaces can be described by the so-called linearized Korteweg-de Vries equation $u_t + u_{xxx} = 0$. Show that the associated dispersion relation is $\omega = -k^3$.

11.4 Fourier Transform and Solution of Initial Value Problem

Consider a general linear wave equation in one-dimension, associated with the dispersion relation

$$\omega = \omega(k)\,. \tag{11.8}$$

The corresponding evolution equation, which is a linear partial differential equation in one space and one time dimensions, can be obtained by the replacements,

$$\omega \longrightarrow \mathrm{i}\frac{\partial}{\partial t}\,, \quad k \longrightarrow -\mathrm{i}\frac{\partial}{\partial x} \tag{11.9a}$$

so that we have the linear dispersive wave equation

$$\mathrm{i}\frac{\partial}{\partial t}u(x,t) = \omega\left(-\mathrm{i}\frac{\partial}{\partial x}\right)u(x,t)\ . \tag{11.9b}$$

Exercise:

4. Apply the transformations (11.9a) to (11.6) and obtain (11.5).

Then the *initial value problem (IVP)* is the following: *Given the initial condition* $u(x,0)$, *find the general solution* $u(x,t)$ *for all* $t>0$ *for specified boundary conditions (say* $u \overset{|x|\to\infty}{\longrightarrow} 0$*).*

The IVP of finding $u(x,t)$ from $u(x,0)$ for linear dispersive systems of the form (11.9b) can be solved through a three step algorithmic procedure, involving direct and inverse Fourier transforms. The method is as follows.

(i) *Direct Fourier Transform (DFT)*

Considering the general solution $u(x,t)$ of (11.9b), its direct Fourier transform is

$$\widehat{u}(k,t) = \int_{-\infty}^{\infty} u(x,t)\mathrm{e}^{-\mathrm{i}kx}\mathrm{d}x\ . \tag{11.10}$$

In particular from the given initial data $u(x,0)$, the associated Fourier coefficient can be obtained as

$$\widehat{u}(k,0) = \int_{-\infty}^{\infty} u(x,0)\mathrm{e}^{-\mathrm{i}kx}\mathrm{d}x\ . \tag{11.11}$$

(ii) *Time Evolution*

How do the Fourier coefficients $\widehat{u}(k,t)$ evolve in time? It is quite easy to see.

Taking the time derivative of (11.10) on both sides, then using (11.9b) into the right hand side, carrying out partial integrations for a sufficient number of times and applying suitable boundary conditions, for example $u(x,t)\to 0$ as $x\to\pm\infty$, we can easily check that the Fourier coefficients $\widehat{u}(k,t)$ evolve as

$$\mathrm{i}\frac{\mathrm{d}\widehat{u}(k,t)}{\mathrm{d}t} = \omega(k)\widehat{u}(k,t)\ . \tag{11.12}$$

The equation (11.12) is a simple, first order, *linear ordinary* differential equation with a constant coefficient. It can be trivially integrated to give

$$\widehat{u}(k,t) = \widehat{u}(k,0)\mathrm{e}^{-i\omega(k)t}\ . \tag{11.13}$$

This is how the Fourier coefficient for any k, $-\infty<k<\infty$, evolves in time.

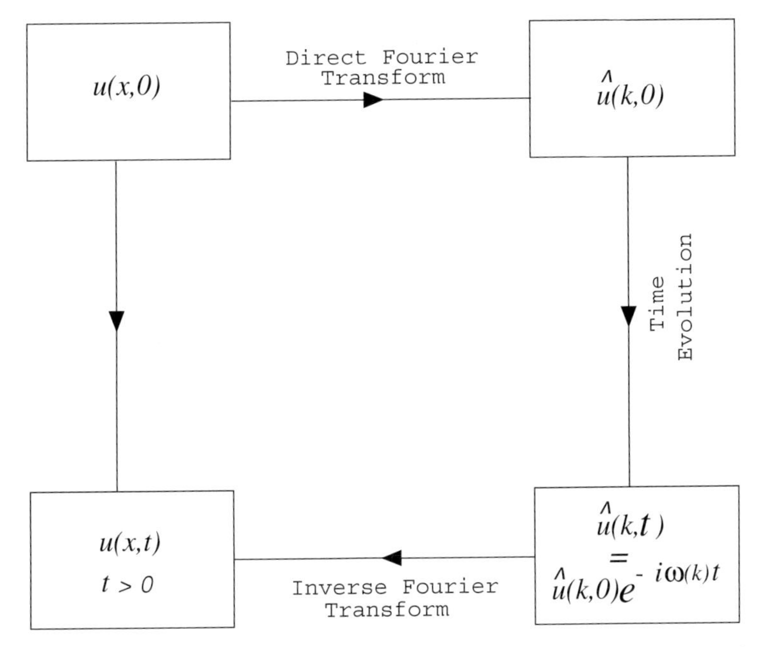

Fig. 11.4. Schematic diagram of the Fourier transform method to solve the initial value problem of linear dispersive wave equations

(iii) *Inverse Fourier Transform (IFT)*

Knowing the Fourier coefficients $\widehat{u}(k,t)$ at an arbitrary time, t, by starting from the initial data $\widehat{u}(k,0)$ obtained from (11.11), we use them in (11.10) and invert them using the inverse Fourier transform to obtain

$$
\begin{aligned}
u(x,t) &= \frac{1}{2\pi}\int_{-\infty}^{\infty} \widehat{u}(k,t)\, \mathrm{e}^{ikx}\mathrm{d}k \\
&= \frac{1}{2\pi}\int_{-\infty}^{\infty} \widehat{u}(k,0)\, \mathrm{e}^{\mathrm{i}(kx-\omega(k)t)}\mathrm{d}k\ .
\end{aligned}
\tag{11.14}
$$

The solution (11.14) may be considered as the most general solution of the IVP of the linear dispersive equation (11.9b). The three step algorithmic Fourier transform method is schematically illustrated in Fig. 11.4.

Exercise:

5. Carry out a similar analysis as above for the dispersion relation $\omega^2 = \omega^2(k)$, and verify that the above approach holds good for linear partial differential equations of second-order in the time variable also, such as (11.5).

11.5 Wave Packet and Dispersion

The general solution (11.14) can be obviously considered as a linear superposition of elementary wave solutions of (11.9b) of the form

$$u(x,t) = A(k)\mathrm{e}^{i(kx-\omega t)}\ , \tag{11.15}$$

over a continuous range of values of k, $-\infty < k < \infty$ (note that integration can be considered as a limiting process of summation). Or in other words, the solution (11.14) corresponds to a *wave packet* or a *wave group* of a large number of waves of the form (11.15). Figure (11.5) gives a schematic representation of superposition of progressing waves, giving rise to a wave packet. While any elementary wave of the form (11.15) moves with a *wave* or

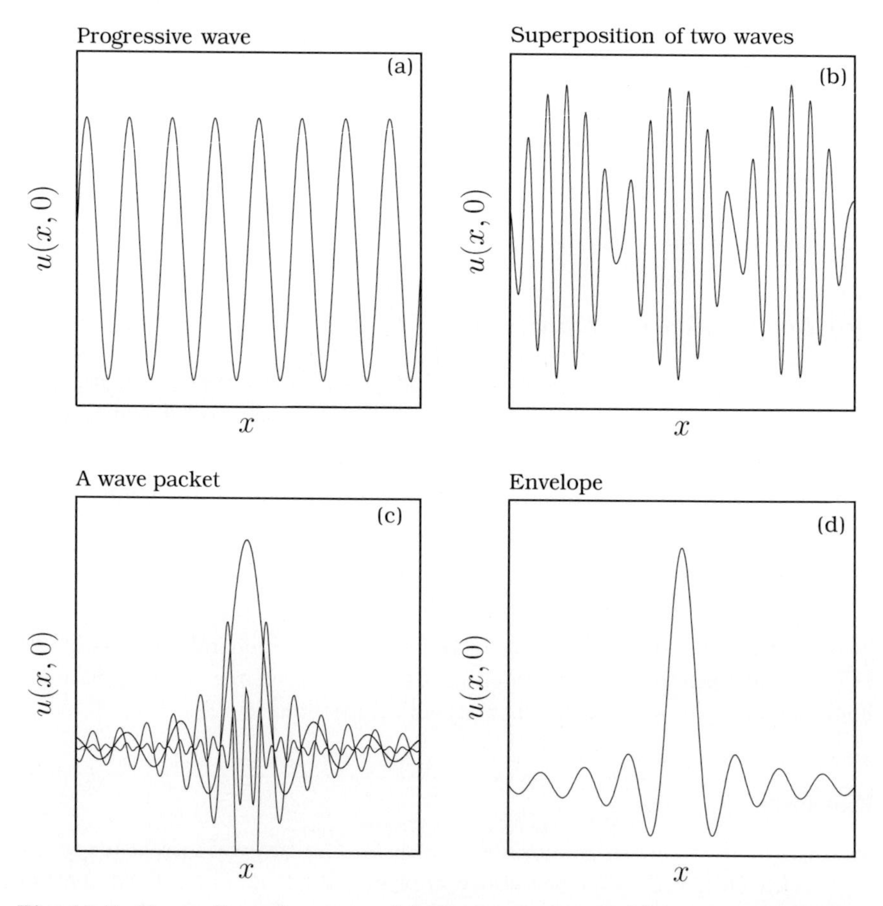

Fig. 11.5. Linear dispersive waves. (**a**) Progressive wave, (**b**) superposition of two linear waves, (**c**) wave packet, superposition of large number of progressing waves and (**d**) envelope of the wave packet at a fixed time (say $t = 0$)

phase velocity $v_p = \omega/k$, how does the wave group (11.14) itself move? Does it move with the same velocity v_p or with a different velocity? The answer to the above question can be obtained through a straight-forward asymptotic analysis (recall such analysis of quantum mechanical wave packets in elementary quantum mechanics to establish whether the wave function described by the Schrödinger equation can be ascribed a particle-like localization property or not, see for example [3]).

In order to realize the nature of the wave packet, let us assume the simple case in which the major contribution to the integral for the wave packet in (11.14) comes from groups of waves in a small neighbourhood of $k = k_0$, $(k_0 - \triangle k < k < k_0 + \triangle k)$. Then the frequency can be expanded as a Taylor series about $k = k_0$,

$$\begin{aligned}\omega(k) &= \omega\left(k - k_0 + k_0\right) = \omega\left(k_0 + \triangle k\right) \\ &= \omega\left(k_0\right) + \mathrm{d}\omega/\mathrm{d}k|_{k=k_0}\triangle k + \text{ higher orders in } (\triangle k)\,,\end{aligned} \tag{11.16}$$

so that the solution (11.14) can be closely approximated as

$$\begin{aligned}u(x,t) &\cong A\left(k_0\right)\,\mathrm{e}^{-\mathrm{i}[\omega(k_0)t - k_0 x]}\int_{-\triangle k}^{\triangle k} \mathrm{e}^{\mathrm{i}\left(x-\omega' t\right)(k-k_0)}\mathrm{d}k \\ &= 2A\left(k_0\right)\,\mathrm{e}^{-\mathrm{i}[\omega(k_0)t - k_0 x]}\,\frac{\sin\left[\triangle k\left(x - \omega' t\right)\right]}{\left(x - \omega' t\right)}\,, \\ \omega' &= \mathrm{d}\omega/\mathrm{d}k|_{k=k_0}\,.\end{aligned} \tag{11.17}$$

A sketch of the wave packet at a given instant is shown in Fig. 11.5, which clearly shows that the wave packet is localized. In particular at $t = 0$, the initial instance, the localization is around $x = 0$ in the form of a nearly localized wave packet with a half-width $\pi/\triangle k$. This packet then moves with a *group velocity*

$$v_g = \frac{\mathrm{d}x}{\mathrm{d}t} = \frac{\mathrm{d}\omega}{\mathrm{d}k}\,, \tag{11.18}$$

which is in general not equal to v_p. It turns out that $v_g = v_p$ only when $\mathrm{d}\omega/\mathrm{d}k$ is independent of k (or ω is proportional to k), that is in the *dispersionless* case (11.2)–(11.3). In the latter case the arbitrary wave profile moves without change of shape. In all other cases, that is in the case of any linear dispersive system, $v_g \neq v_p$, and so the individual constituents of the wave packet start to move with their own wave velocities v_p, different from that of the velocity of the packet of which the individual waves are members. In other words, the wave packet spreads, disperses, diminishes and dies down in due course.

As a result the waves generated in the pond settles, the melody of the veena fades and the radio signal weakens. In short, linear dispersive systems cannot admit localized packets of waves over long distances and times and they have no character of permanence.

Exercise:

6. Illustrate graphically the formation of a wave packet, by combining (a) 2 linear waves, (b) 5 linear waves, (c) 10 linear waves, (d) 50 linear waves and (e) 100 linear waves with same amplitudes but nearby wave numbers at a fixed time, say $t = 0$.

11.6 Nonlinear Dispersive Systems

The waves we discussed so far are linear in nature. They are solutions of linear (dispersive)partial differential equations. Also they are less permanent in the presence of dispersion. They disperse or spread and hence diminish over distance. However, in nature not all waves are so gentle having less permanence. There are many cases of fairly permanent and powerful waves, which may correspond to both constructive and destructive natures. These waves have the capacity to travel extraordinary distances without virtually diminishing in size or shape. All these waves are generally described by *nonlinear* wave equations of dispersive type.

Examples:

Cyclonic waves, tsunami waves, earthquakes, tidal waves, electromagnetic waves in nonlinear optical fibres, solitary waves on shallow water surfaces and so on.

11.6.1 An Illustration of the Wave of Permanence

In 1960 a powerful earthquake struck southern Chile [1]. After levelling 400,000 homes, the earthquake has run its course. Along 1000 kms of coastline whole streets lay submerged, buildings were demolished and boats were lost. Fifteen hours after the Chilean quake, the tsunami – a monstrous wave–hit Hawaii, killing 61 people. Seven hours after, the Japanese islands of Honshu and Hokkaido were struck by a wall of water 21-feet high and 199 people drowned. The tsunami of 1960 was by no means unprecedented. Since ancient times historians have recorded more than two hundred similar waves, some of them more than 100ft high.

11.6.2 John Scott Russel's Great Wave of Translation

Inspite of all their destructive power, such waves have rarely been studied in detail earlier. Their main physical property, namely the capacity to travel extraordinary distances without diminishing in size has often been dismissed as an idle curiosity.

In 1830's the Scottish Naval Architect, John Scott Russel, was carrying out investigations on the shapes of the hulls of ships and speed and forces needed to propel them for Union Canal Company so that they can make safe steam navigation. In August 1834, riding on a horse back, Scott Russel observed the "Great Wave of Translation" in the Union canal connecting the Scottish cities of Edinburgh and Glasgow, where he was carrying out his experiments. He reported his observations to the British Association in his 1844 "Report on Waves" in the following delightful description.

"I believe I shall best introduce the phenomenon by describing the circumstances of my own first acquaintance with it. I was observing the motion of a boat which was rapidly drawn along the narrow canal by a pair of horses, when the boat suddenly stopped not so the mass of water in the channel which it had put in motion; it accumulated round the prow of the vessel in a state of violent agitation, then suddenly leaving it behind, rolled forward with great velocity, assuming the form of a large solitary elevation, a rounded, smooth and well-defined heap of water, which continued its course along the canal apparently without change of form or diminution of speed. I followed it on horse back, and overtook it still rolling on at a rate of some eight or nine miles an hour, preserving its original figure some thirty feet long and a foot to a foot and a half in height. Its height gradually diminished and after a chase of one or two miles I lost it in the windings of the canal..."

– J. Scott Russel

Like a tsunami, this rolling pile of water, a solitary wave, also somehow maintained its shape and speed much larger than the conventional wave which we discussed in the earlier sections. Scott Russel immediately realized that their distinct feature is their longevity and that they have so much staying power that he could use them to pump water uphill, which is not possible ordinarily.

Scott Russel also performed some laboratory experiments generating solitary waves by dropping a weight at one end of a water channel (see Fig. 11.6). He was able to deduce empirically that the volume of water in the wave is equal to the volume of water displaced and further that the speed, c, of the solitary wave is obtained from the relation

$$c^2 = g(h + a) \ , \tag{11.19}$$

where a is the amplitude of the wave, h is the undisturbed depth of water and g is the acceleration due to gravity. A consequence of (11.19) is that taller waves travel faster!

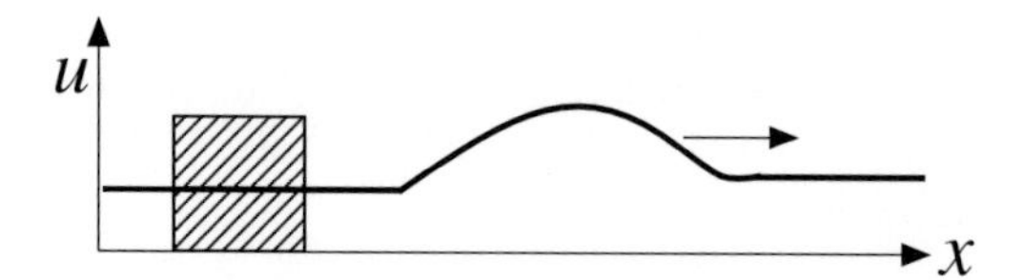

Fig. 11.6. Scott Russel's laboratory experiment: solitary wave in a tank

11.7 Cnoidal and Solitary Waves

To put Russel's formula on a firmer footing, both Boussinesq and Lord Rayleigh [4] assumed that a solitary wave has a length scale much greater than the depth of the water. They deduced from the equations of motion for an inviscid and incompressible fluid, Russel's formula, (11.19), for c. In fact they also showed that the wave profile is given by

$$u(x,t) = a \operatorname{sech}^2[\beta(x-ct)] \,, \tag{11.20a}$$

where

$$\beta^{-2} = \frac{4h^3 g(h+a)}{3a} \tag{11.20b}$$

for any $a > 0$, although sech^2 profile is strictly correct only if $a/h \ll 1$.

11.7.1 Korteweg–de Vries Equation and the Solitary Waves and Cnoidal Waves

The ultimate explanation of the Scott Russel phenomenon was provided by two Dutch physicists Korteweg and de Vries in 1895 [5]. They deduced the famous wave equation responsible for the phenomenon, which now goes by their names. The Korteweg–de Vries (KdV) equation is a simple nonlinear dispersive wave equation (the actual derivation is given in the next chapter). In its modern version it reads as

$$u_t + 6uu_x + u_{xxx} = 0 \,. \tag{11.21}$$

Let us look for elementary wave solutions of (11.21) in the form

$$u = 2f(x-ct) \tag{11.22a}$$

$$= 2f(\xi) \,, \quad \xi = x - ct \,. \tag{11.22b}$$

In view of (11.22b),

$$\frac{\partial}{\partial t} = -c\frac{\partial}{\partial \xi}, \quad \frac{\partial}{\partial x} = \frac{\partial}{\partial \xi}. \tag{11.23}$$

Therefore (11.21) can be reduced to an ordinary differential equation (ode)

$$-cf_\xi + 12ff_\xi + f_{\xi\xi\xi} = 0. \tag{11.24}$$

Equation (11.24) can also be written as

$$-cf_\xi + 6\left(f^2\right)_\xi + f_{\xi\xi\xi} = 0. \tag{11.25}$$

Integrating (11.25), we obtain

$$f_{\xi\xi} + 6f^2 - cf + d = 0, \tag{11.26}$$

where d is the integration constant. Multiplying (11.26) by f_ξ, we obtain

$$f_{\xi\xi}f_\xi + 6f^2f_\xi - cff_\xi + df_\xi = 0. \tag{11.27}$$

Equation (11.27) can be rewritten as

$$\left(\frac{1}{2}f_\xi^2\right)_\xi + \left(2f^3\right)_\xi - \left(\frac{c}{2}f^2\right)_\xi + df_\xi = 0. \tag{11.28}$$

Integrating (11.28) once, we obtain

$$\frac{1}{2}f_\xi^2 + 2f^3 - \frac{c}{2}f^2 + df + b = 0, \tag{11.29}$$

where b is the second integration constant. Equation (11.29) can be rewritten as

$$f_\xi^2 = -4f^3 + cf^2 - 2df - 2b \tag{11.30}$$

or

$$f_\xi^2 = -4\left(f-\alpha_1\right)\left(f-\alpha_2\right)\left(f-\alpha_3\right), \tag{11.31}$$

where α_1, α_2 and α_3 are the three real roots of the polynomial

$$P(f) = -4f^3 + cf^2 - 2df - 2b. \tag{11.32}$$

The solution of (11.31) can be expressed in terms of Jacobian elliptic function (see appendix A) as

$$f(\xi) = f(x-ct) = \alpha_3 - (\alpha_3-\alpha_2)\mathrm{sn}^2\left[\sqrt{\alpha_3-\alpha_1}\,(x-ct), m\right], \tag{11.33a}$$

where

$$(\alpha_1+\alpha_2+\alpha_3) = \frac{c}{4}, \quad m^2 = \frac{\alpha_3-\alpha_2}{\alpha_3-\alpha_1}. \tag{11.33b}$$

Equation (11.33) represents in fact the so-called *cnoidal wave* for obvious reasons.

Special Cases:

(i) $m \approx 0$: *Harmonic wave*

When $m \approx 0$, (11.33a) leads to elementary progressing harmonic wave solutions. This can be verified by taking the limit $m \to 0$ in (11.33). It can be compared with the wave solutions of the linearised version of (11.21) (see appendix A).

(ii) $m = 1$: *Solitary wave*

When $m = 1$, we have

$$f = \alpha_3 - (\alpha_3 - \alpha_2)\tanh^2\left[\sqrt{\alpha_3 - \alpha_1}\,(x - ct)\right] , \tag{11.34}$$

that is,

$$f = \alpha_2 + (\alpha_3 - \alpha_2)\ \mathrm{sech}^2\left[\ \sqrt{\alpha_3 - \alpha_1}\,(x - ct)\right] . \tag{11.35}$$

Choosing now $\alpha_2 = 0$, $\alpha_1 = 0$, we have

$$f = \frac{\alpha_3}{4}\ \mathrm{sech}^2\left[\sqrt{\alpha_3}\,(x - ct)\right] . \tag{11.36}$$

Using (11.33b), (11.36) can be written as

$$f = \frac{c}{4}\ \mathrm{sech}^2\left[\frac{\sqrt{c}}{2}\,(x - ct)\right] . \tag{11.37}$$

Substituting (11.37) into (11.22a), the solution can be written as

$$u(x,t) = 2f = \frac{c}{2}\ \mathrm{sech}^2\left[\frac{\sqrt{c}\,(x - ct)}{2}\right] . \tag{11.38}$$

This is of course the Scott Russel solitary wave (Fig. 11.7), as can be seen after suitable rescaling (see Sect. 12.5).

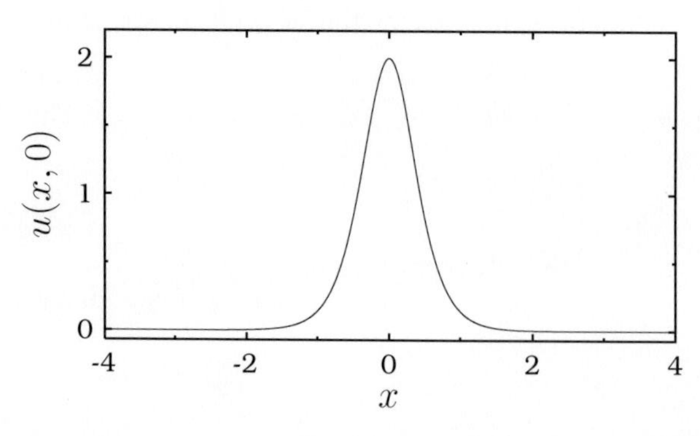

Fig. 11.7. Solitary wave solution (11.38) of the KdV equation (11.21). Here $c = 4$

Exercise:

7. Substitute (11.38) into (11.21) and make sure that (11.38) indeed is a solution of (11.21).

The characteristic feature of the above solitary wave is that the velocity of the wave ($v = c$) is directly proportional to the amplitude ($a = c/2$): the larger the wave, the faster it moves. Unlike the progressing wave, it is fully localized, decaying exponentially fast as $x \to \pm\infty$ (see Fig. 11.7). We will find in the next chapters that the above solitary wave is a remarkably stable entity so as to ascribe particle property to it. It is a purely nonlinear effect.

(iii) $0 < m < 1$: *Cnoidal waves*

When $0 < m < 1$, we have amplitude dependent elliptic function solution (11.33). Rewriting it in the form of $f = f(\omega t - kx)$, we can easily check from (11.33) that the dispersion relation is now amplitude-dependent:

$$\omega = \omega(k, a) \ . \tag{11.39}$$

For example, from the solution (11.33), we can identify $f = f(\omega t - kx)$, with

$$k = \sqrt{\alpha_3 - \alpha_1} \ , \quad \omega = \sqrt{\alpha_3 - \alpha_1}\, c = ck \ , \quad c = 4\,(\alpha_1 + \alpha_2 + \alpha_3) \ . \tag{11.40}$$

Using the relations in (11.33), one can establish a relation of the form (11.39).

11.8 Conclusions

Thus we see that when nonlinearity is present, we have in general nonlinear dispersive waves. The nature of the solution, though regular, is highly dependent on the amplitude or initial conditions. For suitable choices, it may even be possible to obtain solitary wave solutions (but *not necessarily in all systems!*). This will have profound implications on the dynamics. We will consider these aspects in the next chapters. But before this, we wish to see how the KdV equation itself can arise in actual physical contexts so as to understand the historical role it has played in the development of the concept of soliton. This we will take up in the next chapter.

Problems

1. Obtain the cnoidal wave solution of
   a) the (double well) ϕ^4 equation,
   $$\phi_{tt} - \phi_{xx} - m^2\phi + \lambda\phi^3 = 0, \quad m^2, \quad \lambda > 0$$
   and

b) the sine-Gordon equation,
$$\phi_{tt} - \phi_{xx} + m^2 \sin\phi = 0 \, .$$
Find the amplitude-dependent dispersion relations for these cases. Deduce the solitary waves and discuss their characteristic features.

2. Obtain the elliptic function wave solution for the (single well) ϕ^4 case,
$$\phi_{tt} - \phi_{xx} + m^2\phi + \lambda\phi^3 = 0, \ \ m^2, \ \ \lambda > 0$$
and obtain the dispersion relation. Show that no solitary wave exists in this case.

3. Find the travelling wave solution for the following nonpolynomial type field equation:
$$\left(1 + \lambda\phi^2\right)\left(\phi_{tt} - \phi_{xx}\right) - 2\lambda\left(\phi_t^2 - \phi_x^2\right)\phi + m^2\phi = 0 \, .$$
Show that the wave solution is given in terms of trigonometric (sine/cosine) function with a dispersion relation
$$\omega^2 - k^2 = m^2/\left(1 + \lambda A^2\right) \, .$$
Compare the result with the nonpolynomial oscillator equation (2.8).

4. Verify that the solitary wave solution of the modified KdV equation
$$u_t + 6u^2 u_x + u_{xxx} = 0$$
is
$$u(x,t) = k \ \mathrm{sech}\left(kx - k^3 t\right) \, .$$

5. Show that the travelling wave solution of the Boussinesq equation
$$u_{tt} - c^2 u_{xx} - \frac{3c^2}{2} u_{xx} - \frac{c^3}{3} u_{xxxx} = 0$$
is
$$u(x,t) = A \ \mathrm{sech}^2\left[\sqrt{3A/4c}\left(x - c\sqrt{1+A}\, t\right)\right] \, .$$

6. The model equation for the nonlinear modulation of stable plane waves in an unstable medium is given the nonlinear Schrödinger equation (see Chap. 14 also)
$$u_{tt} + \mathrm{i} u_x + 2u|u|^2 = 0 \, .$$
Show that its solitary wave solution is
$$u(x,t) = 2\beta \exp\left[-2\mathrm{i}\left\{\alpha x + 2\left(\alpha^2 - \beta^2\right)t + \alpha\delta_2\right\}\right] \\ \times \mathrm{sech}\left[2\beta\left(x + 4\alpha t + \delta_1\right)\right] \, ,$$
where α, β, δ_1 and δ_2 are constants.

7. Verify that the solitary wave solution of the fifth-order modified KdV equation

$$u_t + \frac{45}{2}\delta^2 u^2 u_x - u_{xxxxx} = 0$$

is

$$u(x,t) = \pm\sqrt{3k/\delta}\ \operatorname{sech}^2\left[A\left(\xi - \xi_0\right) - 1\right] ,$$

where

$$\xi = x - \frac{9}{2}\delta k t\ , \quad A = \frac{\sqrt{\delta}\left(e_1 - e_3\right)}{2}\ , \quad e_1 = -2e_3 = 2\sqrt{\frac{k}{3\delta}}$$

and ξ_0 and k are arbitrary constants.

8. Show that the solitary wave solution of the regularised long wave equation (RLW)

$$u_t - u_{xxt} + u_{xxx} + 6uu_x = 0$$

is

$$u(x,t) = \frac{2k^2}{1-4k^2}\operatorname{sech}^2\left[k\left(x - \frac{4k^2}{1-4k^2}t\right)\right],$$

where $k^2 > 0$ is an arbitrary constant.

9. Verify that the solitary wave solution of the higher-order KdV equation

$$u_t + auu_x + bu_{xxx} + cu_{xxxxx} = 0$$

is

$$u(x,t) = -\frac{105b^2}{169ac}\operatorname{sech}^4\left[\pm\frac{1}{2}\sqrt{\frac{-b}{13c}}\left(x + \frac{36b^2}{169c}\right)\right].$$

Sketch the solitary wave solution for (i) $b > 0$, $c < 0$, $a < 0$ and (ii) $b > 0$, $c < 0$ and $a > 0$. Verify that the solution is convex downward for the case (i) and convex upward for the case (ii) [6].

12. Korteweg–de Vries Equation and Solitons

In the previous chapter, we have seen that nonlinear dispersive systems can have characteristically different types of wave solutions compared to linear dispersive systems. In particular, the former can admit solitary wave solutions, and as an example we have demonstrated this for the case of the KdV equation. How does the KdV equation itself arise typically, how stable is the solitary wave and how do the solitary waves interact mutually? We will analyse these aspects in this chapter and bring out the fact that the KdV equation indeed describes the Scott Russel phenomenon. We also point out how the KdV equation recurs in an entirely different physical context, namely wave propagation in the famous Fermi–Pasta–Ulam (FPU) anharmonic lattice. Further we explain how the results on nonenergy sharing among the modes of the FPU lattice led Norman Zabusky and Martin Kruskal to carry out the now famous numerical experiments, which ultimately lead to the notion of *soliton*. Finally, we also point out how the explicit soliton expressions can be obtained through an algorithmic procedure called the *bilinearization method*, introduced by Hirota, from which one can easily understand the basic soliton properties.

12.1 The Scott Russel Phenomenon and KdV Equation

In 1895, Kortweg and de Vries considered the wave phenomenon underlying the observations of Scott Russel, described earlier in Sect. 11.7, from first principles of fluid dynamics. The basic features of their analysis can be summarized as follows [1].

Consider the one-dimensional (x-direction) wave motion of an incompressible and inviscid fluid (water) in a shallow channel of height h, and of sufficient width with uniform cross-section leading to the formation of a solitary wave propagating under gravity. Let the length of the wave be l and the maximum value of its amplitude, $\eta(x,t)$, above the horizontal surface be a (see Fig. 12.1). Then we can introduce two natural small parameters into the problem,

$$\epsilon = \frac{a}{h} \ll 1 , \tag{12.1a}$$

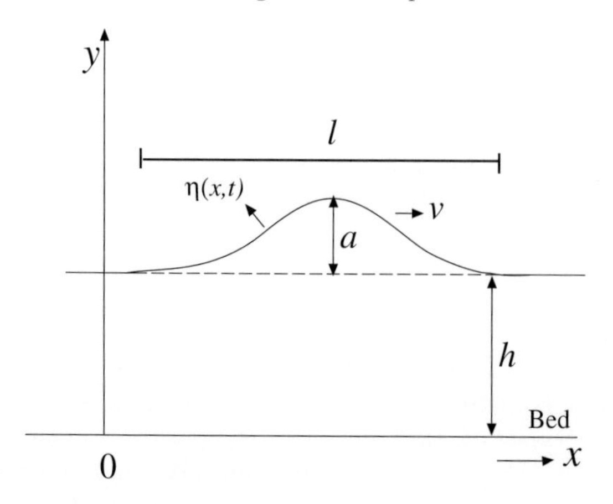

Fig. 12.1. One-dimensional wave motion in a shallow channel

$$\delta = \frac{h}{l} \ll 1 . \tag{12.1b}$$

Now we can proceed with the analysis as follows.

A. *Equation of Motion*

The fluid motion can be described by the velocity vector

$$\boldsymbol{V}(x,y,t) = u(x,y,t)\boldsymbol{i} + v(x,y,t)\boldsymbol{j} , \tag{12.2}$$

where $\boldsymbol{i}$ and $\boldsymbol{j}$ are the unit vectors along the horizontal and vertical directions, respectively. As the motion is irrotational, we have

$$\boldsymbol{\nabla} \times \boldsymbol{V} = 0 . \tag{12.3}$$

Consequently, we can introduce the velocity potential $\phi(x,y,t)$ by the relation

$$\boldsymbol{V} = \boldsymbol{\nabla}\phi . \tag{12.4}$$

(i) *Conservation of Density*

The system obviously admits the following conservation law for the mass density $\rho(x,\, y,\, t)$ of the fluid,

$$\frac{\mathrm{d}\rho}{\mathrm{d}t} = \rho_t + \boldsymbol{\nabla} \cdot (\rho \boldsymbol{V}) = 0 , \tag{12.5}$$

where $\boldsymbol{V}(x,y,t)$ is the velocity vector of the fluid. As ρ is a constant, from (12.5) we have

$$\boldsymbol{\nabla} \cdot \boldsymbol{V} = 0 . \tag{12.6}$$

Then using (12.4) in (12.6), we find that ϕ obeys the Laplace equation

$$\nabla^2\phi(x,y,t) = 0 . \tag{12.7}$$

(ii) *Euler's Equation*

As the density of the fluid $\rho = \rho_0$ =constant, using Newton's law for the rate of change of momentum, we can write

$$\begin{aligned}\frac{\mathrm{d}\boldsymbol{V}}{\mathrm{d}t} &= \frac{\partial \boldsymbol{V}}{\partial t} + (\boldsymbol{V} \cdot \boldsymbol{\nabla})\, \boldsymbol{V} \\ &= -\frac{1}{\rho_0}\boldsymbol{\nabla} p - g\boldsymbol{j}\,,\end{aligned} \tag{12.8}$$

where $p = p(x, y, t)$ is the pressure at the point (x, y) and g is the acceleration due to gravity, which is acting vertically downwards (here $\boldsymbol{j}$ is the unit vector along the vertical direction). Making use of (12.4) in (12.8), we obtain (after one integration)

$$\phi_t + \frac{1}{2}(\boldsymbol{\nabla}\phi)^2 + \frac{p}{\rho_0} + gy = 0\,. \tag{12.9}$$

(iii) *Boundary Conditions*

The above two equations (12.7) and (12.8) or (12.9) for the velocity potential $\phi(x, y, t)$ of the fluid have to be supplemented by appropriate boundary conditions, by taking into account the fact (see Fig. 12.1) that

(a) the horizontal bed at $y = 0$ is hard and
(b) the upper boundary $y = y(x, t)$ is a free surface .

As a result
(a) the vertical velocity at $y = 0$ vanishes,

$$v(x, 0, t) = 0\,, \tag{12.10a}$$

which implies (using (12.2) and (12.4))

$$\phi_y(x, 0, t) = 0\,. \tag{12.10b}$$

(b) As the upper boundary is free, let us specify it by $y = h + \eta(x, t)$ (see Fig. 12.1). Then at the point $x = x_1$, $y = y_1 \equiv y(x, t)$, we can write

$$\frac{\mathrm{d}y_1}{\mathrm{d}t} = \frac{\partial \eta}{\partial t} + \frac{\partial \eta}{\partial x} \cdot \frac{\mathrm{d}x_1}{\mathrm{d}t} = \eta_t + \eta_x u_1 = v_1\,. \tag{12.11a}$$

Since $v_1 = \phi_{1y}$, $u_1 = \phi_{1x}$, the last two parts of (12.11a) can be rewritten as

$$\phi_{1y} = \eta_t + \eta_x \phi_{1x}\,. \tag{12.11b}$$

(c) Similarly at $y = y_1$, the pressure $p_1 = 0$. Then from (12.9), it follows that

$$u_{1t} + u_1 u_{1x} + v_1 v_{1x} + g\eta_x = 0\,. \tag{12.12}$$

Thus the motion of the surface of water wave is essentially specified by the Laplace equation (12.7) and (12.9) along with one fixed boundary condition (12.10b) and two *variable nonlinear* boundary conditions (12.11b) and (12.12). One has to then solve the Laplace equation subject to these boundary conditions.

(iv) *Taylor Expansion of $\phi(x,y,t)$ in y*

Making use of the fact $\delta = h/l \ll 1$, $h \ll l$, we assume $y(= h+\eta(x,t))$ to be small to introduce the Taylor expansion

$$\phi(x,y,t) = \sum_{n=0}^{\infty} y^n \phi_n(x,t) . \tag{12.13}$$

Substituting the above series for ϕ into the Laplace equation (12.7), solving recursively for $\phi_n(x,t)$'s and making use of the boundary condition (12.10b), $\phi_y(x,0,t) = 0$, one can show that

$$u_1 = \phi_{1x} = f - \frac{1}{2} y_1^2 f_{xx} + \text{ higher order in } y_1 , \tag{12.14a}$$

$$v_1 = \phi_{1y} = -y_1 f_x + \frac{1}{6} y_1^3 f_{xxx} + \text{ higher order in } y_1 , \tag{12.14b}$$

where $f = \partial\phi_0/\partial x$. We can then substitute these expressions into the nonlinear boundary conditions (12.11) and (12.12) to obtain equations for f and η.

Exercise:

1. Deduce equations (12.14).

(v) *Introduction of Small Parameters ϵ and δ*

So far the analysis has not taken into account fully the shallow nature of the channel $(a/h = \epsilon \ll 1)$ and the solitary nature of the wave $(a/l = a/h \cdot h/l = \epsilon\delta \ll 1$, $\epsilon \ll 1$, $\delta \ll 1)$, which are essential to realize the Scott Russel phenomenon. For this purpose we stretch the independent and dependent variables in the defining (12.11), (12.12) and (12.14) through appropriate scale changes, but retaining the overall form of the equations. To realize this we can introduce the natural scale changes

$$x = lx' , \quad \eta = a\eta' \tag{12.15}$$

along with

$$t = \frac{l}{c_0} t' , \tag{12.16}$$

where c_0 is a parameter to be determined. Then in order to retain the form of (12.14) we require

$$u_1 = \epsilon c_0 u_1' , \quad v_1 = \epsilon\delta c_0 v_1' , \quad f = \epsilon c_0 f' . \tag{12.17}$$

We also have

$$y_1 = h + \eta(x,t) = h\left(1 + \epsilon\eta'\left(x',t'\right)\right) . \tag{12.18}$$

Substituting the transformations (12.15)–(12.18) into (12.14a), we obtain

$$\begin{aligned} u'_1 &= f' - \frac{1}{2}\delta^2 \left(1 + \epsilon\eta'\right)^2 f'_{x'x'} \\ &= f' - \frac{1}{2}\delta^2 f'_{x'x'} \,, \end{aligned} \tag{12.19}$$

where we have omitted terms proportional to $\delta^2\epsilon$ as small compared to terms of the order δ^2. Similarly from (12.14b), we obtain

$$v'_1 = -\left(1 + \epsilon\eta'\right) f'_{x'} + \frac{1}{6}\delta^2 f'_{x'x'x'} \,. \tag{12.20}$$

Now considering the nonlinear boundary condition (12.11b) in the form

$$v_1 = \eta_t + \eta_x u_1 \,, \tag{12.21}$$

it can be rewritten, after making use of the transformations (12.15)–(12.18) and neglecting terms involving $\epsilon\delta^2$, as

$$\eta'_{t'} + f'_{x'} + \epsilon\eta' f'_{x'} + \epsilon f'\eta'_{x'} - \frac{1}{6}\delta^2 f'_{x'x'x'} = 0 \,. \tag{12.22}$$

Similarly considering the other boundary condition (12.12) and making use of the above transformations, it can be rewritten, after neglecting terms of the order $\epsilon^2\delta^2$, as

$$f'_{t'} + \epsilon f' f'_{x'} + \frac{ga}{\epsilon c_0^2}\eta'_{x'} - \frac{1}{2}\delta^2 f'_{x'x't'} = 0 \,. \tag{12.23}$$

Now choosing the arbitrary parameter c_0 as

$$c_0^2 = gh \tag{12.24}$$

so that η'_x term is of order unity, (12.23) becomes

$$f'_{t'} + \eta'_{x'} + \epsilon f' f'_{x'} - \frac{1}{2}\delta^2 f'_{x'x't'} = 0 \,. \tag{12.25}$$

For notational convenience we will hereafter omit the prime symbol in all the variables, however remembering that all the variables hereafter correspond to rescaled quantities. Then the evolution equation for the amplitude of the wave and the function related to the velocity potential reads

$$\eta_t + f_x + \epsilon\eta f_x + \epsilon f\eta_x - \frac{1}{6}\delta^2 f_{xxx} = 0 \,, \tag{12.26a}$$

$$f_t + \eta_x + \epsilon f_x - \frac{1}{2}\delta^2 f_{xxt} = 0 \,. \tag{12.26b}$$

Note that the small parameters ϵ and δ^2 have occurred in a natural way in (12.26).

Exercise:

2. Deduce (12.19), (12.20), (12.22) and (12.25).

(vi) *Perturbation Analysis*

Since the parameters ϵ and δ^2 are small in (12.26), we can make a perturbation expansion of f in these parameters:

$$f = f^{(0)} + \epsilon f^{(1)} + \delta^2 f^{(2)} + \text{ higher order terms} , \qquad (12.27)$$

where $f^{(i)}$, $i = 0, 1, 2, \ldots$ are functions of η and its spatial derivatives. Substituting this into (12.26) and regrouping, we obtain

$$\eta_t + f_x^{(0)} + \epsilon \left[f_x^{(1)} + \eta f_x^{(0)} + \eta_x f^{(0)} \right] + \delta^2 \left[f_x^{(2)} - \frac{1}{6} f_{xxx}^{(0)} \right] + \text{ higher order terms in } (\epsilon, \delta^2) = 0 , \qquad (12.28a)$$

$$\eta_x + f_t^{(0)} + \epsilon \left[f_t^{(1)} + f^{(0)} f_x^{(0)} \right] + \delta^2 \left[f_t^{(2)} - \frac{1}{2} f_{xxt}^{(0)} \right] + \text{ higher order terms in } (\epsilon, \delta^2) = 0 . \qquad (12.28b)$$

In order that (12.28a) and (12.28b) are self consistent as evolution equations for an one-dimensional wave propagating to the right, we can choose

$$f^{(0)} = \eta + O\left(\epsilon, \delta^2\right) , \qquad (12.29)$$

where $O(\epsilon, \delta^2)$ stands for terms proportional to ϵ and δ^2. Then (12.28) become

$$\eta_t + \eta_x + \epsilon \left[f_x^{(1)} + 2\eta\eta_x \right] + \delta^2 \left[f_x^{(2)} - \frac{1}{6}\eta_{xxx} \right] = 0 , \qquad (12.30a)$$

$$\eta_t + \eta_x + \epsilon \left[f_t^{(1)} + \eta\eta_x \right] + \delta^2 \left[f_t^{(2)} - \frac{1}{2}\eta_{xxt} \right] = 0 , \qquad (12.30b)$$

where higher order terms in ϵ and δ^2 are neglected. Since $f^{(1)}$ and $f^{(2)}$ are functions of η (and its spatial derivatives)

$$f_x^{(1)} = f_\eta^{(1)}\eta_x, \quad f_t^{(1)} = f_\eta^{(1)}\eta_t = -f_\eta^{(1)}\eta_x + O\left(\epsilon, \delta^2\right) = -f_x^{(1)} , \qquad (12.31a)$$

where in the last relation, (12.28) have been used for η_t and η_x. Similarly, we can argue that

$$f_x^{(2)} = f_\eta^{(2)}\eta_x, \quad f_t^{(2)} = -f_\eta^{(2)}\eta_x + O\left(\epsilon, \delta^2\right) = -f_x^{(2)} , \qquad (12.31b)$$

Substituting (12.31) into (12.30), we obtain

$$\eta_t + \eta_x + \epsilon \left[f_x^{(1)} + 2\eta\eta_x \right] + \delta^2 \left[f_x^{(2)} - \frac{1}{6}\eta_{xxx} \right] = 0 , \qquad (12.32a)$$

$$\eta_t + \eta_x + \epsilon[-f_x^{(1)} + \eta\eta_x] + \delta^2 \left[-f_x^{(2)} + \frac{1}{2}\eta_{xxx} \right] = 0 . \qquad (12.32b)$$

Compatibility of these two equations require that

$$f_x^{(1)} = -\frac{1}{2}\eta\eta_x, \quad f_x^{(2)} = \frac{1}{3}\eta_{xxx} . \qquad (12.33)$$

Integrating, we find

$$f^{(1)} = -\frac{1}{4}\eta^2, \quad f^{(2)} = \frac{1}{3}\eta_{xx} \,. \tag{12.34}$$

Substituting $f^{(1)}$ and $f^{(2)}$ into (12.32), we ultimately obtain the KdV equation in the form

$$\eta_t + \eta_x + \frac{3}{2}\epsilon\eta\eta_x + \frac{\delta^2}{6}\eta_{xxx} = 0 \,, \tag{12.35}$$

describing the unidirectional propagation of shallow water waves.

(vii) *The Standard (Contemporary) Form of KdV Equation*

Finally, changing to a moving frame of reference,

$$\xi = x - t, \quad \tau = t \tag{12.36a}$$

so that

$$\frac{\partial}{\partial x} = \frac{\partial}{\partial \xi}\,, \quad \frac{\partial}{\partial t} = -\frac{\partial}{\partial \xi} + \frac{\partial}{\partial \tau}\,, \tag{12.36b}$$

(12.35) can be rewritten as

$$\eta_\tau + \frac{3}{2}\epsilon\eta\eta_\xi + \frac{1}{6}\delta^2\eta_{\xi\xi\xi} = 0 \,. \tag{12.37}$$

Then introducing the new variables

$$u = \frac{3\epsilon}{2\delta^2}\eta, \quad \tau' = \frac{6}{\delta^2}\tau \,, \tag{12.38}$$

(12.37) can be expressed as

$$u_{\tau'} + 6uu_\xi + u_{\xi\xi\xi} = 0 \,. \tag{12.39}$$

Redefining the variables τ' as t and ξ as x, again for notational convenience, we finally arrive at the ubiquitous form of the KdV equation as

$$u_t + 6uu_x + u_{xxx} = 0 \,. \tag{12.40}$$

As noted in the previous chapter, Sect. 11.7, the KdV equation (12.40) admits an exact solitary wave solution, besides the cnoidal wave. The solitary wave is actually the limiting case of the cnoidal wave, having the same empirical relation as that of Scott Russel connecting the velocity and amplitude of the wave, (11.19).

Thus Korteweg and de Vries gave a sound basis for the Scott Russel's Great Wave of Translation. However the ensuing years saw little progress in the study of such solitary waves or the KdV equation itself or any other type of nonlinear waves. One might even say that nonlinear physics and mathematics were playing a rather secondary role probably with the advent of quantum mechanics and other successful linear theories.

12.2 The Fermi–Pasta–Ulam Numerical Experiments on Anharmonic Lattices

Not until more than half a century later, the KdV equation and its solitary wave received their rightful recognition by physicists and mathematicians. The breakthrough came in an entirely different context – this time in the study of wave propagation in nonlinear lattices. For fuller details, see for example J. Ford [2].

12.2.1 The FPU Lattice

In early 1950s, E. Fermi, J. Pasta, and S. Ulam were set to make use of the MANIAC-I analog computer at Los Alamos Laboratory, U.S.A on important problems in physics. Particularly, they were interested in checking the widely held concepts of ergodicity[1] and equipartition[2] of energy in irreversible statistical mechanics. They considered for this purpose, the dynamics of a chain of weakly coupled nonlinear oscillators. The chain contains 32 (or 64) mass points, which interacted through nonlinear forces (see Fig. 12.2).

Then the equation of motion of the lattice for the displacements y_i, $i = 0, 1, 2, ..., N$, can be written (compare with the case of longitudinal vibrations of the elastic rod in Sect. 11.1) as

$$m\frac{\mathrm{d}^2 y_i}{\mathrm{d}t^2} = f\left(y_{i+1} - y_i\right) - f\left(y_i - y_{i-1}\right) , \quad i = 1, 2, ..., N - 1, \tag{12.41}$$

with $y_0 = 0$ and $y_N = 0$, where Fermi, Pasta and Ulam (FPU) assumed the following specific forms for $f(y)$:

$$\begin{aligned}
&\text{(a) quadratic nonlinearity:} && f(y) = y + \alpha y^2 \\
&\text{(b) cubic nonlinearity:} && f(y) = y + \beta y^3 \\
&\text{(c) broken(piecewise) linearity:} && f(y) = \gamma_1 y, \quad |y| < d \\
& && = \gamma_2 y + \delta, \quad |y| > d \\
& && (\alpha, \beta, \delta, d, \gamma_1, \gamma_2 \text{ are constants}).
\end{aligned} \tag{12.42a,b,c}$$

When there is no nonlinearity (for example, $\alpha = 0$ in case (a) or $\beta = 0$ in case (b) above), it is easy to check that the equation of motion (12.41) is separable into their *linear normal modes* and there will be no energy sharing

[1] The ergodic hypothesis states that the time average of some property of an assembly in equilibrium will be the same as the instantaneous ensemble average.

[2] According to the principle of equipartition of energy, the total energy of a molecule is equally distributed among all degrees of freedom of the molecule and the average energy per degree of freedom at temperature T is $kT/2$, where k is the Boltzmann's constant.

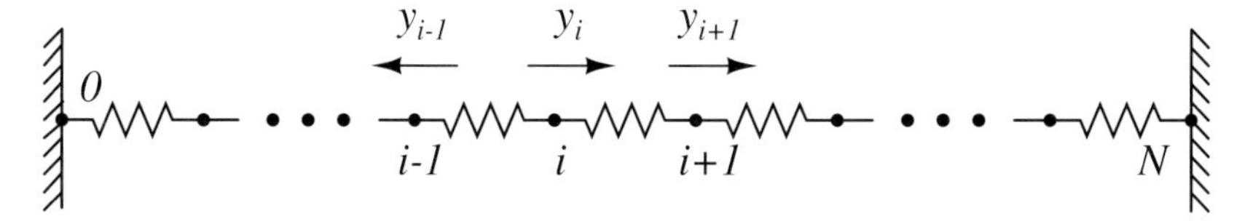

Fig. 12.2. Fermi–Pasta–Ulam nonlinear lattice

among them. (For details of normal mode decomposition, see for example Ref. [3]).

Proof:

Let $f(y) = ky$, $k > 0$. Then the equation of motion is

$$m\frac{\mathrm{d}^2 y_i}{\mathrm{d}t^2} = k\left(y_{i+1} - 2y_i + y_{i-1}\right), \quad i = 1, 2, ..., N-1\ . \tag{12.43}$$

Introducing the normal mode decomposition

$$y_i = \sum_{j=1}^{N-1} a_j(t) \sin(ij\pi/N), \quad i = 1, 2, ..., N-1\ , \tag{12.44}$$

where the coefficients a_j's are functions of t only, into (12.43), we obtain

$$\sum_j \ddot{a}_j(t) \sin(ij\pi/N) + 2\omega_0^2 \sum_j a_j(t) \sin(ij\pi/N)\left[1 - \cos(j\pi/N)\right] = 0\ , \tag{12.45}$$

where $\omega_0^2 = (k/m)$. Since the above equation is true for all values of i, we can easily see that

$$\ddot{a}_j + \Omega_j^2 a_j = 0, \quad \Omega_j^2 = 2\omega_0^2\left[1 - \cos(j\pi/N)\right], \quad j = 1, 2, ..., N-1. \tag{12.46}$$

Note that $a_0 = a_N = 0$. Thus, individual modes oscillate independently in a simple harmonic way and the total energy of the system,

$$E = \frac{m}{2}\sum_{j=1}^{N-1} \dot{y}_j^2 + \frac{k}{2}\sum_{j=2}^{N-1}\left(y_j - y_{j-1}\right)^2 = \frac{1}{2}\sum_{j=1}^{N-1}\left(\dot{a}_j^2 + \Omega_j^2 a_j^2\right) = \sum_{j=1}^{N-1} E_j\ , \tag{12.47}$$

can be written as the sum of the energies of the individual modes. Thus, a given initial energy distribution among the normal modes continues to remain so as there is no coupling between them and no energy sharing will take place among the constituent modes.

12.2.2 FPU Recurrence Phenomenon

However, when one of the weakly nonlinear interaction (12.42) is switched on, the modes get coupled and one would *expect* the energy to flow among the original normal modes back and forth and eventually equipartition of energy among the modes to occur. Numerical analysis should confirm this expectation. The results of FPU are contained in the Los Alamos Report Number 1940 of the year 1955. To their great surprise, FPU found no equipartition of energy to occur. When the energy was assigned to the lowest mode, as time went on only the first few modes were excited and even this energy was given back to the lowest mode after a characteristic time called *recurrence time*.

Figure 12.3 contains the results of FPU for the case $N = 32$ with $\alpha = 0.25$ in (12.42a). Starting with an initial shape at $t = 0$ in the form of a half of a sine wave given by $y_j = \sin(j\pi/32)$, so that only the fundamental harmonic mode was excited (compare with (12.44)) with an initial amplitude $a_1 = 4$ and energy $E_1 = 0.077...$, the figure depicts the evolution of the first four normal mode energies, E_k, $k = 1, 2, 3, 4$. During the time interval $0 \leq t \leq 16$ in Fig. 12.3, where t is measured in periods of the fundamental mode, modes 2,3 and 4 etc. sequentially begin to absorb energy from the initially dominant first mode, as one would expect from a standard analysis. After this the pattern of energy sharing undergoes a dramatic change. Energy is now exchanged primarily only among modes 1 through 6 with all the

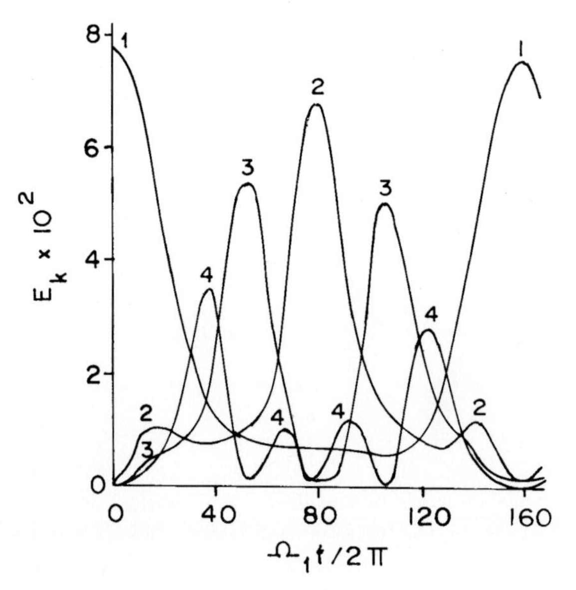

Fig. 12.3. A plot of the normal mode energies $E_k = \frac{1}{2}\left(\dot{a}_k^2 + \Omega_k^2 a_k^2\right)$ for $N = 32$ and $\alpha = 0.25$ in (12.47) [2]. The numbers on the curves represent the modes

higher modes getting very little energy. In fact the motion is almost periodic, with a recurrence period (the so-called *FPU recurrence*) at about $t = 157$ fundamental periods. The energy in the fundamental mode returns to within three percentage of its value at $t = 0$.

The unexpected recurrence phenomenon in the FPU experiments stimulated a great variety of research towards

(1) the statistical behaviour of nonlinear oscillators, mixing, ergodicity, etc.,

(2) the theory of normal mode coupling,

(3) the investigation of nonlinear normal modes, and integrable systems,

and so on. In fact, the FPU experiment is considered to be the trend setter of modern era of nonlinear dynamics.

12.3 The KdV Equation Again!

The entirely unexpected results of FPU experiments motivated many scientists to try to understand nonlinear phenomena more deeply. Martin Kruskal and Norman Zabusky from Princeton Plasma Physics Laboratory set out to understand the FPU recurrence phenomenon through combined analytical and numerical investigations. Their approach, which is based on an asymptotic analysis, is as follows.

12.3.1 Asymptotic Analysis and the KdV Equation

Consider the equation of motion (12.41) of the nonlinear lattice with the combined nonlinear force

$$f = y + \alpha y^2 + \beta y^3 \ . \tag{12.48}$$

Then, consider the continuous limit

$$y_n(t) \stackrel{a \to 0}{\longrightarrow} y(na, t) = y(x, t) \ , \tag{12.49}$$

where a is the lattice parameter. We may also write

$$\begin{aligned} y_{n\pm 1}(t) &\stackrel{a \to 0}{\longrightarrow} y((n \pm 1)a, t) \\ &= y(x \pm a, t) \\ &= y \pm a\frac{\partial y}{\partial x} + \frac{a^2}{2!}\frac{\partial^2 y}{\partial x^2} \pm \frac{a^3}{3!}\frac{\partial^3 y}{\partial x^3} + \frac{a^4}{4!}\frac{\partial^4 y}{\partial x^4} + \ldots \end{aligned} \tag{12.50}$$

Substituting (12.50) into (12.41) with f in the form of (12.48), one can obtain the equation of motion for the continuous case (by retaining terms upto order a^4) as

$$\frac{1}{c^2}\frac{\partial^2 y}{\partial t^2} = \frac{\partial^2 y}{\partial x^2}\left[1 + 2\alpha a\frac{\partial y}{\partial x} + 3\beta a^2\left(\frac{\partial y}{\partial x}\right)^2\right] + \text{higher order terms}, \tag{12.51}$$

where $c^2 = a^2/m$. Then we can consider the following three cases.

(i) *Linear case:* $\alpha = 0,\ \beta = 0$:

Equation (12.51) is the linear dispersionless wave equation discussed in Sect. 11.1.

(ii) $\alpha \neq 0,\ \beta \neq 0$:

This is a hyperbolic equation (when higher order terms are omitted). By using the method of characteristics, one can show that the solution develops multivaluedness–shocks!

(iii) *Addition of a fourth derivative term:*

Physically one does not expect shocks to occur in a nonlinear lattice (certainly the experiments did not reveal any!). So Zabusky and Kruskal added a fourth derivative term $(1/12)\partial^4 y/\partial x^4$ to the right hand side of (12.51) so as to obtain the final form of equation of motion as

$$\frac{1}{c^2}\frac{\partial^2 y}{\partial t^2} = \frac{\partial^2 y}{\partial x^2}\left[1 + 2\alpha a\frac{\partial y}{\partial x} + 3\beta a^2\left(\frac{\partial y}{\partial x}\right)^2\right] + \frac{a^2}{12}\frac{\partial^4 y}{\partial x^4}\,. \tag{12.52}$$

Now considering unidirectional waves (moving to the right), one can make a change of variables,

$$\xi = x - ct, \quad \tau = a^2 ct, \quad y = v/a\,. \tag{12.53}$$

Then

$$\frac{\partial}{\partial t} = a^2 c\frac{\partial}{\partial \tau} - c\frac{\partial}{\partial \xi}, \quad \frac{\partial}{\partial x} = \frac{\partial}{\partial \xi}\,, \tag{12.54}$$

and similarly higher derivatives can be transformed. Equation (12.52) can then be rewritten as

$$\frac{\partial^2 v}{\partial \xi \partial \tau} + \alpha\frac{\partial v}{\partial \xi}\frac{\partial^2 v}{\partial \xi^2} + \frac{3}{2}\beta\frac{\partial^2 v}{\partial \xi^2}\left(\frac{\partial v}{\partial \xi}\right)^2 + \frac{1}{24}\frac{\partial^4 v}{\partial \xi^4} = \frac{a^2}{2}\frac{\partial^2 v}{\partial \tau^2}\,, \tag{12.55}$$

after redefining (α/a^2) as α and (β/a^2) as β. Then in the continuous limit, $a \to 0$, and with the redefinition

$$u = \frac{\partial v}{\partial \xi}\,, \tag{12.56}$$

one finally obtains

$$\frac{\partial u}{\partial \tau} + \alpha u \frac{\partial u}{\partial \xi} + \frac{3}{2}\beta u^2 \left(\frac{\partial u}{\partial \xi}\right) + \frac{1}{24}\frac{\partial^3 u}{\partial \xi^3} = 0 . \tag{12.57}$$

When $\beta = 0$, $\alpha \neq 0$ and if τ and ξ are replaced by the standard notation $t = \sqrt{24}\,\tau$ and $x = \sqrt{24}\,\xi$, we have

$$u_t + \alpha u u_x + u_{xxx} = 0 , \tag{12.58}$$

which is nothing but the KdV equation (12.40), for suitable choice of α but which has now occurred in an entirely new context! Similarly for $\alpha = 0$, $\beta \neq 0$, we have

$$u_t + \frac{3}{2}\beta u^2 u_x + u_{xxx} = 0 , \tag{12.59}$$

which we may call as the *modified KdV equation* or briefly as the *MKdV equation*.

(iv) *Scale change:*

Note that the KdV equation can always be written in the form

$$u_t + p u u_x + q u_{xxx} = 0 \tag{12.58a}$$

under a suitable scale change. Or in other words, making a change of scales of the variables t, x and u, and redefining the variables, (12.58a) can always be written in the form (12.58). We will use this freedom to choose the coefficients p and q as per convenience in our further analysis. This is also true for the MKdV equation.

Exercises:

3. Rewrite (12.58a) through appropriate scale changes of the variables into the forms $u_t \pm 6uu_x + u_{xxx} = 0$.
4. Transform the MKdV equation (12.59) into the form $u_t - 6u^2u_x + u_{xxx} = 0$ through suitable scale changes.
5. Find the solitary wave solution of the MKdV equation (12.59) (see also Chap. 14).

Thus one may conclude that the wave propagation in the FPU lattice with quadratic nonlinear force may be described in a nontrivial way by the KdV equation and the lattice with cubic nonlinear force may be described by the MKdV equation. Then what do these equations have to do with the FPU recurrence phenomenon reported by Fermi, Pasta and Ulam in their famous experiments?

12.4 Numerical Experiments of Zabusky and Kruskal: The Birth of Solitons

Zabusky and Kruskal recalled that the KdV equation admits solitary wave solution (see Sect. 11.7), which has a distinct nonlinear character. Further, if nonlinear normal modes exist leading to recurrence and nonenergy sharing phenomena, then the solitary wave should play an important role. So they initiated a many faceted and deep numerical study of the KdV equation, the results of which was reported in the year 1965 [4]. The KdV equation which Zabusky and Kruskal considered in their numerical analysis was of the form

$$u_t + uu_x + \delta^2 u_{xxx} = 0 \,. \tag{12.58b}$$

Their study mainly focussed on the following two aspects among other things:

(1) What will be the type of solution the system admits for a chosen initial condition, particularly for a spatially periodic initial condition, $u(x,0) = \cos \pi x$, $0 \le x \le 2$, so that u, u_x, u_{xx} are periodic on $[0,2]$ with $u(x,t) = u(x+2,t)$, etc.?

(2) How do solitary waves of the KdV equation interact mutually, particularly when two such waves of differing amplitudes (and so of different velocities) interact?

For their numerical analysis, Zabusky and Kruskal converted the KdV equation (12.58b) into a difference equation, and integrated it keeping the energy as a constant,

$$\begin{aligned} u_i^{j+1} = u_i^{j-1} & - \frac{1}{3}\frac{k}{h}\left(u_{i+1}^j + u_i^j + u_{i-1}^j\right)\left(u_{i+1}^j - u_{i-1}^j\right) \\ & - \frac{\delta^2 k}{h^3}\left(u_{i+2}^j - 2u_{i+1}^j + 2u_{i-1}^j - u_{i-2}^j\right) \,, \\ & i = 0, 1, ..., 2N-1, \end{aligned} \tag{12.60}$$

where a rectangular mesh has been used with temporal and spatial intervals of k and $h = 1/N$, respectively. That is, $u(x,t)$ is approximated by $u_i^j = u(ik, jh)$. The periodic boundary conditions $u_i^j = u_{i+2N}^j$ were used in the calculations. The numerical solutions conserve momentum and almost conserve energy. For further details of numerical calculations see Ref. [4]. The outcome of these numerical experiments may be summarized as follows.

(i) *Periodic Boundary Conditions*

(1) As δ^2 is small, the nonlinearity dominates over the third derivative term. As a consequence the wave steepens in regions where it has a negative slope.

(2) As the wave steepens, $\delta^2 u_{xxx}$ term becomes important and balances the nonlinear term uu_x.

(3) At later time the solution develops a train of eight well-defined (solitary) waves with different amplitudes each like sech^2 functions, with the faster (taller) waves catching up and overtaking the slower (short) waves (Fig. 12.4). These nonlinear waves interact strongly and then continue thereafter almost as if there had been no interaction at all.

(4) Each of the solitary wave pulse moves uniformly at a rate which is linearly proportional to its amplitude. Thus, the solitons spread apart. Because of the periodic boundary condition, two or more solitons eventually overlap spatially and interact nonlinearly (Fig. 12.4). Shortly after interaction, they reappear virtually unaffected in size and shape.

(5) There exists a period T_R, the so-called *recurrence time* at which all the solitons arrive almost in the same phase and almost reconstruct the initial state through nonlinear interaction, thereby explaining the FPU recurrence phenomenon qualitatively.

(6) The persistence of the solitary waves led Zabusky and Kruskal to coin the name *soliton* (after the names such as photon, phonon, etc.) to emphasize the particle-like character of these waves which seem to retain their identities in a collision.

(ii) *Initial Condition With Just Two Solitary Waves*

The above observations can be better understood by considering an initial condition consisting of just two solitary waves of differing amplitudes as

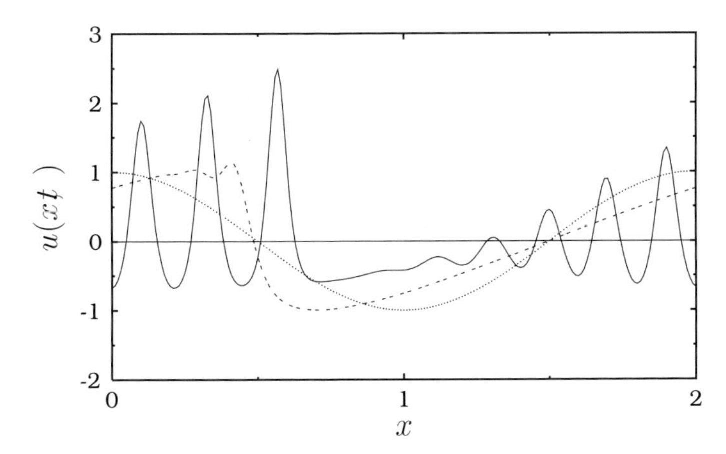

Fig. 12.4. Zabusky–Kruskal's numerical experimental results (Ref. [4]): Solution of the KdV equation (12.58b) with $\delta = 0.022$ and $u(x,0) = \cos \pi x$ for $0 \leq x \leq 2$. The dotted curve represents u at $t = 0$. The dashed curve is the solution at $t = 1/\pi$. The continuous curve gives u at $t = 3.6/\pi$

shown in Fig. 12.5. Suppose that at time $t \to -\infty$ (say $t = -800$ units), two such waves are given, which are well separated and with the bigger one to the right as in Fig. 12.5. Then as the system evolves as per the KdV equation (12.40) or (12.58), after a sufficient time the waves overlap and interact (the bigger one catches up with the smaller one). Following the process still longer, one finds that the bigger one separates from the smaller one, after overtaking it, and asymptotically (as $t \to \infty$) the wave solution regains its initial shape and hence the two solitary waves regain their velocities also. The only effect of the interaction is a phase shift, that is the center of each wave is at a different position than where it would have been if each one of them were travelling alone (Fig. 12.6). Again, because of the analogy with the elastic collision property of particles, Zabusky and Kruskal referred to these solitary waves as *solitons* .

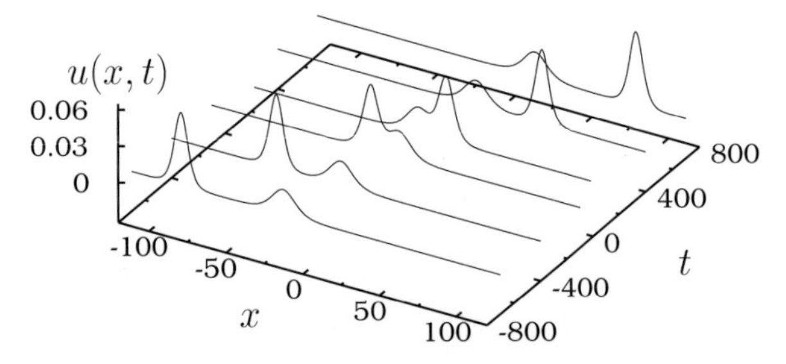

Fig. 12.5. Two-soliton interaction of the KdV equation. The parameters in (12.73) below are fixed as $k_1 = 0.2$, $k_2 = \sqrt{3}\,k_1$, $\omega_1 = k_1^3$, $\omega_2 = k_2^3$ and $\eta_1^{(0)} = 0$, $\eta_2^{(0)} = 0$

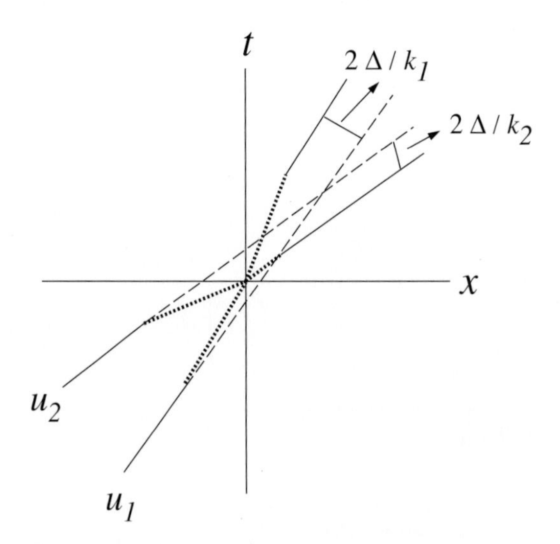

Fig. 12.6. Phase shifts of two interacting solitons

Thus, the major new concepts that emerge from the Zabusky–Kruskal experiments are that

(1) when the nonlinearity suitably balances the linear dispersion as in the KdV equation, solitary waves can arise;
(2) these solitary waves in appropriate nonlinear systems can interact elastically like particles without changing their shapes or velocities and
(3) the solitons can constitute the general solution of the initial value problem of a class of nonlinear dispersive wave equations like the KdV equation.

Naturally the next obvious question arises as to whether exact analytical forms of soliton solutions of the KdV equation, beyond the solitary wave solution, can be obtained which can correspond to all the above numerical results. In fact Martin Kruskal and his coworkers went on further to completely integrate the initial value problem (IVP) of the KdV equation, and in this process also invented a new method to solve the IVP of a class of nonlinear evolution equations. The method is now called the *inverse scattering transform (IST)* method, which may be considered as a natural generalization of the Fourier transform method that is applicable to the linear dispersive systems as discussed earlier in Sect. 11.2.

Before introducing the IST method (in the next chapter) for the KdV equation from a historical perspective and also to know what other interesting nonlinear evolution equations can admit soliton solutions, we wish to introduce a straightforward algorithmic method to obtain the soliton solutions. It is called the *direct* or *bilinearization method*, which was introduced by R. Hirota in 1971 [5]. By using this method, in the next section, we will obtain explicitly the two-soliton solution of the KdV equation which in fact corresponds to the Zabusky–Kruskal's numerical experimental result on two-solitary wave scattering as depicted in Fig. 12.5. We will also indicate the method to obtain more general soliton solutions.

Exercises:

6. Verify that (12.60) is indeed a suitable discretization of the KdV equation (12.59) by making use of the standard numerical methods.
7. Carry out a numerical analysis of the MKdV equation along the lines of Zabusky and Kruskal for the KdV equation.

12.5 Hirota's Direct or Bilinearization Method for Soliton Solutions of KdV Equation

Let us consider the KdV equation (12.40). If it is the aim to obtain soliton solutions alone and not to solve the much more general task of solving the IVP, then we can use the algorithmic *bilinearization method* of Hirota mentioned above. The main ingredient in this method is to introduce a *bilinearizing transformation* so that the given evolution equation can be written in the so-called *bilinear form*: Each term in the transformed equation has a total degree two. Thus with the transformation

$$u = 2\frac{\partial^2}{\partial x^2}\log F\ , \tag{12.61}$$

the KdV equation (12.40) takes the form

$$F_{xt}F - F_xF_t + F_{xxxx}F - 4F_{xxx}F_x + 3F_{xx}^2 = 0\ . \tag{12.62}$$

(Note: How to find the transformation (12.61) beforehand? Initially it was done by inspection, but now sophisticated tools are available in the literature to identify the transformations, which are beyond the scope of this book. See for example Ref. [6] for some details and also appendix J).

Now expanding F in a formal power series in terms of a small parameter ϵ (which one can always introduce into equation (12.62)) as

$$F = 1 + \epsilon f^{(1)} + \epsilon^2 f^{(2)} + ..., \tag{12.63}$$

we substitute it into (12.62). Equating each power of ϵ separately to zero, we get a system of linear partial differential equations (pdes). Upto $O(\epsilon^3)$ we can write them as

$$O\left(\epsilon^0\right):\ \ 0 = 0 \tag{12.64a}$$

$$O(\epsilon)\ \ :\ \ f_{xt}^{(1)} + f_{xxxx}^{(1)} = 0 \tag{12.64b}$$

$$O\left(\epsilon^2\right):\ \ f_{xt}^{(2)} + f_{xxxx}^{(2)} = f_x^{(1)}f_t^{(1)} + 4f_{xxx}^{(1)}f_x^{(1)} - 3\left(f_{xx}^{(1)}\right)^2 \tag{12.64c}$$

$$\begin{aligned} O\left(\epsilon^3\right):\ \ f_{xt}^{(3)} + f_{xxxx}^{(3)} = {} & f_x^{(1)}f_t^{(2)} + f_x^{(2)}f_t^{(1)} - f_{xt}^{(1)}f^{(2)} - f_{xt}^{(2)}f^{(1)} \\ & - f_{xxxx}^{(1)}f^{(2)} - f_{xxxx}^{(2)}f^{(1)} + 4f_{xxx}^{(1)}f_x^{(2)} \\ & + 4f_{xxx}^{(2)}f_x^{(1)} - 6f_{xx}^{(1)}f_{xx}^{(2)}\ . \end{aligned} \tag{12.64d}$$

We can then successively solve this set of linear pdes. To start with, we can easily write the solution of (12.64b) as

$$f^{(1)} = \sum_{i=1}^{N} \mathrm{e}^{\eta_i},\ \ \eta_i = k_ix - \omega_it + \eta_i^{(0)},\ \ \omega_i = k_i^3,\ \ \eta_i^{(0)} = \text{constant}. \tag{12.65}$$

Substituting this into the right hand side of (12.64c), we can solve for $f^{(2)}$. This procedure can be repeated further to find $f^{(3)}$, $f^{(4)}$,... successively. In practice one finds the solution for $N = 1, 2, 3$ and then hypothesize it for arbitrary N which should be then proved by induction. It turns out that for every given value of N in the summation in (12.65), we get a soliton of order N as discussed below.

The task of obtaining the forms of the right hand sides of (12.64) can be simplified enormously by introducing the so-called Hirota's bilinear D-operator. The associated algebra has been developed by Hirota. However, we will not introduce them here and the interested reader may refer, for example, to Ref. [7].

(a) *One-Soliton Solution*

For example, for $N = 1$,

$$f^{(1)} = \mathrm{e}^{\eta_1}, \quad \eta_1 = k_1 x - \omega_1 t + \eta_1^{(0)} \tag{12.66}$$

with $\omega_1 = k_1^3$ and

$$f_{xt}^{(2)} + f_{xxxx}^{(2)} = 0 \,. \tag{12.67}$$

So we can choose $f^{(2)} = 0$. Then one can easily prove that all $f^{(i)} = 0$, $i \geq 3$. Thus the solution to (12.62) becomes

$$F_1 = 1 + \mathrm{e}^{\eta_1}, \quad \eta_1 = k_1 x - k_1^3 t + \eta_1^{(0)} \,. \tag{12.68}$$

Substituting this into the transformation (12.61), we finally obtain the one-soliton solution

$$u(x,t) = \frac{k_1^2}{2} \operatorname{sech}^2 \frac{1}{2} \left(k_1 x - k_1^3 t + \eta_1^{(0)} \right) , \tag{12.69}$$

which is the same as the solitary wave solution (11.38), with the identification $k_1 = \sqrt{c}$.

(b) *2-Soliton Solution*

Proceeding in a similar way for $N = 2$, one has

$$f^{(1)} = \mathrm{e}^{\eta_1} + \mathrm{e}^{\eta_2}, \; \eta_1 = k_1 x - \omega_1 t + \eta_1^{(0)}, \; \eta_2 = k_2 x - \omega_2 t + \eta_2^{(0)} \,. \tag{12.70}$$

Substituting (12.70) into the right hand side of (12.64c), and solving, we obtain

$$f^{(2)} = \mathrm{e}^{\eta_1 + \eta_2 + A_{12}}, \quad \mathrm{e}^{A_{12}} = \left[(k_1 - k_2) \,/\, (k_1 + k_2) \right]^2 \,. \tag{12.71}$$

Using this in (12.63), we then obtain

$$F = 1 + \mathrm{e}^{\eta_1} + \mathrm{e}^{\eta_2} + \mathrm{e}^{\eta_1 + \eta_2 + A_{12}} \,. \tag{12.72}$$

Substituting this in (12.61), we ultimately obtain the two-soliton solution

$$u = \frac{1}{2} \left(k_2^2 - k_1^2 \right) \left[\frac{k_2^2 \operatorname{cosech}^2 (\eta_2/2) + k_1^2 \operatorname{sech}^2 (\eta_1/2)}{\left(k_2 \coth (\eta_2/2) - k_1 \tanh (\eta_1/2) \right)^2} \right] . \tag{12.73}$$

The solution (12.73) when plotted has exactly the same form as in Fig. 12.5, thereby showing the soliton nature of the solitary wave as discussed in the previous section.

(c) *N-Soliton Solutions*

One can proceed as before for the general case, with the choice

$$f^{(1)} = \mathrm{e}^{\eta_1} + \mathrm{e}^{\eta_2} + ... + \mathrm{e}^{\eta_N} \tag{12.74}$$

and then solve successively for $f^{(2)}$, ..., $f^{(N)}$ and finally obtain F and u. Explicit expressions can be written down with some effort, which we desist from doing so here due to its somewhat complicated nature. For more details, see for example Ref. [7]. Further details can be found in the next chapter.

(d) *Asymptotic Analysis*

Let us consider the two-soliton solution (12.73) and analyse the limits $t \to -\infty$ and $t \to +\infty$ separately, so as to understand the interaction of two one-solitons centered around $\eta_1 \approx 0$ or $\eta_2 \approx 0$. Without loss of generality let us assume that $k_2 > k_1$. Then we can see that in the limits $t \to \pm\infty$, $\eta_1 = k_1 x - k_1^3 t + \eta_1^{(0)}$ and $\eta_2 = k_2 x - k_2^3 t + \eta_2^{(0)}$ take the following limiting values:

(i) $t \to -\infty$:

$$\eta_1 \approx 0, \quad \eta_2 \to \infty$$
$$\eta_2 \approx 0, \quad \eta_1 \to -\infty$$

(ii) $t \to +\infty$:

$$\eta_1 \approx 0, \quad \eta_2 \to -\infty$$
$$\eta_2 \approx 0, \quad \eta_1 \to \infty$$

Substituting these limiting values into the two-soliton expression (12.73) and with simple algebra, we can easily show that we have the following solutions for $t \to +\infty$ and $t \to -\infty$:

(i) $t \to -\infty$:

Soliton $1(\eta_1 \approx 0)$:

$$u(x,t) = \frac{1}{2} k_1^2 \operatorname{sech}^2\left(\frac{\eta_1 - \Delta}{2}\right), \quad \Delta = \log\left(\frac{k_2 + k_1}{k_2 - k_1}\right). \tag{12.75}$$

Proof:

Since here $\eta_1 \approx 0$, $\eta_2 \to +\infty$, the two-soliton expression (12.73) becomes

$$
\begin{aligned}
u &\approx \frac{1}{2}\left(k_2^2 - k_1^2\right) \frac{k_1^2 \operatorname{sech}^2(\eta_1/2)}{\left(k_2 - k_1 \tanh\eta_1/2\right)^2} \\
&= 2\left(k_2^2 - k_1^2\right) \frac{k_1^2}{\left[(k_2 - k_1)\, \mathrm{e}^{\eta_1/2} + (k_2 + k_1)\, \mathrm{e}^{-\eta_1/2}\right]^2} \\
&= 2 \frac{k_1^2}{\left[\sqrt{(k_2 - k_1)/(k_2 + k_1)}\, \mathrm{e}^{\eta_1/2} + \sqrt{(k_2 + k_1)/(k_2 - k_1)}\, \mathrm{e}^{-\eta_1/2}\right]^2} \\
&= \frac{2k_1^2}{\left(\mathrm{e}^{(\eta_1 - \Delta)/2} + \mathrm{e}^{-(\eta_1 - \Delta)/2}\right)^2}
\end{aligned}
$$

where

$$
\Delta = \log\left(\frac{k_2 + k_1}{k_2 - k_1}\right) .
$$

Thus (12.75) follows.

Soliton 2 ($\eta_2 \approx 0$) :

Following a procedure similar to the above, we obtain

$$
u(x,t) = \frac{1}{2} k_2^2 \operatorname{sech}^2\left[(\eta_2 + \Delta)/2\right] . \tag{12.76}
$$

(ii) $t \to \infty$:

Soliton 1($\eta_2 \approx 0$) :

$$
u(x,t) = \frac{1}{2} k_1^2 \operatorname{sech}^2\left[(\eta_1 + \Delta)/2\right] . \tag{12.77}
$$

Soliton 2($\eta_1 \approx 0$) :

$$
u(x,t) = \frac{1}{2} k_2^2 \operatorname{sech}^2\left[(\eta_2 - \Delta)/2\right] . \tag{12.78}
$$

Using the above analysis, we can readily interpret the two-soliton solution of the KdV equation given by (12.73) in the following way. Two individual solitary waves (one-solitons) of differing amplitudes, $k_1^2/2$ and $k_2^2/2$ ($k_2 > k_1$), with the smaller one positioned right of the larger one, travel to the right with speeds k_1 and k_2 respectively. The larger one soon catches up with the smaller one, undergoes nonlinear interaction while overtaking it and ultimately the soliton 1 and the soliton 2 get interchanged. The net effect is merely a total *phase shift* 2Δ suffered by the solitons without any change in shape, amplitude or speed. Or in other words, solitary waves of the KdV equation undergo elastic collisions, reminiscent of particle collisions, as demonstrated by the numerical experiments of Zabusky and Kruskal. So they are indeed solitons of the KdV equation.

The analysis can be extended to N-soliton solutions also, but we will refrain from doing so here.

Exercises:

8. Deduce the limiting expressions (12.76), (12.77) and (12.78).
9. Obtain the three-soliton solution of the KdV equation using the Hirota method and carry out an asymptotic analysis.
10. Apply Hirota's method to MKdV equation and obtain the one- and two-soliton solutions.
11. Apply the Hirota method to the following equation and obtain the solitary wave solution. Investigate whether two-soliton solution exists:

$$u_t + u^n u_x + u_{xxx} = 0, \quad n = 3, 4 .$$

12. Obtain the solitary wave solution of the fifth-order MKdV equation given in problem 7 at the end of Chap. 11 through the Hirota method. Does the system possess a two-soliton solution?
13. Using Hirota method, deduce the solitary wave solution of the RLW equation (problem 8 at the end of Chap. 11)

$$u_t - u_{xxt} + u_{xxx} + 6uu_x = 0 .$$

Is there a two-soliton solution?
14. Bilinearize the nonlinear wave equation

$$w_{tt} - w_{xx} - 6(w^2)_{xx} - w_{xxxx} = 0 ,$$

which describes shallow water under gravity and long waves in one-dimensional nonlinear lattices. Using the transformation

$$w = [\log f(x,t)]_{xx} ,$$

bilinearize the equation. Obtain the one- and two-soliton solutions and investigate the collision property [8].

12.6 Conclusions

In this chapter, we have pointed out the KdV equation is quite ubiquitous and that it can appear in very many physical contents. Its solitary wave solution possesses remarkable stability: Under collision with another solitary wave it retains its shape and speed and so it is a soliton. Explicit analytical forms to represent all the soliton aspects can be deduced using the Hirota bilinearization method in the form of N-soliton solutions. In the following chapter, we will show that an even more rigorous formalism in the form of IST method can be developed to solve the Cauchy IVP of KdV equation, which also allows one to prove that it is a completely integrable infinite dimensional nonlinear dynamical system.

13. Basic Soliton Theory of KdV Equation

We have seen in the earlier chapters that the KdV equation is a ubiquitous nonlinear dispersive wave equation occurring in a diverse range of physical phenomena. It admits solitary wave solution, which indeed represents a soliton that undergoes particle-like elastic collision with another soliton, as shown by the numerical experiments of Zabusky and Kruskal. In addition, the long time asymptotic state consists of a series of soliton pulses of differing amplitudes in the background of small amplitude oscillatory tail. As pointed out in the previous chapter, the exact form of the N-soliton solution can also be found. The question is then can one solve the initial value problem of the KdV equation fully and obtain the above results rigorously. In this chapter, we wish to show that an elegant mathematical formalism can be developed, which can be considered as a nonlinear counterpart of the Fourier transform method applicable to linear dispersive systems (see Chap. 11), and it is now popularly called the *inverse scattering transform (IST)* method.

The fact that the Cauchy IVP of the KdV equation can be solved completely by the IST method ensures that it is an integrable system. In fact, it can be proved that the KdV equation is a completely integrable infinite dimensional Hamiltonian nonlinear dynamical system in the Liouville sense. For this purpose, we will also bring out the Hamiltonian structure underlying the KdV equation in this chapter, point out the existence of infinite number of conserved quantities and conservation laws and prove the fact that the scattering data associated with the IST problem provide the required action and angle variables to establish the complete integrability. There also exist other important features associated with the integrable KdV equation, namely the so-called Bäcklund transformation, Painlevé property (see appendix J), etc. We will also briefly point out these aspects in this chapter. In short, KdV equation is a prototypical integrable soliton system, paving the way for the development of the field of 'Integrable Systems/Completely Integrable Systems' in recent times.

13.1 The Miura Transformation and Linearization of KdV: The Lax Pair

13.1.1 The Miura Transformation

It is known that the familiar nonlinear Burgers equation, which is a nonlinear heat equation (see for example, Ref. [1]),

$$u_t + uu_x = \nu u_{xx} \ , \tag{13.1}$$

can be transformed into the standard linear heat equation (see also Chap. 15)

$$v_t = \nu v_{xx} \tag{13.2}$$

under the so-called Cole–Hopf transformation

$$u = -2\nu \frac{v_x}{v} \ . \tag{13.3}$$

In (13.1), ν is a constant parameter.

Exercise:

1. Verify the above.

In the year 1968 R.M. Miura [2], who was working in the group of Martin Kruskal at Princeton, noted that the KdV equation

$$u_t - 6uu_x + u_{xxx} = 0 \tag{13.4}$$

and the modified KdV equation

$$v_t - 6v^2 v_x + v_{xxx} = 0 \tag{13.5}$$

are related to each other through the transformation

$$u = v^2 + v_x \ . \tag{13.6}$$

Note the change in sign in the second term of the KdV equation (13.4) compared to the form given in Chaps. 11 and 12. This is introduced purely for convenience. As noted in Sect. 12.3, (13.4) can be transformed into the form (12.40) by the trivial scale change $u \to -u$ and so also all the further results.

Proof:

We have $u = v^2 + v_x$ and so

$$\begin{aligned} u_t &= 2vv_t + v_{xt} \ , \\ u_x &= 2vv_x + v_{xx} \ , \\ u_{xxx} &= 6v_x v_{xx} + 2vv_{xxx} + v_{xxxx} \ , \end{aligned}$$

so that (13.4) becomes

$$2v(v_t - 6v^2 v_x + v_{xxx}) + (v_t - 6v^2 v_x + v_{xxx})_x = 0 \,, \tag{13.7}$$

which implies the MKdV equation (13.5).

The transformation (13.6) is now called the *Miura transformation* in the literature, which essentially relates two nonlinear equations to each other.

Now in analogy with the Cole–Hopf transformation (13.3) for the Burgers equation, one can think of a transformation

$$v = \frac{\psi_x}{\psi} \tag{13.8}$$

for the MKdV equation (13.5). Then in view of the Miura transformation (13.6), we have the following transformation for the KdV equation:

$$u = \frac{\psi_{xx}}{\psi} \,. \tag{13.9}$$

(Verify (13.9)).

13.1.2 Galilean Invariance and Schrödinger Eigenvalue Problem

At this point we note that the KdV equation (13.4) is invariant (that is its form remains unchanged) under the Galilean transformation

$$x' = x - \lambda t, \;\; t' = t, \;\; u' = u + \lambda \,. \tag{13.10}$$

Under this transformation, we also have

$$\frac{\partial}{\partial x} = \frac{\partial}{\partial x'} \,, \quad \frac{\partial}{\partial t} = \frac{\partial}{\partial t'} - \lambda \frac{\partial}{\partial x'} \tag{13.11}$$

and so the KdV equation (13.4) becomes

$$u'_{t'} - 6u'u'_{x'} + u'_{x'x'x'} = 0 \,. \tag{13.12}$$

Or in other words, the KdV equation is form invariant under the Galilean transformation (13.10). In the new frame of reference, the last of the transformation in (13.10), after using (13.9) and (13.11), can be rewritten as

$$u' = \frac{\psi_{x'x'}}{\psi} + \lambda \,. \tag{13.13}$$

Equation (13.13) can be reexpressed as

$$\psi_{x'x'} + (\lambda - u')\psi = 0 \,. \tag{13.14}$$

Omitting the primes for convenience hereafter, we finally obtain the time independent Schrödinger type linear eigenvalue problem

$$\psi_{xx} + (\lambda - u)\psi = 0 \tag{13.15}$$

in which the unknown function $u(x, t)$ of the KdV equation appears as a "potential", while (13.15) defining the transformation function $\psi(x, t)$ itself is *linear*.

13.1.3 Linearization of the KdV Equation

Treating (13.15) as a linearizing transformation for the KdV equation (13.4), we can obtain an evolution equation for the function $\psi(x,t)$, which reads

$$\psi_t = -4\psi_{xxx} + 6u\psi_x + 3u_x\psi \ . \tag{13.16}$$

Proof:

Let $\psi_t = A\psi + B\psi_x$, where A and B are functions to be determined. Then $(\psi_t)_{xx} = (\psi_{xx})_t$ implies

$$u_t = A_{xx} + 2B_x(u-\lambda) + Bu_x \ , \quad B_{xx} + 2A_x = 0 \ .$$

By expanding A and B in powers of λ, say,

$$A = A_1\lambda + A_0 \ , \quad B = B_1\lambda + B_0$$

and substituting in the above and equating powers of λ to zero, we can obtain (13.16) by finding A and B and making use of (13.15).

Exercise:

2. Deduce (13.16) using the above procedure.

Thus, one can conclude that the nonlinear KdV equation (13.4) is equivalent to *two linear* differential equations, namely the Schrödinger type eigenvalue equation (13.15) and the associated linear time evolution equation (13.16) for the eigenfunction $\psi(x,t)$. Note that in both these linear equations the unknown function $u(x,t)$ (and its derivatives) of the KdV equation occurs as a coefficient. One can also easily check the converse, namely given the two linear systems (13.15) and (13.16), they are equivalent to the nonlinear KdV equation.

Proof:

Consider the compatibility condition

$$(\psi_{xx})_t = (\psi_t)_{xx} \ . \tag{13.17}$$

Substituting for ψ_{xx} on the left hand side from (13.15) and for ψ_t on the right hand side from (13.16), we obtain

$$(\psi_{xx})_t = -uu_x\psi + 2u^2\psi_x + 2\lambda u\psi_x + u_x\lambda\psi - 4\lambda^2\psi_x + u_t\psi \ , \tag{13.18a}$$

$$(\psi_t)_{xx} = 5uu_x\psi + u_x\lambda\psi - u_{xxx}\psi + 2u^2\psi_x - 4\lambda^2\psi_x + 2u\lambda\psi_x \ . \tag{13.18b}$$

Then from the compatibility condition (13.17), we obtain the KdV equation (13.4).

13.1.4 Lax Pair

The above possibility of linearization of the nonlinear KdV equation in terms of the linear systems (13.15) and (13.16) can be rephrased in the following elegant way, as formulated by P.D. Lax [3]. Consider the linear eigenvalue problem

$$L\psi = \lambda\psi\ , \tag{13.19a}$$

where in the present problem the linear differential operator is

$$L = -\frac{\partial^2}{\partial x^2} + u(x,t)\ , \tag{13.19b}$$

and $\lambda = \lambda(t)$ is the eigenvalue at time t. Let the eigenfunction $\psi(x,t)$ evolves as

$$\psi_t = B\psi\ , \tag{13.20a}$$

where in the case of KdV equation the second linear differential operator is

$$B = -4\frac{\partial^3}{\partial x^3} + 3\left(u\frac{\partial}{\partial x} + \frac{\partial}{\partial x}u\right)\ . \tag{13.20b}$$

Then with the requirement that the eigenvalue λ does not change with time, that is

$$\lambda(t) = \lambda(0) = \text{constant}\ , \tag{13.21}$$

the compatibility of (13.19) and (13.20) leads to the *Lax equation* or *Lax condition*

$$L_t = [B, L] = (BL - LB)\ . \tag{13.22}$$

For the specific forms of L and B given by (13.19b) and (13.20b), the Lax equation is indeed equivalent to the KdV equation (13.4), provided λ is unchanged in time. For any other suitable choice of L and B, a different nonlinear evolution equation will be obtained.

One may say that the eigenvalue problem is *isospectral*. The Lax condition (13.22) is the isospectral condition for the Lax pair L and B.

Exercises:

3. Deduce the Lax equation (13.22) using (13.19a), (13.20a) and (13.21).
4. Show that with the choices (13.19b) and (13.20b) for L and B respectively, the Lax equation (13.21) is equivalent to the KdV equation.
5. Find the nonlinear evolution equation which arises for

$$L = -\frac{\partial^2}{\partial x^2} + u(x,t) \,,$$

$$B = \frac{\partial^5}{\partial x^5} - \frac{5}{4} u \frac{\partial^3}{\partial x^3} - \frac{5}{4} \frac{\partial^3}{\partial x^3}(u) + \frac{5}{4}\left(u_{xx} + 3u^2\right) \frac{\partial}{\partial x} + \frac{5}{16} \frac{\partial}{\partial x}\left(u_{xx} + 3u^2\right)$$

in (13.22).

The Lax condition has indeed played a very important role in soliton theory, not only for the KdV equation but also for the other soliton systems as well. In fact it is now an accepted fact that existence of a Lax pair is indeed a decisive hallmark of integrable systems. We will confirm this for a few other soliton systems in the next chapter.

13.2 Lax Pair and the Method of Inverse Scattering: A New Method to Solve the Initial Value Problem

We are interested in solving the *initial value problem* of the KdV equation. That is, "*Given the initial value* $u(x,0)$ *at* $t = 0$, *how does the solution of the KdV equation (13.4) evolve for the given boundary conditions, say* $u(x,t) \overset{x \to \pm\infty}{\longrightarrow} 0$*?*". Also how does the linearization property discussed in the previous section help in this regard? In the following we briefly point out that indeed the linearization property leads to a new method of integrating the nonlinear evolution equation (13.4) through a three steps process, which is a generalization of the method discussed earlier in Chap. 11 for linear dispersive wave equations. The procedure was originally developed by Gardner, Greene, Kruskal and Miura [4]. This method, now called the *inverse scattering transform* (IST) method, may be considered as a nonlinear Fourier transform method. It will be now described as applicable to the KdV equation. The same method can be adopted for other integrable systems as well, which is indicated in the next chapter.

13.2.1 The Inverse Scattering Transform (IST) Method for KdV Equation

The analysis proceeds in three steps similar to the case of Fourier transform method applicable for linear dispersive systems (Chap. 11):

(i) Direct scattering transform analysis.
(ii) Analysis of time evolution of scattering data.
(iii) Inverse scattering transform (IST) analysis.

The method is schematically shown in Fig. 13.1. The details are as follows.

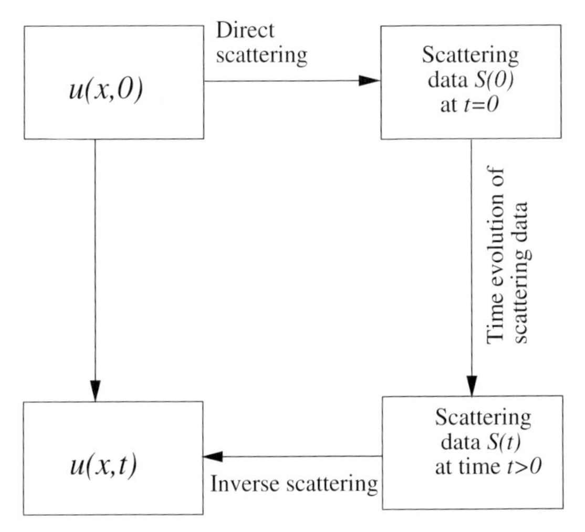

Fig. 13.1. Schematic diagram of the inverse scattering transform method

(i) *Direct Scattering Analysis and Scattering Data at* $t = 0$

The given information is the initial data $u(x,0)$, which has the property that it vanishes sufficiently fast as $x \to \pm\infty$. Now considering the Schrödinger spectral problem (13.19) at $t = 0$,

$$\psi_{xx} + [\lambda - u(x,0)]\psi = 0, \quad u \overset{|x|\to\infty}{\longrightarrow} 0\,, \tag{13.23}$$

it is well known from linear spectral theory (and one-dimensional quantum mechanics) that the system (13.23) admits

(1) a finite number of bound states with eigenvalues

$$\lambda = -\kappa_n^2\,, \quad n = 1, 2, ..., N \tag{13.24a}$$

and normalization constants $C_n(0)$ of the associated eigenstates and
(2) a continuum or scattering states with the continuous eigenvalues,

$$\lambda = k^2\,, \quad -\infty < k < \infty\,. \tag{13.24b}$$

They are further characterized by the reflection coefficient $R(k,0)$ and transmission coefficient $T(k,0)$ such that

$$|R(k,0)|^2 + |T(k,0)|^2 = 1\,. \tag{13.25}$$

Thus, from the given potential (initial data) $u(x,0)$ with the boundary conditions $u \to 0$ as $x \to \pm\infty$, one can carry out a direct scattering analysis of (13.23) to obtain the scattering data at $t = 0$:

$$S(0) = \{\kappa_n, C_n(0), R(k,0), T(k,0),\ \ n = 1, 2, ..., N,\ \ -\infty < k < \infty\}\,. \tag{13.26}$$

In the appendix F, we illustrate these results by solving the model potential $u(x) = -A\mathrm{sech}^2\alpha x$. The general direct scattering analysis for (13.23) is carried out in appendix G.

(ii) *Time Evolution of Scattering Data*

Now as the potential $u(x,t)$ evolves from its initial value $u(x,0)$ so that it satisfies the KdV equation, how does the corresponding scattering data given by (13.26) evolve from $S(0)$ to $S(t)$? In order to understand this one can use the time evolution equation of the eigenfunction (13.16),

$$\psi_t = -4\frac{\partial^3\psi}{\partial x^3} + 3\left(u\frac{\partial}{\partial x} + \frac{\partial}{\partial x}u\right)\psi \,. \tag{13.27}$$

Since the scattering data is intimately associated with the asymptotic ($x \to \pm\infty$) behaviour of the eigenfunction, where the potential $u(x,t) \to 0$, it is enough if we confine the analysis to this region. Thus as $x \to \pm\infty$, (13.27) can be written as

$$\psi_t = -4\frac{\partial^3\psi}{\partial x^3}\,, \qquad x \to \pm\infty \,. \tag{13.28}$$

(a) Scattering States

Without loss of generality, we can write the asymptotic form of the general solution to (13.28) as

$$\psi(x,t) \overset{x\to-\infty}{\longrightarrow} a_+(k,t)\mathrm{e}^{\mathrm{i}kx} + a_-(k,t)\mathrm{e}^{-\mathrm{i}kx} \tag{13.29a}$$

$$\overset{x\to+\infty}{\longrightarrow} b_+(k,t)\mathrm{e}^{\mathrm{i}kx} + b_-(k,t)\mathrm{e}^{-\mathrm{i}kx} \,. \tag{13.29b}$$

Substituting these asymptotic forms into (13.28), we obtain

$$\frac{\mathrm{d}a_\pm}{\mathrm{d}t} = \pm 4\mathrm{i}k^3 a_\pm \,, \quad \frac{\mathrm{d}b_\pm}{\mathrm{d}t} = \pm 4\mathrm{i}k^3 b_\pm \,. \tag{13.30}$$

On integration, we get

$$a_\pm(k,t) = a_\pm(k,0)\mathrm{e}^{\pm 4\mathrm{i}k^3 t} \,, \tag{13.31a}$$

$$b_\pm(k,t) = b_\pm(k,0)\mathrm{e}^{\pm 4\mathrm{i}k^3 t} \,. \tag{13.31b}$$

Consider now the standard type of scattering solutions (involving incident, reflected and transmitted waves)

$$\frac{1}{a_+(k,t)}\psi \overset{x\to-\infty}{\longrightarrow} \mathrm{e}^{\mathrm{i}kx} + R(k,t)\mathrm{e}^{-\mathrm{i}kx} \,, \tag{13.32a}$$

$$\overset{x\to+\infty}{\longrightarrow} T(k,t)\mathrm{e}^{\mathrm{i}kx} \,. \tag{13.32b}$$

Comparing (13.29) and (13.32) and making use of (13.31), it is straightforward to see that

$$R(k,t) = \frac{a_-(k,t)}{a_+(k,t)} = R(k,0)\mathrm{e}^{-8ik^3t} , \tag{13.33a}$$

$$T(k,t) = \frac{b_+(k,t)}{a_+(k,t)} = T(k,0) , \tag{13.33b}$$

with $b_-(k,t)$ taken as zero.

(b) Bound States

By construction, the eigenvalues $\lambda = -\kappa_n^2$, $n = 1, 2, ..., N$, do not change with time:

$$\lambda(t) = \lambda(0) \Longrightarrow \kappa_n(t) = \kappa_n(0) . \tag{13.34}$$

Correspondingly, the eigenfunctions of these discrete states satisfy the time evolution equation

$$\psi_{n,t} = -4\psi_{n,xxx}, \quad x \to \pm\infty . \tag{13.35}$$

Then we have

$$\psi_n(x,t) \longrightarrow \mathrm{e}^{\kappa_n x} , \quad x \to -\infty , \tag{13.36a}$$

$$\longrightarrow C_n(t)\mathrm{e}^{-\kappa_n x}, \quad x \to +\infty , \tag{13.36b}$$

with the normalization condition

$$\int_{-\infty}^{\infty} |\psi_n(x,t)|^2 \mathrm{d}x = 1 . \tag{13.36c}$$

Substituting (13.36) into (13.35), we see that the normalization constants $C_n(t)$ evolve as

$$\frac{\mathrm{d}C_n}{\mathrm{d}t} = 4\kappa_n^3 C_n \tag{13.37a}$$

so that

$$C_n(t) = C_n(0)\mathrm{e}^{4\kappa_n^3 t} . \tag{13.37b}$$

Thus, at an arbitrary future instant of time 't', the scattering data $S(t)$ corresponding to the potential $u(x,t)$ evolves from $S_n(0)$ of the initial data $u(x,0)$:

$$S(t) = \{\kappa_n(t) = \kappa_n(0), \; C_n(t) = C_n(0)\mathrm{e}^{4\kappa_n^3 t}, \; n = 0, 1, 2, ..., N,$$
$$R(k,t) = R(k,0)\mathrm{e}^{-8ik^3 t}, \; -\infty < k < \infty\} . \tag{13.38}$$

(iii) *Inverse Scattering Analysis*

Now given the scattering data $S(t)$ as in (13.38) at time 't', can one invert the data and obtain uniquely the potential $u(x,t)$ of the Schrödinger spectral problem (13.19), in which the time variable 't' enters only as a parameter?

The answer is yes, and it can be done by solving a *linear*, Volterra type singular, *integral equation* called *Gelfand–Levitan–Marchenko integral equation.* The scattering data $S(t)$, given by (13.38), is given as input into this integral equation, which when solved gives the solution $u(x,t)$ of the KdV equation. The linear integral equation reads

$$K(x,y,t) + F(x+y,t) + \int_x^\infty F(y+z,t)K(x,z,t)\mathrm{d}z = 0\ , \quad y > x \tag{13.39a}$$

where

$$F(x+y,t) = \sum_{n=1}^{N} C_n^2(t)\mathrm{e}^{-\kappa_n(x+y)} + \frac{1}{2\pi}\int_{-\infty}^{\infty} R(k,t)\mathrm{e}^{\mathrm{i}k(x+y)}\mathrm{d}k\ . \tag{13.39b}$$

Note that in the above equation the time variable 't' enters only as a parameter and that all the informations about the scattering data are contained in the function $F(x+y,t)$. Solving (13.39), we finally obtain the potential

$$u(x,t) = -2\frac{\mathrm{d}}{\mathrm{d}x}K(x,x+0,t)\ . \tag{13.40}$$

Brief details of the derivation are given in appendix G. Thus the initial value problem of the KdV equation stands solved.

13.3 Explicit Soliton Solutions

Now, we are in a position to obtain all the soliton solutions and the properties associated with them as discussed in the previous chapter. As seen above, for solving the general initial value problem, one has to solve the Gelfand–Levitan–Marchenko integral equation (13.39), with the full set of scattering data $S(t)$. Though this is possible in principle, in practice this may not be completely feasible analytically. However for the special, but important, class of the so-called *reflectionless potentials*, characterized by the condition

$$R(k,t) = 0\ , \tag{13.41}$$

it is possible to solve fully the Gelfand–Levitan–Marchenko integral equation.

An example of the reflectionless case is again the potential $u(x) = -A\mathrm{sech}^2\alpha x$, $A > 0$ (see appendix F). Then if there are N bound states, the corresponding solution, $u(x,t)$, turns out to be the N-soliton solution. First, let us obtain the one and two-soliton solutions explicitly and then generalize the results to N-soliton solution.

13.3.1 One-Soliton Solution ($N = 1$)

Consider the special case of reflectionless potential ($R(k,t) = 0$) with only one bound state, $N = 1$, specified by (example: $u(x) = -2\mathrm{sech}^2 x$ has got one bound state, $\lambda = -1$)

$$\kappa_1 = \kappa, \quad C_1(t) = C(t) = C(0)\mathrm{e}^{+4\kappa^3 t} \tag{13.42}$$

so that in (13.39)

$$F(x+y,t) = C^2(0)\mathrm{e}^{8\kappa^3 t - \kappa(x+y)} = C_0^2 \mathrm{e}^{8\kappa^3 t - \kappa(x+y)} \tag{13.43}$$

and the Gelfand–Levitan–Marchenko integral equation becomes

$$\begin{aligned} K(x,y,t) &+ C_0^2 \mathrm{e}^{8\kappa^3 t - \kappa(x+y)} \\ &+ C_0^2 \mathrm{e}^{8\kappa^3 t} \int_x^\infty \mathrm{e}^{-\kappa(y+z)} K(x,z,t)\mathrm{d}z = 0\,. \end{aligned} \tag{13.44}$$

Then it is straightforward to check that

$$\frac{\partial K}{\partial y} = -\kappa K\,. \tag{13.45}$$

On solving, we have

$$K(x,y,t) = \mathrm{e}^{-\kappa y} h(x,t)\,, \tag{13.46}$$

where the function $h(x,t)$ is to be determined. Substituting (13.46) back into (13.44) and simplifying, we can find that

$$h(x,t) = \frac{-C_0^2 \mathrm{e}^{8\kappa^3 t - \kappa x}}{\left[1 + (C_0^2/2\kappa)\mathrm{e}^{8\kappa^3 t - 2\kappa x}\right]}\,. \tag{13.47}$$

Then from (13.46), we have

$$K(x,y,t) = \frac{-C_0^2 \mathrm{e}^{8\kappa^3 t - \kappa(x+y)}}{\left[1 + (C_0^2/2\kappa)\,\mathrm{e}^{8\kappa^3 t - 2\kappa x}\right]}\,. \tag{13.48}$$

So the corresponding solution to the KdV equation can be obtained from (13.40) as

$$\begin{aligned} u(x,t) &= -2\frac{\mathrm{d}}{\mathrm{d}x}K(x,x+0,t) \\ &= -2\partial K(x,y,t)/\partial x|_{y=x} - 2\partial K(x,y,t)/\partial y|_{y=x} \\ &= -2\kappa^2 \frac{\mathrm{e}^{-2\kappa(x-4\kappa^2 t)-2\delta}}{[1+\mathrm{e}^{-2\kappa(x-4\kappa^2 t)-2\delta}]^2}\,, \qquad \delta = \frac{1}{2}\log\left(2\kappa/C_0^2\right) \\ &= -2\kappa^2 \operatorname{sech}^2\left[\kappa\left(x - 4\kappa^2 t\right) + \delta\right]\,. \end{aligned} \tag{13.49}$$

Expression (13.49) is indeed the one-soliton solution (12.69) obtained by the Hirota method in Sect. 12.5. (Note the scale change and negative sign in (13.49), due to the difference in the coefficients in front of the nonlinear term in (13.4) and (12.40), and also a redefinition of the parameters).

13.3.2 Two-Soliton Solution

Again, let us consider a reflectionless potential such that $R(k,t) = 0$, but now with two bound states (example: $u(x) = -6\mathrm{sech}^2 x$ has two bound states with $\lambda_1 = -4$, $\lambda_2 = -1$), specified by the discrete values κ_1 and κ_2 and the corresponding normalization constants $C_1(t)$ and $C_2(t)$. Then we have the Gelfand–Levitan–Marchenko integral equation in the form

$$\begin{aligned} K(x,y,t) &+ C_{10}^2 \mathrm{e}^{8\kappa_1^3 t} \mathrm{e}^{-\kappa_1(x+y)} + C_{20}^2 \mathrm{e}^{8\kappa_2^3 t} \mathrm{e}^{-\kappa_2(x+y)} \\ &+ C_{10}^2 \mathrm{e}^{8\kappa_1^3 t} \int_x^\infty \mathrm{e}^{-\kappa_1(y+z)} K(x,z,t)\mathrm{d}z \\ &+ C_{20}^2 \mathrm{e}^{8\kappa_2^3 t} \int_x^\infty \mathrm{e}^{-\kappa_2(y+z)} K(x,z,t)\mathrm{d}z = 0 \,, \end{aligned} \tag{13.50}$$

where C_{10} and C_{20} are the normalization constants corresponding to the two bound states. Let

$$K(x,y,t) = \mathrm{e}^{-\kappa_1 y} h_1(x,t) + \mathrm{e}^{-\kappa_2 y} h_2(x,t) \,. \tag{13.51}$$

Using (13.51) in (13.50) and equating the coefficients of $\mathrm{e}^{-\kappa_1 y}$ and $\mathrm{e}^{-\kappa_2 y}$ to zero separately, we can obtain two algebraic equations for h_1 and h_2. Solving them, one obtains

$$h_1(x,t) = \frac{\det \alpha}{\det \beta} \,, \quad h_2(x,t) = \frac{\det \delta}{\det \beta} \,, \tag{13.52a}$$

with

$$\alpha = \begin{pmatrix} \alpha_{11} & \alpha_{12} \\ \alpha_{21} & \alpha_{22} \end{pmatrix}, \quad \beta = \begin{pmatrix} \beta_{11} & \beta_{12} \\ \beta_{21} & \beta_{22} \end{pmatrix}, \quad \delta = \begin{pmatrix} \delta_{11} & \delta_{12} \\ \delta_{21} & \delta_{22} \end{pmatrix}, \tag{13.52b}$$

where

$$\begin{aligned} \alpha_{11} &= -C_{10}^2 \exp\left(8\kappa_1^3 t - \kappa_1 x\right) \,, \\ \alpha_{12} &= \left[C_{10}^2/\left(\kappa_1+\kappa_2\right)\right] \exp\left[8\kappa_1^3 t - \left(\kappa_1+\kappa_2\right)x\right] \,, \\ \alpha_{21} &= -C_{20}^2 \exp\left(8\kappa_2^3 t - \kappa_2 x\right) \,, \\ \alpha_{22} &= 1 + \left[C_{20}^2/\left(2\kappa_2\right)\right] \exp\left(8\kappa_2^3 t - \kappa_2 x\right) \,, \\ \beta_{11} &= 1 + \left[C_{20}^2/\left(2\kappa_1\right)\right] \exp\left(8\kappa_1^3 t - 2\kappa_1 x\right) \,, \\ \beta_{12} &= \left[C_{10}^2/\left(\kappa_1+\kappa_2\right)\right] \exp\left[8\kappa_1^3 t - \left(\kappa_1+\kappa_2\right)x\right] \,, \\ \beta_{21} &= \left[C_{20}^2/\left(\kappa_1+\kappa_2\right)\right] \exp\left[8\kappa_2^3 t - \left(\kappa_1+\kappa_2\right)x\right] \,, \\ \beta_{22} &= 1 + \left[C_{20}^2/\left(2\kappa_2\right)\right] \exp\left(8\kappa_2^3 t - 2\kappa_2 x\right) \,, \\ \delta_{11} &= 1 + \left[C_{10}^2/\left(2\kappa_1\right)\right] \exp\left(8\kappa_1^3 t - 2\kappa_1 x\right) \,, \\ \delta_{12} &= -C_{10}^2 \exp\left(8\kappa_1^3 t - \kappa_1 x\right) \,, \\ \delta_{21} &= \left[C_{20}^2/\left(\kappa_1+\kappa_2\right)\right] \exp\left[-\left(\kappa_1+\kappa_2\right)x + 8\kappa_2^3 t\right] \,, \\ \delta_{22} &= -C_{20}^2 \exp\left(8\kappa_2^3 t - \kappa_2 x\right) \,. \end{aligned} \tag{13.52c}$$

Then

$$u(x,t) = -2\frac{\mathrm{d}}{\mathrm{d}x}\left[\mathrm{e}^{-\kappa_1 y}h_1(x,t) + \mathrm{e}^{-\kappa_2 y}h_2(x,t)\right]$$
$$= -2\left(\kappa_2^2 - \kappa_1^2\right)\frac{\kappa_2^2\,\mathrm{cosech}^2\gamma_2 + \kappa_1^2\,\mathrm{sech}^2\gamma_1}{\left(\kappa_2\,\coth\gamma_2 - \kappa_1\,\tanh\gamma_1\right)^2}\,, \tag{13.53a}$$

where

$$\gamma_i = \kappa_i x - 4\kappa_i^3 t - \delta_i,\;\; \delta_i = \frac{1}{2}\log\left(\frac{C_{i0}^2}{2\kappa_i}\frac{(\kappa_2-\kappa_1)}{(\kappa_2+\kappa_1)}\right),\quad i = 1,2. \tag{13.53b}$$

Note that in the above $\mathrm{d}/\mathrm{d}x$ implies $\partial/\partial x|_{y=x} + \partial/\partial y|_{y=x}$. One can easily check that the form (13.53) is indeed the two-soliton solution of the KdV equation discussed in Sect. 12.7.3, with appropriate scale change and redefinition of parameters.

13.3.3 *N*-Soliton Solution

Considering now reflectionless potentials with N-bound states, we can write the expression (13.39b) as

$$F(x+y,t) = \sum_{n=1}^{N} C_n^2(t)\mathrm{e}^{-\kappa_n(x+y)}$$
$$= \sum_{n=1}^{N} C_n\mathrm{e}^{-\kappa_n x}C_n\mathrm{e}^{-\kappa_n y}$$
$$= \sum g_n(x,t)g_n(y,t)\,, \tag{13.54a}$$

where

$$g_n(x,t) = C_n(t)\mathrm{e}^{-\kappa_n x}\,. \tag{13.54b}$$

Then defining

$$K(x,y) = \sum_{n=1}^{N}\omega_n(x)g_n(y) \tag{13.55}$$

(here the t dependence is suppressed for convenience) and substituting into the Gelfand–Levitan–Marchenko integral equation (13.39a), we obtain

$$\omega_m(x) + g_m(x) + \sum_{n=1}^{N}\omega_n(x)\int_x^\infty g_m(z)g_n(z)\mathrm{d}z = 0\,. \tag{13.56}$$

Defining now the matrices

$$P_{mn}(x) = \delta_{mn} + \int_x^\infty g_m(z)g_n(z)\mathrm{d}z\,, \tag{13.57a}$$
$$\boldsymbol{\omega}(x) = \left(\omega_1(x), \omega_2(x), ..., \omega_N(x)\right)^{\mathrm{T}}\,, \tag{13.57b}$$
$$\mathbf{g}(x) = \left(g_1(x), g_2(x), ..., g_N(x)\right)^{\mathrm{T}}\,, \tag{13.57c}$$

(13.56) can be rewritten as the matrix equation

$$\mathbf{P}(x)\boldsymbol{\omega}(x) = -\mathbf{g}(x) . \tag{13.58a}$$

Or

$$\boldsymbol{\omega}(x) = -\mathbf{P}^{-1}(x)\mathbf{g}(x) . \tag{13.58b}$$

Using (13.55), we have

$$K(x,x) = \mathbf{g}^{\mathrm{T}}(x)\boldsymbol{\omega}(x) = -\mathbf{g}^{\mathrm{T}}(x)\mathbf{P}^{-1}(x)\mathbf{g}(x) . \tag{13.59}$$

Also from (13.57), we have

$$\frac{\mathrm{d}P_{mn}}{\mathrm{d}x} = -g_m(x)g_n(x) . \tag{13.60}$$

Then

$$\begin{aligned} K(x,x) &= -\mathrm{tr}\left(g_m P_{mn}^{-1} g_n\right) \\ &= \mathrm{tr}\left(\mathbf{P}^{-1}\frac{\mathrm{d}\mathbf{P}}{\mathrm{d}x}\right) \\ &= \sum_l \sum_m \frac{P_{ml}}{|\mathbf{P}|}\frac{\mathrm{d}P_{lm}}{\mathrm{d}x} \\ &= \frac{1}{|\mathbf{P}|}\frac{\mathrm{d}|\mathbf{P}|}{\mathrm{d}x} = \frac{\mathrm{d}}{\mathrm{d}x}\log|\mathbf{P}| , \end{aligned} \tag{13.61}$$

where P_{ml} is the cofactor matrix and $|\mathbf{P}|$ is the determinant of $\mathbf{P}$. In (13.61), standard properties of matrices and determinants have been used.

Finally, we can write

$$u(x,t) = -2\frac{\mathrm{d}}{\mathrm{d}x}K(x,x+0,t) = -2\frac{\mathrm{d}^2}{\mathrm{d}x^2}\log|\mathbf{P}| \tag{13.62}$$

as the required N-soliton solution of the KdV equation. It can also be obtained by the Hirota method as described in Chap. 12. One may notice the similarity in the form of (13.62) and the bilinearizing transformation, (12.61).

13.3.4 Soliton Interaction

As described in Chap. 12, the solitons of the KdV equation undergo only elastic collisions without any change in shape or speed, except for the phase shifts. We have seen in Sect. 12.5 from the two-soliton solution expression (12.73) that the larger and smaller solitons undergo phase shifts $\triangle^+$ and $\triangle^-$ respectively given by

$$\triangle^+ = -\triangle^- = \log\left(\frac{\kappa_1-\kappa_2}{\kappa_1+\kappa_2}\right) < 0 . \tag{13.63}$$

Extending this analysis to the N-soliton case, (13.62), assuming that $\kappa_1 > \kappa_2 > ... > \kappa_N > 0$, then for fixed γ_n, for $t \to \pm\infty$

$$u(x,t) \sim -2\kappa_n^2 \operatorname{sech}^2\left(\gamma_n + \Delta_n^{\pm}\right), \quad \gamma_n = \kappa_n\left(x - 4\kappa_n^2 t + \delta_n\right) \tag{13.64a}$$

so that the nth soliton undergoes a phase shift given by

$$\begin{aligned}\Delta_n &= \Delta_n^+ - \Delta_n^- \\ &= \sum_{m=n+1}^{N} \log\left(\frac{\kappa_n - \kappa_m}{\kappa_n + \kappa_m}\right) - \sum_{m=1}^{n-1} \log\left(\frac{\kappa_m - \kappa_n}{\kappa_m + \kappa_n}\right) .\end{aligned} \tag{13.64b}$$

13.3.5 Nonreflectionless Potentials

Let us now consider the case in which the initial state of the KdV equation is such that the potential is nonreflectionless, that is $R(k,0) \neq 0$. Then as discussed in Sect. 13.2, the reflection coefficient evolves according to (13.33a). Correspondingly, the contribution to $F(x+y)$ in (13.39) comes from both the bound states and continuum states. Solving the Gelfand–Levitan–Marchenko integral equation exactly in this case becomes impossible. However, it is possible to carry out a perturbative analysis for sufficiently large t and to show that asymptotically the solution of the KdV equation consists of N individual solitons in the background of small amplitude dispersive propagating waves, which in due course disperse and die down (see for example Ref. [5]).

To see the nature of the dispersive waves, let us consider the case in which there is no bound state at all so that (13.39b) becomes

$$F(x,t) = \frac{1}{2\pi}\int_{-\infty}^{\infty} R(k,0)\mathrm{e}^{-8\mathrm{i}k^3 t}\mathrm{e}^{\mathrm{i}kx}\mathrm{d}k . \tag{13.65}$$

Substituting this into the Gelfand–Levitan–Marchenko equation, one can solve it and show that the solution of the KdV equation for sufficiently large t is

$$u(x,t) \approx \frac{1}{2\pi}\int_{-\infty}^{\infty} 4\mathrm{i}kR(k,0)\mathrm{e}^{-8\mathrm{i}k^3 t}\mathrm{e}^{-2\mathrm{i}kx}\mathrm{d}k \tag{13.66a}$$

so that

$$u(x,t) = \frac{1}{2\pi}\int_{-\infty}^{\infty} F(k)e^{-\mathrm{i}(\omega t - \hat{k}x)}\mathrm{d}k , \quad \omega = -\hat{k}^3,\ \hat{k} = -2k . \tag{13.66b}$$

Equation (13.66) is nothing but the wave packet solution (11.14) of the linearized KdV equation (see Sect. 11.4). We have already discussed the dispersive nature of the solution in Chap. 11.

13.4 Hamiltonian Structure of KdV Equation

Having solved the IVP of the KdV equation, we now turn our attention to establish the integrability aspects of it. As a prelude, in this section we wish to bring out the Hamiltonian nature of the KdV equation and obtain appropriate Hamiltonian form for it. We will also point out the fact that the KdV equation may be considered as an *infinite dimensional completely integrable dynamical system* in the Liouville sense. (In fact all soliton possessing systems belong to the class of such completely integrable dynamical systems. Few other examples will be given in the next chapter). For this purpose, we will first summarize the standard Lagrangian and Hamiltonian description applicable to continuous dynamical systems in the next subsection. For details see for example, Ref. [6].

13.4.1 Dynamics of Continuous Systems

Starting from the Lagrange's or Hamilton's equations of motion (see Table 1.1 of Chap. 1) of an N-particle system, one can deduce the equation of motion of a continuous system by a suitable $N \to \infty$ limit as indicated in Sect. 11.1 (see for example Ref. [6]). However, one can develop an independent Lagrangian or Hamiltonian approach for continuous systems directly.

A. Lagrange Equation of Motion

Let us consider an one-dimensional continuous system described by the real field variable $\phi(x,t)$. Then one can define the Lagrangian of the continuous system by

$$L = \int \mathcal{L}\left(\phi, \phi_t, \phi_x, \phi_{xx}, ..., x, t\right) \mathrm{d}x \;, \tag{13.67}$$

where $\mathcal{L}$ is the Lagrangian density, which depends on the field ϕ and its derivatives. Then the Euler–Lagrange equation of motion of the system becomes

$$\frac{\partial}{\partial t}\left(\frac{\partial \mathcal{L}}{\partial \phi_t}\right) + \frac{\partial}{\partial x}\left(\frac{\partial \mathcal{L}}{\partial \phi_x}\right) - \frac{\partial^2}{\partial x^2}\left(\frac{\partial \mathcal{L}}{\partial \phi_{xx}}\right) + ... - \frac{\partial \mathcal{L}}{\partial \phi} = 0 \;. \tag{13.68}$$

B. Hamilton's Equations of Motion

Similarly one can define a Hamiltonian (functional) of the continuous system by

$$H = \int \mathcal{H}\left(\phi, \pi, \phi_t, \phi_x, ...\right) \mathrm{d}x \;, \tag{13.69}$$

where the Hamiltonian density $\mathcal{H}$ is given by

$$\mathcal{H} = \pi \phi_t - \mathcal{L} \;. \tag{13.70}$$

Here the canonically conjugate field momentum is defined as

$$\pi = \frac{\partial \mathcal{L}}{\partial \phi_t} \tag{13.71}$$

in analogy with canonically conjugate momentum for finite dimensional systems. Then the Hamiltonian equation of motion can be written as

$$\phi_t = \frac{\delta \mathcal{H}}{\delta \pi} , \tag{13.72a}$$

$$\pi_t = -\frac{\delta \mathcal{H}}{\delta \phi} . \tag{13.72b}$$

Here the functional derivative $\delta/\delta\phi$ stands for

$$\frac{\delta}{\delta\phi} = \frac{\partial}{\partial\phi} - \frac{\partial}{\partial x}\left(\frac{\partial}{\partial\phi_x}\right) + \frac{\partial^2}{\partial x^2}\left(\frac{\partial}{\partial\phi_{xx}}\right) + \cdots \tag{13.72c}$$

and $\delta/\delta\pi$ is also similarly defined.

C. Poisson Brackets and Integrals of Motion

Defining now the Poisson bracket between two functionals U and V as

$$\{U, V\} = \int \left[\frac{\delta U}{\delta\phi}\frac{\delta V}{\delta\pi} - \frac{\delta U}{\delta\pi}\frac{\delta V}{\delta\phi}\right] \mathrm{d}x , \tag{13.73a}$$

where

$$U = \int \mathcal{U}(\phi, \pi, \cdots)\mathrm{d}x , \quad V = \int \mathcal{V}(\phi, \pi, \cdots)\mathrm{d}x , \tag{13.73b}$$

one can easily check that

$$\frac{\mathrm{d}U}{\mathrm{d}t} = \frac{\partial U}{\partial t} + \{U, H\} . \tag{13.74}$$

If U is independent of time and its Poisson bracket with the Hamiltonian H vanishes, then

$$\frac{\mathrm{d}U}{\mathrm{d}t} = \{U, H\} = 0 \tag{13.75}$$

and so U is a *constant of motion* or a *conserved quantity* or it is an *integral of motion*. In particular if H is independent of time, the Hamiltonian is a conserved quantity. Similarly any other quantity which does not explicitly depend on time is also a constant of motion if the Poisson bracket of it with the Hamiltonian vanishes.

D. Completely Integrable Hamiltonian System

Suppose given a conservative system, whose Hamiltonian is constant in time, and if there exists another functionally independent integral or conserved quantity, F_1, then by (13.75) it is in involution with H

$$\{F_1, H\} = 0 . \tag{13.76a}$$

For an N-degrees of freedom Hamiltonian system if there are N functionally independent integrals of motion, including the Hamiltonian, which are in involution with each other, then one can say that the system is completely integrable in the Liouville sense. (See Chap. 10 for further discussions on this point). That is we demand that

$$\{F_n, F_m\} = 0 \, , \quad n, m = 1, 2, ..., N. \tag{13.76b}$$

One can extend this notion of complete integrability to continuous systems by extending n, m to 1,2,...,∞.

One way to show the complete integrability of a Hamiltonian system with N-degrees of freedom defined by the Hamiltonian

$$H = H\left(q_1, q_2, ..., q_N, p_1, p_2, ..., p_N\right) \tag{13.77a}$$

is to introduce a suitable canonical transformation to action and angle variables from the set $(q_1, q_2, ..., q_N, p_1, p_2, ..., p_N)$ to $(\theta_1, \theta_2, ..., \theta_N, J_1, J_2, ..., J_N)$. (Recall that the $N = 2$ case was briefly discussed in Sect. 7.2). Then one can write the Hamiltonian in terms of the momentum variables or the action variables $(J_1, J_2, ..., J_N)$ alone as

$$H = H\left(J_1, J_2, ..., J_N\right) \, . \tag{13.77b}$$

Then the resultant equations of motion take the simple form

$$\dot{\theta}_i = \frac{\partial H}{\partial J_i} = \nu_i \, , \quad \dot{J}_i = \frac{\partial H}{\partial \theta_i} = 0 \, , \quad i = 1, 2, ..., N, \tag{13.77c}$$

which can be integrated completely to give the solution

$$\theta_i(t) = \nu_i t + \beta_i \, , \quad J_i(t) = J_i(0) \, . \tag{13.78}$$

In this Liouville sense the dynamical system can be integrated completely. The action integrals J_i's can be related to the involutive integrals F_i, $i = 1, 2, ..., N$, suitably. Finally, the concept can also be extended to continuous systems also appropriately.

13.4.2 KdV as a Hamiltonian Dynamical System

Let us consider the Lagrangian density

$$\mathcal{L} = \left[\frac{1}{2}\psi_x\psi_t - \psi_x^3 - \frac{1}{2}\psi_{xx}^2\right] \, . \tag{13.79}$$

Then, from (13.68), the equation of motion for the ψ field becomes

$$\psi_{xt} - 6\psi_x\psi_{xx} + \psi_{xxxx} = 0 \, . \tag{13.80}$$

Defining

$$u = \psi_x \, , \tag{13.81}$$

(13.80) is seen to reduce to the KdV equation (13.4). Thus (13.79) may be considered as the Lagrangian of the KdV equation through the potential field function $\psi(x, t)$.

Defining now the canonically conjugate momentum

$$\pi = \frac{\partial \mathcal{L}}{\partial \psi_t} = \frac{\psi_x}{2} , \tag{13.82}$$

the Hamiltonian density becomes (see (13.70))

$$\mathcal{H} = \frac{1}{2}\psi_{xx}^2 + \psi_x^3 = \left[\pi_x^2 + \frac{1}{4}\psi_{xx}^2 + 2\pi^2\psi_x + \pi\psi_x^2\right] . \tag{13.83}$$

Then the Hamiltonian of the KdV equation is

$$H = \int \left[\pi_x^2 + \frac{1}{4}\psi_{xx}^2 + 2\pi^2\psi_x + \pi\psi_x^2\right] \mathrm{d}x . \tag{13.84}$$

Now using the expression (13.83) for the Hamiltonian density into the Hamilton's equation of motion (13.72), we can readily derive

$$\psi_t = \psi_x^2 + 4\psi_x\pi - 2\pi_{xx} , \tag{13.85a}$$

$$\pi_t = 4\pi\pi_x + 2\pi_x\psi_x + 2\pi\psi_{xx} - \frac{1}{2}\psi_{xxxx} . \tag{13.85b}$$

Then with the substitution $\pi = \psi_x/2$ ((13.82)), one can easily check that the evolution equation for ψ_x or π is identical from both (13.85a) and (13.85b). It also coincides with the KdV-ψ field equation (13.80) as it should be. One can thus give both a Lagrangian and Hamiltonian description for the KdV equation and conclude that it is a Hamiltonian continuous system in the dynamical sense.

One can also give an alternative Hamiltonian description, by writing (13.83) into terms of the KdV field function $u = \psi_x$:

$$\mathcal{H} = \frac{1}{2}u_x^2 + u^3 , \quad \widehat{\mathcal{H}} = \int \left(\frac{1}{2}u_x^2 + u^3\right) \mathrm{d}x . \tag{13.86}$$

Then writing the Hamiltonian equation of motion for a single field in the form (for further details see for example, Refs. [5,7,8])

$$u_t = \frac{\partial}{\partial x}\frac{\delta \widehat{\mathcal{H}}}{\delta u} , \tag{13.87}$$

we obtain the KdV equation (13.4), after using the definition (13.72c) for the functional derivative.

13.4.3 Complete Integrability of the KdV Equation

In order to understand the complete integrability property of the KdV equation, it is more convenient to use the Hamiltonian (13.87) than the standard form (13.72). Correspondingly the definition of the Poisson bracket between two functionals U and V (for the KdV equation) can be redefined[1] as

[1] It may be pointed out that in recent years certain modifications have been suggested to the Poisson bracket structure given by (13.88) as there exist some

$$\{U, V\} = \int \mathrm{d}x \frac{\delta U}{\delta u(x)} \frac{\partial}{\partial x} \frac{\delta V}{\delta u(x)} . \tag{13.88}$$

We know that any transformation from one set of canonical variables (p, q) to a new set (P, Q) is *canonical* provided the Poisson brackets of the new set satisfy the relations

$$\{P, P\} = 0 , \quad \{Q, Q\} = 0 , \quad \{P, Q\} = \delta(x - x') . \tag{13.89}$$

Or in other words, if such a transformation exists, then the relation (13.89) ensures that P and Q are indeed canonical variables.

It so happens that for the KdV equation one can find a suitable canonical transformation from the continuous field variable $u(x)$ to a new set of canonical variables (P_i, Q_i) and $(P(k), Q(k))$, $i = 1, 2, ..., N$ and $-\infty < k < \infty$, so that the later are infinite in number. More interestingly one can prove that the P's and Q's are not only canonical variables but also the action and angle variables, respectively, of the KdV system. Consequently, the Hamiltonian (13.86) can be written purely as a function of the action variables, P_i's and $P(k)$'s, alone. The resulting equation of motion can be obviously integrated trivially and in this Liouville sense the KdV system becomes a completely integrable, but infinite dimensional (or degrees of freedom), nonlinear dynamical system.

Now, how can one find such a canonical transformation (CT)? Indeed one finds that the direct scattering transform, which we discussed in Sect. 13.2, does correspond to such a CT and that the scattering data $S(t)$ provides the necessary coordinates to construct the required action and angle variables. Though the analysis is somewhat involved, but direct, it has been successfully performed by Zakharov and Faddeev in 1981 [8], see also Ablowitz and Clarkson [5] for fuller details. In the following, we give essential details only.

We have seen in Sect. 13.2.1, that the scattering data $S(t)$, as given in (13.38), consists of the discrete eigenvalues $\kappa_n(t) = \kappa_n(0)$, $n = 0, 1, ..., N$, the normalization constants $C_n(t) = C_n(0)\mathrm{e}^{4\kappa_n^3 t}$ and the reflection coefficient $R(k, t) = R(k, 0)\, \mathrm{e}^{-8\mathrm{i}k^3 t}$, $-\infty < k < \infty$.

Considering the linear eigenvalue problem

$$\phi_{xx} + \left[k^2 - u(x)\right] \phi(x) = 0 , \tag{13.90}$$

it can be converted into an integral equation

$$\phi(x, k) = \mathrm{e}^{-\mathrm{i}kx} \left[1 - \frac{1}{2\mathrm{i}k} \int_{-\infty}^{x} \left[1 - \mathrm{e}^{2\mathrm{i}k(x-\xi)}\right] u(\xi)\phi(\xi, k)\mathrm{e}^{\mathrm{i}k\xi}\mathrm{d}\xi\right] . \tag{13.91}$$

(For details see for example, Ablowitz and Clarkson [5], also appendix G). Then one can define the quantities

---

technical difficulties in connection with the satisfaction of Jacobi identity for certain class of functionals. However, these are outside the scope of this book and the interested reader may refer to the original works on the topic, see for example Ref. [7].

$$a(k) = 1 - \frac{1}{2\mathrm{i}k} \int_{-\infty}^{\infty} u(\xi)\phi(\xi, k) \mathrm{e}^{\mathrm{i}k\xi} \mathrm{d}\xi \,, \tag{13.92a}$$

$$b(k) = \frac{1}{2\mathrm{i}k} \int_{-\infty}^{\infty} u(\xi)\phi(\xi, k) \mathrm{e}^{2\mathrm{i}k\xi} \mathrm{d}\xi \,, \tag{13.92b}$$

which are related to $R(k)$ and $T(k)$. Considering now the new variables

$$P(k) = k\pi^{-1} \log |a(k)|^2 \,, \quad Q(k) = -\frac{1}{2}\mathrm{i} \log \left[\frac{b(k)}{\bar{b}(k)}\right] \,, \tag{13.93}$$

one can evaluate the Poisson brackets

$$\{P(k), Q(k')\} = \int_{-\infty}^{\infty} \frac{\delta P(k)}{\delta u(x)} \frac{\partial}{\partial x} \frac{\delta Q(k)}{\delta u(x)} \mathrm{d}x \,, \tag{13.94}$$

and show that

$$\begin{aligned} &\{P(k), Q(k')\} = \delta(k - k') \,, \quad \{P(k), P(k')\} = 0 \,, \\ &\{Q(k), Q(k')\} = 0 \,. \end{aligned} \tag{13.95}$$

Similarly with

$$P_j = \kappa_j^2 \,, \quad Q_j = -2 \log |C_j| \,, \quad j = 1, 2, ..., N, \tag{13.96}$$

one can show that

$$\{P_i, Q_j\} = \delta_{ij} \,, \quad \{P_i, P_j\} = 0 \,, \quad \{Q_i, Q_j\} = 0 \tag{13.97}$$

and also that the Poisson brackets of P_i's and Q_i's with that of $P(k)$'s and $Q(k)$'s vanish. Thus, the set $\{P_i, Q_i, P(k), Q(k)\}$ indeed form a complete canonical set and that the direct scattering transformation is indeed a canonical transformation.

Further, one can also show that in terms of these new canonical variables the Hamiltonian $\widehat{\mathcal{H}}$ given by (13.86) becomes

$$\widehat{\mathcal{H}} = -\frac{32}{5} \sum_{j=1}^{N} P_j^{5/2} + 8 \int_{-\infty}^{\infty} k^3 P(k) \mathrm{d}k \,. \tag{13.98}$$

Thus, the Hamiltonian is a function of the action variables (momenta) only. So the resultant equations of motion can be trivially integrated and solved for the variables $Q_i(t)$, $P_i(t)$, $Q(k,t)$, and $P(k,t)$ $i = 1, 2, ..., N$, in terms of their initial values. As a result the KdV equation can be considered as a completely integrable infinite dimensional dynamical system.

Exercise:

6. Write down the Hamilton's equation of motion for the action and angle variables P_i, Q_i, $P(k)$, $Q(k)$ and integrate them.

13.5 Infinite Number of Conserved Densities

Using the KdV equation (13.4), one can write the following conservation laws easily:

$$u_t + \left(-3u^2 + u_{xx}\right)_x = 0\,, \tag{13.99a}$$

$$\left(u^2\right)_t + \left(-4u^3 + 2uu_{xx} - u_x^2\right)_x = 0\,, \tag{13.99b}$$

$$\left(\frac{1}{2}u_x^2 + u^3\right)_t + \left(3u^2 u_{xx} + u_x u_{xxx} - \frac{9}{2}u^4 - 6uu_x^2 - \frac{1}{2}u_{xx}^2\right)_x = 0\,. \tag{13.99c}$$

Note that (13.99b) is obtained by multiplying with u throughout the KdV equation (13.4). Similarly (13.99c) can be obtained. These equations are in the so-called conservative form and they correspond to conservation laws, because they can be written as

$$\frac{\partial P}{\partial t} + \frac{\partial Q}{\partial x} = 0\,, \tag{13.100}$$

where P and Q are functions of u, u_x, ... such that they vanish at $x \to \pm\infty$, since $u \overset{|x|\to\infty}{\longrightarrow} 0$ sufficiently fast. If P and Q are connected by a gradient relationship, that is $P = F_x$, then (13.100) gives $Q = -F_t$. Then integrating (13.100), we have

$$\frac{\partial}{\partial t}\int_{-\infty}^{\infty} P\mathrm{d}x = 0\,. \tag{13.101}$$

In other words, each of

$$I = \int_{-\infty}^{\infty} P\mathrm{d}x \tag{13.102}$$

constitutes a conserved quantity of the KdV equation. In particular from (13.99) we see that

$$I_1 = \int_{-\infty}^{\infty} u\mathrm{d}x\,,\quad I_2 = \int_{-\infty}^{\infty} u^2\mathrm{d}x\,,\quad I_3 = \int_{-\infty}^{\infty}\left(\frac{1}{2}u_x^2 + u^3\right)\mathrm{d}x \tag{13.103}$$

are specific integrals of motion. Note that $I_3 = \widehat{\mathcal{H}}$ is the Hamiltonian of the system, see (13.86).

Interestingly, the KdV equation possesses many more conservation laws and constants of motion: in fact they are infinitely many in number. In order to realize them we can proceed as follows.

Introducing the so-called Gardner transformation [9]

$$u = \omega + \epsilon\omega_x + \epsilon^2\omega^2 \,, \tag{13.104a}$$

where ϵ is a small parameter, and substituting it into the KdV equation (13.4), one can show that u is a solution of the KdV equation provided

$$\omega_t - 6\left(\omega + \epsilon^2\omega^2\right)\omega_x + \omega_{xxx} = 0 \,. \tag{13.104b}$$

Expressing ω now formally as a power series in ϵ,

$$\omega(x,t;\epsilon) = \sum_{n=0}^{\infty} \epsilon^n \omega_n(x,t) \,, \tag{13.104c}$$

and substituting it into (13.104b), one can equate each power of ϵ separately to zero. Then one obtains the following,

$$O\left(\epsilon^0\right) : \left(\omega_0\right)_t = \left(3\omega_0^2 - \omega_{0xx}\right)_x \,, \tag{13.105a}$$

$$O\left(\epsilon^1\right) : \left(\omega_1\right)_t = \left(6\omega_0\omega_1 - \omega_{1xx}\right)_x \,, \tag{13.105b}$$

$$O\left(\epsilon^2\right) : \left(\omega_2\right)_t = \left(3\omega_1^2 + 6\omega_0\omega^2 + 2\omega_0^2 - \omega_{2xx}\right)_x \,, \tag{13.105c}$$

and so on. However from (13.104a), again comparing the coefficients of powers of ϵ, one finds

$$\omega_0 = u \,, \quad \omega_1 = u_x \,, \quad \omega_2 = u_{xx} + u^2 \,, \text{ etc.} \tag{13.106}$$

Substituting (13.106) into the conservation laws (13.105), one can obtain the previous conservation laws (13.99) again, as well as further conservation laws which are infinite in number.

Finally, one can also check that the infinite number of integrals of motion arising above are functionally independent and involutive as the Poisson brackets among them vanish:

$$\{I_n, I_m\} = \int_{\infty}^{\infty} \frac{\delta I_n}{\delta u(x)} \frac{\partial}{\partial x} \frac{\delta I_m}{\delta u(x)} \mathrm{d}x = 0 \,. \tag{13.107}$$

This is yet another property indicative of the complete integrability of the KdV equation.

13.6 Bäcklund Transformations

Finally, we point out one more important property of the KdV equation, namely that it admits the so called (auto)*Bäcklund transformation (BT)*. A BT is a transformation which connects the solutions of two differential equations. If the transformation connects two distinct solutions of the same equation, then it is called an *auto-Bäcklund* transformation (see also the next chapter). Existence of such an auto-Bäcklund transformation is indicative of the existence of soliton solutions and in some sense integrability of the system as well. There are several ways of obtaining such Bäcklund transformations, but we will consider the BT for KdV equation without its actual derivation. For some details, see for example Ref. [5,10].

Now introducing the transformation $u = \psi_x$, the KdV equation (13.4) becomes (13.80). Integrating it with respect to x and taking the integration 'constant' to be zero without loss of generality, we obtain

$$\psi_t - 3\psi_x^2 + \psi_{xxx} = 0 \, . \tag{13.108}$$

Equation (13.108) is often called the *potential KdV* equation. If ω and $\overline{\omega}$ are any two solutions of the potential KdV equation (13.108), then the auto-Bäcklund transformation of it is

$$\omega_x + \overline{\omega}_x + 2\kappa^2 + \frac{1}{2}\left(\omega - \overline{\omega}\right)^2 = 0 \, , \tag{13.109a}$$

$$\omega_t + \overline{\omega}_t - 3\left(\omega_x - \overline{\omega}_x\right)\left(\omega_x + \overline{\omega}_x\right) + \omega_{xxx} - \overline{\omega}_{xxx} = 0 \, , \tag{13.109b}$$

where κ is a real parameter. Then the two equations are compatible with (13.108).

The use of such a Bäcklund transformation is immediately obvious. If $\omega = 0$, the trivial solution to (13.108), then solving (13.109), we obtain the one-soliton solution

$$\overline{\omega}(x,t) = -2\kappa \tanh\{\kappa\left(x - 4\kappa^2 t\right) + \delta\} \, , \tag{13.110a}$$

so that

$$\overline{u}(x,t) = \overline{\omega}_x = -2\kappa^2 \operatorname{sech}^2\left\{\kappa\left(x - 4\kappa^2 t\right) + \delta\right\} \, , \tag{13.110b}$$

which is nothing but the one-soliton solution of the KdV equation. One can then use the one-soliton solution in (13.109) as the new 'seed' solution and obtain two-soliton solution and the process can be continued to obtain higher order solitons. For more details we refer to Ref. [5].

Exercises:

7. Deduce the Lagrangian associated with the modified KdV equation.
8. Obtain the Hamiltonian form of the modified KdV equation.
9. Using the general soliton form (13.62), deduce the one, two and three soliton solutions of the KdV equation.
10. Use the Bäcklund transformation (13.109) and the one-soliton solution (13.110) to obtain the two-soliton solution.
11. Using (13.107), show that I_1, I_2 and I_3 are mutually in involution.
12. Obtain the form of the conserved quantities I_4 and I_5, using (13.104).

13.7 Conclusions

We have seen conclusively that the KdV equation, deduced originally by Korteweg and de Vries in 1895 to explain the Scott Russel phenomenon, is indeed a completely integrable infinite dimensional nonlinear dynamical system. It possesses many remarkable properties such as N-solitons, IST solvability, Hirota bilinearization, Hamiltonian structure, Liouville integrability, infinite number of conservation laws and constants of motion and Bäcklund transformation, among others. It also possesses the Painlevé property in a generalized sense (see appendix J). It is indeed a remarkable dynamical system, exhibiting very coherent and localized structures namely solitons, even though it is a nonlinear dispersive system. Is KdV equation a unique one or does there exist other systems as well possessing similar properties? We will investigate this question in the next chapter.

14. Other Ubiquitous Soliton Equations

In the previous chapter, we have seen that the KdV equation is a completely integrable, infinite-dimensional, nonlinear dynamical system. It possesses exact soliton solutions exhibiting remarkable particle-like collision properties. Its Cauchy initial value problem is completely solvable through the Inverse Scattering Transform (IST) procedure by making use of the associated Lax pair of operators. Using these linear operators the KdV equation can be written as the consistency condition between two linear differential equations; that is the KdV equation is linearizable. Further, it has infinite number of conservation laws and possesses a Bäcklund transformation (BT) and so on. Also, the KdV equation is a ubiquitous system in that it occurs not only in the description of shallow water waves and wave propagation in anharmonic lattices discussed earlier, but also in many other physical systems involving unidirectional dispersive wave propagation such as ion-acoustic plasma waves, stratified internal waves, collision-free hydromagnetic waves and so on [1–3]. The KdV equation is a remarkable nonlinear dynamical system indeed!

But how ubiquitous or how common are the properties exhibited by the KdV equation (including its soliton nature)? Are they specific to the KdV equation alone or do other interesting dynamical systems share with the KdV equation the same general properties? If not, the KdV equation will have to be considered as just another interesting dynamical system. Indeed, it happens that there exist many other important nonlinear dynamical systems which are also soliton possessing nonlinear evolution equations of dispersive type, having the same general properties as the KdV equation. They are also completely integrable through the IST procedure and bilinearizable, possess Lax pairs, infinite conservation laws, BTs and so on. Some standard examples are

(1) the modified KdV equation, (12.59) already discussed in Chap. 12,

$$u_t \pm 6u^2 u_x + u_{xxx} = 0\ , \tag{14.1}$$

(2) the sine-Gordon (sG) equation, (in light-cone coordinates)

$$u_{xt} + m^2 \sin u = 0\ , \quad m^2 = \text{constant}\ , \tag{14.2}$$

(3) the nonlinear Schrödinger equation,

$$\mathrm{i}q_t + q_{xx} \pm 2|q|^2 q = 0\ , \quad q \in C \tag{14.3}$$

(4) the continuous Heisenberg ferromagnetic spin equation,

$$\boldsymbol{S}_t = \boldsymbol{S} \times \boldsymbol{S}_{xx}\,, \quad \boldsymbol{S} = (S_1, S_2, S_3)\,, \quad \boldsymbol{S}^2 = 1 \tag{14.4}$$

(5) the Toda lattice equation,

$$\ddot{u}_n = \exp\left[-\left(u_n - u_{n-1}\right)\right] - \exp\left[-\left(u_{n+1} - u_n\right)\right], \quad i = 1, 2, .., N, \tag{14.5}$$

and so on, to name a few. The list is very large indeed and it is an active area of research to identify and to isolate such integrable systems [1,2]. Even in higher spatial dimensions, particularly in two spatial dimensions, several integrable nonlinear evolution equations such as the Kadomstev-Petviashvile, Davey–Stewartson and Ishimori equations [1,2] have been identified which possess interesting solutions and properties. These equations can be considered as higher dimensional generalizations of integrable soliton equations in one-dimension, namely the KdV, NLS and Heisenberg spin equations, respectively.

In the following, in Sect. 14.1, we will briefly explain specific physical situations (for illustrative purpose) where some of the above mentioned evolution equations in one spatial dimension arise naturally. Then in Sect. 14.2 it will be pointed out how these equations can be considered as the consistency conditions between two sets of linear equations: a (2×2) matrix linear eigenvalue problem, namely the so-called Zakharov–Shabat (ZS) and Ablowitz–Kaup–Newell–Segur (AKNS) linear eigenvalue problem, and the associated time evolution equation. Next, in Sect. 14.3, the explicit forms of the basic solitary wave solutions are obtained. Then, in Sect. 14.4, we point out how Hirota's bilinearization method can be applied to all the above equations to show that the solitary waves discussed above are also soliton solutions by obtaining multisoliton solutions. Next in Sect. 14.5, it is shown how the IST method can be applied to the AKNS type soliton equations to solve their IVPs and in particular to obtain soliton solutions. Then, in Sect. 14.6, the Bäcklund transformations associated with the above soliton equations are discussed, while the infinite number of conservation laws and associated constants of motion are briefly introduced in Sect. 14.7. Finally in Sect. 14.8, the Hamiltonian structure and complete integrability aspects are discussed, while in Sect. 14.9 a brief future outlook is given.

14.1 Identification of Some Ubiquitous Nonlinear Evolution Equations from Physical Problems

In this section, we will point out briefly how some of the soliton possessing nonlinear evolution equations arise in important physical contexts. The MKdV equation was already shown in Chap. 12 to arise in the study of wave propagation in anharmonic lattices. Here we will point out that the NLS equation arises in the study of light wave propagation in optical fibers, the sine-Gordon equation arises in the study of flux propagation in Josephson junctions, the Heisenberg spin equation appears in the study of the classical dynamics of a ferromagnetic lattice while the Toda equation arises in the study of anharmonic lattices. These situations should be considered only as prototypical as these equations are ubiquitous and occur in varied physical situations.

14.1.1 The Nonlinear Schrödinger Equation in Optical Fibers

The study of optical wave propagation in a nonlinear dispersive (dielectric) fiber (see Fig. 14.1) has been receiving considerable attention in recent times as the fiber can support under suitable circumstances a stable pulse called *optical soliton* [3,4]. It arises essentially due to a compensation of the effect of dispersion of the pulses by the nonlinear response of the medium.

The analysis of such pulse propagation naturally starts from the Maxwell's equations for electromagnetic wave propagation in a dielectric medium,

$$\nabla^2 \boldsymbol{E} - \frac{1}{c^2}\frac{\partial^2 \boldsymbol{E}}{\partial t^2} = -\mu_0 \frac{\partial^2 \boldsymbol{P}}{\partial t^2}, \tag{14.6}$$

where the induced polarization for silica fibers is

$$\boldsymbol{P}(\boldsymbol{r},t) = \epsilon_0 \left[\chi^{(1)} \cdot \boldsymbol{E} + \chi^{(3)} \vdots\, \boldsymbol{EEE} + \cdots \right]. \tag{14.7}$$

In (14.6), $\boldsymbol{E}$ represents the electric field, c is the velocity of light, μ_0 and ϵ_0 are the permeability and permittivity of free space, respectively, and $\chi^{(m)}$ is the mth order susceptibility tensor.

In order to analyse (14.6) it is necessary to make several simplifying assumptions:

(1) The nonlinear part of the induced polarization is treated as a small perturbation to the linear part.
(2) The optical field is assumed to maintain its polarization along the fiber.
(3) Fiber loss is assumed to be very small.
(4) The nonlinear response of the fiber is assumed to be instantaneous.

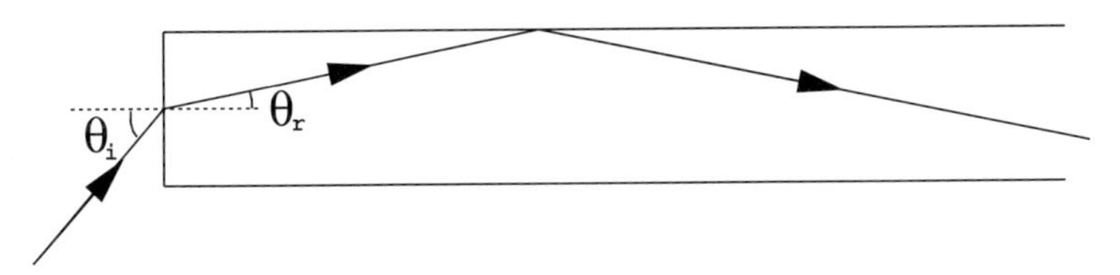

Fig. 14.1. Wave propagation in an optical fiber

(5) In a slowly varying envelope approximation for pulse propagation along the fiber, the electric field can be written as

$$\boldsymbol{E}(\boldsymbol{r},t)=\frac{1}{2}\boldsymbol{e}\left[F(x,y)E(z,t)\mathrm{e}^{\mathrm{i}(k_0z-\omega_0t)}+\mathrm{c.c}\right], \tag{14.8}$$

where c.c stands for complex conjugate, $\boldsymbol{e}$ is the unit polarization vector of the light assumed to be linearly polarized, $E(z,t)$ is the slowly varying electric field, $F(x,y)$ is the mode distribution function in the (x,y) plane, while k_0 and ω_0 denote the propagation constant and central frequency of the optical pulse, respectively.

Under the above assumptions, rewriting (14.6) by using the method of separation of variables and introducing the coordinate system, $T=t-z/v_g$, moving with the pulse at the group velocity $v_g=\partial k/\partial\omega$, one can obtain a wave equation for the evolution of E as

$$\mathrm{i}\frac{\partial E}{\partial z}-\frac{k''}{2}\frac{\partial^2E}{\partial T^2}+\gamma_0|E|^2E=0, \tag{14.9}$$

where $\gamma_0=n_2\omega_0/(cA_{\mathrm{eff}})$. Here A_{eff} denotes the effective core area of the single-mode fiber, n_2 represents the nonlinear refractive index coefficient and the parameter $k''=\partial^2k/\partial\omega^2|_{\omega=\omega_0}=-1/v_g^2\,(\mathrm{d}v_g/\mathrm{d}\omega)$ accounts for the group velocity dispersion (GVD). After normalizing (14.9) and using the transformation, $q=(\gamma_0T_0^2/|k''|)^{1/2}\times E$, $\xi=z|k''|/T_0^2$, $\tau=T/T_0$ and then redefining ξ as z and τ as t we get the ubiquitous nonlinear Schrödinger (NLS) equation

$$\mathrm{i}q_z-\mathrm{sgn}\,(k'')\,q_{tt}+2|q|^2q=0, \tag{14.10}$$

in which T_0 represents the width of the incident pulse, z and t are the normalized distance and time along the direction of propagation and q is the normalized envelope. Interchanging t and x, then one obtains the standard form of NLS equation (14.3).

14.1.2 The Sine-Gordon Equation in Long Josephson Junctions

Let us consider a simple model for the propagation of transverse electromagnetic waves on a superconducting 'strip-line' transmission system, where the propagation occurs in the x-direction while the electric field (E) is oriented in the y-direction and the magnetic field (H) is in the z-direction (Fig. 14.2a). Following the discussion by Scott in Ref. [5], the equivalent circuit for transverse electromagnetic mode (TEM) can be given as in Fig. 14.2b. Denoting V as the voltage across the insulating region and J as the longitudinal current and making use of the Kirchhoff's voltage and current laws respectively, we can write

$$\frac{\partial V}{\partial x}=-L\frac{\partial I}{\partial t}, \tag{14.11a}$$

$$\frac{\partial I}{\partial x}=-C\frac{\partial V}{\partial t}-I_0\sin\phi. \tag{14.11b}$$

(a)

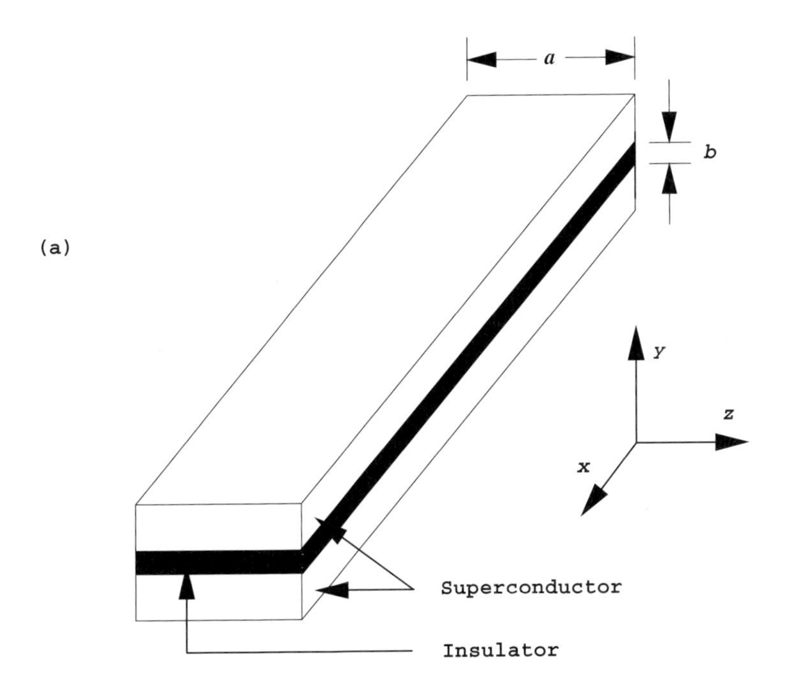

(b)

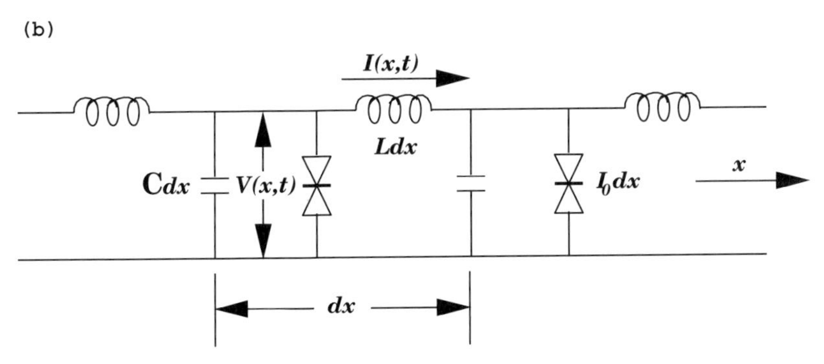

Fig. 14.2. (**a**) A superconducting strip-line transmission system, supporting transverse electromagnetic propagation in the x-direction. (**b**) Equivalent electrical circuit. Adapted from Ref. [5]

Here L is the inductance per unit length of the strip-line and C is the capacitance per unit length of the insulating region, while $I_0 \sin\phi$ term represents Josephson tunnelling of superconducting electrons through the insulating barrier as in Fig. 14.2. The quantity ϕ represents the change in phase of the superconducting wave function across the barrier. It can be shown to be related to the voltage V by

$$\frac{\partial \phi}{\partial t} = \frac{2e}{\hbar} V , \tag{14.12}$$

where e is the electronic charge and $\hbar = h/2\pi$, h being the Planck constant.

Now using (14.12) in (14.11a), it can be rewritten as

$$\frac{\partial^2 \phi}{\partial x \partial t} = -L\frac{\partial I}{\partial t} , \tag{14.13}$$

which on integration over t can be written as

$$\frac{\partial \phi}{\partial x} = -2\frac{eL}{\hbar} I . \tag{14.14}$$

Making use of (14.12) and (14.14) for V and I, respectively, in (14.11b), we can obtain the evolution equation for the flux density ϕ as

$$\frac{1}{c^2}\frac{\partial^2 \phi}{\partial t^2} - \frac{\partial^2 \phi}{\partial x^2} + m^2 \sin\phi = 0 , \tag{14.15a}$$

where

$$c = 1/\sqrt{LC} \quad \text{and} \quad m^2 = 2eLI_0/\hbar . \tag{14.15b}$$

Changing to a new frame of reference, called the light-cone coordinates, $\xi = \frac{1}{2}(x+ct)$ and $\eta = \frac{1}{2}(x-ct)$, the form (14.2) can be readily obtained.

14.1.3 Dynamics of Ferromagnets: Heisenberg Spin Equations

Spin excitations in ferromagnets are effectively expressed in terms of the Heisenberg's nearest neighbour spin-spin exchange interaction with additional anisotropy and external field dependent forces [2,6]. For the simplest isotropic case, the Hamiltonian for quasi-one-dimensional ferromagnets is given by

$$\mathcal{H} = -J \sum_{(i,j)} \boldsymbol{S}_i \cdot \boldsymbol{S}_j , \tag{14.16}$$

where the spin operator $\boldsymbol{S}_i = (S_i^x, S_i^y, S_i^z)$ and J is the exchange integral. The summation in (14.16) extends over nearest neighbours only. The Heisenberg equation of motion,

$$\frac{\mathrm{d}\boldsymbol{S}_i}{\mathrm{d}t} = [\boldsymbol{S}_i, \mathcal{H}] \equiv \boldsymbol{S}_i\mathcal{H} - \mathcal{H}\boldsymbol{S}_i , \quad i = 1, 2, ..., N, \tag{14.17}$$

in the long wavelength and low temperature (semiclassical $\hbar \to 0$) limit, can be expressed in terms of classical unit vectors as

$$\frac{\mathrm{d}\boldsymbol{S}_i}{\mathrm{d}t} = \{\boldsymbol{S}_i, \mathcal{H}\}_{\mathrm{PB}} , \quad \boldsymbol{S}_i{}^2 = 1 , \tag{14.18}$$

where the Poisson brackets between two functions of spin can be defined as

$$\{A, B\}_{\mathrm{PB}} = \sum_i \epsilon_{\alpha\beta\gamma} \frac{\partial A}{\partial S_i^\alpha} \frac{\partial B}{\partial S_i^\beta} S_i^\gamma , \quad \alpha, \beta, \gamma = 1, 2, 3, \tag{14.19}$$

where $\epsilon_{\alpha\beta\gamma}$ is the Levi–Civita tensor (see (14.195) below also). Correspondingly, the classical equation of motion for the spins associated with the Hamiltonian (14.16) can be written as

$$\frac{\mathrm{d}\boldsymbol{S_i}}{\mathrm{d}t} = J\boldsymbol{S_i} \times (\boldsymbol{S}_{i+1} + \boldsymbol{S}_{i-1}) \ . \tag{14.20}$$

Additional interactions can also be included in the same way. For example with a uniaxial anisotropy and external magnetic field along the z-direction, the Hamiltonian is

$$\mathcal{H} = -J \sum_{(i,j)} \boldsymbol{S_i} \cdot \boldsymbol{S_j} + A \sum_i (S_i^z)^2 - \mu \boldsymbol{B} \cdot \sum \boldsymbol{S_i} \ , \tag{14.21}$$

so that the equation of motion for the spins becomes

$$\frac{\mathrm{d}\boldsymbol{S}_i}{\mathrm{d}t} = \boldsymbol{S}_i \times \{J(\boldsymbol{S}_{i+1} + \boldsymbol{S}_{i-1}) - 2AS_i^z \boldsymbol{n} + \mu \boldsymbol{B}\} \ , \tag{14.22}$$

where $\boldsymbol{B} = (0,0,B)$ and $\boldsymbol{n} = (0,0,1)$.

In the continuum limit in a quasi one-dimensional ferromagnetic system (compare with the analysis of FPU lattice in Sect. 12.2.1),

$$\boldsymbol{S}_i(t) \to \boldsymbol{S}(x,t), \quad \boldsymbol{S}_{i\pm1}(t) = \boldsymbol{S}(x,t) \pm a \frac{\partial \boldsymbol{S}}{\partial x} + \frac{a^2}{2} \frac{\partial^2 \boldsymbol{S}}{\partial x^2} \pm \dots \ , \tag{14.23}$$

where a is the lattice parameter. Then the equation of motion (14.20) in the limit $a \to 0$ and after a suitable rescaling becomes

$$\frac{\partial \boldsymbol{S}}{\partial t} = \boldsymbol{S} \times \frac{\partial^2 \boldsymbol{S}}{\partial x^2} \ . \tag{14.24}$$

Similarly (14.22) in the continuum limit becomes

$$\frac{\partial \boldsymbol{S}}{\partial t} = \boldsymbol{S} \times \left(\frac{\partial^2 \boldsymbol{S}}{\partial x^2} + 2AS^z \boldsymbol{n} + \mu \boldsymbol{B} \right) \ . \tag{14.25}$$

Spin equations of the type (14.24) or (14.25) are special cases of the so-called Landau–Lifshitz equation, which were derived originally by Landau and Lifshitz from phenomenological arguments. It was not until 1977 that the complete integrable nature of many of these systems was realized [6]. In fact, by mapping (14.24) on a moving space curve with curvature [6]

$$\kappa(x,t) = \left[\frac{\partial \boldsymbol{S}}{\partial x} \cdot \frac{\partial \boldsymbol{S}}{\partial x} \right]^{1/2} \ , \tag{14.26}$$

and torsion

$$\tau(x,t) = \kappa^{-2} \left(\boldsymbol{S} \cdot \frac{\partial \boldsymbol{S}}{\partial x} \times \frac{\partial^2 \boldsymbol{S}}{\partial x^2} \right) \ , \tag{14.27}$$

which are respectively related to energy density and current density, (14.24) gets mapped onto the ubiquitous nonlinear Schrödinger equation

$$\mathrm{i}q_t + q_{xx} + 2|q|^2 q = 0 \ , \tag{14.28}$$

where the complex variable

$$q(x,t) = \frac{1}{2}\kappa(x,t)\exp\left(\mathrm{i}\int_{-\infty}^{x}\tau \mathrm{d}x'\right) . \tag{14.29}$$

14.1.4 The Lattice with Exponential Interaction: The Toda Equation

We have seen in Chap. 12 that the study of wave propagation in anharmonic lattices by Fermi, Pasta and Ulam has led to path-breaking developments, including the identification of the soliton concept. The original FPU lattice with quadratic/cubic/broken nonlinearities is a nonintegrable, and even chaotic, system. However, we have also seen that in suitable limits the integrable KdV equation can arise. This led scientists to search for explicit integrable lattice models (see also Sect. 10.10). The first and the most famous of such models as briefly pointed out in Sect. 10.10 is the one introduced by Morikazu Toda in 1967 and it is now called the Toda lattice [7]. Toda's original idea was to look for lattice models which admit elliptic function travelling waves so that in the appropriate limit solitary waves can arise as in the case of the KdV equation (see Sect. 12.3). Toda achieved this by making use of the generalized addition theorem for Jacobian elliptic functions (see also appendix A). For further details see for example, Ref. [7] and appendix I.

Considering thus the equation of motion of the nonlinear lattice as in the case of the FPU lattice in Sect. 12.2, equation (12.41),

$$m\frac{\mathrm{d}^2 y_i}{\mathrm{d}t^2} = f\left(y_{i+1}-y_i\right) - f\left(y_i - y_{i-1}\right) \ , \quad i = 1,2,...,N-1, \tag{14.30}$$

Toda arrived at the potential

$$f(r) = a\frac{\mathrm{e}^{-br}}{r} + br \ , \tag{14.31}$$

where a and b are constants. One may note that in the limit $a \to 0$ equation (14.31) is the harmonic oscillator potential, while for $b \to 0$ it is the hard sphere potential.

14.2 The Zakharov–Shabat (ZS)/ Ablowitz–Kaup–Newell–Segur (AKNS) Linear Eigenvalue Problem and NLEES

Following the invention of the IST technique applicable to the KdV equation by Kruskal and his coworkers in 1967 (as discussed earlier in Chap. 13), wherein the KdV equation is associated with the linear Schrödinger spectral problem, Zakharov and Shabat in a remarkable work [8] in 1972 took a further decisive step by showing that the nonlinear Schrödinger equation (14.3) is also IST solvable and possesses soliton solutions. This was achieved by associating the NLS equation with a Lax pair L and B, but which are now (2×2) matrix linear differential operators. It immediately showed that the IST procedure is not restricted to the KdV equation alone, but it has wider applicability. More importantly, the analysis of Zakharov and Shabat clearly established that the soliton property is a more encompassing phenomenon possessed by a class of nonlinear dispersive systems. It gave a tremendous impetus to other scientists to look for general prescriptions to identify integrable soliton equations. In this connection, Ablowitz, Kaup, Newell and Segur (AKNS) in 1974 [9] developed a sound formalism to identify a class of soliton possessing NLEEs including the KdV, MKdV, NLS and sG equations. We briefly indicate the procedure in this section.

14.2.1 The AKNS Linear Eigenvalue Problem and AKNS Equations

As mentioned above, Zakharov and Shabat in their seminal 1972 paper [8] gave the Lax pair for the nonlinear Schrödinger equation (14.3) as

$$L = \mathrm{i}\begin{bmatrix} 1-p & 0 \\ 0 & 1+p \end{bmatrix}\frac{\partial}{\partial x} + \begin{bmatrix} 0 & q \\ q^* & 0 \end{bmatrix} \tag{14.32a}$$

and

$$B = -\mathrm{i}\begin{bmatrix} -p\frac{\partial^2}{\partial x^2} - \frac{|q|^2}{1-p} & -\mathrm{i}q_x \\ \mathrm{i}q_x^* & -p\frac{\partial^2}{\partial x^2} + \frac{|q|^2}{1+p} \end{bmatrix}, \tag{14.32b}$$

where

$$1 - p^2 = 4\,, \tag{14.32c}$$

so that the Lax equation $L_t = [B, L]$ is equivalent to the NLS equation. Note that in the present case L and B are (2×2) matrix linear differential operators in contrast to the scalar linear differential operators needed for the linearization of the KdV equation (see Sect. 13.2). The nature of the linear differential operators L and B in (14.32) prompted AKNS [9] to propose a general (2×2) matrix linear eigenvalue problem for the complex potentials $q(x,t)$ and $r(x,t)$ in the form

$$v_{1x} + \mathrm{i}\zeta v_1 = q v_2\,, \tag{14.33a}$$

$$v_{2x} - \mathrm{i}\zeta v_2 = r v_1 \tag{14.33b}$$

or equivalently

$$V_x = MV\,, \tag{14.34a}$$

where the matrices

$$V = \begin{bmatrix} v_1 \\ v_2 \end{bmatrix}, \quad M = \begin{bmatrix} -\mathrm{i}\zeta & q \\ r & \mathrm{i}\zeta \end{bmatrix}. \tag{14.34b}$$

Correspondingly the time evolution of the eigenfunction $V = (v_1, v_2)^{\mathrm{T}}$ can be written in a general sense as the set of linear differential equations

$$v_{1t} = Av_1 + Bv_2, \tag{14.35a}$$

$$v_{2t} = Cv_1 + Dv_2 . \tag{14.35b}$$

Here A, B, C and D are yet undetermined functions of ζ, t, x, q, r and their spatial derivatives. Again (14.35) can be written equivalently as

$$V_t = NV , \quad N = \begin{pmatrix} A & B \\ C & D \end{pmatrix} . \tag{14.36}$$

Note that there is no x-derivative term of V in (14.35), as they can be eliminated by using (14.33).

It is very easy to check that the compatibility of (14.34) and (14.36) (that is $V_{xt} = V_{tx}$) gives rise to the matrix evolution equation

$$M_t - N_x + [M, N] = 0 \tag{14.37}$$

or equivalently

$$D = -A , \tag{14.38a}$$

$$A_x = qC - rB , \tag{14.38b}$$

$$q_t = 2Aq + B_x + 2\mathrm{i}\zeta B , \tag{14.38c}$$

$$r_t = -2Ar + C_x - 2\mathrm{i}\zeta C . \tag{14.38d}$$

In the above derivation we have assumed that the eigenvalue parameter is *isospectral*, that is it does not change with time. (Note that this condition can be relaxed in certain cases leading to inhomogeneous evolution equations. However, such a study is outside the scope of the present book and we do not consider them here).

Now, how can one obtain meaningful evolution equations from (14.38)? We note that in (14.38), the eigenvalue parameter ζ appears explicitly. As a result, we can assume that the quantities A, B,C and D are well defined functions of ζ and they can be expanded as power series in ζ or ζ^{-1}. Then truncating such power series at finite powers in ζ or ζ^{-1} so that (14.37) or (14.38) are consistent, one can obtain many interesting nonlinear evolution equations. It is also possible to proceed with the analysis for very general forms of A, B and C. For details, see the original paper of AKNS [9].

14.2.2 The Standard Soliton Equations

Now, let us assume that the functions A, B, and C can be expressed as polynomials in ζ and as an example are of degree 3. Then we may write

$$A = A_1\zeta^3 + A_2\zeta^2 + A_3\zeta + A_4 , \tag{14.39a}$$

$$B = B_1\zeta^3 + B_2\zeta^2 + B_3\zeta + B_4 , \tag{14.39b}$$

$$C = C_1\zeta^3 + C_2\zeta^2 + C_3\zeta + C_4 , \tag{14.39c}$$

where A_i, B_i, C_i are functions of t, x, q, r and their x derivatives to be determined. Substituting (14.39) in the AKNS equations (14.38) and collecting the coefficients of various powers of ζ and equating them to zero individually, we obtain determining equations for the quantities A_i, B_i and C_i. Solving them recursively, we can obtain [9]

$$A_1 = a_3\,, \quad A_2 = a_2\,, \quad A_3 = \frac{1}{2}\left(a_3 qr + a_1\right)\,,$$

$$A_4 = -\frac{\mathrm{i}}{4}\left(qr_x - q_x r\right) a_3 + a_0\,, \tag{14.39d}$$

$$B_1 = 0\,, \quad B_2 = \mathrm{i}a_3 q\,, \quad B_3 = \left(\mathrm{i}a_2 q - \frac{1}{2}a_3 q_x\right)\,,$$

$$B_4 = \left(\mathrm{i}a_1 q + \frac{\mathrm{i}}{2}a_3 q^2 r - \frac{1}{2}a_2 q_x - \frac{\mathrm{i}}{4}a_3 q_{xx}\right) \tag{14.39e}$$

and

$$C_1 = 0\,, \quad C_2 = \left(\mathrm{i}a_2 r + \frac{1}{2}a_3 r_x\right)\,,$$

$$C_3 = \left(\mathrm{i}a_1 r + \frac{\mathrm{i}}{2}a_3 q r^2 + \frac{1}{2}a_2 r_x - \frac{\mathrm{i}}{4}a_3 r_{xx}\right)\,. \tag{14.39f}$$

Using the expressions (14.39) in (14.38), the resultant NLEE is

$$q_t + \frac{\mathrm{i}}{4}a_3\left(q_{xxx} - 6qrq_x\right) + \frac{1}{2}a_2\left(q_{xx} - 2q^2 r\right) - \mathrm{i}a_1 q_x - 2a_0 q = 0, \tag{14.40a}$$

$$r_t + \frac{\mathrm{i}}{4}a_3\left(r_{xxx} - 6qrr_x\right) - \frac{1}{2}a_2\left(r_{xx} - 2qr^2\right) - \mathrm{i}a_1 r_x + 2a_0 r = 0, \tag{14.40b}$$

where a_0, a_1, a_2 and a_3 are integration constants. Many interesting nonlinear evolution equations of contemporary interest can be obtained by choosing these constants appropriately. Some of the important evolution equations are as follows.

(1) **KdV Equation**
For the choice $a_0 = a_1 = a_2 = 0$, $a_3 = -4\mathrm{i}$ and $r = -1$ in (14.40), we have the KdV equation

$$q_t + 6qq_x + q_{xxx} = 0\,. \tag{14.41}$$

(2) **MKdV Equation**
Again for the same choice, $a_0 = a_1 = a_2 = 0$, $a_3 = -4\mathrm{i}$ but with $r = \mp q$, we have the MKdV equation

$$q_t \pm 6q^2 q_x + q_{xxx} = 0\,. \tag{14.42}$$

(3) **The Nonlinear Schrödinger Equation**
For $a_0 = a_1 = a_3 = 0$, $a_2 = -2\mathrm{i}$ and $r = \mp q^*$, we obtain the nonlinear Schrödinger equation

$$\mathrm{i}q_t + q_{xx} \pm 2|q|^2 q = 0\,. \tag{14.43}$$

(4) **Hirota Equation**
With $a_0 = a_1 = 0$, $a_2 = -\mathrm{i}$ and $a_3 = 4\,\epsilon\,\mathrm{i}^3$, $r = -q^*$, one obtains the so-called Hirota equation

$$\mathrm{i}q_t + \frac{1}{2}q_{xx} + |q|^2q^* + \mathrm{i}\epsilon(q_{xxx} + 6|q|^2q_x) = 0\,. \tag{14.44}$$

Similarly, by developing A, B and C in the AKNS equations (14.38) in inverse powers of ζ, we get another set of interesting equations. For example, with the following choice for A, B and C, one can obtain the sine-Gordon equation (14.2), after suitable scalings:

$$A = -\frac{\mathrm{i}}{4\zeta}\cos u\,, \quad B = C = -\frac{\mathrm{i}}{4\zeta}\sin u\,, \quad q = -r = -\frac{1}{2}u_x\,. \tag{14.45}$$

Many other interesting nonlinear evolution equations can also obtained for suitable choices of A, B and C.

Finally, the Heisenberg spin equation and the Toda lattice equation can also be shown to fall within the Lax formalism/AKNS type formalism. For some details, see for example, the problems at the end of the chapter and also appendix I.

Exercises:

1. Verify that the Lax pair given in (14.32) lead to the NLS equation.
2. Obtain explicitly the forms for A, B and C given by (14.39).

14.3 Solitary Wave Solutions and Basic Solitons

Just as the KdV equation possesses solitary wave (Sect. 11.7) and soliton solutions, the nonlinear evolution equations deduced from the AKNS equations (14.38) also possess a variety of solitary waves and soliton solutions. First, let us indicate the nature of the solitary waves of some of the ubiquitous evolution equations in this section, before investigating their solitonic nature in the subsequent sections.

14.3.1 The MKdV Equation: Pulse Soliton

Proceeding as in the case of the KdV equation to obtain its solitary wave solution (Sect. 11.7), we can look for the elementary wave solution of the MKdV equation and obtain elliptic function (cnoidal) wave solutions. Then taking appropriate limiting form the solitary wave solution can be obtained.

As in Sect. 11.7, we look for wave solutions of the MKdV equation (14.1),

$$u(x,t) = f(\zeta)\,, \quad \zeta = (x - ct)\,. \tag{14.46}$$

Then (14.1) with positive sign can be reduced to the form

$$-c\frac{\mathrm{d}f}{\mathrm{d}\zeta}+6f^2\frac{\mathrm{d}f}{\mathrm{d}\zeta}+\frac{\mathrm{d}^3f}{\mathrm{d}\zeta^3}=0\,. \tag{14.47}$$

This can be integrated once to give

$$-cf+2f^3+\frac{\mathrm{d}^2f}{\mathrm{d}\zeta^2}=A\,, \tag{14.48}$$

where A is an arbitrary constant. As usual this equation can again be integrated to give (see Sect. 11.5)

$$-\frac{c}{2}f^2+\frac{f^4}{2}+\frac{1}{2}\left(\frac{\mathrm{d}f}{\mathrm{d}\zeta}\right)^2=Af+B\,, \tag{14.49}$$

where B is the new integration constant. On integrating this once again, we can obtain the elliptic function solution. However, in the limiting case $A=0$, $B=0$, we have

$$\int\frac{\mathrm{d}f}{\sqrt{f^2(c-f^2)}}=\int\mathrm{d}\zeta\,. \tag{14.50}$$

In explicit form, this can be expressed as

$$\frac{1}{\sqrt{c}}\,\mathrm{sech}^{-1}\left(\frac{f}{\sqrt{c}}\right)=(\zeta-\zeta_0)\,, \tag{14.51}$$

where ζ_0 is the integration constant. Rewriting, we obtain the solitary wave of the MKdV equation in the form

$$f=\sqrt{c}\,\mathrm{sech}\left[\sqrt{c}\,(x-ct)\right]\,. \tag{14.52}$$

Note again the fact that the velocity of the solitary wave is proportional to the square of the amplitude (compare with the KdV equation case, Sect. 11.5). We will verify later on that this solitary wave is a soliton indeed.

What about the possibility of solitary wave in the case of negative sign in (14.1)? In this case (14.47) should be replaced by

$$-c\frac{\mathrm{d}f}{\mathrm{d}\zeta}-6f^2\frac{\mathrm{d}f}{\mathrm{d}\zeta}+\frac{\mathrm{d}^3f}{\mathrm{d}\zeta^3}=0\,. \tag{14.53}$$

It can be shown that (14.53) does not lead to a solitary wave of the form (14.52). What else does one get? (See problem 7 below).

14.3.2 The sine-Gordon Equation: Kink, Antikink and Breathers

Let us consider the normalized form of the sine-Gordon equation (14.2) in the laboratory coordinates (see (14.15)) in the form

$$u_{xx}-u_{tt}=\sin u\,. \tag{14.54}$$

In terms of the wave variable $\zeta=(x-ct)$, (14.54) reduces to the ode,

$$\left(1-c^2\right)u_{\zeta\zeta} = \sin u\ . \tag{14.55}$$

Integrating (14.55) in the standard way one can obtain elliptic function solution. Choosing the integration constant suitably so that $u \to 0$ (mod 2π) and $u_\zeta \to 0$ as $\zeta \to \pm\infty$, the solitary wave solution can be obtained. Its form is

$$u(x,t) = 4\tan^{-1}\left[\exp\left(\pm\frac{(x-ct-x_0)}{\sqrt{1-c^2}}\right)\right]\ . \tag{14.56}$$

Note that in the right hand side of the solution (14.56) inside the bracket both the '+' and '−' signs are admissible. Then we can distinguish two kinds of solitary waves which actually turn out to be solitons as well [5].

(i) Kink Soliton

Consider the '+' sign in (14.56). It represents a monotonic change in the value of u from 0 to 2π as ζ increases from $-\infty$ to $+\infty$ (Fig. 14.3a). Eventhough it is not a true pulse, note that the changes in its amplitude occur in a localized region. However, the derivative of u is indeed a true pulse (see below). It is a *kink* type solitary wave.

(ii) Antikink Soliton

The solution (14.56) with '−' sign corresponds to a monotonic decrease of u from the value 2π at $x=-\infty$ to 0 at $x=+\infty$ (see Fig. 14.3b). We have now an antikink type solitary wave.

In both the above cases, the derivatives of u, namely

$$u_x = \pm\frac{2}{\sqrt{1-c^2}}\,\mathrm{sech}\left(\frac{x-ct-x_0}{\sqrt{1-c^2}}\right) \tag{14.57a}$$

and

$$u_t = \mp\frac{2c}{\sqrt{1-c^2}}\,\mathrm{sech}\left(\frac{x-ct-x_0}{\sqrt{1-c^2}}\right) \tag{14.57b}$$

are both localized solitary pulses.

(iii) Breathers

Yet another class of localized solutions is admitted by the sine-Gordon equation, which is not present in the KdV equation. It is the so-called breather solution. To identify such a solution straightforwardly, one can apply the method of separation of variables. Let us look for a solution of (14.54) in the form

$$u(x,t) = 4\tan^{-1}\left[X(x)T(t)\right]\ . \tag{14.58}$$

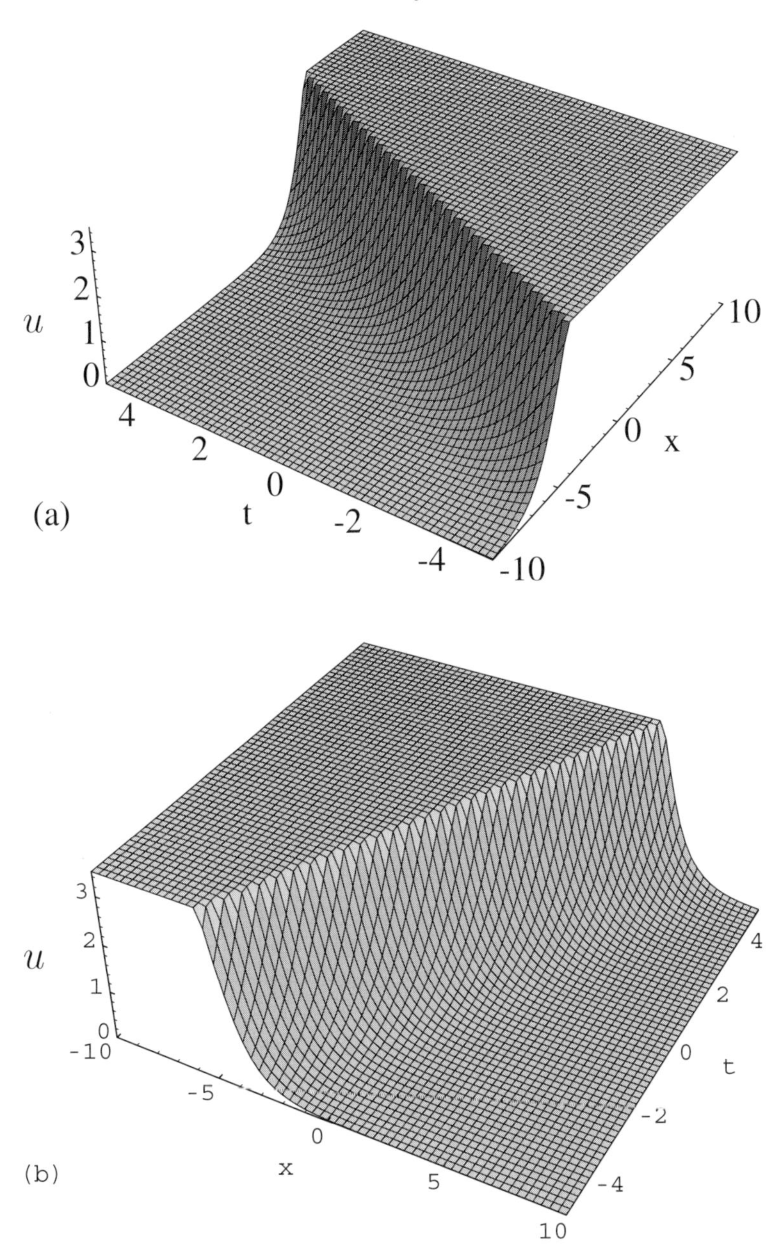

Fig. 14.3. (**a**) A kink solution of the sine-Gordon equation. (**b**) An antikink solution of the sine-Gordon equation

Then substituting (14.58) into (14.54) one obtains (see Ref. [5] for fuller details)

$$\left(1 + X^2T^2\right)\left(X_{xx}T - XT_{tt} - XT\right)$$
$$= 2XT\left[\left(X_x\right)^2 T^2 - \left(T_t\right)^2 X^2 - X^2T^2\right] . \tag{14.59}$$

If we now make the assumption that

$$X_x^2 = a_1X^4 + a_2X^2 + a_3 \tag{14.60a}$$

and

$$T_t^2 = b_1T^4 + b_2T^2 + b_3 , \tag{14.60b}$$

where a_1, a_2, a_3, b_1, b_2 and b_3 are constants, then we can force a separation of the equations for X and T:

$$X_{xx} = 2a_1X^3 + a_2X , \tag{14.61a}$$

and

$$T_{tt} = 2b_1T^3 + b_2T . \tag{14.61b}$$

In this case, using (14.60) and (14.61) in (14.59) and requiring that the terms proportional to X^5T^3 and X^3T^5 vanish, one can show that the following conditions on the constants have to be satisfied:

$$b_1 = -a_3 , \quad b_2 = a_2 - 1 \text{ and } b_3 = -a_1 . \tag{14.62}$$

Consequently, when $X(x)$ satisfies (14.60a), $T(t)$ obeys the equation

$$\left(T_t\right)^2 = -a_3T^4 + \left(a_2 - 1\right)T^2 - a_1 . \tag{14.63}$$

As before (as in the case of KdV and MKdV equations) the solutions for $X(x)$ and $T(t)$ can be expressed in terms of suitable Jacobian elliptic functions (see appendix A). Such a solution, which may be considered as the bound oscillation of a kink and an antikink, is

$$u(x,t) = 4\tan^{-1}\left\{A \operatorname{dn}\left[\beta\left(x - x_0\right); k_x\right] \operatorname{sn}\left(\omega t; k_t\right)\right\} , \tag{14.64}$$

where the moduli of the elliptic functions are given by

$$k_x^2 = 1 - \left[\frac{1 - \beta^2\left(1 + A^2\right)/A^2}{\beta^2\left(1 + A^2\right)}\right] \tag{14.65a}$$

and

$$k_t^2 = \frac{A^2\left[1 - \omega^2\left(1 + A^2\right)\right]}{\omega^2\left(1 + A^2\right)} , \tag{14.65b}$$

where $\beta = \omega A$. One can show that in the limit $k_x^2 \to 1$, $k_t^2 \to 0$, so that $\omega^2(1 + A^2) = 1$ and $\omega^2 + \beta^2 = 1$, we have the *breather* or *bion* solution (see Fig. 14.4)

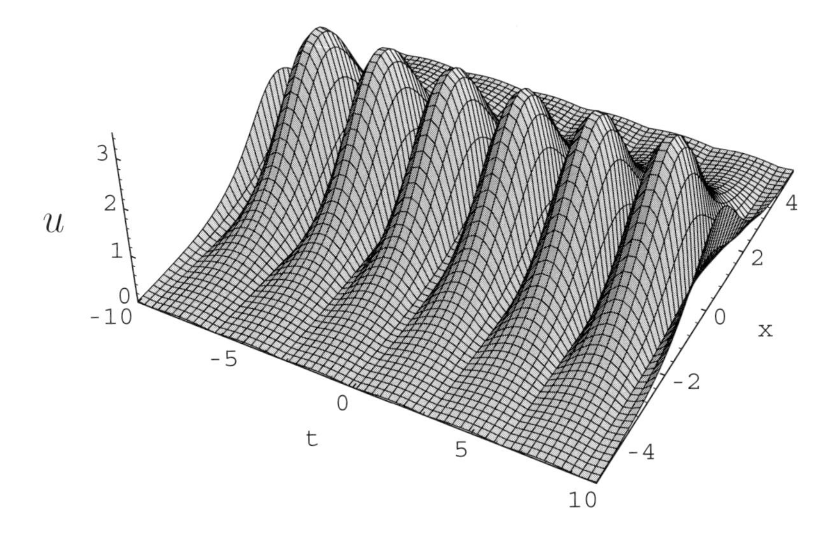

Fig. 14.4. A breather solution of the sine-Gordon equation

$$u(x,t) = 4 \ \tan^{-1}\left[\frac{\beta \sin \omega t}{\omega \ \cosh \beta \, (x - x_0)}\right] . \tag{14.66}$$

Other possible solutions which one may obtain along these lines are plasma oscillations, kink-antikink oscillations and so on (see problems 11 and 12 at the end of the chapter for some details).

Finally one can also obtain a two-soliton solution. For example, a kink of velocity $+v$ colliding at the origin with a kink of velocity $-v$ can be given as

$$u(x,t) = 4 \tan^{-1}\left[\frac{v \ \sinh\left(x/\sqrt{1-v^2}\ \right)}{\cosh\left(vt/\sqrt{1-v^2}\ \right)}\right] . \tag{14.67}$$

In a similar way the kink-antikink collision can be represented by the solution

$$u(x,t) = 4 \ \tan^{-1}\left[\frac{\sinh\left(vt/\sqrt{1-v^2}\ \right)}{v \ \cosh\left(x/\sqrt{1-v^2}\ \right)}\right] . \tag{14.68}$$

There are several ways to derive the above two-soliton solutions. They can be obtained, for example, from (14.58) for suitable choice of parameters as in the case of breathers (see problem 12 at the end of the chapter). They can also be obtained by the IST method, Hirota method or the Bäcklund transformation method. For more details see the following sections.

Exercises:

3. Integrate (14.55) explicitly and obtain the solitary wave solution (14.56).

4. Solve the equations (14.61) using (14.63) and obtain the solution (14.64).
5. Obtain explicitly the breather solution (14.66) from (14.64) by taking the indicated limits. For details, see for example Ref. [5].

14.3.3 The Nonlinear Schrödinger Equation: Envelope Soliton

Considering the nonlinear Schrödinger equation (14.3), we look for wave solution of the form

$$q(x,t) = \widehat{q}(\tau)\exp\left(\mathrm{i}v_e\xi/2\right), \quad \tau = (x - v_e t), \quad \xi = (x - v_c t), \tag{14.69}$$

where v_e and v_c are arbitrary constants. Substituting (14.69) into the NLS equation (14.3), we obtain the ode

$$\widehat{q}_{\tau\tau} + \frac{1}{4}\left(2v_e v_c - v_e^2\right)\widehat{q} + 2\widehat{q}^3 = 0 . \tag{14.70}$$

Defining the constant 'a' through the relation

$$4a^2 = v_e^2 - 2v_c v_e , \tag{14.71a}$$

(14.70) can be rewritten as

$$\widehat{q}_{\tau\tau} - a^2\widehat{q} + 2\widehat{q}^3 = 0 . \tag{14.71b}$$

Integrating once it becomes

$$\widehat{q}_\tau^2 = a^2\widehat{q}^2 - \widehat{q}^4 + c , \tag{14.71c}$$

where c is an arbitrary constant. Taking c to be zero and integrating once again we get

$$\int \mathrm{d}\tau = \int \frac{\mathrm{d}\widehat{q}}{\sqrt{\widehat{q}^2\left(a^2 - \widehat{q}^2\right)}} . \tag{14.72}$$

Carrying out the integration and rewriting, we obtain

$$\widehat{q} = a \operatorname{sech} a\left(x - v_e t - x_0\right) , \quad (x_0 = \text{arbitrary constant}) \tag{14.73}$$

so that the solitary wave solution to the NLS equation (14.3) can be written as

$$q(x,t) = a \exp\left[\mathrm{i}\frac{v_e x}{2} + \mathrm{i}\left(a^2 - \frac{v_e^2}{4}t\right)\right] \operatorname{sech}\left[a\left(x - v_e t - x_0\right)\right] . \tag{14.74}$$

In fact, it can be shown that the solitary wave solution (14.74) is indeed the one-soliton solution of the NLS equation, which is also called the *envelope soliton solution*, due to the presence of the envelope term in the right hand side of (14.71). One can also choose $v_e = 0$ without loss of generality so that the stationary form of the solitary wave solution becomes (see also problem 10 at the end of the chapter)

$$q(x,t) = a\exp\left(\mathrm{i}a^2 t\right) \operatorname{sech} a\left(x - x_0\right) . \tag{14.75}$$

Exercise:

6. Plot (i) Re$q(x,t)$ versus x and (ii) Im$q(x,t)$ versus x, for any fixed t and identify the envelope nature of the solution (14.74). Here Re and Im correspond to real and imaginary parts.

14.3.4 The Heisenberg Spin Equation: The Spin Soliton

Expressing the Heisenberg spin equation (14.4) in terms of the polar coordinates $\theta(x,t)$ and $\phi(x,t)$ defined by the relation

$$\boldsymbol{S} = (S_1,\ S_2,\ S_3) \equiv (\sin\theta\sin\phi,\ \sin\theta\cos\phi,\ \cos\theta)\ , \tag{14.76}$$

the equation of motion can be written as

$$\theta_t = -2\theta_x\phi_x\cos\theta - \phi_{xx}\sin\theta\ , \tag{14.77a}$$

$$\phi_t\sin\theta = \theta_{xx} - \phi_x^2\sin\theta\cos\theta\ . \tag{14.77b}$$

Now let us look for the wave solution of the form

$$\theta(x,t) = \theta(\xi),\quad \phi(x,t) = \overline{\phi}(\xi) + \Omega t,\quad \xi = (x - ct)\ , \tag{14.78}$$

where Ω is a parameter. Defining the new variable

$$\chi(\xi) = \overline{\phi}_\xi\sin\theta(\xi)\ , \tag{14.79}$$

(14.77) can be rewritten as

$$\chi_\xi = \theta_\xi(c - \chi\cot\theta)\ , \tag{14.80a}$$

$$\theta_{\xi\xi} = -\chi(c - \chi\cot\theta) + \Omega\sin\theta\ . \tag{14.80b}$$

It follows immediately from the above equation that

$$\theta_\xi^2 + \chi^2 + 2\Omega\cos\theta = 2K = \text{constant} \tag{14.81}$$

so that the energy density

$$\varepsilon(x,t) = \frac{1}{2}\left(\frac{\partial \boldsymbol{S}}{\partial x}\right)^2 = \frac{1}{2}\boldsymbol{S}_\xi^2 = \frac{1}{2}\left[\theta_\xi^2 + \chi^2\right] = -\Omega\cos\theta(\xi) + K\ . \tag{14.82}$$

Now defining yet another set of new variables

$$z(\xi) = \cos\theta(\xi)\ ,\quad y(\xi) = \overline{\phi}_\xi\sin^2\theta(\xi)\ , \tag{14.83}$$

(14.80) can be rewritten as

$$y_\xi = -cz_\xi\ , \tag{14.84a}$$

$$z_{\xi\xi} - cy + \Omega\left(1 - z^2\right) + \frac{z\left(z_\xi^2 + y^2\right)}{(1 - z^2)} = 0\ . \tag{14.84b}$$

Now making use of the relation (14.81) in (14.84), one can solve the coupled equations (14.84) and obtain the general solution. In particular, in the special case,

$$K = \Omega = \frac{1}{2}c^2 , \qquad (14.85)$$

one can write down the spin solitary wave solution as

$$\phi(x,t) = \tan^{-1}\{\tanh[c(x-ct)/2]\} + \frac{1}{2}cx \qquad (14.86)$$

and

$$\cos\theta(u) = \{\tanh[c(x-ct)/2]\}^2 . \qquad (14.87)$$

Correspondingly the energy density takes the form of a solitary wave,

$$\varepsilon(u) = \frac{1}{2}c^2 \operatorname{sech}^2(c(x-ct)/2) \qquad (14.88)$$

and that the total energy is finite:

$$E = \int_{-\infty}^{\infty} \varepsilon(u)\mathrm{d}u = 2c . \qquad (14.89)$$

For details of the above derivation, see the reference given below under exercise 9. Again it has been shown in the literature [6] that the above solitary wave solution is a soliton indeed.

Exercises:

7. Deduce the polar form (14.77) of the Heisenberg spin equation (14.4).
8. Verify the relations (14.81) and (14.82).
9. Solve explicitly (14.84) and obtain the solutions given in (14.86) and (14.87) [10].
10. Obtain the form of $\boldsymbol{S}$ using (14.76) and (14.86)–(14.87) (see Sect. 14.4.4 below).

14.3.5 The Toda Lattice: Discrete Soliton

The equation of motion for Toda lattice, which is now a system of coupled nonlinear odes, can also be treated suitably and the existence of elliptic function wave, solitary wave and soliton solutions can be proved. Details are given in appendix I.

14.4 Hirota's Method and Soliton Nature of Solitary Waves

It may be recalled that we have introduced the so-called Hirota's bilinearization method in Sect. 12.5 in order to prove the soliton nature of the KdV solitary wave by obtaining the two-soliton and in principle N-soliton solutions. We also saw that using the resultant solutions the elastic nature of the collision of the KdV solitons can be easily proved. We will now demonstrate that using the same general method that each of the solitary wave solutions discussed in the specific examples considered in the previous section is in fact a soliton and it possesses the property that under collision with another solitary wave its shape and velocity remain unchanged. Thus, the presently considered evolution equations are also indeed soliton possessing nonlinear evolution equations solvable by the AKNS–ZS type (2×2) inverse scattering transformation (IST) method (see below) or its generalization, thereby showing the genericity of the soliton phenomenon. We will illustrate briefly this property for each of the cases by making use of the Hirota's bilinearization method.

14.4.1 The Modified KdV Equation

The MKdV equation (14.1) under the transformation

$$u(x,t) = \frac{G(x,t)}{F(x,t)} , \tag{14.90}$$

where G and F are real functions, can be easily shown to be bilinearized as

$$(G_t F - G F_t) + (G_{xxx} F - 3G_{xx} F_x + 3G_x F_{xx} - G F_{xxx}) = 0 , \tag{14.91a}$$

$$F F_{xx} - F_x^2 = G^2 . \tag{14.91b}$$

Now introducing the following series expansions in (14.91),

$$F = 1 + \epsilon^2 F_2 + \epsilon^4 F_4 + \ldots = \sum_{m=0}^{\infty} \epsilon^{2n} F_{2n} , \quad F_0 = 1 , \tag{14.92a}$$

$$G = \epsilon G_1 + \epsilon^3 G_3 + \ldots , \tag{14.92b}$$

where ϵ is an arbitrary but small parameter, we equate equal powers of ϵ to obtain the following equations:

$$O\left(\epsilon^0\right) \; : \; 0 = 0 \tag{14.93a}$$

$$O\left(\epsilon\right) \; : \; \left(\partial_t + \partial_x^3\right) G_1 = 0 , \tag{14.93b}$$

$$\partial_x^2 F_1 = G_1^2 \tag{14.93c}$$

$$O\left(\epsilon^2\right) \; : \; \left(\partial_t + \partial_x^3\right) G_3 = -\left(\partial_t + \partial_x^3\right) G_1 F_2 , \tag{14.93d}$$

$$2\partial_x^2 F_4 = 4 G_1 G_3 - \partial_x^3 \left(F_2^2\right) \tag{14.93e}$$

and so on.

As in the case of the KdV equation, from (14.93b,c), if we assume

$$G_1 = \mathrm{e}^{\eta_1} ,$$
$$\eta_1 = k_1 x - k_1^3 t + \eta_1^{(0)}, \quad k, \eta_1^{(0)} : \text{ real constant parameters} \tag{14.94}$$

then

$$F_2 = \frac{\mathrm{e}^{2\eta_1}}{4k_1^2} , \tag{14.95a}$$

so that

$$F = 1 + F_2 = 1 + \frac{\mathrm{e}^{2\eta_1}}{4k_1^2} . \tag{14.95b}$$

Then, from (11.90) we get the *one-soliton* solution, with $G_j = 0 = F_i$, $j > 1$, $i > 2$,

$$u(x,t) = k_1 \,\mathrm{sech}\eta_1 . \tag{14.96}$$

Note that the form (14.96) is in agreement with the solitary wave solution (14.52) obtained earlier by elementary means.

Similarly with the form

$$G_1 = \mathrm{e}^{\eta_1} + \mathrm{e}^{\eta_2}, \quad \eta_i = k_i x - k_i^3 t + \eta_i^{(0)} , \quad i = 1,2 \tag{14.97a}$$

we have

$$F_2 = \frac{\mathrm{e}^{2\eta_1}}{4k_1^2} + \frac{\mathrm{e}^{2\eta_2}}{4k_2^2} . \tag{14.97b}$$

Making use of the above forms in (14.93d–e), we can obtain G_3 and F_4 with $G_j = 0$, $F_i = 0$, $j \geq 3$, $i > 4$ and write the two-soliton solution of the MKdV equation as

$$u(x,t) = \frac{2\left[k_1 \cosh(\eta_2 + A_{12}/2) + k_2 \cosh(\eta_1 + A_{21}/2)\right]}{\alpha_1 \left[\cosh\left(\eta_1 + \eta_2 + \frac{R_4}{2}\right) + \alpha_2 \cosh\left(\eta_1 - \eta_2 + \frac{R_1}{2} - \frac{R_2}{2}\right) + \alpha_3\right]} , \tag{14.98a}$$

where

$$A_{12} = \log \frac{\alpha_1^2}{4k_2^2} , \quad A_{21} = \log \frac{\alpha_1^2}{4k_1^2} , \tag{14.98b}$$

$$\alpha_1 = \frac{k_1 - k_2}{k_1 + k_2} , \quad \alpha_2 = \left(\frac{k_1 + k_2}{k_1 - k_2}\right)^2 , \quad \alpha_3 = \frac{4k_1 k_2}{(k_1 + k_2)^2} , \tag{14.98c}$$

$$R_1 = \log \frac{1}{4k_1^2} , \quad R_2 = \log \frac{1}{4k_2^2} , \quad R_4 = \log \frac{\alpha_1^4}{16k_1^2 k_2^2} . \tag{14.98d}$$

One can carry out the asymptotic ($t \to \pm\infty$) analysis of the two-soliton solution as in the case of the KdV equation in Sect. 12.5 and verify the elastic nature of the soliton collision. $N \geq 3$ soliton solutions can also be obtained in a similar manner. For a proof for the existence of N-soliton solution (N: arbitrary), see for example Ref. [11].

Exercises:

11. Obtain the bilinearized form (14.91) using (14.90) in (14.1).
12. Obtain the explicit form of 2-soliton solution (14.98).
13. Carry out an asymptotic analysis of the 2-soliton solution (14.98) to verify the soliton nature.

14.4.2 The NLS Equation

Again with the transformation $u = G(x,t)/F(x,t)$, where now G is a complex function while F is real, the NLS equation (14.3) can be bilinearized as

$$\mathrm{i}\,(G_t F - G F_t) + (G_{xx}F + GF_{xx} - 2G_x F_x) = 0\,, \tag{14.99a}$$

$$2FF_{xx} - 2F_x^2 - 2GG^* = 0\,. \tag{14.99b}$$

Assuming the series expansion,

$$F = 1 + \epsilon^2 F_2 + \epsilon^4 F_4 + \ldots\,, \qquad G = \epsilon G_1 + \epsilon^3 G_3 + \ldots\,, \tag{14.100}$$

(14.99) can be recast in the form

$$O(\epsilon)\ :\ \left(\mathrm{i}\partial_t + \partial_x^2\right) G_1 = 0 \tag{14.101a}$$

$$O\left(\epsilon^2\right)\ :\ \partial_x^2\,(2F_2) = 2G_1G_1^* \tag{14.101b}$$

$$O\left(\epsilon^3\right)\ :\ \left(\mathrm{i}\partial_t + \partial_x^3\right) G_3 = -\left(\mathrm{i}\partial_t + \partial_x^2\right) G_1F_2 \tag{14.102a}$$

$$O\left(\epsilon^4\right)\ :\ \partial_x^2\,(2F_4) = -\partial_x^2\left(F_2^2\right) + 2\left(G_1G_3^* + G_3G_1^*\right) \tag{14.102b}$$

and so on. Solving recursively one can then obtain soliton solutions:

(i) One-Soliton Solution (Envelope Soliton Solution):

Assuming

$$G = G_1 = \mathrm{e}^{\eta_1}\,, \tag{14.103a}$$

we get

$$F = 1 + F_2 = 1 + \frac{\mathrm{e}^{\eta_1+\eta_1^*}}{\left(p_1 + p_1^*\right)^2}\,, \tag{14.103b}$$

where $\eta_1 = p_1 x + \mathrm{i}p_1^2 t + \eta_1^{(0)}$ and p_1 and $\eta_1^{(0)}$ are now complex constants. Thus, the one-soliton solution can be written as

$$u(x,t) = \frac{G}{F} = \frac{\mathrm{e}^{\eta_1}}{1 + \left(\mathrm{e}^{2\eta_{1R}}/(4p_{1R}^2)\right)} = p_{1R}\,\mathrm{e}^{\mathrm{i}z_1}\,\mathrm{sech} z_2\,, \tag{14.104a}$$

where

$$z_1 = \left[p_{1I}x + \left(p_{1R}^2 - p_{1I}^2\right)t + \eta_{1I}^{(0)}\right],$$
$$z_2 = \left[p_{1R}\left(x - 2p_{1I}t\right) + \eta_{1R}^{(0)} + \Delta\right] \tag{14.104b}$$

and

$$\Delta = \log\frac{1}{2p_{1R}}. \tag{14.104c}$$

This form is obviously in agreement with the envelope solitary wave solution (14.74) obtained earlier.

(ii) Two-Soliton Solution:

Now assuming

$$G_1 = \mathrm{e}^{\eta_1} + \mathrm{e}^{\eta_2}, \tag{14.105}$$

and proceeding as above we can obtain the two-soliton solution as

$$u(x,t) = \frac{f_1}{f_2}, \tag{14.106a}$$

where

$$f_1 = \mathrm{e}^{\eta_1} + \mathrm{e}^{\eta_2} + \mathrm{e}^{\eta_1+\eta_1^*+\eta_2+\delta_1} + \mathrm{e}^{\eta_1+\eta_2+\eta_2^*+\delta_2}, \tag{14.106b}$$

$$f_2 = 1 + \mathrm{e}^{\eta_1+\eta_1^*+R_1} + \mathrm{e}^{\eta_1+\eta_2^*+\delta_0} + \mathrm{e}^{\eta_1^*+\eta_2+\delta_0^*} + \mathrm{e}^{\eta_2+\eta_2^*+\eta_2+R_2} + \mathrm{e}^{\eta_1+\eta_1^*+\eta_2+\eta_2^*+R_3} \tag{14.106c}$$

$$\mathrm{e}^{\delta_0} = \frac{1}{(k_1+k_2^*)^2}, \quad \mathrm{e}^{R_1} = \frac{1}{(k_1+k_1^*)^2}, \quad \mathrm{e}^{R_2} = \frac{1}{(k_2+k_2^*)^2}, \tag{14.106d}$$

$$\mathrm{e}^{\delta_1} = \frac{(k_1-k_2)^2}{(k_1+k_1^*)^2(k_1^*+k_2)^2}, \tag{14.106e}$$

$$\mathrm{e}^{\delta_2} = \frac{(k_2-k_1)^2}{(k_2+k_2^*)^2(k_1+k_2^*)^2}, \tag{14.106f}$$

$$\mathrm{e}^{R_3} = \frac{|k_1-k_2|^4}{(k_1+k_1^*)^2(k_2+k_2^*)^2|k_1+k_2^*|^4}. \tag{14.106g}$$

Exercises:

14. Deduce the bilinearized form (14.99).
15. Obtain the two-soliton solution (14.106) using (14.105) in (14.101) and (14.102).
16. Establish the elastic collision property of the envelope soliton of the NLS equation by carrying out an asymptotic ($t \to \pm\infty$) analysis of (14.106).

14.4.3 The sine-Gordon Equation

The sine-Gordon equation (14.2) can be bilinearized by using the transformation,

$$u = 2\,\mathrm{i}\log\frac{f^*}{f}\ . \qquad (14.107)$$

Then (14.2) can be rewritten as

$$2\left(f_x f_t - f f_{xt}\right) + m^2\left(f^2 - f^{*2}\right) = 0\ . \qquad (14.108)$$

With the substitution

$$f = 1 + \epsilon F_1 + \epsilon^2 F_2 + \dots\ , \qquad (14.109)$$

(14.108) can be rewritten as

$$O(\epsilon)\ :\ 2F_{1xt} + (F_1 - F_1^*) = 0\ , \qquad (14.110a)$$

$$O\left(\epsilon^2\right)\ :\ 2F_{2xt} + (F_2 - F_2^*) = -2\left(F_{1xt}F_1 - F_{1t}F_{1x}\right) - \frac{1}{2}\left(F_1^2 - F_1^{*2}\right) \qquad (14.110b)$$

and so on.

(i) One-Soliton (Kink) Solution

If we start with

$$f = 1 + \mathrm{e}^{\eta_1 + \mathrm{i}\pi/2},\quad \eta_1 = k_1 x - \frac{t}{k_1} + \eta_1^{(0)}\ , \qquad (14.111)$$

where k_1 and $\eta_1^{(0)}$ are now real constants, then the kink solution follows:

$$u = 2\,\mathrm{i}\log\frac{f}{f^*} = 4\tan^{-1}\left(\mathrm{e}^{\eta_1}\right)\ . \qquad (14.112)$$

Note that in the laboratory coordinates the solution (14.112) is the same as (14.56).

(ii) Two-Kink Solution

Proceeding as in the previous cases, we can consider terms upto $O(\epsilon^2)$ in (14.109) and obtain

$$f = 1 + \mathrm{e}^{\eta_1 + \mathrm{i}\pi/2} + \mathrm{e}^{\eta_2 + \mathrm{i}\pi/2} + \frac{(k_1 - k_2)^2}{(k_1 + k_2)^2}\mathrm{e}^{\eta_1 + \eta_2 + \mathrm{i}\pi}\ . \qquad (14.113a)$$

Then the two-soliton solution can be written from (14.107) as

$$u(x,t) = 4\tan^{-1}\left[\left(\frac{k_1 - k_2}{k_1 + k_2}\right)\frac{\sinh\frac{1}{2}(\eta_1 + \eta_2 + A_{12})}{\cosh\frac{1}{2}\left(\eta_1 - \eta_2\right)}\right]\ , \qquad (14.113b)$$

where

$$A_{12} = \log \left(\frac{k_1 - k_2}{k_1 + k_2} \right)^2 . \tag{14.113c}$$

Exercise:

17. Obtain the breather solution of the sine-Gordon equation using the Hirota method.

14.4.4 The Heisenberg Spin System

Under the stereographic transformation

$$S_1 = \frac{2\mathrm{Re}(\omega)}{(1 + \omega\omega^*)}, \quad S_2 = \frac{2\mathrm{Im}(\omega)}{(1 + \omega\omega^*)}, \quad S_3 = \frac{1 - \omega\omega^*}{(1 + \omega\omega^*)}, \tag{14.114}$$

where $\omega = \omega(x,t)$ is a complex function, the Heisenberg spin equation (14.4) can be rewritten as

$$(1 + \omega\omega^*)(\mathrm{i}\omega_t + \omega_{xx}) - 2\omega^*\omega_x^2 = 0 . \tag{14.115}$$

Under the transformation

$$\omega = \frac{g}{f} , \tag{14.116}$$

(14.4) in its component form can be bilinearized as

$$\mathrm{i}\,(f_t^* g - f^* g_t) - (f_{xx}^* g - 2 f_x^* g_x + f^* g_{xx}) = 0 , \tag{14.117a}$$

$$\begin{aligned} \mathrm{i}\,(f_t^* f - f^* f_t - g_t^* g + g^* g_t) - [(f_{xx}^* f - 2f_x^* f_x + f^* f_{xx}) \\ - (g_{xx}^* g - 2g_x^* g_x + g^* g_{xx})] = 0 , \end{aligned} \tag{14.117b}$$

$$f_x^* f - f^* f_x + g_x^* g - g^* g_x = 0 . \tag{14.117c}$$

Expanding now the functions g and f in the series

$$g = \epsilon g_1 + \epsilon^3 g_3 + ... \tag{14.118a}$$

$$f = 1 + \epsilon^2 f_2 + \epsilon^4 f_4 + ... , \tag{14.118b}$$

substituting them in (14.117) and equating the coefficients of various powers of ϵ to zero, we obtain the following system of equations:

$$\mathrm{i}\, g_{1t} + g_{1xx} = 0 , \tag{14.118c}$$

$$\begin{aligned} \mathrm{i}\,(f_{2t}^* - f_{2t}) - (f_{2xx}^* + f_{2xx}) = \mathrm{i}\,(g_{1t}^* g_1 - g_1^* g_{1t}) \\ - (g_{1xx}^* g_1 - 2g_{1x}^* g_{1x} + g_1^* g_{1xx}) , \end{aligned} \tag{14.118d}$$

$$f_{2x}^* - f_{2x} = g_1^* g_{1x} - g_{1x}^* g_1 , \tag{14.118e}$$

and so on.

Now, (14.118c) admits solution of the form

$$g_1 = N\mathrm{e}^{\chi_1}\ , \quad \chi_1 = a_1 x + \mathrm{i}a_1^2 t\ , \quad N, a_1 : \text{ complex constants}\ . \tag{14.119}$$

Substituting g_1 in (14.118d) we get

$$f_2 = A\exp\left(\chi_1 + \chi_1^*\right)\ , \quad A = \frac{-|N|^2 a_1^2}{\left(a_1 + a_1^*\right)^2}\ . \tag{14.120}$$

We get the one-soliton solution for spin components by introducing the quantities,

$$\Delta = \log\left(\frac{|N|\ |a|}{2a_{1R}}\right)\ , \quad \delta = \text{phase of } \left(N a_1^*\right)\ . \tag{14.121}$$

We have

$$\omega = \frac{-a_{1R}\exp\left[\,\mathrm{i}\left(\chi_{1I} + \delta\right)\right]}{a_{1R}\sinh\left(\chi_{1R} + \Delta\right) + \mathrm{i}\, a_{1I}\cosh\left(\chi_{1R} + \Delta\right)} \tag{14.122a}$$

and

$$\omega^* = \frac{-a_{1R}\exp\left[-\mathrm{i}\left(\chi_{1I} + \delta\right)\right]}{a_{1R}\sinh\left(\chi_{1R} + \Delta\right) - \mathrm{i}\, a_{1I}\cosh\left(\chi_{1R} + \Delta\right)}\ , \tag{14.122b}$$

$$\chi_{1R} = a_{1R}\left(x - 2a_{1I}t\right)\ , \quad \chi_{1I} = a_{1I}x + \left(a_{1R}^2 - a_{1I}^2\right)t\ . \tag{14.122c}$$

Substituting ω and ω^* in (14.114) we obtain the one-soliton solution for the spin components as

$$\begin{aligned} S^+ &= S_1 + \mathrm{i}\, S_2 \\ &= \frac{2a_{1R}}{a_{1R}^2 + a_{1I}^2}\left[\exp\left(\mathrm{i}\chi_{1I}\right)\right]\left[\mathrm{i}a_{1I} - a_{1R}\tanh\left(\chi_{1R}\right)\right]\mathrm{sech}\left(\chi_{1R}\right)\ , \end{aligned} \tag{14.123}$$

and

$$S_3 = 1 - \frac{2a_{1R}^2}{\left(a_{1R}^2 + a_{1I}^2\right)}\mathrm{sech}^2\left(\chi_{1R}\right)\ . \tag{14.124}$$

Multisoliton solutions can be generated in a similar way.

Note that the solitary wave solution (14.87) follows from (14.124) with $S_3 = \cos\theta$ and the special choice $a_{1R} = a_{1I} = c/2$.

Exercises:

18. Obtain the stereographic version (14.115) for (14.4) using the transformations (14.114).
19. Obtain the two-soliton solution for the spin components of (14.4).

14.5 Solutions via IST Method

We have pointed out earlier in Sect. 14.2 that a large number of soliton possessing nonlinear evolution equations are obtainable as the compatibility conditions for the ZS–AKNS linear eigenvalue problem (14.34) and its time evolution (14.36). The resulting class of equations included the MKdV, NLS and sine-Gordon equations among others. As in the case of the KdV equation, the existence of a linear eigenvalue problem (and its time evolution) associated with these NLEEs allows one to develop an inverse scattering transform (IST) analysis for a given set of initial data $q(x,0)$ and $r(x,0)$ for the boundary conditions $q(x,t)$, $r(x,t) \overset{|x|\to\infty}{\longrightarrow} 0$ (see for example Refs. [1,5,11] and also appendix H). Consequently a set of coupled Gelfand–Levitan–Marchenko linear integral equations can be solved to obtain the general solution. In short, the three step (IST) procedure advocated for the KdV equation in Chap. 13 (see also the schematic diagram, Fig. 13.1) to solve its IVP can be again used in the present case as well.

14.5.1 Direct and Inverse Scattering

In appendix H, we have given a detailed discussion on the direct and inverse scattering analysis associated with the linear eigenvalue problem (14.33). In particular, it has been pointed out that direct scattering analysis associated with the given 'potentials' $q(x)$, $r(x)$ identifies the scattering data ((H.20))

$$S = \left\{C_k, \overline{C}_k, \zeta_k, \overline{\zeta}_k, R(\xi), \overline{R}(\xi)\ ,\ \ k = 1,2,...,N(\overline{N})\right\}, \tag{14.125}$$

where C_k, $\overline{C}_k$ are the normalization constants, ζ_k, $\overline{\zeta}_k$ are the discrete eigenvalues and $R(\xi)$ and $\overline{R}(\xi)$ are the reflection coefficients.

Making use of the scattering data $\{S\}$, one can identify the functions

$$F(x) = \sum_{k=1}^{N} C_k^2 \mathrm{e}^{\mathrm{i}\zeta_k x} + \frac{1}{2\pi}\int_{-\infty}^{\infty} R(\xi)\mathrm{e}^{\mathrm{i}\xi x}\mathrm{d}\xi \tag{14.126a}$$

and

$$\overline{F}(x) = \sum_{k=1}^{\overline{N}} \overline{C}_k^2 \mathrm{e}^{-\,\mathrm{i}\overline{\zeta}_k x} + \frac{1}{2\pi}\int_{-\infty}^{\infty} \overline{R}(\xi)\mathrm{e}^{-\,\mathrm{i}\xi x}\mathrm{d}\xi\ . \tag{14.126b}$$

Then the 'potentials' $q(x)$ and $r(x)$ can be obtained by solving the associated Gelfand–Levitan–Marchenko linear integral equations (see (H.24))

$$\overline{K}(x,y) + F(x+y)\begin{pmatrix}0\\1\end{pmatrix} + \int_x^{\infty} K(x,z)F(z+y)\mathrm{d}z = 0\ , \quad x < y, \tag{14.127a}$$

$$K(x,y) + \overline{F}(x+y)\begin{pmatrix}1\\0\end{pmatrix}$$

$$+\int_x^\infty \overline{K}(x,z)\overline{F}(z+y)\mathrm{d}z = 0\,, \quad x<y, \tag{14.127b}$$

where

$$K(x,y) = \begin{pmatrix} K_1(x,y) \\ K_2(x,y) \end{pmatrix} \text{ and } \overline{K}(x,y) = \begin{pmatrix} \overline{K}_1(x,y) \\ \overline{K}_2(x,y) \end{pmatrix}. \tag{14.127c}$$

Then

$$q(x) = 2K_1(x,x+0), \quad r(x) = -2K_2(x,x+0) \tag{14.128}$$

and thus the inverse problem of constructing the potentials $q(x)$ and $r(x)$ stands solved.

One can also easily check from (14.127) that

$$\begin{aligned} K_1(x,y) &= -\overline{F}(x+y) \\ &\quad + \int_x^\infty \mathrm{d}z \int_x^\infty \mathrm{d}z' K_1(x,z)F(z+z')\,\overline{F}(z'+y)\,, \end{aligned} \tag{14.129}$$

where F and $\overline{F}$ are as given by (14.126). Similarly, an integral of equation for K_2 can also be given. Solving them, $q(x)$ and $r(x)$ can be determined using (14.128). Particularly, restricting to the reflectionless case, $R(\xi) = 0 = \overline{R}(\xi)$, one can obtain soliton solutions as in the case of the KdV equation.

14.5.2 Time Evolution of the Scattering Data

The time dependence of the eigenfunction of the linear eigenvalue problem (14.33) is given by (14.35); that is (using (14.38a)),

$$v_{1t} = Av_1 + Bv_2\,, \tag{14.130a}$$

$$v_{2t} = Cv_1 - Av_2\,. \tag{14.130b}$$

Since $q(x,t), r(x,t) \overset{|x|\to\infty}{\longrightarrow} 0$, typically the quantities $A(q,r,\zeta)$, $B(q,r,\zeta)$ and $C(q,r,\zeta)$ have the asymptotic behaviour (see the various examples given in Sect. 14.3.2)

$$A \overset{|x|\to\infty}{\longrightarrow} A_\infty(\zeta)\,, \quad B \overset{|x|\to\infty}{\longrightarrow} 0\,, \quad C \overset{|x|\to\infty}{\longrightarrow} 0\,, \tag{14.131}$$

where $A_\infty(\zeta)$ is a constant depending only on the eigenvalue ζ. For example, it can be verified from Sect. 14.2.2 that in the case of the NLS equation, $A_\infty = -2\mathrm{i}\zeta^2$, for the MKdV equation, $A_\infty = -4\mathrm{i}\zeta^3$ and for the sG equation, $A_\infty = \mathrm{i}/(4\zeta)$. Consequently, in the asymptotic regimes, $x \to \pm\infty$,

$$v_1(x,t) = v_1(x,0)\mathrm{e}^{A_\infty t}\,, \quad |x| \to \infty \tag{14.132a}$$

$$v_2(x,t) = v_2(x,0)\mathrm{e}^{-A_\infty t}\,, \quad |x| \to \infty\,. \tag{14.132b}$$

Then using the fact that while obtaining the compatibility condition (14.37) the eigenvalue ζ is required to be constant in time and the relations (H.6)–(H.7), (H.15), (H.19) in appendix H, one can easily prove from (14.132) that the evolution of the scattering data S given in (H.20) or (14.125) satisfy the relations

$$\zeta_k(t) = \zeta_k(0), \quad \overline{\zeta}_k(t) = \overline{\zeta}_k(0), \quad k = 1, 2, ..., N(\overline{N}) , \tag{14.133a}$$

$$C_k^2(t) = C_k^2(0)\mathrm{e}^{-2A_\infty t}, \quad \overline{C}_k^2(t) = \overline{C}_k^2(0)\mathrm{e}^{2A_\infty t} \tag{14.133b}$$

and

$$R(\xi, t) = R(\xi, 0)\mathrm{e}^{-2A_\infty t}, \quad \overline{R}(\xi, t) = \overline{R}(\xi, 0)\mathrm{e}^{2A_\infty t} . \tag{14.133c}$$

Using these expressions into the integral equations (14.126)–(14.129) and solving them, the Cauchy initial value problem associated with the AKNS evolution equation can in principle be solved.

14.5.3 Soliton Solutions

As in the case of the KdV equation discussed in Sect. 13.2, the pure soliton solutions for the various nonlinear evolution equations solvable by the AKNS formalism can be obtained by restricting to the case of reflectionless 'potentials', corresponding to $R(\xi, 0) = \overline{R}(\xi, 0) = 0$. We will demonstrate here how the single soliton solution follows in typical cases. Extension to more general solitons is straightforward in principle as in the case of the KdV equation.

A. Envelope Soliton of the NLS Equation

Assuming the number of bound states to be $N = 1$ ($\overline{N} = 1$) so that $\zeta_1 = \zeta$ and $C_1 = C$ in (14.126) and $R(\xi) = 0 = \overline{R}(\xi)$, one has

$$F(x) = C^2\mathrm{e}^{\mathrm{i}\zeta x}, \quad \zeta = \xi + \mathrm{i}\,\eta \tag{14.134}$$

and since (from (14.39d) and (14.43) for the case of the NLS equation)

$$A = -2\mathrm{i}\zeta^2 + \mathrm{i}|q|^2 , \quad r = -q* \tag{14.135}$$

in (14.130), we have

$$A_\infty = -2\mathrm{i}\zeta^2 . \tag{14.136}$$

Then

$$C^2(t) = C^2(0)\mathrm{e}^{4\mathrm{i}\zeta^2 t} . \tag{14.137}$$

Defining

$$C^2(0) = \alpha + \mathrm{i}\beta , \tag{14.138}$$

we have (from (14.126a))

$$F(x+z;t) = (\alpha + \mathrm{i}\beta)\exp\Big[\,\mathrm{i}\,(\xi + \mathrm{i}\eta)\,(x+z) - 8\xi\eta t + 4\mathrm{i}\,(\xi^2 - \eta^2)\,t\Big] . \tag{14.139}$$

Substituting this into (14.129) and noting that

$$K_1(x,z,t) \propto \exp\left[-\mathrm{i}(\xi - \mathrm{i}\eta)z\right], \quad \overline{F} = -F^* , \tag{14.140}$$

from the relations (14.128) one can obtain envelope soliton solution as

$$q(x,t) = 2\eta \exp\left[-2\mathrm{i}\xi x + 4\mathrm{i}\left(\xi^2 - \eta^2\right)t + \mathrm{i}\delta_1\right] \\ \times \operatorname{sech}\left[2\eta(x - 4\xi t) + \delta_2\right] , \tag{14.141a}$$

where

$$\exp(\mathrm{i}\delta_1) = (\beta + \mathrm{i}\alpha)/\sqrt{\beta^2 + \alpha^2} , \quad \exp(\delta_2) = 2\eta/\sqrt{\beta^2 + \alpha^2} . \tag{14.141b}$$

Equation (14.141) is nothing but the envelope soliton expression obtained by direct and Hirota methods earlier.

B. Sine-Gordon Equation

In this case (see (14.45))

$$A_\infty = \frac{\mathrm{i}}{4\zeta} , \tag{14.142}$$

and so

$$C^2(t) = C^2(0)\mathrm{e}^{-\mathrm{i}t/2\zeta} \tag{14.143}$$

so that

$$F(x+z,t) = \pm 2k\mathrm{e}^{-k(x+z)-t/2k} , \quad \zeta_1 = \mathrm{i}k . \tag{14.144}$$

Substituting into the Gelfand–Levitan–Marchenko integral equation one can obtain

$$K(x,z,t) = \pm \frac{2k \exp\left[-k(x+z) - t/2k\right]}{1 + \exp(-4kx - t/k)} . \tag{14.145}$$

Then

$$\frac{\partial u}{\partial x} = \pm 4K_1(x,x,t) \\ = \pm 4k \operatorname{sech}(2kx + t/2k) \tag{14.146}$$

so that

$$u(x,t) = 4\tan^{-1}\left\{\exp\left[\pm(2kx + t/2k)\right]\right\} , \tag{14.147}$$

which in the laboratory coordinates correspond to the standard kink (antikink) solution (14.56).

Generalizing the above analysis the soliton solutions of other nonlinear evolution equations as well as multisoliton solutions as in the case of the KdV equation can be obtained.

Exercises:

20. Show that the scattering data for the AKNS eigenvalue problem evolve in time according to (14.133) by using the asymptotic form of the eigenfunction (14.132) and the results in appendix H.
21. Solve the Gelfand–Levitan–Marchenko integral equation to obtain the envelope soliton solution for the NLS equation by using (14.140) in (14.129).
22. Obtain the expression (14.145) explicitly by solving the Gelfand–Levitan–Marchenko equation for the sine-Gordon equation.

14.6 Bäcklund Transformations

In the earlier chapter (Sect. 13.6), we have pointed out that the KdV equation admits the so-called (auto)Bäcklund transformation (BT) in that the BT connects its two different solutions. As a result the BT provides an efficient tool to construct multisoliton solutions; in particular, starting from the trivial solution as the seed solution, one can construct the one-soliton solution using the BT and then reusing the one-soliton solution as the new seed solution, one can obtain the two-soliton solution and so on.

The existence of a BT is not confined to the KdV equation alone: This property is shared by other IST solvable soliton equations as well. In fact, historically it was for the sine-Gordon equation the auto-BT was originally derived by Bäcklund in the year 1880 in connection with the study of surfaces with constant negative Gaussian curvature [12]. Generally, they arose in the theories of differential geometry and differential equations as a generalization of contact transformation (see for example, Ref. [12]). A classical example is the Cauchy–Riemann conditions for an analytic function

$$f(z) = u(x,y) + \mathrm{i}v(x,y) , \quad z = (x + \mathrm{i}y) \tag{14.150a}$$

so that

$$u_x = v_y , \quad v_x = -u_y . \tag{14.150b}$$

Then the above relation may be considered as a BT connecting the Laplace equation into itself:

$$\Delta u = u_{xx} + u_{yy} = 0 , \quad \Delta v = 0 . \tag{14.151}$$

Thus a BT for a partial differential equation in two independent variables can be essentially defined [11] as a pair of partial differential equations involving two dependent variables and their derivatives which together imply that each one of the dependent variables satisfies separately a partial differential equation, usually of higher order than the BT. Thus, for example, the BT

$$v_x = F(u, v, u_x, u_t, x, t) , \tag{14.152a}$$
$$v_t = F(u, v, u_x, u_t, x, t) \tag{14.152b}$$

will imply that u and v satisfy pdes of the general form

$$P(u) = 0 , \quad Q(v) = 0 . \tag{14.153}$$

If P and Q are of the same form then the transformation (14.152) constitutes an auto-BT.

From the above kind of considerations, the BT for the sine-Gordon equation turns out to be

$$\left(\frac{u+v}{2}\right)_x = k \sin\left(\frac{u-v}{2}\right) \tag{14.154a}$$
$$\left(\frac{u-v}{2}\right)_t = \frac{1}{k} \sin\left(\frac{u+v}{2}\right) , \tag{14.154b}$$

where k is a parameter, so that u and v satisfy the sG equation $u_{xt} = \sin u$ in the light cone coordinates. Similarly, BTs can be worked out for all other soliton equations also. For the various methods available to obtain BTs, see for example Refs. [5,11,12].

BTs can in principle be integrated to generate higher-order solitons, though in practice they are difficult to solve in general. The usual way to get around this difficulty is to utilize the permutability property of the BT, which states that if we make two successive BTs from a given initial solution u_0, we will end up with the same final solution no matter what sequential order of the two BTs we take. In other words, if k_1 and k_2 represent the parameters of the two BTs and if

$$u_0 \xrightarrow{k_1} u_1 \xrightarrow{k_2} u_{12} , \quad u_0 \xrightarrow{k_2} u_2 \xrightarrow{k_1} u_{21} ,$$

then $u_{12} = u_{21}$. From such permutability one can derive a nonlinear superposition formula, expressing u_{12} algebraically in terms of u_0, u_1 and u_2. For example for the sine-Gordon equation the following superposition law can be derived:

$$u_{12} = u_0 + 4 \tan^{-1}\left[\frac{k_2+k_1}{k_2-k_1} \tan\left(\frac{u_1-u_2}{4}\right)\right] . \tag{14.155}$$

The superposition law can be repeatedly applied to construct soliton solutions of higher and higher order and one can make use of symbolic computation packages such as MAPLE effectively for this purpose.

Exercises:

23. Prove (14.151) for the variables u and v.
24. Show that the BT (14.154) implies sG equation for both the variables u and v.
25. Find u_{12} from (14.155) by choosing $u_0 = 0$ and u_1 and u_2 as the kink and antikink solutions respectively of the sG equation.

14.7 Conservation Laws and Constants of Motion

We have shown in the earlier chapter (Sect. 13.5), that the integrable KdV equation admits infinite number of independent conservation laws and so admits infinite number of involutive integrals motion so that its complete integrability can be established. Is this property true in the present case of NLEEs solvable by ZS–AKNS eigenvalue problem as well? In the case of the KdV equation, we used the so-called Gardner transformation to obtain the infinite conservation laws. In this section, we show that by a straightforward analysis of the linear equations associated with a given integrable NLEE, the conservation laws can be deduced in a recursive way. To see this we proceed as below in the case of the (2×2) ZS–AKNS scattering problem (14.33)–(14.36). This can be done for the case of the Schrödinger spectral problem as well (see problem 29 below) in order to reproduce the conservation laws for the KdV equation.

A. Infinite Constants of Motion

Now considering the linear eigenvalue problem (14.33) for the eigenfunction $(v_1, v_2)^T$, using the asymptotic forms of the Jost functions $\phi(x,\zeta)$, $\overline{\phi}(x,\zeta)$, $\psi(x,\zeta)$, $\overline{\psi}(x,\zeta)$ as $x \to \pm\infty$ discussed in the appendix H, see (H.4), one can easily check from (H.6) that

$$\phi(x,\zeta) = \begin{pmatrix} \phi_1(x,\zeta) \\ \phi_2(x,\zeta) \end{pmatrix} \overset{x\to\infty}{\longrightarrow} \begin{pmatrix} a(\zeta)\mathrm{e}^{-\mathrm{i}\zeta x} \\ b(\zeta)\mathrm{e}^{\mathrm{i}\zeta x} \end{pmatrix} \tag{14.156}$$

so that

$$a(\zeta) = \lim_{x\to\infty} \phi_1 \mathrm{e}^{\mathrm{i}\zeta x} \,. \tag{14.157}$$

Now assuming $(\phi_1, \phi_2)^T$ as the solution of the linear eigenvalue equation (14.33), one can eliminate ϕ_2 in favour of ϕ_1 and obtain an equation for ϕ_1 in the form

$$q\left(\phi_{1xx} + \mathrm{i}\zeta\phi_{1x}\right) - \left(\phi_{1x} + \mathrm{i}\zeta\phi_1\right) q_x - \mathrm{i}\zeta q\left(\phi_{1x} + \mathrm{i}\zeta\phi_1\right) = rq^2\phi_1 \,. \tag{14.158}$$

The above form suggests that with the substitution (see for example, Ref. [11])

$$\phi_1 = \exp\left[-\mathrm{i}\zeta x + \Gamma(x,t)\right] \,, \tag{14.159}$$

where Γ is some function of x and t, (14.158) can be simplified as

$$q\Gamma_{xx} - 2\mathrm{i}\zeta q\Gamma_x - q_x\Gamma_x + \Gamma_x^2 = rq^2 \,. \tag{14.160}$$

Defining

$$\chi = \Gamma_x \,, \quad \Gamma = \int_{-\infty}^{x} \chi \mathrm{d}x' \,, \tag{14.161}$$

one can easily check that equation (14.160) above can be recast in the form of a Riccati equation:

$$q\left(\frac{\chi}{q}\right)_x - 2\mathrm{i}\zeta\chi + \chi^2 = qr \tag{14.162}$$

or

$$2\mathrm{i}\zeta\chi = \chi^2 - qr + q\left(\frac{\chi}{q}\right)_x . \tag{14.163}$$

Then, from equation (H.21) in the appendix H, we can check that $\phi_1(x,\zeta)$ has the asymptotic form:

$$\phi_1(x,\zeta) \overset{|\zeta|\to\infty}{\longrightarrow} \mathrm{e}^{-\mathrm{i}\zeta x} . \tag{14.164}$$

Consequently,

$$\chi(x,\zeta) \overset{|\zeta|\to\infty}{\longrightarrow} 0 . \tag{14.165}$$

Then we can expand χ in increasing powers of ζ^{-1} as

$$\chi = \sum_{n=0}^{\infty} \frac{\chi_n(x,t)}{(2\mathrm{i}\zeta)^{n+1}} . \tag{14.166}$$

Substituting the series expansion (14.166) into (14.163) and equating equal powers of $(1/\zeta)$, we can obtain the recursion relation [11]

$$\begin{aligned} &\chi_0 = -qr , \quad \chi_1 = -qr_x , \\ &\chi_{n+1} = q\left(\frac{\chi_x}{q}\right)_x + \sum_{k=0}^{n-1} \chi_k \chi_{n-k-1} , \quad n \geq 1. \end{aligned} \tag{14.167}$$

Now from the fact that $\phi_1 \to \mathrm{e}^{-\mathrm{i}\zeta x}$ as $x \to -\infty$, equation (H.4a), we can conclude that $\chi \to 0$ as $x \to -\infty$. Then from (14.157), taking log on both sides, we can easily check that

$$\begin{aligned} \log a(\zeta) &= \Gamma(x \to +\infty) \\ &= \int_{-\infty}^{\infty} \chi(x,t)\mathrm{d}x \\ &= \sum_{n=0}^{\infty} \int_{-\infty}^{\infty} \frac{\chi_n(x,t)}{(2\mathrm{i}\zeta)^{n+1}} \mathrm{d}x \\ &= \sum_{n=0}^{\infty} \frac{C_n}{(2\mathrm{i}\zeta)^{n+1}} , \end{aligned} \tag{14.168}$$

where

$$C_n = \int_{-\infty}^{\infty} \chi_n(x,t)\mathrm{d}x . \tag{14.169}$$

The χ_k's are explicitly given by the recursion relation (14.167).

From the time evolution equations for the eigenfunctions, one can easily check that $a(\zeta,t)$ is independent of time for all ζ ($\mathrm{Im}\zeta \geq 0$), and so $\log a(\zeta,t)$ is independent of time as well (see also Sect. 14.5.2). As a result, all the C_n's must be independent of time, for all $n \geq 0$. Consequently, there exists an infinite number of global constants of motion, C_n, $n = 0, 1, 2, \ldots$ The first few of them are

$$C_0 = \int_{-\infty}^{\infty} (-qr)\mathrm{d}x \,, \tag{14.170a}$$

$$C_1 = \int_{-\infty}^{\infty} (-qr_x)\,\mathrm{d}x \,, \tag{14.170b}$$

$$C_2 = \int_{-\infty}^{\infty} \left[-qr_{xx} + q^2 r^2\right] \mathrm{d}x \,, \tag{14.170c}$$

$$C_3 = \int_{-\infty}^{\infty} \left[-qr_{xxx} + 4qrr_x + rqq_x\right] \mathrm{d}x \,. \tag{14.170d}$$

B. Conservation Laws

As we have done earlier, let us make the substitution for ϕ_1 from (14.159) both in the linear eigenvalue problem (14.33) and the time evolution equation (14.35) for $v = (\phi_1, \phi_2)^T$ and eliminate ϕ_2 as before. Then (14.35) can be written as

$$\phi_{1t} = A\phi_1 + \frac{B}{q}\left(\phi_{1x} + \mathrm{i}\zeta\phi_1\right) \,. \tag{14.171}$$

Substituting now the expression (14.159), one obtains

$$\Gamma_t = A + \frac{B}{q}\Gamma_x \,. \tag{14.172}$$

Taking an x-derivative on both sides, and using the definition $\chi = \Gamma_x$ ((14.161)), we get

$$\partial_t \chi = \partial_x \left(A + \frac{B}{q}\chi\right) \,. \tag{14.173}$$

Then making use of the asymptotic expansion (14.166) for χ on both sides, we can write the above equation as

$$\partial_t \left[\sum_{n=0}^{\infty} \frac{\chi_n}{(2\mathrm{i}\zeta)^{n+1}}\right] = \partial_x \left[A + \frac{B}{q}\sum_{n=0}^{\infty} \frac{\chi_n}{(2\mathrm{i}\zeta)^{n+1}}\right] \,. \tag{14.174}$$

As before, one can equate equal powers of ζ^{-1} and solving recursively one can obtain infinite conservation laws. One may note that these conservation laws will be different sets for different prescriptions of A and B corresponding to different nonlinear evolution equations.

Example 1: NLS Equation

Considering the NLS equation

$$\mathrm{i}q_t + q_{xx} + 2|q|^2 q = 0 \,, \tag{14.175}$$

we have

$$A = -2\mathrm{i}\zeta^2 + \mathrm{i}|q|^2 \,, \quad B = 2\zeta q + \mathrm{i}q_x \,. \tag{14.176}$$

Then (14.174) becomes

$$\partial_t \left[\sum_{n=0}^{\infty} \frac{\chi_n}{(2\mathrm{i}\zeta)^{n+1}} \right] + \mathrm{i}\partial_x \left[2\zeta^2 - |q|^2 + \left(2\mathrm{i}\zeta - \frac{q_x}{q} \right) \sum_{n=0}^{\infty} \frac{\chi_n}{(2\,\mathrm{i}\zeta)^{n+1}} \right] = 0 \,. \tag{14.177}$$

Comparing the coefficients of ζ^{-n} for $n \geq 0$ (for $n < 0$, they are trivial) one can obtain the required infinite number of conservation laws:

$$n = 0 \;:\quad \frac{\partial}{\partial t}|q|^2 + \frac{\partial}{\partial x}\mathrm{i}\,(q q_x^* - q^* q_x) = 0 \tag{14.178}$$

$$n = 1 \;:\quad \frac{\partial}{\partial t}\,(-q q_x^*) + \frac{\partial}{\partial x}\mathrm{i}\left(|q_x|^2 - q q_{xx}^* - |q|^4 \right) = 0 \tag{14.179}$$

$$n = 2 \;:\quad \frac{\partial}{\partial t}\left(q q_{xx}^* + |q|^4 \right) + \frac{\partial}{\partial x}\mathrm{i}\left[q q_{xxx}^* + 4|q|^2 q q_x^* - q_x q_{xx}^* \right] = 0 \tag{14.180}$$

and so on. One can check that the constants of motion given in (14.170) for $r = -q^*$ follow from the above conservation laws straightforwardly.

Example 2: sine-Gordon Equation

For the sine-Gordon equation

$$u_{xt} = \sin u \,, \tag{14.181}$$

we have from equation (14.45) that

$$A = -\frac{\mathrm{i}}{4\zeta}\cos u \,, \quad B = -\frac{\mathrm{i}}{4\zeta}\sin u \tag{14.182}$$

and so (14.174) can be written as

$$\partial_t \left[\sum_{n=0}^{\infty} \frac{\chi_n}{(2\mathrm{i}\zeta)^{n+1}} \right] + \mathrm{i}\partial_x \left[-\frac{\mathrm{i}}{4\zeta}\cos u + \frac{B}{u} \sum_{n=0}^{\infty} \frac{\chi_n}{(2\mathrm{i}\zeta)^{n+1}} \right] = 0 \,. \tag{14.183}$$

Consequently, the first few of the conservation laws follow:

$$(1 - \cos u)_t + \left(-\frac{1}{2}\phi_t^2 \right)_x = 0 \,. \tag{14.184}$$

$$\left(-\frac{1}{2}u_x^2 \right)_t + (1 - \cos u)_x = 0 \,. \tag{14.185}$$

$$\left(\frac{1}{4}u_x^4 - u_{xx}^2\right)_t + \left(u_x^2 \cos u\right)_x = 0 . \tag{14.186}$$

$$\left(\frac{1}{6}u_x^6 - \frac{2}{3}u_x^2 u_{xx}^2 + \frac{8}{5}u_x^3 u_{xxx} + \frac{4}{3}u_{xxx}^2\right)_t + \left[\left(\frac{1}{9}u_x^4 - \frac{4}{3}u_{xx}^2\right)\cos u\right]_x = 0. \tag{14.187}$$

14.8 Hamiltonian Structure and Integrability

We have seen in Chap. 13 that the KdV equation can be considered as a Hamiltonian dynamical system under appropriate Poisson bracket relations (see Sect. 13.4). It was also pointed out that the scattering data $S(t)$ associated with the Schrödinger spectral operator $L(t)$ provides the set of canonical coordinates of action and angle type to show that the KdV equation is an infinite dimensional completely integrable nonlinear dynamical system in the Liouville sense. In this section, we wish to point out briefly that a similar conclusion can be made for all the soliton possessing equations of AKNS type solvable by the IST method in the sense that all these systems are also Hamiltonian systems and they are again completely integrable infinite dimensional dynamical systems.

14.8.1 Hamiltonian Structure

Following our discussion on the dynamics of continuous systems in Sect. 13.4.1, we can define suitable canonical Poisson brackets and Hamiltonians for the soliton equations solvable by the AKNS formalism also. We mention below some of these cases.

A. The sine-Gordon Equation

Defining the Hamiltonian

$$H = \frac{1}{2}\int_{-\infty}^{\infty} \left[\pi^2 + u_x^2 + 2m^2(1 - \cos u)\right] \mathrm{d}x \tag{14.188}$$

along with the Poisson bracket

$$\{u(x), \pi(y)\}_{\mathrm{PB}} = \delta(x - y) , \tag{14.189}$$

the sG equation in the laboratory coordinates can be written equivalently in the Hamiltonian from

$$u_t = \pi , \tag{14.190a}$$

$$\pi_t = \left(u_{xx} - m^2 \sin u\right) . \tag{14.190b}$$

B. The Nonlinear Schrödinger Equation

Writing the Hamiltonian defined by

$$H = \int_{-\infty}^{\infty} \left[-q_x^* q_x + (qq^*)^2 \right] , \tag{14.191}$$

and using the Poisson bracket

$$\{q(x), q^*(y)\}_{\mathrm{PB}} = \mathrm{i}\delta(x - y) , \tag{14.192}$$

one can deduce the nonlinear Schrödinger equation (14.3):

$$\mathrm{i}q_t + q_{xx} + 2|q|^2 q = 0 . \tag{14.193}$$

C. The Heisenberg Ferromagnetic Spin Equation

For the spin Hamiltonian

$$H = \frac{1}{2} \int_{-\infty}^{\infty} \left(\frac{\partial \boldsymbol{S}}{\partial x} \right)^2 \mathrm{d}x, \quad \boldsymbol{S} = (S_1, S_2, S_3), \quad \boldsymbol{S}^2 = 1, \tag{14.194}$$

one can define the Poisson bracket relations for the spin components as

$$\{S_\alpha(x), S_\beta(y)\}_{\mathrm{PB}} = \epsilon_{\alpha\beta\gamma} S_\gamma(x) \delta(x - y), \quad \alpha,\ \beta,\ \gamma = 1, 2, 3 \tag{14.195}$$

where $\epsilon_{\alpha\beta\gamma}$ is the Levi–Civita tensor ($\epsilon_{123} = \epsilon_{231} = \epsilon_{312} = +1$, $\epsilon_{213} = -1$, etc. and $\epsilon_{112} = 0$, etc.). Then the equation of motion for the Heisenberg spin system, equation (14.4), follows,

$$\boldsymbol{S}_t = \boldsymbol{S} \times \boldsymbol{S}_{xx} . \tag{14.196}$$

Similar Hamiltonian formulation can be given for any other soliton equation solvable by the AKNS formalism. As a result, following the analysis for the KdV equation in Sect. 13.4.3, one can identify the scattering data $S(t)$ associated with the AKNS spectral problem to provide suitable infinite sets of canonical coordinates of the action and angle type so that the Hamiltonian can be written purely in terms of action (momenta) variables alone. Consequently, the resultant Hamilton's equation of motion stand trivially solved and thereby proving the complete integrability of the underlying nonlinear evolution equation. In the following we will briefly demonstrate this for the case of the NLS equation.

14.8.2 Complete Integrability of the NLS Equation

Making use of the results in appendix H, Sect. H.3, and specializing to the case of the NLS equation, that is $r = -q^*$, we know that the scattering coefficient $a(\zeta)$ has the following properties:

(i) $a(\zeta)$ is analytic in the upper half plane, $\mathrm{Im}\zeta > 0$

(ii) $a(\zeta) \to 1$ as $|\zeta| \to \infty$, $\mathrm{Im}\zeta > 0$, and

(iii) $a(\zeta)$ has finite number of (simple) zeros at $\zeta = \zeta_m$, $m = 1, 2, \cdots, N$, $\mathrm{Im}\zeta > 0$.

We also assume that for real ξ, $\xi^n \log a(\xi) \to 0$ as $|\xi| \to \infty$ for all $n \geq 0$. Following the discussion in Ref. [11], and keeping the above properties of $a(\zeta)$ in mind, we can now define a function

$$\nu(\zeta) = a(\zeta) \prod_{m=1}^{N} \frac{(\zeta - \zeta_m^*)}{(\zeta - \zeta_m)} . \tag{14.197}$$

Then it is easy to see that $\nu(\zeta)$ has the same properties as that of $a(\zeta)$, except that it has no zeros for $\mathrm{Im}\zeta \geq 0$. Similarly $\nu^*(\zeta)$ is analytic in the lower half ζ-plane and

$$\nu^*(\zeta) = a^*(\zeta) \prod_{m=1}^{N} \frac{(\zeta - \zeta_m)}{(\zeta - \zeta_m^*)} \tag{14.198}$$

is analytic with no zeros for $\mathrm{Im}\zeta \leq 0$, where $\nu^*(\zeta) \to 1$ as $|\zeta| \to \infty$.

With the above properties, applying Cauchy's integral formula, for $\mathrm{Im}\zeta > 0$, we have

$$\log \nu(\zeta) = \frac{1}{2\pi \mathrm{i}} \int_{-\infty}^{\infty} \frac{\log \nu(\xi)}{\xi - \zeta} \mathrm{d}\xi \tag{14.199a}$$

and

$$0 = \frac{1}{2\pi \mathrm{i}} \int_{-\infty}^{\infty} \frac{\log \nu^*(\xi)}{\xi - \zeta} \mathrm{d}\xi . \tag{14.199b}$$

Adding the above two equations and making use of (14.197), we can write

$$\log a(\zeta) = \sum_{i=1}^{N} \log \left[\frac{\zeta - \zeta_m}{\zeta - \zeta_m^*} \right] + \frac{1}{2\pi \mathrm{i}} \int_{-\infty}^{\infty} \frac{\log aa^*(\xi)}{\xi - \zeta} \mathrm{d}\xi . \tag{14.200}$$

Now expressing $\log a(\zeta)$ as a power series in ζ^{-1} for large ζ ($\mathrm{Im}\zeta > 0$), we can write

$$\log a(\zeta) = \sum_{n=1}^{\infty} \frac{\widehat{C}_n}{\zeta^n} . \tag{14.201}$$

Comparing (14.201) with (14.200) for large ζ, the coefficients of expansion C_n can be expressed as

$$\widehat{C}_n = -\frac{1}{\mathrm{i}\pi} \int_{-\infty}^{\infty} \log |a(\zeta)| \zeta^{n-1} \mathrm{d}\zeta - \sum_{l=1}^{N} \frac{1}{n} (\zeta_l^n - \zeta_l^{*n}) . \tag{14.202}$$

Alternatively as indicated in Sect. 14.8, equations (14.170), by expressing the AKNS eigenvalue problem in the form of a Riccati equation, $\log a(\zeta)$ can be expressed as in (14.201), where now the $\widehat{C}_n$'s are

$$\widehat{C}_1 = \frac{\mathrm{i}}{2}\int_{-\infty}^{\infty} q^* q \mathrm{d}x \ , \tag{14.203a}$$

$$\widehat{C}_2 = \frac{1}{8}\int_{-\infty}^{\infty} \left(q_x^* q - q^* q_x\right) \mathrm{d}x \ , \tag{14.203b}$$

$$\widehat{C}_3 = -\frac{\mathrm{i}}{8}\int_{-\infty}^{\infty} \left(-|q_x|^2 + |q|^4\right) \mathrm{d}x \ , \tag{14.203c}$$

and so on. It may be noted that $\widehat{C}_1$, $\widehat{C}_2$ and $\widehat{C}_3$ are related respectively to important physical observables, namely number of particles, momentum and energy. Comparing (14.202) with (14.203), one can express the physical quantities in terms of the scattering data.

Specifically, the Hamiltonian $H(= 8\mathrm{i}\widehat{C})$ can be expressed as

$$\begin{aligned} H &= \int_{-\infty}^{\infty} \left[-|q_x|^2 + |q|^4\right] \mathrm{d}x \\ &\equiv -\frac{8}{\pi}\int_{-\infty}^{\infty} \zeta^2 \log|a(\zeta)|\mathrm{d}\zeta - \frac{8\mathrm{i}}{3}\sum_l \left(\zeta_l^3 - \zeta_l^{*3}\right) \ . \end{aligned} \tag{14.204}$$

Equation (14.204) provides the desired expression for the Hamiltonian in terms of scattering data.

Now making the canonical transformations to the new set of variables involving the scattering data (see appendix H)

$$P(\zeta) = -\frac{2}{\pi}\log|a(\zeta)|, \quad Q(\zeta) = \arg b(\zeta), \quad -\infty < \zeta < \infty \ , \tag{14.205a}$$

$$\begin{aligned} \xi_l &= 4 \ \mathrm{Re} \ \zeta_l, \quad \eta_l = -\log|C_l|, \\ \alpha_l &= 4 \ \mathrm{Im} \ \zeta_l, \quad \beta_l = \arg C_l, \quad l = 1, 2, ..., N, \end{aligned} \tag{14.205b}$$

one can show that the following canonical Poisson brackets are satisfied [13]:

$$[Q(\zeta), P(\zeta)]_{PB} = \delta\left(\zeta - \zeta'\right), \ [\eta_l, \xi_{l'}]_{PB} = \delta_{ll'}, \ [\beta_l, \alpha_{l'}]_{PB} = \delta_{ll'} . \tag{14.206}$$

Consequently, the Hamiltonian can be written as

$$\begin{aligned} H &= \int_{-\infty}^{\infty} \left[-|q_x|^2 + |q|^4\right] \mathrm{d}x \\ &= \int_{-\infty}^{\infty} 4\zeta^2 P(\zeta)\mathrm{d}\zeta + \sum_l \left(\frac{1}{4}\xi_l^2\alpha_l - \frac{1}{12}\alpha_l^3\right) \end{aligned} \tag{14.207}$$

in terms of the 'momenta' variables alone.

Consequently, the new variables $(P(\zeta), Q(\zeta))$, (ξ_l, η_l) and (α_l, β_l) constitute three sets of action and angle variables such that the equations of motion become trivial:

$$\frac{\mathrm{d}P(\zeta)}{\mathrm{d}t} = 0 \ , \quad \frac{\mathrm{d}\xi_l}{\mathrm{d}t} = 0 \ , \quad \frac{\mathrm{d}\alpha_l}{\mathrm{d}t} = 0 \ , \tag{14.208a}$$

$$\frac{\mathrm{d}Q(\zeta)}{\mathrm{d}t} = 4\zeta^2 \ , \quad \frac{\mathrm{d}\eta_l}{\mathrm{d}t} = \xi_l\alpha_l \ , \quad \frac{\mathrm{d}\beta_l}{\mathrm{d}t} = -\frac{1}{4}\alpha_l^2 \ . \tag{14.208b}$$

As a consequence, (14.208) can be trivially integrated and so the nonlinear Schrödinger equation turns out to be also a completely integrable infinite dimensional nonlinear dynamical system similar to the KdV equation as discussed in Chap. 13.

One can make a similar analysis for all other soliton systems solvable by the AKNS eigenvalue problem as well as for the Heisenberg ferromagnetic spin system and the Toda lattice. For further details see for example Refs. [1,11,13].

Exercises:

26. For two functionals A and B associated with the sG field, obtain the Poisson bracket $\{A,B\}_{\text{PB}}$ and hence deduce (14.190).
27. Carry out the same as above for the NLS field and deduce the Hamilton's equation of motion.
28. Write down the Poisson brackets form for two functionals of classical spin vectors A and B and hence deduce the Heisenberg spin equation (14.196).
29. Obtain the expression (14.204) for the Hamiltonian of the NLS equation in terms of the scattering coefficients.
30. Deduce the Poisson bracket relation (14.204), using (14.192) and the results in appendix H. (For details, see for example Refs. [11,13]).
31. Deduce the form of (14.205) for the Hamiltonian of the NLS system, using the relations given in (14.204).

14.9 Conclusions

In this chapter, we have seen the remarkable fact that a large class of nonlinear dispersive systems of differing complexity are completely integrable soliton possessing nonlinear dynamical systems, provided they can be included in the AKNS formalism. Evidently, the (2×2) matrix AKNS formalism itself can be extended in several directions, for example to $(n \times n)$ matrix eigenvalue problems, various discretized versions, periodic problems, higher dimensional problems and so on, encompassing many other integrable nonlinear evolution equations. However, these topics are outside the scope of this book and the interested reader may refer to references such as [1,11,14,15]. But the important fact is that all these systems may be considered as completely integrable systems and they are characterized by Lax pairs, Bäcklund transformations, Hirota bilinearizations, conservation laws, Hamiltonian structures, infinite symmetries, Painlevé property and so on.

In spite of the above developments, there remains many challenging problems: Not many physically interesting nonlinear evolution equations are known to be integrable in higher spatial dimensions. For example, the physically interesting two-spatial dimensional versions of sG, NLS and Heisenberg spin equations whose form read, respectively, as

$$u_{tt} - u_{xx} - u_{yy} + m^2 \sin u = 0 , \tag{14.209}$$

$$\mathrm{i}q_t + q_{xx} + q_{yy} + \frac{1}{2}|q|^2 q = 0 , \tag{14.210}$$

$$\boldsymbol{S}_t = \boldsymbol{S} \times (\boldsymbol{S}_{xx} \times \boldsymbol{S}_{yy}) , \quad \boldsymbol{S} = (S_1, S_2, S_3) , \quad \boldsymbol{S}^2 = 1 . \tag{14.211}$$

None of these important systems are known to be integrable; but still they should possess many interesting dynamical phenomena. Nothing much is known about them. On the other hand, certain modified equations [1] of the above systems are found to be integrable and possess exponentially localized solutions and interesting collision properties, including soliton type interactions. For example, the well known (2+1)-dimensional integrable versions [1] of the NLS and Heisenberg spin equations are, respectively, the following:

Davey–Stewartson equation:

$$\begin{aligned} \mathrm{i}q_t + q_{xx} + q_{yy} + 2|q|^2 q + qu = 0 , \\ u_{xx} - u_{yy} = 4\left(|q|^2\right)_{xx} . \end{aligned} \tag{14.212}$$

Ishimori equation:

$$\begin{aligned} \boldsymbol{S}_t + \boldsymbol{S} \times (\boldsymbol{S}_{xx} + \boldsymbol{S}_{yy}) + u_x \boldsymbol{S}_y + u_y \boldsymbol{S}_x = 0 , \\ u_{xx} - u_{yy} = -2\boldsymbol{S} \cdot (\boldsymbol{S}_x \times \boldsymbol{S}_y) , \end{aligned} \tag{14.213}$$

where $\boldsymbol{S}^2 = 1$, $\boldsymbol{S} = (S_1, S_2, S_3)$. The above systems are known to be integrable and possess many properties discussed in this chapter. They also possess exponentially localized solutions called 'dromions', whose interaction property can be more general than solitons. (For details see for example, Refs. [1,15]). It is a challenging question as to how to understand the properties of systems such as (14.209)–(14.211). Naturally, extension to three spatial dimensions is even more harder and not much progress has been made in this direction.

From another point of view multicomponent nonlinear evolution equations may exhibit shape changing collision properties of solitons in contrast to the elastic (shape preserving) collision of solitons encountered in this and earlier chapters. For example, the coupled NLS equations (Manakov system), describing two mode (birefringence) propagation in optical fibers [3],

$$\mathrm{i}q_{1t} + q_{1xx} + 2\left(|q_1|^2 + |q_2|^2\right) q_1 = 0 , \tag{14.214a}$$

$$\mathrm{i}q_{2t} + q_{2xx} + 2\left(|q_1|^2 + |q_2|^2\right) q_2 = 0 \tag{14.214b}$$

admits such shape changing collisions [16]. In Figs. 14.5 and 14.6, we have given the plot of one- and two-soliton solution of the Manakov model for a typical choice of soliton parameters which show the shape changing property in the two modes of (14.214) (see also Chap. 16 for further discussions). This type of behaviour again indicates further possible generalization of the standard soliton property in integrable systems. On may look for many further important progress in integrable systems in future, the possibility of which makes the field very exciting.

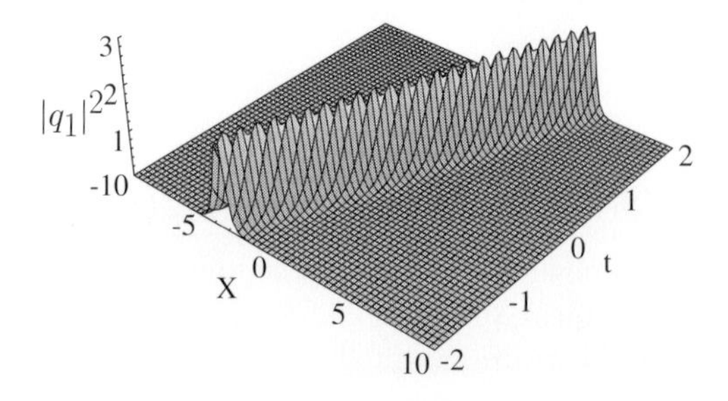

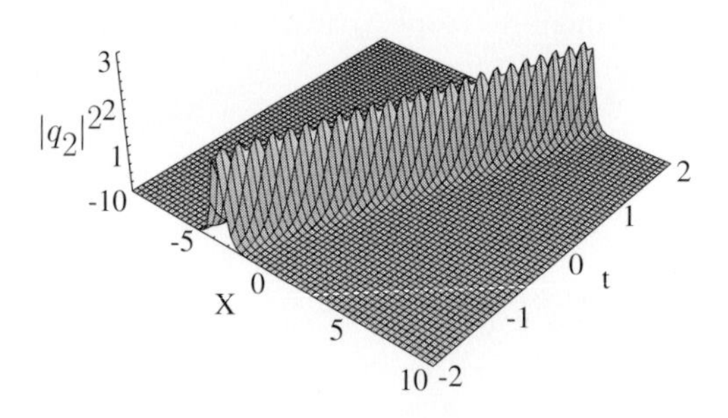

Fig. 14.5. One-soliton solution of the Manakov system (14.214) for the two components q_1 and q_2. (See Chap. 16 also)

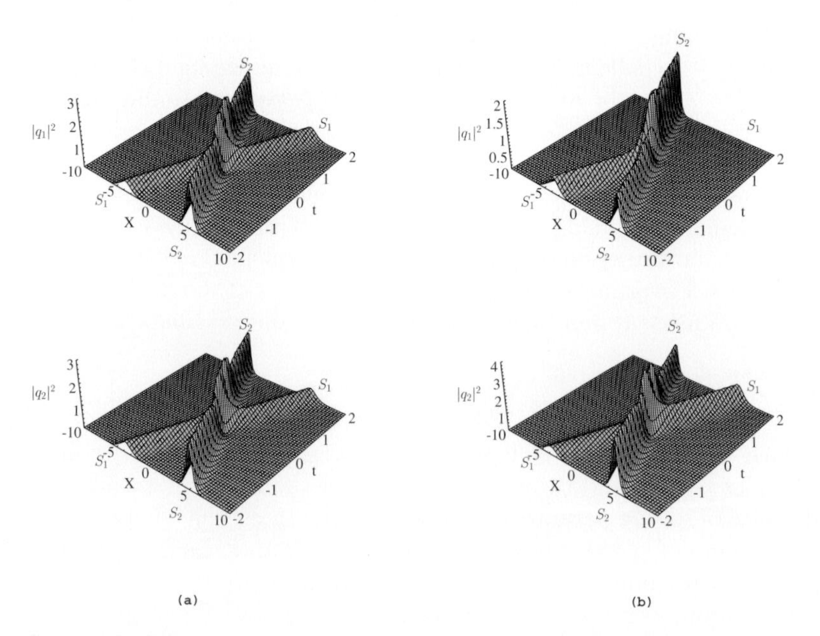

Fig. 14.6. Soliton interaction in the Manakov system (14.214). (**a**) Elastic collision for specific parametric choice. (**b**) Shape changing collision for general parameter choice

Problems

1. Obtain the NLS equation (14.9) for light wave propagation in an optical fiber using the substitution (14.8) in (14.6) under suitable approximations. (See Ref. [4] for details).
2. Analyse the equivalence between the spin evolution equation (14.24) and the NLS equation (14.28) by mapping the spin system on a moving space curve, defined by the Serret–Frenet equations in elementary differential geometry [6, 10].
3. Obtain the basic elliptic function solution of the Toda lattice defined by equations (14.30) and (14.31) using addition theorem of elliptic functions (for details see Ref. [7]), and also appendix I).
4. Consider A, B and C in (14.39) as fifth degree polynomials in ζ and deduce their form. Obtain the resultant NLEEs and identify the resulting equations.
5. (a) Defining the spin matrix $S = \boldsymbol{S}\cdot\boldsymbol{\sigma}$, where $\boldsymbol{\sigma} \equiv (\sigma_x, \sigma_y, \sigma_z)$ is the Pauli spin matrix vector, show that the equation of motion of the Heisenberg ferromagnetic spin system (14.4) can be reexpressed as

$$S_t = -\frac{\mathrm{i}}{2}\left[S,\ S_{xx}\right] .$$

[Hint: Make use of the relation $(\boldsymbol{\sigma}\cdot\boldsymbol{A}) = (\boldsymbol{A}\cdot\boldsymbol{B}) + \mathrm{i}\boldsymbol{\sigma}\cdot(\boldsymbol{A}\times\boldsymbol{B})$].
(b) Show that the Lax pair for the above spin matrix evolution equation is

$$L = \mathrm{i}S\frac{\partial}{\partial x}, \qquad B = -2\mathrm{i}S\frac{\partial^2}{\partial x^2} - \mathrm{i}S_x\frac{\partial}{\partial x} .$$

6. Show that the Toda lattice equation is the consistency condition for the system of linear equations (see Ref. [1])

$$\alpha_n v_{n+1} + \alpha_{n-1}v_{n-1} + \beta_n v_n = \frac{1}{2}\left(\zeta + \zeta^{-1}\right)v_n ,$$

$$\frac{\partial v_n}{\partial t} = 2\alpha_n v_{n+1} + \left(\sum_{k=-\infty}^{n}\left(\ln\alpha_{k-1}\right)_t - \zeta^{-1}\right)v_n ,$$

where

$$\alpha_n = \frac{1}{2}\exp\left[-\left(u_n - u_{n-1}\right)\right], \qquad \beta_n = -\frac{1}{2}u_{n-1,t}$$

and ζ is a constant. (See also appendix I).
7. Obtain the explicit wave solutions admitted by the negative sign MKdV equation (14.1) and investigate whether solitary wave solution exists.
8. Find the elliptic function wave solution of the MKdV equation, by solving equation (14.49) for the constants A, $B \neq 0$, and show that in a suitable limit the solitary wave solution (14.52) follows.

9. a) Show that the sine-Gordon equation (14.54) (in one space and one time dimension) is Lorentz invariant. Hence show that in a frame of reference moving with a velocity c, the kink (antikink) solution (14.56) can be written as $u = 4\tan^{-1}\exp(\pm X)$, where X is the transformed spatial coordinate.
   b) Deduce the elliptic function wave solution of the sine-Gordon equation.
10. a) Show that the NLS equation (14.3) is Galilean invariant under the transformation $x \to X = (x - v_0 t)$, $t \to T = t$. Hence show that the envelope soliton (14.74) in a moving frame can be expressed as $q(X,T) = a\exp\left(\mathrm{i}a^2 T\right)\ \mathrm{sech}aX$.
    b) Obtain the wave solution of the NLS equation in terms of Jacobian elliptic functions.
11. The sG equation (14.54) admits 'plasma oscillations' of the form

$$u(x,t) = 4\tan^{-1}\left[A\mathrm{cn}\left(\beta x; k_x\right)\ \mathrm{cn}\left(\omega t; k_t\right)\right]$$

where

$$k_x^2 = \frac{A^2\left[\beta^2\left(1+A^2\right)+1\right]}{\beta^2\left(1+A^2\right)^2}, \quad k_t^2 = \frac{A^2\left[\omega^2\left(1+A^2\right)-1\right]}{\omega^2\left(1+A^2\right)^2},$$

$$\omega^2 - \beta^2 = \frac{\left(1-A^2\right)}{\left(1+A^2\right)},$$

for suitable choice of parameters in (14.63), see Ref. [5].
12. Show that for yet another choice of parameters in (14.62), the sG equation (14.54) admits the kink-antikink oscillations given by

$$u(x,t) = 4\tan^{-1}\left[A\mathrm{dn}\left(\beta x; k_x\right)\frac{\mathrm{sn}\left(\omega t; k_t\right)}{\mathrm{cn}\left(\omega t; k_x\right)}\right],$$

where

$$k_x^2 = 1 - \left[\frac{\omega^2\left(A^2-1\right)-1}{\beta^2\left(A^2-1\right)}\right], \quad k_t^2 = 1 - \left[\frac{A^2\left\{\omega^2\left(A^2-1\right)-1\right\}}{\omega^2\left(A^2-1\right)}\right]$$

with $\beta = \omega A$ (again see Ref. [5]).
13. Investigate what kind of solution results for the Heisenberg spin system, if the parameter Ω is chosen as zero in equation (14.81).
14. Write down the Hamiltonian for the field system (14.54). Show that the values of the total energy of the sG system (14.54) corresponding to kink, antikink and breather solutions are finite.
15. Obtain the value of the total energy $E = \frac{1}{2}\int_{-\infty}^{\infty}\left[-|q_x|^2 + |q|^4\right]\mathrm{d}x$ of the NLS system (14.3) for the envelope soliton and show that it is finite.
16. For the MKdV equation (14.1), instead of the transformation (14.87) try $u(x,t) = \mathrm{i}\log\left(f^*/f\right)$. Show that one can still bilinearize the MKdV equation and obtain its soliton solutions.

17. For the sG equation (14.2) try the substitution

$$u = 2\mathrm{i} \log \left[(F - \mathrm{i}G)/(F + \mathrm{i}G) \right] = 4 \tan^{-1}(F/G)$$

to bilinearize it. Obtain the kink soliton solution for this case.
18. Obtain the three-soliton solution of the MKdV equation using the Hirota method and carry out an asymptotic analysis to verify the elastic nature of collisions.
19. Deduce the three envelope soliton solution of the NLS equation through the Hirota method and verify that asymptotically this solution separates into three individual envelope solitons.
20. Carry out a similar procedure as in problems 18 or 19 for the kink solution of the sine-Gordon equation.
21. Generalize the calculations in problems 18–20 to the case of N-solitons (see Refs. [1,11] for details).
22. Bilinearize the Hirota equation (14.44) through the transformation $u(x,t) = G/F$ and obtain its soliton solutions.
23. Carry out the inverse scattering analysis for the MKdV equation to obtain its soliton solution.
24. Obtain the soliton (solitary wave) solution of the so-called Sasa–Satsuma equation [17]

$$\mathrm{i}q_t + q_{xx} + 2|q|^2 q + \mathrm{i}\left(q_{xxx} + 6|q|^2 q_x + 3|q|^2 q\right) = 0 \, .$$

25. Generalize the form of F in (14.134) to the two-bound state case and obtain the two-soliton solution for the NLS equation by solving the associated Gelfand–Levitan–Marchenko equation.
26. Obtain the kink-antikink bound state solution for the sG equation using the IST method.
27. Develop the IST formalism for the Heisenberg spin system, using the Lax pair given in problem 5. Obtain the expression for the spin soliton in this formalism (for details, see for example [18]).
28. Deduce the autoBäcklund transformation for the NLS equation (see for example, Ref. [11]).
29. Work out the autoBäcklund transformation for the MKdV equation [11].
30. Deduce the conservation laws associated with the KdV equation using the Schrödinger spectral problem by converting the latter into a Riccati equation.
31. Using (14.174) and (14.39), find the first few conservation laws and constants of motion for the MKdV equation.
32. Identify the Poisson bracket relation for the MKdV field and set up the Hamiltonian formulation for the equation of motion.
33. What is the relevant Poisson bracket relation for the Toda chain? Identify the Hamiltonian formulation of the system.
34. Work out the canonical transformations for the sG system and identify the action and angle variables. Prove the complete integrability of the system (see for example, Ref. [13]).

35. Show that the Sasa–Satsuma equation given in problem 24 can be recast in the form

$$u_t + \epsilon \left(u_{xxx} + 6|u|^2 u_x + 3u|u|_x^2 \right) = 0$$

under a combined Galilei and gauge transformation. Then show that the Lax pair for this equation can be written as

$$M = \begin{pmatrix} -\mathrm{i}\lambda & 0 & u \\ 0 & -\mathrm{i}\lambda & u^* \\ -u^* & -u & 0 \end{pmatrix}$$

and

$$\begin{aligned} N = &-4\mathrm{i}\epsilon\lambda^3 \begin{pmatrix} 1 & 0 & 0 \\ 0 & 1 & 0 \\ 0 & 0 & -1 \end{pmatrix} + 4\epsilon\left(\lambda^2 - |u|^2\right) \begin{pmatrix} 0 & 0 & u \\ 0 & 0 & u^* \\ -u^* & -u & 0 \end{pmatrix} \\ &+2\mathrm{i}\epsilon\lambda \begin{pmatrix} |u|^2 & u^2 & u_x \\ (u^*)^2 & |u|^2 & u_x^* \\ u_x^* & u_x & -2|u|^2 \end{pmatrix} - \epsilon \begin{pmatrix} 0 & 0 & u_{xx} \\ 0 & 0 & u_{xx}^* \\ -u_{xx}^* & -u_{xx} & 0 \end{pmatrix} \\ &+\epsilon\left(uu_x^* - u_x u^*\right) \begin{pmatrix} 1 & 0 & 0 \\ 0 & -1 & 0 \\ 0 & 0 & 0 \end{pmatrix} . \end{aligned}$$

36. Bilinearize the Manakov equation (14.214) and obtain the one-soliton solution of the system (see Ref. [16] and also Chap. 16).
37. Show that with the (3×3) extension of the matrices in the AKNS formalism (see equations (14.34) and (14.36)) such that

$$M = \begin{pmatrix} -\mathrm{i}\zeta & q_1 & q_2 \\ -q_1^* & \mathrm{i}\zeta & 0 \\ -q_2^* & 0 & \mathrm{i}\zeta \end{pmatrix}$$

and

$$N = \begin{pmatrix} -2\mathrm{i}\zeta^2 + \mathrm{i}\left(|q_1|^2 + |q_2|^2\right) & 2\zeta q_1 + \mathrm{i}q_{1x} & 2\zeta q_2 + \mathrm{i}q_{2x} \\ -2\zeta q_1^* + \mathrm{i}q_{1x}^* & 2\mathrm{i}\zeta^2 - \mathrm{i}|q_1|^2 & -\mathrm{i}q_1^* q_2 \\ -2\zeta q_2^* + \mathrm{i}q_{2x}^* & -\mathrm{i}q_1 q_2^* & 2\mathrm{i}\zeta^2 - \mathrm{i}|q_2|^2 \end{pmatrix},$$

the compatibility (14.37) is equivalent to the Manakov equation (14.214).
38. Deduce the infinite conservation laws and constants of motion of the Manakov system (14.214) by generalizing the procedure outlined in Sect. 14.7.

15. Spatio-Temporal Patterns

Patterns abound in nature. If we look around us, we can see a myriad of interesting patterns ranging from uniform to very complex types. They occur in varied phenomena encompassing physics, chemistry, biology, social dynamics, economics and so on. Essentially they are distinct structures on a space-time scale which arise as a collective and cooperative phenomenon due to the underlying large number of constituent subsystems. The latter could be aggregates of particles, atoms, molecules, circuits, cells, bacteria, defects, dislocations and so on. When these aggregates can move and/or interact, they give rise to various patterns. A small select set of patterns occurring in nature and in physical and chemical systems is shown in Figs. 15.1 and 15.2, respectively. Such patterns tell us much about the dynamics, both at the macroscopic as well as at the microscopic levels of the underlying systems. Naturally, when the interactions among the constituents are nonlinear, one might expect novel and unexpected patterns.

Patterns could be simple as well as complex. They could be stationary or changing with time. They could also tend towards a goal or target asymp-

Fig. 15.1. A select band of patterns in natural systems: (**a**) zebra (**b**) panther (**c**) snow leopard (**d**) peacock (**e**) giraffe (**f**) pillar coral and (**g**) daisy coral

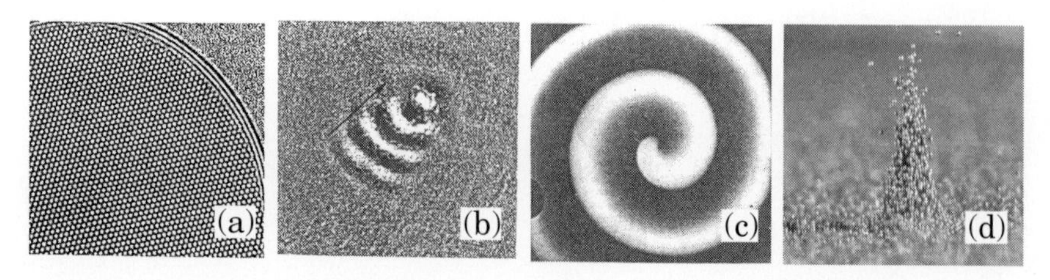

Fig. 15.2. A select band of patterns in physical and chemical systems: (**a**) Hexagonal pattern in Rayleigh–Bénard system (**b**) localized propagating wave in binary fluid convection (**c**) spiral wave in the Belousov–Zhabotinsky reaction and (**d**) localized standing wave in vertically vibrated layer of bronze balls

totically. Homogeneous or uniform patterns, though trivial, are important basic structures. One can often have travelling wave patterns, especially in dispersive systems as we have seen earlier. On the other hand, the different soliton waves which we discussed earlier are novel type of spatio-temporal patterns, retaining their identity for ever. Perturbations of them can also give rise to further interesting structures. But even more novel structures, which can mimic naturally occurring patterns, arise when one considers nonlinear diffusive (especially the so-called reaction-diffusion) systems and dissipative systems. Their analysis is much more intricate and difficult. However, many of the ideas we have already discussed in the earlier chapters on nonlinear dynamical systems will be helpful in dealing with reaction-diffusion equations as well. In this chapter, we wish to give a rudimentary outline of these systems and the various spatio-temporal patterns which they can give rise to.

When large aggregates of microstructures consisting of particles, atoms, molecules, defects, dislocations, etc. are able to move and/or interact, the evolution of the concentrations of the species can be shown to obey nonlinear *diffusion equations of reactive type*. They may be deduced from the underlying mass, energy, momentum, etc. balance equations. Before going into the details of the nonlinear regime, let us consider briefly the derivation of the linear diffusion equation.

15.1 Linear Diffusion Equation

In an assembly of particles such as atoms, molecules, cells, bacteria, chemicals, defects and so on mentioned above the individual constituents move in a complicated and random way depending upon the various interactions and environment. However, when the microscopic irregular behaviour gives rise to some macroscopic or gross regular motion of the assemblies, one can talk of a *diffusion* process. In such a situation one tries to obtain a continuum model equation for the global behaviour in terms of a particle density or concentration for the whole process based on suitable balancing laws rather than obtaining them from the motion of individual microscopic particles. There are several phenomenological ways to obtain the necessary evolution equation, for example using random walk process or Fokker–Planck equation and so on [1].

The classical or the standard approach [1] to diffusion is to use the Fick's law of diffusion. According to this law, the flux, $\boldsymbol{J}$, of a material of the kind mentioned above is proportional to the gradient of the concentration of the material, that is,

$$\boldsymbol{J} = -D\boldsymbol{\nabla}C\,, \tag{15.1}$$

where $C(\boldsymbol{r}, t)$ is the concentration of the species and D is its diffusivity. The minus sign shows that due to diffusion the matter is transported from regions of high concentration to regions of low concentration.

To start with, for simplicity, let us consider diffusion in one dimension and then generalize it to three dimensions. The relevant conservation equation states that *the rate of change of the amount of material in a region is equal to the rate of flow across the boundary plus any that is created within this boundary.* If the bounded region corresponds to $x_0 < x < x_1$ and no material is created within this region, then

$$\frac{\partial}{\partial t}\int_{x_0}^{x_1} C(x,t)\mathrm{d}x = J(x_0,t) - J(x_1,t)\,. \tag{15.2}$$

In the continuous limit, $x_1 = x_0 + \Delta x$. Taking $\Delta x \to 0$, and using (15.1), we obtain the classical diffusion equation in one dimension, namely,

$$\frac{\partial C}{\partial t} = -\frac{\partial J}{\partial x} = \frac{\partial}{\partial x}\left(D\frac{\partial C}{\partial x}\right)\,. \tag{15.3}$$

If D, the so-called diffusion coefficient, is a constant, then (15.3) takes the standard form

$$\frac{\partial C}{\partial t} = D\frac{\partial^2 C}{\partial x^2}\,. \tag{15.4}$$

Generalizing to three dimensions, the classical diffusion equation can be written in the general form (also see below, Sect. 15.3)

$$\frac{\partial C(\boldsymbol{r},t)}{\partial t} = D\nabla^2 C\,. \tag{15.5}$$

This is the linear diffusion equation. However, if D is also a function of $\boldsymbol{r}$ and t, then (15.5) should be modified to the form

$$\frac{\partial C}{\partial t} = \boldsymbol{\nabla}\cdot(D\boldsymbol{\nabla}C)\,. \tag{15.6}$$

Equations (15.4)–(15.6) all belong to the category of parabolic, linear, partial differential equations. For the general properties of such equations, see for example Ref. [2].

Exercise:

1. If we release an amount of Q of bacteria per unit length at the instant $x = 0$ at $t = 0$, that is

$$C(x, 0) = Q\delta(x) ,$$

where $\delta(x)$ is the Dirac delta function, then one can prove that

$$C(x, t) = \frac{Q}{\sqrt{4\pi Dt}} \mathrm{e}^{-x^2/(4Dt)} , \quad t > 0 .$$

Verify this. (See for example, Ref. [1]). Hint: Use Laplace transform of $C(x, t)$ with respect to t and Fourier transform of the resultant function with respect to x.

15.2 Nonlinear Diffusion and Reaction-Diffusion Equations

So far we have assumed that there is no source of particle production in a given region. On the other hand if such a source is present, which we designate as f, then from the general conservation theorem stated in the previous section we have

$$\frac{\partial}{\partial t} \int_V C(\boldsymbol{r}, t) \mathrm{d}V = - \int_S \boldsymbol{J} \cdot \mathrm{d}\boldsymbol{S} + \int_V f \mathrm{d}V . \tag{15.7}$$

Here f can in general be a function of the concentration $C(\boldsymbol{r}, t)$ as well as $\boldsymbol{r}$ and t. Applying divergence theorem to the surface integral and assuming that $C(\boldsymbol{r}, t)$ is continuous, equation (15.7) can be simplified to a volume integral,

$$\int_V \left[\frac{\partial C}{\partial t} + \boldsymbol{\nabla} \cdot \boldsymbol{J} - f(C, \boldsymbol{r}, t) \right] \mathrm{d}V = 0 . \tag{15.8}$$

Since the volume is arbitrary, the integrand must be zero and so we obtain the generalized rate equation

$$\frac{\partial C}{\partial t} + \boldsymbol{\nabla} \cdot \boldsymbol{J} = f(C, \boldsymbol{r}, t) . \tag{15.9}$$

If the process is again the classical Fick's diffusion of the form

$$\boldsymbol{J} = -D\boldsymbol{\nabla} C , \tag{15.10}$$

then (15.9) takes the form

$$\frac{\partial C}{\partial t} = \boldsymbol{\nabla} \cdot (D\boldsymbol{\nabla} C) + f(C, \boldsymbol{r}, t) . \tag{15.11}$$

In (15.11), we can even consider D to be a function of C also. Since f is also in general a nonlinear function of C, one may call equation (15.11) as the general *nonlinear diffusion equation* for one species.

Example: Fisher's equation

Suppose C in (15.11) stands for the population density of a particular species, and f represents the birth-death process. In the logistic population growth (recall the logistic map, Chap. 4), $f = aC(1 - \delta C)$, where a is the linear reproduction rate and δ is the inhibition rate. If D is a constant, then the resulting equation is the Fisher's equation

$$\frac{\partial C}{\partial t} = D\nabla^2 C + aC(1 - \delta C) . \tag{15.12}$$

Originally Fisher proposed the one-dimensional version of this equation in 1937 for the spread of an advantageous gene in a population (see Ref. [1] for details).

15.2.1 Nonlinear Reaction-Diffusion Equations

Let us consider a situation in which there are two or more number of species such as different chemicals, bacteria, populations and so on which interact and evolve. In this case the concentration density C of the single species in (15.11) can be replaced by a density vector $\boldsymbol{C} = (C_1, C_2, ..., C_n)^{\mathrm{T}}$, where the components stand for individual concentration densities. Correspondingly, the diffusion coefficient D will be replaced by a diffusion matrix $\underline{D}$ and the scalar source term f by a vector $\boldsymbol{f} = (f_1, f_2, ..., f_n)$. Examples include the Belousov–Zhabotinsky reaction discussed in Chap. 1, Sect. 4, the Brusselator model, the Gierer–Meinhardt model for biological pattern formation, the FitzHugh–Nagumo nerve conduction model and so on. Brief details of them are given below. All the reaction-diffusion equations given below are subject to suitable initial and boundary conditions.

A. The Oregonator Model

One simple model which tries to explain the various features of Belousov–Zhabotinski reaction is the Oregonator model introduced by Field, Körös and Noyes of University of Oregon, U.S.A., in 1972. In its simplest version, it retains only the concentration u_1 of the autocatalytic species $HBrO_2$ and the concentration u_2 of the transition ion catalyst in the oxidized state Ce^{3+} or Fe^{3+}. In dimensionless units, a simplified version of this model reads (Ref. [3])

$$\frac{\partial u_1}{\partial t} = D_1\nabla^2 u_1 + \eta^{-1}\left[u_1(1 - u_1) - \frac{bu_2(u_1 - a)}{(u_1 + a)}\right] , \tag{15.13a}$$

$$\frac{\partial u_2}{\partial t} = D_2\nabla^2 u_2 + u_1 - u_2 . \tag{15.13b}$$

Here η, a and b are parameters.

B. Lotka–Volterra Predator-Prey Model

Dynamics of interacting population of species including diffusional effects can be modelled in some situations by the coupled nonlinear differential equations [4]

$$\frac{\partial S_1}{\partial t} = D_1 \frac{\partial^2 S_1}{\partial x^2} + a_1 S_1 - b_1 S_1 S_2 \,, \tag{15.14a}$$

$$\frac{\partial S_2}{\partial t} = D_2 \frac{\partial^2 S_2}{\partial x^2} + a_2 S_2 - b_2 S_1 S_2 \,, \tag{15.14b}$$

where $S_1(x,t)$ and $S_2(x,t)$ are the population densities of prey and predator, while D_1 and D_2 are the diffusivities of the two populations, respectively. The parameter a_1 and a_2 are respectively the linear rates of birth and death of the individual species and b_1 and b_2 are respectively the linear decay and growth factors due to interactions.

C. Gierer–Meinhardt Model

The following set of nonlinear partial differential equations describes a possible interaction between an activator a and its rapidly diffusing antagonist h (see for example Ref. [5])

$$\frac{\partial a}{\partial t} = D_a \nabla^2 a + \rho_a \frac{a^2}{(1 + k_a a^2)\, h} - \mu_a a + \sigma_a \,, \tag{15.15a}$$

$$\frac{\partial h}{\partial t} = D_h \nabla^2 h + \rho_h a^2 - \mu_h h + \sigma_h \,, \tag{15.15b}$$

where D_a, D_b are the two diffusion coefficients, μ_a, μ_b are the removal rates and ρ_a, ρ_b are the cross-reaction coefficients. Also σ_a, σ_h are basic production terms and k_a is a saturation constant.

D. FitzHugh–Nagumo Nerve Conduction Equation

The propagation of the electrical impulses along the axonal membrane of a nerve fibre is represented by the so called Hodgkin–Huxley model. A simplified version of it is the FitzHugh–Nagumo nerve conduction equation [6,7]. It reads as

$$\frac{\partial V}{\partial t} = \frac{\partial^2 V}{\partial x^2} + V - V^3 + R + I(x,t) \,, \tag{15.16a}$$

$$\frac{\partial R}{\partial t} = c(V + a - bR) \,. \tag{15.16b}$$

Here $V(x,t)$ is the membrane potential, R is the (lumped) recovery variable, $I(x,t)$ is the externally injected current and the parameters a and b are positive constants, while c stands for the temperature factor.

E. The Brusselator Model [8]

The formation of chemical patterns has been studied theoretically by means of a number of reaction-diffusion type model systems. One of the best studied model among them is the Brusselator model, introduced by Lefever, Nicolis and Prigogine from Brussels. It is based on the following chemical reactions

$$\begin{aligned} A &\to X, \\ B + X &\to Y, \\ 2X + Y &\to 3X, \\ X &\to E, \end{aligned} \qquad (15.17)$$

where the concentrations of the species A, B, and E are maintained constant and are thus real constant parameters of the system. After appropriate scaling, the evolution of the active species X and Y can be described by the following set of equations:

$$\frac{\partial X}{\partial t} = A - (B+1)X + X^2Y + D_X\nabla^2 X = f(X,Y) , \qquad (15.18a)$$

$$\frac{\partial Y}{\partial t} = BX - X^2Y + D_Y\nabla^2 Y = g(X,Y) , \qquad (15.18b)$$

where D_X and D_Y are the diffusion coefficients.

The above types of diffusion equations are naturally called *reaction-diffusion equations*, where in general there can be more than one diffusion coefficient. Such forms were also proposed as models for the chemical basis of morphogenesis (a term representing the development of structure during the growth of organism). For more details, see for example Ref. [1]. One finds that the study of reaction-diffusion systems reveals considerable amount of information about the formation and structure of a variety of patterns in nonlinear systems. We will describe briefly the occurrence and nature of some of these patterns in the next section.

15.2.2 Dissipative Systems

We have already discussed the effect of dissipation or damping on nonlinear systems in earlier chapters (see Chap. 5). The dissipative systems have the property that any arbitrary subvolume in phase space shrinks to zero asymptotically. In other words, the phase trajectories eventually end up in low dimensional attractors. In the case of higher dimensional (extended) systems, dissipation plays an important role in bringing out more novel patterns. Particularly in systems which are driven away from equilibrium due to diffusion, dissipation gives rise to more interesting spatio-temporal patterns and they are termed as *dissipative structures*. The best example is the *Turing instability* which leads to the formation of many of the naturally occurring heterogeneous patterns (see below). We will be discussing such patterns in the following sections. Typical examples of such pattern forming dissipative systems are the Brusselator model equation (15.18), the Gierer–Meinhardt model equation (15.15), the Oregonator model equation (15.13) and so on. Another interesting system which is often studied in the literature as a model for dissipative structures is the so-called Ginzburg–Landau equations [3] for the complex amplitude $q(\boldsymbol{r}, t)$,

$$\frac{\partial q}{\partial t} = (1 + \mathrm{i}c_1)\,\nabla^2 q + (1 - \mathrm{i}c_2)\,|q|^2 q\,, \tag{15.19}$$

where c_1 and c_2 are constant parameters. Note that in the limit c_1^{-1}, $c_2^{-1} \to 0$, (15.19) reduces to ubiquitous nonlinear Schrödinger equation (in three dimensions) discussed in Chap. 14.

15.3 Spatio-Temporal Patterns in Reaction-Diffusion Systems

We have seen in the earlier sections that reaction-diffusion equations occur naturally in many physical, chemical and biological problems. Then the question arises as to what kind of distinct space-time structures or patterns they give rise to in contrast to the dispersive systems considered in the earlier chapters. In fact one finds that eventhough no rigorous analytical tools as in the case of soliton systems of the earlier chapters are available in the present case, combined local analysis and numerical investigations have lead to the identification of a number of important spatio-temporal patterns in nonlinear reaction-diffusion systems. Some of them are as follows:

(1) uniform or steady states (trivial)
(2) autowaves, including travelling wavefronts and pulses
(3) ring waves, spiral waves and scroll waves
(4) Turing patterns
    a) rolls
    b) stripes
    c) hexagons
    d) rhombus and so on
(5) localized structures
(6) spatio-temporal chaos

and so on. We will now discuss briefly the occurrence of the above types of patterns in the present and in the following sections.

Considering the dynamics of upto two species, whose concentrations are represented by $X(\boldsymbol{r}, t)$ and $Y(\boldsymbol{r}, t)$ (generalization to more number of species can in principle be carried out without much difficulty), we can write the general form of the reaction-diffusion equations as

$$\frac{\partial X}{\partial t} = D_X \nabla^2 X + F(X, Y)\,, \tag{15.20a}$$

$$\frac{\partial Y}{\partial t} = D_Y \nabla^2 Y + G(X, Y)\,, \tag{15.20b}$$

where F and G contain the nonlinear interacting terms, while D_X and D_Y are the corresponding diffusion coefficients. Specific examples are already given in the previous section. The single species equations like the Fisher's equation are then simply special cases of (15.20). Then we may identify the following types of structures or patterns in the present as well as in the following sections.

15.3.1 Homogeneous Patterns

The trivial, but important, class of patterns is the homogeneous or uniform steady state patterns (X_0, Y_0). These are obtained from the relations

$$F\left(X_0, Y_0\right) = G\left(X_0, Y_0\right) = 0\,. \tag{15.21}$$

The set of solutions to (15.21) constitutes the allowed homogeneous states. They are the states in which all the constituents of the system have attained uniform or equilibrium values, stable or not. Their stability against small perturbations can be analysed easily by linearizing equations (15.20) about the steady state solutions.

15.3.2 Autowaves: Travelling Wave Fronts, Pulses, etc.

Nonlinear reaction-diffusion systems often admit the so-called *autowaves* or autonomous waves, which propagate in an active excitable medium at the expense of energy stored in the medium, even in the *absence* of external driving forces. We have seen in the earlier chapters that dispersive systems admit different kinds of propagating waves such as sinusoidal travelling waves, wavepackets, solitary waves and solitons. However all these waves carry energy and information and do not consume any energy associated with the system. The medium may be considered as *passive* in such a case. Dispersion and nonlinearity are sufficient to sustain the waves as they belong to conservative systems. In contrast, the waves that propagate in nonlinear reaction-diffusion as well as dissipative systems do so in a self-sustained manner by inducing a local release of energy and use it to trigger the same process in adjacent regions. The medium in this case is said to be *active*. Typical examples include waves of combustion (for example, a flame propagating along the cord of a cracker), waves of phase transition, concentration waves in chemical reactions, propagation of nerve impulses, excitation waves in the cardiac tissue, epidemic waves in ecological communities and so on. In the following we consider some specific examples.

A. Travelling Wavefronts

In order to realize the nature of travelling wavefronts which can arise in reaction diffusion systems, let us first consider the Fisher's equation (15.12) (which is a nonlinear diffusion equation) in one spatial dimension in a normalized form as an example,

$$\frac{\partial u}{\partial t} = \frac{\partial^2 u}{\partial x^2} + u(1-u) \; . \tag{15.22}$$

Exercise:

2. Show that under the transformation $t \to t^* = at$, $x \to x^* = (a/D)^{\frac{1}{2}}x$, $C \to u^* = u/\delta$, (15.12) in the one-dimensional case reduces to the dimensionless form (15.22).

We look for a wave solution of the form

$$u = f(x - ct) = f(\zeta) \; , \quad \zeta = x - ct \; , \tag{15.23}$$

where c is the wave speed, which we assume to be nonnegative. Using (15.23), equation (15.22) reduces to the form

$$f'' + cf' + f(1-f) = 0 \; , \tag{15.24}$$

where prime denotes differentiation with respect to the wave variable ζ. Since (15.24) admits two stationary state or constant solutions

$$f = f_0 = 0 \;\text{ and }\; f = f_1 = 1 \; , \tag{15.25}$$

one can look for a wavefront solution satisfying the condition

$$\lim_{\zeta\to\infty} f(\zeta) = 0 \; , \qquad \lim_{\zeta\to-\infty} f(\zeta) = 1 \; . \tag{15.26}$$

Rewriting (15.24) as a system of two first-order coupled differential equations

$$f' = g \; , \tag{15.27a}$$

$$g' = -cf - f(1-f), \qquad \left({}' = \frac{\mathrm{d}}{\mathrm{d}\zeta}\right) \tag{15.27b}$$

one can draw the phase trajectories in the $(f, g = f')$ plane around the two fixed points $(0,0)$ and $(1,0)$. It can be easily shown by carrying out a linear stability analysis (refer Chap. 3) that the nature of the fixed points are as follows: The phase trajectories for a typical value of c with $c^2 > 4$ is depicted

Fixed points (F.P.)	Condition	Nature of F.P.
$(0,0)$	$c^2 \geq 4$	stable node
$(0,0)$	$c^2 < 4$	stable focus
$(1,0)$	all c values	saddle

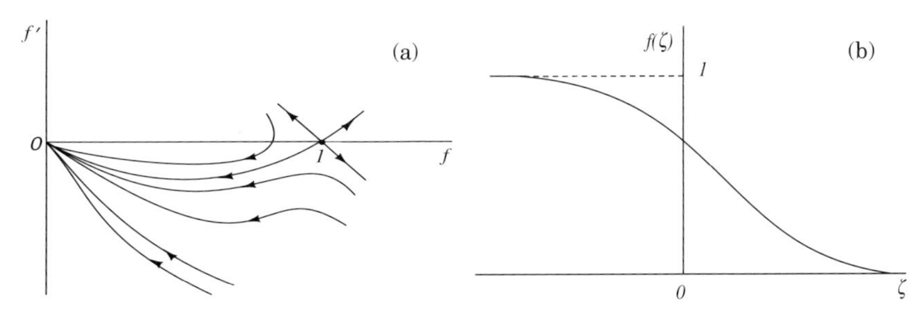

Fig. 15.3. (**a**) Phase trajectories for equation (15.27) with $c^2 > 4$ and (**b**) travelling wavefront solution of (15.24)

in Fig. 15.3a. The corresponding travelling wavefront obtained by numerically solving the equation (15.27) for the boundary conditions (15.26) is depicted in the Fig. 15.3b.

From the nature of the trajectories in Fig. 15.3a one can argue that there is a trajectory from $(1, 0)$ to $(0, 0)$ lying entirely in the fourth quadrant $f \geq 0$, $f' < 0$ in the $(f - f')$ plane for all wave speeds $c \geq c_{\text{min}} = 2$ (see Ref. [1]).

Do all initial conditions $u(x, 0)$ for the Fisher's equation always evolve into the above type of travelling wavefront and, if such a solution exists, what is its wave speed c? This is an intricate question and several analytic studies exist in the literature on this question [1]. It has been shown that the solution critically depends on the nature of the initial condition. Kolmogoroff, Petrovsky and Piscounoff (see Ref. [1]) have proved that if $u(x, 0)$ has the form

$$u(x,0) = u_0(x) \geq 0$$
$$u_0(x) = \begin{cases} 1 & x \leq x_1 \\ 0 & x \geq x_2 \, , \end{cases} \tag{15.28}$$

where $x_1 < x_2$ and $u_0(x)$ is continuous, then the solution $u(x, t)$ of the Fisher's equation (15.22) evolves to a travelling wavefront solution $f(\zeta)$ with $\zeta = x - 2t$ and with a wave speed $c = 2$.

Exercise:

3. Carry out the linear stability analysis for the fixed points of equation (15.26) and identify their nature.

How stable are these travelling fronts, particularly to small perturbations? One can carry out a lineal stability analysis in this situation, with reference to the full Fisher's equation (15.22). It has been shown that the travelling wave solutions are stable for small finite domain perturbations in a moving frame of reference and as a result typical numerical simulations of the Fisher's equation result in stable wavefront solutions with speed $c = 2$. For details see Ref. [1].

B. An explicit solution

Let us consider the ansatz

$$f(\zeta) = 1/\left(1 + a\mathrm{e}^{b\zeta}\right)^s , \tag{15.29}$$

where a, b and s are positive constants to be determined. Substituting (15.29) into (15.24), one can find a nontrivial solution which is obtained for the specific choice

$$s = 2, \quad a = \sqrt{2} - 1, \quad b = 1/\sqrt{6}\,, \quad c = 5/\sqrt{6}\,. \tag{15.30}$$

Exercises:

4. Prove that $f(\zeta) = \left[1 + \left(\sqrt{2} - 1\right)\mathrm{e}^{(\zeta/\sqrt{6})}\right]^{-2}$ is indeed a travelling wavefront solution of the Fisher's equation (15.22) with $\zeta = x - \left(5/\sqrt{6}\,\right)t$.
5. Solve the system (15.27) numerically for $c^2 > 4$ and obtain the travelling wavefront solution given in Fig. 15.3b.

A. Travelling Pulses

Earlier we have seen that travelling wavefronts can occur in scalar diffusive wave equations such as the Fisher's equation. They can also occur in reaction-diffusion equations such as the Belousov–Zhabotinsky system, equations (15.13). More interestingly, reaction-diffusion systems with more than one variable can admit even more novel and richer structures. Particularly, these structures can arise when the system such as (15.13) has a single stable steady state but which under suitable perturbation can exhibit a threshold behaviour. In this case the perturbation can undergo an appreciable growth before eventually dying down. So, for a finite time the situation will appear to be like that of the initiation of a travelling wavefront varying between two steady states. Thus, the existence of such a threshold capability is to provide a travelling pulse wave.

A typical example for the onset of such travelling wave pulses is the FitzHugh–Nagumo nerve conduction model for the action potential of a neuronal axon (see also equation (15.16)):

$$\frac{\partial u}{\partial t} = f(u) - v + D\frac{\partial^2 u}{\partial x^2} , \tag{15.31a}$$

$$\frac{\partial v}{\partial t} = bu - sv , \tag{15.31b}$$

Here $f(u) = u(a - u)(u - 1)$. The variable u is related to the membrane potential $V(x,t)$ given in equation (15.16) and v is related to the recovery variable $R(x,t)$. The quantities D, b, s and a are all positive parameters ($0 < a < 1$) associated with the axon. A typical travelling wave pulse is given in Fig. 15.4 for equation (15.31). The figure is self explanatory.

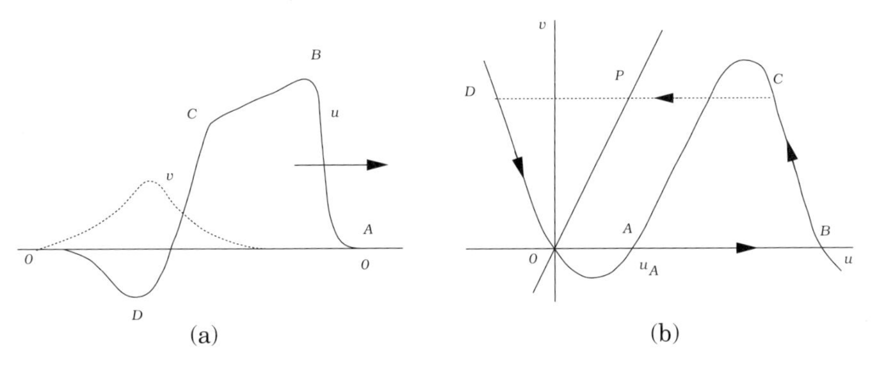

Fig. 15.4. (**a**) Typical travelling wave pulse of the system (15.31) and (**b**) phase trajectory in the (u, v) plane (adapted from Ref. [1])

B. Travelling Wavetrains

Typical reaction-diffusion systems, which exhibit limit cycles in the absence of diffusion, can exhibit travelling wave solutions. A particularly illustrative model is the so-called $(\lambda - \omega)$ system introduced by Kopell and Howard in 1973 (see Ref. [1]). It is a two species reaction-diffusion model whose mechanism is described by the set of equations for the variables u and v in the form

$$\frac{\partial u}{\partial t} = \lambda(r)u - \omega(r)v + D\frac{\partial^2 u}{\partial x^2} , \tag{15.32a}$$

$$\frac{\partial v}{\partial t} = \lambda(r)u + \omega(r)v + D\frac{\partial^2 v}{\partial x^2} , \tag{15.32b}$$

where $r^2 = u^2 + v^2$. Here $\omega(r)$ and $\lambda(r)$ are functions of r. It can be checked that if r_0 is an isolated zero of $\lambda(r)$ for some $r_0 > 0$ and $\lambda'(r_0) < 0$ and $\omega(r_0) \neq 0$, then the spatially homogeneous system with $D = 0$ in (15.32) has a limit cycle solution (see Ref. [1]). More interestingly, the system (15.32) admits travelling wave solutions also. To see this clearly, we can proceed as follows.

Let us look for wave solutions of the form

$$u = \alpha \cos(\nu t - kx) , \tag{15.33a}$$

and

$$v = \alpha \sin(\nu t - kx) . \tag{15.33b}$$

Substituting these into equation (15.32) and separating out the coefficients of $\sin(kx - \omega t)$ and $\cos(kx - \omega t)$, one can readily obtain

$$r = \omega(\alpha) , \tag{15.34a}$$

and

$$k^2 = \lambda(\omega)/D . \tag{15.34b}$$

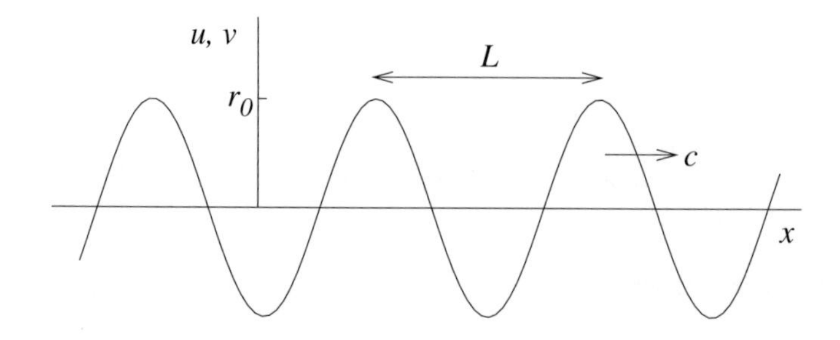

Fig. 15.5. A typical small amplitude travelling wave solution for the $\lambda - \omega$ system (15.32)

Then, we have a one parameter family of travelling wave solutions to (15.32) in the form

$$u = \alpha \cos\left(\omega t - \sqrt{\lambda/D}\, x\right) , \quad v = \alpha \sin\left(\omega t - \sqrt{\lambda/D}\, x\right) , \tag{15.35}$$

where ω and λ depend on α as given by (15.34). The wave speed is

$$v = \frac{\nu}{k} = \frac{\omega(\alpha)}{\sqrt{\lambda(\alpha)/D}} \tag{15.36}$$

and it depends on the amplitude α, which is of course a characteristic feature of nonlinear systems. A typical small amplitude travelling wave solution for the $\lambda - \omega$ system (15.32) is given in Fig. 15.5.

Exercise:

6. Verify that the $(\lambda - \omega)$ system (15.32) in the case $D = 0$ admits a limit cycle solution for suitable conditions on $\lambda(r)$ and $\omega(r)$.

15.3.3 Ring Waves, Spiral Waves and Scroll Waves

In two and three spatial dimensions, reaction diffusion systems characteristically admit patterns such as ring waves, spiral waves and scroll waves [8,9]. Such waves have been observed in nature as well as in many chemical, biological and physical systems. We will give a very brief account of them in the following.

A. Ring Waves

Ring waves are a class of autowaves which arise in certain two-dimensional reaction-diffusion systems. These waves are sometimes called concentric waves and they are for example responsible for the normal functioning of cardiac muscle [9].

Ring waves were originally observed by Zaikin and Zhabotinsky in chemical reactions and by Parker–Rhodes in fields of mushrooms [9]. In this case the solutions of the reaction-diffusion equations do not depend upon the azimuthal angle, but only on the radial coordinate. For example considering the two-dimensional version of the FitzHugh–Nagumo equation (15.31) in the form

$$\frac{\partial u}{\partial t} = D\left(\frac{\partial^2 u}{\partial x^2} + \frac{\partial^2 u}{\partial y^2}\right) + f(u) - v\,, \tag{15.37a}$$

$$\frac{\partial v}{\partial t} = bu - sv\,, \tag{15.37b}$$

where D is the diffusion constant associated with the potential variable u, while the recovery variable v does not diffuse. As before b and s are constant parameters and $f(u)$ represents the nonlinearity. One of the interesting type of waves admitted by this system is the *ring waves*. To obtain them, let us look for circularly symmetric solutions $u(x, y, t) \equiv u(r, t)$ so that the FitzHugh–Nagumo equation (15.37) becomes

$$\frac{\partial u}{\partial t} = D\left(\frac{\partial^2 u}{\partial r^2} + \frac{1}{r}\frac{\partial u}{\partial r}\right) + f(u) - v\,, \tag{15.38a}$$

$$\frac{\partial v}{\partial t} = bu - sv\,, \quad \left(r = \sqrt{x^2 + y^2}\right)\,. \tag{15.38b}$$

Then asymptotically (that is for $r \gg D/v_0$) one can look for plane wave solutions of the form $u = u(\xi)$, $\xi = r - vt$, when a point of constant amplitude on the wave front moves outward with a radial velocity $v = v_0 - D/r$.

An interesting feature of the ring waves is that when rings meet, their intersected portions get destroyed since the recovery variable v requires a time of the order $1/b$ to relax back to its resting value, thereby permitting the propagation of a new ring [9]. Thus the rings do not pass through each other but instead form cusps in the wave fronts (see Fig. 15.6).

B. Spiral Waves

Another important class of waves which occurs in reaction-diffusion systems with more than one spatial dimension is the *spiral waves* in two spatial dimensions and its extension in three dimensions, namely the *scroll waves*. In order to appreciate the possibility of spiral waves, let us consider the following qualitative picture of a single wave propagating around a circular obstacle [10]. The wave repeatedly travels along the same path at a frequency given by the wave velocity divided by the circumference of the obstacle. If the radius is gradually decreased the frequency increases until the wavefront of a new wave catches up with the tail of the previous wave. At this point the rate at which the wave travels around the obstacle cannot increase further because the new wave cannot reexcite regions that are still recovering from the previous wave.

Fig. 15.6. Ring waves (adapted from Ref. [9])

If the radius is made even smaller, the wave is forced to adopt a spiral shape that continues to rotate around a central core (Fig. 15.6). The spiral cannot enter the central core because the region is still in the so-called refracting (or recovering) state and cannot be reexcited. Typical example where such spiral waves can arise is the case of cardiac arrhythmia in the heart during irregularities in the heart beat. The phenomenon is known to be modelled by the so-called Hodgkin–Huxley model, which is a system of five coupled first-order nonlinear partial differential equations (see for example, Ref. [7]). Spiral wave solutions are generated in several reaction-diffusion equations in two-dimensions by a variety of special initial conditions. These include the Belousov–Zhabotinsky reaction system, the $\lambda - \omega$ system, the Brusselator model and so on [1]. Spiral waves in fact do not require an obstacle to be present and can be produced and maintained in media that can be considered to be entirely homogeneous. In a homogeneous medium the spiral can either rotate, or the tip of the spiral can *meander* around the central core in a motion that can be biperiodic, quasiperiodic and perhaps even chaotic.

Mathematically, spiral waves can be analysed as follows. Let us assume that a spiral is rotating around the central core. If one stands at a fixed position in the medium, it will appear locally as if a periodic wavetrain is passing the person because every time the spiral turns a wavefront will move past the person.

To represent such a rotating spiral wave one can use polar coordinates $r(=\sqrt{x^2+y^2}\ )$ and $\theta(=\tan^{-1}(y/x))$ and look for periodic waves of the forms

$$X(x,y,t) \equiv \widehat{X}(r)\cos\phi\,, \tag{15.39a}$$

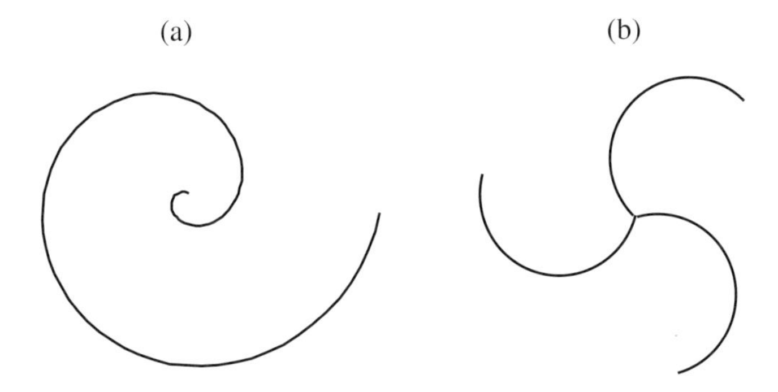

Fig. 15.7. Typical (**a**) 1-armed spiral and (**b**) 3-armed spiral

$$Y(x, y, t) \equiv \widehat{Y}(r) \sin\phi \,. \tag{15.39b}$$

for the reactants X and Y in equation (15.20), where the phase ϕ may be chosen as

$$\phi = \Omega t \pm m\theta + \psi(r) \,, \tag{15.40}$$

where Ω is the frequency, m is the number of spirals and $\psi(r)$ is a function which describes the spiral. The $\pm$ sign in front of $m\theta$ determines the sense of rotation. Figures 15.7 represent 1-armed and 3-armed spirals.

Example:

A steady state spiral is obtained for $\theta = 0$, $\phi = \psi(r)$. Then one gets Archimedean spiral (Fig. 15.7a) for $\theta = ar$, $a > 0$ and logarithmic spiral (Fig. 15.7b) for $\theta = a \ln r$ with the core at $r = 0$.

C. Scroll Waves

Scroll waves are the extension of spiral waves in three dimensions [9]. They can be realized by extending the two-dimensional spiral waves in the vertical direction. The properties of these scroll waves are analogous to spiral waves.

15.3.4 Turing Instability and Turing Patterns

As we have noted in the previous section, reaction-diffusion systems often exhibit diffusion driven instability or Turing instability. It occurs when a homogeneous steady state which is stable due to small perturbations in the absence of diffusion becomes unstable in the presence of diffusion. In the following by considering the Brusselator model equation (15.18) as an example, we derive the conditions for the diffusion driven instability and also deduce for illustrative purpose the mechanism for the pattern or mode selection.

To arrive at the conditions for Turing instability we shall investigate, for simplicity, the system (15.18) in one dimension in a domain $\mathcal{D} \in \{\boldsymbol{r} = (x, 0, 0) \mid 0 \leq |x| \leq L\}$ with zero flux boundary conditions. Note that by zero flux boundary condition we mean here that the spatial derivatives of the dependent variable $X(x,t)$ and $Y(x,t)$ vanish at $x = 0$ and at $x = L$.

A. Stability of Homogeneous Steady State in the Absence of Diffusion

To start with, first we consider equation (15.18) with $D_X = D_Y = 0$ (in the absence of diffusion). Then the homogeneous steady state solution (X^*, Y^*) of (15.18) is

$$(X^*, Y^*) = (A,\ B/A)\ . \tag{15.41}$$

Our aim is to find the linear stability of the above steady state (X^*, Y^*) in the absence of diffusion, that is, in the absence of spatial dependence, and to arrive at the conditions for which it becomes unstable when the diffusion is present. The stability of the steady states is determined by the eigenvalues of the Jacobian matrix associated with the system (15.18). The Jacobian matrix is given by

$$M = \begin{pmatrix} \frac{\partial f}{\partial X}\big|_{X^*,Y^*} & \frac{\partial f}{\partial Y}\big|_{X^*,Y^*} \\ \\ \frac{\partial g}{\partial X}\big|_{X^*,Y^*} & \frac{\partial g}{\partial Y}\big|_{X^*,Y^*} \end{pmatrix} = \begin{pmatrix} B-1 & A^2 \\ \\ -B & -A^2 \end{pmatrix}\ . \tag{15.42}$$

The stability determining eigenvalues are the roots of the characteristic equation

$$\lambda^2 + \left(A^2 - B + 1\right)\lambda + A^2 = 0 \tag{15.43}$$

of the above Jacobian matrix. From (15.42) the eigenvalues are obtained as

$$\lambda_{1,2} = \frac{1}{2}\left[-\left(A^2 - B + 1\right) \pm \sqrt{\left(A^2 - B + 1\right)^2 - 4A^2}\,\right]\ . \tag{15.44}$$

For the homogeneous state (X^*, Y^*) to be stable, both the above eigenvalues must have negative real parts (refer Sect. 3.4.). This is guaranteed if

$$B < 1 + A^2\ . \tag{15.45}$$

Thus the homogeneous state (X^*, Y^*) is stable for $B < 1 + A^2$.

Exercise:

7. Show that the stability determining Jacobian matrix for the homogeneous solution (15.41) of the Brusselator equation (15.18) in the absence of diffusion has the form (15.42).

B. Stability of Homogeneous Steady State in the Presence of Diffusion

Now let us consider the full reaction-diffusion system (15.18). We perturb the steady states (X^*, Y^*) into $(X^* + \delta X, Y^* + \delta Y)$, δX, $\delta Y \ll 1$. Then the linearized equations are written as

$$\frac{\partial}{\partial t}(\delta X) = (B-1)\delta X + A^2 \delta Y + D_X \nabla^2 \delta X \,, \tag{15.46a}$$

$$\frac{\partial}{\partial t}(\delta Y) = -B\delta X - A^2 \delta Y + D_Y \nabla^2 \delta Y \,. \tag{15.46b}$$

The solution to the above equations can be assumed in the following form,

$$\delta X(\boldsymbol{r}, t) = \delta X_0(\boldsymbol{r}) \mathrm{e}^{\lambda t} \,, \tag{15.47a}$$

$$\delta Y(\boldsymbol{r}, t) = \delta Y_0(\boldsymbol{r}) \mathrm{e}^{\lambda t} \,, \tag{15.47b}$$

where $\delta X_0(\boldsymbol{r})$ and $\delta Y_0(\boldsymbol{r})$ represent the time independent solution of (15.46) and λ is the eigenvalue determining the temporal growth.

To solve equation (15.46) subjected to the boundary conditions mentioned earlier, we first consider the spatial eigenvalue problem (the time independent part) defined by

$$\nabla^2 \delta X_0 + k^2 \delta X_0 = 0 \,, \tag{15.48a}$$

$$\nabla^2 \delta Y_0 + k^2 \delta Y_0 = 0 \,, \tag{15.48b}$$

where k^2 is the eigenvalue to be determined. One can write the solutions of the above time independent part (keeping in mind the boundary conditions) as

$$\delta X_0(\boldsymbol{r}) \propto \cos(\boldsymbol{k} \cdot \boldsymbol{r}) \,, \quad \delta Y_0(\boldsymbol{r}) \propto \cos(\boldsymbol{k} \cdot \boldsymbol{r}) \,. \tag{15.49}$$

The eigenvalue k in this case is nothing but the wave number. It is also clear from the above solutions that due to the form of the solution (15.49) and the zero flux boundary conditions, k takes only discrete values, $k = n\pi/L$, where n is an integer.

Now substituting (15.47) along with (15.49) into (15.46), one can find the stability determining eigenvalues λ. They are the roots of the characteristic polynomial

$$\begin{vmatrix} \lambda - B + 1 + D_X k^2 & -A^2 \\ B & \lambda + A^2 + D_Y k^2 \end{vmatrix} = 0 \,, \tag{15.50}$$

or

$$\lambda^2 + \lambda \left[k^2 (D_X + D_Y) + (A^2 - B + 1) \right] + h(k^2) = 0 \,, \tag{15.51}$$

where

$$h(k^2) = D_X D_Y k^4 + \left[D_X A^2 + D_Y (1 - B) \right] k^2 + A^2 \,. \tag{15.52}$$

C. Turing Instability

The steady state (X^*, Y^*) is linearly stable in the presence of diffusion if both the solutions of the quadratic equation (15.51) have negative real parts. Even if one of the roots has a positive real part, the steady state is unstable. On the other hand, we have already imposed the constraint that the steady state is stable in the absence of diffusion in equation (15.44), that is, Re $\lambda(k^2 = 0) < 0$. Now, in order to have *Turing instability* we require that at least one of the roots of the above equation (15.51) should have a positive real part, or in other words, Re $\lambda > 0$ for some $k \neq 0$. This happens when either one of the coefficients of λ in (15.51) is negative or $h(k^2) < 0$ for some $k \neq 0$ (see Fig. 15.8). Since we have already imposed the condition that $B < 1 + A^2$, and by definition $k^2 (D_X + D_Y)$ is also positive, for all $k \neq 0$, the coefficient of λ, namely,

$$k^2 (D_X + D_Y) + A^2 - B + 1 > 0 \,. \tag{15.53}$$

So the only way to have Re $\lambda > 0$ is that $h(k^2) < 0$ for some $k \neq 0$. That is,

$$D_X D_Y k^4 + \left[D_X A^2 + D_Y (1 - B)\right] k^2 + A^2 < 0 \,. \tag{15.54}$$

The *necessary condition* to satisfy the above requirement is that

$$D_X A^2 + D_Y (1 - B) < 0 \,. \tag{15.55}$$

This implies that $D_X \neq D_Y$ (since $B < 1 + A^2$) and more specifically we require that

$$D_X < D_Y \,. \tag{15.56}$$

Thus, the variable Y should diffuse faster than X for Turing instability to occur.

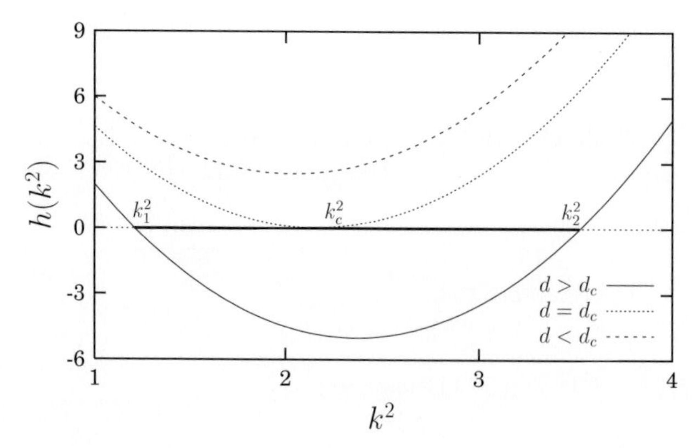

Fig. 15.8. Plot of $h\left(k^2\right)$ as given in equation (15.51) for the Brusselator system (15.18) versus wave number $\left(k^2\right)$ for $A = 0.4$, $B = 10$ and $D_X = 1$

Note that the above inequality is only the necessary condition but not a sufficient one for Re $\lambda > 0$ for $k \neq 0$. To identify the latter we proceed as follows. For $h\left(k^2\right)$ to be negative for some nonzero k, the minimum of $h(k^2)$, namely h_{min}, must be negative. From (15.51), one can show that

$$h_{\text{min}} = A^2 - \frac{\left[(B-1)D_Y - A^2 D_X\right]^2}{4D_X D_Y} < 0 \tag{15.57}$$

at

$$k^2 = k_m^2 = \frac{(B-1)D_Y - A^2 D_X}{2D_X D_Y} \; . \tag{15.58}$$

At the bifurcation point, that is at the value of k^2 where the stability changes, $h_{\text{min}} = 0$. From (15.57), one can easily check that at $h_{\text{min}} = 0$,

$$(B-1)^2 d_c^2 - 2A^2(B+1)d_c + A^4 = 0 \; , \tag{15.59}$$

where $d = d_c = D_Y/D_X$. Thus the critical wave number k_c is given by

$$k_c^2 = \frac{1}{D_X}\left(\frac{(B-1)d_c - A^2}{2d_c}\right) \; . \tag{15.60}$$

Exercise:

8. Verify the above minimization of $h\left(k^2\right)$ given by equation (15.57).

D. Mode Selection

Summarizing the conditions for Turing or diffusion driven instability for the system (15.18), one can write,

$$\begin{aligned} &B < 1 + A^2, \;\; (B-1)D_Y - D_X A^2 < 0 \; , \\ &\left[(B-1)D_Y - D_X A^2\right]^2 - 4D_X D_Y A^2 < 0 \; . \end{aligned} \tag{15.61}$$

Figure 15.8 shows the variation of $h\left(k^2\right)$ as a function of k^2. The variation of thc largest eigenvalue (Re λ) as a function of k^2 is depicted in Fig. 15.9. It is clear from the figures that for $d < d_c$ all the modes are stable and for $d > d_c$ there will be a range of k^2 values for which the system grows. This range of k^2 values for $d > d_c$ can given by

$$\begin{aligned} k_1^2 &= \frac{(B-1)d - A^2 - \sqrt{[(B-1)d - A^2]^2 - 4dA^2}}{2dD_X} \\ &< k^2 < \frac{(B-1)d - A^2 + \sqrt{[(B-1)d - A^2]^2 - 4dA^2}}{2dD_X} = k_2^2 \; . \end{aligned} \tag{15.62}$$

It may be observed that the modes which lie in the range $k_1 \leq k \leq k_2$ will grow as t increases, while all other modes tend to zero asymptotically.

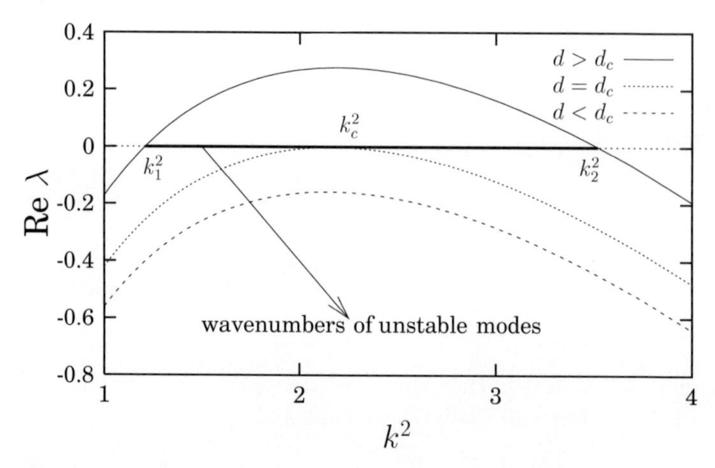

Fig. 15.9. Plot of the largest eigenvalue (Re λ) as a function of the wave number for $A = 0.4$, $B = 10$ and $D_X = 1$

Consequently, a spatially heterogenous solution emerges from the sum of the unstable modes and equation (15.47) can be written as (for large t)

$$\delta X(\boldsymbol{r}, t) \approx \sum_{k_1}^{k_2} c_{1k} \exp\left[\lambda\left(k^2\right)t\right] \delta X_0^k(\boldsymbol{r}) , \qquad (15.63a)$$

$$\delta Y(\boldsymbol{r}, t) \approx \sum_{k_1}^{k_2} c_{2k} \exp\left[\lambda\left(k^2\right)t\right] \delta Y_0^k(\boldsymbol{r}) , \qquad (15.63b)$$

where $\delta X_0^k(\boldsymbol{r})$ and $\delta Y_0^k(\boldsymbol{r})$ are the solutions of the time independent part as in equation (15.49) with $k = n\pi/L$ and c_{1k} and c_{2k} are constants. $\lambda\left(k^2\right)$ is the positive solution of the quadratic equation (15.51).

Let us consider the one-dimensional domain $x \in (0, p)$. Then the solution can be written as

$$\delta X(x, t) \approx \sum_{n_1}^{n_2} c_{1k} \exp\left[\lambda\left(\frac{n^2\pi^2}{p^2}\right)t\right] \cos\frac{n\pi x}{p} , \qquad (15.64a)$$

$$\delta Y(x, t) \approx \sum_{n_1}^{n_2} c_{2k} \exp\left[\lambda\left(\frac{n^2\pi^2}{p^2}\right)t\right] \cos\frac{n\pi x}{p} , \qquad (15.64b)$$

and finally one can write the solution to the system (15.18) as (here we are considering the X variable only for illustrative purpose)

$$X(x, t) \approx X_0 + \sum_{n_1}^{n_2} c_{1k} \exp\left[\lambda\left(\frac{n^2\pi^2}{p^2}\right)t\right] \cos\frac{n\pi x}{p} . \qquad (15.65)$$

If we assume the domain size such that the range of unstable wave numbers admits only the $k = 1$ mode then equation (15.65) becomes

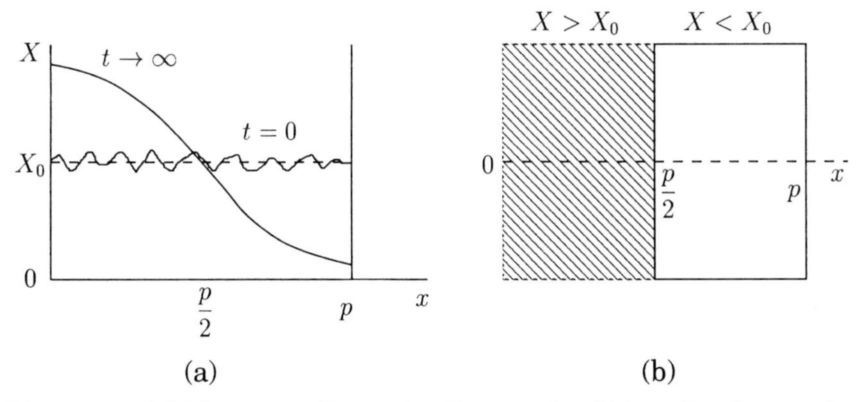

Fig. 15.10. (**a**) The temporally growing linear mode which evolves from random initial conditions at $t = 0$ into a finite amplitude spatial pattern at $t \to \infty$ and (**b**) The actual pattern showing the region of higher concentration (shaded region) compared to the steady state X_0 and the region (unshaded) of lower concentration in comparison with the steady state (adapted from Ref. [1])

$$X(x,t) \approx X_0 + \epsilon \exp\left[\lambda\left(\frac{\pi^2}{p^2}\right)t\right] \cos\frac{\pi x}{p} . \tag{15.66}$$

Here $\epsilon = c_{11}$ and is assumed to be very small. The above solution corresponds to the unstable mode which becomes dominant when t increases. The development of the corresponding pattern is shown in Fig. 15.10. However this growth cannot continue indefinitely as otherwise there will be unbounded increase in the dynamical variables.

At certain stage the nonlinearity in the problem will take over and dampen the growth, leading to an eventual pattern as shown in Fig. 15.10 in this simple case. More complex patterns can obviously be obtained when one considers situations where more number of modes are unstable and also when the spatial dimension is greater than one. For more detailed analysis of such patterns, the readers may refer to Refs. [1,3,8].

15.3.5 Localized Structures

Another interesting phenomenon observed recently is the existence of *soliton-like* localized structures in driven dissipative systems. These structures have been observed in a variety of pattern forming systems experimentally as well as from numerical simulations. Very recently Umbanhowar, Melo and Swinney [11] have discovered localized particle like excitations called *oscillons* in granular layers (see Fig. 15.2d). They find that when a thin 'sand-like' layer of minute brass balls are excited into motion by the vertical vibration of their container, they exhibit well defined localized structures at critical excitation value. Such localized excitations have been found to oscillate at half the driving frequency.

Similar localized structures have been also observed in highly dissipative two and three-dimensional systems like the convection of binary mixtures (see Fig. 15.2b) and a flowing viscous film [12]. Further, theoretical studies on the Ginzburg–Landau equations show the existence of localized breathing oscillations [13].

It has also been realized that dissipation is a necessary condition to have localized excitations. Essentially these localized structures arise due to a tendency of certain nonlinear systems which localize dissipation in the presence of external forces.

15.3.6 Spatio-Temporal Chaos

The study of spatio-temporal or extensive chaos has also been receiving considerable interest in recent times. The phenomenon of chaos in low dimensional systems, for example in systems described by ordinary differential equations, has been well analysed (as we have seen in Chaps. 2–9). However, investigations are still under progress to understand chaos in systems with infinite number of degrees of freedom. Examples include partial differential equations, arrays of coupled ordinary differential equations, coupled map lattices, cellular automata and so on. The associated attractors in the case of extended systems are typically multidimensional sets with fractal structure. For such systems, it has been found that mathematically one can define a smooth set called the *inertial manifold* which is an invariant and contains the attractor. The existence of such inertial sets has been shown for models such as the complex Ginzburg–Landau and the Kuramoto–Sivashinsky equations [13,14].

Spatio-temporal chaos can be defined by both finite time correlation as well as correlation length (space). There is another term *turbulence* which is widely used in disordered flow but sometimes people refer turbulence as synonym of spatio-temporal chaos. The chaotic attractor in extended systems is characterized using the dimension of the invariant measure, inertial manifold of the attractor as a function of the control parameters and system size. Chaos correlation length, Lyapunov exponents, fractal dimensions, information functions and so on have been used to understand the underlying phenomena of the spatio-temporal chaotic systems [15,16]. Particularly, in spatially extended systems, the fractal dimension increases linearly with the system size. The connection between the above measures and the system size is an important topic of current research [3]. In the following we briefly discuss the different types of spatio-temporal chaos observed in a class of dynamical systems.

(a) Complex Ginzburg–Landau (CGL) Equation

The complex Ginzburg–Landau equation (15.19) has been used as a prototype model to study spatio-temporal patterns including spatio-temporal chaos. It was pointed out in Sect. 15.2.2 that in the limit c_1^{-1}, $c_2^{-1} \to 0$, the complex term dominates and the CGL equation (15.19) reduces to the nonlinear Schrödinger (NLS) equation, which is a Hamiltonian system with infinitely many conservation laws (in one space dimension) and so is integrable as we have seen in Chap. 14. On the other hand, in the limit when the coefficients are real (15.19) shows a rich variety of complex dynamics. Since the CGL equation has different behaviours in the two extreme limits, the dynamics in the region in between is extremely complex. Not only are there new types of coherent structures, like stable pulse solutions, sources, sinks, homoclinic orbit solutions, etc., the equation has regimes where the behaviour is intrinsically chaotic.

(b) Spatio-Temporal Chaos in Rayleigh–Bénard Convection

When a fluid is heated from below, the hot fluid will swell up from the bottom forming spatio-temporal patterns. Such space-time structures are often called convection patterns. In experiments, convection roll patterns are observed when a thin layer of fluid is heated from below and cooled from above. This kind of patterns is also referred as *spiral chaos*. Spiral chaos state is explained in terms of *invasive defects* that expands throughout the system via their intrinsic dynamics.

Theoretically spiral chaos which arises in convection experiments have been studied using *Swift–Hohenberg* equation [3], an equation for a single real field ψ that is a function of two space variables (representing the horizontal coordinates in a convection system) and time of the following form

$$\dot{\psi} = \epsilon\psi - \left(\nabla^2 + 1\right)^2 \psi - \psi^3 \,. \tag{15.67}$$

Here ϵ is the control parameter and $\psi(x, y, t)$ is a real field. Figure 15.11 shows a typical spiral and spiral chaos patterns observed in computer simulation of Rayleigh–Bénard convection experiment.

In the convection experiment, if the convection apparatus is rotated about a vertical axis, the Coriolis force changes the fluid motion. For large enough rotation rates the stationary stripe pattern (steady rolls) becomes unstable to a dynamic pattern of carnivorous domains. This type of unstable patterns are referred as *domain chaos* [18]. The steady roll state occurs in convection due to instability at the critical system parameter value, namely, Rayleigh number[1] $R = R_c$. The experimentally observed domain chaos consists of differently oriented rolls with a persistent dynamics of domains. The dynamical equation that exhibits domain chaos can be written as [18]

$$\begin{aligned}\dot{\psi} = \epsilon\psi - \left(\nabla^2 + 1\right)^2 \psi - g_1\psi^3 + g_2\hat{z}\cdot\nabla\times\left[(\nabla\psi)^2\nabla\psi\right] \\ + g_3\nabla\cdot\left[(\nabla\psi)^2\nabla\psi\right] \,.\end{aligned} \tag{15.68}$$

[1] Rayleigh number refers to a dimensionless measure of the temperature difference across the fluid.

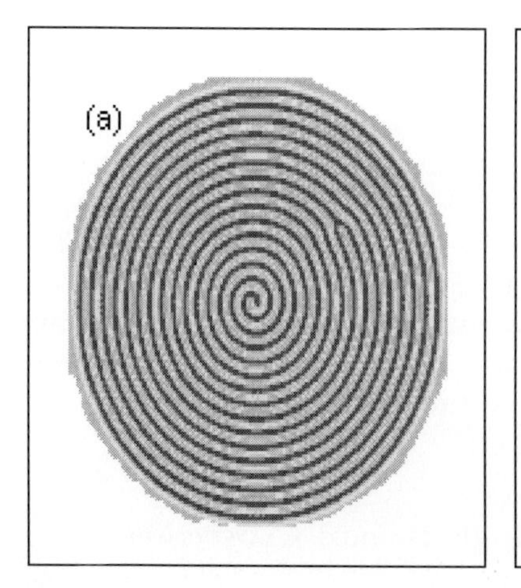

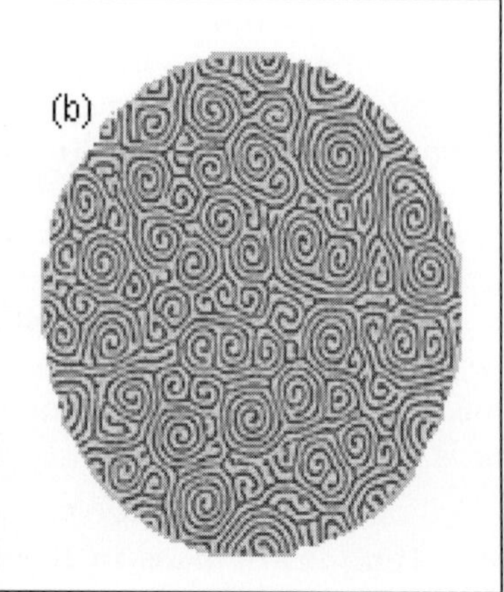

Fig. 15.11. (**a**) Spiral and (**b**) spiral chaos patterns in Rayleigh–Bénard convection (as simulated on the CRAY C90 by James Gunton and coworkers), adapted from Ref. [17]

When $g_2 = g_3 = 0$, the above equation reduces to Swift–Hohenberg equation. Numerical simulation of the above equation shows domain chaos for selected values of parameters. A typical domain chaos pattern of (15.68) is depicted in Fig. 15.12.

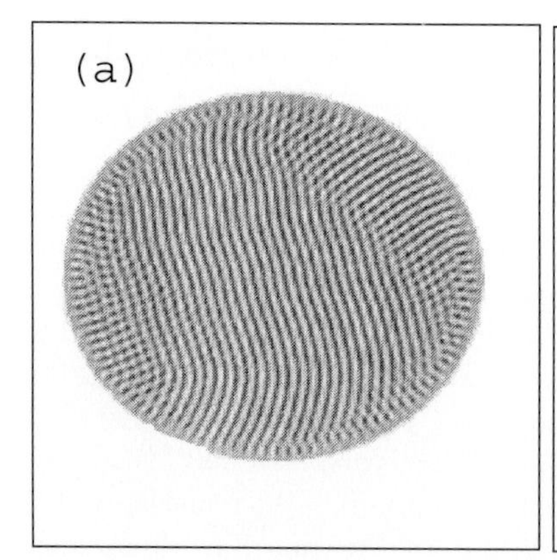

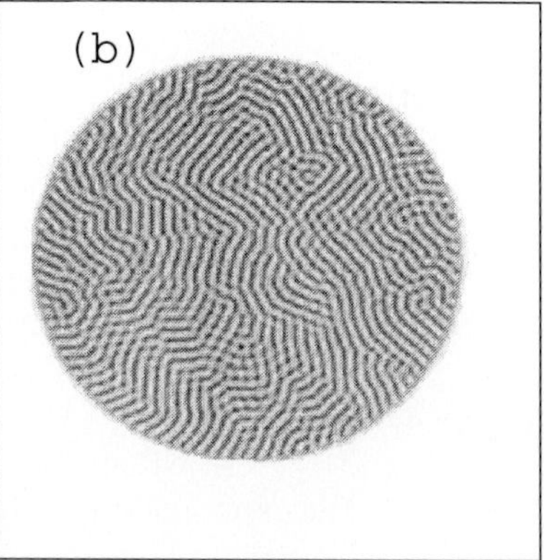

Fig. 15.12. Snapshots of domain configuration for two different values of control parameter (**a**) $\epsilon = 0.01$ and (**b**) $\epsilon = 0.3$ (taken from Ref. [18])

(c) Coupled Map Lattices

Another interesting simple model that exhibits spatio-temporal chaos is the so-called coupled map lattice (CML). Apart from the fact that coupled map lattices are relatively easy to analyse, these systems are useful to explain various characteristic properties of higher dimensional chaotic systems. In certain occasions CML is considered as coarse-grained version of continuum systems. A typical model of CML exhibiting spatio-temporal chaos can be represented by the following equations [19]

$$x_{n+1}^{j} = (1-\epsilon) f\left(x_{n}^{j}\right) + \frac{\epsilon}{2}\left[f\left(x_{n}^{j-1}\right) + f\left(x_{n}^{j+1}\right)\right], \quad j = 1, 2, ..., N. \quad (15.69)$$

Here $f(x) = \mu x(1-x)$, represents logistic map function. Figure 15.13 shows the pattern of the spatio-temporal chaos observed in the above one-dimensional lattice. Another property of the CML is that it shows interesting transition from a low dimensional (temporal) chaos to spatio-temporal chaos through size instability.

There exists a large number of systems in various areas of science and applied science which exhibit spatio-temporal chaos though the details are different. Considerable efforts are being made at present for a full understanding of spatio-temporal chaos.

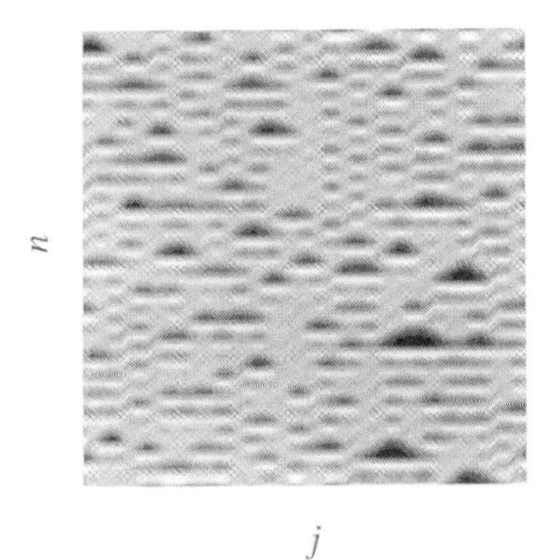

Fig. 15.13. Spatio-temporal chaos in CML (15.69) with $\mu = 4$ and $N = 50$

15.4 Cellular Neural/Nonlinear Networks (CNNs)

The most common feature in the autowave process and other pattern formations in reaction-diffusion systems is the presence of an active nonlinear medium. We have already discussed the role of active medium in Sect. 15.3 for the autowave process. However there are physical situations where propagation of waves is inhibited beyond certain spatial distance in such a medium, example: the failure of electrical impulses in the nerves of patients suffering from *multiple sclerosis*. Interestingly there are theorems which clearly show that continuous models cannot in general exhibit such *propagation failure*. In order to simulate this kind of new phenomenon, in addition to the existing several kinds of patterns noted above, it becomes necessary to consider discrete versions of diffusively coupled nonlinear dynamical systems such as nonlinear oscillators, electronic circuits, etc., to mimic reaction-diffusion processes [20]. It is very often possible to treat the total system as an assembly of a large number of identical local systems which are coupled (through diffusion) to each other [21]. Here the local systems are defined as those obeying the diffusionless part of the equations, for example the Belousov–Zhabotinsky chemical reaction. In many situations, the discrete systems are modelled by appropriate nonlinear electronic circuits, for example as in the case of impulse propagation along nerve fiber, propagation of action potential in cardiac tissues and so on.

Under these circumstances, it is of great interest to investigate the dynamics of arrays of diffusively coupled nonlinear oscillators and systems. In such cases one can often consider an array of interconnected locally coupled cells such as neurons, nonlinear circuits, nonlinear oscillators and so on. Such aggregates of cells may be called *cellular neural networks* (CNNs) in the case of neurons and *cellular nonlinear networks* (again CNNs) in the case of oscillators and circuits [21].

15.4.1 Cellular Nonlinear Networks (CNNs)

In general, a CNN is defined mathematically by the dynamics of the constituent subsystems (state equations of the individual oscillators or cells) and a *synaptic law* which specifies the interaction with their neighbours. Such cellular neural networks, namely, interconnections of sufficiently large number of simple dynamical units can exhibit extremely complex, synergetic and self organizing behaviours. Theoretically one can consider a system of interconnected sets of coupled ordinary differential equations (odes) to represent a macroscopic system, in which each of the set of coupled odes corresponds to the evolution of the individual subsystem. Such collection or aggregates could represent the *discrete reaction-diffusion system*. In other words, in reaction-diffusion systems, the CNNs have linear synaptic law which approximates the spatial Laplacian operator (nearest neighbour coupling). A CNN can be represented by the following four specifications:

(1) The cell dynamics given by

$$\dot{u}_j = g_j(u_j) + I_j(u_1, u_2, \ldots, u_{j-1}, u_j, u_{j+1}, \ldots, u_m), \; j = 1, 2, \ldots, m, \tag{15.70}$$

where $u_j \in \mathcal{R}^n$ and I_j represents the interaction between the jth and the remaining cells.

(2) A synaptic law representing the interaction between the cells. For the reaction-diffusion CNN this can be given in terms of the discrete spatial Laplacian operator, $I_j = \tilde{D}(u_{j+1} - 2u_j + u_{j-1})$, where $\tilde{D}$ is the diffusion coefficient matrix.

(3) Appropriate boundary conditions.

(4) Initial conditions.

Specific examples are the coupled array of anharmonic oscillators, Josephson junctions, continuously stirred tank reactors (CSTR) exhibiting travelling waves, propagation of nerve impulses (action potential) along the neuronal axon, propagation of cardiac action potential in the cardiac tissues, etc. During the past few years several investigations have been carried out to understand the spatio-temporal behaviour of these coupled nonlinear oscillators and systems. The studies on these systems include the travelling wave phenomenon, Turing patterns, spatio-temporal chaos and synchronization. Of particular interest among coupled arrays is the study of *diffusively coupled driven systems* as they represent diverse topics like Faraday instability, granular hydrodynamics, self organized criticality and so on. Identification of localized structures in these systems has also been receiving considerable attention very recently.

A rather powerful and practical way of studying CNN systems is to model the constituent cells in terms of suitable nonlinear electronic circuits, which are then interconnected through appropriate linear resistors (so as to give rise to diffusive coupling). The advantage of such arrays of nonlinear electronic circuits is that they are quite flexible, that is they can mimic real systems but also can be studied on their own merit, they are easy to produce and easy to study experimentally and numerically. From this point of view, one can consider the dynamics of very simple diffusively coupled driven nonlinear electronic circuits to realize novel spatio-temporal patterns. As we have seen in Chap. 6, the Murali–Lakshmanan–Chua circuit may be considered as the simplest second order nonlinear nonautonomous dissipative circuit consisting of a single nonlinear element, namely, the Chua's diode. This simple circuit can exhibit a variety of interesting bifurcations, chaos and so on when driven by external periodic force as discussed in Chap. 6. Therefore it will be of considerable interest to study the dynamics of one and two dimensional arrays of coupled MLC circuits, as a prototype of CNNs for illustrative purpose. In this section, we will briefly indicate the type of patterns admitted by these arrays.

15.4.2 Arrays of MLC Circuits: Simple Examples of CNN

We will now consider one and two-dimensional arrays of MLC circuits [20], where the intercell couplings are effected by linear resistors. Such resistive couplings are then equivalent to diffusive couplings. Let us deduce the underlying dynamical equations for such CNNs.

A. One-Dimensional Array

Figure 15.14 shows a schematic representation of an one-dimensional chain of *resistively* coupled MLC circuits. Note that the individual cells correspond to isolated MLC circuits. Using the standard Kirchhoff's laws, the dynamics of the one-dimensional chain can be easily shown to be governed by the following system of equations, in terms of the rescaled variables (see Chap. 6),

$$\dot{x}_i = y_i - h(x_i) + D(x_{i+1} + x_{i-1} - 2x_i) , \tag{15.71a}$$

$$\dot{y}_i = -\sigma y_i - \beta x_i + F \sin \omega t , \quad i = 1, 2, ..., N. \tag{15.71b}$$

Or equivalently in the CNN form (15.70), we have

$$\dot{u}_i = g(u_i) + \tilde{D}(u_{i+1} - 2u_i + u_{i-1}) , \quad i = 1, 2, \ldots, N, \tag{15.72a}$$

where

$$u_i = \begin{pmatrix} x_i \\ y_i \\ z_i \end{pmatrix} , \; g = \begin{pmatrix} y_i - h(x_i) \\ -\sigma y_i - \beta x_i + F \sin \omega t \\ \omega \end{pmatrix} , \; \tilde{D} = \begin{pmatrix} D & 0 & 0 \\ 0 & 0 & 0 \\ 0 & 0 & 0 \end{pmatrix} . \tag{15.72b}$$

Here the diffusion coefficient D is related to the resistive coupling strength,

$$D = \frac{R_c}{G} , \tag{15.72c}$$

N is the chain length and $h(x)$ is the three segment piecewise linear function (see Sect. 6.3.1),

$$h(x) = \begin{cases} \epsilon + m_2 x + (m_0 - m_1) , & x \geq x_2 \\ \epsilon + m_0 x, & x_1 \leq x \leq x_2 \\ \epsilon + m_1 - (m_0 - m_1) , & x \leq x_1 \end{cases} \tag{15.72d}$$

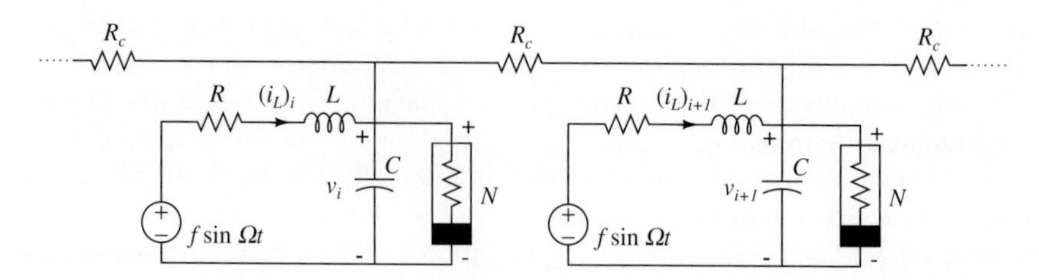

Fig. 15.14. Circuit realization of an one-dimensional array of coupled MLC circuits

where ϵ is the dc offset included for convenience. Naturally, depending on the choice of the slopes m_0, m_1 and m_2, one can fix the characteristic curve of the Chua's diode. For typical forms of $h(x)$, the coupled MLC circuit exhibits interesting dynamics such as wave propagation and its failure, Turing patterns, spatio-temporal chaos and so on. In the following discussion, we will fix N at $N = 100$.

B. Two-Dimensional Array

As in the case of the one-dimensional array introduced above, one can also consider a two-dimensional square array with each cell in the array being coupled to four of its nearest neighbours, for example, with linear resistors. The state equations can be now written in dimensionless form as

$$\begin{aligned} \dot{x}_{i,j} &= y_{i,j} - h\left(x_{i,j}\right) + D_1\left(x_{i+1,j} + x_{i-1,j} + x_{i,j+1} + x_{i,j-1} - 4x_{i,j}\right) \\ &\equiv f\left(x_{i,j}, y_{i,j}\right), \qquad (15.73a) \\ \dot{y}_{i,j} &= -\sigma y_{i,j} - \beta x_{i,j} + D_2\left(y_{i+1,j} + y_{i-1,j} + y_{i,j+1} + y_{i,j-1} - 4y_{i,j}\right) \\ &\quad +F\sin\omega t \\ &\equiv g\left(x_{i,j}, y_{i,j}\right), \quad i,j = 1,2,...,N. \qquad (15.73b) \end{aligned}$$

This two-dimensional array has $N \times N$ cells arranged in a square lattice. In our numerical study we will again take $N = 100$.

Exercise:

9. Obtain the dynamical equations (15.71) and (15.73) for the one and two-dimensional arrays respectively of MLC circuits using Kirchhoff's laws.

15.4.3 Active Wave Propagation and its Failure in One-Dimensional CNNs

To illustrate the wave propagation failure let us consider the CNN model described by the array of coupled MLC circuits in one dimension, equations (15.71). For this purpose we have numerically integrated equations (15.71) using fourth-order Runge–Kutta method with fixed step size. In this analysis we fix the parameters at $\{\beta, \sigma, m_0, m_1, m_2, \epsilon, F\} = \{1.0, 1.0, -2.25, 1.5, .25, 0, 0\}$ so that the system admits bistability which is a necessary condition to observe a wave front. Zero flux boundary conditions are used in the numerical computations, which in this context mean setting $x_0 = x_1$ and $x_{N+1} = x_N$ at each integration step; similar choice has been made for the variable y also. To start with, we will study in this subsection the autonomous case ($F = 0$) and extend our studies to the nonautonomous case ($F \neq 0$) in the next subsection.

The choice of the values of the parameters guarantees the existence of two stable equilibrium points $P_i^+ = \{\sigma(m_1 - m_0 - \epsilon)/(\beta + m_2\sigma),\ \beta(m_0 -$

$m_1 + \epsilon)/(\beta + m_2\sigma)\}$ and $P_i^- = \{\sigma(m_1 - m_0 - \epsilon)/(\beta + m_1\sigma),\ \beta(m_0 - m_1 + \epsilon)/(\beta + m_1\sigma)\}$ for each cell, $i = 1, 2, \ldots, 100$, along with the unstable equilibrium point $P_i^0 = (-\sigma\epsilon/(\beta + m_0\sigma),\ \beta\epsilon/(\beta + m_0\sigma))$. In the particular case corresponding to the above parametric choice, each cell in the array has three equilibrium points $P_i^+ = (3.0, -3.0)$, $P_i^- = (-1.5, 1.5)$ and $P_i^0 = (0, 0)$. Out of these three equilibrium points, P_i^+ and P_i^- are stable while P_i^0 is unstable. Due to the asymmetry in the function $h(x)$ for the present parametric choices, the basin of attraction of the point P_i^+ is much larger than that of P_i^- and it is harder to steer a trajectory back into the basin P_i^- once it is in the basin of P_i^+.

Now we choose an initial condition such that the first few cells in the array are excited to the P_i^+ state (having a larger basin of attraction compared to that of P_i^-) and the rest are set at the P_i^- state. In other words a wave front in the array is initiated by means of the two stable steady states. On actual numerical integration of equations (15.71) with $N = 100$ and with the diffusion coefficient chosen at a higher value, $D = 2.0$, a motion of the wave front towards right (see Fig. 15.15a) is observed, that is a travelling wave front is found. After about 80 time units the wave front reaches the 100th cell so that all the cells are now settled at the more stable state (P_i^+) as demonstrated in Fig. 15.15a. When the value of D is decreased in steps and the analysis is repeated, the phenomenon of travelling wavefronts continues to be present.

However, below a critical value of the diffusion coefficient $(D = D_c)$ a failure in the propagation has been observed, which in the present case turns out to be $D = D_c = 0.4$. This means that the initiated wavefront is unable to move as time progresses and Fig. 15.15b shows the propagation failure for $D = 0.4$.

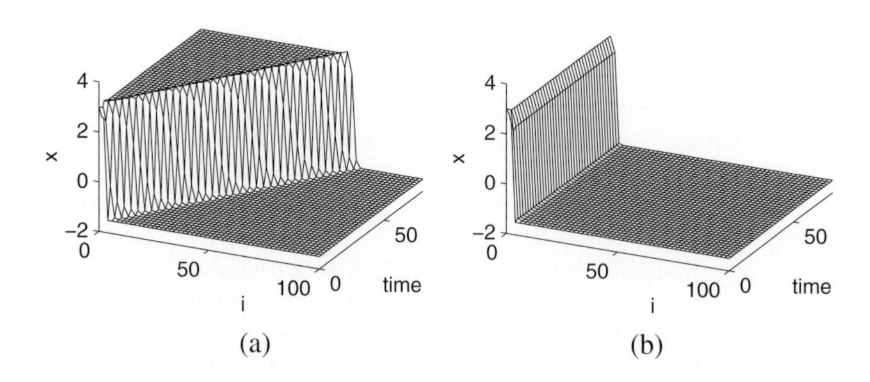

Fig. 15.15. Space-time plot (**a**) showing propagation of wave front in one-dimensional array (100 cells) of MLC circuits for $D = 2.0$ and (**b**) propagation failure for $D = 0.4$

It is possible to develop an analysis based on the linear stability of various steady states present in the system in order to explain the initiation of the wavefront and its propagation. However, the details are beyond the scope of this book and for relevant references see for example Ref. [20].

Exercise:

10. Obtain the explicit form of the steady states P_i^+, P_i^- and P_i^0 and investigate their linear stability with reference to equations (15.71) for $F = 0$.

15.4.4 Turing Patterns

Another interesting dynamical phenomenon in the coupled arrays is the formation of Turing patterns. As noted earlier in Sect. 15.3.4, these patterns are observed in many reaction-diffusion systems when a homogeneous steady state which is stable due to small perturbations in the absence of diffusion becomes unstable in the presence of diffusion [1]. To be specific, the Turing patterns can be observed in a two variable reaction-diffusion system when one of the variables diffuses faster than the other and undergoes Turing bifurcation, that is, when diffusion driven instability (Sect. 15.3) is present.

Treating the coupled array of MLC circuits as a discrete version of a reaction-diffusion system, one can as well observe the Turing patterns in this model also. For clarity, we will consider the two-dimensional array given by the dynamical equations (15.73). In order to realize Turing patterns, one has to study the linear stability of system (15.73) near the steady state. In continuous systems, the linear stability analysis is necessary to arrive at the conditions for diffusion driven instability as we have done earlier in Sect. 15.3.4 for the case of the Brusselator system. For discrete cases also one can follow the same derivation as in the case of continuous systems by considering perturbations of the form $\exp[\mathrm{i}(kj - \lambda t)]$ [1]. Here k and λ are considered to be independent of the position j ($j = 1, 2, ..., N$). For equations (15.73), the criteria for the diffusion driven instability can be derived by finding the conditions for which the steady states in the absence of diffusive coupling are linearly stable and become unstable when the coupling is present. One can easily show the eigenvalues that guarantee the linear stability in the absence of coupling are the roots of the characteristic equation [20]

$$\lambda^2 - (f_x + g_y)\,\lambda + f_x g_y - f_y g_x = 0\,, \tag{15.74}$$

where f_x, f_y, g_x and g_y are the partial derivatives of the quantities f and g given in equations (15.73) without the coupling coefficients ($D_1 = D_2 = 0$) and evaluated at the steady state. It can be further seen easily that the steady state is stable in the absence of coupling if and only if the roots of (15.74) (λ_1 and λ_2) have negative real parts.

Apart from the above condition, in order to satisfy the instability in the presence of coupling (Turing instability), at least one of the roots of the characteristic equation,

$$\lambda_s^2 - \left[k^2 (D_1 + D_2) - (f_x + g_y)\right] \lambda_s + m\left(k^2\right) = 0 \tag{15.75a}$$

with

$$m\left(k^2\right) = D_1 D_2 k^4 - (D_2 f_x + D_1 g_y) k^2 + f_x g_y - f_y g_x \;, \tag{15.75b}$$

should have positive real part.

A straightforward calculation shows that the following conditions should be satisfied for the general reaction-diffusion system of the form given by (15.73):

$$\begin{aligned} &f_x + g_y < 0, \;\; f_x g_y - f_y g_x > 0, \;\; f_x D_2 - g_y D_1 > 0 \;, \\ &(f_x D_2 - g_y D_1)^2 - 4 D_1 D_2 (f_x D_2 - g_y D_1) > 0 \;. \end{aligned} \tag{15.76}$$

The critical wave number for the discrete system (15.73) can be obtained (compare with (15.60)) as

$$\cos(k_c) = 1 - \frac{f_x D_2 - g_y D_1}{4 D_1 D_2} \;. \tag{15.77}$$

Combining equations (15.76)–(15.77), one obtains the condition for the Turing instability as

$$\frac{f_x D_2 - g_y D_1}{8 D_1 D_2} \le 1 \;. \tag{15.78}$$

Now one can apply these conditions to the coupled oscillator system of the present study. For this purpose, we have fixed the parameters for the two-dimensional model (15.73) as

$$\{\beta, \sigma, \epsilon, m_0, m_1, m_2\} = \{0.8, 0.92, 0.1, -0.5, 0.5, 0.5\}$$

and verified that this choice satisfies the conditions (15.76) to (15.78). The numerical simulations have been carried out using an array of size 100×100 and random initial conditions near the steady states have been chosen for the x and y variables. Figures 15.16a–d show how the diffusion driven instability leads to stable hexagonal pattern (Fig. 15.16d) after passing through intermediate stages (Figs. 15.16a–c). Further, the spontaneously formed patterns are fairly uniform hexagonal patterns having a penta-hepta defect pair. These defects are inherent and very stable.

15.4.5 Spatio-Temporal Chaos

Next, we move on to a study of the spatio-temporal chaotic dynamics of the array of coupled MLC circuits when individual cells are driven by external periodic force. The motivation is that over a large domain of (F, ω) values the individual MLC circuits typically exhibit various bifurcations and transition

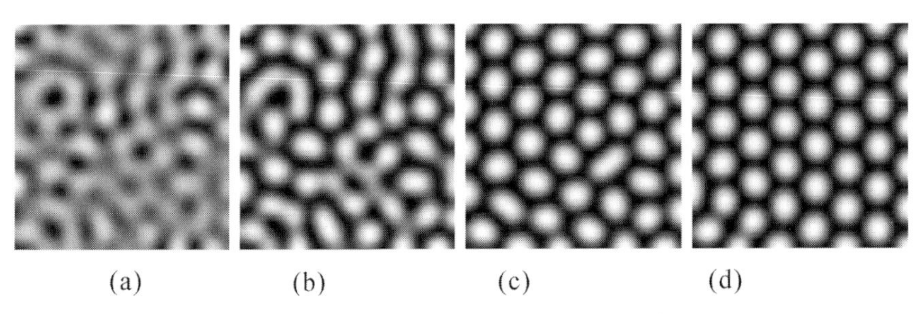

(a) (b) (c) (d)

Fig. 15.16. The spontaneous formation of Turing patterns in an array of 100×100 oscillators for the parameters $\beta = 0.8$, $\sigma = 1$, $m_0 = -0.5$, $m_1 = 0.5$, $m_2 = 0.5$, $\epsilon = 0.1$, $F = 0.0$, $\omega = 0.75$, $D_1 = 1$ and $D_2 = 10$ in equations (15.73) at various time units (**a**) $T = 50$, (**b**) $T = 100$, (**c**) $T = 500$ and (**d**) $T = 2000$

to chaotic motion. So one would like to know how the coupled array behaves collectively in such a situation, for fixed values of the parameters. For this purpose, we set the parameters at $\{\beta, \sigma, \epsilon, m_0, m_1, m_2, \omega\} = \{1.0,\ 1.015,\ 0,\ -1.02,\ -0.55,\ -0.55,\ 0.75\}$. The uncoupled systems exhibit period doubling bifurcations and chaotic dynamics in the presence of external force. In our numerical simulations, here we have mainly considered the one-dimensional array specified by (15.71) and assumed periodic boundary conditions.

A. Spatio-Temporal Regular and Chaotic Motion

Numerical simulations were performed by considering 50 cells (for convenience) and random initial conditions using fourth order Runge–Kutta method for six choices of F values. The coupling coefficient in (15.71) was chosen as $D = 1.0$. Out of these, the first three lead to period-T, period-$2T$, period-$4T$ oscillations, respectively, and the remaining choices correspond to chaotic dynamics of the single MLC circuit. Figures 15.17a–g show the space-time plots for $F = 0.05$, $F = 0.08$, $F = 0.09$, $F = 0.12$, $F = 0.13$ and $F = 0.15$, respectively. From the Figs.15.17a–c, it can be observed that for $F = 0.05$, 0.08 and 0.09 the MLC array also exhibits regular periodic behaviour with periods T, $2T$ and $4T$, respectively, in time alone as in the case of the single MLC circuit. However, for $F = 0.12$ and 0.13 (Figs. 15.17d and 15.17e) one obtains *space-time periodic* oscillations even though each of the individual uncoupled MLC circuits for the same parameters exhibits chaotic dynamics. We may say that a kind of controlling of chaos occurs due to the coupling, though the coupling strength is large here. From the above analysis it can be seen that the macroscopic system shows regular behaviour in spite of the fact that the microscopic subsystems will oscillate chaotically when the coupling is removed.

Finally for $F = 0.15$ the coupled system shows spatio-temporal chaotic dynamics (Fig. 15.17f) and this was confirmed by calculating the Lyapunov exponents using standard algorithms. For example, we calculated the Lyapunov

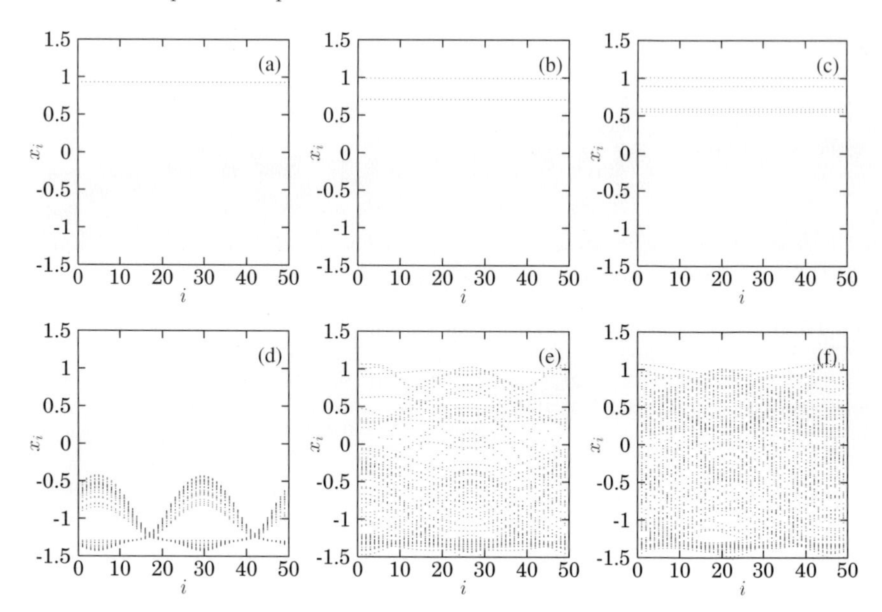

Fig. 15.17. Space-amplitude plot showing the spatio-temporal periodic and chaotic oscillations in 50 coupled MLC circuits for various values of external periodic forcing strength: (**a**) $F = 0.05$, (**b**) $F = 0.08$, (**c**) $F = 0.091$, (**d**) $F = 0.12$, (**e**) $F = 0.13$ and (**f**) $F = 0.15$

exponents for the $N = 50$ coupled oscillators and we find the largest three exponents have the values $\lambda_{\max} = \lambda_0 = 0.1001$, $\lambda_1 = 0.0776$, $\lambda_2 = 0.0092$ and the rest are negative (see next subsection for further analysis).

B. Size Instability, Chaos Synchronization and Suppression of STC

Since the above study of spatio-temporal chaos involves a large number of coupled chaotic oscillators, it is of great interest to analyse the size dependence of the dynamics of these systems. For this purpose we consider the case of 10 coupled oscillators with periodic boundary conditions and numerically solve the system with the other parameters chosen as in above. The value of F is chosen in the range $(0.12, 0.15)$. We find that this set up shows a different behaviour as compared to the 50 cells case. Actually the system gets synchronized to a chaotic orbit rather than showing periodic behaviour or spatio-temporal chaos as in the case of 50 cells described above.

To start with we analyse the dynamics for $F = 0.12$ by slowly increasing the system size from $N = 10$. It has been found that the coupled system (15.71) shows synchronized motion for $N \leq 42$. This was confirmed by calculating the Lyapunov spectrum which shows only one positive exponent (corresponding to the synchronized chaotic motion) with the rest being negative. For example, for $N = 42$, $\lambda_{\max} = 0.1162$, with the rest of the exponents being negative. The existence of only one positive exponent is a necessary condition to have chaos synchronization. Figure 15.18a shows the dynamics of the 5th cell in the array. The system shows entirely different behaviour when we increase the system size to $N = 43$. As noted in the previous Sect. 15.5.6, there occurs a kind of suppression of spatio-temporal chaos. Figure 15.18b shows the resultant periodic orbit in the 5th cell of the array. The maximal Lyapunov exponent is found to be negative in this case ($\lambda_{\text{max}} = -0.001474$). Similar phenomenon has been observed for $F = 0.13$ also.

Next we consider the case of $F = 0.15$ in which the coupled system in the previous subsection showed spatio-temporal chaos. From numerical simulations, we again observed a synchronized motion for $N \leq 31$ and the corresponding Lyapunov spectrum shows one positive exponent only with all the other exponents being negative. But for $N > 31$, the coupled system shows spatio-temporal chaos. The Lyapunov spectrum in this case (for $N > 31$) possesses multiple positive exponents. For example, for $N = 43$, ($\lambda_{\text{max}} = \lambda_0 = 0.0997$, $\lambda_1 = 0.0633$, $\lambda_2 = 0.0038$).

The above type of size instability behaviour has also been found in a coupled Rössler system as well which undergoes the so called short wavelength bifurcation (see for more details Ref. [20]). In such cases, one can find the exact value of the size below which stable synchronous oscillations occur. However, we find that the coupled MLC circuits do not show short wavelength bifurcations. To arrive at a criterion for the size instability one can compute the so-called Lyapunov dimension. It can be shown that the fractal dimension per unit size, namely the dimension density, is an appropriate quantity for characterizing such spatio-temporal chaotic systems [20]. The topic is an evolving subject and one can expect much progress in understanding such systems in future.

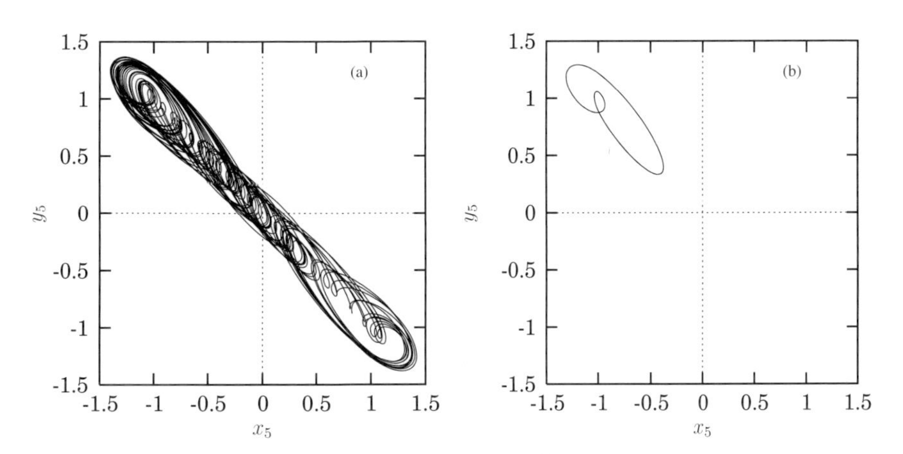

Fig. 15.18. (**a**) The chaotic attractor at the 5th cell for the synchronized state ($N = 42$) and (**b**) the periodic orbit in the 5th cell for the controlled state ($N = 43$)

15.5 Some Exactly Solvable Nonlinear Diffusion Equations

We have seen in the earlier sections that nonlinear reaction-diffusion systems give rise to very many complex patterns and their analysis becomes extremely complicated. Combined numerical and analytical approaches are needed to elucidate the underlying structures. However, it will be very valuable to identify and isolate integrable/solvable physically interesting nonlinear diffusion systems. In this section, we will point out few such model systems, which can then be used for further study of reaction-diffusion systems.

15.5.1 The Burgers Equation

We have already seen in Chap. 13 that the Burgers equation given in the form

$$u_t + uu_x = u_{xx} \tag{15.79}$$

is linearizable in the sense that under the Cole–Hopf transformation

$$u = \frac{v_x}{v} \tag{15.80}$$

it can be transformed into the linear heat equation

$$v_t - v_{xx} = 0\,. \tag{15.81}$$

The Burgers equation was suggested as a model to describe the structure of shock waves in gas dynamics. Later, inhomogeneous Burgers equation was used to model turbulence, propagation of acoustic waves and so on [22]. For suitable initial and boundary conditions, knowing the solution of the linear heat equation, one can obtain the solution of the Burgers equation through the Cole–Hopf transformation (15.80).

15.5.2 The Fokas–Yortsos–Rosen Equation

Consider a homogeneous permeable solid filled with a fluid which contains a chemical substance of concentration $c = c(t, \boldsymbol{r})$. For a prescribed adsorption-desorption equilibrium maintained locally, the concentration of the chemical substance is governed by the diffusion-convection equation $\partial u/\partial t = u^2\left(d\nabla^2 u - \boldsymbol{v}\cdot\nabla u\right)$ in which $u = 1/\left(1 + k^{-1}c\right)$ is the dimensionless dependent variable and $\boldsymbol{v}$ is the local convective flow velocity of the fluid. By assuming the one-dimensional uniform-flow conditions $u = u(x,t)$, $\boldsymbol{v}\cdot\nabla u = v\mathrm{d}u/\mathrm{d}x$, $v \neq 0$, one obtains the evolution equation in the form [23]

$$u_t = du^2 u_{xx} - u^2 u_x\,. \tag{15.82}$$

Under the variable transformation

$$x = -d\ \mathrm{ln}v, \quad u = -d\frac{v_x}{v}\,, \tag{15.83}$$

(15.82) reduces to the linear heat equation (15.81).

15.5.3 Generalized Fisher's Equation

As a final example, consider the model for bacterial colony growth based upon a reaction-diffusion equation scheme which includes two reaction terms to take care of the growth that can occur locally and growth that occurs non-locally by reorganization and expansion of compact colonies through the division and growth of cells within the colony. Thus, in an unstirred medium, the growth in microbial number occurs by cell division and the spatio-temporal growth kinetics in the absence of fluctuations is given by

$$\frac{\partial n(\boldsymbol{r},t)}{\partial t} = D\nabla^2 n(\boldsymbol{r},t) + \lambda(\nabla n(\boldsymbol{r},t))^2 + \lambda n(\boldsymbol{r},t)G(n(\boldsymbol{r},t),n_m)\ , \quad (15.84)$$

where D is the diffusion coefficient, ∇^2 is the Laplacian operator and $n(\boldsymbol{r},t)$ is the local microbial number density with $N(t) = \int \mathrm{d}\boldsymbol{r} n(\boldsymbol{r},t)$, where the integral is taken over the sample volume. The second term on the right hand side of (15.84) corresponds to the nonlocal growth occurring at concentration gradients most notably at the surface of the compact microcolony while the third term represents the local growth of cell number upto a maximum cell concentration n_m characterized by a local rate λ and local growth function $G(n,n_m)$. A typical case of equation (15.84) can be written as

$$u_t - \Delta u - \frac{m}{1-u}(\nabla u)^2 - u(1-u) = 0 \quad (15.85)$$

with $m = 2$. This equation can be reduced to the linear heat equation via the transformation [24]

$$u = 1 - \frac{1}{1+\chi}\ . \quad (15.86)$$

Equation (15.84) in $(1+1)$ dimensions reads as

$$u_t - u_{xx} - \frac{m}{1-u}u_x^2 - u + u^2 = 0 \quad (15.87)$$

whose exact solution is of the form

$$u = 1 - a\Big/\left[a + A_1 \mathrm{e}^{m_1(ax-bt)} + A_2 \mathrm{e}^{m_2(ax-bt)}\right]\ ,$$

$$m_{1,2} = \frac{-b \pm \sqrt{b^2 - 4a^2}}{2a^2}\ , \quad (15.88)$$

where A_1 and A_2 are integration constants and a, b are arbitrary constants for $m = 2$. The (2+1) dimensional case of equation (15.84) is found to contain a large class of interesting solutions of various types including travelling wave solutions and static patterns [25].

It will be interesting and useful to identify more such explicitly solvable nonlinear diffusive systems. These models can then be profitably used to investigate more complex patterns.

15.6 Conclusion

It is clear from the above analysis that the problem of understanding spatio-temporal patterns in nonlinear reaction-diffusion systems is only in recent times being taken up for analysis in a comprehensive way. Much more indepth analysis remains to be carried out to fully understand these systems. We have only tried to present a rudimentary view point of the multitude of spatio-temporal patterns present in nonlinear diffusion systems. Obviously, the area is a challenging one and one can expect much progress in the coming years.

Problems

1. Show that the linear diffusion equation (15.5) is parabolic in nature (see for example, Ref. [26]).
2. Carry out a Painlevé analysis of the waveform equation of the Fisher's equation (15.22) and show that it satisfies the P-property only for $c = 5/\sqrt{6}$.
3. Show that under the transformations
$$f = -6 \exp\left[-\sqrt{2/3}\,\zeta\right] W(Z) + 1\,,$$
$$Z = -\sqrt{6}\,\exp\left(-\sqrt{6}\,\zeta\right)\,,\quad \zeta = \left(x - \frac{5}{\sqrt{6}}t\right)\,,$$
equation (15.22) reduces to the ode whose solution can be given in terms of elliptic function. Show that the limiting form of such a function gives the solution (15.29) and (15.30).
4. The classical Fisher's equation (with a different nonlinear term)
$$u_t - u_{xx} - u + u^3 = 0$$
reduces to the form
$$f'' + cf' + f - f^3 = 0, \qquad \left(' = \frac{d}{d\xi}\right)$$
for $u = f(x-ct) = f(\xi)$. Find the transformations under which the above ode can be reduced to
$$\ddot{W} + W^3 = 0$$
and find the corresponding elliptic function solution.
5. Show that the Fisher's equation
$$u_t - u_{xx} + u(u-a)(u-1) = 0$$
admits travelling wavefront solution of the form
$$u(x,t) = 1\Big/\left[1 + \exp(\pm x - vt)/\sqrt{2}\,\right], \quad v = \pm(1-2a)/\sqrt{2}\,.$$

6. Carry out an analysis of the travelling wave form of the FitzHugh–Nagumo equation (15.31) and prove the existence of travelling wave pulses given in Fig. 15.4.
7. Obtain the steady states of the Lotka–Volterra model equations (15.14) and investigate the Turing instability therein.
8. Extend the Turing instability analysis of the Brusselator model equation (15.18) discussed in Sect. 15.3.4 to two-dimensional problem and discuss the pattern selection nature.
9. Find the stationary solution of the Landau–Ginzburg equation (15.19) and investigate its Turing instability in the one-dimensional case.
10. Show that the so-called Landau–Lifshitz equation with Gilbert damping (in the special case of Heisenberg ferromagnetic spin equation with damping),

$$\boldsymbol{S}_t = \boldsymbol{S} \times \boldsymbol{S}_{xx} + \lambda \boldsymbol{S} \times (\boldsymbol{S} \times \boldsymbol{S}_{xx}) ,$$

where λ is the damping parameter, under stereographic projection (14.114) can be written as [27]

$$\mathrm{i}\omega_t + (1 - \mathrm{i}\lambda) \left[\omega_{xx} - 2\omega_x^2 \omega^* / (1 + \omega\omega^*)\right] = 0 .$$

Note that when $\lambda = 0$, the above equation is the isotropic Heisenberg ferromagnetic spin equation. Obtain the stationary state solution of the above equation and discuss the Turing instability associated with it.
11. For an one-dimensional array of Chua's circuits write down the CNN form of equations of motion. Investigate the nature of steady states and travelling wave phenomenon therein.
12. Show that the generalized Fisher's equation

$$u_t - u_{xx} + \frac{(\beta + 1)}{u} u_x^2 - u\left(u^\beta - 1\right) = 0$$

can be transformed into the linear heat equation

$$v_{tt} - v_{xx} = \beta(v - 1)$$

by letting $u = v^{-1/\beta}$.
13. Show that the nonlinear diffusion equation $u_{xx} = u_x^2 u_t$ can be transformed into the linear heat equation $\omega_{z_1 z_1} = \omega_{z_2}$ through the transformation via $z_1 = u$, $z_2 = t$ and $\omega = x$.

16. Nonlinear Dynamics: From Theory to Technology

Starting from the basic definition of nonlinearity and nonlinear dynamical systems, we have passed through several stages of developing novel concepts such as bifurcations, chaos, controlling, synchronization, integrability, solitons, patterns and so on in the previous fifteen chapters. We have also seen from time to time various physical applications in which nonlinear dynamical systems naturally arise and exhibit one or other or many of the above properties. In this final and concluding chapter, we proceed further and try to point out that the various theoretical notions we have introduced so far are also developing into exciting applications in frontier or cutting edge technologies. We have already indicated a few such possibilities like optical solitons for efficient communication, spread spectrum secure communication through chaos synchronization, stochastic resonance for medical applications, to name a few. In the following, we wish to highlight some of the exciting technological possibilities/real world applications such as

1. chaotic cryptography
2. chaos based internet protocol
3. chaos in communications
4. chaos in financial marketing
5. optical soliton based communication system
6. soliton based optical computers and
7. nonlinear dynamics of magnetic recording

just to indicate the potentialities and to give a flavour of the developing technologies. The list is by no means exhaustive and it is only indicative of the numerous possibilities.

In any case it is highly satisfying and appealing to realize that already many of the theoretical notions of nonlinear dynamics are finding their way into applications in current technologies which can lead to revolutionary developments in our daily life. We will briefly indicate in the following some of these potential technological applications.

16.1 Chaotic Cryptography

In Chap. 9 (Sect. 9.5), we have seen that identical chaotically evolving systems can be synchronized under suitable coupling. This has led to the possibility of spread spectrum secure communication, which is developing into a powerful technological tool. In addition, the notion of chaos synchronization can also be profitably used to develop a safe and reliable cryptographic system (transmission of secretive messages) for private use as shown by He and Vaidya [1]. In particular such secure methods are very much in need because cryptography is extremely useful in this age of information revolution not only for military and diplomatic purposes but also for remote log-in protocols, electronic voting, electronic banking, distributed management of data bases and so on. In order to appreciate these possibilities, let us first introduce some of the basic notions of classical cryptography [1].

16.1.1 Basic Idea of Cryptography

The basic method of classical cryptography is first to break the intended message into units which are then coded as the *plain text*. The basic unit could be a single letter such as a member of the alphabets or equivalent numerals or punctuation marks and so on. For example, the letters $(A, ..., Z)$ can be coded into the numerals $(0, ..., 25)$. Then the process of encryption is that the plain text p is disguised into a *cipher text* c through a *secret key* k by a suitable operation, for example $c = p + k \bmod(26)$, and k is a set of nonrepeating integers $k \in (k_1, k_2, \ldots, k_{26})$. A reliable and preferable cryptographic system can be obtained when the secret key k is

(1) as random as possible
(2) not smaller than the size of the message and
(3) available to both the sender and receiver simultaneously.

It is interesting to note that all these criteria are satisfied to a great extent by a chaotic signal (as the key k) when the underlying chaotic system is synchronized with another identical system through appropriate coupling. The idea can be further extended to nonidentical systems through the notion of generalized synchronization. Let us briefly illustrate the method of He and Vaidya [1] in the following.

16.1.2 An Elementary Chaotic Cryptographic System

Let us assume that a sender and a receiver want to use a cryptographic system based on a chaotic signal as the key. Let them choose a chaotic system such as the Lorenz system, the autonomous Duffing–van der Pol (ADVP) oscillator or a Murali–Lakshmanan–Chua (MLC) circuit or any other similar dynamical system discussed earlier in the book. The control parameters are chosen such that the signal is chaotic. Starting from a predetermined initial time after leaving out the transients, let the sender and receiver agree to make measurements at suitable intervals of time so as to minimize correlation between signals at successive instants.

As an example, we choose the ADVP oscillator [2] satisfying the following dynamical equations as the drive system

$$\dot{x}_1 = -\nu\left(x_1^3 - \alpha x_1 - y_1\right) , \tag{16.1a}$$

$$\dot{y}_1 = x_1 - y_1 - z_1 , \tag{16.1b}$$

$$\dot{z}_1 = \beta y_1 , \tag{16.1c}$$

which was shown to be synchronizable with an identical system with one way coupling in the form

$$\dot{x}_2 = -\nu\left(x_2^3 - \alpha x_2 - y_2\right) + \nu\epsilon\left(x_1 - x_2\right) , \tag{16.2a}$$

$$\dot{y}_2 = x_2 - y_2 - z_2 , \tag{16.2b}$$

$$\dot{z}_2 = \beta y_2 , \tag{16.2c}$$

when the parameters are fixed at $\alpha = 0.45$, $\nu = 100$, $\beta = 300$ and $\epsilon = 1.35$.

The possibility of synchronization can be proved by showing that there exits a Lyapunov function associated with systems (16.1) and (16.2) in the form

$$E = (\beta/2)\left(x_1 - x_2\right)^2 + (\nu\beta/2)\left(y_1 - y_2\right)^2 + (\nu/2)\left(z_1 - z_2\right)^2 . \tag{16.3}$$

Now we make a set up as shown in Fig. 16.1. While the coupling between the sender and receiver is made through x coupling, the secret key is derived from the z-component, for example. A more complex key can be obtained by compounding x, y and z signals in a desired fashion. For example one can assign the key as

$$k = 1000\left|z_1\right| \tag{16.4}$$

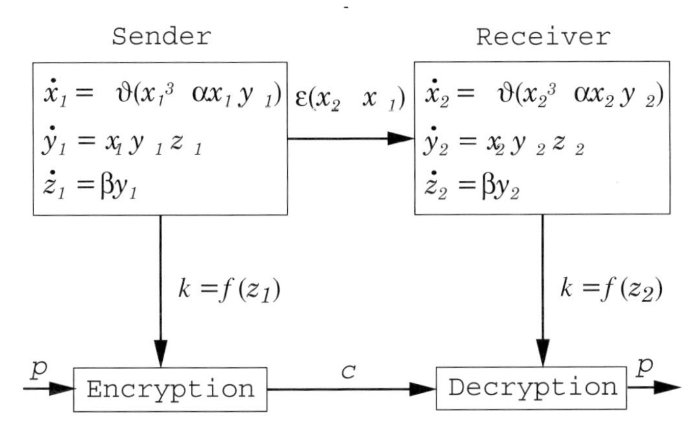

Fig. 16.1. Schematic diagram of chaotic cryptography using ADVP oscillator

by the sender. It can be simultaneously obtained by the receiver as the systems are in synchronization and so the later will also obtain the key as

$$k = 1000\,|z_2| = 1000\,|z_1| \ . \tag{16.5}$$

Then encryption can be made at the sender's end to obtain the cipher text as

$$c = p + 1000\,|z_1| \ . \tag{16.6}$$

At the receiver's end, the plain text can be recovered through the decryption

$$p = c - 1000\,|z_2| = c - 1000\,|z_1| \ . \tag{16.7}$$

Now an eavesdropper who tries to intercept cannot in general hope to synchronize with the sending chaotic system as the structure of the system is secret. Even if he succeeds to construct the attractor from the intercepted chaotic signal, as long as the key is uncorrelated with the x-signal, the key is secure and so deciphering is almost impossible.

In Tables 16.1 and 16.2, we demonstrate how the text '*START OPERATION*' is first converted into a cipher text (Table 16.1) by means of the chaotic signal of the ADVP oscillator used as the key from a predetermined time $t = 25$ units. The measurements are made at each time unit leading to a different key to disguise a different message unit. In Table 16.2 using the synchronized chaotic signal, the key is used by the receiver to decrypt the signal to receive the message securely.

The security of the key can be further enhanced by additional modifications to the notion of synchronization of chaos such as generalized synchronization, compounding of signals, multi-step parameter modulation, etc. and much further research is being carried on the engineering aspects in connection with the actual implementation of the method.

Exercises:

1. Investigate chaos cryptography using synchronization of (a) Lorenz system and (b) Chua's circuit for secure transmission of a picture.
2. Investigate how internet protocols can be developed using chaotic cryptography for secure communication (see for example, Ref. [3]).

16.2 Using Chaos (Controlling) to Calm the Web

In recent times it has been increasingly realized that traffic flow in congested motor-ways shows self-similar, fractal pattern. Several mathematical models have been devised to represent chaotic traffic flow. Interestingly, following such studies internet trafficking and video trafficking have been modelled on similar lines.

Table 16.1. Chaotic cryptography (transmitter)

Time t	$z_1(t)$	Keys k	Plain text p	Cipher text $c = p + k \bmod(26)$
25.0	0.7397	0739	$18(S)$	$29(18+11)$
25.1	−3.0795	3079	$19(T)$	$30(19+11)$
25.2	−1.3162	1316	$00(A)$	$16(00+16)$
25.3	2.1797	2179	$17(R)$	$38(17+21)$
25.4	−0.5157	0515	$19(T)$	$40(19+21)$
25.5	−3.5087	3508	$14(O)$	$38(14+24)$
25.6	0.0924	0092	$15(P)$	$29(15+14)$
25.7	2.8198	2819	$04(E)$	$15(04+11)$
25.8	0.2256	0225	$17(R)$	$34(17+17)$
25.9	−1.4845	1484	$00(A)$	$02(00+02)$
26.0	1.6979	1697	$19(T)$	$26(19+07)$
26.1	2.4876	2487	$08(I)$	$26(08+18)$
26.2	−0.9716	0971	$14(O)$	$23(14+09)$
26.3	−0.8177	0817	$13(N)$	$24(13+11)$

An interesting report by M. Sincell [4] which has appeared in the journal 'Physics World', published by the Institute of Physics, U.K., claims that physicists use chaos to calm the Web. By this it is meant that bottlenecks created by increased flow of information in the Internet can be controlled and cleared and the flow can be speeded up just like the way in which efforts have been made to clear chaotic traffic jams. The report presents some qualitative details as to how this is technologically possible. In the following we essentially make use of this report to demonstrate another emerging technology using chaos.

Consider the flow of traffic in a motor-way. By analysing aerial photographs of traffic flows, scientists have identified chaotic oscillations in congested parts of the motor-way [5]. By detailed analysis of the underlying dynamics, it can be shown that by separating lanes of the motor-way slip road during heavy traffic flow one can control the chaos that has occurred on the road and increase the speed of the traffic flow.

Table 16.2. Chaotic cryptography (receiver)

Time t	$z_2(t)$	Keys k	Cipher text c	Plain text $p = c - k \bmod(26)$
25.0	0.7397	0739	29	18(S)
25.1	−3.0795	3079	30	19(T)
25.2	−1.3162	1316	16	00(A)
25.3	2.1797	2179	38	17(R)
25.4	−0.5157	0515	40	19(T)
25.5	−3.5087	3508	38	14(O)
25.6	0.0924	0092	29	15(P)
25.7	2.8198	2819	15	04(E)
25.8	0.2256	0225	34	17(R)
25.9	−1.4845	1484	02	00(A)
26.0	1.6979	1697	26	19(T)
26.1	2.4876	2487	26	08(I)
26.2	−0.9716	0971	23	14(O)
26.3	−0.8177	0817	24	13(N)

In the case of Internet also the idea is that a similar proposal can work for Internet traffic congestion. The point is that exponential increase in the volume of Internet traffic is causing very complex problems in the information transmission: Bottlenecked information in the Internet simply disappears. This means in the modern world of high technology, information loss is equivalent to loss of money. As considerable information indeed is lost in the Internet, a huge financial loss does occur.

One can try to compare the Internet traffic with that of traffic of cars and vehicles in a motor-way/highway. Like those on the motor-way, it is claimed that Internet traffic jams are chaotic and not random. This is a very important distinction, because we have seen in Chap. 9 (Sect. 9.4) that chaos can be controlled by different methods. One particular way of controlling is to identify the unstable periodic orbits (UPOs) associated with the chaotic attractor and then stabilize these orbits by appropriate intervention. One example is the OGY (Ott–Grebogi–Yorke) method of controlling chaos as briefly discussed in Sect. 9.4.

Let us look for a moment into the way the Internet traffic occurs. It is shown schematically in Fig. 16.2. The figure is self evident. Now chaotic Internet traffic causes much of its havoc in the so-called router (see the schematic Fig. 16.2). As kilobyte-sized information packets, which can be called as the cars of the information superhighway, zip to and from computers in the network, they occasionally pass through the router, which redirects packets towards their final destination. It takes time for a router to decide where to send a packet, and while making the decision, routers store packets in memory buffers as Sincell points out in his Physics World report [4].

During an Internet rush hour, packets start coming in faster than the router can redirect them and send them on their way, so the buffers fill up. To make room for incoming traffic, the router starts deleting, or dropping from the buffer. All dropped packets must be retransmitted. Consequently an overloaded router can really slow down response times. Though technical details are not available, several software companies have made use of chaos control techniques like in the motor-way traffic problems so that a significant reduction in retransmission for congested links occur.

Exercise:

3. Develop a suitable nonlinear model for chaotic traffic flow and analyse its dynamics (see for example, Refs. [6,7]).

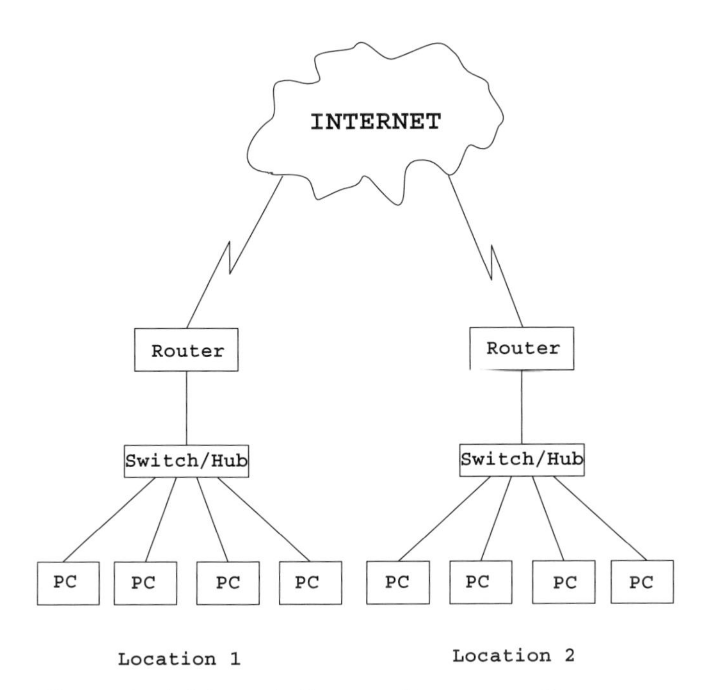

Fig. 16.2. A Schematic diagram of internet traffic

16.3 Some Other Possibilities of Using Chaos

Using chaos in a purposeful way is gaining momentum in different directions. These range from applications in communications to financial markets. Without going into the technical details, we quote in the following two reports which appeared in the recent 'Physics World' issues [8,9] which are essentially self explanatory. We include these reports merely to show the many technological possibilities which can be thought of among a variety of different applications.

16.3.1 Communicating by Chaos

The Physics World report of September 2001 (p.6) by J. Moore states the following. Microelectronic engineers at University College, Cork in Ireland, have invented a new type of communications receiver that uses chaos to overcome signal interference. The receiver, which has been developed by a team led by Peter Kennedy, uses a technique known as differential chaos shift keying (DCSK). It could enable much cleaner signals to be transmitted in "multi-path" channels – where multiple digital signals coexist from different electronic sources.

The use of chaotic signals to enhance mobile communication has attracted growing interest in recent years because they can suppress unwanted signals in a multi-path environment. Stray or source signals that reflect off walls or buildings, for example, can interfere with a signal and cause it to fade, which is why mobile-phone signals are sometimes so weak. Unlike conventional–"coherent"–communications systems, where the signal consists of weighted sums of sinusoidal wave functions, non-coherent–"chaotic"–communications systems send segments that are spread over a broad range of frequencies and are at relatively low signal densities.

The fact that chaotic communications systems send so many different chaotic segments is the key to why they work better in multi-channel environments. A particular chaotic segment that reflects off a wall, for example, may try to interfere with the original signal at the DCSK-enabled receiver, which could be built into a mobile phone. However, it will only interfere destructively with the small component of the original signal that has the same chaotic characteristics. All other segments of original signal remain unaffected.

It appears that this kind of work is a very important step towards the use of chaos in real communications systems. Kennedy's group developed and tested their system over the experimental Industrial, Scientific, Medical (ISM) band, which operates at about 2.4GHz. The band uses Bluetooth technology, in which low power transmitters are allowed to operate without a governmental licence. It is claimed that such a system has very low sensitivity to Doppler effects, low sensitivity to impulsive interference, and is tolerant to transmitter and receiver nonlinearities and is also a potentially low cost system.

16.3.2 Chaos and Financial Markets

In recent times both academic and applied researchers studying financial market and other economic indicators are trying to apply ideas from chaotic dynamics [10]. In fact some economists and management scientists believe that nonlinear dynamics concepts may prove to be as powerful ideas in finance as they have been in biology. Typically one tries to understand commonalities of things that are driven by selection and innovation, which are basic to nonlinear evolution of dynamical systems as we have seen in earlier chapters. Such evolutions are taking place all the time in financial markets, social changes in societies, etc. In particular, there exists considerable empirical evidence for chaos in financial markets and macroeconomic indicators.

One way of looking at the above connection is the following. Our earlier discussions in Chaps. 4–9 clearly indicated that chaotic behaviour is very much related to disorder and hence disorder in different aspects of world can be related to chaos. Multidimensional nonlinear dynamical systems exhibiting chaos have no possibility of long term predictability, but they do have a small exploitable amount of short term order. (Recall the fact that evolution of nearby trajectories have correlation over a short period initially before diverging). For example, the Physics World report of P. Guynne indicates that J. Doyne Farmer, presently at Santa Fe Institute, has applied chaos theory to roulette wheels and tested chaos concepts to predict in real time where the ball would land on the real roulette wheels in the Casinos of Las Vegas. Such ideas lead to the possibility of beating the standard patterns in gambling.

Consequently attempts are being made to forecast market behaviour by using the short-term predictability inherent in chaos theory. Actually what one does is to create algorithms which can identify islands of predictability in the market and put money on those predictions. Typically one looks through historical data for patterns and identifies them. These are then implemented in computer modules, which integrates considerable amount of various informations together to make trading decisions.

Exercises:

4. There has been efforts to use chaotic lasers for communications purpose. Develop a project on such a possibility (see for example, Ref. [11]).
5. Carry out a time series analysis of the daily return data of any of the leading stock exchange markets of the world for the past five years and identify the nonlinear and chaotic characteristics in these data (see for example for such an analysis on Athens Stock Exchange Market, Ref. [12]).

16.4 Optical Soliton Based Communications

Earlier in Chap. 14 (Sect. 14.1), we have introduced how optical solitons arise in connection with the propagation of intense electromagnetic waves in optical fibres. In the present section, we present a capsule summary of the technological applications of these optical solitons which brought the studies on them to the verge of the modern cutting edge technology. The optical soliton, which forms due to a delicate balance between the group velocity dispersion and the Kerr nonlinearity (intensity dependent refractive index), is capable of propagating over long distances without change in its waveform and preserves shape and energy during collision with other solitons. This property has led scientists and engineers to choose optical soliton as the right candidate for optical transmission which will fetch a high speed and long distance global network. A detailed analysis of the historical background, various developments and the present state of the technology is discussed by various experts in a special issue of the September 2000 (Vol.10, No.3) issue of the journal 'Chaos' (American Institute of Physics). Some basic ideas alone are given below.

It was in the year 1973 that Hasegawa and Tappert [13] have theoretically demonstrated that pico-second (10^{-12} sec.) pulse propagation through single mode optical fiber is governed by NLS equation (as pointed out in Sect. 14.1) and proved the existence of optical soliton. Following their work Mollenauer, Stolen and Gordon [14] have experimentally succeeded in propagating optical solitons with a duration of 10 pico-seconds with carrier wavelength 1.5μ m in optical fibers. These theoretical predictions and their experimental confirmations (observations) attracted serious interest in the optical communication community and provided a way to construct all-optical transmission system.

The aforementioned soliton propagation through optical fiber suffers a major impediment due to fiber loss, an inevitable loss arising due to fiber property, and hence requires repeaters to be installed periodically in order to regenerate them. In the year 1983 Hasegawa [15] made the imaginative proposal of repeaterless communications which uses the Raman gain of fiber itself to amplify the optical solitons. Making use of this idea in the year 1988 Mollenauer and Smith [16] demonstrated experimentally the first long distance all-optical soliton transmission, showing soliton propagation over more than 4000kms, using a recirculating loop. In the mean time fiber amplifiers have been also developed to overcome the difficulties arising in Raman fiber amplifiers. To be specific, the invention of erbium doped fiber amplifiers enhanced the study on the optical soliton regeneration. In 1989 Nakazawa, Kimura and Suzuki [17] demonstrated experimentally the reshaping of optical solitons. Following this an immense amount of research work is being carried out at present in this direction. It has also been observed that light amplification in erbium-doped glass fibers allows intercontinental communication at 10 billion bits per second and opens new avenues of data transmission via optical solitons.

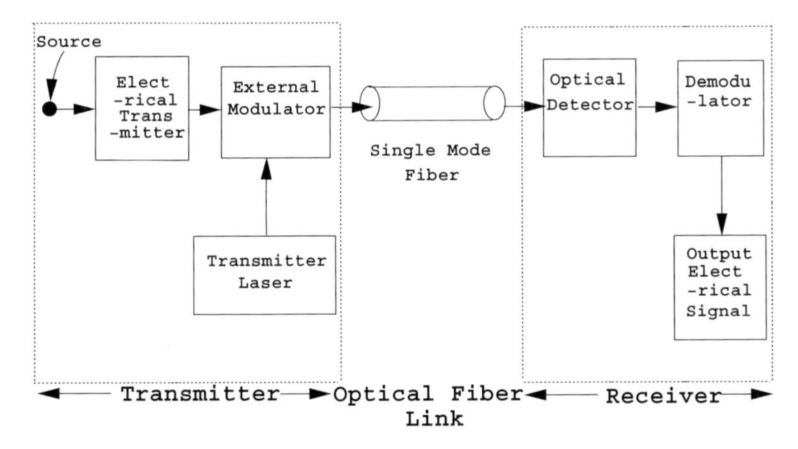

Fig. 16.3. Schematic diagram of the fiber-based coherent optical communications system

Another important area in which the optical soliton propagation has potential application is the wavelength division multiplexing (WDM) system, which is a consequence of the wide bandwidth of optical transmission systems. Usually in linear regime only one channel can be transmitted, as its wavelength must be at the zero dispersion point of the fiber. But as we have seen earlier soliton propagation requires nonzero dispersion and as a result of this several channels can be transmitted at different wavelengths. Further the robustness of these solitons to fiber nonlinearities other than the Kerr effect makes it advantageous over multiple-channel linear system.

For illustrative purpose, we have given below in Fig. 16.3 the schematic diagram of the fiber-based coherent optical communications system.

Not withstanding the above positive aspects, optical soliton transmission length is limited by Gordon–Haus timing jitter and the spontaneous emission-noise [18, 19]. These length limits can be overcome by using various soliton transmission control schemes, like the use of sliding frequency-guiding filters [20], periodic synchronous modulation with optical filtering [21], etc. Further, in real optical communication lines the varying dispersion will degrade the soliton pulse quality. In order to overcome this the concept of dispersion compensation has been introduced. Here by changing dispersion proportional to the soliton power the dispersive wave radiation and collision induced frequency shift have been reduced. For more detailed analysis regarding such dispersion management technique see the various articles in the special issue of 'Chaos' mentioned above.

Exercises:

6. Investigate how the nonlinear Schrödinger equation model for pulse propagation in optical fibers has to be modified to include the effects of fiber loss and Raman gain [13,19].

7. Analyse the various technological issues involved in the use of optical solitons for efficient communications taking into account the positive and negative aspects.
8. Investigate the effect of Gordon–Haus timing jitter and spontaneous emission noise [18,19] on pulse propagation in optical fibers.

16.5 Soliton Based Optical Computing

Towards the end of Chap. 14, we mentioned that the Manakov model represented by the set of two-coupled nonlinear Schrödinger equations (CNLS),

$$\mathrm{i}q_{1t} + q_{1xx} + 2\mu\left(\left|q_1^2\right| + \left|q_2^2\right|\right)q_1 = 0\,, \tag{16.8a}$$

$$\mathrm{i}q_{2t} + q_{2xx} + 2\mu\left(\left|q_1^2\right| + \left|q_2^2\right|\right)q_2 = 0\,, \tag{16.8b}$$

possesses certain novel properties, which are not present in the standard soliton equations in (1+1) dimensions. This is concerned with the shape changing nature of solitons under collision, in contract to the pure shape preserving elastic collision possessed by the standard system (of Chap. 14). In this section, we will point out briefly how this property can be profitably used to develop an all optical computer without interconnecting discrete components.

16.5.1 Photo-Refractive Materials and the Manakov Equation

The above type of CNLS equations are not only useful to describe soliton propagation in optical fibers, but can also be shown to describe incoherent soliton beam propagation in the so-called photo-refractive materials. These materials can often exhibit high nonlinearity with very low intensity optical pulses, even with milliwatt power lasers or incandescent bulbs. Examples of such photo-refractive materials include lithium niobate ($LiNiO_3$) and strontium barium niobate ($SrBaNiO_3$) crystals.

It can be shown that beam propagation in a photo-refractive medium is governed by the scalar (single component) nonlinear Schrödinger equation, where the beam is propagating along the z-axis and is allowed to diffract only in the x-direction. The soliton formation here arises due to a delicate balance between the diffraction and the nonlinearity which arise in the medium [22,23]. One can also show that two component beam propagation in such photo-refractive medium is governed by the 2-CNLS equation (16.8). Generalizing this, it can be shown that the governing equation of N-self trapped mutually incoherent wave packets in such a medium is given by the following N-CNLS equations [24]

$$\mathrm{i}q_{jz} + q_{jxx} + 2\mu\sum_{p=1}^{N}\left|q_p\right|^2 q_j = 0\,, \quad j = 1, 2, ..., N. \tag{16.9}$$

Here q_j represents the jth component of the beam and $\sum_{p=1}^{N} |q_p|^2$ represents the change in the refractive index profile created by all the incoherent components of the light beam and 2μ represents the strength of nonlinearity.

16.5.2 Soliton Solutions and Shape Changing Collisions

The Manakov system (16.8) is completely integrable: It possesses a Lax pair of (3×3) matrix form as given in problem 36 below Chap. 14. It can be bilinearized with the substitution

$$q_1 = \frac{g_1}{f}, \quad q_2 = \frac{g_2}{f}. \tag{16.10}$$

Proceeding as in the case of the NLS equation in Chap. 14 (Sect. 14.3.3), one can obtain the one-soliton solution [25] as

$$\begin{pmatrix} q_1 \\ q_2 \end{pmatrix} = \begin{pmatrix} A_1 \\ A_2 \end{pmatrix} k_{1R} \mathrm{e}^{\mathrm{i}\eta_{1I}} \operatorname{sech}\left(\eta_{1R} + R/2\right), \tag{16.11}$$

where $\eta_j = k_j\left(x + \mathrm{i}k_j t\right)$, $j = 1$, $A_1 = \alpha_1^{(1)}/\Delta$, $A_2 = \alpha_1^{(2)}/\Delta$, $\Delta = \sqrt{\mu\left(\left|\alpha_1^{(1)}\right|^2 + \left|\alpha_1^{(2)}\right|^2\right)}$, $\mathrm{e}^R = \Delta^2/\left(k_1 + k_1^*\right)^2$, $\alpha_1^{(1)}$, $\alpha_1^{(2)}$, and $k_1 \left(= k_{1R} + \mathrm{i}k_{1I}\right)$ are complex constants. Here $\sqrt{\mu}\left(A_1, A_2\right)^{\mathrm{T}}$ represents the unit polarization vector, $\left(k_{1R}A_1, k_{1R}A_2\right)$ gives the amplitudes of the two modes q_1 and q_2, respectively, of the Manakov one-soliton, subject to the condition $|A_1|^2 + |A_2|^2 = \mu^{-1}$ (see Fig. 14.5).

The two-soliton solution is a bit more complicated expression, though it can be explicitly written down using the Hirota bilinear form of (16.8) using the transformation (16.10) mentioned above. For details see Ref. [25]. The general form of the two-soliton solution is

$$q_1 = \left(\alpha_1^{(1)}\mathrm{e}^{\eta_1} + \alpha_2^{(1)}\mathrm{e}^{\eta_2} + \mathrm{e}^{\eta_1+\eta_1^*+\eta_2+\delta_1} + \mathrm{e}^{\eta_1+\eta_2^*+\eta_2+\delta_2}\right)\Big/D, \tag{16.12a}$$

$$q_2 = \left(\alpha_1^{(2)}\mathrm{e}^{\eta_1} + \alpha_2^{(2)}\mathrm{e}^{\eta_2} + \mathrm{e}^{\eta_1+\eta_1^*+\eta_2+\delta_1'} + \mathrm{e}^{\eta_1+\eta_2^*+\eta_2+\delta_2'}\right)\Big/D, \tag{16.12b}$$

where

$$\begin{aligned} D = 1 &+ \mathrm{e}^{\eta_1+\eta_1^*+R_1} + \mathrm{e}^{\eta_1+\eta_2^*+\delta_0} + \mathrm{e}^{\eta_1^*+\eta_2+\delta_0^*} + \mathrm{e}^{\eta_2+\eta_2^*+R_2} \\ &+ \mathrm{e}^{\eta_1+\eta_1^*+\eta_2+\eta_2^*+R_3}, \end{aligned}$$

$$\mathrm{e}^{\delta_0} = \frac{\kappa_{12}}{k_1 + k_2^*}, \quad \mathrm{e}^{R_1} = \frac{\kappa_{11}}{k_1 + k_1^*}, \quad \mathrm{e}^{R_2} = \frac{\kappa_{22}}{k_2 + k_2^*},$$

$$\mathrm{e}^{\delta_1} = \frac{k_1 - k_2}{\left(k_1 + k_1^*\right)\left(k_1^* + k_2\right)}\left(\alpha_1^{(1)}\kappa_{21} - \alpha_2^{(1)}\kappa_{11}\right),$$

$$\mathrm{e}^{\delta_2} = \frac{k_2 - k_1}{\left(k_2 + k_2^*\right)\left(k_1 + k_2^*\right)}\left(\alpha_2^{(1)}\kappa_{12} - \alpha_1^{(1)}\kappa_{22}\right),$$

$$\mathrm{e}^{\delta_1'} = \frac{k_1 - k_2}{\left(k_1 + k_1^*\right)\left(k_1^* + k_2\right)}\left(\alpha_1^{(2)}\kappa_{21} - \alpha_2^{(1)}\kappa_{11}\right),$$

$$
\begin{aligned}
\mathrm{e}^{\delta_2'} &= \frac{k_2 - k_1}{(k_2 + k_2^*)(k_1 + k_2^*)} \left(\alpha_2^{(2)} \kappa_{12} - \alpha_1^{(2)} \kappa_{22} \right) , \\
\mathrm{e}^{R_3} &= \frac{|k_1 - k_2|^2}{(k_1 + k_1^*)(k_2 + k_2^*)|k_1 + k_1^*|^2} (\kappa_{11}\kappa_{22} - \kappa_{12}\kappa_{21}) , \\
\kappa_{ij} &= \left(\mu \sum_{n=1}^{2} \alpha_i^{(n)} \alpha_j^{(n)*} \right) \Big/ (k_i + k_j^*) , \quad i, j = 1, 2.
\end{aligned} \tag{16.12c}
$$

An asymptotic ($t \to \pm\infty$) analysis of the above two-soliton solution reveals the dramatic fact that the amplitude of the solitons can change under collision, even though their velocities remain unchanged (see Fig. 14.6). The standard elastic collisions occur under very special circumstances. In fact one obtains asymptotically the following.

(i) Limit $t \to -\infty$:

(a) S_1 **(**$\eta_{1R} \sim 0,\ \eta_{2R} \to -\infty$**):**

$$
\begin{aligned}
\begin{pmatrix} q_1 \\ q_2 \end{pmatrix} &\to \begin{pmatrix} q_1^{(1)}(t \to -\infty) \\ q_2^{(1)}(t \to -\infty) \end{pmatrix} \\
&= \begin{pmatrix} \alpha_1^{(1)} \\ \alpha_1^{(2)} \end{pmatrix} \left(\mathrm{e}^{\eta_1} \Big/ \left(1 + \mathrm{e}^{\eta_1 + \eta_1^* + R_1} \right) \right) \\
&= \begin{pmatrix} A_1^{1-} \\ A_2^{1-} \end{pmatrix} k_{1R} \mathrm{e}^{\mathrm{i}\eta_{1I}} \operatorname{sech}(\eta_{1R} + R_1/2) ,
\end{aligned} \tag{16.13a}
$$

where

$$
\left(A_1^{1-}, A_2^{1-}\right) = \left[\mu \left(\alpha_1^{(1)} \alpha_1^{(1)*} + \alpha_1^{(2)} \alpha_1^{(2)*} \right) \right]^{-1/2} \left(\alpha_1^{(1)}, \alpha_1^{(2)} \right) . \tag{16.13b}
$$

Here superscript 1^- denotes S_1 at the limit $t \to -\infty$ and subscripts 1 and 2 refer to the modes q_1 and q_2, respectively.

(b) S_2 **(**$\eta_{2R} \sim 0,\ \eta_{1R} \to \infty$**):**

$$
\begin{aligned}
\begin{pmatrix} q_1 \\ q_2 \end{pmatrix} &\to \begin{pmatrix} q_1^{(2)}(t \to -\infty) \\ q_2^{(2)}(t \to -\infty) \end{pmatrix} \\
&= \begin{pmatrix} \mathrm{e}^{\delta_1 - R_1} \\ \mathrm{e}^{\delta_1' - R_1} \end{pmatrix} \left(\mathrm{e}^{\eta_2} \Big/ \left(1 + \mathrm{e}^{\eta_2 + \eta_2^* + R_3 - R_1} \right) \right) \\
&= \begin{pmatrix} A_1^{2-} \\ A_2^{2-} \end{pmatrix} k_{2R} \mathrm{e}^{\mathrm{i}\eta_{2I}} \operatorname{sech}(\eta_{2R} + (R_3 - R_1)/2) ,
\end{aligned} \tag{16.14a}
$$

where

$$A_l^{2-} = (a_1/a_1^*)\ c\left[\mu\left(\alpha_2^{(1)}\alpha_2^{(1)*} + \alpha_2^{(2)}\alpha_2^{(2)*}\right)\right]^{-1/2} \times \left[\alpha_1^{(l)}\kappa_{11}^{-1} - \alpha_2^{(l)}\kappa_{21}^{-1}\right], \quad l = 1, 2 \tag{16.14b}$$

in which

$$a_1 = (k_1 + k_2^*)\left[(k_1 - k_2)\left(\alpha_1^{(1)*}\alpha_2^{(1)} + \alpha_1^{(2)*}\alpha_2^{(2)}\right)\right]^{1/2} \tag{16.14c}$$

and

$$c = \left(\frac{1}{|\kappa_{12}|^2} - \frac{1}{\kappa_{11}\kappa_{22}}\right)^{-1/2}. \tag{16.14d}$$

(ii) Limit $t \to -\infty$:

(a) $S_1\ (\eta_{1R} \sim 0,\ \eta_{2R} \to \infty)$**:**

$$\begin{pmatrix} q_1 \\ q_2 \end{pmatrix} \to \begin{pmatrix} q_1^{(1)}(t \to +\infty) \\ q_2^{(1)}(t \to +\infty) \end{pmatrix} = \begin{pmatrix} e^{\delta_2 - R_2} \\ e^{\delta_2' - R_2} \end{pmatrix}\left(e^{\eta_1} \Big/ \left(1 + e^{\eta_1 + \eta_1^* + R_3 - R_2}\right)\right) = \begin{pmatrix} A_1^{1+} \\ A_2^{1+} \end{pmatrix} k_{1R} e^{i\eta_{1I}} \operatorname{sech}\left(\eta_{1R} + (R_3 - R_2)/2\right), \tag{16.15a}$$

where

$$A_1^{1+} = (a_2/a_2^*)\, c\left[\mu\left(\alpha_1^{(1)}\alpha_1^{(1)*} + \alpha_1^{(2)}\alpha_1^{(2)*}\right)\right]^{-1/2} \times \left[\alpha_1^{(1)}\kappa_{12}^{-1} - \alpha_2^{(1)}\kappa_{22}^{-1}\right], \quad l = 1, 2 \tag{16.15b}$$

in which

$$a_2 = (k_2 + k_1^*)\left[(k_1 - k_2)\left(\alpha_1^{(1)}\alpha_2^{(1)*} + \alpha_1^{(2)}\alpha_2^{(2)*}\right)\right]^{1/2}. \tag{16.15c}$$

(b) $S_2\ (\eta_{2R} \sim 0,\ \eta_{1R} \to -\infty)$**:**

$$\begin{pmatrix} q_1 \\ q_2 \end{pmatrix} \to \begin{pmatrix} q_1^{(2)}(t \to +\infty) \\ q_2^{(2)}(t \to +\infty) \end{pmatrix} = \begin{pmatrix} \alpha_2^{(1)} \\ \alpha_2^{(2)} \end{pmatrix}\left(e^{\eta_2} \Big/ \left(1 + e^{\eta_2 + \eta_2^* + R_2}\right)\right) = \begin{pmatrix} A_1^{2+} \\ A_2^{2+} \end{pmatrix} k_{2R} e^{i\eta_{2I}} \operatorname{sech}\left(\eta_{2R} + R_2/2\right), \tag{16.16a}$$

where

$$\left(A_1^{2+}, A_2^{2+}\right) = \left[\mu \left(\alpha_2^{(1)}\alpha_2^{(1)*} + \alpha_2^{(2)}\alpha_2^{(2)*}\right)\right]^{-1/2} \left(\alpha_2^{(1)}, \alpha_2^{(2)}\right) . \qquad (16.16b)$$

The complex parameters $A_i^{j\pm}$ satisfy the relation $|A_1^{j\pm}|^2 + |A_2^{j\pm}|^2 = \mu^{-1}$. It has been shown in Ref. [25] that the polarization components after collision, A_i^{j+}, are related to the components A_i^{j-} before collision through the relation $A_i^{j+} = T_i^j A_i^{j-}$, $i, j = 1, 2$, where

$$\begin{aligned} |T_j^1|^2 &= \left|1 - \lambda_2 \left(\alpha_2^{(j)}/\alpha_1^{(j)}\right)\right|^2 \Big/ |1 - \lambda_1\lambda_2| , \\ |T_j^2|^2 &= |1 - \lambda_1\lambda_2| \Big/ \left|1 - \lambda_1 \left(\alpha_1^{(j)}/\alpha_2^{(j)}\right)\right|^2 , \quad j = 1, 2, \\ \lambda_1 &= \kappa_{21}/\kappa_{11} \ \text{ and } \ \lambda_2 = \kappa_{12}/\kappa_{22} . \end{aligned} \qquad (16.17)$$

The above expressions clearly show that there is an intensity redistribution among the solitons S_1 and S_2. In fact, from the equations (16.13) to (16.17) we observe that the initial amplitudes of the two modes of the two-solitons $\left(A_1^{1-}k_{1R}, A_2^{1-}k_{1R}\right)$ and $\left(A_1^{2-}k_{2R}, A_2^{2-}k_{2R}\right)$, undergo a redistribution among themselves and the solitons emerge with amplitudes $\left(A_1^{1+}k_{1R}, A_2^{1+}k_{1R}\right)$ and $\left(A_1^{2+}k_{2R}, A_2^{2+}k_{2R}\right)$, respectively, where $A_i^{l\pm}$, $i, l = 1, 2$, are given above. The changes in the amplitudes due to collision are essentially given by the transition matrix T_i^j, as mentioned earlier which is equal to one only when $\alpha_1^{(1)}/\alpha_2^{(1)} = \alpha_1^{(2)}/\alpha_2^{(2)}$ and is in general different for all other values of $\alpha_i^{(j)}$'s, $i, j = 1, 2$.

The above shape changing interaction can also be specified in terms of a complex parameter ρ, which is the ratio of the expressions for the two modes in a particular soliton, that is

$$\rho_j^{\pm} = \frac{q_1^j(t \to \pm\infty)}{q_2^j(t \to \pm\infty)} = \frac{A_1^{j\pm}}{A_2^{j\pm}} , \quad j = 1, 2 \qquad (16.18)$$

which is a complex number that defines the change in the soliton state, besides the other unchanging complex number k_j, $j = 1, 2$. Let us represent the general two-soliton collision in the Manakov system schematically as in Fig. 16.4, where the complex numbers ρ_1, ρ_L, ρ_2 and ρ_R define the variable soliton states, while k_1 and k_2 denote the constant soliton parameters. From the two-soliton solutions presented by Radhakrishnan, Lakshmanan and Hieterinta [25] above and its asymptotic analysis, recently Jakubowski, Steigliz and Squier [26] have noted that equivalently the transition in the soliton states due to collision can be represented through a Möbius transformation or bilinear transformation, which is a linear fractional transformation (LFT) for the change in the state variables ρ_1 and ρ_L,

$$\rho_2 = \frac{a(\rho_1)\rho_L + b(\rho_1)}{c(\rho_1)\rho_L + d(\rho_1)}, \quad (ad - bc) \neq 0, \qquad (16.19a)$$

where

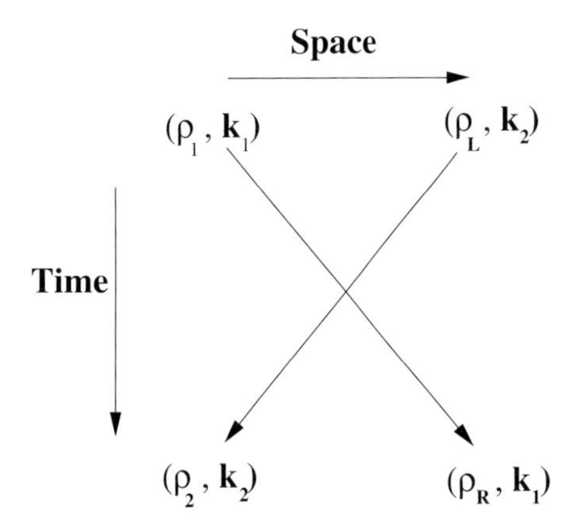

Fig. 16.4. Collision of two-CNLS solitons

$$a = (1-g)/\rho_1^* + \rho_1 \ , \quad b = g\rho_1/\rho_1^* \ , \quad c = g \ , \tag{16.19b}$$

$$d = (1-g)\rho_1 + 1/\rho_1^* \ , \quad g\,(k_1, k_2) = \frac{k_1 + k_1^*}{k_2 + k_1^*} \tag{16.19c}$$

and

$$\rho_R = \frac{a'\,(\rho_L)\,\rho_1 + b'\,(\rho_L)}{c'\,(\rho_L)\,\rho_1 + d'\,(\rho_L)} \ , \quad (a'd' - b'c') \neq 0, \tag{16.20a}$$

where

$$a' = (1-h^*)\,/\rho_L^* + \rho_L \ , \quad b' = h^*\rho_L/\rho_L^* \ , \quad c' = h^* \ , \tag{16.20b}$$

$$d' = (1-h^*)\,/\rho_L + 1/\rho_L^* \ , \quad h^* = h^*\,(k_1, k_2) = g\,(k_2, k_1) \ . \tag{16.20c}$$

In the above it is assumed, without loss of generality, that k_{1R}, $k_{2R} > 0$. The LFT's posses many interesting properties. Some of the important properties of the above transformations include the following:

1. Existence of inverse transformations
2. Existence of fixed points
3. Existence of implicit forms.

In particular, when viewed as an operator every soliton has an inverse, which will undo the effect of the operator on state, provided k remains unchanged. This property can then be used profitably to design logic gates as shown below.

16.5.3 Optical Soliton Based Computation

We have seen in the previous section that the optical solitons under energy sharing pairwise collisions undergo a state change represented by a LFT for the state variables ρ_1 and ρ_L of the two solitons into ρ_R and ρ_2, respectively, given by (16.20) and (16.19). Then one can treat the right moving soliton states as corresponding to data (particles) and left moving solitons as operators (or vice versa), such that

$$\rho_R = T_{\rho_L}(\rho_1) \ , \tag{16.21}$$

$$\rho_2 = T_{\rho_1}(\rho_L) \ . \tag{16.22}$$

Then, it is easy to see that for every operator T_{ρ_L} or T_{ρ_1}, there exists an inverse $T_{\rho_L}^{-1}$ or $T_{\rho_1}^{-1}$ such that the successive operations by the operator and its inverse restore the original data.

Example:

Let $\rho_L = 0$. Then using the LFT (16.20), we have

$$\rho_R = T_0(\rho_1) = (1-h^*)\,\rho_1 \ . \tag{16.23}$$

Let a second operator $\rho'_L = \infty$ operates on ρ_R (see Fig. 16.5). The new state ρ'_R is then

$$\rho'_R = \frac{1}{(1-h^*)}\rho_R = \frac{1}{(1-h^*)}(1-h^*)\,\rho_1 = \rho_1 \tag{16.24}$$

so that the initial state ρ_1 is restored. Thus 0 and ∞ are inverses to each other. Similarly, for every operator corresponding to the state ρ an inverse can be obtained. What is the consequence of this for computation?

The existence of an inverse for any given operator so that the data is restored on successive operation by the operator and its inverse allows one to assign logical, binary, 0 and 1 or TRUE and FALSE states in terms of the

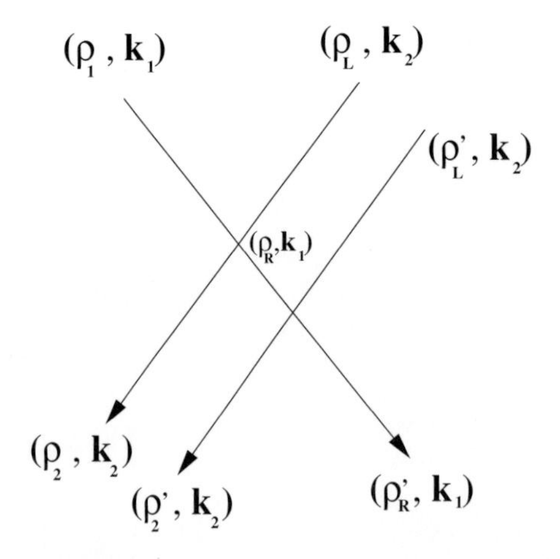

Fig. 16.5. State restoring property of shape changing solitons under collision

complex 0 and 1 states of the parameter ρ. In particular, one can use the actuator as 0 and the inverse as ∞ (see Fig. 16.5).

Using the above facts, very recently Steiglitz [27] has shown that one can construct the various logic gates such as COPY, FANOUT, NOT, ONE and finally the universal NAND gate also so as to develop a Turing equivalent computing machine (at least in a theoretical sense) purely based on optical soliton interactions which do not use any interconnecting discrete components in a bulk nonlinear medium like the photo-refractive materials.

For obtaining the above gates let us rotate the figure of the scattering process suitably so as to treat data as solitons travelling vertically downwards and the operators as solitons travelling horizontally (Fig. 16.6). With this picture one can design the various logic gates as follows.

A. The COPY Gate and FANOUT Gate

Let us consider the collision of three down moving solitons with a horizontal soliton as in Fig. 16.7 as studied by Steiglitz [27]. Let the input state be represented by ρ_1, while the actuator state is taken as $\rho_L = 0$. The other two solitons down moving are in the arbitrary states z and y. Then using the transformations (16.19) and (16.20), we can easily check that the horizontally moving soliton transforms after successive collisions respectively into

$$\rho_2 = T_{\rho_1}(0) , \tag{16.25}$$

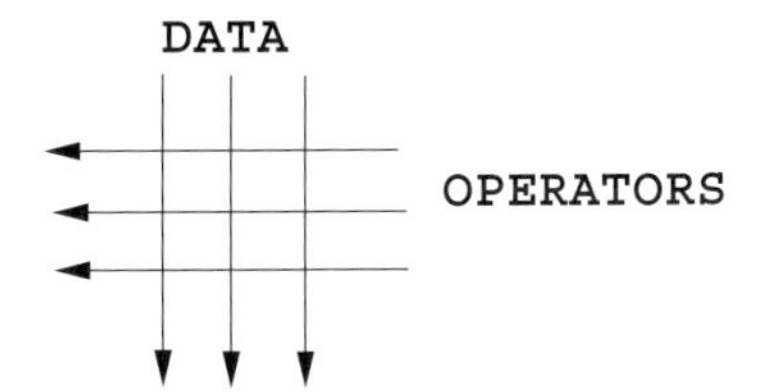

Fig. 16.6. The soliton collision in terms of data and operators [27]

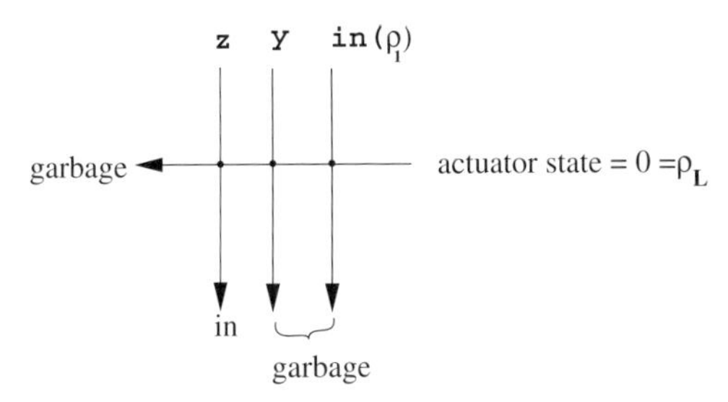

Fig. 16.7. Steiglitz's construction of COPY gate [27]

$$\rho_2' = T_y(\rho_2) = T_y[T_{\rho_1}(0)] \tag{16.26}$$

and finally the vertical down moving soliton z after collision becomes

$$\begin{aligned} \text{output} &= \rho_R' = T'_{\rho_2}(z) , \\ &= T_{T_y[T_{\rho_1(0)}]}(z) . \end{aligned} \tag{16.27}$$

Now, if we assign for the input state the logical values 0 or 1 corresponding $\rho_1 = 0$ or $\rho_1 = \infty$, respectively, and demand that in=out, we obtain two complex equations for the two complex arbitrary parameters y and z,

$$0 = T_{T_y[T_0(0)]}(z) , \tag{16.28}$$

$$\infty = T_{T_y[T_\infty(0)]}(z) . \tag{16.29}$$

These equations take the explicit form

$$\begin{aligned} &\{(1-h^*)[(1-g)yy^*+1][(1-g^*)yy^*+1]+gg^*yy^*\}z \\ &\quad +h^*g[(1-g^*)yy^*+1]y = 0 , \end{aligned} \tag{16.30}$$

$$\begin{aligned} &h^*z[g^*(1-g)]y^* + (1-h^*)[g(1-g^*)]y \\ &\quad +(1-g)(1-g^*)yy^* = 0 . \end{aligned} \tag{16.31}$$

Solving (16.30) and (16.31) one can obtain a set of solution (z_c, y_c), if it exists, which then copies the input at one site and places it at the output site, giving rise to the COPY gate.

In the above collision process, several unutilized solitons emerge after collision, which are named as 'garbage' solitons. However, one can use them also profitably to generate copies of inputs as outputs in a 'time gated' manner by colliding appropriate 'inverse' solitons with them. Thus one can obtain the FANOUT gate (Fig. 16.8).

B. The NOT and ONE Gates

As one has obtained the COPY gate by requiring the state out=in for both the logical values 0 and 1 (or the ρ values 0 and ∞), We can require that

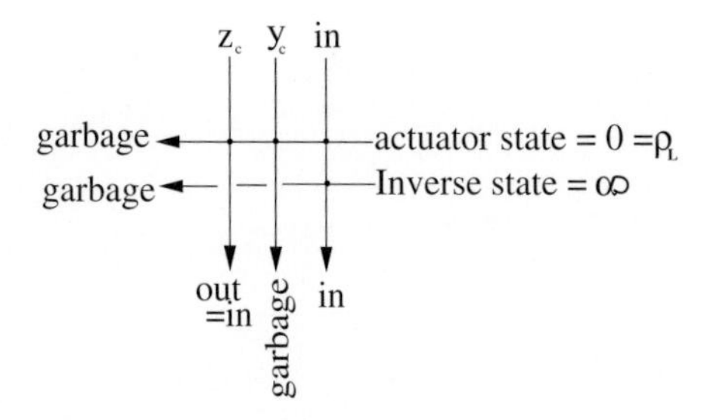

Fig. 16.8. FANOUT gate [27]

when the input state is 0 the output is 1 and when the input is 1 the output is 0. This requirement again gives two complex equations for the two unknowns z and y:

$$\infty = T_{T_y[T_0(0)]}(z)\ , \tag{16.32}$$

$$0 = T_{T_y[T_\infty(0)]}(z)\ . \tag{16.33}$$

The resultant solution gives the set (z_n, y_n), if it exists, giving rise to the NOT gate.

Similarly, if we require that for both the inputs 0 and 1, the output should be 1, the resultant two complex equations solve for (z_1, y_1) giving rise to the ONE gate.

C. The NAND Gate

The existence of FANOUT, NOT and ONE gates are sufficient to design a NAND gate [27]. Using two outputs as equivalent to a given input as in the FANOUT gate (Fig. 16.8), one can form a z converter and y converter such that the following values are chosen:

Value of input to the converter	Value of z converter	Value of y converter
0	z_1	y_1
1	z_n	y_n

In the above (z_1, y_1) are the values of (z, y) in the ONE gate, while (z_n, y_n) are the values of (z, y) in the NOT gate.

Thus, using one output from a given gate as the input of a FANOUT gate to produce the z and y converter to the required value as the above, while the other output as the second input of the present gate, one can have a standard three collision arrangement as discussed earlier. Then the following output results for the gate depicted in Fig. 16.9:

Left input	Right input	Value of z	Value of y	Output
0	0	z_1	y_1	1
0	1	z_1	y_1	1
1	0	z_n	y_n	1
1	1	z_n	y_n	0

As a result, one finds that in this two input one output arrangement one essentially obtains the universal NAND gate binary operations. It may be noted that all the above results can be obtained with the help of explicit

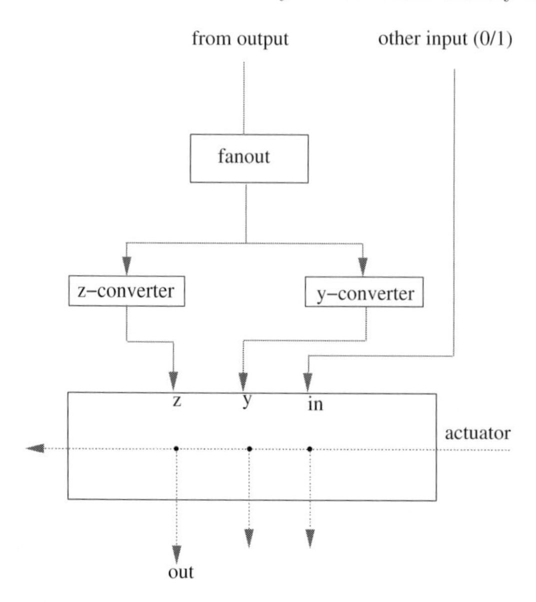

Fig. 16.9. Construction of the NAND gate (adapted from Ref. [27])

multisoliton solutions. However, such studies are beyond the scope of the book and so we do not discuss them here.

The importance of the NAND gate is that it is universal [27]. Using suitable interconnects and FANOUTS, it can lead to other logical functions. One can also implement 'wiring' through the light-light collisions, implying that one can implement any logic using the Manakov model.

Exercises:

9. Write down the bilinearized form of the Manakov system (16.8) using the transformation (16.10) and obtain the one-soliton solution (16.11) (see Ref. [25]).
10. Obtain the two-soliton solution (16.12) using the bilinearized version of (16.8).
11. Deduce the transition matrix T_i^j (see (16.17)) for collision of two Manakov solitons.
12. Using the transition matrices deduce the forms of the LFTs given by (16.19) and (16.20).
13. Numerically investigate the COPY gate equations (16.32) and (16.33) and identify suitable solutions.
14. Write down the explicit form of the NOT gate equation for the soliton parameters and numerically investigate the nature of the solution.
15. Deduce the equation of the ONE gate for the relevant soliton parameters and analyse it.

16.6 Micromagnetics and Magnetoelectronics

Micromagnetics is the subject which is concerned with the study of detailed magnetization reversal process in magnetic materials [28]. Particularly, it encompasses the study of ferromagnets and ferromagnetic thin films (used in magnetic thin film sensors and devices). The theory considers the ferromagnetic free energy in the ferromagnetic material to consist of

(i) the ferromagnetic exchange energy,
(ii) the magnetic anisotropy energy,
(iii) the magnetoelastic energy,
(iv) the magnetostatic energy and
(v) the magnetic potential energy due to external magnetic fields.

The magnetization orientation $\boldsymbol{M}(\boldsymbol{r}, t)$ follows the Landau–Lifshitz–Gilbert equation [29,30]

$$\frac{\partial \boldsymbol{M}}{\partial t} = -\gamma \boldsymbol{M} \times \boldsymbol{F}_{\text{eff}} - \frac{\lambda}{M} \boldsymbol{M} \times \left(\boldsymbol{M} \times \boldsymbol{F}_{\text{eff}}\right) , \tag{16.34}$$

where γ is the gyromagnetic ratio and λ is the damping constant and $\boldsymbol{F}_{\text{eff}}$ is the effective magnetic field. The first term in the equation describes the gyromagnetic motion (precession of $\boldsymbol{M}$ about $\boldsymbol{F}_{\text{eff}}$) and the second describes the rotation of the effective field. Note that $M = |\boldsymbol{M}|$ is a constant. The Heisenberg ferromagnet equations discussed earlier in Chap. 14 (Sect. 14.1) are then essentially special cases of the above Landau–Lifshitz equation when the damping vanishes and $\boldsymbol{F}_{\text{eff}}$ takes special forms. The complex magnetization patterns and the detailed spin structures within the domain boundaries are obtained by solving (16.34). The structures so obtained can then be used for different applications (see also the special issue on 'Magnetoelectronics' – Physics Today, April 1995). Some of them are discussed below briefly.

A. Magnetoresistive Recording

Over a century now, for magnetic recording most systems have used an inductive head for writing and reading which employ coils to both induce a magnetic field (write mode) and sense a recorded area (read mode). Recently, a more powerful reading head called the *magnetoresistance* (MR) head, has been introduced into disk products which employs a sensor whose resistance changes in the presence of a magnetic field. Its performance gain has enhanced the density of storage by up to 50% commercial conversion to MR heads is only just starting.

B. Magneto-Optical Recording

Again instead of using the conventional inductive head for recording, optical pulses or lasers have become usage. Optical recording is expected to increase in capacity and transfer rate by a factor about 20 over the next decade. Rewritable systems will be based largely on magneto optical technologies that exploit the similar mark sizes made possible by new short wavelength lasers.

In the above two models, it is mainly the interaction between the magnetization of the medium and the electrical field of the head in the case of magnetoresistive head and the electromagnetic (optic or laser) field in the case of magneto optical head that play important roles. For efficient storage, excitation of the magnetization of the medium (which may be due to thermal or due to other external disturbances) if any should be localized. Theoretically speaking, these different magnetic interactions can be accommodated in appropriate spin Hamiltonian models and the explicit form of (16.35) may then be considered for investigation.

C. Magnetic Films for Better Recording

The fundamental magnetization process in thin films can be characterized by the formation, motion and annihilation of magnetization vortices. When a sufficiently strong external magnetic field is applied, magnetization reversal takes place and these vortices move across the film. If the intergranular exchange coupling in magnetic films is large, the size of the vortices will be larger and travel more freely over larger distances. Thus the intergranular exchange coupling has a significant impact on the properties of recorded bits in thin films because in the case of large vortices the recording noise is large and in the case of two vortices the noise is low.

Vortices form an interesting class of solutions to multidimensional nonlinear evolution equations. Hence here also solving higher-dimensional Landau-Lifshitz equations of the form (16.35) with intergranular exchange coupling and interaction with large external fields for vortex like solutions is an important task for future.

D. Study of Single Domain Magnets

The behavior of individual magnetic domains has become important for technology. The problem of media noise which is one of the fundamental present day problems of magnetic storage can be avoided if we use individual magnetic domains to store each bit. The study of single domain magnets (mesoscopic magnets) has accelerated the development of new theoretical approaches to magnetic dynamics. Thus for the next few years attention has to be paid as to how quantum mechanical effects influence the properties of all these small systems. All these require intense study of the nonlinear dynamics of magnetic systems.

Exercise:

16. Prepare project reports on each of the issues indicated above in magnetoelectronics and magneto-optics.

16.7 Conclusions

We have tried to present in this book a unified and rather comprehensive picture of nonlinear dynamics at an elementary level. Starting from the basic notions of nonlinearity and nonlinear dynamical systems, we have introduced the concepts of attractors, bifurcations, chaos, controlling, synchronization, integrability, solitons, spatio-temporal patterns and so on through typical examples, analytical descriptions, numerical investigations, nonlinear circuits and exact analysis. We have also emphasized that the developed notions are not only of theoretical and conceptual relevance but also have great application potentialities even at cutting-edge technologies. There is no doubt that further great advances are bound to take place in the coming years in understanding more complex and complicated nonlinear dynamical systems. Nonlinear dynamics will continue to be a challenging and exciting topic of study to unravel the myriad of complex and regular behaviours of nature around us.

A. Elliptic Functions and Solutions of Certain Nonlinear Equations

In this appendix we give a brief account of the properties of Jacobian elliptic functions and obtain the solutions of certain physically important nonlinear differential equations in terms of them.

Consider the nonlinear differential equation

$$\ddot{x} = A + Bx + Cx^2 + Dx^3 \,, \quad (^{\cdot} = \mathrm{d}/\mathrm{d}t) \tag{A.1}$$

where A, B, C and D are real constants. Integration of (A.1) once, after multiplying it by $\dot{x}$, gives

$$(\dot{x})^2 = a + 2Ax + Bx^2 + \frac{2C}{3}x^3 + \frac{D}{2}x^4 \,, \tag{A.2}$$

where a is the integration constant. We rewrite (A.2) in the convenient form

$$(\dot{x})^2 = \left(\frac{\mathrm{d}x}{\mathrm{d}t}\right)^2 = a + bx + cx^2 + dx^3 + ex^4 \,, \tag{A.3}$$

where $b = 2A$, $c = B$, $d = 2C/3$ and $e = D/2$. If $(\alpha, \beta, \gamma, \delta)$ are the four real roots of the equation

$$x^4 + (d/e)x^3 + (c/e)x^2 + (b/e)x + (a/e) = 0 \,, \tag{A.4}$$

then (A.3) can be rewritten as

$$\left(\frac{\mathrm{d}x}{\mathrm{d}t}\right)^2 = e(x-\alpha)(x-\beta)(x-\gamma)(x-\delta) \,. \tag{A.5}$$

For the special choice, namely, $e = k^2$, $\alpha = 1$, $\beta = -1$, $\gamma = 1/k$ and $\delta = -1/k$, (A.5) takes the canonical form

$$\left(\frac{\mathrm{d}x}{\mathrm{d}t}\right)^2 = \left(1 - x^2\right)\left(1 - k^2x^2\right) \,. \tag{A.6}$$

The solution of (A.6) is related to the *elliptic integral of the first kind* [1,2] given by

$$t(x,k) = \int_0^x \frac{\mathrm{d}x'}{\sqrt{(1-x'^2)\,(1-k^2x'^2)}} \,, \qquad k^2 < 1. \tag{A.7}$$

When the transformation $x = \sin\phi$ is used then (A.7) becomes

$$t = \int_0^{\phi} \frac{d\phi}{\sqrt{(1 - k^2 \sin^2 \phi)}}, \quad k^2 < 1. \tag{A.8}$$

The parameter k is called the *modulus of the elliptic integral*. The elliptic functions are then the inverse of elliptic integrals.

The various Jacobian elliptic functions are defined as

$$\begin{aligned} \mathrm{sn}(t,k) &= \sin\phi\,, \\ \mathrm{cn}(t,k) &= \cos\phi\,, \\ \mathrm{dn}(t,k) &= \sqrt{1 - k^2 \sin\phi}\,, \\ \mathrm{am}(t,k) &= \phi\,, \\ \mathrm{tn}(t,k) &= \tan\phi\,. \end{aligned}$$

Often $\mathrm{sn}(t,k)$ is written as $\mathrm{sn}t$, $\mathrm{cn}(t,k)$ as $\mathrm{cn}t$ and so on. Some of the important relationships between the Jacobian elliptic functions and the identities obeyed by them are as follows [1]:

$$\begin{aligned} &\mathrm{sn}^2 t + \mathrm{cn}^2 t = 1\,, \\ &\mathrm{dn}^2 t - k^2 \mathrm{cn}^2 t = 1 - k^2\,, \\ &k^2 \mathrm{sn}^2 t + \mathrm{dn}^2 t = 1\,, \\ &\mathrm{sn}(t,0) = \sin t\,, \\ &\mathrm{cn}(t,0) = \cos t\,, \\ &\mathrm{dn}(t,0) = 1\,, \\ &\mathrm{sn}(t,1) = \tanh t\,, \\ &\mathrm{cn}(t,1) = \mathrm{dn}(t,1) = \mathrm{sech} t\,. \end{aligned}$$

The derivatives of the elliptic functions are

$$\begin{aligned} &\frac{d}{dt}\mathrm{sn}t = \mathrm{cn}t\,\mathrm{dn}t\,, \\ &\frac{d}{dt}\mathrm{cn}t = -\mathrm{sn}t\,\mathrm{dn}t\,, \\ &\frac{d}{dt}\mathrm{dn}t = -k^2 \mathrm{sn}t\,\mathrm{cn}t\,, \\ &\frac{d}{dt}\mathrm{am}t = \mathrm{dn}t\,, \end{aligned}$$

and so on. Solutions of some simple nonlinear equations can be expressed in terms of elliptic functions. In the following we give a few examples of them.

(1) *Pendulum Equation*

The undamped pendulum equation is given by

$$\ddot{\theta} + \frac{g}{L}\sin\theta = 0\,. \tag{A.9}$$

As in the case of (A.1), integrating once with respect to t, we obtain

$$\dot{\theta} = \frac{\mathrm{d}\theta}{\mathrm{d}t} = \sqrt{2g/L}\,\sqrt{\cos\theta - \cos\bar{\theta}}\,, \tag{A.10}$$

where we have assumed $\dot{\theta} = 0$, when θ takes its maximum value $\theta = \bar{\theta}$. Rewriting (A.10) as

$$\mathrm{d}t = \sqrt{L/2g}\,\frac{\mathrm{d}\theta}{\left(\cos\theta - \cos\bar{\theta}\right)^{1/2}}\,, \tag{A.11}$$

we introduce the change of variable

$$\cos\theta = 1 - 2k^2\sin^2\phi\,, \qquad k = \sin\left(\bar{\theta}/2\right)\,. \tag{A.12}$$

Now from (A.12) we have

$$\begin{aligned}\cos\theta &= 1 - 2k^2\left(1 - \cos^2\phi\right)\\ &= 1 - 2k^2 + 2k^2\cos^2\phi\\ &= \cos\bar{\theta} + 2k^2\cos^2\phi\,.\end{aligned}$$

That is,

$$\cos\theta - \cos\bar{\theta} = 2k^2\cos^2\phi\,. \tag{A.13}$$

Further, we find that

$$\sin\theta = 2k\sin\phi\left(1 - k^2\sin^2\phi\right)^{1/2} \tag{A.14}$$

and

$$\mathrm{d}\theta = \frac{2k\cos\phi}{\sqrt{1 - k^2\sin^2\phi}}\mathrm{d}\phi\,. \tag{A.15}$$

Substituting (A.13) and (A.15) in (A.11), we obtain

$$\mathrm{d}t = \sqrt{L/g}\,\frac{\mathrm{d}\phi}{\sqrt{1 - k^2\sin^2\phi}}\,. \tag{A.16}$$

Integration of the above equation gives

$$t = \sqrt{L/g}\int_0^{\phi}\frac{\mathrm{d}\phi}{\sqrt{1 - k^2\sin^2\phi}}\,, \tag{A.17}$$

where we have assumed for simplicity that $\phi = 0$ at $t = 0$. Using (A.8), (A.17) can be written as

$$\mathrm{sn}(t\sqrt{g/L}\,, k) = \sin\phi = \frac{1}{k}\sin(\theta/2)\,.$$

We can thus write the solution of (A.9) as

$$\theta = 2\,\sin^{-1}\left[k\,\mathrm{sn}\left(t\sqrt{g/L}\,, k\right)\right], \quad k = \sin(\bar{\theta}/2)\,. \tag{A.18}$$

(2) *Pure Cubic Anharmonic Oscillator*

Consider the anharmonic oscillator with pure cubic potential. Its equation of motion is

$$\ddot{x} = 6x^2 \,. \tag{A.19}$$

The solution of (A.19) can be easily shown to be of the form

$$x = A^2 \left[\frac{-k^2}{1+k^2} + \frac{1}{\mathrm{sn}^2\{A\,(t-t_0)\,,k\}} \right] , \tag{A.20}$$

where A and t_0 are arbitrary constants and k^2 is the root of the equation $1 - k^2 + k^4 = 0$. (Verify).

(3) *Undamped Cubic Anharmonic Oscillator*

The solution of the equation

$$\ddot{x} + \omega_0^2 x + \beta x^3 = 0 \tag{A.21}$$

can be given [3] in the following four forms by following the procedure indicated in the beginning of this appendix and depending upon the signs of ω_0^2 and β:

(i) ω_0^2, $\beta > 0$:

$$x(t) = A\mathrm{cn}\left[\Omega\,(t-t_0)\,,k\right] \,, \tag{A.22a}$$

where A and t_0 are arbitrary constants and

$$\Omega = \sqrt{\omega_0^2 + \beta A^2}, \quad k^2 = \frac{\beta A^2}{2\,(\omega_0^2 + \beta A^2)} . \tag{A.22b}$$

(ii) $\omega_0^2 > 0$, $\beta < 0$:

$$x(t) = A\mathrm{sn}\left[\Omega\,(t-t_0)\,,k\right] \,, \tag{A.23a}$$

with

$$\Omega = \sqrt{\omega_0^2 - (|\beta| A^2/2)} \,,$$

$$k^2 = \frac{|\beta| A^2}{2\omega_0^2 - |\beta| A^2} \,, \quad 0 \le A \le \sqrt{\omega_0^2/|\beta|} \,. \tag{A.23b}$$

(iii) $\omega_0^2 < 0$, $\beta > 0$:

$$x(t) = A\mathrm{dn}\left[\Omega\left(t - t_0\right), k\right] \, , \tag{A.24a}$$

with

$$\begin{aligned} \Omega &= \frac{\beta A^2}{2} \, , \\ k^2 &= \frac{2\left(\beta A^2 - |\omega_0^2|\right)}{\beta A^2} \, , \quad \sqrt{|\omega_0^2|/\beta} \le A \le \sqrt{2|\omega_0^2|/\beta} \, . \end{aligned} \tag{A.24b}$$

Another form of the solution is

$$x(t) = A\mathrm{cn}\left[\Omega\left(t - t_0\right), k\right] \, , \tag{A.25a}$$

with

$$\begin{aligned} \Omega &= \sqrt{-|\omega_0^2| + \beta A^2} \, , \\ k^2 &= \frac{\beta A^2}{2\left(-|\omega_0|^2 + \beta A^2\right)} \, , \quad \sqrt{2|\omega_0^2|/\beta} \le A < \infty \, . \end{aligned} \tag{A.25b}$$

(4) *The Damped Cubic Anharmonic Oscillator*

First we show that [4] the equation

$$\ddot{x} + d\dot{x} + \alpha x + \beta x^3 = 0 \tag{A.26}$$

in the case $\alpha = 2d^2/9$ can be reduced to the equation

$$\ddot{W} + W^3 = 0 \tag{A.27}$$

which is similar to (A.21) and then obtain its solution. We introduce the change of variables

$$x = kW(Z)\mathrm{e}^{at} \, , \quad Z = b\mathrm{e}^{ct} \, , \tag{A.28}$$

where k, a, b, c are constants to be determined. To determine them we substitute (A.28) in (A.26) and obtain

$$\begin{aligned} & k\, b^2c^2\mathrm{e}^{(a+2c)t}W'' + kbc\mathrm{e}^{(a+c)t}W'(2a + c + d) \\ & \quad + k\mathrm{e}^{at}W(a^2 + ad + \alpha) + k^3\beta\mathrm{e}^{3at}W^3 = 0 \, , \end{aligned} \tag{A.29}$$

where prime denotes differentiation with respect to Z. Since the terms W' and W are absent in (A.27), we equate their coefficients separately to zero in (A.29) and obtain

$$2a + c + d = 0 \, , \tag{A.30}$$

$$a^2 + ad + \alpha = 0 \, . \tag{A.31}$$

Now (A.29) becomes

$$b^2c^2ke^{(a+2c)t}W'' + \beta k^3 e^{3at}W^3 = 0 \,. \tag{A.32}$$

To bring this equation into the form of (A.27), we choose the arbitrary constants a and c such that

$$c = a \,. \tag{A.33}$$

Then from (A.30) and (A.31) we get

$$a = -d/3 \,, \quad \alpha = 2d^2/9 \,. \tag{A.34}$$

Now (A.32) can be rewritten as

$$W'' + \frac{9\beta k^2}{b^2d^2}W^3 = 0 \,, \tag{A.35}$$

which reduces to (A.27) when we set

$$k = \sqrt{2d^2/(9\beta)} \,. \tag{A.36}$$

where b^2 is chosen as 2.

To summarize, (A.26) reduces to (A.27) under the change of variables

$$x = \sqrt{2d^2/(9\beta)}\, W(Z)e^{-dt/3} \,, \quad Z = -\sqrt{2}\, e^{-dt/3} \,. \tag{A.37}$$

Case 1: $\beta > 0$

From (A.21) and (A.23), the solution of (A.27) is obtained as

$$W = \lambda \mathrm{cn}(\lambda v; k) \,, \quad v = Z - Z_0 \,, \quad k^2 = 1/2 \,, \tag{A.38}$$

where λ and Z_0 are integration constants. Using the transformation (A.37), the solution of (A.26) is written as

$$x(t) = \sqrt{2d^2/(9\beta)}\, e^{-dt/3}\lambda \mathrm{cn}(\lambda v; k) \,, \quad v = -\sqrt{2}\, e^{-dt/3} - Z_0 \,. \tag{A.39}$$

Case 2: $\beta < 0$

For $\beta < 0$, (A.26) can be rewritten as

$$\ddot{x} + d\dot{x} + \alpha x - |\beta|x^3 = 0 \,. \tag{A.40}$$

Again, using the same transformation (A.37) with the replacement $\beta = -|\beta|$, (A.40) becomes

$$\ddot{W} = W^3 \,. \tag{A.41}$$

Integrating the above equation once we obtain

$$(\dot{W})^2 = \frac{W^4}{2} + C_0 \,, \tag{A.42}$$

where C_0 is an integration constant. We treat the cases $C_0 = 0$ and $C_0 \neq 0$ separately.

For $C_0 \neq 0$, with the initial condition $W(0) = W_0 = (-2C_0)^{1/4}$ and $\dot{W}(0) = 0$, real solution of (A.41) exists only for $C_0 < 0$ and is given by

$$W = \frac{W_0}{\mathrm{cn}(W_0 v; k)}, \quad k^2 = 1/2, \; ' \; v = Z - Z_0, \tag{A.43}$$

where Z_0 is a second integration constant so that the solution of (A.40) becomes

$$\begin{aligned} x(t) &= \sqrt{2d^2/(9|\beta|)}\, W_0\, \mathrm{e}^{-dt/3} \mathrm{cn}\,(W_0 v; k)^{-1}, \\ v &= -\sqrt{2}\, E^{-dt/3} - Z_0. \end{aligned} \tag{A.44}$$

For $C_0 = 0$, straightforward integration of (A.42) gives the solution

$$x(t) = \pm\sqrt{2d^2/(9|\beta|)}\, \left[1 \pm C_1 \mathrm{e}^{dt/3}\right]^{-1}, \tag{A.45}$$

where C_1 is the second integration constant.

(5) *Undamped Quadratic Anharmonic Oscillator*

Finally, we give the solution of the equation

$$\ddot{x} = A + Bx + Cx^2. \tag{A.46}$$

Multiplying (A.46) by $\dot{x}$ and integrating once we obtain

$$(\dot{x})^2 = a + bx + cx^2 + dx^3, \tag{A.47a}$$

where

$$b = 2A, \quad c = B, \quad d = 2C/3 \tag{A.47b}$$

and a is an integration constant. Defining α_1, α_2 and α_3 ($\alpha_1 < \alpha_2 < \alpha_3$) as the real roots of the polynomial

$$x^3 + (c/d)x^2 + (b/d)x + (a/d) = 0 \tag{A.48}$$

(A.47a) can be rewritten as

$$\left(\frac{\mathrm{d}x}{\mathrm{d}t}\right)^2 = d\,(x - \alpha_1)\,(x - \alpha_2)\,(x - \alpha_3). \tag{A.49}$$

The relationship between α_1, α_2 and α_3 and the parameters in (A.48) are obtained by comparing the coefficients of various powers of x in the right side of (A.47a) and (A.49) which gives

$$\begin{aligned} &\alpha_1 + \alpha_2 + \alpha_3 = -c/d = -3B/(2C) \\ &\alpha_1\alpha_2 + \alpha_2\alpha_3 + \alpha_3\alpha_1 = b/d = 3A/C \\ &\alpha_1\alpha_2\alpha_3 = -a/d = -3a/(2C). \end{aligned}$$

When the change of variable

$$z^2 = (x - \alpha_3)\,/\,(\alpha_2 - \alpha_3) \tag{A.50}$$

is introduced, (A.49) becomes

$$\left(\frac{\mathrm{d}z}{\mathrm{d}t}\right)^2 = \lambda^2\left(1-z^2\right)\left(1-m^2z^2\right) , \tag{A.51a}$$

where

$$m^2 = (\alpha_3-\alpha_2)/(\alpha_3-\alpha_1) , \quad \lambda^2 = -d(\alpha_3-\alpha_1)/4 . \tag{A.51b}$$

One can easily verify that the solution of (A.51) is

$$z = \mathrm{sn}(\lambda t, m) . \tag{A.52}$$

Now the solution of (A.46) is obtained from (A.50) and (A.52) as

$$x(t) = \alpha_3 - (\alpha_3-\alpha_2)\,\mathrm{sn}^2(\lambda t, m) . \tag{A.53}$$

For other forms of solutions of (A.46) one may refer to Ref. [1]. Such solutions will also be useful in writing the cnoidal wave solutions of the Korteweg–de Vries equation as discussed in Chap. 11.

(6) *The Toda Lattice*

One of the most exciting applications of elliptic functions to dynamical systems is the possibility of finding propagating wave solutions of the Toda lattice. The details are given in appendix I, Sect. 4.

Problems

1. Starting from (A.7), show that $\mathrm{sn}(t,1) = \tanh t$ and $\mathrm{cn}(t,1) = \mathrm{dn}(t,1) = \mathrm{sech} t$.
2. For the pendulum equation (A.9) show that if the maximum displacement of the pendulum is $\bar{\theta} = \pi$ then the solution is

$$\theta(t) = 2\tan^{-1}\left(\mathrm{e}^{\sqrt{(g/L)}\,t}\right) - \frac{\pi}{2} .$$

3. Prove the following addition theorems of Jacobian elliptic functions [1,2]

$$\mathrm{sn}(u\pm v) = \frac{\mathrm{sn}u\mathrm{cn}v\mathrm{dn}v \pm \mathrm{cn}u\mathrm{sn}v\mathrm{dn}u}{1-k^2\mathrm{sn}^2u\mathrm{sn}^2v} ,$$

$$\mathrm{cn}(u\pm v) = \frac{\mathrm{cn}u\mathrm{cn}v \mp \mathrm{sn}u\mathrm{sn}v\mathrm{dn}u\mathrm{dn}v}{1-k^2\mathrm{sn}^2u\mathrm{sn}^2v} ,$$

$$\mathrm{dn}(u\pm v) = \frac{\mathrm{dn}u\mathrm{dn}v \mp k^2\mathrm{sn}u\mathrm{sn}v\mathrm{cn}u\mathrm{cn}v}{1-k^2\mathrm{sn}^2u\mathrm{sn}^2v} .$$

4. Show that for $m=1$ the solution (A.53) reduces to $x(t) = \alpha_2 + (\alpha_3 - \alpha_2)\mathrm{sech}^2\lambda t$.

5. For the anharmonic oscillator equation $\ddot{x} + d\dot{x} + \omega_0^2 x + \alpha x^2 = 0$ obtain the change of variable which reduce it to the form $\ddot{W} + 6W^2 = 0$ for $d = \pm\sqrt{(25/6)\,\omega_0^2}$, α-arbitrary. Then find the solution of the anharmonic oscillator equation.
6. Obtain the change of variable and the parametric restrictions for which the system

$$\ddot{x} + d\dot{x} + \omega_0^2 x + \alpha x^2 + \beta x^3 = 0$$

can be reduced to the form $\ddot{W} + W^3 = 0$. Also construct the solution of the corresponding system.

B. Perturbation and Related Approximation Methods

In this appendix, we will present brief details of some of the important approximation methods available in the literature to obtain approximate analytic solutions of weakly perturbed nonlinear dynamical systems. In particular we will discuss (i) perturbation and related methods to obtain periodic solutions of nonintegrable equations of motions approximately and (ii) canonical perturbation theory to investigate weakly perturbed conservative systems.

B.1 Approximation Methods for Nonlinear Differential Equations

Consider the nonlinear oscillator system governed by the equation of motion

$$\ddot{x} + \omega_0^2 x = \epsilon f(x, \dot{x}, t) , \quad \epsilon \ll 1, \quad \dot{} = \frac{\mathrm{d}}{\mathrm{d}t} , \tag{B.1}$$

where the function f contains the nonlinearity. Several perturbative methods have been developed [1,2] to obtain approximate periodic solutions valid for small strengths of the parameter $\epsilon(\ll 1)$. These include

(1) the Lindstedt–Poincaré perturbation method,

(2) Averaging methods,

(3) Harmonic balance method and

(4) Multiple-scale perturbation method.

Each one of these procedures has its own advantages and disadvantages as well as range of validity and is suitable to deal with specific situations. We illustrate the method of Lindstedt–Poincaré with an example and then briefly summarize the salient features of the other methods.

(1) *Lindstedt–Poincaré Perturbation Method*

This well known method essentially develops a series expansion to the solution $x(t)$ of (B.1) in the small parameter ϵ, but avoids the notorious secular terms present in the standard series expansion through a renormalization procedure. One looks for a solution of the form

$$x(t) = x(\tau, \epsilon) = \sum_{n=0}^{\infty} \epsilon^n x_n(\tau) \ , \quad \tau = \omega t \ , \tag{B.2}$$

where the new independent variable τ is an unspecified function of ϵ. The system of governing recursive set of differential equations for the $x_n(\tau)$'s, which are linear in nature, will contain ω in the coefficient of the derivatives, and this permits the frequency and amplitude to interact. Then one can choose the function ω in such a way as to eliminate the secular terms.

Let us apply the method to the van der Pol equation (see Chap. 3, Sect. 3.5)

$$\ddot{x} - \epsilon \dot{x} \left(1 - x^2\right) + \omega_0^2 x = 0 \ . \tag{B.3}$$

Following (B.2), we rewrite (B.3) as

$$\omega^2 x'' - \epsilon \omega x' \left(1 - x^2\right) + \omega_0^2 x = 0 \ , \quad \tau = \omega t \ , \tag{B.4}$$

where prime stands for differentiation with respect to τ. For $\epsilon \neq 0$ we assume the solution to be of the form

$$x(\tau) = x_0(\tau) + \epsilon x_1(\tau) + \epsilon^2 x_2(\tau) + \ldots \tag{B.5}$$

We expand ω also in a power series in ϵ,

$$\omega = \omega_0 + \epsilon \omega_1 + \epsilon^2 \omega_2 + \ldots \ , \tag{B.6}$$

and substitute the two series expressions for x and ω in (B.4), and equate like powers of ϵ. This gives rise to a system of second-order linear differential equations for the x_i's. Let the initial conditions be such that

$$x_0(0) = a, \ \ x_0'(0) = 0, \ \ x_i(0) = x_i'(0) = 0, \ \ i \neq 0 \ . \tag{B.7}$$

Substitution of the series expansions (B.5) and (B.6) into (B.3) gives

$$\begin{aligned} &\left(\omega_0^2 + 2\epsilon\omega_0\omega_1 + \epsilon^2\omega_1^2 + 2\epsilon^2\omega_0\omega_2 + \ldots\right)\left[x_0'' + \epsilon x_1'' + \epsilon^2 x_2'' + \ldots\right] \\ &\quad - \epsilon\left[1 - \left(x_0 + \epsilon x_1 + \ldots\right)^2\right]\left(x_0' + \epsilon x_1' + \ldots\right)\left(\omega_0 + \epsilon\omega_1 + \ldots\right) \\ &\quad + \omega_0\left(x_0 + \epsilon x_1 + \ldots\right) = 0 \ . \end{aligned} \tag{B.8}$$

Equating the coefficients of like powers of ϵ we obtain,

$$\epsilon^0 : \ x_0'' + x_0 = 0 \ , \tag{B.9a}$$

$$\epsilon^1 : \ x_1'' + x_1 = -\frac{2}{\omega_0}\left[\omega_1 x_0'' - \frac{1}{2} x_0' \left(1 - x_0^2\right)\right] \ , \tag{B.9b}$$

$$\begin{aligned} \epsilon^2 : \ x_2'' + x_2 = &-\frac{1}{\omega_0^2}\left[\left(\omega_1^2 + 2\omega_0\omega_2\right)x_0'' + 2\omega_0\omega_1 x_1'' - \omega_1 x_0'\left(1 - x_0^2\right)\right. \\ &\left. - \omega_0 x_1'\left(1 - x_0^2\right) - 2\omega_0 x_0' x_0 x_1\right] \ , \end{aligned} \tag{B.9c}$$

and so on. The solution of (B.9a), subject to (B.7), is

$$x_0(t) = a \cos \omega_0 t \ . \tag{B.10}$$

Substituting (B.10) for x_0 in (B.9b), we obtain

$$x_1'' + x_1 = \frac{a}{4\omega_0}\left(a^2 - 4\right)\sin\tau + \frac{2\omega_1 a}{\omega_0}\cos\tau + \frac{a^3}{4\omega_0}\sin 3\tau \,. \qquad \text{(B.11)}$$

Equation (B.11) is an inhomogeneous linear second-order differential equation and it can be easily integrated [3] following the procedure indicated in Chap. 1, exercise 4. It can be easily checked that the first two terms on the right hand side of (B.11) give rise to the so-called secular terms (proportional to τ here which grow uncontrollably as $t \to \infty$) on integration. To avoid them we can renormalize the frequency and amplitude by choosing

$$\omega_1 = 0 \quad \text{and} \quad a = 2 \,. \qquad \text{(B.12)}$$

The resultant solution subject to the initial conditions (B.7) can be easily written as

$$x_1(t) = -\frac{1}{4\omega_0}\left(\sin 3\omega_0 t - 3\sin\omega_0 t\right) \,. \qquad \text{(B.13)}$$

Now the solution upto the first-order terms in ϵ after simplification is given by

$$x(t) = 2\cos\omega_0 t + \frac{\epsilon}{4\omega_0}\left(3\sin\omega_0 t - \sin 3\omega_0 t\right) \,. \qquad \text{(B.14)}$$

The procedure can be continued at higher orders in ϵ as well. For example, solving the second-order linear homogeneous differential equation (B.9c) as above subject to the initial conditions (B.7), we can deduce the second-order correction to the solution as

$$x_2(t) = \frac{1}{16\omega_0^2}\left[\frac{19}{6}\cos\omega t - 3\cos 3\omega t - \frac{1}{6}\cos 5\omega t\right] \,, \qquad \text{(B.15)}$$

where now the frequency is modified as

$$\omega = \omega_0 + \epsilon^2 \frac{7}{16\omega_0} \,. \qquad \text{(B.16)}$$

Using (B.10), (B.13), (B.15) and (B.16) in (B.5) and (B.6), we can readily obtain the approximate periodic solution to the van der Pol equation upto order ϵ^2.

(2) *Averaging Methods*

Another useful technique to deal with nonlinear oscillator equations of the type given by (B.1) is the method of averaging. There are various versions of it, including the Krylov–Bogoliubov method, the Krylov–Bogoliubov–Mitropolsky technique, the generalised method of averaging, averaging using canonical variables, averaging using Lie series and transforms and averaging using Lagrangians, and so on [1,2].

Most of the averaging procedures start with the variation of parameters to transform the dependent variable from x to $a(t)$ and $\phi(t)$, where

$$x = a(t) \cos(\omega_0 t + \phi(t)) , \tag{B.17a}$$

$$\dot{x} = -a(t)\omega_0 \sin(\omega_0 t + \phi(t)) . \tag{B.17b}$$

Using (B.17) in (B.1), one obtains a set of differential equations for the slowly changing amplitude $a(t)$ and phase $\phi(t)$ in the form

$$\dot{a} = -\frac{\epsilon}{\omega_0} f(a\cos\theta, -a\omega_0 \sin\theta) \sin\theta , \tag{B.18a}$$

$$\dot{\phi} = -\frac{\epsilon}{a\omega_0} f(a\cos\theta, -a\omega_0 \sin\theta) \cos\theta , \tag{B.18b}$$

where $\theta = \omega_0 t + \phi$. Now we can assume that $a(t)$ and $\phi(t)$ are *slowly* varying so that they remain nearly constant during a time interval of duration $T_0 = 2\pi/\omega_0$. As a result one obtains the autonomous system of *averaged* equations

$$\dot{a} = -\frac{\epsilon}{2\pi\omega_0} \int_0^{2\pi} f(a\cos\theta, -a\omega_0 \sin\theta) \sin\theta \mathrm{d}\theta , \tag{B.19a}$$

$$\dot{\phi} = -\frac{\epsilon}{2\pi a\omega_0} \int_0^{2\pi} f(a\cos\theta, -a\omega_0 \sin\theta) \cos\theta \mathrm{d}\theta . \tag{B.19b}$$

(3) *Harmonic Balance Method*

Here the idea is to express the periodic solution of an equation of the form (B.1) as a finite Fourier series,

$$x(t) = \sum_{m=0}^{M} A_m \cos m(\omega t + \delta) . \tag{B.20}$$

Substituting the series (B.20) into the given ordinary differential equation and equating each of the lowest $(m+1)$ harmonics to zero, one can obtain a system of $(m+1)$ algebraic equations relating ω and A_m. Usually these equations are solved for A_0, A_2, A_3,..., A_m and ω in terms of A_1. The accuracy of the resulting solution will then depend on the value of A_1 and the number of harmonics in the assumed solution (B.20).

(4) *Multiple-Scale Perturbation Method*

We look for a periodic solution of (B.1) in the form

$$x(t,\epsilon) = x_0(T_0, T_1, ...) + \epsilon x_1(T_0, T_1, ...) + ... , \tag{B.21}$$

where the multiple time scales are $T_0 = t$, $T_1 = \epsilon t$, ... and

$$\frac{\mathrm{d}}{\mathrm{d}t} = \sum_{n=0}^{\infty} \epsilon^n D_n , \quad D_n = \frac{\partial}{\partial T_n} \tag{B.22}$$

from which higher order derivatives can also be calculated. The introduction of the new time scales is not a new expansion and t, ϵt, $\epsilon^2 t$, etc. measure the same time on different scales. Using (B.21) in (B.1), one can set up a recursive set of linear differential equations which can be solved exactly. Though the method is little more involved, it has definite advantages in that different orders of contributions can be treated through different scales.

B.2 Canonical Perturbation Theory for Conservative Systems

If a given nonlinear dynamical system is associated with a Hamiltonian function $H = H(p, q)$, then the equation of motion can be expressed in the form of Hamilton's canonical equations [4]. One can develop a canonical perturbation theory in the case in which the Hamiltonian can be split up into two parts, namely an unperturbed dominant part and a weak perturbation. In particular, if there exists a canonical transformation to action and angle variables which makes the unperturbed part to depend solely on the action variables, while the perturbation part can depend both on the action and angle variables, irrespective of the dimension of the system, a suitable perturbation theory [5–7] can be developed as long as the perturbation is sufficiently weak. In the following, we provide brief details of the procedure first for a one degree of freedom system and then extend the analysis to two degrees of freedom case, which can then be generalised to the N-degrees of freedom systems.

B.2.1 One Degree of Freedom Hamiltonian Systems

Let the given Hamiltonian system with one degree of freedom under the action and angle variables transformation be given by

$$H(I, \theta) = H_0(I) + \epsilon H_1(I, \theta) , \tag{B.23}$$

where ϵ is small parameter. A simple example is the one-dimensional linear harmonic oscillator as discussed in Sect. 3.7.1. The canonical equations of motion associated with (B.23) are

$$\dot{I} = -\epsilon \frac{\partial}{\partial \theta} H_1(I, \theta), \tag{B.24a}$$

$$\dot{\theta} = \omega_0(I) + \epsilon \frac{\partial}{\partial I} H_1(I, \theta), \tag{B.24b}$$

where $\omega_0(I) = (\partial/\partial I) H_0(I)$. The essential idea behind the canonical perturbation theory is to find a new set of action and angle variables (J, ϕ) for the perturbed system (B.23) such that the new Hamiltonian is a function of J only, that is $H(I, \theta) \to \widehat{H}(J)$.

In order to obtain such a transformation one tries to find a type-II generating function $S = S(\theta, J)$, where θ is the old angle coordinate and J is the new action variable. Then from standard results on canonical transformations [4], one can write the familiar relations

$$I = \frac{\partial}{\partial \theta} S(\theta, J), \tag{B.25a}$$

$$\phi = \frac{\partial}{\partial J} S(\theta, J). \tag{B.25b}$$

Expanding now S in a power series in ϵ,

$$S = S_0 + \epsilon S_1 + \epsilon^2 S_2 + ..., \tag{B.26}$$

where $S_0 = J\theta$, and substituting (B.25) and (B.26) into (B.24), and equating equal powers of ϵ, one obtains

$$O\left(\epsilon^0\right) : H_0(J) = \widehat{H}_0(J) \tag{B.27a}$$

$$O\left(\epsilon^1\right) : \frac{\partial S_1}{\partial \theta}\frac{\partial H_0}{\partial J} + H_1(J,\theta) = \widehat{H}_1(J) \tag{B.27b}$$

$$O\left(\epsilon^2\right) : \frac{1}{2}\left[\left(\frac{\partial S_1}{\partial \theta}\right)^2 \frac{\partial^2 H_0}{\partial J^2}\right] + \frac{\partial S_2}{\partial \theta}\frac{\partial H_0}{\partial J} + \frac{\partial S_1}{\partial \theta}\frac{\partial H_1}{\partial J} + H_2(J,\theta)$$
$$= \widehat{H}_2(J) \tag{B.27c}$$

and higher-order terms in ϵ. Assuming H_1 and S to be periodic in θ, integrating both sides of (B.27b) on both sides [5], we have

$$\widehat{H}_1(J) = \frac{1}{2\pi}\int_0^{2\pi} H_1(J,\theta)\mathrm{d}\theta\ . \tag{B.28}$$

Then we have

$$\frac{\partial}{\partial \theta} S_1(\theta, J) = \frac{1}{\omega_0(J)}\left[\widehat{H}_1(J) - H_1(J,\theta)\right]\ . \tag{B.29}$$

Expanding now both $S_1(\theta, J)$ and $[\widehat{H}_1(J) - H_1(J,\theta)]$ as Fourier series in θ, one can write

$$\left[\widehat{H}_1(J) - H_1(J,\theta)\right] = -\sum_{k=-\infty}^{\infty} A_k(J)\mathrm{e}^{\mathrm{i}k\theta}\ , \tag{B.30a}$$

$$S_1(J,\theta) = \sum_{k=-\infty}^{\infty} B_k(J)\mathrm{e}^{\mathrm{i}k\theta}\ , \tag{B.30b}$$

one can easily check from (B.29) that

$$S_1(J,\theta) = \sum_k \frac{\mathrm{i}A_k(J)}{k\omega_0(J)}\mathrm{e}^{\mathrm{i}k\theta}. \tag{B.31}$$

Then to first-order in ϵ, we have the action and angle variables,

$$\phi = \theta + \epsilon\frac{\partial}{\partial J} S_1(J,\theta)\ , \tag{B.32a}$$

$$J = I - \epsilon\frac{\partial}{\partial \theta} S_1(J,\theta)\ , \tag{B.32b}$$

so that the corrected frequency is

$$\omega(J) = \omega_0(J) + \epsilon \frac{\partial}{\partial J} \widehat{H}_1(J) . \tag{B.33}$$

Example:

Let

$$H(p, q) = H_0(p, q) + \epsilon H_1(p, q) , \tag{B.34}$$

where

$$H_0(q, p) = \frac{1}{2} \left(p^2 + \omega^2 q^2 \right) , \quad H_1(p, q) = q^3 . \tag{B.35}$$

The Hamiltonian (B.34) in terms of the action and angle variables is

$$H(I, \theta) = I\omega_0 + \epsilon \left(\frac{2I}{\omega_0} \right)^{3/2} \sin^3 \theta , \tag{B.36}$$

where $\theta = \omega_0 t + \delta$. Following the above procedure, one can obtain $\widehat{H}_1(J, \theta) = 0$, that is, there is no first-order energy correction. Equation (B.29) gives

$$S_1 = \frac{1}{\omega_0} \left(\frac{2J}{\omega_0} \right)^{3/2} \left[\frac{1}{3} \sin^2 \theta \cos \theta + \frac{2}{3} \cos \theta \right] . \tag{B.37}$$

Using (B.37) one can find second-order energy correction as

$$\widehat{H}_2(J) = -\frac{15\pi J^2}{2\omega_0^4} , \tag{B.38}$$

and hence the new frequency is

$$\omega(J) = \frac{\partial}{\partial J} H(J) = \omega_0 - \epsilon^2 \frac{15\pi J}{\omega_0^4} . \tag{B.39}$$

B.2.2 Two Degrees of Freedom Systems

Let us consider the two degrees of freedom Hamiltonian system in action and angle variables,

$$H = H_0 \left(I_1, I_2 \right) + \epsilon H_1 \left(I_1, I_2, \theta_1, \theta_2 \right) . \tag{B.40}$$

The canonical equations of motion are

$$\dot{I}_i = -\epsilon \frac{\partial H_1}{\partial \theta_i} , \tag{B.41a}$$

$$\dot{\theta}_i = \omega_{0i} \left(I_1, I_2 \right) + \epsilon \frac{\partial H_1}{\partial I_i} , \quad i = 1, 2. \tag{B.41b}$$

Under the basic idea of canonical perturbation theory as demonstrated for the one degree of freedom case, one tries to find a new Hamiltonian $\widehat{H}(J_1, J_2)$, by identifying a type-II generating function $S = S(J_1, J_2, \theta_1, \theta_2)$, so that

$$I_i = \frac{\partial S}{\partial \theta_i}\,, \tag{B.42a}$$

$$\phi_i = \frac{\partial S}{\partial J_i}\,, \quad i = 1, 2\,. \tag{B.42b}$$

Assuming again a power series expansion

$$S\left(J_1, J_2, \theta_1, \theta_2\right) = S_0 + \epsilon S_1\left(J_1, J_2, \theta_1, \theta_2\right) + \epsilon^2 S_2 + ..., \tag{B.43a}$$

where $S_0 = J_1\theta_1 + J_2\theta_2$, one can obtain the following equations:

$$O\left(\epsilon^0\right) : H_0\left(J_1, J_2\right) = \widehat{H}_0\left(J_1, J_2\right) \tag{B.43b}$$

$$O\left(\epsilon^1\right) : \frac{\partial S_1}{\partial \theta_1}\frac{\partial H_0}{\partial J_1} + \frac{\partial S_1}{\partial \theta_2}\frac{\partial H_0}{\partial J_2} + H_1\left(J_1, J_2, \theta_1, \theta_2\right) = \widehat{H}_1\left(J_1, J_2\right) \tag{B.43c}$$

$$\begin{aligned} O\left(\epsilon^2\right) : & \frac{1}{2}\left[\left(\frac{\partial S_1}{\partial \theta_1}\right)^2 \frac{\partial^2 H_0}{\partial J_1^2} + \left(\frac{\partial S_1}{\partial \theta_2}\right)^2 \frac{\partial^2 H_0}{\partial J_2^2}\right] + \frac{\partial S_2}{\partial \theta_1}\frac{\partial H_0}{\partial J_1} \\ & + \frac{\partial S_2}{\partial \theta_2}\frac{\partial H_0}{\partial J_2} + \frac{\partial S_1}{\partial \theta_1}\frac{\partial H_1}{\partial J_1} + \frac{\partial S_1}{\partial \theta_2}\frac{\partial H_1}{\partial J_2} + H_2\left(J_1, J_2, \theta_1, \theta_2\right) \\ & = \widehat{H}_2\left(J_1, J_2\right) \end{aligned} \tag{B.43d}$$

and so on.

Following the the procedure mentioned in Sect. B.2.1, we can obtain the relation

$$\widehat{H}_1\left(J_1, J_2\right) = \int_0^{2\pi}\int_0^{2\pi} H_1\left(J_1, J_2, \theta_1, \theta_2\right) \mathrm{d}\theta_1 \mathrm{d}\theta_2\,. \tag{B.44}$$

Further expanding both S_1 and $[H_1(J_1, J_2, \theta_1, \theta_2) - \widehat{H}(J_1, J_2)]$ in a two-dimensional Fourier series, that is,

$$S_1\left(J_1, J_2, \theta_1, \theta_2\right) = \sum_k B_k\left(J_1, J_2\right) \mathrm{e}^{\mathrm{i}\boldsymbol{k}\cdot\boldsymbol{\theta}}\,, \tag{B.45a}$$

$$\left[H_1 - \widehat{H}_1\right] = \sum_k A_k\left(J_1, J_2\right) \mathrm{e}^{\mathrm{i}\boldsymbol{k}\cdot\boldsymbol{\theta}}\,, \tag{B.45b}$$

where $\boldsymbol{k}\cdot\boldsymbol{\theta} = k_1\theta_1 + k_2\theta_2$ and $\boldsymbol{k} = (k_1, k_2)$, one can easily obtain

$$S_1\left(J_1, J_2, \theta_1, \theta_2\right) = \mathrm{i}\sum_k \frac{A_k \mathrm{e}^{\mathrm{i}\boldsymbol{k}\cdot\boldsymbol{\theta}}}{\boldsymbol{k}\cdot\boldsymbol{\omega}_0\left(J_1, J_2\right)}\,, \tag{B.46}$$

where $\boldsymbol{k}\cdot\boldsymbol{\omega}_0 = k_1\omega_{01} + k_2\omega_{02}$, $\omega_{0i} = \partial H_0(J_1, J_2)/\partial J_i$, $i = 1, 2$. Thus knowing H_1 and S_1 from (B.45) and (B.46), respectively, one can proceed to solve (B.43d) to obtain H_2 and S_2, and so on.

It may be immediately noted that the presence of the term $(\boldsymbol{k}\cdot\boldsymbol{\omega}_0)$ in the denominator of (B.46) gives rise to the problem of small denominators or the phenomenon of resonance. The corresponding invariant torus in the phase space is called the *resonant surface*. If the fundamental frequencies $\omega_{0i}(J_1, J_2)$, $i = 1, 2$, are *commensurable* so that their ratio is a rational, the small denominators in (B.46) will prevent the series from converging. When the two frequencies are *incommensurable*, it may so happen that $\boldsymbol{k}\cdot\boldsymbol{\omega}_0(J_1, J_2)$ is arbitrarily close to to zero. This phenomenon represents a physical as well as a mathematical difficulty. It arises from real resonances that change the topology of the phase space trajectories.

From the above analysis, it is clear that there are two problems associated with the canonical perturbation theory: (i) the problem of convergence of the power series in (ϵ) and (ii) the problem of small denominators. However, the celebrated KAM theorem discussed in Sect. 7.6 asserts that for sufficiently *irrational* $\omega_{0i}(J_1, J_2)$, $i = 1, 2$, there is a convergent perturbation in ϵ leading to a new action-angle set so that the tori are only slightly deformed but not destroyed.

What is the nature of the dynamics near a resonance? It can be shown by straightforward but careful analysis [6,7] that near a resonance surface the dynamics of any perturbed Hamiltonian system is just the dynamics of a pendulum. Consequently the effect of any additional perturbation may be analysed as pointed out in Sect. 7.2.3.

As examples of the above analysis of the two degrees of freedom Hamiltonian system one can study the model systems of Walker and Ford [8] discussed in Sect. 7.2.3 and draw the conclusions discussed therein.

Problems

1. Verify the second-order correction to the solution $x(t)$ ((B.15)) of the van der Pol equation (B.3).
2. Show that the approximate solution of the perturbed harmonic oscillator equation
$$\ddot{x} + \omega_0^2 x + \epsilon\left(x^3 + \dot{x}\right) = 0$$
by the method of Krylov–Bogoliubov is
$$x(t) = a(0)\mathrm{e}^{-\epsilon t/2}\sin\left[\omega_0 t + \phi(0) + \frac{3}{8\omega_0}a^2(0)\left(1 - \mathrm{e}^{-\epsilon t}\right)\right],$$
where $a(0)$ and $\phi(0)$ are arbitrary constants.
3. Construct an approximate solution of the van der Pol equation (B.3) by the Krylov–Bogoliubov and multiple scale methods.
4. Show that the approximate solution of the undamped anharmonic oscillator equation
$$\ddot{x} + x + \epsilon x^3 = 0, \quad 0 < \epsilon \ll 1, \quad \dot{x}(0) = 0$$
by the method of multiple scale is
$$x(t) = a\cos\left[\left(1 + \epsilon\frac{3}{8}a^2\right)t\right], \quad t < 1/\epsilon$$
where a is a constant.

5. Obtain the approximation periodic solution of the periodically driven Duffing oscillator equation

$$\ddot{x} + \omega_0^2 x + \epsilon(\dot{x} + x^3) = f \cos \Omega t$$

by the various perturbation methods. Compare the approximate solutions with the numerical solution for specific values of the parameters of the system.
6. Investigate the effect of perturbation to the one-dimensional pendulum system near the resonance for the Hamiltonian

$$H = \frac{p^2}{2} - \omega_0^2 \cos x + \epsilon \omega_0^2 x \sin \omega t \ .$$

7. Apply the canonical perturbation theory to the example of Walker and Ford [8] with the Hamiltonian

$$H = I_1 + I_2 - I_1^2 - 3I_1 I_2 + I_2^2 + 2I_1 I_2 + I_2^2 + 2I_1 I_2 \cos(2\theta_1 - 2\theta_2)$$

and find the energy correction upto second-order. See also the discussions in Chap. 7, Sect. 2.

C. A Fourth-Order Runge–Kutta Integration Method

The goal of an integration algorithm is to approximate the behaviour of a continuous-time system on a digital computer. Since digital computers are discrete-time in nature, an integration algorithm essentially models a continuous system by a discrete system. Several numerical schemes are available in the literature to solve given differential equations. The most widely used and simple algorithm is the Runge–Kutta method [1].

For simplicity and illustrative purpose, let us consider a first-order differential equation of the form

$$\frac{\mathrm{d}x}{\mathrm{d}t} = F(x,t)\ . \tag{C.1}$$

Let the value of x at time t_i is known and denote it as x_i. Suppose we wish to calculate the value of x at time $t_{i+1} = t_i + h$, where h is the increment in time. The Runge–Kutta family of algorithms has its origin in the idea of approximating $\mathrm{d}F/\mathrm{d}t$ by its Taylor series approximation. We here present only the fourth-order algorithm. The term K-th order means that K terms of the Taylor expansion are used in the approximation. To get more accuracy it is not necessary to go to higher order approximations but instead a smaller step size can be used. The truncation error in the fourth-order algorithm is $O(h^5)$. The formulae for the fourth-order Runge–Kutta algorithm for (C.1) are

$$x_1 = hF\left(x_i, t_i\right)\ , \tag{C.2a}$$

$$x_2 = hF\left(x_i + x_1/2, t_i + h/2\right)\ , \tag{C.2b}$$

$$x_3 = hF\left(x_i + x_2/2, t_i + h/2\right)\ , \tag{C.2c}$$

$$x_4 = hF\left(x_i + x_3, t_i + h\right)\ , \tag{C.2d}$$

$$t_{i+1} = t_i + h\ , \tag{C.3}$$

$$x_{i+1} = x_i + \frac{1}{6}\left(x_1 + 2x_2 + 2x_3 + x_4\right)\ . \tag{C.4}$$

Starting with the value of $F\left(x_i, t_i\right)$ at (x_i, t_i), x_1 may be computed from (C.2a). x_2 is obtained with values of (x_i, t_i, x_1), using (C.2b). Similarly x_3 and x_4 are obtained from (C.2c) and (C.2d) respectively. Equation (C.4) yields the value of x_{i+1}. The entire process is repeated to find $x_{i+2}, x_{i+3}, \ldots$ In (C.4), each of the four x_1, x_2, x_3 and x_4 is an approximate value of x. x_1

is the value of x at t_i, x_2 is the approximation value a half-step later, at time $t_i + h/2$. x_3 is also the value at time $t_i + h/2$ but it is calculated using the slope x_2. x_4 is the value of x at t_{i+1} and is calculated by the most recent value x_3. Finally, these four values are averaged together to give the approximation to x which is used to predict x_{i+1}. The above Runge–Kutta algorithm can be extended straightforwardly to integrate m-coupled first-order equations as well.

We now give the Runge–Kutta algorithm for the Duffing oscillator equation, whose dynamics is numerically studied in Sect. 5.1,

$$\frac{\mathrm{d}^2x}{\mathrm{d}t^2} + d\frac{\mathrm{d}x}{\mathrm{d}t} + ax + bx^3 = f \sin \omega t \ . \tag{C.5}$$

To apply the fourth-order Runge–Kutta method we rewrite the above equation as a system of two first-order coupled equations, namely,

$$\dot{x} = y = FX(y) \ , \tag{C.6a}$$

$$\dot{y} = -dy - ax - bx^3 + f \sin \omega t = FY(x, y, t) \ . \tag{C.6b}$$

Let (x_0, y_0) are the known values at time $t = t_0$. The Runge–Kutta algorithm to find (x, y) at time $t = t_0 + h$ is given below.

Fourth-order Runge–Kutta algorithm for Duffing oscillator equation

$FX(y) = y$
$FY(x, y, t, d, a, b, f, \omega) = -d * y - a * x - b * x * x * x + f * \sin(\omega * t)$
$x_1 = h * FX(y_0)$
$y_1 = h * FY(x_0, y_0, t_0, d, a, b, f, \omega)$
$x_2 = h * FX(y_0 + y_1/2)$
$y_2 = h * FY(x_0 + x_1/2, y_0 + y_1/2, t_0 + h/2, d, a, b, f, \omega)$
$x_3 = h * FX(y_0 + y_2/2)$
$y_3 = h * FY(x_0 + x_2/2, y_0 + y_2/2, t_0 + h/2, d, a, b, f, \omega)$
$x_4 = h * FX(y_0 + y_3)$
$y_4 = h * FY(x_0 + x_3, y_0 + y_3, t_0 + h, d, a, b, f, \omega)$
$t = t_0 + h$
$x = x_0 + (x_1 + 2 * x_2 + 2 * x_3 + x_4)/6$
$y = y_0 + (y_1 + 2 * y_2 + 2 * y_3 + y_4)/6$

For other standard methods such as Runge–Kutta–Fehlberg method, Adams–Bashforth–Moulton multistep method, etc. the readers may refer to Ref. [1].

Problems

1. Find the numerical solution of the equation $\dot{x} = 1 - x^2$ for $t \in [0, 2]$ with the initial condition $x(0) = 1$ for the time steps $h = 0.2$, 0.1, 0.05 and 0.01. Compare the numerical solutions with the exact solution $x(t) = \tan(t + \tan^{-1} x(0))$.
2. Compare the numerical solution of the equation $\ddot{x} + 3\dot{x} + x = 0$, $x(0) = 2$, $\dot{x}(0) = 1$ obtained by the Runge–Kutta method over the time interval $[0, 10]$ for the time steps $h = 0.3$, 0.2, 0.1 and 0.01 with the exact solution.
3. Analyse the problem of step size versus error for the fourth-order Runge–Kutta method [1].
4. Solve the Lorenz equations (5.5) numerically and verify the results discussed in Sect. 5.2.
5. Obtain the results on Hénon–Heiles system (Sect. 7.3) using Runge–Kutta fourth-order method.

D. Nature of Phase Space Trajectories for λ_1, $\lambda_2 < 0$ and $\lambda_1 < 0 < \lambda_2$ (Sect. 3.4.2)

In Sect. 3.4.2 we studied the classification of equilibrium points and pointed out that for a node and saddle the trajectories in the phase space are parabolic and hyperbolic curves respectively. In this appendix we explicitly show the results. Let us now consider the equation (3.8). We introduce the linear transformation [1]

$$u = \alpha\xi + \beta\eta \,, \tag{D.1a}$$

$$v = \gamma\xi + \delta\eta \,, \tag{D.1b}$$

where α, β, γ and δ are to be determined so that we obtain the equation of the form

$$\frac{dv}{du} = m\frac{v}{u} \,, \tag{D.2}$$

where m is a real constant. Differentiating (D.1) with respect to t and then substituting for $\dot{\xi}$ and $\dot{\eta}$ from (3.8), we get

$$\dot{u} = S_1 u + \eta(\alpha b + \beta(d - S_1)) \,, \tag{D.3a}$$

$$\dot{v} = S_2 v + \xi(\delta c + \gamma(a - S_2)) \,, \tag{D.3b}$$

where S_1 and S_2 are obtained from

$$\beta c + \alpha\,(a - S_1) = 0 \,, \tag{D.4a}$$

$$\gamma b + \delta\,(d - S_2) = 0 \,. \tag{D.4b}$$

Setting the coefficients of η and ξ separately to zero in (D.3b), that is,

$$\alpha b + \beta\,(d - S_1) = 0 \,, \tag{D.5a}$$

$$\delta c + \gamma\,(a - S_2) = 0 \tag{D.5b}$$

we obtain the equations

$$\dot{u} = S_1 u \,, \tag{D.6a}$$

$$\dot{v} = S_2 v \,. \tag{D.6b}$$

The four unknowns α, β, γ and δ can be found by solving the set of equations (D.4). For nontrivial solution we require that S_1 and S_2 are the roots of the equation

$$\begin{vmatrix} a-S & c \\ b & d-S \end{vmatrix} = S^2 - (a+d)S + (ad - bc) = 0 \qquad \text{(D.7)}$$

which is same as the characteristic equation (3.11) and therefore we can write $S_{1,2} = \lambda_{1,2}$. From (D.5), we obtain (D.2) with $m = S_2/S_1 = \lambda_2/\lambda_1$. The solution of (D.2) is

$$v = C_1 |u|^m , \qquad \text{(D.8)}$$

where C_1 is the integration constant. For λ_1, $\lambda_2 < 0$, m is positive and (D.7) represent parabolic curves. Thus, in a suitably transformed coordinate axes the trajectories approach the equilibrium point along parabolic paths as shown in Fig. 3.1b.

When $\lambda_1 < 0 < \lambda_2$ (corresponding to the saddle point) we have $m < 0$ and (D.7) represent hyperbolic curves. In this case the trajectories deviate from the equilibrium point along hyperbolic paths.

Problems

1. For $b = c = 0$ in (3.6) show that after shifting $x \to x - x_0$, $y \to y - y_0$ the solution curve is $y = C_1|x|^m$, where C_1 is an integration constant and $m = d/a$.
2. Show that for the system $\dot{x} = x$, $\dot{y} = ay$ the solution curve can be described by $y = C|x|^a$, where C is a constant. From the above solution determine the nature of the equilibrium point $(x_0, y_0) = (0, 0)$ for (i) $0 < a < 1$, (ii) $a > 1$, (iii) $a = 1$ and (iv) $a < 0$ [1].

E. Fractals and Multifractals

In Chap. 4, we have illustrated an important characteristic property of the chaotic attractors, namely, the self-similar structure with special reference to the Hénon attractor. An object or a mathematical set with a self-similar pattern is called a *fractal*, which is a term coined by B.B. Mandelbrot [1–3]. A fractal has a sprawling and tenuous pattern which when magnified reveals repetitive levels of detail so that similar structure exists at all scales. Some general characteristic properties of fractal sets are their fine structures, that is details on arbitrary small scales, self-similar patterns and noninteger dimensions. Many of the fractal descriptions go back to classical mathematics and mathematicians of the past, like Georg Cantor (1883), Helge von Koch (1904), Waclaw Sierpinski (1916), Gaston Julia (1918) and Felix Hausdorff (1919).

A simple example of a natural structure with self-similar property is the familiar cauliflower. It contains branches or parts which when removed and compared with the whole are very much the same, except that they are smaller in size. These clusters which again look very similar to the whole as well as to the first generation branches. This self-similarity can be seen for about 3 to 4 stages. After that the structures are too small for further dissection. Some other examples are: Black spleen wort fern, sponge, growth pattern of some crystals, random branching pattern of blood vessels, random patterns of clouds and coastlines, electrical discharges, chaotic attractors, mathematical sets such as Cantor set, Koch curve, Julia set and Mandelbrot set.

I. FRACTALS

We now describe the construction of some simple fractals and point out some of their characteristic properties [2–4].

1. *Cantor Set*

The basic Cantor set is an infinite set of points in the unit interval $[0, 1]$ and its existence was published in 1883 by the German mathematician Georg Cantor. The middle third Cantor set is one of the best known fractal sets and is constructed from a unit interval by a sequence of deletion operation. Let S_0 be the interval $[0, 1]$ and delete its middle third and call the new set as S_1. It consists of two intervals namely $[0, 1/3]$ and $[2/3, 1]$. Next, delete the middle thirds of each of these intervals and obtain the set S_2. Repeating this process k times we obtain the set S_k which is obtained by deleting the middle thirds of each interval in S_{k-1}. In the limit $k \to \infty$ we get the Cantor set S. Figure E.1 depicts the construction of the middle third Cantor set. The set S has a self-similar property. It is clear that the parts $[0, 1/3]$ and $[2/3, 1]$ in S_1 are geometrically similar to S_0 except they are scaled by a factor 1/3. Again the parts of S_2 in each of four intervals are similar to S_0 but scaled by a factor $(1/3)^2$ and so on.

What is the dimension of the Cantor set constructed as above? From Fig. E.1 and (8.34) its fractal dimension is computed as

$$
\begin{aligned}
D_F &= \lim_{\epsilon \to 0} -\frac{\ln M(\epsilon)}{\ln \epsilon} \\
&= -\frac{\ln 2^k}{\ln 3^{-k}} \\
&= \frac{\ln 2}{\ln 3} = 0.631 \, .
\end{aligned}
$$

What is the significance of this number? To understand it, let us calculate the total length of the Cantor set. At step 0 the length is 1; at step 1 the length is $2/3 = 0.666$; at step 2 the length is 0.444; at step 3 the length is 0.296 and so on. That is, though the number of segments diverges as the construction process continues, the total length of the set (as well as the length of individual segments) decreases and approaches zero. From (8.35) we can say that the dimension of the set simply gives the relation between the number of segments M and the length ϵ of each segment. We conclude that the Cantor set is a fractal with a noninteger dimension $D_F = 0.631$.

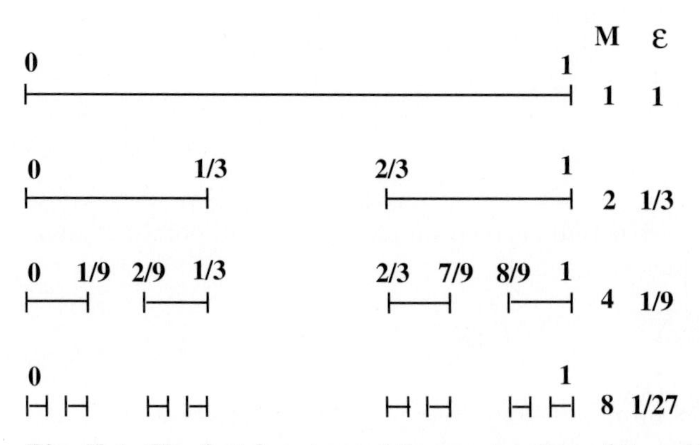

Fig. E.1. The first few steps of the construction of the middle third Cantor set

2. *Sierpinski Triangle*

The Sierpinski triangle is another fractal introduced by the Polish mathematician Waclaw Sierpinski in 1916. Its construction process starts with an equilateral triangle and consists of the following iterative rules:

(1) connect the mid points of the sides with line segments and

(2) remove the middle triangle of the 4 triangles formed.

Continuation of these iteration rules lead to a very porus, open figure much like a triangular filter. Figure E.2 shows the object obtained at few successive steps. The completed Sierpinski triangle naturally decomposes into three triangular parts, an upper portion and two lower portions. These parts are exact replica of the whole original figure. Likewise, each of these parts itself can be decomposed into three smaller ones again exact replica of the original figure. The fractal dimension of Sierpinski triangle is determined as

$$D_F = \lim_{\epsilon \to 0} - \frac{\ln 3^{k+1}}{\ln(1/2)^k} \approx 1.585 \, .$$

3. *Few More Examples*

Other interesting mathematical constructions exhibiting self-similar structure include Koch curve, snowflake, Julia set and Mandelbrot set [1–3]. They may be constructed as follows.

(i) Koch Curve: One begins with a straight-line segment. It is partitioned into three equal segments. Then replace the middle-one by an equilateral triangle and remove its base. Now there are four line segments of equal size. Repeat the above scheme infinitely large number of times to each of the segments formed at every stage. The resulting curve is called the Koch curve.

(ii) Snowflake: Snowflake is composed of three congruent parts, each of which is a Koch curve.

(iii) Julia Set: It is the basin boundary of points in the $x - y$ plane which under iteration of a complex map $z_{n+1} = f(z_n) + c$, where $z(= x + \mathrm{i}y)$ and c are complex, do not escape to infinity. Some forms of $f(z)$ and values of c giving beautiful Julia sets are: $f(z) = z^2$, $c = -1, 0.25, 0.2+\mathrm{i}0.3, -0.9+\mathrm{i}0.12$; $f(z) = \lambda \sin z$, $c = 0$, $\lambda = 1, 1+\mathrm{i}0.1, 1+\mathrm{i}0.4$. Obviously Julia set is described in the plane of the state variables (x, y).

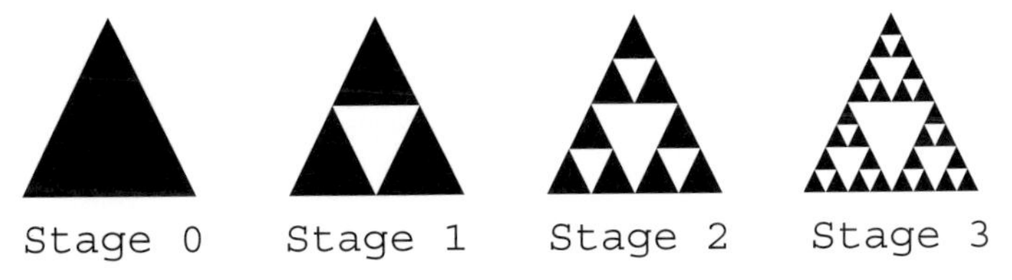

Fig. E.2. The first few successive steps of constructing a Sierpinski triangle

(iv) Mandelbrot Set: It is described in the parameter space (a, b) of the control parameter $c = a+ib$ in the above mentioned Julia map. It is the basin boundary of points in the (a, b) plane for fixed initial conditions $z_0 = x_0 + iy_0$ which under the iterations of the map do not diverge to infinity.

II. MULTIFRACTALS

The Cantor set discussed above is homogeneous. That is, the points are uniformly distributed in the nonempty regions and hence the probability distribution p of points in the all nonempty subintervals are identical. There are inhomogeneous fractals in which the points are not uniformly distributed. Some regions of the object may have relatively large values of p while in other regions they are small. In the calculation of the fractal dimension all the nonempty regions contribute equally. Therefore, the quantity D_F is insufficient to characterize the nonuniformity or inhomogeneity of the object. Generally, chaotic attractors are inhomogeneous [2–4]. Such an inhomogeneous object or set is called a *multifractal* and is characterized by *generalised dimensions*.

A multifractal means essentially a multiscale fractal. A uniform fractal, for example, the middle third Cantor set, is constructed by iterating a single first generation length l repeatedly, so that after k iterations the length of each segments is l^k. Multifractal is constructed by the repeated iteration of two or more first generation length scales $l_1, l_2, \ldots$ and the result is a nonuniform object.

Let the given object consisting of m pieces be covered with N number of n-dimensional boxes of equal size ϵ. For the inhomogeneous object an important quantity which can be determined is the sum of the q-th power of p_i, that is

$$\tau(q) = \sum_{i=1}^{N} p_i^q \tag{E.1}$$

for $-\infty < q < \infty$. Here p_i's, $i = 1, 2, \ldots, N$ are the probability of finding a piece in an ith box. For $q = 0$, eq.(E.1) gives $N(\epsilon)$ and therefore

$$\tau(0) = N(\epsilon) = \epsilon^{-D_0} , \tag{E.2}$$

where D_0 is now the fractal dimension. Because of the complexity of the distributions the scaling for $\tau(q)$ for $\epsilon \to 0$ depends on q as

$$\tau(q) \approx \epsilon^{(q-1)D_q} , \tag{E.3}$$

where D_q is the so-called *order-q generalised dimension*. D_q can be calculated from the relation [1]

$$D_q = \frac{1}{q-1} \lim_{\epsilon \to 0} \frac{\ln \sum_i p_i^q}{\ln(1/\epsilon)} . \tag{E.4}$$

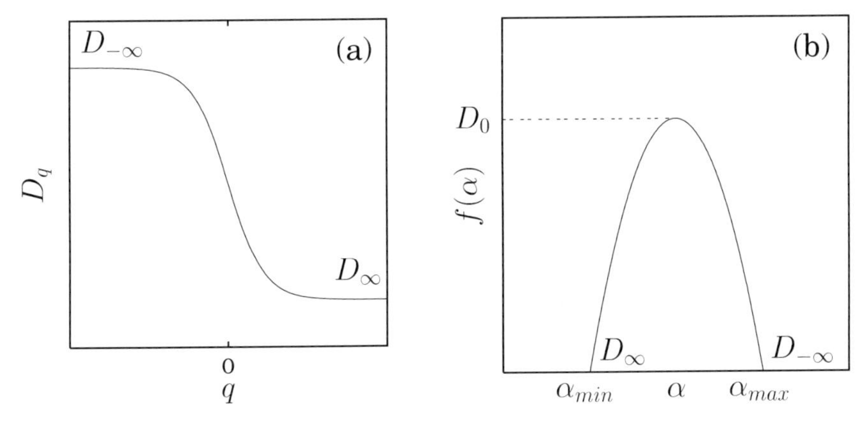

Fig. E.3. Typical plots of (**a**) q versus D_q and (**b**) α versus $f(\alpha)$

For all q, we have $D_q > 0$. It can be shown that D_q decreases monotonically with increase in q. Given D_q, one can calculate two other quantities $\alpha(q)$ and $f(\alpha)$ as

$$\alpha(q) = \frac{\mathrm{d}}{\mathrm{d}q}(q-1)D_q\,, \quad f(\alpha) = q\alpha(q) - (q-1)D_q\,. \tag{E.5}$$

Figure E.3a shows a typical plot of q versus D_q. Here D_∞ corresponds to the region in the set where the measure is most concentrated, while $D_{-\infty}$ corresponds to that where the measure is most rarified. In this sense D_q describes the nonuniformity property of a fractal object. Similarly, certain features of α versus $f(\alpha)$ are quit general. A typical form of α versus $f(\alpha)$ curve is depicted in Fig. E.3b. It may be noted that $f(\alpha)$ is a convex function and is increasing for $q > 0$ and decreasing for $q < 0$. The maximum of the $f(\alpha)$ curve is the fractal dimension D_0. The slope $\partial f/\partial\alpha$ is unity for $q = 1$. As $q \to -\infty$, $\alpha \to \alpha_{max}$. These properties imply that the information about the most intense region is obtained from the quantities at $q \to \infty$ and about the more rarified region is obtained at $q = -\infty$.

The theory of fractals has considerable applications in the study of diffusion-limited-aggregation (DLA), diffusion and discharge experiments, viscous fingering, percolation, image process and so on. For more details about fractals the reader may refer to Refs. [2–4], for example.

Problems

1. Investigate the nature of simple fractals such as Koch curve, snowflake, Julia set and Mandelbrot set [2–4] discussed above.

2. Show that for $q \to 1$ the (information)dimension D_1 becomes

$$D_1 = \frac{\sum p_i \ln p_i}{\ln(1/\epsilon)} \ .$$

3. Compute D_q and $f(\alpha)$ spectrum and verify their general properties for the Hénon attractor (Fig. 4.18a) and the chaotic attractor (Fig. 5.6b) of the Duffing oscillator equation.

F. Spectrum of the sech$^2 \alpha x$ Potential

Let us consider the time-independent Schrödinger equation

$$-\frac{\hbar^2}{2m}\frac{\mathrm{d}^2\psi}{\mathrm{d}x^2} + V(x)\psi(x) = E\psi(x) \tag{F.1a}$$

for the potential

$$V(x) = -\frac{\hbar^2\alpha^2}{2m}\lambda(\lambda-1)\mathrm{sech}^2\alpha x \,. \tag{F.1b}$$

Here α and λ are real parameters with $\lambda > 1$. It can be rewritten as

$$\frac{\mathrm{d}^2\psi}{\mathrm{d}x^2} + \left[k^2 + \alpha^2\lambda(\lambda-1)\mathrm{sech}^2\alpha x\right]\psi(x) = 0\,, \tag{F.2}$$

where $k^2 = 2mE/\hbar^2$. The potential (F1.b) is usually called the modified Pöschl–Teller potential in the quantum mechanics literature (see for example Ref. [1]).

With the introduction of the new variable y =cosh$^2\alpha x$ and writing the eigenfunction in the form

$$\psi = y^{\lambda/2}\, v(y)\,, \tag{F.3}$$

we arrive at the hypergeometric differential equation [1]

$$y(1-y)v'' + \left[(\lambda+\frac{1}{2}) - (\lambda+1)y\right]v' - \frac{1}{4}\left(\lambda^2 + \frac{k^2}{\alpha^2}\right)v = 0\,. \tag{F.4}$$

The solution of (F.4) can be easily found to be

$$\begin{aligned} v(y) &= A\,{}_2F_1\left(a, b, \frac{1}{2}; 1-y\right) \\ &\quad + B\sqrt{(1-y)}\,{}_2F_1\left(a+\frac{1}{2}, b+\frac{1}{2}, \frac{3}{2}; 1-y\right)\,, \end{aligned} \tag{F.5a}$$

where

$$a = \frac{1}{2}\left(\lambda + \mathrm{i}\frac{k}{\alpha}\right), \quad b = \frac{1}{2}\left(\lambda - \mathrm{i}\frac{k}{\alpha}\right) \tag{F.5b}$$

and ${}_2F_1(a, b, c; x)$ is the hypergeometric function with the usual notation. We look for a fundamental system of two real standard solutions in which one is odd and the other is even in x.

The even standard solution can be obtained from (F.5) with $B = 0$ and $A = 1$ as

$$\psi_e(x) = \cosh^\lambda \alpha x \; {}_2F_1\left(a, b, \frac{1}{2}; -\sinh^2 \alpha x\right) \tag{F.6a}$$

and the following odd standard solution can be deduced for $A = 0$, $B = -\mathrm{i}$,

$$\psi_0(x) = \cosh^\lambda \alpha x \sinh^\lambda \alpha x \; {}_2F_1\left(a + \frac{1}{2}, b + \frac{1}{2}, \frac{3}{2}; -\sinh^2 \alpha x\right) . \tag{F.6b}$$

We can consider the cases (i)$E > 0$ and (ii)$E < 0$ separately.

Case (i): Nonlocalized States ($E > 0$)

From the asymptotic form of solutions (F.6), see for example Ref. [1], we can infer the following. When E is positive, a and b become complex conjugates. Then the general solution for $|x| \to \infty$ can be written as

$$\psi = \frac{\widetilde{A}}{2} C_e \left(\mathrm{e}^{\mathrm{i}\phi_e} \mathrm{e}^{-\mathrm{i}kx} + \mathrm{e}^{-\mathrm{i}\phi_e} \mathrm{e}^{\mathrm{i}kx}\right) - \frac{\widetilde{B}}{2} C_0 \left(\mathrm{e}^{\mathrm{i}\phi_0} \mathrm{e}^{-\mathrm{i}kx} + \mathrm{e}^{-\mathrm{i}\phi_0} \mathrm{e}^{\mathrm{i}kx}\right), \quad x \to -\infty \tag{F.7a}$$

and

$$\psi = \frac{\widetilde{A}}{2} C_e \left(\mathrm{e}^{\mathrm{i}\phi_e} \mathrm{e}^{\mathrm{i}kx} + \mathrm{e}^{-\mathrm{i}\phi_e} \mathrm{e}^{-\mathrm{i}kx}\right) + \frac{\widetilde{B}}{2} C_0 \left(\mathrm{e}^{\mathrm{i}\phi_0} \mathrm{e}^{\mathrm{i}kx} + \mathrm{e}^{-\mathrm{i}\phi_0} \mathrm{e}^{-\mathrm{i}kx}\right), \quad x \to +\infty \tag{F.7b}$$

where the constants C_e,C_0, ϕ_e and ϕ_0 can be obtained by making an asymptotic analysis of the solutions (F.6) at $x \to \pm\infty$. The constants $\widetilde{A}$ and $\widetilde{B}$ are arbitrary, while the constants ϕ_e and ϕ_0 can be shown to be of the form [1]

$$\phi_e = \arg \frac{\Gamma(\mathrm{i}k/\alpha) \exp[-\mathrm{i}(k/\alpha) \log 2]}{\Gamma\left[(\lambda/2) + \mathrm{i}k/(2\alpha)\right] \Gamma\left[(1-\lambda)/2 + \mathrm{i}k/(2\alpha)\right]} , \tag{F.7c}$$

$$\phi_0 = \arg \frac{\Gamma(\mathrm{i}k/\alpha) \exp[-\mathrm{i}(k/\alpha) \log 2]}{\Gamma\left[(\lambda-1)/2 + \mathrm{i}k/(2\alpha)\right] \Gamma\left[(1-\lambda)/2 + \mathrm{i}k/(2\alpha)\right]} . \tag{F.7d}$$

Now we look for the specific asymptotic form ($|x| \to \infty$) of the solution

$$\psi = \begin{cases} \mathrm{e}^{\mathrm{i}kx} + R\mathrm{e}^{-\mathrm{i}kx}, & x \to -\infty \\ T\mathrm{e}^{\mathrm{i}kx}, & x \to +\infty \end{cases} \tag{F.8}$$

where R and T are reflection and transmission coefficients. Comparing (F.8) with (F.7), we can obtain R and T as

$$R = \frac{1}{2} \left[(\cos 2\phi_e + \cos 2\phi_0) + \mathrm{i}(\sin 2\phi_e + \sin 2\phi_0)\right] , \tag{F.9a}$$

and

$$T = \frac{1}{2}\left[(\cos 2\phi_e - \cos 2\phi_0) + \mathrm{i}\,(\sin 2\phi_e - \sin 2\phi_0)\right] \,. \tag{F.9b}$$

With the introduction of the parameter

$$p = \frac{\sinh(\pi k/\alpha)}{\sin(\pi\lambda)} \,, \tag{F.10a}$$

one can show that the reflection and transmission coefficients satisfy the relations

$$|R|^2 = \frac{1}{1+p^2}, \qquad |T|^2 = \frac{p^2}{1+p^2} \,, \tag{F.10b}$$

satisfying the conservation law $|T|^2 + |R|^2 = 1$.

Case (ii): Bound States ($E < 0$)

For the given potential (F.1b), the bound states exist only for negative energies. Therefore we write $k = i\kappa$, $\kappa > 0$, so that

$$E = -\frac{\hbar^2}{2m}\kappa^2 \,. \tag{F.11}$$

Again making an asymptotic analysis of (F.6), the eigenvalues for even eigenstates can be obtained from the relation

$$\frac{\kappa}{\alpha} = \lambda - 1 - 2n, \qquad n = 0, 1, 2, \ldots \tag{F.12a}$$

and for the odd states it follows that

$$\frac{\kappa}{\alpha} = \lambda - 2 - 2n, \qquad n = 0, 1, 2, \ldots \tag{F.12b}$$

As a result the energy spectrum can be given by

$$E_n = -\frac{\hbar^2\alpha^2}{2m}(\lambda - 1 - n)^2, \qquad n \leq \lambda - 1 \tag{F.13}$$

where $n = 0, 1, 2, \ldots$ Since $n \leq \lambda - 1$ for a given λ, there exists only a finite number (N) of bound states.

Problems

1. Derive the relations (F.7), (F.9) and (F.10b).
2. For the potentials $V = -2\mathrm{sech}^2 x$ and $-6\mathrm{sech}^2 x$ obtain the reflection and transmission coefficients for nonlocalized states and the energy spectrum for bound states.

G. Inverse Scattering Transform for the Schrödinger Spectral Problem

In this appendix, we will briefly explain how the scattering data for a given potential can be obtained for the case of Schrödinger spectral problem based on the analysis of Ref. [1]. We also point out how the inverse problem, namely reconstructing the potential uniquely from the given scattering data, can be tackled. The results are then applicable to solve the initial value problem of the KdV equation through the inverse scattering method as described in Chap. 13. For details see Refs. [1,2].

G.1 The Linear Eigenvalue Problem

Let us consider the KdV equation (13.1). The associated linear spectral problem is the time-dependent Schrödinger-like eigenvalue equation

$$\psi_{xx} - u(x)\psi = -\zeta^2\psi \,, \tag{G.1a}$$

where ζ is the eigenvalue, which we assume to be complex in general. The potential $u(x)$ is assumed to satisfy the following condition

$$\int_{-\infty}^{\infty} (1+|x|)|u(x)|\mathrm{d}x \;<\; \infty \,. \tag{G.1b}$$

This is to ensure that the potential $u(x)$ is a rapidly decaying function and approaches zero sufficiently fast as $|x| \to \infty$. (More stringent conditions can also be introduced which are however not of interest to us at present). Thus, as $x \to \pm\infty$, $u(x) \to 0$ and so in this limit $\psi(x,\zeta)$ can be taken to have the form of a linear combination of the functions $\mathrm{e}^{\mathrm{i}\zeta x}$ and $\mathrm{e}^{-\mathrm{i}\zeta x}$. We thus choose the two Jost functions $\phi_\pm(x,\zeta)$ satisfying (G.1) such that

$$\phi_+(x,\zeta) \sim \mathrm{e}^{\mathrm{i}\zeta x} \quad \text{as} \quad x \to +\infty \,, \tag{G.2a}$$

$$\phi_-(x,\zeta) \sim \mathrm{e}^{-\mathrm{i}\zeta x} \quad \text{as} \quad x \to -\infty \,. \tag{G.2b}$$

Formally by considering the Jost function ϕ_+ along with the boundary conditions in the regime $x \to \infty$, ψ can be written as

$$\psi = \mathrm{e}^{-\mathrm{i}\zeta x}\phi_+(x,\zeta) \,. \tag{G.3}$$

Substitution of this into (G.1) gives

$$\phi_{+xx} - 2\mathrm{i}\zeta\phi_{+x} = u(x)\phi_+(x,\zeta)\,. \tag{G.4}$$

The Green's function associated with this problem satisfies the differential equation

$$\frac{\mathrm{d}^2G(x,y)}{\mathrm{d}y^2} - 2\mathrm{i}\zeta G(x,y) = -\delta(x-y)\,. \tag{G.5}$$

The Green's function $G(x,y)$ can be obtained by substituting the Fourier transform of it in (G.5) and evaluating the resulting integral by contour integration. It is found to be

$$G(x,y) = \frac{1}{2\mathrm{i}\zeta} - \frac{\mathrm{e}^{2\mathrm{i}\zeta(x-y)}}{2\mathrm{i}\zeta}\,. \tag{G.6}$$

Now the solution of (G.1) can be written as

$$\mathrm{e}^{-\mathrm{i}\zeta x}\phi_+(x,\zeta) = G(\infty) + \int_x^\infty G(x,y)u(y)\phi_+(y,\zeta)\mathrm{e}^{-\mathrm{i}\zeta y}\mathrm{d}y\,. \tag{G.7}$$

Identifying $G(\infty)$ as 1, the singular Volterra integral equation corresponding to (G.1) is

$$\mathrm{e}^{-\mathrm{i}\zeta x}\phi_+(x,\zeta) = 1 + \int_x^\infty M(x,y,\zeta)\phi_+(y,\zeta)\mathrm{e}^{-\mathrm{i}\zeta y}\mathrm{d}y \tag{G.8}$$

with the kernel $M(x,y,\zeta)$ having the form

$$M(x,y,\zeta) = \begin{cases} -\sin\zeta(x-y)\mathrm{e}^{-\mathrm{i}\zeta(x-y)}u(y)/\zeta & \text{for } \zeta \neq 0 \\ -(x-y)u(y) & \text{for } \zeta = 0\,. \end{cases} \tag{G.9}$$

One can prove that the Neumann series for $\mathrm{e}^{-\mathrm{i}\zeta x}\phi_+(x,\zeta)$ [1,2] converges. It then follows that (G.8) has a unique solution for each ζ with $\mathrm{Im}\zeta \geq 0$ for u obeying the condition (G.1b). Since $M(x,y,\zeta)$ is regular for $\mathrm{Im}\zeta > 0$ and vanishes as $|\zeta| \to \infty$, $\mathrm{e}^{-\mathrm{i}\zeta x}\phi_+(x,\zeta)$ is regular for $\mathrm{Im}\zeta > 0$. Similarly, it can be proved that $\mathrm{e}^{\mathrm{i}\zeta x}\phi_-(x,\zeta)$ is analytic for $\mathrm{Im}\zeta > 0$. One can also check that $\mathrm{e}^{-\mathrm{i}\zeta x}\phi_+(x,\zeta) \to 1$ as $\zeta \to \infty$ for $\mathrm{Im}\zeta > 0$, which is also valid for $\mathrm{e}^{\mathrm{i}\zeta x}\phi_-(x,\zeta)$.

G.2 The Direct Scattering Problem

On the real axis $\zeta = \xi$ (that is, the eigenvalues are real), from (G.2) we have two linearly independent solutions $\phi_+(x,\xi)$ and $\phi_-(x,\xi)$ (except for $\xi = 0$). So any other solution to (G.1) will be a linear combination of these two functions, for example the solution

$$\phi_-(x,\xi) = a(\xi)\phi_+(x,-\xi) + b(\xi)\phi_+(x,\xi)\,, \tag{G.10}$$

where $a(\xi)$ and $b(\xi)$ are coefficients depending on ξ only. One can then define the reflection coefficient as

$$R(\xi) = \frac{b(\xi)}{a(\xi)}\,. \tag{G.11}$$

Now defining the Wronskian of any two functions $\psi_1(x)$ and $\psi_2(x)$,

$$W[\psi_1, \psi_2] = \begin{vmatrix} \psi_1 & \psi_2 \\ \psi_{1x} & \psi_{2x} \end{vmatrix} = \psi_1\psi_{2x} - \psi_2\psi_{1x} , \tag{G.12}$$

we find that for (G.1) $W[\psi_1, \psi_2]$ is independent of x and $\mathrm{d}W/\mathrm{d}x = 0$, that is $W[\psi_1, \psi_2]$ is a constant. We normalize the wave function (see also (G.2)) such that

$$W[\psi_1, \psi_2] = 1 . \tag{G.13}$$

Now using (G.10) in (G.12) along with the asymptotic form (G.2) we can easily establish that

$$|a(\xi)|^2 = 1 + |b(\xi)|^2 . \tag{G.14}$$

With the help of (G.12) and the Jost functions we find $a(\xi)$ on the real axis as

$$a(\xi) = \frac{1}{2i\xi} W[\phi_-(x,\xi), \phi_+(x,\xi)] . \tag{G.15}$$

Since u obeys the condition (G.1b) and $\phi_+ \mathrm{e}^{-\mathrm{i}\zeta x}$, $\phi_- \mathrm{e}^{\mathrm{i}\zeta x}$ are analytic in the upper half plane, $a(\zeta)$ is also analytic and hence we can extend $a(\xi)$ into the upper half ζ-plane by

$$a(\zeta) = \frac{1}{2\mathrm{i}\zeta} W[\phi_-(x,\zeta), \phi_+(x,\zeta)] . \tag{G.16}$$

Further from the asymptotic form (G.2), it is easy to see that

$$a(\zeta) \to 1 \text{ as } \zeta \to \infty . \tag{G.17}$$

Thus, for the given potential one can construct the Jost functions $\phi_\pm$ and the scattering parameters a and b. Computing these quantities and deducing their analytic properties constitute the direct scattering problem.

Now, let us look at the nature of $a(\zeta)$. What are the values ζ_1, ζ_2, ..., ζ_N for which $a(\zeta) = 0$ in the upper half plane? By recalling the property of $a(\zeta)$ given in (G.17) and applying the fact that $a(\zeta)$ is continuous for $\mathrm{Im}\zeta \geq 0$, we can check that the values ζ_i must be finite. If $u(x)$ is real then they cannot lie on the real axis because of (G.14). Thus the function $a(\zeta)$ can have only a finite number of simple zeros at ζ_1, ζ_2, ..., ζ_N which should lie on the imaginary axis of the upper half plane.

Since $a(\zeta)$ vanishes at ζ_1, ζ_2, ..., ζ_N, from (G.16), the Wronskian of ϕ_- and ϕ_+ also vanishes at $\zeta = \zeta_k$, $k = 1, 2, ..., N$. So ϕ_- and ϕ_+ must be linearly dependent at these points:

$$\phi_-(x, \zeta_k) = b_k \phi_+(x, \zeta_k) . \tag{G.18}$$

It is to be noted that $\phi_\pm(x, \zeta_k) \to 0$ as $x \to \pm\infty$, because $\zeta_k = i\eta_k$, $\eta_k > 0$, by construction. Thus, we have bound states for $\zeta = \zeta_k$, $k = 1, 2, ..., N$. We can define the normalized bound states as

$$\phi_k(x) = C_k \phi_+(x, \zeta_k) \ , \quad C_k^2 = -\mathrm{i}\, b_k / a'(\zeta_k) \ , \tag{G.19}$$

so that

$$\mathrm{e}^{-\mathrm{i}\zeta_k x} \phi_k(x) \to C_k \ \text{ as } \ x \to \infty \ . \tag{G.20}$$

Thus, the scattering data for a given potential $u(x)$ is

$$S = \{R(\xi), \bar{R}(\xi), \ \ -\infty < \xi < \infty, \ \ \zeta_k, C_k^2, \ \ k = 1, 2, ..., N\} \ . \tag{G.21}$$

G.3 The Inverse Scattering Problem

The potential $u(x)$ can be reconstructed uniquely as follows. In order to do this we consider the function $\Phi(x, \zeta)$ as identified from the above analysis,

$$\Phi(x, \zeta) = \begin{cases} \mathrm{e}^{\mathrm{i}\zeta x} \phi_-(x, \zeta)/a(\zeta) & \text{for } \mathrm{Im}\zeta > 0 \\ \mathrm{e}^{\mathrm{i}\zeta x} \phi_+(x, -\zeta) & \text{for } \mathrm{Im}\zeta < 0. \end{cases} \tag{G.22}$$

It has the following properties:

(1) $\Phi(x, \zeta) \to 1$ as $\zeta \to \infty$.

(2) It has a discontinuity on the real axis

$$\Phi(x, \xi + \mathrm{i}0) - \Phi(x, \xi - \mathrm{i}0) = R(\xi) \mathrm{e}^{\mathrm{i}\xi x} \phi_+(x, \xi) \ .$$

(3) It has simple poles at $\zeta = \zeta_k$, $k = 1, 2, ..., N$, in the upper half ζ-plane with residues

$$\frac{b_k}{a'(\zeta_k)} \mathrm{e}^{\mathrm{i}\zeta_k(x)} \phi_+(x, \zeta_k) = \mathrm{i} C_k^2 \mathrm{e}^{\mathrm{i}\zeta_k x} \phi_+(x, \zeta_k) \ .$$

Now let us apply the residue theorem to $(\Phi(x, \zeta) - 1)/(\zeta - \zeta_0)$ $(\mathrm{Im}\zeta_0 < 0)$ around the two semi-circles as shown in Fig. G.1. We get

$$\int_{\Gamma_1 + \Gamma_2} \frac{\Phi(x, \zeta) - 1}{\zeta - \zeta_0} \mathrm{d}\zeta$$
$$= 2\pi\mathrm{i} \left[\sum_{k=1}^{N} (\text{residue at } \zeta_k) + (\text{residue at } \zeta_0) \right] \ . \tag{G.23}$$

As $R \to \infty$, the integral round the curved part tends to 0 and so we get

$$\int_{-\infty}^{\infty} \frac{R(\xi) \mathrm{e}^{\mathrm{i}\xi x} \phi_+(x, \xi)}{\xi - \zeta_0} \mathrm{d}\xi$$
$$= 2\pi\mathrm{i} \left[\sum_{k=1}^{N} \mathrm{i}\, C_k^2 \frac{\mathrm{e}^{\mathrm{i}\zeta_k x} \phi_+(x, \zeta_k)}{\zeta_k - \zeta_0} + \mathrm{e}^{\mathrm{i}\zeta_0 x} \phi_+(x, -\zeta_0) - 1 \right] \ . \tag{G.24}$$

After rearranging the terms and letting $\zeta_0 \to -\xi'$ from the lower half plane the following expression can be obtained,

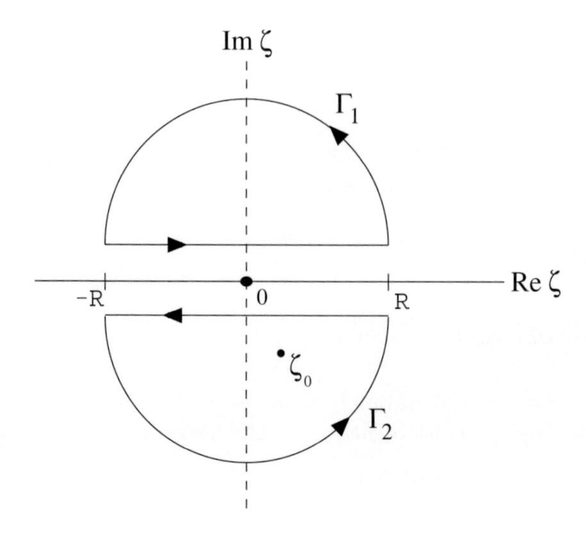

Fig. G.1. Path of integration for the function $\Phi(x,\zeta)$, see (G.23)

$$\mathrm{e}^{-\mathrm{i}\xi' x}\phi_+(x,\xi') - 1 = -\mathrm{i}\sum_{k=1}^{N} C_k^2 \frac{\mathrm{e}^{\mathrm{i}\zeta_k x}}{\zeta_k+\xi'}\phi_+(x,\zeta_k) + \frac{1}{2\pi\mathrm{i}}\int_{-\infty}^{\infty}\frac{R(\xi)\mathrm{e}^{\mathrm{i}\xi x}\phi_+(x,\xi)}{\xi+\xi'+\mathrm{i}0}\,\mathrm{d}\xi\,. \qquad (G.25)$$

Taking the Fourier transform of the function on the left hand side of the above equation, we can write the expression

$$\phi_+(x,\zeta) - \mathrm{e}^{-\mathrm{i}\zeta x} = \int_{-\infty}^{\infty} K(x,y)\mathrm{e}^{\mathrm{i}\zeta y}\,\mathrm{d}y\,. \qquad (G.26)$$

From this $K(x,y)$ can be obtained as

$$K(x,y) = \frac{1}{2\pi}\int_{-\infty}^{\infty}\left(\phi_+(x,\xi) - \mathrm{e}^{\mathrm{i}\xi x}\right)\mathrm{e}^{-\mathrm{i}\xi y}\,\mathrm{d}\xi\,. \qquad (G.27)$$

By substituting the expression for $\phi_+(x,\xi) - \mathrm{e}^{\mathrm{i}\xi x}$ from (G.25), one can find

$$K(x,y) = -\theta(y-x)\left[\sum C_k^2\mathrm{e}^{\mathrm{i}\zeta_k y}\phi_+(x,\zeta_k) + \frac{1}{2\pi}\int_{-\infty}^{\infty} R(\xi)\mathrm{e}^{\mathrm{i}\xi y}\phi_+(x,\xi)\,\mathrm{d}\xi\right], \qquad (G.28)$$

where θ is the Heaviside step function. For $x<y$, $\theta(x,y)=1$, (G.28) reduces to

$$K(x,y) + F(x+y) + \int_x^{\infty} K(x,z)F(z+y)\mathrm{d}z = 0 \quad \text{for } x<y\,, \qquad (G.29a)$$

where

$$F(x) = \sum_{k=1}^{N} C_k^2 e^{i\zeta_k x} + \frac{1}{2\pi} \int_{-\infty}^{\infty} R(\xi) e^{i\xi x} \, d\xi \, . \tag{G.29b}$$

The linear integral equation given by (G.29) is known as the *Gelfand–Levitan–Marchenko equation*.

G.4 Reconstruction of the Potential

By taking Fourier transform of the Volterra equation (G.8) we get

$$K(x,y) = \frac{1}{2\pi} \int_{-\infty}^{\infty} e^{i\xi(x-y)} \int_{x}^{\infty} M(x,z,\xi) e^{-i\xi z} \times \left\{ e^{i\xi z} + \int_{-\infty}^{\infty} K(z,\omega) e^{i\xi\omega} d\omega \right\} dz d\xi \, . \tag{G.30}$$

Evaluating the integral

$$\int_{-\infty}^{\infty} e^{i\xi(x-y-z+\omega)} M(x,z,\xi) d\xi = -\pi \left\{ \theta(x-y-z+\omega) - \theta(-x-y+z+\omega) \right\} u(z) \tag{G.31}$$

and using it in (G.30) we get

$$K(x,y) = \frac{1}{2} \int_{(x+y)/2}^{\infty} u(z) \, dz + \frac{1}{2} \int_{x}^{\infty} \int_{x+y-z}^{-x+y+z} K(z,\omega) u(z) \, d\omega dz \quad \text{for } x < y \, . \tag{G.32}$$

After the substitution $y = x + 0$, the above equation becomes

$$K(x, x+0) = \frac{1}{2} \int_{x}^{\infty} u(z) \, dz \, . \tag{G.33}$$

Replacing z by x and using the condition $u(x) \to 0$ as $x \to \infty$, we can find the potential as

$$u(x) = -2 \frac{d}{dx} K(x, x+0) \, . \tag{G.34}$$

Problems

1. Deduce the Green's function (G.6).
2. Write down the scattering data S for the Pöschl–Teller potential using the results of appendix F.

H. Inverse Scattering Transform for the Zakharov–Shabat Eigenvalue Problem

H.1 The Linear Eigenvalue Problem

We consider the following scattering problem [1,2] which is the generalized Zakharov–Shabat system (Sect. 14.2):

$$v_{1x} + \mathrm{i}\zeta v_1 = q(x) v_2 \,, \tag{H.1a}$$

$$v_{2x} - \mathrm{i}\zeta v_2 = r(x) v_1 \,, \tag{H.1b}$$

where the potentials $q(x)$ and $r(x)$ are assumed to satisfy the condition

$$\int_{-\infty}^{\infty} \{|q(x)| + |r(x)|\} \,\mathrm{d}x \;<\; \infty \,, \tag{H.2}$$

and the eigenvalue ζ can be complex, in general. The condition (H.2) is invoked to ensure that the potentials $q(x)$ and $r(x)$ are rapidly decaying functions and approach zero sufficiently fast as $|x| \to \infty$. Then it is obvious that the solutions

$$v(x,\zeta) = \begin{pmatrix} v_1(x,\zeta) \\ v_2(x,\zeta) \end{pmatrix} \tag{H.3}$$

are oscillatory functions asymptotically (as $|x| \to \infty$) and can be written as linear combination of $\begin{pmatrix} \mathrm{e}^{-\mathrm{i}\zeta x} \\ 0 \end{pmatrix}$ and $\begin{pmatrix} 0 \\ \mathrm{e}^{\mathrm{i}\zeta x} \end{pmatrix}$. This leads one to choose the Jost functions (eigenfunctions of the given problem) $\phi(x,\zeta)$, $\bar{\phi}(x,\zeta)$, $\psi(x,\zeta)$ and $\bar{\psi}(x,\zeta)$ which satisfy the following boundary conditions,

$$\phi(x,\zeta) \sim \begin{pmatrix} \mathrm{e}^{-\mathrm{i}\zeta x} \\ 0 \end{pmatrix}, \quad \bar{\phi}(x,\zeta) \sim \begin{pmatrix} 0 \\ \mathrm{e}^{\mathrm{i}\zeta x} \end{pmatrix} \quad \text{as } x \to -\infty \tag{H.4a}$$

and

$$\psi(x,\zeta) \sim \begin{pmatrix} 0 \\ \mathrm{e}^{\mathrm{i}\zeta x} \end{pmatrix}, \quad \bar{\psi}(x,\zeta) \sim \begin{pmatrix} \mathrm{e}^{-\mathrm{i}\zeta x} \\ 0 \end{pmatrix} \quad \text{as } x \to \infty \,. \tag{H.4b}$$

As in the case of the KdV problem (Schrödinger spectral problem, appendix G) here also by using Green's function associated with (H.1), one can construct the coupled singular-Volterra integral equations as

$$\phi_1(x,y) = \mathrm{e}^{-\mathrm{i}\zeta x} + \int_{-\infty}^{x} \mathrm{e}^{-\mathrm{i}\zeta(x-y)} q(y)\phi_2(y,\zeta)\,\mathrm{d}y\,, \tag{H.5a}$$

$$\phi_2(x,y) = \int_{-\infty}^{x} \mathrm{e}^{\mathrm{i}\zeta(x-y)} r(y)\phi_1(y,\zeta)\,\mathrm{d}y\,. \tag{H.5b}$$

One can show that for $\mathrm{Im}(\zeta) \geq 0$, the Neumann series for $\mathrm{e}^{\mathrm{i}\zeta x}\phi_1(x,\zeta)$ is convergent. Thus it follows that (H.5) has unique solutions for each ζ with $\mathrm{Im}(\zeta) \geq 0$ for the potentials obeying the condition (H.2) and is regular for $\mathrm{Im}\zeta > 0$. Similar results apply for $\psi(x,\zeta)$ in the upper half ζ-plane and for $\bar{\phi}(x,\zeta)$ and $\bar{\psi}(x,\zeta)$ in the lower half ζ-plane.

H.2 The Direct Scattering Problem

On the real axis $\zeta = \xi$, by the linear independence of solutions, we obtain the following relationships between the eigenfunctions

$$\phi(x,\xi) = a(\xi)\bar{\psi}(x,\xi) + b(\xi)\psi(x,\xi)\,, \tag{H.6a}$$
$$\bar{\phi}(x,\xi) = \bar{b}(\xi)\bar{\psi}(x,\xi) + \bar{a}(\xi)\psi(x,\xi)\,, \tag{H.6b}$$

where $a(\xi)$, $b(\xi)$, $\bar{a}(\xi)$ and $\bar{b}(\xi)$ are constants so that the reflection coefficients are defined as

$$R(\xi) = \frac{b(\xi)}{a(\xi)} \quad \text{and} \quad \bar{R}(\xi) = \frac{\bar{b}(\xi)}{\bar{a}(\xi)}\,. \tag{H.7}$$

Now defining the Wronskian of two functions $u(x)$ and $v(x)$ as

$$W[u(x), v(x)] = u_1(x)v_2(x) - v_1(x)u_2(x)\,, \tag{H.8}$$

we find that if u and v are solutions of (H.1), then the Wronskian is independent of x. From the Wronskian $W[\phi(x,\zeta),\ \bar{\phi}(x,\zeta)]$, the following condition can be obtained,

$$a(\xi)\bar{a}(\xi) - b(\xi)\bar{b}(\xi) = 1\,. \tag{H.9}$$

It follows from the above equation that (H.6) is invertible and hence

$$\psi(x,\xi) = a(\xi)\bar{\phi}(x,\xi) - \bar{b}(\xi)\phi(x,\xi)\,, \tag{H.10a}$$
$$\bar{\psi}(x,\xi) = -b(\xi)\bar{\phi}(x,\xi) + \bar{a}(\xi)\phi(x,\xi)\,. \tag{H.10b}$$

Then $a(\xi)$ and $\bar{a}(\xi)$ can be expressed as

$$a(\xi) = W[\phi(x,\xi), \psi(x,\xi)]\,, \tag{H.11a}$$
$$\bar{a}(\xi) = -W[\bar{\phi}(x,\xi), \bar{\psi}(x,\xi)]\,. \tag{H.11b}$$

These relations can be extended into the upper half ζ-plane, as in the case of KdV equation, and become

$$a(\zeta) = W[\phi(x,\zeta), \psi(x,\zeta)]\,, \tag{H.12a}$$
$$\bar{a}(\zeta) = -W[\bar{\phi}(x,\zeta), \bar{\psi}(x,\zeta)]\,. \tag{H.12b}$$

In addition to (H.12), the following analytic properties follow from the Wronskian relations:

(1) $e^{i\zeta x}\phi$, $e^{-i\zeta x}\psi$ are analytic in the upper half plane ($\mathrm{Im}\zeta > 0$).

(2) $e^{-i\zeta x}\bar{\phi}$, $e^{i\zeta x}\bar{\psi}$ are analytic in the lower half plane ($\mathrm{Im}\zeta < 0$).

(3) $a(\zeta)$ is regular in the upper half plane and $\bar{a}(\zeta)$ is regular in the lower half plane.

(4) $a(\zeta)$ and $\bar{a}(\zeta) \to 1$ as $\zeta \to \infty$ in their respective planes.

(5) $b(\zeta)$ and $\bar{b}(\zeta)$ are in general are not analytic.

Thus for given potentials $q(x)$ and $r(x)$, we can construct the quantities ϕ, $\bar{\phi}$, ψ, $\bar{\psi}$, a, $\bar{a}$, b and $\bar{b}$ and deduce their analytic properties which constitute a part of the direct scattering problem. In the upper half plane, $a(\zeta)$ has zeros at $\zeta = \zeta_k$, $k = 1, 2, ..., N$. For simplicity we shall assume that these zeros are simple and do not lie on the real axis. The Wronskian of the functions ϕ and ψ vanishes at these zeros. Henceforth we can say that $\phi(\zeta_k)$ is proportional to $\psi(\zeta_k)$, that is,

$$\phi(x, \zeta_k) = b_k \psi(x, \zeta_k) \tag{H.13}$$

and we have bound states or discrete states.

We can define the normalized bound states

$$\phi_k(x) = C_k \psi(x, \zeta_k) \tag{H.14}$$

so that

$$e^{-i\zeta_k x}\phi_k(x) \to \begin{pmatrix} 0 \\ C_k \end{pmatrix} \quad \text{as } x \to \infty , \tag{H.15}$$

where

$$C_k^2 = -i b_k / a'(\zeta_k) . \tag{H.16}$$

Similar arguments can be applied to $\bar{a}(\zeta)$ in the lower half plane. This has zeros at $\zeta = \bar{\zeta}_k$, $k = 1, 2, ..., \bar{N}$ and at these points

$$\bar{\phi}(x, \bar{\zeta}_k) = \bar{b}_k \bar{\psi}(x, \bar{\zeta}_k) . \tag{H.17}$$

Here the normalized bound states are defined as

$$\bar{\phi}_k(x) = \bar{C}_k \bar{\psi}(x, \zeta_k) \tag{H.18}$$

such that

$$e^{i\bar{\zeta}_k x}\bar{\phi}_k(x) \to \begin{pmatrix} \bar{C}_k \\ 0 \end{pmatrix} \quad \text{as } x \to \infty . \tag{H.19}$$

After determining the eigenvalues, the reflection coefficients and normalization constants, we can reconstruct the potentials $q(x)$, $r(x)$ as discussed below. These data are then the scattering data for the above direct scattering problem:

$$\begin{aligned} S = \{ & R(\xi), \bar{R}(\xi), \ -\infty < \xi < \infty, \ \zeta_k, C_k^2, \ k = 1, 2, ..., N, \\ & \bar{\zeta}_k, \bar{C}_k^2, \ k = 1, 2, ..., \bar{N} \} . \end{aligned} \tag{H.20}$$

H.3 Inverse Scattering Problem

In order to reconstruct the potentials we consider a function $\Phi(x,\zeta)$ defined by

$$\Phi(x,\zeta) = \begin{cases} \frac{1}{a(\zeta)} \mathrm{e}^{\mathrm{i}\zeta x}\phi(x,\zeta) & \text{for } \mathrm{Im}\zeta > 0, \\ \\ \mathrm{e}^{\mathrm{i}\zeta x}\bar{\psi}(x,\zeta) & \text{for } \mathrm{Im}\zeta < 0. \end{cases} \tag{H.21}$$

It has the following properties,

(1) $\Phi(x,\zeta) \to \begin{pmatrix} 1 \\ 0 \end{pmatrix}$ as $\zeta \to \infty$.

(2) It has a discontinuity on the real axis,

$$\Phi(x,\xi + \mathrm{i}\,0) - \Phi(x,\xi - \mathrm{i}\,0) = R(\xi)\mathrm{e}^{\mathrm{i}\xi x}\psi(x,\xi) \,.$$

(3) It has simple poles at $\zeta = \zeta_k$, $k = 1, 2, ..., N$ in the upper half plane. They have the following residues,

$$\frac{b_k}{a'(\zeta_k)}\mathrm{e}^{\mathrm{i}\zeta_k(x)}\psi(x,\zeta_k) = \mathrm{i}C_k^2\mathrm{e}^{\mathrm{i}\zeta_k x}\psi\left(x,\zeta_k\right) \,.$$

Now let us apply the residue theorem to $\left[\Phi(x,\zeta) - \begin{pmatrix} 1 \\ 0 \end{pmatrix}\right] \Big/ (\zeta - \zeta_0)$, ($\mathrm{Im}\zeta_0 < 0$), around the two semi-circles as shown in Fig. G.1. Again following the steps given in appendix G, we obtain

$$\begin{aligned} \mathrm{e}^{\mathrm{i}\xi' x}\bar{\psi}(x,\xi') - \begin{pmatrix} 1 \\ 0 \end{pmatrix} &= -\mathrm{i}\sum_{k=1}^{N}\left[C_k^2\mathrm{e}^{\mathrm{i}\zeta_k x}\psi\left(x,\zeta_k\right) / \left(\zeta_k - \xi'\right)\right] \\ &+ \frac{1}{2\pi\mathrm{i}}\int_{-\infty}^{\infty}\left[R(\xi)\mathrm{e}^{\mathrm{i}\xi x}\psi\left(x,\xi\right) / \left(\xi - \xi' + \mathrm{i}\,0\right)\right]\mathrm{d}\xi \,. \end{aligned} \tag{H.22}$$

By taking the Fourier transform of the function

$$\psi(x,\zeta) - \mathrm{e}^{\mathrm{i}\zeta x}\begin{pmatrix} 0 \\ 1 \end{pmatrix} = \int_{-\infty}^{\infty} K(x,y)\mathrm{e}^{\mathrm{i}\zeta y}\,\mathrm{d}y \,, \tag{H.23a}$$

where

$$K(x,y) = \begin{pmatrix} K_1(x,y) \\ K_2(x,y) \end{pmatrix} \tag{H.23b}$$

and

$$\bar{\psi}(x,\zeta) - \mathrm{e}^{-\mathrm{i}\zeta x}\begin{pmatrix} 1 \\ 0 \end{pmatrix} = \int_{-\infty}^{\infty}\left(\bar{K}(x,y)\mathrm{e}^{-\mathrm{i}\zeta y}\right)\,\mathrm{d}y \,, \tag{H.23c}$$

then finding their inverse Fourier transforms and making use of (H.22) and proceeding along the lines in appendix G, we obtain

$$\bar{K}(x,y) + F(x+y)\begin{pmatrix}0\\1\end{pmatrix} + \int_x^\infty K(x,z)F(z+y)\mathrm{d}z = 0,\ x < y. \quad \text{(H.24a)}$$

Similarly

$$K(x,y) + \bar{F}(x+y)\begin{pmatrix}1\\0\end{pmatrix} + \int_x^\infty \bar{K}(x,z)\bar{F}(z+y)\mathrm{d}z = 0,\ x < y, \quad \text{(H.24b)}$$

where

$$F(x) = \sum_{k=1}^{N} C_k^2 \mathrm{e}^{\mathrm{i}\zeta_k x} + \frac{1}{2\pi}\int_{-\infty}^{\infty} R(\xi)\mathrm{e}^{\mathrm{i}\xi x}\,\mathrm{d}\xi \quad \text{(H.25a)}$$

and

$$\bar{F}(x) = \sum_{k=1}^{\bar{N}} \bar{C}_k^2 \mathrm{e}^{-\mathrm{i}\bar{\zeta}_k x} + \frac{1}{2\pi}\int_{-\infty}^{\infty} \bar{R}(\xi)\mathrm{e}^{-\mathrm{i}\xi x}\,\mathrm{d}\xi\ . \quad \text{(H.25b)}$$

Further from (H.24a), we can write

$$\begin{aligned}\bar{K}(x,y) &= -\theta(y-x)\left[\sum_{k=1}^{N} C_k^2 \mathrm{e}^{\mathrm{i}\zeta_k y}\psi(x,\zeta_k)\right] \\ &\quad + \frac{1}{2\pi}\int_{-\infty}^{\infty}\left\{F(\xi)\mathrm{e}^{\mathrm{i}\xi y}\psi(x,\xi)\right\}\mathrm{d}\xi \quad \text{(H.26a)} \\ &= -\theta(y-x)\Bigg[\int_{-\infty}^{\infty} K(x,z)F(z+y)\,\mathrm{d}z \\ &\quad + F(x+y)\begin{pmatrix}0\\1\end{pmatrix}\Bigg], \quad \text{(H.26b)}\end{aligned}$$

where θ is the step-function, $\theta(x) = 1$ for $x > 0$ and $\theta(x) = -1$ for $x < 0$.

H.4 Reconstruction of the Potentials

After expanding $\mathrm{e}^{\mathrm{i}\zeta x}\bar{\psi}(x,\zeta)$ asymptotically ($\zeta \to \infty$), we find from (H.22) and (H.26) that

$$\begin{aligned}\mathrm{e}^{\mathrm{i}\zeta x}\bar{\psi}(x,\zeta) &= \begin{pmatrix}1\\0\end{pmatrix} + \frac{1}{\zeta}\Bigg[\mathrm{i}\sum_{k=1}^{N} C_k^2 \mathrm{e}^{\mathrm{i}\zeta_k x}\psi(x,\zeta_k) \\ &\quad - \frac{1}{2\pi\mathrm{i}}\int_{-\infty}^{\infty} R(\xi)\mathrm{e}^{\mathrm{i}\xi x}\psi(x,\xi)\mathrm{d}\xi + O(1/\zeta)\Bigg] \\ &= \begin{pmatrix}1\\0\end{pmatrix} + \frac{\mathrm{i}}{\zeta}\Bigg[\int_{-\infty}^{\infty} K(x,y)F(x+y)\,\mathrm{d}y\end{aligned}$$

$$+ F(2x) \begin{pmatrix} 0 \\ 1 \end{pmatrix} \Bigg] + O\left(1/\zeta^2\right)$$

$$= \begin{pmatrix} 1 \\ 0 \end{pmatrix} - \frac{\mathrm{i}}{\zeta} \bar{K}(x, x+0) + O\left(1/\zeta^2\right) . \tag{H.27}$$

From the original scattering problem (H.1) with (H.2), we get

$$\mathrm{e}^{\mathrm{i}\zeta x} \bar{\psi}(x, \zeta) = \begin{pmatrix} 1 \\ 0 \end{pmatrix} - \frac{\mathrm{i}}{2\zeta} \begin{pmatrix} \int_x^\infty q(y) r(y) \, \mathrm{d}y \\ -r(x) \end{pmatrix} + O(1/\zeta^2). \tag{H.28}$$

Comparing eq.(H.28) with eq.(H.27), we get

$$r(x) = -2\bar{K}_2(x, x+0) \tag{H.29a}$$

and

$$\int_x^\infty q(y) r(y) \, \mathrm{d}y = 2\bar{K}_1(x, x+0) . \tag{H.29b}$$

Applying a similar treatment to $\mathrm{e}^{-\mathrm{i}\zeta x} \psi(x, \zeta)$, we get

$$q(x) = 2K_1(x, x+0) \tag{H.30a}$$

and

$$\int_x^\infty q(y) r(y) \, \mathrm{d}y = -2K_2(x, x+0) . \tag{H.30b}$$

Thus, for the Zakharov–Shabat 2×2 problem we have reconstructed the scattering potentials $q(x)$ and $r(x)$ by using the IST technique.

Problems

1. Prove the convergence of the Neumann series for $\mathrm{e}^{\mathrm{i}\zeta x} \phi_1(x, \zeta)$ using (H.5).
2. Obtain the explicit form of the potential $q(x)$ for the reflectionless case, $R(\xi) = 0 = \bar{R}(\xi)$, for $N = 1 = \bar{N}$ and $N = 2 = \bar{N}$ bound states using (H.24) and (H.25).

I. Integrable Discrete Soliton Systems

In Chap. 10, we have seen that there exists a large number of finite dimensional integrable nonlinear dynamical systems both of Hamiltonian and non-Hamiltonian types, which can be identified by a Painlevé type singularity structure analysis. The integrals of motion can then be obtained through a generalized symmetry approach. Towards the end of the chapter, we also pointed out there exists a number of interesting lattice type integrable discrete dynamical systems such as the Toda lattice, Calogero–Moser type systems, discretized nonlinear Schrödinger equations and so on which can be either finite dimensional (for example periodic lattice) or infinite dimensional but discrete. These systems also possess soliton properties very often. In this Appendix, we briefly point out their integrability property and existence of Lax pairs (discussed in more detail in Chaps. 13 and 14 in connection with continuous soliton systems).

I.1 Integrable Finite Dimensional N-Particles System on a Line: Calogero–Moser System

Consider the finite dimensional lattice system described by the Hamiltonian

$$H = \frac{1}{2}\sum_{i=1}^{N} p_i^2 + \frac{1}{2}g^2 \sum_{i\neq j} V(q_i - q_j) \,, \tag{I.1}$$

where (q_i, p_i) $i = 1, 2, \ldots, N$ are the canonical coordinates and momenta respectively of a N particle system on an one-dimensional lattice with the potential V depending on the relative displacement $(q_i - q_j)$. The equation of motion is then

$$\ddot{q}_i = -g^2 \sum_{j\neq i} \frac{\partial V(q_i - q_j)}{\partial q_i}, \quad i = 1, 2, ..., N \tag{I.2}$$

The Calogero–Moser (CM) systems essentially correspond to the following four choices of V [1]:

$$\text{(a)}\;\; V(x) = \frac{1}{x^2} \tag{I.3}$$

$$\text{(b)}\ \ V(x) = \frac{a^2}{\sinh^2 ax} \tag{I.4}$$

$$\text{(c)}\ \ V(x) = \frac{a^2}{\sin^2 ax} \tag{I.5}$$

$$\text{(d)}\ \ V(x) = a^2 \rho(ax) \tag{I.6}$$

Here ρ denotes the elliptic Weierstrass function (related to the Jacobian elliptic functions) and a is a parameter. Though potentials (a)–(c) are in some sense limiting cases of the potential (d), it is customary to treat all the four cases separately.

Let us first consider the CM system corresponding to the inverse square potential (I.3). The Hamiltonian is

$$H = \frac{1}{2}\left[\sum_{i=1}^{N} p_i^2 + g^2 \sum_{i \neq j} \frac{1}{(q_i - q_j)^2}\right] \tag{I.7}$$

and the equation of motion is

$$\ddot{q}_i = 2g^2 \sum_{j \neq i} \frac{1}{(q_i - q_j)^3}, \qquad i = 1, 2, ..., N\ . \tag{I.8}$$

Define now the matrices L and M,

$$(L)_{jk} = \delta_{jk} p_j + \mathrm{i}g \frac{(1 - \delta_{jk})}{(q_j - q_k)} \tag{I.9}$$

and

$$(M)_{jk} = \mathrm{i}g\delta_{jk} \sum_{l \neq j} \frac{1}{(q_j - q_l)^2} - \mathrm{i}g \frac{(1 - \delta_{jk})}{(q_j - q_k)^2}\ . \tag{I.10}$$

Note that the matrix M is antihermitian, while L is hermitian. Then it can be shown that the so-called Lax equation (see also Chaps. 13 and 14)

$$\frac{\mathrm{d}L}{\mathrm{d}t} = [L, M] \equiv (LM - ML) \tag{I.11}$$

is equivalent to the equation of motion (I.8).

It can be proved that any equation of the form (I.11) for given arbitrary $(N \times N)$ matrices L and M implies the existence of N time-independent involutive integrals of motion of the form

$$I_k = \mathrm{Tr}\left(L^k\right), \quad \frac{\mathrm{d}}{\mathrm{d}t} I_k = 0\ , \quad k = 1, 2, ..., N \tag{I.12}$$

provided $\mathrm{Tr}(LM) = \mathrm{Tr}(ML)$. Here $\mathrm{Tr}A (= \mathrm{trace}A)$ is the sum of all diagonal elements of the matrix.

Proof:

Since the matrix M is antisymmetric, the matrix $U(t)$ defined by $\mathrm{d}U/\mathrm{d}t = MU$, $U(0) = I$, is unitary ($UU^{-1} = U^{-1}U = I$). Then, the quantities I_k's are obviously invariant under the unitary transformation

$$L \longrightarrow \widehat{L}(t) = U(t)L(t)U^{-1}(t) , \tag{I.13}$$

as may be verified directly. We can then interpret $U(t)$ as the unitary evolution matrix which evolves $L(t)$ so that $\widehat{L} = L(0)$. Thus $\widehat{L}(t)$ is time-independent.

Now, let

$$L(t)\phi(t) = \lambda(t)\phi(t) , \tag{I.14}$$

where $\lambda(t)$ and $\phi(t)$ are the eigenvalues and eigenvectors, respectively, of $L(t)$. Then using (I.13), (I.14) can be written as

$$\widehat{L}(t)U(t)\phi(t) = U(t)L(t)\phi(t) = \lambda(t)U(t)\phi(t) . \tag{I.15}$$

Thus, $\lambda(t)$ is an eigenvalue of both $\widehat{L}(t)$ and $L(t)$ and so must be independent of time, that is $\lambda(t) = \lambda =$ constant. Let λ_i, $i = 1, 2, \ldots, N$, be the N independent eigenvalues of the time dependent matrix $L(t)$. Then we can write I_i's as a linear function of the N independent eigenvalues (which are constants in time) of L. Extending this argument, one can easily check that I_i's are again functions of the N independent constant eigenvalues of L and so are conserved quantities.

Further, if the I_k's are involutive in the Poisson bracket sense, then one can identify I_k's as the new momenta

$$P_k = I_k , \quad k = 1, 2, ..., N \tag{I.16}$$

and choose a new set of canonical coordinates Q_k as the conjugate coordinates. Consequently, the equations of motion become

$$\frac{\mathrm{d}P_k}{\mathrm{d}t} = 0 , \quad \frac{\mathrm{d}Q_k}{\mathrm{d}t} = \frac{\partial H}{\partial P_k} = \nu_k(\text{constants}), \quad k = 1, 2, ..., N , \tag{I.17}$$

which can be obviously trivially integrated. Normally in the above cases $H = I_2/2$. For more details, see Refs. [1–3].

I.2 The Toda Lattice

Let us consider the one-dimensional lattice of N-particles whose dynamics is governed by [4]

$$H = \frac{1}{2}\sum_{i=1}^{N} p_i^2 + \sum_{i=0}^{N} \exp(q_i - q_{i+1}) . \tag{I.18}$$

Then, the equations of motion read

$$\dot{q}_i = p_i \;, \tag{I.19a}$$
$$\dot{p}_i = \mathrm{e}^{-(q_i - q_{i-1})} - \mathrm{e}^{-(q_{i+1} - q_i)} \;, \qquad i = 1, 2, ..., N \;. \tag{I.19b}$$

Let us introduce the canonical transformation

$$a_i = \frac{1}{2}\mathrm{e}^{-(q_{i+1} - q_i)/2} \;, \tag{I.20a}$$
$$b_i = \frac{1}{2}p_i \tag{I.20b}$$

in (I.19). Then the equation of motion becomes

$$\dot{a}_i = a_i\left(b_i - b_{i+1}\right) \;, \tag{I.21a}$$
$$\dot{b}_i = 2\left(a_{i-1}^2 - a_i^2\right) \;. \tag{I.21b}$$

One may introduce now the periodic boundary conditions so that we have a periodic lattice of N particles satisfying the relations

$$a_{i+N} = a_i \;, \quad b_{i+N} = b_i \quad i = 1, 2, ..., N \;. \tag{I.22}$$

Now let us introduce $(N \times N)$ matrices

$$L = \begin{bmatrix} b_1 & a_1 & & & & & & a_N \\ a_1 & b_2 & & & & & & \\ & & \cdot & & 0 & & & \\ & & & \cdot & & & & \\ & & & b_{n-1} & a_{n-1} & & & \\ & & & a_{n-1} & b_{n+1} & & & \\ & & 0 & & & \cdot & & \\ & & & & & & \cdot & \\ & & & & & & b_{N-1} & a_{N-1} \\ a_N & & & & & & a_{N-1} & b_N \end{bmatrix} \tag{I.23a}$$

and

$$M = \begin{bmatrix} 0 & a_1 & & & & & & -a_N \\ -a_1 & 0 & & & & & & \\ & & \cdot & & 0 & & & \\ & & & \cdot & & & & \\ & & & 0 & a_{n-1} & & & \\ & & & -a_{n-1} & 0 & & & \\ & & 0 & & & \cdot & & \\ & & & & & & \cdot & \\ & & & & & & 0 & a_{N-1} \\ a_N & & & & & & -a_{N-1} & 0 \end{bmatrix} \;. \tag{I.23b}$$

Then, one can easily check that the Lax equation (I.11) is equivalent to the equation of motion (I.21) of the Toda lattice. Again the arguments as in the previous case will then prove that there exists N involutive integrals of motion of the form (I.12) to show the complete integrability of the periodic Toda lattice.

I.3 Other Discrete Lattice Systems

In recent times many other interesting integrable discrete lattice systems have been identified in the literature (see for example, Ref. []5). In fact with every integrable soliton partial differential equation one can identify integrable lattice systems. To realize the existence of such possibilities, it is more advantageous to consider them as arising from discretized spectral problems of Schrödinger and AKNS types in analogy with continuous systems considered in Chaps. 13 and 14. In fact, it can be easily checked that the existence of Lax pair for Toda lattice as discussed above is equivalent to the existence of discrete Schrödinger scattering problem [5]:

$$a_i\phi_{i+1} + a_{i-1}\phi_i - b_i\phi_i = k\phi_i \; , \tag{I.24}$$

where a_i and b_i are as defined in (I.21) and k is the spectral parameter. With this identification, one can develop an IST formalism for the discretized Schrödinger spectral problem (I.24) to obtain soliton solutions and establish complete integrability. For details see for example Ref. [5].

In a similar fashion, one can consider the discretized (2×2) AKNS type scattering problem

$$v_{1,n+1} = kv_{1,n} \mp u_n^* v_{2,n} \; , \tag{I.25a}$$

$$v_{2,n+1} = k^{-1}v_{2,n} + u_n v_{1,n} \; . \tag{I.25b}$$

Along with the time evolution of the eigenfunctions,

$$\frac{\partial v_{1,n}}{\partial t} = A_n v_{1,n} + B_n v_{2,n} \; , \tag{I.26a}$$

$$\frac{\partial v_{2,n}}{\partial t} = C_n v_{1,n} + D_n v_{2,n} \; , \tag{I.26b}$$

where

$$A_n = -\mathrm{i}\left(1 - k^2 \mp u_n^* u_{n-1}\right) \; , \quad B_n = \mp\mathrm{i}\left(ku_n^* - k^{-1}u_{n-1}^*\right) \; , \tag{I.26c}$$

$$C_n = \mathrm{i}\left(ku_{n-1} - k^{-1}u_n\right) \; , \quad D_n = \mathrm{i}\left(1 - k^{-2} \mp u_n u_{n-1}^*\right) \; , \tag{I.26d}$$

(I.25) leads to the discretized integrable NLS equation:

$$\mathrm{i}\frac{\partial u_n}{\partial t} = (u_{n+1} + u_{n-1} - 2u_n) \pm u_n u_n^* (u_{n+1} + u_{n-1}) \; . \tag{I.27}$$

One can also develop the relevant IST formalism [5].

Some of the other similarly investigated integrable discretized soliton equations include the following:

(1) <u>the discrete KdV (or nonlinear ladder) equation:</u>

$$\frac{\partial u_n}{\partial t} = \exp(u_{n+1}) - \exp(u_{n-1}) \; . \tag{I.28}$$

(2) the discrete sine-Gordon equation:

$$\frac{\partial u_{n+1}}{\partial t} - \frac{\partial u_n}{\partial t} = \sin(u_{n+1} + u_n) . \tag{I.29}$$

(3) the discrete MKdV equation:

$$\frac{\partial u_n}{\partial t} = \left(1 \pm h^2 u_n^2\right)(u_{n+1} - u_n) . \tag{I.30}$$

(4) the discrete Heisenberg spin lattice:

$$\frac{\partial \boldsymbol{S}_n}{\partial t} = \frac{\boldsymbol{S}_n \times \boldsymbol{S}_{n+1}}{[1 + \boldsymbol{S}_n \cdot \boldsymbol{S}_{n+1}]} + \frac{\boldsymbol{S}_n \times \boldsymbol{S}_{n-1}}{[1 + \boldsymbol{S}_n \cdot \boldsymbol{S}_{n-1}]} , \tag{I.31a}$$

$$\boldsymbol{S}_n = (S_{n,x}, S_{n,y}, S_{n,z}) , \quad S_{n,x}^2 + S_{n,y}^2 + S_{n,z}^2 = 1 . \tag{I.31b}$$

There exists many such integrable discretized lattice systems for soliton equations and it is an interesting research topic to identify and investigate such integrable systems.

I.4 Solitary Wave (Soliton) Solution of the Toda Lattice

Let us consider the equation of motion of the Toda lattice (I.18) in the form

$$\frac{\mathrm{d}^2 q_n}{\mathrm{d}t^2} = \mathrm{e}^{-(q_n - q_{n-1})} - \mathrm{e}^{-(q_{n+1} - q_n)} . \tag{I.32}$$

Defining now the new set of variables [4]

$$r_n = q_{n+1} - q_n \tag{I.33}$$

(I.32) can be rewritten as

$$\frac{\mathrm{d}^2 r_n}{\mathrm{d}t^2} = \left(2\mathrm{e}^{-r_n} - \mathrm{e}^{-r_{n-1}} - \mathrm{e}^{-r_{n+1}}\right) . \tag{I.34}$$

Defining now another new variable

$$\mathrm{e}^{-r_n} = 1 + \dot{s}_n \ \text{ or } \ r_n = -\ln(1 + \dot{s}_n) , \quad \left(\cdot = \frac{\mathrm{d}}{\mathrm{d}t}\right) \tag{I.35}$$

(I.34) can be rewritten as

$$\frac{\mathrm{d}}{\mathrm{d}t} \ln(1 + \dot{s}_n) = (s_{n-1} - 2s_n + s_{n+1}) . \tag{I.36}$$

Now from the properties of the Jacobian elliptic functions (see appendix A), it can be shown that the function

$$\epsilon(u) = \int_0^u \mathrm{dn}^2 u' \mathrm{d}u' \tag{I.37}$$

satisfies the addition rule

$$\epsilon(u+v)+\epsilon(u-v)-2\epsilon(u)=\frac{\epsilon''(u)}{(1/\mathrm{sn}^2 v)-1+\epsilon'(u)}\,,\quad\left({}'=\frac{\mathrm{d}}{\mathrm{d}u}\right). \tag{I.38}$$

Further defining the new function

$$Z(u)=\epsilon(u)-\frac{E}{K}u\,, \tag{I.39}$$

where $K(k)$ and $E(k)$ are the complete elliptic integrals of the first and second kind respectively, one can show that $Z(u)$ is a periodic function with period $4K$. In terms of this new function, the addition rule (I.38) can be recast as

$$Z(u+v)+Z(u-v)-2Z(u)=\frac{\mathrm{d}}{\mathrm{d}u}\ln\left[1+\frac{Z'(u)}{(1/\mathrm{sn}^2 v)-1+(E/K)}\right]. \tag{I.40}$$

Considering now wave propagation in the Toda lattice such that

$$u=2\left(\nu t\pm\frac{n}{\lambda}\right)\,,\quad \nu=\frac{2K}{\lambda}\,, \tag{I.41}$$

where λ (wavelength) and ν (frequency) are constants and then comparing (I.36) and (I.41), we have

$$\mathrm{e}^{-r_n}=1+(2K\nu)^2\left[\mathrm{dn}^2\left\{2\left(\frac{n}{\lambda}\pm\nu t\right)K\right\}-\frac{E}{K}\right]\,, \tag{I.42a}$$

where the wavelength λ and frequency ν are related by

$$2K\nu=\sqrt{\frac{1}{\mathrm{sn}^2(2K/\lambda)}-1+\frac{E}{K}}\,. \tag{I.42b}$$

A limiting form of this solution is obtained by allowing the modulus $k\to 1$. In this case we have

$$\left(\mathrm{e}^{-r_n}-1\right)=\beta^2\mathrm{sech}^2(\kappa n\pm\beta t)\,, \tag{I.43a}$$

with

$$\beta=\sinh\kappa\,. \tag{I.43b}$$

Thus, we have a solitary pulse like wave with the speed

$$c=\frac{\beta}{\kappa}=\frac{\sinh\kappa}{\kappa}\,, \tag{I.44}$$

where the lattice spacing is chosen as the unit length. In fact, by applying the IST procedure to the Toda lattice, one can show [4] that the above solitary wave is a soliton indeed and multisoliton solutions can also be found as in the case of other soliton systems. For more details, see for example Refs. [4,5].

The above procedure can be extended to other discrete integrable systems mentioned above to obtain soliton solutions.

Problems

1. Obtain the Lax pair for the CS system with the potential $V(x) = a^2/\sin^2 ax$.
2. Verify that the Lax equation (I.11) is equivalent to the equation of motion (I.8).
3. Verify the correctness of the Lax pair (I.23) for the Toda lattice.
4. Deduce the addition rule (I.38), using the standard properties of Jacobian elliptic functions (see appendix A).
5. Show that the function $Z(u)$ given by (I.39) is a periodic function with period $4K(k)$.
6. Deduce the solitary wave solution (I.43) of the Toda lattice.

J. Painlevé Analysis for Partial Differential Equations

In Chap. 10, we have seen in detail that the Painlevé singularity structure analysis in which the solution of the given equation of motion of a dynamical system is analysed for the presence or absence of movable critical singular points can be used as a powerful test to identify and isolate integrable nonlinear dynamical systems. In Chaps. 11–14, it has been pointed out that there exists a large number of soliton possessing completely integrable infinite dimensional dynamical systems. The question naturally arises as to whether the Painlevé property can be associated with integrable partial differential equations (pdes) also and whether an algorithmic procedure of Painlevé analysis can be developed for these systems as well. The answer is in the affirmative and in fact the generalized Painlevé analysis, namely the Weiss–Tabor–Canevale (WTC) procedure, has been very successful not only in identifying integrable pdes in $(1+1)$ and $(2+1)$ dimensions, but also in capturing many of the integrability indicators such as the linear eigenvalue problem, bilinear transformation, Bäcklund transformation and other properties. In this appendix, we will briefly indicate the procedure and apply them to a few interesting nonlinear evolution equations to demonstrate the usefulness of the method.

J.1 The Painlevé Property for PDEs

In analogy with odes, it is now well recognized that a systematic approach to determine whether a nonlinear pde is integrable or not is to investigate the singularity structure of the solutions, namely the Painlevé property. This approach, which was originally suggested by Weiss, Tabor and Carnevale [1] (WTC), aims to determine the presence or absence of movable noncharacteristic critical singular manifolds (of branching type, both algebraic and logarithmic, and essential singular type). When the system is free from movable critical manifolds so that the solution is single-valued, the Painlevé property holds, suggesting its integrability. Otherwise, the system is nonintegrable.

The essential idea [1] is that there exists a major difference between analytic functions of one complex variable and analytic functions of several complex variables in the sense that the singularities of a function of several complex variables cannot be isolated. If $f = f(z_1, z_2, \cdots, z_n)$ is a meromorphic function of N complex variables ($2N$ real variables), the singularities of f occur along analytic manifold of (real) dimension $2N-2$. These manifolds are determined by the conditions

$$\phi\,(z_1, z_2, ..., z_n) = 0\,, \tag{J.1}$$

where ϕ is an analytic function of $(z_1, z_2, ..., z_n)$ in a neighbourhood of the manifold. Further we require that the singularity manifold be noncharacteristic,

$$\phi_{z_i} \neq 0\,, \quad i = 1, 2, ..., N \tag{J.2}$$

so that Cauchy initial data can be prescribed unambiguously.

In the above situation, we say that a pde has the Painlevé property when the solutions of the pde is 'single-valued' about the movable singular manifold. From a practical point of view, we can proceed as follows. If the singularity manifold is determined by (J.1) and $u = u(z_1, z_2, ..., z_n)$ is a solution of the pde, then one can express the solution in the form of a Laurent series around the movable singular manifold,

$$u = u\,(z_1, z_2, ..., z_n) = \phi^p \sum_{j=1}^{\infty} u_j \phi^j = \sum_{j=1}^{\infty} u_j \phi^{j+p}\,, \tag{J.3}$$

where $\phi = \phi(z_1, z_2, ..., z_n)$ and $u_j = u_j(z_1, z_2, ..., z_n)$ are analytic functions of z_1, z_2, $\ldots$, z_n in the neighbourhood of the manifold (J.1). We require p to be a negative integer, in general. Substitution of (J.3) into the given pde determines the possible values of p and defines the recursion relations for u_j, $j = 0, 1, 2, ...$ Then the identification of sufficient number N of arbitrary functions, depending upon the nature of the pde, by assuming the 'formal' Laurent series will ensure that the Painlevé property is satisfied. Note that the above procedure is analogous to that of the algorithmic procedure developed for odes in Chap. 10.

J.1.1 Painlevé Analysis

Let us consider a nonlinear evolution equation of the form

$$u_t + K(u) = 0\,, \tag{J.4}$$

where $K(u)$ is a nonlinear functional of $u(x_1, x_2, \cdots, x_M, t) = u(X, t)$ and its derivatives up to order N, so that (J.4) is an Nth order nonlinear pde. Then one may say that (J.4) possesses the Painlevé or P-property if the following conditions are satisfied.

(A) The solutions of (J.4) are single-valued about the noncharacteristic movable singular manifold. More precisely, if the singular manifold is determined by

$$\phi(X,t) = 0, \quad \phi_{x_i}(X,t) \neq 0, \quad \phi_t(X,t) \neq 0, \quad i = 1,2,...,M, \tag{J.5}$$

and $u(X,t)$ is a solution of the pde (J.4), then we have the Laurent expansion

$$u(X,t) = [\phi(X,t)]^{-m} \sum_{j=0}^{\infty} u_j(X,t)\phi^j(X,t)\,, \tag{J.6}$$

where $\phi(X,t)$ and $u_j(X,t)$ are analytic functions of (X,t) in a deleted neighbourhood of the singular manifold (J.2), and m is an integer.

(B) By Cauchy–Kovalevskaya theorem the solution (J.6) contains N arbitrary functions, one of them being the singular manifold ϕ itself and the others coming from the u_j's.

Then the WTC procedure to test the given pde for its P-property essentially consists of the following three steps [1]:

(i) Determination of leading-order behaviours,
(ii) Identification of powers j (resonances) at which the arbitrary functions can enter into the Laurent series expansion (J.6), and
(iii) Verifying that at the resonance values j a sufficient number of arbitrary functions exist without the introduction of movable critical singular manifold.
(iv) Identifying integrability properties.

An important feature of the WTC formalism is that the generalized Laurent series expansion can not only reveal the singularity structure aspects of the solution and integrability nature of a given pde, but can also provide an effective algorithm which in most cases successfully captures many of its properties, namely the linearization, symmetries, BTs and so on.

J.2 Examples

We now apply the WTC procedure to some ubiquitous soliton equations and point out how the Painlevé property can be investigated.

J.2.1 KdV Equation

Let us consider the KdV equation (13.1) in the form

$$u_t - 6uu_x + u_{xxx} = 0\,. \tag{J.7}$$

Let us expand the solution $u(x,t)$ about the noncharacteristic singularity manifold,

$$\phi(x,t) \approx 0, \quad \phi_x, \phi_t \neq 0\,, \tag{J.8}$$

as the Laurent series

$$u(x,t) = \sum_{j=0}^{\infty} u_j(x,t)\phi^{j+p} \,, \tag{J.9}$$

where $u_j(x,t)$'s are coefficient functions to be determined subject to the KdV equation (J.7). Here p is the exponent of the leading order. Then the WTC analysis proceeds as follows.

(i) Leading Order Analysis

Considering the leading order behaviour

$$u \approx u_0(x,t)\phi^p \tag{J.10}$$

and substituting it into equation (J.7), we can easily check that

$$p = -2, \quad u_0 = 2\phi_x^2 \,. \tag{J.11}$$

(ii) Identification of Resonances

Next, we investigate the powers in the right hand side of (J.9) at which arbitrary functions enter. For this purpose we write down the recursion relation obeyed by the u_j's, which is obtained by substituting (J.9) into (J.7):

$$\begin{aligned} &u_{j-3,t} + (j-4)\phi_t u_{j-2} - 6\sum_{l=0}^{j} u_{j-l}\left(u_{l-1,x} + (l-2)u_l\phi_x\right) \\ &\quad + u_{j-3,xxx} + 3(j-4)u_{j-2,xx}\phi_x + 3(j-3)(j-4)\phi_x^2 u_{j-1,x} \\ &\quad + 3(j-4)u_{j-2,x}\phi_{xx} + (j-2)(j-3)(j-4)u_j\phi_x^3 \\ &\quad + 3(j-3)(j-4)\phi_x\phi_{xx}u_{j-1} + (j-4)u_{j-2}\phi_{xxx} = 0 \,. \end{aligned} \tag{J.12}$$

For various values of j, these relations read as follows:

$$j=0: \quad u_0 = 2\phi_x^2 \tag{J.13}$$

$$j=1: \quad u_1 = -2\phi_{xx} \tag{J.14}$$

$$j=2: \quad \phi_x\phi_t - 6\phi_x^2 u_2 + 4\phi_x\phi_{xxx} - 3\phi_{xx}^2 = 0 \tag{J.15}$$

$$j=3: \quad \phi_{xt} - 6\phi_{xx}u_2 + 6\phi_x^2 u_3 + \phi_{xxxx} = 0 \tag{J.16}$$

$$j=4: \quad \frac{\partial}{\partial x}\left(\phi_{xt} + \phi_{xxxx} - 6\phi_{xx}u_2 + 6\phi_x^2 u_3\right) = 0 \tag{J.17}$$

$$\begin{aligned} j=5: \quad &6u_1u_{4x} + 6u_2u_{3x} - u_{3xx} - u_{3t} - 2u_4\phi_{xxx} - 6u_{4x}\phi_{xx} \\ &\quad - 2u_4\phi_t - 6u_{4xx}\phi_x - 18u_{5x}\phi_x^2 + 6u_3^2\phi_x \\ &\quad + 6u_3u_{2x} + 6u_4u_{1x} + 6u_5u_{0x} + 6u_0u_{5x} \\ &\quad - 18u_5\phi_{xxx}\phi_x + 12u_1u_5\phi_x + 12u_2u_4\phi_x = 0 \end{aligned} \tag{J.18}$$

$$j = 6: \quad \frac{\partial}{\partial x}\{\text{L.H.S. of } (J.18)\} = 0 \,. \tag{J.19}$$

For $j > 6$, one can deduce the coefficients u_j in terms of coefficients upto $j = 6$ by using the recursion relation (J.12) in a similar way.

(iii) Arbitrary Functions

It may be noted from the above that by (J.16) the compatibility condition (J.17) is identically satisfied. Similarly by (J.18) equation (J.19) is satisfied identically. As a consequence the functions u_4 and u_6 in the Laurent expansion (J.9) can be chosen as arbitrary. In addition the manifold $\phi(x,t)$ itself is arbitrary. Thus, we can say that the Laurent series (J.8) representing the general solution of the third-order pde (J.7) in a deleted neighbourhood of the singularity manifold admits three arbitrary functions $u_2(x,t)$, $u_4(x,t)$ and $\phi(x,t)$ at the powers $j = 4$, 6 and -1, respectively, (the so-called *resonance* values) without the introduction of any movable critical manifold. Thus, the solution is single-valued about the singularity manifold and so the KdV equation satisfies the Painlevé property. Then it can be expected to be integrable. To verify this assertion, we proceed as follows.

(iv) Integrability Properties

Since the functions u_2 and u_4 are arbitrary, let us choose them as

$$u_2 = u_4 = 0 \,. \tag{J.20}$$

Furthermore, by requiring

$$u_3 = 0 \,, \tag{J.21}$$

it can be easily seen from the above that

$$u_j = 0, \quad j \geq 3 \,. \tag{J.22}$$

In addition, let us require that u_2 satisfies the KdV equation itself,

$$u_{2t} - 6u_2 u_{2x} + u_{2xxx} = 0 \,, \tag{J.23}$$

so that from (J.9) we can write

$$u = \frac{u_0}{\phi^2} + \frac{u_1}{\phi} + u_2 \,. \tag{J.24}$$

This means that the Laurent series has been cut off at the 'constant' level term (in ϕ). Now using the relations (J.13) and (J.14) in (J.24), we have

$$\begin{aligned} u(x,t) &= 2\frac{\phi_x^2}{\phi^2} - 2\frac{\phi_{xx}}{\phi} + u_2(x,t) \\ &= -2\frac{\partial^2}{\partial x^2}\log\phi + u_2(x,t) \,, \end{aligned} \tag{J.25}$$

where $\phi(x,t)$ satisfies (J.15) and (J.16).

Equation (J.25) can be further interpreted as follows. If u and u_2 solve the KdV equation (J.7), they are related by (J.25) and so it is an auto-Bäcklund transformation, provided (J.15) and (J.16) are consistent. Moreover, if we choose the trivial solution, $u_2 = 0$, then (J.15) corresponds to the bilinearized Hirota form of the KdV equation (see Sect. 12.5) and so (J.25) in this case corresponds to the bilinearizing transformation.

Finally, considering (J.16) and (J.17) and making the change of variable

$$\phi_x = \psi^2 \,, \tag{J.26}$$

one obtains

$$\psi_{xx} + (\lambda - u_2)\,\psi = 0 \tag{J.27a}$$

and

$$\psi_t - 3u_2\psi_x + 3\lambda\psi_x + \psi_{xxx} = 0 \,, \tag{J.27b}$$

where λ is a constant of integration. We may immediately note that (J.27) constitute the linearization of the KdV equation in terms of the Lax pair discussed in Chap. 14, which formed the basis for showing the complete integrability of the KdV equation. Thus we have seen that the basic properties of integrability, namely the BT, bilinearization and Lax pair follow straightaway from the Painlevé analysis. This shows clearly the power and applicability of the Painlevé property to nonlinear dynamical systems, particularly for soliton possessing nonlinear partial differential equations.

J.2.2 The Nonlinear Schrödinger Equation

Next, we apply the Painlevé analysis to the nonlinear Schrödinger equation

$$\mathrm{i}q_t + q_{xx} + \frac{1}{2}|q|^2 q = 0 \,, \quad q \in C \,, \tag{J.28}$$

for which we consider its conjugate form also together:

$$-\mathrm{i}q_t^* + q_{xx}^* + \frac{1}{2}|q|^2 q^* = 0 \,. \tag{J.29}$$

Defining formally

$$q(x,t) = a(x,t) \,, \quad q^*(x,t) = b(x,t) \,, \tag{J.30}$$

we have

$$\mathrm{i}a_t + a_{xx} + \frac{1}{2}a^2 b = 0 \,, \tag{J.31a}$$

$$-\mathrm{i}b_t + b_{xx} + \frac{1}{2}ab^2 = 0 \,. \tag{J.31b}$$

Then we can proceed as in the case of the KdV equation for the present two component equation also.

(i) Leading Order Analysis

Let

$$a \approx a_0\phi^{\alpha}\,, \quad b \approx b_0\phi^{\beta}\,. \tag{J.32}$$

Then using them in (J.31), we find

$$\alpha = \beta = -1\,, \quad a_0 b_0 = -\frac{1}{4}\phi_x^2\,. \tag{J.33}$$

(ii) Resonances and Arbitrary Functions

Starting with the Laurent series

$$a = \sum_j a_j \phi^{j-1}\,, \quad b = \sum_j b_j \phi^{j-1}\,, \tag{J.34}$$

we get the recursion relation

$$\begin{aligned} \mathrm{i}\,(a_{j-2,t} &+ (j-2)\phi_t a_{j-1}) + a_{j-2,xx} + 2(j-2)\phi_x a_{j-1,x} \\ &+ (j-1)(j-2)\phi_x^2 a_j + (j-2)\phi_{xx} a_{j-1} \\ &+ \frac{1}{2}\sum_k \sum_l a_k a_l b_{j-k-l} = 0\,, \end{aligned} \tag{J.35a}$$

$$\begin{aligned} -\mathrm{i}\,(b_{j-2,t} &+ (j-2)\phi_t b_{j-1}) + b_{j-2,xx} + 2(j-2)\phi_x b_{j-1,x} \\ &+ (j-1)(j-2)\phi_x^2 b_j + (j-2)\phi_x b_{j-1} \\ &+ \frac{1}{2}\sum_k \sum_l b_k b_l a_{j-k-l} = 0\,. \end{aligned} \tag{J.35b}$$

Then the coefficients at various powers are as follows.

$$j = 0: \quad \frac{1}{2}a_0 b_0 = -2\phi_x^2\,. \tag{J.36}$$

Note that for two unknown functions $a_0(x,t)$ and $b_0(x,t)$, we have only one relation (J.36). So one of the functions can be taken as arbitrary.

$$\begin{aligned} j = 1: \quad a_1 &= (1/12\phi_x^4)\,(-8a_{0x}\phi_x^3 - 6\mathrm{i}a_0\phi_x^2\phi_t \\ &\qquad -2a_0\phi_{xx}\phi_x^2 - a_0^2 b_{0x}\phi_x)\,, \end{aligned} \tag{J.37a}$$

$$\begin{aligned} b_1 &= (1/12\phi_x^4)\,(-8b_{0x}\phi_x^3 + 6\mathrm{i}b_0\phi_x^2\phi_t \\ &\qquad -2b_0\phi_{xx}\phi_x^2 - b_0^2 a_{0x}\phi_x)\,. \end{aligned} \tag{J.37b}$$

$$\begin{aligned} j = 2: \quad a_2 = (4/3a_0^2b_0^2)\Bigg(&-\mathrm{i}a_0b_0a_{0t} - \frac{1}{2}a_0^2b_0a_1b_1 - \frac{1}{2}a_0b_0^2a_1^2 \\ &-a_0b_0a_{0xx} - \frac{1}{2}\mathrm{i}a_0^2b_{0t} + \frac{1}{4}b_1^2a_0^3 \\ &+\frac{1}{2}a_0^2b_{0xx}\Bigg)\,, \end{aligned} \tag{J.38a}$$

$$b_2 = (4/3a_0^2b_0^2)\left(\mathrm{i}a_0b_0b_{0t} - \frac{1}{2}a_0b_0^2a_1b_1 - \frac{1}{2}a_0^2b_0b_1^2 \right.$$
$$-a_0b_0b_{0xx} + \frac{1}{2}\mathrm{i}b_0^2a_{0t} + \frac{1}{4}a_1^2b_0^3$$
$$\left. + \frac{1}{2}b_0^2a_{0xx}\right) . \quad \text{(J.38b)}$$

$$j = 3: \quad b_0a_3 + a_0b_3 = -(2/b_0)\,(2b_{2x}\phi_x + b_2\phi_{xx} + b_0b_1a_2$$
$$+ b_1b_2a_0 + (1/2)a_1\left(2b_0b_2 + b_1^2\right)$$
$$- \mathrm{i}\,(b_2\phi_t + b_{1t}) + b_{1xx}) , \quad \text{(J.39a)}$$
$$b_0a_3 + a_0b_3 = -(2/a_0)\,(2a_{2x}\phi_x + a_2\phi_{xx} + a_0a_1b_2$$
$$+ a_1a_2b_0 + (1/2)b_1\left(2a_0a_2 + a_1^2\right)$$
$$+ \mathrm{i}(a_2\phi_t + a_{1t}) + a_{1xx}) . \quad \text{(J.39b)}$$

Substituting (J.36–38) into (J.39) we obtain a single equation and hence one of the functions a_3 or b_3 becomes arbitrary.

$$j = 4: \quad -b_0a_4 + a_0b_4 = -\ (2/a_0)\,(2a_3\phi_{xx} + \mathrm{i}\,(a_{2t} + 2a_3\phi_t)$$
$$+ a_0a_1b_3 + (1/2)b_2\left(2a_0a_2 + a_1^2\right)$$
$$+ (1/2)b_1\,(2a_0a_3 + 2a_1a_2)$$
$$+ (1/2)b_0\left(2a_1a_3 + a_2^2\right)$$
$$+ 4a_{3x}\phi_x + a_{2xx}\ ) , \quad \text{(J.40a)}$$
$$-b_0a_4 + a_0b_4 = (2/b_0)\,(2b_3\phi_{xx} - \mathrm{i}\,(b_{2t} + 2b_3\phi_t) + b_0b_1a_3$$
$$+ (1/2)a_2\left(2b_0b_2 + b_1^2\right)$$
$$+ (1/2)a_1\,(2b_0b_3 + 2b_1b_2)$$
$$+ (1/2)a_0\left(2b_1b_3 + b_2^2\right)$$
$$+ 4b_{3x}\phi_x + b_{2xx}\ ) . \quad \text{(J.40b)}$$

Similarly, here also we get a single equation and hence one of the function a_4 or b_4 becomes arbitrary.

One can proceed further to higher values of j and check that all other coefficients a_j, b_j, $j > 4$, can be given in terms of coefficients upto $j = 4$. Thus the Laurent series (J.34) contains four arbitrary functions at $j = -1$ (the arbitrary singular manifold, $j = 0$ (a_0 or b_0), $j = 3$ (a_3 or b_3) and $j = 4$ (a_4 or b_4). Considering the coupled equations (J.28) and (J.29), then we can conclude that the solutions are free from movable critical manifolds and so the Painlevé property is satisfied and the system is expected to be integrable as it should be.

In fact one can proceed as in the case of KdV equation and obtain the (2×2) AKNS linear eigenvalue problem associated with the NLS equation (Chap. 14), see Ref. [2]. However, here we demonstrate briefly the Hirota bilinearization and BT.

As in the case of KdV equation, we cut off the Laurent series (J.34) at the constant level term, that is

$$a = \frac{a_0}{\phi} + a_1 \ , \quad b = \frac{b_0}{\phi} + b_1 \ ; , \tag{J.41}$$

by choosing the coefficients $a_j, b_j, j \geq 2$, all to be zero, provided the resultant conditions on the manifold $\phi(x,t)$ are consistent. Let us now demand that both the set (a,b) and (a_1,b_1) are solutions of the NLS equations (J.31), so that (J.41) constitute the auto-BT. Considering now the trivial solution $a_1 = b_1 = 0$, one has

$$a = \frac{a_0}{\phi} \ , \quad b = \frac{b_0}{\phi} \tag{J.42}$$

as the bilinearizing transformation. Indeed with the choice

$$q = \frac{g(x,t)}{f(x,t)} \ , \quad q^* = \frac{h(x,t)}{f(x,t)} \ , \tag{J.43}$$

where g and h are complex functions and f is a real function, as we have seen in Sect. 14.6, the NLS equation gets bilinearized.

Similar analysis can be performed for all other integrable nonlinear pdes, whether they are of dispersive or diffusive type. Thus Painlevé analysis has come to stay as a very useful working tool [3] to check whether a given equation is integrable or not (though many basic questions regarding its meaning and applicability remain to be answered completely).

Problems

1. Show that the Burgers equation $u_t + uu_x = \nu u_{xx}$ satisfies the Painlevé property and that the Bäcklund transform deduced from the Laurent series leads to the linearizing Cole–Hopf transformation (see Ref. [1]).
2. Show that the MKdV equation possesses the Painlevé property (see again Ref. [1]).
3. For the sG equation $u_{xt} = \sin u$, prove the Painlevé property after transforming the variable $v = \mathrm{e}^{iu}$ (see also Ref. [1]).
4. Apply the Painlevé analysis to the equation

   $$\phi_{tt} - \phi_{xx} + m^2\phi + \lambda\phi^3 = 0$$

   and investigate its integrability property.
5. Investigate the Painlevé property for the evolution equation $u_x = -auv$, $v_t = bu_x$ [4] and obtain its general solution [5].
6. Prove the Painlevé property of the Liouville equation $u_{xt} = \mathrm{e}^u$, by using the transformation $u = \log v$ and hence obtain the general solution [5].

Glossary

Ablowitz–Ramani–Segur (ARS) algorithm: A systematic algorithmic procedure to identify whether a given nonlinear ordinary differential equation is of Painlevé type or not.

Analog simulation circuit: An electronic circuit designed to mimic the dynamics of a system modelled by a nonlinear evolution equation.

Attractor: It is a bounded region of phase space of a dynamical system towards which nearby trajectories asymptotically approach. The attractor may be a point or a closed curve or an unclosed but bounded orbit (even very complex one). Stable equilibrium points, stable limit cycles and chaotic orbits are some examples.

Autonomous system: A system with no explicit time-dependent term in its equation of motion.

Bäcklund transformation: A transformation connecting the solutions of two differential equations. If it connects the solution of the same differential equation then the transformation is an auto-Bäcklund transformation.

Band-merging bifurcation: Merging of two or more bands of a m-band chaotic attractor at a critical value of a control parameter.

Basin of attraction (of an orbit): It is the collection of initial conditions in phase space such that the trajectories started from them asymptotically approach the attracting orbit in the limit $t \to \infty$.

Belousov–Zhabotinsky reaction: A chemical reaction in which an organic molecule (malonic acid) is oxidized by bromate ions and catalyzed by a Ce^{4+} or Ce^{3+} ion. The basic reactants are $Ce_2(SO_4)_3$, $NaBrO_3$, $CH_2(COOH)_2$ and H_2SO_4.

Bifurcation: A sudden/abrupt qualitative change in the dynamics of a system at a critical value of a control parameter when it is varied smoothly.

Bifurcation diagram: A plot illustrating qualitative changes in the dynamical behaviour of a system as a function of a control parameter.

Breather: A localized oscillatory solitary wave solution admitted by the sine-Gordon equation.

Brusselator: It describes a hypothetical three-molecular chemical reaction with autocatalytic step under far-from-equilibrium conditions. It has the form $\dot{X} = A + X^2Y - BX - X$, $\dot{Y} = BX - X^2Y$ in the absece of diffusion.

Butterfly effect: The term coined by Lorenz to describe high sensitive dependence of chaotic solution on small uncertainty in the initial condition.

Center: It is a a neutral type equilibrium point in the neighbourhood of which trajectories form closed orbits (in the phase space) about it without ever approaching it. It is also called an elliptic equilibrium point.

Chaos: A phenomenon or process of occurrence of bounded nonperiodic evolution in deterministic nonlinear systems with high sensitive dependence on initial conditions. Consequently nearby chaotic orbits diverge exponentially in phase space.

Chaotic masking: A method for information processing based on chaos in which the information carrying signal is combined with a chaotic signal suitably.

Chaotic scattering: A type of scattering process in which scattering functions (for example scattering angle) are sensitive to changes in initial conditions.

Chua's circuit: A simple, third-order, autonomous electronic circuit consisting of two linear capacitors, a linear inductor, a linear resistor, and only one nonlinear element (resistor) namely, Chua's diode, exhibiting chaotic dynamics.

Chua's diode: A five-segment piecewise-linear (nonlinear) resistor, whose effective nontrivial operating region is the three middle segments region (negative-resistance region) with two break points and two slopes. The current-voltage ($v_R - i_R$) characteristic for Chua's diode is represented as

$$i_R = h(v_R) = m_0 v_R + 0.5(m_1 - m_0)[|v_R + B_P| - |v_R - B_P|] ,$$

where m_1 and m_0 are the two negative slopes and B_P is the break point voltage.

Cobweb diagram: A graphical representation of iterations of a map.

Conservative map: A map which preserves phase space area/volume (as the case may be) on each iteration. Example: Standard map.

Conservative system: A system whose total mechanical energy is conserved. A system of the form $\dot{X} = \boldsymbol{F}(X)$, $X = (x_1, x_2, ..., x_n)$ is said to be conservative if $\nabla \cdot \boldsymbol{F} = 0$.

Controlling of chaos: Conversion of chaotic motion to desired periodic/regular motion of a nonlinear system through minimal perturbations/preassigned changes.

Correlation dimension: A quantitative measure used to describe geometric and probabilistic features of attractors. It is an integer for regular attractors such as fixed point, limit cycle and quasiperiodic orbit. It is noninteger for strange attractor.

Correlation function: A statistical measure used to characterize regular and chaotic motions. For periodic motion it oscillates while for chaotic motion it decays to zero.

Crisis: A sudden qualitative change in the dynamics caused by the collision of an attractor with an unstable periodic orbit. Intermittency, sudden widening bifurcation and band-merging bifurcation are some of the crises.

Cryptography: The art of hiding information in a string of bits that are meaningless to any unauthorized party.

Dispersive wave: A propagating travelling wave, satisfying a dispersion relation $\omega = \omega(k)$, such that ω/k is not a constant, where ω is the frequency and k is the wave number. A dispersive wave is said to be nonlinear if the dispersion relation also depends on the amplitude of the wave, otherwise, it is a linear dispersive wave.

Dissipative system: A system whose total mechanical energy is not conserved and decreases in time. A system of the form $\dot{X} = \boldsymbol{F}(X)$ is said to be dissipative if $\nabla \cdot \boldsymbol{F} < 0$.

Duffing oscillator: The second-order, damped and driven nonlinear oscillator with cubic restoring force satisfying the ordinary differential equation $\ddot{x} + \alpha\dot{x} + \omega_0^2 x + \beta x^3 = f \sin \omega t$, where α is the strength of damping, ω_0 is the natural frequency, β is the strength of nonlinear restoring force. It may be single-well, double well or double-hump type for ($\omega_0^2 > 0$, $\beta > 0$), ($\omega_0^2 < 0$, $\beta > 0$) or ($\omega_0^2 > 0$, $\beta < 0$), respectively.

Duffing–van der Pol oscillator: A combination of Duffing and van der Pol equations: $\ddot{x} - p(1 - x^2)\dot{x} + \omega_0^2 x + \beta x^3 = f \cos \omega t$.

Equilibrium point: An admissible solution of $F(X) = 0$ for a system $\dot{X} = F(X)$. It is also called fixed point or singular point of the system.

Fast Fourier transform: An efficient algorithm for the calculation of power spectrum of a signal $x(t)$. The number of sample points required in this algorithm is an integral power of 2. If N is the number of sample points in the signal $x(t)$ then the fast Fourier transform requires about $N \log_2 N$ arithmetic operations only compared to N^2 operations in conventional methods.

Feigenbaum constant: A universal property of certain chaotic (dissipative) dynamical systems related to the period doubling bifurcation sequence. The ratio of the successive differences in the control parameter during period doubling bifurcations approaches the universal number $4.6692 \cdots$ for dissipative systems.

Fermi–Pasta–Ulam (FPU) recurrence phenomenon: The recurrence of initial state of a nonlinear lattice after a characteristic recurrence time as demonstrated originally 1955 by Fermi, Pasta and Ulam in an one-dimensional anharmonic lattice of finite size.

Fixed point: See equilibrium point.

Flip bifurcation: See period doubling bifurcation.

Fractal: An object or a mathematical set or an attractor of a dynamical system with self-similar property. It is characterized by noninteger dimension.

Fractal dimension: A quantitative measure of a set of points in an n-dimensional space that characterizes its space-filling properties, leading to the definition of objects with noninteger dimensions.

Frequency-response equation: An equation connecting the frequency of the external periodic force and amplitude of oscillation of a nonlinear oscillator.

Frequency-response curve: See resonance curve.

Heisenberg ferromagnetic spin equation: A mathematical model equation describing the classical version of spin excitations in ferromagnets. The equation in the isotropic continuum case is: $\boldsymbol{S}_t = \boldsymbol{S} \times \boldsymbol{S}_{xx}$, $\boldsymbol{S} = (S_1, S_2, S_3)$, $\boldsymbol{S}^2 = 1$.

Henon–Heiles equation: A simple model for the simulation of a three atom solid and is used as a prototype model to study nonlinear dynamics of conservative continuous time dynamical systems. The equation of motion is: $\dot{x} = p_x$, $\dot{p}_x = -(x + 2xy)$, $\dot{y} = p_y$, $\dot{p}_y = -(y + x^2 - y^2)$. The Hamiltonian of the system is $H = \frac{1}{2}(p_x^2 + p_y^2) + \frac{1}{2}(x^2 + y^2 + 2x^2 y - \frac{2}{3}y^3)$.

Henon map: A two-dimensional map often used to illustrate the self-similar property of strange attractor. The map is: $x_{n+1} = 1 - ax_n^2 + y_n$, $y_{n+1} = bx_n$, $a, b > 0$.

Hirota's bilinearization method: A method introduced by Hirota to construct N-soliton solution of soliton equations wherein given soliton type equation is transformed to the bilinear form and a formal series expansion is effected.

Hopf bifurcation: The birth of a limit cycle from an equilibrium point when a control parameter is varied. If the limit cycle is stable (unstable) then the bifurcation is called supercritical (subcritical).

Hyperbolic equilibrium point: See saddle.

Hyperchaos: Chaotic dynamics with more than one positive Lyapunov exponents.

Integrability: A property related to the existence of sufficient number of independent integrals of motion for a dynamical system so that the equation of motion can in principle be integrated in terms of regular (meromorphic) functions. The system exhibits completely regular behaviour.

Intermittency route to chaos: A route to chaos where regular orbital behaviour is intermittently interrupted by short time irregular bursts. As the control parameter is varied, the duration of the bursts increase, leading to full scale chaos.

Invariant curve: A smooth closed curve formed by the intersecting points of a quasiperiodic orbit on a suitable Poincaré surface of section.

Inverse scattering transform (IST) method: An elegant nonlinear Fourier transform method developed by Gardner, Greene, Kruskal and Miura to solve initial value problem of nonlinear dispersive wave equations.

Island: A closed invariant curve on the Poincaré surface of section of a conservative system. Trajectories from inside and outside the invariant curve cannot cross it as time progresses.

Jump phenomenon: Multivaluedness of stable frequency-response curve of a weakly nonlinear oscillator driven by a sinusoidal force, showing a hysteresis type behaviour.

Kink: A localized solitary wave type solution admitted by the sine-Gordon equation, varying from 0 to 2π as the spatial variable x changes from $-\infty$ to $+\infty$.

Kolmogorov–Arnold–Moser (KAM) theorem: A theorem concerned with the effect of weak Hamiltonian perturbation on periodic and quasiperiodic orbits in Hamiltonian systems. It asserts that for sufficiently weak perturbation for systems whose unperturbed frequencies are suitably irrational, the periodic trajectories (on tori) persist.

Korteweg–de Vries equation: A mathematical model equation proposed by Korteweg and de Vries to account for the Scott Russel phenomenon. The equation of the form $u_t - 6uu_x + u_{xxx} = 0$ describes unidirectional wave propagation in one-dimensional nonlinear dispersive systems such as a shallow channel.

Laurent series: A two way power series expansion (including both negative and positive powers) of an analytic function valid in the neighbourhood of a singular point.

Lax pair/Lax equations: A pair of linear differential operators L and B, associated with a given nonlinear evolution equation (NLEE). If the compatibility of these operators gives rise to the Lax equation or Lax condition, $L_t - B_x + [L, B] = 0$, equivalent to the original NLEE, integrability of the system is ensured.

Limit cycle: An isolated closed orbit in the phase space associated with a dynamical system.

Linear circuit: An electronic circuit only with linear circuit elements like linear resistor, linear capacitor and linear inductance only.

Linear differential equation: It is a differential equation satisfying the linear superposition principle. In practical terms, if each of the terms of a given differential equation after rationalization has a total degree either 1 or 0 in the dependent variables and their derivatives then it is a linear differential equation.

Linear force: A force $\boldsymbol{F}(\boldsymbol{r}, \mathrm{d}\boldsymbol{r}/\mathrm{d}t, t)$ which is directly proportional to $\boldsymbol{r}$ or $\mathrm{d}\boldsymbol{r}/\mathrm{d}t$ or both, but not products of them.

Linear oscillator: An oscillatory system modelled by linear differential equation.

Linear superposition principle: A property associated with linear differential equations. The property is that if u_1 and u_2 are two linearly independent solutions of a linear differential equation then $u = au_1 + bu_2$ is also a solution of it, where a and b are arbitrary (complex)constants.

Linear system: A dynamical system driven by a linear force. The equation of motion of such a system is linear.

Linear wave: A wave solution of a linear dynamical system.

Logistic map: A discrete time analog of the logistic equation for population growth. The map is: $x_{n+1} = ax_n(1 - x_n)$, where a is a parameter with $0 \le a \le 4$ and $0 < x < 1$.

Lorenz system: The paradigmic nonlinear chaotic system originally introduced by E. Lorenz in 1963 in connection with atmospheric convection. The equation is: $\dot{x} = \sigma(y - x)$, $\dot{y} = rx - y - xz$, $\dot{z} = xy - bz$ where σ, r and b are parameters.

Lyapunov exponents: Numbers providing a quantitative average measure of the divergence of nearby trajectories in phase space. All negative exponents represent regular and periodic orbits, while at least one positive exponent signals the presence of chaotic motion.

Miura transformation: A transformation relating two nonlinear equations. Originally it related the Korteweg-de Vries and modified Korteweg-de Vries equations through a Riccati equation.

Morse oscillator: The Morse potential has provided a useful model for interatomic potentials and for fitting the vibration spectra of diatomic molecules. The damped and driven Morse oscillator is: $\ddot{x} + \alpha\dot{x} + \beta \mathrm{e}^{-x}(1 - \mathrm{e}^{-x}) = f\cos\omega t$ where α, β, f and ω are parameters.

Murali–Lakshmanan–Chua (MLC) circuit: The simplest second-order nonautonomous, nonlinear circuit consisting of a linear resistor, a linear inductor, a linear capacitor, a sinusoidal driving force and only one nonlinear element, namely, Chua's diode. Its dynamical equation reads, $\dot{x} = y - h(x)$, $\dot{y} = -\beta(1+\nu)y - \beta x + f \sin \omega t$, where

$$h(x) = \begin{cases} bx + a - b & x \geq 1 \\ ax & \mid x \mid \leq 1 \\ bx - a + b & x \leq -1 \, . \end{cases}$$

Noether's theorem: A theorem establishing the connection between invariance and existence of integrals of motion for Lagrangian systems. It states that for a given Lagrangian L, if the action integral is invariant under a continuous one parameter group of transformations, there exists a conserved quantity.

Nonautonomous system: A system with at least one explicit time-dependent term in its equation of motion.

Nondispersive wave: Any propagating wave travelling without change of form so that the phase velocity is independent of the wave number.

Nonlinear circuit: An electronic circuit containing at least one nonlinear circuit element, like a nonlinear resistor, inductor or capacitor.

Nonlinear differential equation: A differential equation which is not linear is said to be nonlinear. Even if one of the terms of a given differential equation after rationalization has a total degree other than 0 or 1 in the dependent variables and their derivatives then it is a nonlinear differential equation.

Nonlinear dynamics: The field of study of behaviour of dynamical systems acted upon by nonlinear forces.

Nonlinear force: A force which is not linear (see linear force).

Nonlinear oscillator: An oscillatory system modelled by nonlinear differential equation.

Nonlinear Schrödinger equation: A simple ubiquitous model equation describing envelope wave propagation. Example: Wave propagation in optical fibres. The equation is: $iq_t + q_{xx} + 2|q|^2 q = 0$, $q \in C$.

Nonlinear system: A dynamical system driven by nonlinear force. The equation of motion of such a system is nonlinear.

Optical soliton: Formation of solitons when light interacts strongly with matter leading to wave propagation. Examples: Intense laser light propagation in optical fibres, photo-refractive media, etc.

Painlevé property: If the solution of a nonlinear differential equation is free from movable (integration constant/initial condition dependent) critical singular points (branch points and essential singularities) it is said to possess the Painlevé property. In this case the system is expected to be integrable.

Period doubling: Denotes the bifurcation sequence of periodic motions for a nonlinear dynamical system in which the period doubles at each bifurcation as a control parameter is varied. Beyond a critical accumulation parameter value, chaotic motions occur. It is also referred as subharmonic bifurcation and flip bifurcation.

Periodic attractor: A stable limit cycle motion.

Phase point: A point in the phase space representing a state of a system at any instant of time.

Phase portrait: It is a plot of solution of a system in its phase space.

Phase space: An abstract space with each of the variables needed to specify the state of a system representing the orthogonal coordinates.

Pitchfork bifurcation: It is a bifurcation of an equilibrium point into three equilibrium points. If the newly born two equilibrium points are stable (unstable), while the original one becomes unstable (stable) then the bifurcation is called supercritical (subcritical).

Poincaré–Bendixson theorem: A theorem concerned with the existence of a limit cycle of two-dimensional systems.

Poincaré map: Any suitable hyperplane of phase space is a Poincaré surface of section. The relation between the successive intersections of the phase trajectories with this section in a single direction constitutes the Poincaré map. For a periodically driven system it is simply the stroboscopic map taken at every period of the external force.

Poincaré surface of section: See Poincaré map.

Point attractor: A stable equilibrium point in the phase space of a dynamical system.

Power spectrum: The distribution of power in a signal $x(t)$ is most commonly quantified by means of power density spectrum or simply the power spectrum. It is magnitude-square of the Fourier transform of the signal $x(t)$. It can detect the presence of chaos when the spectrum is broad-banded.

Primary resonance: See resonance.

Quantum chaos: Quantum manifestations of classical chaos.

Quasiperiodic attractor: An attracting quasiperiodic orbit.

Quasiperiodic orbit: It is an orbit which is neither periodic nor chaotic but almost periodic and takes place on a n-dimensional torus.

Quasiperiodic route to chaos: Transition from quasiperiodic motion to chaotic motion when a control parameter is varied.

Repeller: An unstable orbit. Examples include saddle, unstable limit cycle, etc.

Resonance: A weakly nonlinear oscillator exhibiting periodic response to an external periodic force of frequency $\omega = k\omega_0$ (where ω_0 is the natural frequency of the oscillator in the absence of the applied force) is said to be in resonance of order k with the external force. When $\omega \approx k\omega_0$ then the system is close to a resonance of order k. The case $\omega = \omega_0$ is referred as primary resonance. $\omega = k\omega_0$ with $k > 1$ is called secondary resonance. An integer value of k corresponds to subharmonic resonance. If $k = 1/n$, where n is an integer then it refers to superharmonic resonance.

Resonance curve: A plot of amplitude of oscillation versus frequency or amplitude of an external periodic force of an oscillator. It is also called frequency-response curve.

Resonance layer: The region of chaotic motion formed near a separatrix in the phase space of a conservative system.

Rössler system: A simple model equation related to chemical reactions in a stirred tank. The equation is: $\dot{x} = -(y + z)$, $\dot{y} = x + ay$, $\dot{z} = b + z(x - c)$, where a, b and c are parameters.

Saddle: An unstable equilibrium point in the neighbourhood of which trajectories approach it along two directions only and deviate along all other directions. It is also called hyperbolic point.

Saddle-node bifurcation: A bifurcation of an equilibrium point into two equilibrium points at a critical value of a control parameter with one being of saddle type and the other being stable node.

Scott Russel phenomenon: Solitary wave propagation observed by John Scott Russel during 1834 in the Union canal connecting Edinburgh and Glasgow, which is related to the soliton solution of the Korteweg–de Vries equation.

Secondary resonance: See resonance.

Self-similar structure: A geometrical structure which is invariant under change of scale. Fractals and strange attractors of nonlinear systems possess self-similar structure.

Separatrix: An orbit separating two different types of orbits in the phase space.

Signal to-noise-ratio: Ratio of the strength of the signal peak to the mean amplitude of the background noise at the input signal frequency. It is used to characterize stochastic resonance phenomenon.

Sine-Gordon equation: A simple model equation for example for the propagation of transverse electromagnetic waves in a superconducting transmission system. The equation in light cone coordinates is $u_{xt} + m^2 \sin u = 0$, where m^2 is a constant.

Singularities/regular points: Points in the complex plane of the independent variables of an ordinary differential equation at which its general solution cease to be analytic. Poles, branch points and essential singularities are singular points.

Solitary wave: A localized nonlinear wave travelling without change of its form and velocity.

Soliton: A solitary wave which preserves its identity even after collision with another solitary wave.

Spatio-temporal chaos: Chaotic dynamics over both space and time scales in a spatially extended nonlinear system. It can be found in certain nonlinear systems described by partial differential equations and coupled map lattices, especially in dissipative and diffusive systems.

Stable focus: It is a stable equilibrium point in the neighbourhood of which trajectories approach it along spiral paths.

Stable node: A stable equilibrium point in the neighbourhood of which trajectories approach it along parabolic paths.

Stable star: It is a stable equilibrium point in the neighbourhood of which trajectories approach it along straight-line paths.

Standard map: A discrete time analogue of the equation of vertical pendulum. It is a prototype two-dimensional map used in the study of various nonlinear phenomena of conservative systems. The map is : $I_{n+1} = I_n + K \sin\theta_n \mod(2\pi)$, $\theta_{n+1} = \theta_n + I_{n+1} \mod(2\pi)$.

Stochastic resonance: A phenomenon of enhancement of response of a weakly periodically or nonperiodically driven nonlinear system by noise.

Strange nonchaotic attractor: A nonperiodic attractor like chaotic attractor with noninteger dimension but insensitive to initial conditions.

Subharmonic bifurcation: See period doubling.

Subharmonic resonance: See resonance.

Sudden widening bifurcation: An abrupt change in the size of an attractor at a critical value of a control parameter. It is also sometimes dubbed as crisis.

Supercycle/super stable: A periodic attractor with maximum stability.

Superharmonic resonance: See resonance.

Symmetries: Invariance of the dynamical systems (equation of motion) under transformations lead to the identification corresponding symmetries such as space-time translations, rotations, reflections and so on.

Synchronization (of chaotic systems): The possibility of driving two identical chaotic systems to have the same phase and amplitude through appropriate coupling.

Time series: The measured values of a physical variable of a dynamical system at regular intervals of time.

Toda lattice: A model equation introduced by M. Toda in 1967 to study wave propagation in a nonlinear lattice admitting soliton solutions. The equation is:

$$\ddot{u}_n = \exp\left[-(u_n - u_{n-1})\right] - \exp\left[-(u_{n+1} - u_n)\right], \quad n = 1, 2, ..., N$$

Torus: A two-torus is a doughnut shaped surface in three-dimensional space. An n-torus is its generalization in $(n+1)$-dimensional space.

Trajectory plot: A plot of a physically measurable quantity, for example position of a particle, versus time.

Transcritical bifurcation: Exchange or transfer of stability of two equilibrium points at a critical value of a control parameter.

Transient motion: An initial short (time) evolution of a system before confining to its attractor.

Turing instability: A diffusion driven instability in a reaction-diffusion system. It occurs when a homogeneous steady state, which is stable due to small perturbations in the absence of diffusion, becomes unstable in the presence of diffusion.

Turing patterns: Patterns formed due to Turing instability in reaction-diffusion systems.

Unstable star: An unstable equilibrium point in the neighbourhood of which trajectories deviate from it along straight-line paths.

Unstable node: It is an unstable equilibrium point in the neighbourhood of which trajectories deviate from it along parabolic paths.

Unstable focus: An unstable equilibrium point in the neighbourhood of which trajectories deviate from it along spiral paths.

Unstable periodic orbit: An unstable period-1, or $2, \cdots$, or n fixed point or limit cycle. A chaotic orbit contains a large number of unstable periodic orbits.

van der Pol oscillator: A second-order differential equation with nonlinear damping introduced by van der Pol to describe the oscillation in a vacuum tube circuit. The equation is: $\ddot{x} - \epsilon(1 - x^2)\dot{x} + x = 0$, where ϵ is a parameter.

Wave: Any disturbance through a medium due to a physical event. The shape of the disturbance is known as waveform.

Wave equation: Any partial differential equation (linear or nonlinear) satisfied by the wave solution (linear or nonlinear).

References

Chapter 1

1. H. Goldstein: *Classical Mechanics* (Narosa, New Delhi 1990)
2. N.C. Rana, P.S. Joag: *Classical Mechanics* (Tata McGraw Hill, New Delhi 1991)
3. J. Gleick: *Chaos: Making a New Science* (Viking, New York 1987)
4. R.J. Field, L. Györgyi (Eds.): *Chaos in Chemistry and Biochemistry* (World Scientific, Singapore 1993)
5. M.C. Cross, P.C. Hohenberg: Rev. Mod. Phys. **63**, 851 (1993)

Chapter 2

1. T. Davis: *Introduction to Nonlinear Differential and Integral Equations* (Dover, New York 1962)
2. E.L. Ince: *Ordinary Differential Equations* (Dover, New York 1956)
3. P.M. Mathews, M. Lakshmanan: Quart. Appl. Maths. **32**, 315 (1974)
4. S. Parthasarathy, M. Lakshmanan: J. Sound and Vibr. **137**, 523 (1990)
5. N. Minorsky: *Nonlinear Oscillations* (Van Nostrand, New Jersey 1962)
6. A.H. Nayfeh, D.T. Mook: *Nonlinear Oscillations* (Wiley, New York 1979)
7. P. Hagedorn: *Nonlinear Oscillations* (Oxford Sci. Publ., Oxford 1988)

Chapter 3

1. H. Goldstein: *Classical Mechanics* (Narosa Publ. House, New Delhi 1990)
2. P. Glendenning: *Stability, Instability and Chaos* (Cambridge Univ. Press, Cambridge 1994)
3. J. Hale, H. Kocak: *Dynamics and Bifurcations* (Springer, New York 1991)
4. C. Grebogi, E. Ott: Phys. Rev. Lett. **50**, 935 (1983)
5. F. C. Moon, G.X. Li: Phys. Rev. Lett. **55**, 1439 (1985)
6. J. Guckenheimer, P. Holmes: *Nonlinear Oscillations, Dynamical Systems and Bifurcations of Vector Fields* (Springer, New York 1983)
7. G. Schmidt, A. Tondl: *Nonlinear Vibrations* (Cambridge Univ. Press, Cambridge 1986)
8. E.N. Lorenz: J. Atmos. Sci. **20**, 130 (1963)
9. E.A. Jackson: Phys. Rev. A **44**, 4839 (1991)
10. S. Rajasekar, M. Lakshmanan: J. Theor. Biol. **166**, 275 (1994)

11. A.H. Nayfeh, B. Balachandran: *Applied Nonlinear Dynamics* (John–Wiley, New York 1995)
12. L.G. Reichl: *The Transition to Chaos: In Conservative Classical Systems – Quantum Manifestation* (Springer, New York 1992)
13. A.J. Lichtenberg, M.A. Lieberman: *Regular and Stochastic Motion* (Springer, New York 1983)
14. E.A.D. Foster: Inter. J. Bifur. and Chaos **6**, 647 (1996)
15. A.B. Goryachev, A.A. Polezhaev, D.S. Chernavskii: Chaos **6**, 78 (1996)
16. N. Minorsky: *Nonlinear Oscillations* (Van Nostrand, Princeton 1962)
17. F. Arogoul, A. Arneodo, P. Richetti: Phys. Lett. A **120**, 269 (1987)
18. M.K. Stephen Yeung, S.H. Strogatz: Phys. Rev. E **58**, 4421 (1998)

Chapter 4

1. P. Glendenning: *Stability, Instability and Chaos* (Cambridge Univ. Press, Cambridge 1994)
2. J.E. Marsden, M. McCracken: *The Hopf Bifurcation and its Application* (Springer, New York 1976)
3. S. Rajasekar, M. Lakshmanan: Physica D **32**, 146 (1988)
4. Yu.A. Kuznetsov, S. Muratori, S. Rinaldi: Inter. J. Bifur. and Chaos **2**, 117 (1992)
5. J. Guckenheimer, P. Holmes: *Nonlinear Oscillations, Dynamical Systems, and Bifurcations of Vector Fields* (Springer, New York 1983)
6. R.M. May: Nature **26**, 459 (1976)
7. J.T. Sandefure: *Discrete Dynamical Systems Theory and Applications* (Oxford Press, Oxford 1990)
8. M.J. Feigenbaum: J. Stat. Phys. **19**, 25 (1978); Los Alamos Science **1**, 4 (1980)
9. H.G. Schüster: *Deterministic Chaos* (Verlag, Weinheim 1988)
10. E.N. Lorenz: J. Atmos. Science **20**, 130 (1963)
11. J. Gleick: *Chaos: Making A New Science* (Viking, New York 1987)
12. S.N. Rasband: *Chaotic Dynamics of Nonlinear Systems* (Wiley, New York 1990)
13. M. Hénon: Commun. Math. Phys. **50**, 69 (1976)
14. D. Ruelle: Math. Intelligencer **2**, 126 (1980)
15. D. Ruelle: Commun. Math. Phys. **82**, 137 (1981)
16. S. Newhouse, D. Ruelle, F. Takens: Commun. Math. Phys. **64**, 35 (1978)
17. P. Manneville, Y. Pomeau: Phys. Lett. A **75**, 1 (1979)
18. Y. Pomeau, P. Manneville: Commun. Math. Phys. **74**, 189 (1980)
19. J.E. Hirsh, B.A. Huberman, D.J. Scalapino: Phys. Rev. A **25**, 519 (1982)
20. J. San-Martin, J.C. Antoranz: Phys. Lett. A **219**, 69 (1996); Chaos, Solitons and Fractals **10**, 1539 (1999)
21. C. Grebogi, E. Ott, J. Yorke: Phys. Rev. Lett. **50**, 93 (1983); C. Grebogi, E. Ott, F. Romeiras, J. A. Yorke: Phys. Rev. A **36**, 5365 (1987)
22. N. Platt, E.A. Spiegel, C. Tresser: Phys. Rev. Lett. **70**, 279 (1993)
23. T. Kawabe, Y. Kondo: Prog. Theor. Phys. **85** (1991)
24. S. Sinha, P.K. Das: Pramana J. Phys. **48**, 87 (1997)
25. S.P. Dawson, C. Grebogi: Chaos, Solitons and Fractals **1**, 137 (1991)
26. D.R. Chialvo: Chaos, Solitons and Fractals **5**, 461 (1995)
27. M. Bier, T. Bountis: Phys. Lett. A **104**, 239 (1984)
28. A. Venkatesan, M. Lakshmanan: Phys. Rev. E **63**, 026219 (2001)
29. P. Coullet, C. Tresser, A. Arneodo: Phys. Lett. A **77**, 327 (1980)

30. P. Bak: Physics Today, December, 39 (1986)
31. T.T. Chia, B.L. Tan: Phys. Rev. E **54**, 5985 (1996)]
32. R. Kariotis, H. Suhl, J.P. Eckmann: Phys. Rev. Lett. **54**, 1106 (1985)
33. C. Grebogi, E. Ott, F. Romeiras, J.A. Yorke: Phys. Rev. A **36**, 5365 (1987)
34. P. Philominathan, P. Neelamegam, S. Rajasekar: Physica A **242**, 391 (1997)
35. G. Stolovitzky, T.J. Kaper, L. Sirovich: Chaos **5**, 671 (1995)
36. J. Güemez, J.M. Gutierrez, A. Iglesias, M.A. Matias: Phys. Lett. A **190**, 429 (1994)
37. M. Ding, C. Grebogi, E. Ott: Phys. Rev. A **39**, 2593 (1989)
38. K. Stefanski: Chaos, Solitons and Fractals **9**, 83 (1998)

Chapter 5

1. G. Duffing: *Erzwungene Schwingungen bei Veränderlicher Eigenfrequenz* (Braunschweig, 1918)
2. F.C. Moon, P.J. Holmes: J. Sound and Vib. **65**, 285 (1979)
3. M. Lakshmanan, K. Murali: *Chaos in Nonlinear Oscillators – Controlling and Synchronization* (World Scientific, Singapore 1997)
4. A. Prasad, S.S. Negi, R. Ramaswamy: Int. J. Bifur. Chaos **11**, 291 (2001)
5. A. Venkatesan, M. Lakshmanan, A. Prasad, R. Ramaswamy: Phys. Rev. E **61**, 3641 (2000)
6. A. Venkatesan, M. Lakshmanan: Phys. Rev. E **63**, 026219 1–14 (2001)
7. E.N. Lorenz: J. Atmos. Sci. **20**, 130 (1963)
8. B. van der Pol: Phil. Mag. **43**, 700 (1927)
9. U. Parlitz, W. Lauterborn: Phys. Rev. A **36**, 1428 (1987)
10. G.L. Baker, J.P. Gollub: *Chaotic Dynamics: An Introduction* (Cambridge Univ. Press, Cambridge 1990)
11. O.E. Rössler: Phys. Lett. A **71**, 155 (1979)
12. W. Knob, W. Lauterborn: J. Chem. Phys. **93**, 3950 (1990)
13. R.H. Dalling, M.E. Goggin: Amer. J. Phys. **62**, 563 (1994)
14. M.C. Kohn: *Practical Numerical Methods, Algorithms and Programs* (Wiley, New York 1990) p.9
15. C. Grebogi, S. Hammel, J. Yorke, T. Sauer: Phys. Rev. Lett. **65**, 1527 (1990)
16. T. Sauer, J. Yorke: Nonlinearity **4**, 961 (1991)
17. Z. Liu, Z. Zhu: Inter. J. Bifur. and Chaos **7**, 1383 (1996)
18. A. Venkatesan, M. Lakshmanan: Phys. Rev. E **55**, 5134 (1997)
19. T. Kapitaniak, W.H. Steeb: J. Sound and Vibr. **143**, 167 (1990)
20. C. Grebogi, E. Ott, F. Romeiras, J.A. Yorke: Phys. Rev. A **36**, 5365 (1987)
21. S. Rajasekar, M. Lakshmanan: Physica D **32**, 146 (1988)
22. W. Wang: J. Phys. A **22**, L627 (1989)
23. S. Rajasekar: Chaos, Solitons and Fractals **7**, 1799 (1996)
24. T. Kai, K. Tomita: Prog. Theor. Phys. **61**, 54 (1979); J. Stat. Phys. **21**, 65 (1979)
25. B.L. Hao, S.Y. Zhang: J. Stat. Phys. **28**, 769 (1982)
26. P. Wang, J. Dai, H. Zhang: Phys. Rev. A **41**, 3250 (1990)
27. S. Rajasekar: Pramana J. Physics **41**, 509 (1992)
28. G. Rega, A. Salvatori: Inter. J. Bifur. and Chaos **6**, 1529 (1996)
29. P.E. Phillipson, P. Schüster: Inter. J. Bifur. and Chaos **8**, 471 (1998)
30. A. Milik, P. Szmolyan, H. Löffelmann, E. Gröller: Inter. J. Bifur. and Chaos **8**, 505 (1998)

31. L. Flepp, R. Holzner, E. Brun, M. Finardi, R. Badii: Phys. Rev. Lett. **67**, 2244 (1991)
32. T.A. Newton, D. Martin, A.R.B. Leipnik: 'A Double Strange Attractor'. In:*Dynamical Systems Approaches to Nonlinear Problems in Systems and Circuits* ed. by F.M.A. Salam, M.L. Levi (SIAM, Philadelphia 1988) pp.117
33. S. Rajasekar, M. Lakshmanan: J. Theor. Biol. **166**, 275 (1994)
34. P.L. Sachdev, R. Sarathy: Chaos, Solitons and Fractals **4**, 2015 (1994)
35. E.A. Jackson: Inter. J. Bifur. and Chaos **5**, 1255 (1995)
36. B.A. Niyazov, M.V. Sharov, F.R. Sultanova: Inter. J. Bifur. and Chaos **2**, 873 (1992)

Chapter 6

1. L.O. Chua, C.A. Desoer, E.S. Kuh: *Linear and Nonlinear Circuits* (McGraw–Hill, Singapore 1987)
2. L.O. Chua: IEICE Trans. Fund. E **76**, 704 (1993); R.N. Madan, *Chua's Circuit: A Paradigm for Chaos* (World Scientific, Singapore 1993)
3. R.N. Madan (Ed.): *Special Issue on Chua's Circuit: A Paradigm for Chaos*, Part I and II of Journal of Circuit Syst. Comput. **3** (1993)
4. M.P. Kennedy: IEEE Trans. Circuits and Systems–I **40**, 640 (1993); 657
5. K. Murali, M. Lakshmanan, L.O. Chua: Int. J. Bifur. and Chaos **4**, 1511 (1994); IEEE Trans. Circuits and Systems–I **41**, 462 (1994)
6. M. Lakshmanan, K. Murali: *Chaos in Nonlinear Oscillators: Controlling and Synchronization* (World Scientific, Singapore 1996)
7. G. Sarafian, B.Z. Kaplan: Int. J. Bifur. and Chaos **7**, 1665 (1997)
8. P.S. Linsay: Phys. Rev. Lett. **47**, 1349 (1981)
9. E.R. Hunt: Phys. Rev. Lett. **49**, 1054 (1982)
10. J.Y. Huang, J.J. Kim: Phys. Rev. A **36**, 1495 (1987)
11. T.P. Weldon: Amer. J. Phys. **58**, 936 (1990)
12. M.P. Kennedy: IEEE Trans. Circuits and Systems–I **41**, 771 (1994)
13. G.P. King, T. Gaito: Phys. Rev. A **46**, 3092 (1992)
14. A. Tamasevicius, A. Cenys: Chaos, Solitons and Fractals **9**, 115 (1998)
15. M. Itoh, H. Murakami: Inter. J. Bifur. and Chaos **4**, 1023 (1994)
16. A.S. Elwakil, A.M Soliman: Chaos, Solitons and Fractals **8**, 335 (1997)
17. T. Shou, F. Moss, A. Bulsara: Phys. Rev. A **45**, 5394 (1992)
18. Y.H. Kao, C.S. Wang: Phys. Rev. E **48**, 2514 (1993)
19. K. Murali, M. Lakshmanan: IEEE Trans. Circuit Syst.–I **39**, 264 (1992); Int. J. Bifur. and Chaos **2**, 621 (1992)
20. K. Thamilmaran, M. Lakshmanan, K. Murali: Int. J. Bifur. and Chaos **10**, 1175 (2000)
21. K.M. Cuomo, A.V. Oppenheim: Phys. Rev. Lett. **71**, 65 (1993)

Chapter 7

1. M. Hénon, C. Heiles: Astronomical Journal **69**, 73 (1964)
2. A.J. Lichtenberg, M.A. Lieberman: *Regular and Stochastic Motion* (Springer, New York 1983)
3. R.S. MacKay, J.D. Meiss: *Hamiltonian Dynamical Systems* (Adam Hilger, Bristol 1987)

4. A.M. Ozorio de Almeida: *Hamiltonian Systems* (Cambridge University Press, Cambridge 1988)
5. L. Reichl: *The Transition to Chaos: In Conservative Classical Systems – Quantum Manifestations* (Springer, New York 1992)
6. G.M. Zaslavsky, R.Z. Sagdeev, D.A. Usikov, A.A. Cherikov: *Weak Chaos and Quasiregular Patterns* (Cambridge University Press, Cambridge 1991)
7. G.H. Walker, J. Ford: Phys. Rev. **188**, 416 (1969)
8. F.Gustavson: Astron. J. **21**, 670 (1966)
9. B.V. Chirikov: Phys. Rep. **52**, 265 (1979)
10. T.C. Bountis: Physica D **3**, 371 (1981)
11. A.N. Kolmogorov: Dokl. Akad. Nauk. SSSR **98**, 525 (1954)
12. V.I. Arnold: Russ. Math. Surveys **18**, 85 (1963)
13. J. Moser: Nachr. Akad. Wiss. Göttingen Math. Phys. **K1**, 1 (1962)
14. M. Lakshmanan, R. Sahadevan: Phys. Rep. **224**, 1 (1993)
15. Z. Deng, F.T. Hioe: Phys.Rev.Lett. **55**, 1539 (1985)
16. R. Ramaswamy, P. Siders, R.A. Marcus: J. Chem. Phys. **74**, 4418 (1981)
17. K. Ganesan, M. Lakshmanan: Phys. Rev. A **42**, 3940 (1990)
18. K. Ganesan, R. Gebarowski: Pramana J. Phys. **48**, 379 (1997)
19. B. Dey, C.N. Kumar, A. Sen: Int. J. Mod. Phys. A **8**, 1755 (1993)
20. G. Benettin, L. Galgani, A. Giorgilli: IL Nuovo Cimento **89**, 103 (1985)
21. P. Collas, D. Klein, H.P. Schwebler: Chaos **8**, 466 (1998)
22. N. Srivastava, C. Kaufman and G. Müller: Computers in Physics 549 (1990)
23. A.A. Chernikov, M.Ya. Natenzon, B.A. Petrovichev, R.Z. Sagdeev, G.M. Zaslavsky: Phys. Lett. A **129**, 377 (1988)
24. B. Esser, H. Schanz: Chaos, Solitons and Fractals **4**, 2067 (1994)
25. J.P. Yeh: Int. J. Bifur. and Chaos **3**, 733 (1997)
26. N. Bisai, A. Sen, K.K. Jain: J. Plasma Phys. **56**, 209 (1996)
27. B. Malraison, P. Atten: Phys. Rev. Lett. **49**, 723 (1992)
28. X. Tong, F.P.J. Rimrott: Chaos, Solitons and Fractals **1**, 179 (1991)
29. G.M. Zaslavsky, M. Edelman, B.A. Niyazov: Chaos **7**, 159 (1997)
30. A. Banerjee, P.L. Taylor: Phys. Rev. B **30**, 6489 (1984)
31. G. Ananthakrishna, R. Balakrishnan, B.L. Hao: Phys. Lett. A **121**, 407 (1987)
32. A. Becker, P. Eckelt: Chaos **3**, 487 (1993)
33. R. Sankaranarayanan, A. Lakshminarayanan, V.B. Sheorey: Phys. Lett. A **279**, 313 (2001)

Chapter 8

1. J.P. Eckmann, D. Ruelle: Rev. Mod. Phys. **57**, 617 (1985); H.G. Schüster: *Deterministic Chaos* (Physik–Verlag, Weinheim 1984)
2. T.S. Parker, L.O. Chua: *Practical Numerical Algorithms for Chaotic Systems* (Springer, New York 1990)
3. A. Wolf, J.B. Swift, H.L. Swinney, J.A. Vastano: Physica D **16**, 285 (1985)
4. C.R. Wylie, L.C. Barrett: *Advanced Engineering Mathematics* (McGraw–Hill, New York 1995)
5. R.M. May: Nature **261**, 459 (1976)
6. T.S. Bellows: J. Anim. Ecol. **50**, 139 (1981)
7. J.G. Milton, J. Belair: Theor. Pop. Biol. **37**, 273 (1990)
8. T.T. Chia, B.L. Tau: Phys. Rev. E **54**, 5985 (1996)
9. K.P. Harikrishnan, V.M. Nandakumaran: Phys. Lett. A **142**, 483 (1989)
10. M. Ding, C. Grebogi, E. Ott: Phys. Rev. A **39**, 2593 (1989)

11. T. Nishikawa, K. Kaneko: Phys. Rev. E **54**, 6114 (1996)
12. V. Mehra, R. Ramaswamy: Phys. Rev. E **53**, 3420 (1996)
13. H. Hate, T. Horita, H. Mori: Prog. Theor. Phys. **82**, 897 (1989)
14. K. Criest, U. Parlitz, W. Lauterborn: Prog. Theor. Phys. **83**, 875 (1990)
15. G. Rangarajan, S. Habib, R. Ryne: Phys. Rev. Lett. **80**, 3747 (1998)
16. K. Ramasubramanian, M.S. Sriram: Physica D **139**, 283 (2000);
17. A. Adrover, S. Cerbelli, M. Giona: Int. J. Bifur. and Chaos **12**, 353 (2002)
18. W.H. Press, S.A. Teukolsky, W.T. Wellerling, B.P. Flannery: *Numerical Recipes in Fortran* (Cambridge University Press, Cambridge 1993)
19. I. Bull, R. Lincke: Amer. J. Phys. **64**, 906 (1996)
20. C. Grebogi, E. Ott, S. Pelikan, J.A. Yorke: Physica D **13**, 261 (1984)
21. E.G. Gwinn, R.M. Westervelt: Phys. Rev. A **33**, 4143 (1986)
22. P. Grassberger, I. Procaccia: Physica D **9**, 189 (1983)
23. F.C. Moon, G.X. Li: Physica D **17**, 99 (1985)
24. J.L. Kaplan, J.A. Yorke: In *Functional Differential Equations and Approximations of Fixed Points* ed. by H.O. Peitgen, H.O. Walter (Springer, New York 1979) pp.204–240

Chapter 9

1. H.D.I. Abarbanel: *Analysis of Observed Chaotic Data* (Springer, New York 1996)
2. A.R. Bulsara, L. Gammaitoni: Physics Today **49**, 39 (1996); P. Jung: Phys. Rep. **234**, 175 (1993)
3. S. Bleher, C. Grebogi, E.Ott: Physica D **46**, 87 (1990); E.Ott, T. Tel: Chaos **3**, 417 (1993)
4. M. Lakshmanan, K. Murali: *Chaos in Nonlinear Oscillators: Controlling and Synchronization* (World Scientific, Singapore 1996)
5. M.C. Gutzwiller: *Chaos in Classical and Quantum Mechanics* (Springer, New York 1990); F. Haake: *Quantum Signatures of Chaos* (Springer, New York 1991); K. Nakamura: *Quantum Chaos* (Cambridge University Press, Cambridge 1993); *Quantum versus Chaos* (Kluwer Academic Publishers, Dordrecht 1997); L. Reichl: *The Transition to Chaos: In Conservative Classical Systems – Quantum Manifestations* (Springer, New York 1992); H.J. Stöckmann: *Quantum Chaos, An Introduction* (Cambridge University Press, Cambridge 1999)
6. J. Gao, Z. Zheng: Phys. Rev. E **49**, 3807 (1994)
7. M.C. Valsakumar, K.P.N. Murthy, S. Venkadesan: 'Time series analysis: A review'. In: *Proceedings of Nonlinear Phenomena in Materials Science-III: Instabilities and Patterning*. ed. by G. Ananthakrishna, L.P.Kubin, G. Martin: Key Eng. Materials **103**, 141 (1995)
8. B. McNamara, K. Wiesenfeld: Phys. Rev. A **39**, 4854 (1989)
9. S. Fauve, F. Heslot: Phys. Lett. A **97**, 5 (1983)
10. R. Benzi, A. Sutera, A. Vulpiani: J. Phys. A **14**, L453 (1981)
11. G.P. Golubev, I.Kh. Kaufman, D.G. Luchinsky, P.V.E. McClintock, S.M. Soskin, N.D. Stein: Inter. J. Bifur. and Chaos **8**, 843 (1998)
12. V.S. Anishchenko, M.A. Safonova, L.O. Chua: Inter. J. Bifur. and Chaos **2**, 397 (1992)
13. H. Goldstein: *Classical Mechanics* (Narosa Publishing House, New Delhi 1990)
14. B.P. Koch, B. Bruhn: J. Phys. A **25**, 3945 (1992)
15. V. Daniels, M. Vallieres, J.M. Yuan: Chaos **3**, 475 (1993)
16. B. Bruhn, B.P. Koch: Inter. J. Bifur. and Chaos **3**, 999 (1993)

17. A. Becker, P. Eckelt: Chaos **3**, 487 (1993)
18. G. Stolovitzky, T.J. Kaper, L. Sirovich: Chaos **5**, 671(1995)
19. T. Kapitaniak: *Controlling Chaos* (Academic, San Diego 1996)
20. G. Chen, X. Dong: *From Chaos to Order* (World Scientific, Singapore 1998);
21. E. Ott, C. Grebogi, J. Yorke: Phys. Rev. Lett. **64**, 1196 (1990)
22. A. Hubler, E. Luscher: Naturwssenschaften **76**, 67 (1989)
23. B.A. Huberman, L. Lumer: IEEE Trans. Circuits Syst. **37**, 547 (1990)
24. S. Sinha, R. Ramaswamy, J. Subba Rao: Physica D **43**, 118 (1990)
25. R. Lima, M. Pettini: Phys. Rev. A **41**, 726 (1990)
26. J. Singer, Y.Z. Wang, H.H. Bau: Phys. Rev. Lett. **66**, 1123 (1991)
27. Y. Braiman, I. Goldhirsch: Phys. Rev. Lett. **66**, 2545 (1991)
28. E.A. Jackson: Phys. Rev. A **44**, 4839 (1991)
29. K. Pyragas: Phys. Lett A **170**, 421 (1992)
30. G. Chen, X. Dong: Int. J. Bifur. and Chaos **2**, 407 (1992)
31. S. Rajasekar, M. Lakshmanan: Physica D **67**, 282 (1993)
32. S. Parthasarathy, S. Sinha: Phys. Rev. E **51**, 6239 (1995)
33. S. Rajasekar, K. Murali, M. Lakshmanan: Chaos, Solitons and Fractals **8**, 1545 (1997)
34. E. Ott, C. Grebogi, J.A. Yorke: 'Controlling chaotic dynamical systems'. In: *Chaos: Soviet–American Perspectives on Nonlinear Science*. ed. by D.K. Campbell (Amer. Inst. Phys., New York 1990) pp.153–172
35. S. Rajasekar, M. Lakshmanan: J. Theor. Biol. **166**, 275(1994)
36. S. Rajasekar, M. Lakshmanan: Inter. J. Bifur. and Chaos **2**, 201 (1992)
37. S. Rajasekar: Chaos, Solitons and Fractals **5**, 2135 (1995).
38. M. Lakshmanan, S. Rajasekar: Proc. Nat. Acad. Sci. India **66**, 37 (1996)
39. L.M. Pecora, T.L. Carroll: Phys. Rev. Lett. **64**, 821 (1990)
40. L.M. Pecora, T.L. Carroll: Phys. Rev. A **44**, 2374 (1991)
41. K. Murali, M. Lakshmanan: Phys. Rev. E **49**, 4882 (1994)
42. L.M. Pecora, T.L. Carroll, G.A. Johnson, D.J. Mar, J.F. Heagy: Chaos **7**, 520 (1997)
43. R. Brown, L. Kocarev: Chaos **10**, 344 (2000)
44. G. Casati, B.B. Chirikov, I. Guarneri, D.L. Shepelyansky: Phys. Rep. **154**, 77 (1987)
45. H. Friedrich, D. Wintgen: Phys. Rep. **183**, 37 (1989)
46. F.M. Izrailev: Phys. Rep. **196**, 299 (1990)
47. K. Nakamura, M. Lakshmanan: Phys. Rev. Lett. **57**, 1661 (1986)
48. E.J. Heller, S. Tomosovic: Physics Today **46**, 38 (1993)
49. K. Ganesan, M. Lakshmanan: Phys. Rev. A **42**, 3940 (1990)
50. K. Ganesan, M. Lakshmanan: Phys. Rev. A **48**, 964 (1993)
51. K. Ganesan, R. Gebarowski: Pramana J. Phys. **48**, 379 (1997)
52. D. Kaplan, L. Glass: *Understanding Nonlinear Dynamics* (Springer, New York 1995); S.H. Strogatz: *Nonlinear Dynamics and Chaos: With Applications in Physics, Biology, Chemistry and Engineering* (Addison–Wesley, Manchester 1994); S.N. Rasband: *Chaotic Dynamics of Nonlinear Systems* (Wiley, New York 1990); E. Ott: *Chaos in Dynamical Systems* (Cambridge University Press, New York 1993)

Chapter 10

1. A.H. Nayfeh, D.T. Mook: *Nonlinear Oscillatons* (Wiley, New York 1979)
2. J. Kevorkian, J. D. Cole: *Perturbation Methods in Applied Mathematics* (Springer, New York 1981)
3. V.V. Kozlev: Russ. Math. Surveys **38**, 1 (1983)
4. See for example, B. Grammaticos, A. Ramani in *Integrability of Nonlinear Systems* ed. by Y. Kosmann-Schwarzbach, B. Grammaticos, K.M. Tamizhmani (Springer, Berlin 1997) pp.30–94
5. J. Hieterinta: Phys. Rep. **147**, 87 (1987)
6. A. Ramani, B. Grammaticos, T. Bountis: Phys. Rep. **180**, 160 (1989)
7. M. Lakshmanan, R. Sahadevan: Phys. Rep. **224**, 1 (1993)
8. H. Goldstein: *C*lassical Mechanics (Narosa Publishing House, New Delhi 1990)
9. E.I. Ince: *Ordinary Differential Equations* (Dover Publ., New York 1956); see also M.D. Kruskal, P.A. Clarkson: Stud. Appl. Math. **86**, 87 (1992)
10. M. Lakshmanan, P. Kaliappan: J. Math. Phys. **24**, 795 (1983)
11. G. Bluman, S. Kumei: *Symmetries and Differential Equations* (Springer, New York 1989).
12. H. Stephani: *Differential Equations: Their Solutions Using Symmetries* (Cambridge University Press, Cambridge 1989)
13. E.T. Whittaker: *A Treatise on the Analytical Dynamics of Particles and Rigid Bodies* (Cambridge University Press, Cambridge 1937)
14. S. Chandrasekhar: *Principles of Stellar Dynamics* (Dover Publi., New York 1942) Chap. III
15. M.J. Ablowitz, P.A. Clarkson: *Soliton, Nonlinear Evolution Equations and Inverse Scattering* (Cambridge University, Cambridge 1992)
16. J. Hoppe: *Lectures on Integrable Systems* (Springer, Berlin 1992)
17. S. Parthasarathy, M. Lakshmanan: J.Sound and Vib. **137**, 523 (1990)
18. S. Parthasarathy, M. Lakshmanan: J.Phys.A **23**, L1223 (1990)
19. M. Tabor, J.D. Gibbon: Physica D **18**, 180 (1986)
20. M. Lakshmanan, M. Senthilvelan: J. Phys. A **25**, 1259 (1992)
21. K. Ganesan, M. Lakshmanan: Phys. Rev. Lett. **62**, 232 (1989)
22. J.A. Ketoja, I.I. Satija: Physica D **109**, 70 (1997)
23. A. Prasad, R. Ramaswamy, I.I. Satija, V. Shah: Phys. Rev. Lett. **83**, 4530 (1999)

Chapter 11

1. A.C. Scott: *Nonlinear Science: Emergence and Dynamics of Coherent Structures* (Oxford Univ. Press, New York 1999)
2. H. Goldstein: *Classical Mechanics* (Narosa, New Delhi 1990) Chap. 12.
3. L.I. Schiff: *Quantum Mechanics* (McGraw Hill, New York 1988)
4. R.K. Bullough: "The Wave" "Par Excellence", 'The solitary progressive great wave of equilibrium of the fluid: An early history of the solitary wave'. In: *Solitons: Introduction and Applications* ed. by M. Lakshmanan (Springer, New York 1988)
5. D.J. Korteweg, G. de Vries: Phil. Mag. **39**, 422 (1895)
6. A. Jeffrey, M.N.B. Mohamad: Chaos, Solitons and Fractals **1**, 187 (1991)

Chapter 12

1. See for example, R. K. Bullough, "The Wave" "Par Excellence", The solitary progressive great wave of equilibrium of the fluid: An early history of the solitary wave. In *Solitons: Introduction and Applications* ed. by M. Lakshmanan (Springer, New York 1988); G.B. Whitham: *Linear and Nonlinear Waves* (John–Wiley, New York 1974)
2. J. Ford: Phys. Reports **213**, 271 (1992)
3. H. Goldstein: *Classical Mechanics* (Narosa Publishing House, New Delhi 1990)
4. N.J. Zabusky, M.D. Kruskal: Phys. Rev. Lett. **15**, 240 (1965)
5. R. Hirota, Phys. Rev. Lett. **27**, 1192 (1971); J.Phys. Soc. Japan **33**, 1456 (1972); R. Hirota in *Bäcklund Transformations* ed. by R.M. Miura (Springer, New York 1976)
6. M. Lakshmanan: Inter. J. Bifur. and Chaos **3**, 3 (1993)
7. M.J. Ablowitz, P.A. Clarkson: *Solitons, Nonlinear Evolution Equations and Inverse Scattering* (Cambridge Univ. Press, Cambridge 1992)
8. R. Hirota: J. Math. Phys. **14**, 805 (1973)

Chapter 13

1. P.L. Bhatnagar: *Nonlinear Waves in One-Dimensional Dispersive Systems* (Oxford University Press, Bombay 1979)
2. R.M. Miura: J. Math. Phys. **9**, 1202 (1968)
3. P.D. Lax: Commun. Pure Appl. Math. **21**, 467 (1968)
4. C.S. Gardner, J.M. Greene, M.D. Kruskal, R.M. Miura: Phys. Rev. Lett. **19**, 1095 (1967)
5. M.J. Ablowitz, P.A. Clarkson: *Solitons, Nonlinear Evolution Equations and Inverse Scattering* (Cambridge Univ. Press, Cambridge 1992)
6. H. Goldstein: *Classical Mechanics* (Narosa Publishing House, New Delhi 1990)
7. A. Kundu, B. Basu Mallick: J. Phys. Soc. Jpn. **59**, 1560 (1990); J. Phys. A **23**, L709 (1990)
8. V.E. Zakharov, L.D. Faddeev: Funct. Anal. Appl. **5**, 280 (1971)
9. C.S. Gardner: J. Math. Phys. **12**, 1548 (1971)
10. R.M. Miura (Ed.): *Bäcuund Transformations* Lecture Notes in Mathematics 515 (Springer, New York 1976)

Chapter 14

1. M.J. Ablowitz, P.A.Clarkson: *Solitons, Nonlinear Evolution Equations and Inverse Scattering* (Cambridge University Press, Cambridge 1991)
2. M. Lakshmanan (Ed.): *Solitons* (Springer, Berlin 1990)
3. G.P. Agrawal: *Nonlinear Fiber Optics* 2nd edn. (Academic Press, New York 1995)
4. A. Hasegawa, Y. Kodama: *Solitons in Optical Communications* (Oxford University Press, Oxford 1995)
5. A.C. Scott: *Nonlinear Science: Emergence and Dynamics of Coherent Structures* (Oxford University Press, Oxford 1999)
6. M. Lakshmanan: Phys. Lett A **61**, 53 (1977)
7. M. Toda: *Theory of Nonlinear Lattices* (Springer, Berlin 1981)

8. V.E. Zakharov, A.B. Shabat: Sov. Phys. JETP **34**, 62 (1972)
9. M.J. Ablowitz, D.J. Kaup, A.C. Newell, H. Segur: Stud. Appl. Maths. **53**, 294 (1974)
10. M. Lakshmanan, Th.W. Ruijgrok, C.J. Thompson: Physica A **84**, 577 (1976)
11. M.J. Ablowitz, H. Segur: *Solitons and Inverse Scattering Transform* (SIAM, Philadelphia 1981)
12. C. Rogers, W.F. Shadwick: *Bäcklund Transformations and Applications* (Academic Press, New York 1982)
13. L.D. Fadeev: In: *Solitons* ed. by R.M. Bullough, P.J. Caudrey (Springer, Berlin 1980) pp.339–354
14. S. Novikov, S.V. Manakov, L.P. Pitaevskii, V.E. Zakharov: *Theory of Solitons – The Inverse Scattering Method* (Consultants Bureau, New York 1984)
15. B.G. Konopelchenko: *Solitons in Multidimensions* (World Scientific, Singapore 1993)
16. R. Radhakrishnan, M. Lakshmanan, J.Hietarinta: Phys. Rev. E **56**, 2213 (1997); T. Kanna, M. Lakshmanan: Phys. Rev. Lett. **86**, 5043 (2001)
17. N. Sasa, J. Satsuma: J. Phys. Soc. Japan **60**, 409 (1991)
18. H. Fogedby: J. Phys. A **13**, 1467 (1980)

Chapter 15

1. J.D. Murray: *Mathematical Biology* (Springer, New York 1989)
2. E. Taylor: *Partial Differential Equations*, Vol. I–III (Springer, New York 1996)
3. M.C. Cross, P.C. Hohenberg: Rev. Mod. Phys. **63**, 851 (1993)
4. A.Okubo: *Diffusion and Ecological Problems: Mathematical Models* (Springer, New York 1980)
5. A. Koch, H. Meinhardt: Rev. Mod. Phys. **66**, 1481 (1994)
6. R. FitzHugh: Biophys. J. **1**, 445 (1961); R. FitzHugh: In: *Biological Engineering* ed. by H.P. Schwarn: (McGraw–Hill, 1969) pp. 1–85
7. A.C. Scott: Rev. Mod. Phys. **47**, 487 (1975)
8. D. Walgraef: *Spatio-temporal Pattern Formation* (Springer, New York 1996)
9. A.C. Scott: *Nonlinear Science: Emergence and Dynamics of Coherent Structures* (Oxford University Press, Oxford 1999)
10. A.V. Holden: Physics World (1998) No.11, p.29
11. P.B. Umbanhowar, F. Melo, H.L. Swinney: Nature **382**, 793 (1996)
12. J. Fineberg: Nature **382**, 763 (1996)
13. R.J. Deissler, H.R. Brand: Phys. Rev. Lett. **74**, 4847 (1995)
14. R. Temam: Mathematical Intelligencer **12**, 68 (1990)
15. D.A. Egolf, H.S. Greenside: Nature **369**, 124 (1994)
16. T. Bohr, E. Bosch, W. van de Water: Nature **372**, 48 (1994)
17. J. Gunton et al.: URL http://www.psc.edu/science/Gunton/gunton.html
18. M.C. Cross, M. Louie, D. Meiron: Phys. Rev. E **63**, 045201 (2001)
19. K. Kaneko: Physica D **37**, 1 (1989)
20. M. Lakshmanan, P. Muruganandam: Proceedings of Indian National Science Academy PINSA **66A**, 393 (2000)
21. L.O.Chua: *CNN: A Paradigm for Complexity* (World Scientific , Singapore 1998)
22. P.L. Sachdev: *Nonlinear Diffusion Waves* (Cambridge University Press, Cambridge 1987)
23. G.S. Rosen: Phys. Rev. Lett. **49**, 1844 (1982)
24. X.Y. Wang, S. Fan, T. Kyu: Phys. Rev. E **56**, R4931 (1996)

25. P.S. Bindu, M. Senthil Velan, M. Lakshmanan: J. Phys. A **34**, L689 (2001)
26. P.V. O'Neil: *Beginning Partial Differential Equations* (John Wiley, New York 1999)
27. M. Lakshmanan, K. Nakamura: Phys. Rev. Lett. **53**, 2497 (1984)

Chapter 16

1. R. He, P.G. Vaidya: Phys. Rev. E **57**, 1532 (1998)
2. M. Lakshmanan, K. Murali: *Chaos in Nonlinear Oscillators: Controlling and Chaos* (World Scientific, Singapore 1996)
3. M.I. Sobhy, A. Schchata: Inter. J. Bifur. and Chaos **10**, 2831 (2000))
4. M. Sincell: Physics World (March, 2001) p.7
5. W.E. Leland, M.S. Taqqu, W. Willinger, D.V. Wilson: IEEE Transactions in Networking **2**, 1(1994)
6. D. Helbing: Rev. Mod. Phys. **73**, 1067 (2001)
7. T. Nagatani: Phys. Rev. E **60**, 1535 (1999)
8. J. Moore: Physics World (September, 2001) p.6
9. P. Guynne: Physics World (October, 2001) p.9
10. B. Lebaron: Phil. Trans. R. Soc. London A **348**, 397 (1994)
11. G.D. van Wiggerenand, R. Roy: Science **279**, 1198 (1998)
12. G. Papaioannou, A. Karytinos: Inter. J. Bifur. and Chaos **5**, 1557 (1995)
13. A.Hasegawa, F.D. Tappert: Appl. Phys. Lett. **23**, 142 (1973); 171; see also A. Hasegawa, Y. Kodama: *Soliton in Optical Communications* (Oxford University Press, Oxford 1995)
14. L.F. Mollenauer, R.H. Stolen, J.P. Gordon: Phys. Rev. Lett. **45**, 1095 (1980)
15. A. Hasegawa: Opt. Lett. **8**, 650 (1983)
16. L.F.Mollenauer, K. Smith: Opt. Lett. **13**, 675 (1988)
17. M. Nakazawa, Y. Kimura, K. Suzuki: Electron Lett. **25**, 199 (1989)
18. E. Desurvirl: Physics Today **47**, 20 (1994)
19. E. Iannone, F. Matera, A. Mecozzi, M. Settembre: *Nonlinear Optical Communication Networks* (A Wiley–Interscience Publication, New York 1998)
20. See for example, G.P. Agrawal: *Nonlinear Fiber Optics* (Academic Press, New York 1995)
21. A. Hasegawa" *New Trends in Optical Soliton Transmission Systems* (Kluwer Academic, Dordrecht 1998)
22. P.W. Boyd: *Nonlinear Optics* (Academic Press, London 1992)
23. D.N. Christodulides, M.J. Carvalho: J. Opt. Soc. Am. **B12**, 1682 (1995)
24. D.N. Christodulides, T.H.Coskun, M.Mitchell, M.Segev: Phys. Rev. Lett. **78**, 646 (1997)
25. R. Radhakrishnan, M. Lakshmanan, J. Hietarinta: Phys. Rev. E **56**, 2213 (1997)
26. M.H. Jakubowski, K. Steiglitz, R. Squier: Phys. Rev. E **58**, 6752 (1998)
27. K. Steiglitz: Phys. Rev. E **63**, 016608 (2000)
28. W.F. Brown Jr.: *Micromagnetics* (Wiley, New York 1963)
29. M. Lakshmanan, K. Nakamura: Phys. Rev. Lett. **53**, 2497 (1984)
30. R. Urban, G. Woltersdorf, B. Heinrich, Phys. Rev. Lett. **87**, 217204 (2001)

Appendix A

1. F. Bowman: *Introduction to Elliptic Functions* (Wiley, New York 1953)
2. T. Davis: *Introduction to Nonlinear Differential and Integral Equations* (Dover, New York 1962)
3. P.M. Mathews, M. Lakshmanan: Ann. of Phys.(N.Y) **79**, 171 (1973)
4. S. Parthasarathy, M. Lakshmanan: J. Sound and Vibr. **137**, 523 (1990)

Appendix B

1. A.H. Nayfeh, D.T. Mook: *Nonlinear Oscillations* (Wiley, New York 1979)
2. P. Hagedorn: *Nonlinear Oscillations* (Oxford Sci. Publ., Oxford 1988)
3. T.L. Saaty, J. Bram: *Nonlinear Mathematics* (Dover, New York 1964)
4. H. Goldstein: *Classical Mechanics* (Narosa, New Delhi 1990)
5. M.Tabor: *Chaos and Integrability in Nonlinear Dynamics: An Introduction* (John–Wiley, New York 1989)
6. L.G. Reichl: *The Transition to Chaos: In Conservative Classical Systems – Quantum Manifestations* (Springer, New York 1992)
7. G.M. Zaslavsky, R.Z. Sagdeev, D.A. Usikov, A.A. Cherikov: *Weak Chaos and Quasiregular Patterns* (Cambridge University Press, Cambridge 1991)
8. G.H. Walker, J. Ford: Phys. Rev. **188**, 416 (1969)

Appendix C

1. J.H. Mathews: *Numerical Methods for Mathematics, Science and Engineering* (Prentice–Hall of India, New Delhi 1998)

Appendix D

1. N. Minorsky: *Nonlinear Oscillations* (Van Nostrand, New York 1962)

Appendix E

1. T.C. Halsey, M.H. Jenson, L.P. Kadanoff, I. Procaccia, B.I. Shraiman: Phys. Rev. A **33**, 1141 (1986)
2. B.B. Mandelbrot: *The Fractal Geometry of Nature* (Freeman, San Francisco 1982)
3. H.O. Peitgen, P.H. Richter: *The Beauty of Fractals* (Springer, Berlin 1986)
4. H.G. Schüster: *Deterministic Chaos* (Verlag, Weinheim 1988)

Appendix F

1. S. Flügge: *Practical Quantum Mechanics* (Springer, New York 1974)

Appendix G

1. P.J. Caudrey: Unpublished Lecture Notes, Department of Mathematics, University of Manchester Institute of Science and Technology, Manchester, England (1980)
2. M.J. Ablowitz, P.A. Clarkson: *Solitons, Nonlinear Evolution Equations and Inverse Scattering* (Cambridge Univ. Press, Cambridge 1991)

Appendix H

1. P.J. Caudrey: Unpublished Lecture Notes, Department of Mathematics, University of Manchester Institute of Science and Technology, Manchester, England (1980)
2. M.J. Ablowitz, P.A. Clarkson: *Solitons, Nonlinear Evolution Equations and Inverse Scattering* (Cambridge Univ. Press, Cambridge 1991)

Appendix I

1. J. Hoppe: *Lectures on Integrable Systems* (Springer, Berlin 1992)
2. F. Calogero: Lett. Nuovo Cimento **13**, 411 (1975); **16**, 77 (1976)
3. J. Moser: Adv. Math. **16**, 197 (1975)
4. M. Toda: *Theory of Nonlinear Lattices* (Springer, Berlin 1981)
5. M.J. Ablowitz, P.A. Clarkson: *Solitons, Nonlinear Evolution Equations and Inverse Scattering* (Cambridge Univ. Press, Cambridge 1991)

Appendix J

1. J. Weiss, M. Tabor, G. Carnevale: J. Math. Phys. **24**, 522 (1983)
2. A.C. Newell, M. Tabor, Y.B. Zheng: Physica D **29**, 1 (1987)
3. M.J. Ablowitz, P.A. Clarkson: *Solitons, Nonlinear Evolution Equations and Inverse Scattering* (Cambridge Univ. Press, Cambridge 1991)
4. H. Cheng: Stud. Appl. Math. **70**, 183 (1984)
5. K.M. Tamizhmani, M. Lakshmanan: J. Math Phys. **27**, 2257 (1986)

Index

Druck: Strauss Offsetdruck, Mörlenbach
Verarbeitung: Schäffer, Grünstadt